BIOCHEMISTRY

A Comprehensive Review

BIOCHEMISTRY
A Comprehensive Review

N. V. Bhagavan, Ph.D.

Professor of Biochemistry and Medical Technology
University of Hawaii School of Medicine
Honolulu, Hawaii

Consultant Biochemist to the
Kaiser Foundation Hospital
Honolulu, Hawaii

J. B. Lippincott Company
Philadelphia • Toronto

ISBN 0-397-52063-8

Library of Congress Catalog Card Number 74-3317

Printed in the United States of America

1   3   5   4   2

Library of Congress Cataloging in Publication Data

Bhagavan, N   V
    Biochemistry.

   1. Biological chemistry. I. Title.
[DNLM: 1. Biochemistry. QU4 B575b 1974]
QP514.2.B45   574.1'92   74-3317
ISBN 0-397-52063-8

Preface

The purpose of this synopsis is to create an interest and a sustained enthusiasm towards the study of biochemistry for those students pursuing medical and other health-related studies. To most of these students, it is essential to continually relate biochemical principles to the art of healing. It is hoped that this book can guide students in acquiring sufficient information in biochemistry to increase their comprehension of how biochemically determinable constituents vary in normal and abnormal states and to relate metabolic disorders to biochemical lesions.

In order to provide concise and complete summaries, detailed discussions of experimental observations have been omitted. In the process, subtle points and various ramifications of the subject material may not appear in the text. With this in mind, it is suggested that the student refer for greater details to suitable references such as those listed at the end of the book.

I am grateful to the contributions made to various parts of the book by the authors whose names appear along with the chapters on which they have worked. Special mention should be made of the invaluable contribution made by John H. Bloor who reviewed the manuscript in its entirety and supplemented many of the deficiencies.

I am indebted to Janet Kawata for her painstaking typing of the entire manuscript and to Anita G. Bloor for patient proofreading.

I wish to express my sincere thanks to John Smith, Douglas Campbell, and Terry Copperman who reviewed selected sections and suggested changes that have enhanced the value of the synopsis. Thanks are also due to E. Yanagihara, L. Kimura, and S. Perreira for their assistance with the chapter on Immunochemistry.

I am grateful to Dr. Sheldon I. Freedman, Dr. Eduardo A. Porta, and Dr. Stanley W. Hartroft for the many fruitful discussions pertaining to the clinical aspects of laboratory medicine.

N. V. Bhagavan

Honolulu, Hawaii
November 1973

Contents

1
Acids, Bases and Buffers

1. Acids are substances that liberate hydrogen ions (protons) in solution (proton donor).

2. Bases are substances that bind hydrogen ions (protons) and thus remove them from solution (proton acceptor). Note: This is the most useful definition of acids and bases for biological work. There are other definitions: Acid is an electron-pair acceptor and base is an electron-pair donor (known as the Lewis Acid Concept).

3. Strong and weak electrolytes.

   a. Strong electrolytes (acids, bases or salts) are completely ionized in aqueous solutions.

   b. Weak electrolytes (acids, bases or salts) are partially ionized in aqueous solutions. The following equation shows the relationship between an undissociated compound and its ions.

$$HA \rightleftharpoons H^+ + A^-$$
(weak acid)

(ions)

   c. Applying the Law of Mass Action

cation   anion

$$K_a = \frac{(H^+) \quad (A^-)}{(HA) \leftarrow (undissociated\ acid)}$$

where K_a = ionization constant of the acid and parentheses indicate molar concentrations of the species. Similarly, K_b represents the ionization constant of a base. The higher the value of K_a, the greater the number of H^+ ions liberated per mole of acid in solution and hence the stronger the acid. K_a is thus a measure of the <u>strength of an acid</u>.

$$K_a = \frac{(H^+)\ (A^-)}{(HA)}$$

Rearranging $$H^+ = \frac{K_a(HA)}{(A^-)}$$

Taking logarithms (to base 10) on both sides

$$\log (H^+) = \log K_a + \log (HA) - \log (A^-)$$

Multiplying by -1

$$-\log (H^+) = -\log K_a - \log (HA) + \log (A^-)$$

Using the definitions

$$-\log (H^+) = pH \text{ and } -\log K_a = pK_a$$

one gets

$$pH = pK_a + \log \frac{(A^-)}{(HA)}$$

or

$$pH = pK_a + \log \frac{salt}{acid}$$

This relationship is commonly known as the <u>Henderson-Hasselbalch</u> <u>Equation</u>.

<u>Dissociation of Water</u>

1. Considering water as a weak acid

$$H_2O \rightleftharpoons H^+ + OH^-$$

The dissociation constant

$$K_a^{H_2O} = \frac{(H^+) (OH^-)}{(H_2O)}$$

Since in dilute aqueous solutions, the molarity of water is very nearly constant, this is usually rewritten as

$$K_w = K_a^{H_2O} (H_2O) = (H^+)(OH^-)$$

The concentration of pure water is 55.6 M so that $K_w = 55.6 \ K_a^{H_2O}$. K_w, called the dissociation constant or ion product for water, has the value of 1×10^{-14} at $25^\circ C$ (room temperature).

2. In aqueous solution, then $(H^+)(OH^-) = 1 \times 10^{-14}$
    So that, in <u>pure water</u> $(H^+) = 1 \times 10^{-7}$ moles/liter
                                               $(OH^-) = 1 \times 10^{-7}$ moles/liter

    and: pH = 7
          pOH = 7

    For all aqueous solutions $pK_w = pH + pOH = 14$
                          and $pOH = 14 - pH.$

    pOH is not used descriptively (as is pH), but it is useful in
    calculating the pH of an alkaline solution.

<u>Problem 1</u> What is the pH of 0.0004 N NaOH solution? Strong
                 electrolyte, completely dissociated.

    <u>Method 1</u>: $(OH^-) = 0.0004 M = 4 \times 10^{-4} M$

$$pOH = -\log (OH^-) = -\log (4 \times 10^{-4}) = -\log 4 - \log 10^{-4}$$

$$= -(0.602 - 4) = 3.398$$

        pOH + pH = 14 so that pH = 14 - pOH

        Then pH = 14 - 3.398 = <u>10.602</u>

    <u>Method 2</u>: $(H^+)(OH^-) = 10^{-14}$

$$(H^+) = 10^{-14}/(OH^-) = 10^{-14}/(4 \times 10^{-14}) = 2.5 \times 10^{-11}$$

$$pH = -\log (H^+) = -\log (2.5 \times 10^{-11})$$

$$= -\log 2.5 - \log 10^{-11} = -(0.398 - 11)$$

$$pH = \underline{10.602}$$

<u>Problem 2</u> Calculate the pH of 0.005 M HCl solution. Strong
                 electrolyte, completely dissociated.

$$(H^+) = 5 \times 10^{-3} M$$

$$pH = 3 - 0.7 = \underline{2.3}$$

<u>Problem 3</u> What is the pH of 0.1 M solution of CH_3COOH? Given
                 $K_a = 1.86 \times 10^{-5}$. Acetic acid is a weak acid and
                 therefore poorly dissociated.

$$CH_3COOH \rightleftharpoons (CH_3COO^-) + (H^+)$$

$$K_a = \frac{(CH_3COO^-)\ (H^+)}{(CH_3COOH)}$$

Since $(CH_3COO^-) = (H^+)$, define $(CH_3COO^-) = (H^+) = X$ at equilibrium.

In order to get one CH_3COO^- and one H^+, one CH_3COOH has to dissociate. Therefore, if the concentrations of the ions are X, the loss in CH_3COOH concentration must also be X. Since the initial concentration of CH_3COOH is 0.1 M, the concentration at equilibrium is $(CH_3COOH) = 0.1 - X$. Then

$$K_a = \frac{(X)\ (X)}{(0.1 - X)}$$

To make this easier to solve, assume that the loss of CH_3COOH due to dissociation is small compared to 0.1 M. This is acceptable because

$$\frac{K_a}{(CH_3COOH)}$$ is less than approximately 10^{-3}.

Then $(0.1 - X) \approx 0.1$ and $K_a = \dfrac{X^2}{0.1}$.

Note: HAc = acetic acid = CH_3COOH.

$$\left[\text{If } \frac{K_a}{(HAc)} \geq 10^{-3}, \text{ there will be an error of } \geq 3.2\% \text{ in the value of } \left[H^+\right]. \right]$$

$$X^2 = 0.1\ K_a$$

$$X = \sqrt{0.1\ K_a}$$

$$= \sqrt{(0.1)(1.86 \times 10^{-5})} = \sqrt{1.86 \times 10^{-6}}$$

$$= 1.36 \times 10^{-3} = (H^+)$$

Note: The concentration of $X = (1.36 \times 10^{-3}$ M) is small compared to the concentration of free acid = (0.1 M).

$$pH = 3 - 0.13$$

$$= \underline{2.87}$$

Problem 4 What is the pH of 0.1 N NH_4OH?

$$K_b = 2 \times 10^{-5}$$

Answer: pH = <u>11.15</u>

Buffers

1. These are solutions that minimize changes in pH when acids or bases are added to them.

2. They are usually mixtures of a weak acid and a salt of the same acid with a strong base, although they can also be mixtures of a weak base and its salt with a strong acid.

   Example: CH_3COOH and H_2CO_3 are weak acids. When reacted with NaOH, a strong base, the salts $CH_3COO^-Na^+$ and $Na^+HCO_3^-$ are formed.
   Then: ($CH_3COOH + CH_3COO^-Na^+$) and ($H_2CO_3 + Na^+HCO_3^-$) are both buffer systems.

3. A buffer works in the following manner.

   To make a buffer, mix CH_3COOH, $CH_3COO^-Na^+$, and water. The species present in solution are CH_3COOH, CH_3COO^-, Na^+, and H_2O. (There are also small amounts of H^+ and OH^- as there must be in all aqueous solutions. These need not be considered in calculations.)

   Add H^+: $H^+ + CH_3COO^- \rightarrow CH_3COOH$. Almost all of the protons react with acetate ions to produce weakly ionized acetic acid. The H^+ is thereby prevented from changing the pH appreciably.

   Add OH^-: $OH^- + CH_3COOH \rightarrow CH_3COO^- + H_2O$. Almost all of the hydroxyl radicals react with acetic acid molecules to produce more acetate and more water. The OH^- is thus absorbed and does not affect the pH very much.

4. Adding H^+ or OH^- to a buffer causes <u>slight</u> pH changes (since mass action expressions must be obeyed) provided there is enough salt (CH_3COO^-) or acid (CH_3COOH) to react with them. If, for example, all of the acid is converted to a salt by adding a large amount of OH^-, then the solution no longer behaves as a buffer. Adding more OH^- will cause the pH to rise rapidly, just as if the solution contained only the salt. The maximum buffering is said to occur when the molarities of the salt and acid are equal. Then the

buffer has its maximum capacity to absorb <u>either</u> H^+ or OH^-. (Note also that, mathematically, the pK_a corresponds to an inflection point in the titration curve. Hence, the pK_a is the point of minimum slope or minimum change in pH for a given change in $\boxed{H^+}$ or $\boxed{OH^-}$. Since a buffer is intended to give only a small pH change with added H^+ or OH^-, the best buffer is the one which gives the smallest change.) From the Henderson-Hasselbalch Equation, this is seen to occur when the pH of the solution equals the pK_a (or pK_b) of the weak acid (or base) forming the buffer.

$$pH = pK + \log \frac{(salt)}{(acid)}$$

if (salt) = (acid) then $\log \dfrac{(salt)}{(acid)} = \log 1 = 0$ and <u>pH = pK</u>.

5. Buffers are part of the homeostatic mechanisms whereby the neutrality of the body fluids is regulated.

Blood pH: 7.35-7.45 = normal range

7.8 → death
↑ ←——————— alkalosis
7.45
7.35 } normal range
↓ ←——————— acidosis
6.8 → death

Buffer Systems of the Blood

1. The most important buffer of <u>plasma</u> is the bicarbonate-carbonic acid system

$$H_2CO_3 \leftrightarrows HCO_3^- + H^+$$

$$pK_a = 6.1$$

Using the Henderson-Hasselbalch Equation, the ratio of HCO_3^-/H_2CO_3 in blood at physiolocal pH (7.4) is determined as follows:

$$7.4 = 6.1 + \log \frac{salt}{acid}$$

(Note that pH 7.4 is not very close to the pH of maximum buffering for carbonic acid.)

$$\log \frac{salt}{acid} = 1.3$$

Taking antilogarithms on both sides $\dfrac{\text{salt}}{\text{acid}} = \dfrac{20}{1}$

Average normal conc. of bicarbonate found in blood = 27 mEq/L (0.027N).

Average normal conc. of carbonic acid found in blood = 1.35 mEq/L (0.00135N).

So that $\dfrac{27}{1.35} = \dfrac{20}{1}$.

Note that the <u>ratio</u> is quite invariant although the concentrations going into it may vary.

2. The effectiveness of the HCO_3^-/H_2CO_3 buffer is due to HCO_3^- being present in high concentration. Among anions, only chloride (103 mEq/L) is in higher concentration. Chloride, however, has no buffering capacity at all.

3. The continuing ability of blood to buffer is dependent upon various processes involving red blood cells, lungs and kidneys. These are essential in removing catabolites produced by the body. The diagram on page 8 illustrates some of these aspects.

4. Serum Protein Buffer System. There are many proteins in serum and they have many weakly acidic (glutamate, aspartate) and weakly basic (lysine, arginine, histidine) amino acid sidechains which act as components of a buffer system. Such effects are insignificant, however, compared to the buffering capacity of hemoglobin (in erythrocytes) and of the bicarbonate system (in the plasma).

5. The phosphate buffer system

   at pH 7.4 of plasma $\dfrac{HPO_4^=}{H_2PO_4^-} = \dfrac{80}{20}$

   This is a minor buffering system in blood, since the plasma concentration of PO_4^{-3}-containing species is only 0.3-0.4 mEq/L. It is important, however, in raising plasma pH through excretion of $H_2PO_4^-$ by the kidney.

6. The hemoglobin buffer system buffers CO_2 (as carbonic acid) produced during metabolic processes. There are two mechanisms for this.

a. Hemoglobin (Hb) exists as oxyhemoglobin ($HHbO_2$), deoxy-
    hemoglobin (HHb), and as their potassium salts ($KHbO_2$ and KHb).
    Since HHb and $HHbO_2$ are weak acids, the two buffer systems,
    (HHb/KHb) and ($HHbO_2/KHbO_2$), are present in the erythrocyte.
    (This is similar to the buffering of serum proteins in 4 (above).
    The proton released is <u>not</u> the same one as in <u>b</u> below.)

b. More important than <u>a</u> and a very effective buffer is the
    acceptance of H^+ ions by histidine in the hemoglobin molecule
    with the release of O_2 (Bohr proton).

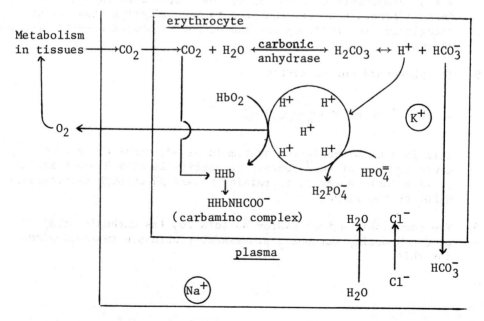

Oxyhemoglobin   Deoxygenated Hemoglobin

or $HbO_2^- + H_2CO_3 \rightleftarrows HHb + O_2 + HCO_3^-$

An increase in P_{CO_2} and a decrease in pH will <u>favor</u> the
dissociation of HbO_2 and thus permit the imidazole group to
accept H^+ ions. In the tissues then

HCO_3^-, produced within the erythrocytes, diffuses into the plasma
and in order to maintain electrical neutrality, the principal plasma
anion Cl^- goes into the erythrocytes. Note that the concentrations
of the principal intracellular cation (K^+) and extracellular
(plasma) cation (Na^+) do not undergo significant changes due to the
relative impermeability of the erythrocyte membrane. The exact
opposite set of changes takes place at the lungs with the release
of CO_2.

The bicarbonate/carbonic acid system participates in buffering
against acids and bases as follows:

$$NaHCO_3^- + HCl \longrightarrow H_2CO_3 + NaCl$$

$$H_2O + CO_2 \text{ (can be removed by the lungs)}$$

$$H_2CO_3 + NaOH \longrightarrow NaHCO_3 + H_2O$$

The above reactions may be summarized as follows:

Hemoglobin actually absorbs about 60% of the hydrogen ions
produced by H_2CO_3 formation from CO_2. This is far more than any
other buffer system in the blood. However, since hemoglobin and
carbonic anhydrase are present only in the erythrocytes, the
HCO_3^-/H_2CO_3 system in the plasma is an indispensable intermediary
in transporting the acid. This is discussed further in Chapters 7
and 12.

Titration Curves

1. The profiles below (Figure 1) are constructed by measuring the pH
    of an acid solution as it is titrated with a base. Titration
    curves were prepared using 25 ml of 1 M acid and titrating with
    1 M NaOH (or any other strong base).

2. Note in the diagram below the differences in curves for a strong
    acid (HCl) and weak acid ($NH_4Cl \rightleftharpoons NH_3 + H^+ + Cl^-$).

3. Calculation of pK_a for unknown weak acids and bases (strong acids
    are completely ionized in H_2O and show no pK_a) is possible with a
    titration curve for the acid or base.

4. One can determine the pK_a of a weak acid used in the preparation
    of a buffer of known pH by observing the shape of the titration
    curve of the buffer.

Figure 1. Titration Profiles of Some Acids*

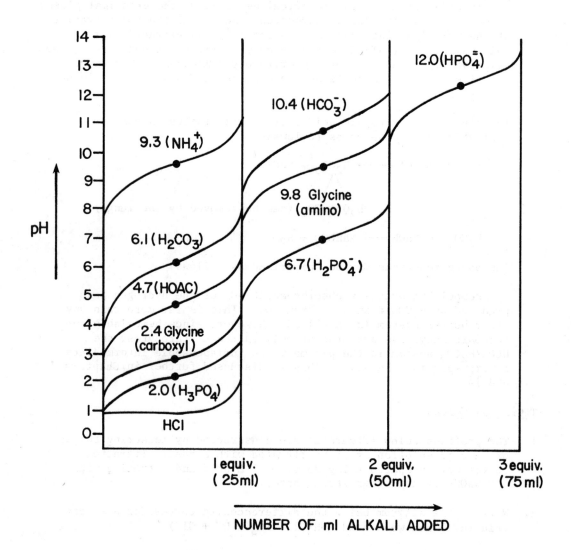

Note: pK_{a_1}: $H_2CO_3^- \rightleftarrows HCO_3^- + H^+$ pK_{a_2}: $HCO_3^- \rightleftarrows CO_3^= + H^+$

Note: The points marked, and the accompanying numbers on the above
        diagram, are the pK_a values of the compound indicated.
        HOAC = acetic acid.

* Reproduced with permission from Raffelson, Max E., Jr., Brinkley,
  Stephen B., and Hayashi, James A.: Basic Biochemistry, ed. 3,
  New York, 1971, The MacMillan Company.

Table 1

pH Values of Human Body Fluids and Secretions

Body Fluid or Secretion	pH
Blood	7.4
Milk	6.6-6.9
Hepatic Bile	7.4-8.5
Gallbladder Bile	5.4-6.9
Urine (normal)	6.0
Gastric Juice (parietal secretion)	0.87
Pancreatic Juice	8.0
Intestinal Juice	7.7
Cerebrospinal Fluid	7.4
Saliva	7.2
Aqueous Humor of Eye	7.2
Tears	7.4
Urine (range in various disease states)	4.8-7.5
Feces	7.0-7.5

Intracellular pH of a resting muscle cell
(at 37°C and at an extracellular pH of 7.4)   about 7.1

2
Amino Acids and Proteins

Chemistry of the Amino Acids

```
         H    α-Carbon
         |   ↙
     R-C-NH₂
         |
       COOH
```

1. Almost all of the naturally occurring amino acids are α-amino acids.
    In these, an amino group, a carboxyl group, and an R-group are
    all attached to the α-carbon.

2. They are classified in accordance with the structure of the R-group.

3. With the exception of glycine (R=H) and β-alanine (amino group
    attached to the β-carbon, the second carbon from the carboxyl group),
    the amino acids have at least one asymmetric carbon atom (the
    α-carbon) and hence are optically active.

4. Most amino acids found in living systems are L-α-amino acids. The
    L indicates the absolute configuration based on the relationship
    of the amino acid to D and L glyceraldehyde. Diagrams showing this
    are given below. Recall, however, that the α-carbon will be
    tetrahedral so that, actually, the -CHO and -CH₂OH groups of
    glyceraldehyde are going back into the page and the -H and -OH
    groups are coming out of the page.

$$
\begin{array}{cccc}
\text{O} & \text{O} & \text{O} & \text{O} \\
\parallel & \parallel & \parallel & \parallel \\
\text{C-OH} & \text{C-H} & \text{C-H} & \text{C-OH} \\
| & | & | & | \\
\text{H}_2\text{N-C-H} & \text{HO-C-H} & \text{H-C-OH} & \text{H-C-NH}_2 \\
| & | & | & | \\
\text{R} & \text{CH}_2\text{OH} & \text{CH}_2\text{OH} & \text{R}
\end{array}
$$

L-amino acid   L-glyceraldehyde   D-glyceraldehyde   D-amino acid

The designation D or L does not indicate the direction in which a
solution of the compound rotates the plane of polarized light.
This is designated d or l and varies from one amino acid to
another.

12

5. All amino acids isolated from plant and animal protein are
    L-α-amino acids. β-alanine is found in the vitamin pantothenic
    acid. Some D-amino acids are found in polypeptide antibiotics
    such as the gramicidins and bacitracins. D-glutamic acid and
    D-alanine are found in bacterial cell walls. Carnosine (β-alanyl
    histidine) and anserine (β-alanyl-1-methyl histidine) are found
    in skeletal muscles (20-30 millimoles per kg) but their exact
    function is not known.

6. Threonine, cystine, isoleucine, 3- and 4-hydroxyproline, and
    hydroxylysine have <u>two</u> asymmetric carbon atoms and therefore have
    <u>four isomeric forms</u>. General rule: Number of possible isomeric
    forms of any compound equals 2^n where n represents the number of
    asymmetric carbon atoms.

    Example:

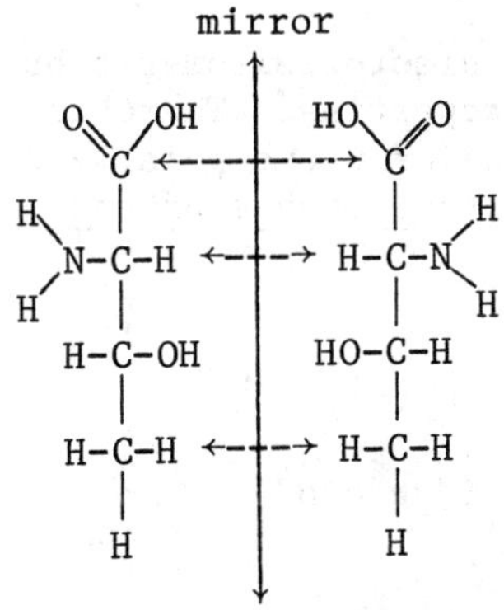

L-threonine   D-threonine

∇

enantiomeric pair

Enantiomeric pairs –
similar chemical properties,
rotate beam of plane polarized
light equally but in opposite
directions.

Note: Identical atoms of each molecule are equal distances on
       either side of the mirror plane, as indicated by the dotted
       arrows in the figure above. The other two isomers are
       designated as L-allothreonine and D-allothreonine
       (Greek-allos=other). The structures of these follow.

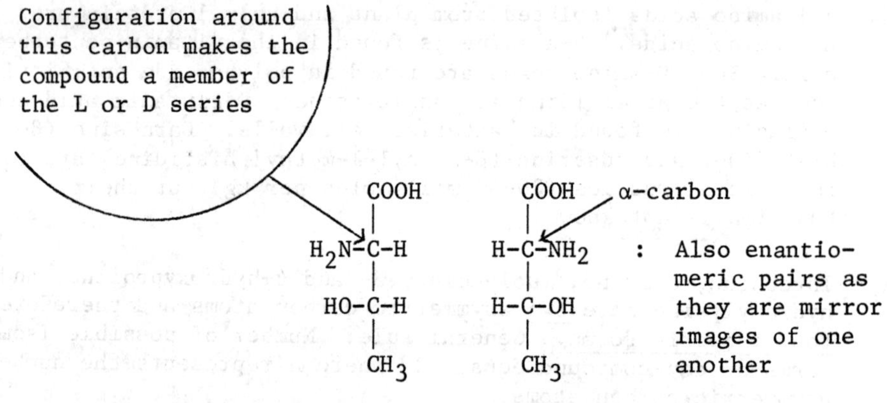

Configuration around
this carbon makes the
compound a member of
the L or D series

```
        COOH              COOH    α-carbon
         |                 |
   H₂N-C-H           H-C-NH₂    :  Also enantio-
         |                 |            meric pairs as
      HO-C-H            H-C-OH          they are mirror
         |                 |            images of one
        CH₃               CH₃           another
```

L-allothreonine   D-allothreonine
                    ▽
           enantiomeric pair

L-allothreonine and L-threonine are a diastereoisomeric pair
(diastereomers) with different chemical properties. There is no
relationship between the degree of rotation of plane polarized
light by these compounds. They are <u>not mirror images of one
another</u>.

7. Amino acids as electrolytes

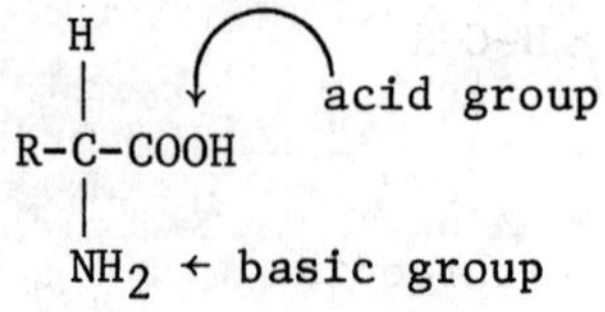

```
        H
        |  ⟶  acid group
   R-C-COOH
        |
       NH₂  ← basic group
```

a. They are ampholytes and contain both acidic and basic groups
    (or proton donating and proton accepting groups).

b. At physiological pH's they exist as dipolar ions or zwitterions

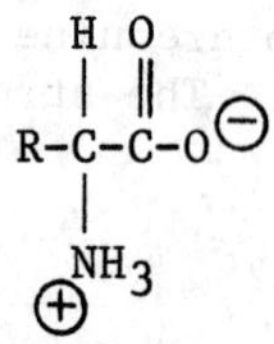

```
        H  O
        |  ||
   R-C-C-O⊖
        |
       NH₃
       ⊕
```

This form is electrically neutral (does not migrate in an electric
field).

In acidic solution (below pH 2) they exist as

$$R-\underset{\underset{\overset{|}{NH_3^+}}{|}}{\overset{\overset{H}{|}}{C}}-\overset{\overset{O}{\parallel}}{C}-OH$$

In basic solution (above pH 9.5) they exist as

$$R-\underset{\underset{\overset{|}{NH_2}}{|}}{\overset{\overset{H}{|}}{C}}-\overset{\overset{O}{\parallel}}{C}-O^-$$

They hardly ever occur as

$$R-\underset{\underset{\overset{|}{NH_2}}{|}}{\overset{\overset{H}{|}}{C}}-\overset{\overset{O}{\parallel}}{C}-OH$$

Isoelectric point - The pH at which a dipolar ion does not migrate in an electric field. (i.e., the pH at which the dipolar ion is the predominant species in solution; also called pI).

Isoionic point - The isoelectric point when the isoelectric pH is determined in water solution only, in the absence of any other solutes. The isoionic point is also defined as the <u>pH at which the number of cations equals the number of anions</u>.

c. Addition of (H^+) and (OH^-) ions: When H^+ is added, the electrically neutral amino acid becomes positively charged (curves below are for glycine; note the directions of the pH scales).

<u>Figure 2</u>. <u>Titration of Glycine with an Acid</u>

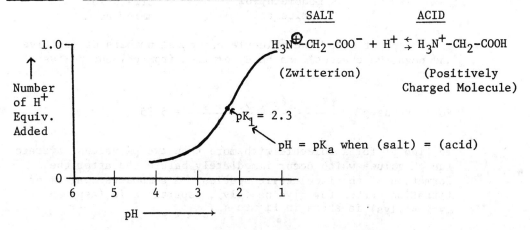

When OH^- is added, the electrically neutral amino acid becomes negatively charged.

Figure 3. Titration of Glycine with a Base

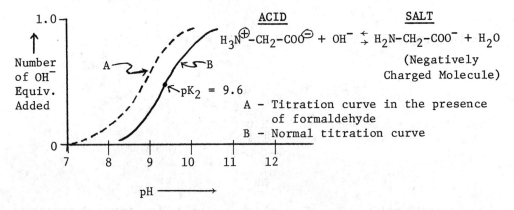

$$H_3\overset{\oplus}{N}-CH_2-COO^{\ominus} + OH^- \underset{\rightarrow}{\leftarrow} H_2N-CH_2-COO^- + H_2O$$

ACID   SALT

(Negatively Charged Molecule)

A - Titration curve in the presence of formaldehyde

B - Normal titration curve

$pK_2 = 9.6$

Number of OH^- Equiv. Added

pH ⟶

Reaction with formaldehyde (HCHO): Formaldehyde combines with the NH_2 groups in neutral or slightly basic solutions and thus alters the pK of the NH_2 group. This is shown by the dashed line in the above graph.

Reaction

$$R-CH-COO^- \underset{-(CH_2O)}{\overset{+(CH_2O)}{\rightleftarrows}} R-CH-COO^- \underset{-(CH_2O)}{\overset{+(CH_2O)}{\rightleftarrows}} R-CH-COO^-$$

NH_2

$HN-CH_2OH$

N HOH_2C CH_2OH

monomethylol amino acid

dimethylol amino acid

pI (the pH at which the molecule has equal numbers of positive and negative charges) can be determined from pK_1 and pK_2 as shown.

For Glycine $pI = \dfrac{pK_1 + pK_2}{2} = \dfrac{2.3 + 9.6}{2} = 5.95$

To find pI for a molecule with more than two pK values, average the pK values which occur immediately before and after the formation of the isoelectric species. A composite diagram of titration curves for glycine (gly), aspartic acid (asp) and lysine (lys) is shown in Figure 4 .

Figure 4. Titration Profiles of Glycine, Lysine and Aspartic Acid*

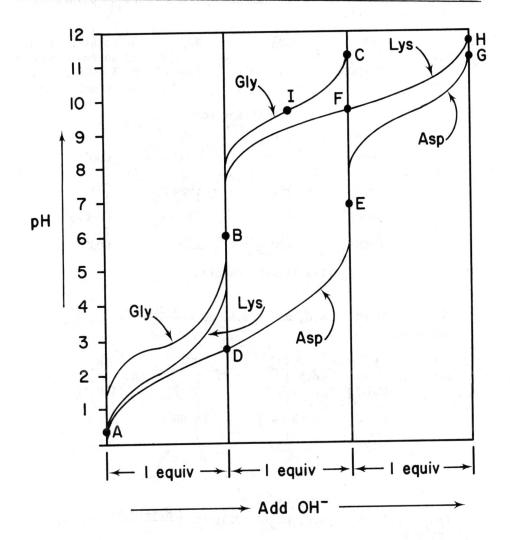

* From Orten, James M. and Neuhaus, Otto W.: Biochemistry, ed. 8,
St. Louis, 1970, The C.V. Mosby Co. used with permission

The following amino acid species are the predominant forms in the solution at the pH's indicated by the lettered points on the above graph.

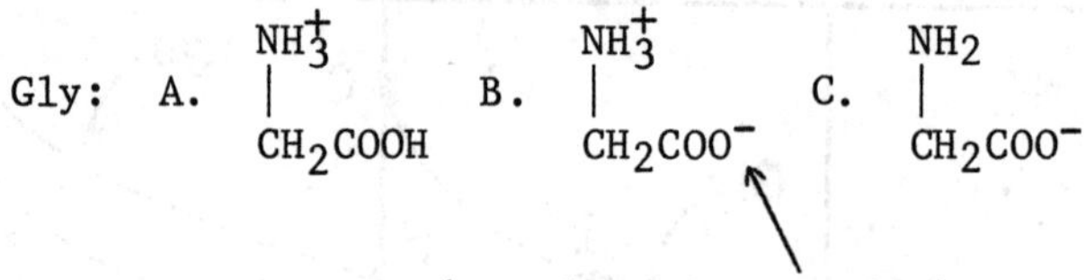

Gly: A. $\underset{\displaystyle CH_2COOH}{\overset{\displaystyle NH_3^+}{|}}$ B. $\underset{\displaystyle CH_2COO^-}{\overset{\displaystyle NH_3^+}{|}}$ C. $\underset{\displaystyle CH_2COO^-}{\overset{\displaystyle NH_2}{|}}$ At point I, there is an equal mixture of B + C; pH = pK_a for $\alpha\text{-}NH_3^+ \leftrightarrows \alpha\text{-}NH_2 + H^+$

(Isoelectric species)

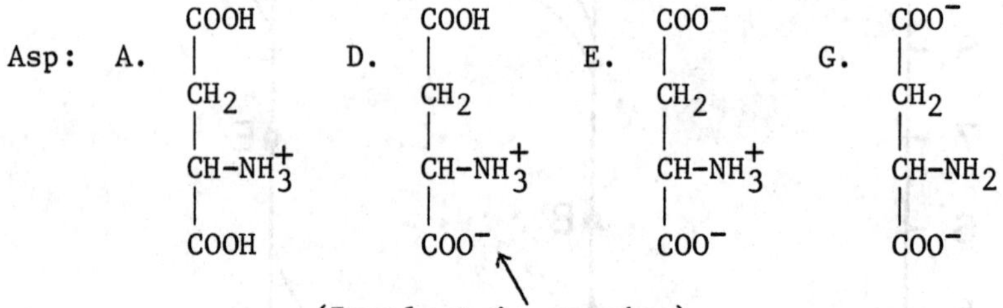

Asp: A. $\begin{matrix} COOH \\ | \\ CH_2 \\ | \\ CH\text{-}NH_3^+ \\ | \\ COOH \end{matrix}$ D. $\begin{matrix} COOH \\ | \\ CH_2 \\ | \\ CH\text{-}NH_3^+ \\ | \\ COO^- \end{matrix}$ E. $\begin{matrix} COO^- \\ | \\ CH_2 \\ | \\ CH\text{-}NH_3^+ \\ | \\ COO^- \end{matrix}$ G. $\begin{matrix} COO^- \\ | \\ CH_2 \\ | \\ CH\text{-}NH_2 \\ | \\ COO^- \end{matrix}$

(Isoelectric species)

(pK_1 = 2.0; pK_2 = 4.0; pK_3 = 9.8; pI = $\dfrac{4.0+2.0}{2}$ = 3.0 for aspartate)

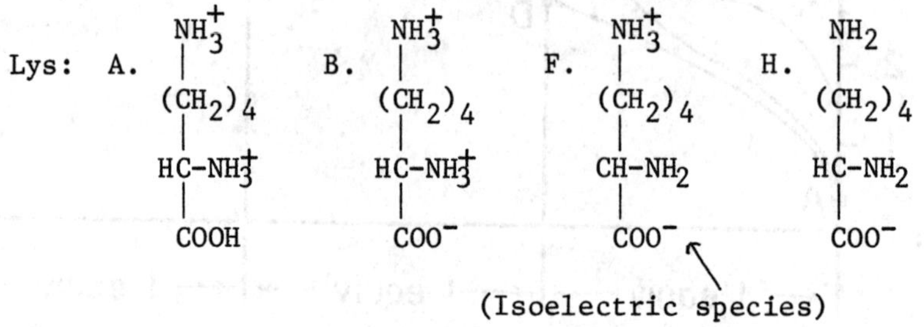

Lys: A. $\begin{matrix} NH_3^+ \\ | \\ (CH_2)_4 \\ | \\ HC\text{-}NH_3^+ \\ | \\ COOH \end{matrix}$ B. $\begin{matrix} NH_3^+ \\ | \\ (CH_2)_4 \\ | \\ HC\text{-}NH_3^+ \\ | \\ COO^- \end{matrix}$ F. $\begin{matrix} NH_3^+ \\ | \\ (CH_2)_4 \\ | \\ CH\text{-}NH_2 \\ | \\ COO^- \end{matrix}$ H. $\begin{matrix} NH_2 \\ | \\ (CH_2)_4 \\ | \\ HC\text{-}NH_2 \\ | \\ COO^- \end{matrix}$

(Isoelectric species)

(pK_1 = 2.2; pK_2 = 8.9; pK_3 = 10.5; pI = $\dfrac{8.9+10.5}{2}$ = 9.7 for lysine)

Structures of Some Amino Acids   (Abbreviations used throughout this text are indicated in parentheses following the names of the amino acids.)

1. Glycine (Gly). Simplest amino acid (not optically active); used in many biosynthetic reactions such as porphyrin and purine biosynthesis.

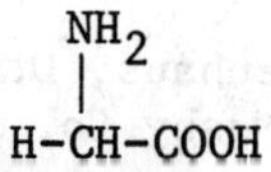

$$\underset{\displaystyle H\text{-}CH\text{-}COOH}{\overset{\displaystyle NH_2}{|}}$$

2. Alanine (Ala). Substrate for GPT (one of the transaminases); other amino acids may be considered derivatives of alanine, with substitutions on the β-carbon. The β-carbon and substitutions attached to it make up the R-group.

$$\beta \searrow \quad \overset{\displaystyle NH_2}{\underset{\displaystyle H}{\overset{\displaystyle |}{\underset{\displaystyle |}{H_3C-C-COOH}}}} \nwarrow \alpha$$

3. Cysteine (Cys). The HS-group is referred to as a sulfhydryl group. It is essential for the activity of many enzymes. Heavy metals inactivate some proteins by combining with HS-groups.

$$\overset{\displaystyle NH_2}{\underset{}{\overset{|}{HS-CH_2-CH-COOH}}}$$

4. Cystine. Oxidized form of cysteine; 2 cysteine $\rightleftarrows$ cystine + 2H; in proteins, this amino acid often links two separate peptide chains together; it is also capable of forming a disulfide bridge within the same peptide chain. This amino acid does not have its own transfer RNA and is not incorporated into a polypeptide during ribosomal protein synthesis. It is, instead, formed after completion of the polypeptide by the oxidation of two cysteine residues. It is responsible for the formation of one kind of kidney stone.

$$\overset{\displaystyle NH_2}{\overset{|}{S-CH_2-CH-COOH}}$$
$$\overset{|}{\underset{|}{S-CH_2-CH-COOH}}$$
$$NH_2$$

5. Methionine (Met). An essential amino acid for humans; responsible for donating methyl groups in biosynthetic reactions involving transmethylation.

$$\overset{\displaystyle NH_2}{\overset{|}{H_3C-S-CH_2-CH_2-CH-COOH}}$$

(Note: 3, 4, and 5 are also referred to as sulfur containing amino acids.)

6. Leucine (Leu). One of the branched chain amino acids; the R-group
    is hydrophobic (water hating) and interacts with other hydrophobic
    substances (including other hydrophobic amino acid R-groups). It
    is an essential amino acid for man.

$$CH_3\!\diagdown\!\!\!\!\!\underset{CH_3\diagup}{CH}\!-\!CH_2\!-\!\overset{\overset{\displaystyle NH_2}{|}}{CH}\!-\!COOH$$

R-Group

7. Isoleucine (Ile). An essential amino acid for man; hydrophobic
    sidechain.

$$CH_3\!\diagdown\!\underset{\underset{CH_3\diagup}{}}{CH_2}\!\diagdown\!CH\!-\!\overset{\overset{\displaystyle NH_2}{|}}{CH}\!-\!COOH$$

8. Valine (Val). An essential amino acid for man; hydrophobic side
    chain.

$$CH_3\!\diagdown\!\!\!\!\underset{CH_3\diagup}{CH}\!-\!\overset{\overset{\displaystyle NH_2}{|}}{CH}\!-\!COOH$$

(Note: 6, 7, and 8 are known as branched chain amino acids with
hydrophobic R-groups. A defect in the metabolism of these three
compounds is the cause of maple syrup urine disease.)

9. Serine (Ser). In phosphoproteins, the phosphate is linked through
    serine; involved in the active center of many enzymes.

$$\underset{\underset{\displaystyle OH}{|}}{CH_2}\!-\!\overset{\overset{\displaystyle NH_2}{|}}{CH}\!-\!COOH$$

10. Threonine (Thr). An essential amino acid for man.

$$H_3C\!-\!\underset{\underset{\displaystyle OH}{|}}{CH}\!-\!\overset{\overset{\displaystyle NH_2}{|}}{CH}\!-\!COOH$$

(Note: 9 and 10 are hydroxyl group-containing amino acids.)

11. Phenylalanine (Phe). An essential amino acid in man; its metabolism is defective in phenylketonuria; hydrophobic sidechain.

$$\text{C}_6\text{H}_5-\text{CH}_2-\overset{\displaystyle \overset{\textstyle NH_2}{|}}{\text{CH}}-\text{COOH}$$

12. Tyrosine (Tyr). Accumulates in tissues in tyrosinosis; analysis for the phenol ring is used to quantitate proteins by Folin's method. It is involved in the synthesis of thyroxine, catecholamines, and melanin.

$$\text{HO}-\text{C}_6\text{H}_4-\text{CH}_2-\overset{\displaystyle \overset{\textstyle NH_2}{|}}{\text{CH}}-\text{COOH}$$

13. Tryptophan (Trp). Essential amino acid in man; indole ring system is involved in the formation of serotonin; metabolites of tryptophan are involved in <u>carcinoid disease</u> (an epithelial growth resembling cancer).

$$\text{indole}-\text{CH}_2-\overset{\displaystyle \overset{\textstyle NH_2}{|}}{\text{CH}}-\text{COOH}$$

(Note: 11, 12, and 13 are known as aromatic amino acids.)

14. Aspartic Acid (Asp). Substrate for GOT (a transaminase enzyme of great clinical significance).

$$\text{HOOC}-\text{CH}_2-\overset{\displaystyle \overset{\textstyle NH_2}{|}}{\text{CH}}-\text{COOH}$$

15. Glutamic Acid (Glu). Substrate for both GOT and GPT transaminases.

$$\text{HOOC}-\text{CH}_2-\text{CH}_2-\overset{\displaystyle \overset{\textstyle NH_2}{|}}{\text{CH}}-\text{COOH}$$

Note: The above amino acids (14 and 15) are known as acidic amino
acids and are often present in proteins as the corresponding
amides, asparagine and glutamine.

$$\overset{O}{\overset{\|}{(H_2N)-C}}-CH_2-\overset{NH_2}{\overset{|}{CH}}-COOH \qquad \overset{O}{\overset{\|}{(H_2N)-C}}-CH_2-CH_2-\overset{NH_2}{\overset{|}{CH}}-COOH$$

Asparagine (Asn)   Glutamine (Gln)

Note: Asn and Gln have their own specific transfer RNA carriers.

16. Arginine (Arg). Involved in urea synthesis; has a guanido group.

$$\overset{H}{\overset{|}{H_2N-C-N}}-(CH_2)_3-\overset{NH_2}{\overset{|}{CH}}-COOH$$
$$\overset{\|}{NH}\leftarrow$$
Guanido group

17. Lysine (Lys).

$$H_2N-\underset{\varepsilon}{\overset{H_2}{\overset{|}{C}}}-\underset{\delta}{\overset{H_2}{\overset{|}{C}}}-\underset{\gamma}{\overset{H_2}{\overset{|}{C}}}-\underset{\beta}{\overset{H_2}{\overset{|}{C}}}-\underset{\alpha}{\overset{NH_2}{\overset{|}{CH}}}-COOH$$

The end NH_2 group is referred to as the epsilon (ε) NH_2 group, since
it is attached to the fifth carbon (ε) from the carboxyl group; an
essential amino acid in man. δ-Hydroxylysine is found in collagen
and it is formed postribosomally.

18. Histidine (His). Has an imidazole ring attached to the β-carbon
of alanine; the decarboxylated product is histamine, which is
produced in mast cells. They are found around blood vessels of
the tissues, in the pleural membrane of the lung, and in the mucosa
of the stomach. Histamine constricts bronchial smooth muscles,
increases gastric secretion rich in hydrochloric acid, and
dilates capillaries (the latter effect causing edema and drop in
blood pressure). The histidine residue in the hemoglobin molecule
functions as a proton acceptor in a reversible manner.

$$\text{HC} \Large=\normalsize \overset{\displaystyle \text{NH}_2}{\underset{}{}} \text{C-CH}_2\text{-CH-COOH}$$

(Note: 16, 17, and 18 are basic amino acids.)

19. Proline (Pro). Present in collagen; a heterocyclic amino acid
containing a pyrrolidine ring.

Proline (and hydroxyproline) residues act as turning or disrupting
points in the formation of α-helix in a protein molecule. However,
the proline ring does not prevent other types of helical
structures. Also note--proline, when present as a part of a
peptide, has no additional H on the N to participate in hydrogen
bond formation.

20. 4-Hydroxyproline. Present in collagen. The isomeric compound
3-hydroxyproline is also found in collagen. Hydroxylation
reactions of some of the proline residues present in collagen
are postribosomal modifications.

(Note: 19 and 20 ⎯NH replaces -NH₂ group, hence the name *imino*
acids. It is actually more similar to a secondary amine than to
an imine, however.)

Naturally Occurring Amino Acids Which Do Not Occur in Proteins

1. β-Alanine. Part of pantotheine and of coenzyme A.

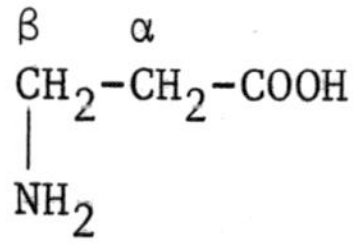

$$\overset{\beta}{CH_2}-\overset{\alpha}{CH_2}-COOH$$
$$|$$
$$NH_2$$

2. Taurine. Comes from metabolism of the S-containing amino acids; conjugated with bile acids in the liver.

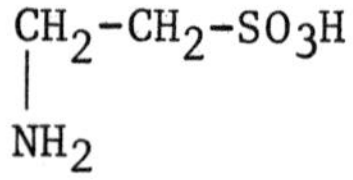

$$CH_2-CH_2-SO_3H$$
$$|$$
$$NH_2$$

3. α-Aminobutyric Acid.

$$NH_2$$
$$|$$
$$H_3C-CH_2-CH-COOH$$

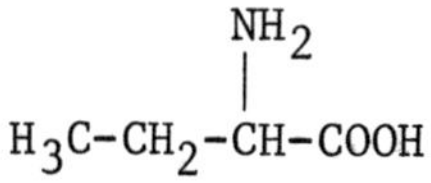

4. γ-Aminobutyric Acid. Present in brain tissue, may be a synaptic transmitter. (see page 428)

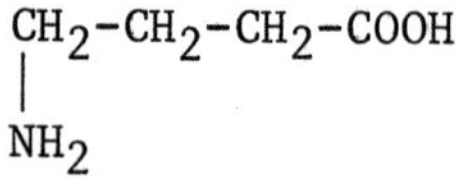

$$CH_2-CH_2-CH_2-COOH$$
$$|$$
$$NH_2$$

5. β-Aminoisobutyric Acid. End-product in pyrimidine metabolism; found in urine of patients with an inherited error in this metabolism.

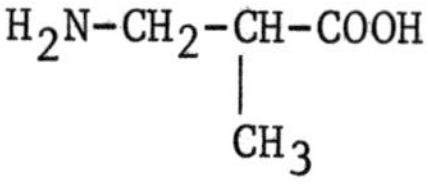

$$H_2N-CH_2-CH-COOH$$
$$|$$
$$CH_3$$

6. Homocysteine. Involved in methionine biosynthesis. Demethylated methionine.

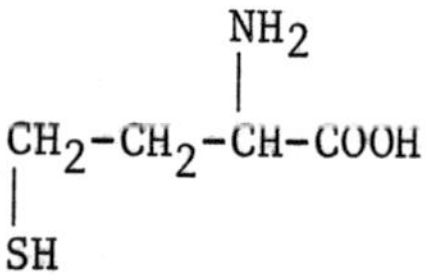

$$NH_2$$
$$|$$
$$CH_2-CH_2-CH-COOH$$
$$|$$
$$SH$$

7. Homoserine. Involved in threonine, aspartate, and methionine metabolism.

$$\underset{\overset{|}{\text{HOCH}_2-\text{CH}_2-\text{CH}-\text{COOH}}}{\overset{\text{NH}_2}{}}$$

8. Cysteinesulfinic Acid. Present in rat brain tissue.

$$\begin{array}{c} \text{NH}_2 \\ | \\ \text{CH}_2-\text{CH}-\text{COOH} \\ | \\ \text{SO}_2\text{H} \end{array}$$

9. Cysteic Acid. Present in wool.

$$\begin{array}{c} \text{NH}_2 \\ | \\ \text{CH}_2-\text{CH}-\text{COOH} \\ | \\ \text{SO}_3\text{H} \end{array}$$

10. Ornithine. A urea cycle intermediate.

$$\begin{array}{c} \text{NH}_2 \\ | \\ \text{CH}_2-\text{CH}_2-\text{CH}_2-\text{CH}-\text{COOH} \\ | \\ \text{NH}_2 \end{array}$$

11. Citrulline. A urea cycle intermediate.

$$\begin{array}{c} \text{H}_2\text{C}-\text{CH}_2-\text{CH}_2-\text{CH}-\text{COOH} \\ | \qquad\qquad\qquad | \\ \text{HN} \qquad\qquad \text{NH}_2 \\ | \\ \text{C}=0 \\ | \\ \text{NH}_2 \end{array}$$

12. Homocitrulline. Present in urine of normal children.

$$H_2C-CH_2-CH_2-CH_2-CH-COOH$$

with HN and NH_2 substituents, HN connected to C=O, which connects to NH_2

13. 5-Hydroxytryptophan. Decarboxylated product is serotonin, or 5-hydroxytryptamine. It is present in the central nervous system as well as in the intestinal mucosa.

Structure: HO-substituted indole ring with $-CH_2-CH(NH_2)-COOH$ side chain; ring nitrogen bears H.

14. Monoiodotyrosine. Present in thyroid tissue and blood serum.

Structure: benzene ring with I substituent and HO substituent, $-CH_2-CH(NH_2)-COOH$ side chain.

15. 3,5-Diiodotyrosine. Found in association with thyroid globulin.

Structure: benzene ring with two I substituents and HO substituent, $-CH_2-CH(NH_2)-COOH$ side chain.

16. 3,5,3'-Triiodothyronine. Designated T_3; thyroid hormone present in thyroid tissue (structure at top of next page).

17. Thyroxine (3,5,3',5'-Tetraiodothyronine). Designated T_4; present in blood, converted to T_3 (#16 above) at the tissue level; found in association with thyroid globulin. T_3 and T_4 are thyroid hormones.

18. Azaserine. A potent inhibitor of tumor growth. It is not a naturally occurring amino acid.

$$N \equiv N = CH-C-O-CH_2-CH-COOH$$

with $\| \atop O$ below the C, and NH_2 below the CH.

Chemical Properties of Amino Acids Due to Carboxyl Group

1. Salt formation and titration (discussed above under electrolytes).

2. Formation of esters and amides

3. Formation of aminoacyl chlorides

4. Decarboxylation

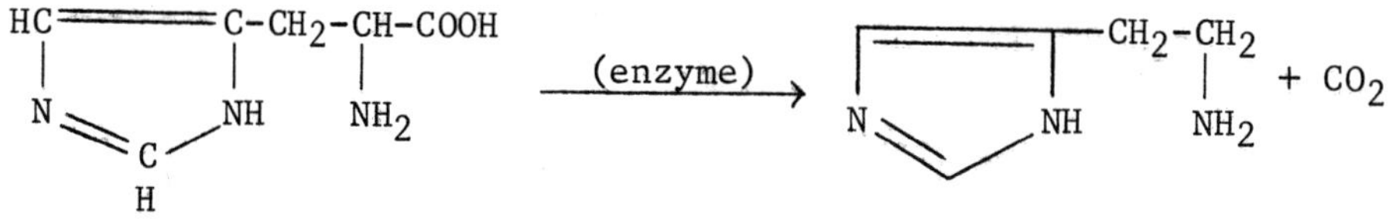

$$HC \!\!=\!\! C\text{-}CH_2\text{-}CH\text{-}COOH \quad \xrightarrow{\text{(enzyme)}} \quad CH_2\text{-}CH_2 \;+\; CO_2$$

Histidine   Histamine

Chemical Properties Due to NH_2 Group

1. Acylation

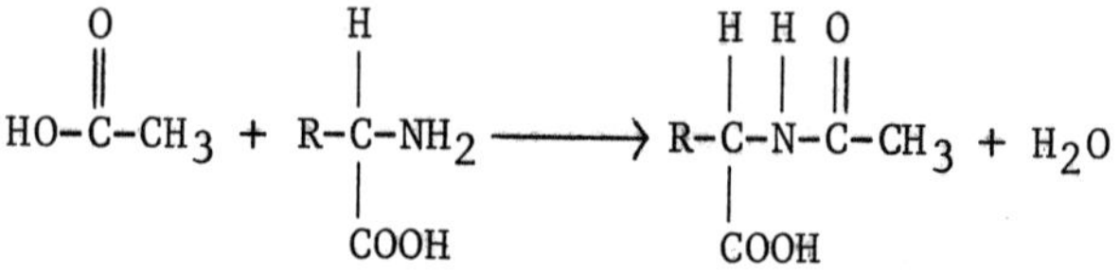

$$HO\text{-}\overset{O}{\overset{\|}{C}}\text{-}CH_3 \;+\; R\text{-}\underset{COOH}{\overset{H}{\underset{|}{\overset{|}{C}}}}\text{-}NH_2 \longrightarrow R\text{-}\underset{COOH}{\overset{H}{\underset{|}{\overset{|}{C}}}}\text{-}\overset{H}{\overset{|}{N}}\text{-}\overset{O}{\overset{\|}{C}}\text{-}CH_3 \;+\; H_2O$$

2. Benzoylation

$$\bigcirc\!\!-COOH \;+\; H_2N\text{-}CH_2\text{-}COOH \xrightarrow{-H_2O} \bigcirc\!\!-\overset{O}{\overset{\|}{C}}\text{-}\overset{H}{\overset{|}{N}}\text{-}CH\text{-}COOH$$

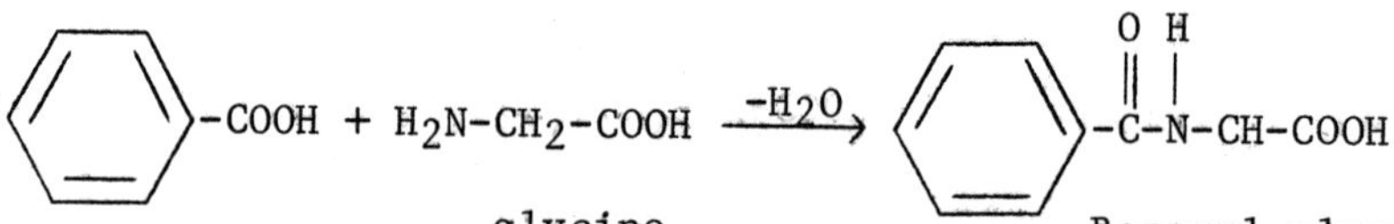

                  glycine   Benzoyl glycine
                                             (Hippuric acid)

3. Methylation

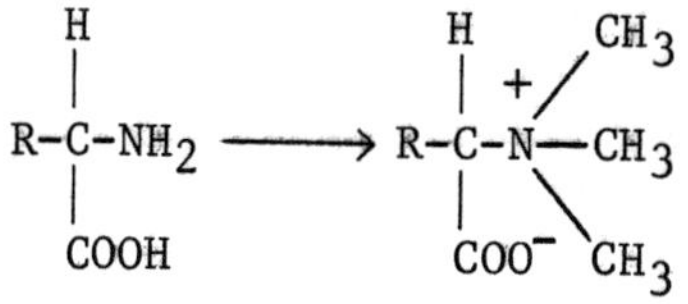

$$R\text{-}\underset{COOH}{\overset{H}{\underset{|}{\overset{|}{C}}}}\text{-}NH_2 \longrightarrow R\text{-}\underset{COO^-}{\overset{H}{\underset{|}{\overset{|}{C}}}}\text{-}\overset{+}{N}\!\!\underset{CH_3}{\overset{CH_3}{<}}$$

Betaine of an amino acid

4. Reaction with Sanger's Reagent (1-fluoro-2,4-dinitrobenzene, FDNB) reacts with free amino end groups to form dinitrophenylamino acids (DNP a.a.) which are yellow. This reaction is used to identify terminal amino acids (see top of next page).

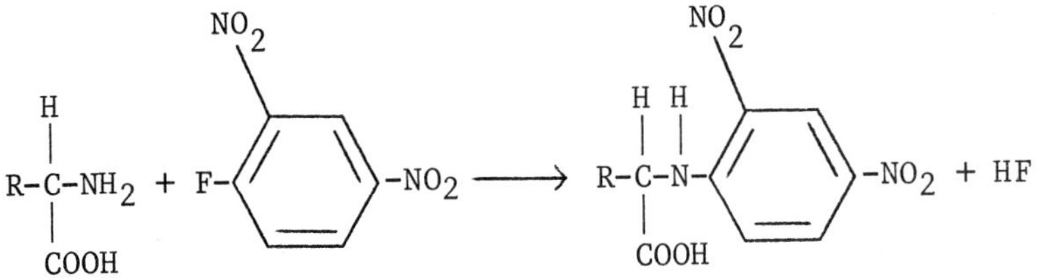

$$R-\overset{\overset{\displaystyle H}{|}}{\underset{\underset{\displaystyle COOH}{|}}{C}}-NH_2 \ + \ F-\langle\overset{NO_2}{\bigcirc}\rangle-NO_2 \ \longrightarrow \ R-\overset{\overset{\displaystyle H}{|}}{\underset{\underset{\displaystyle COOH}{|}}{C}}-\overset{\overset{\displaystyle H}{|}}{N}-\langle\overset{NO_2}{\bigcirc}\rangle-NO_2 \ + \ HF$$

DNP - amino acid

5. Reaction with nitrous acid (HNO_2). This reaction is the basis of the amino nitrogen method of Van Slyke.

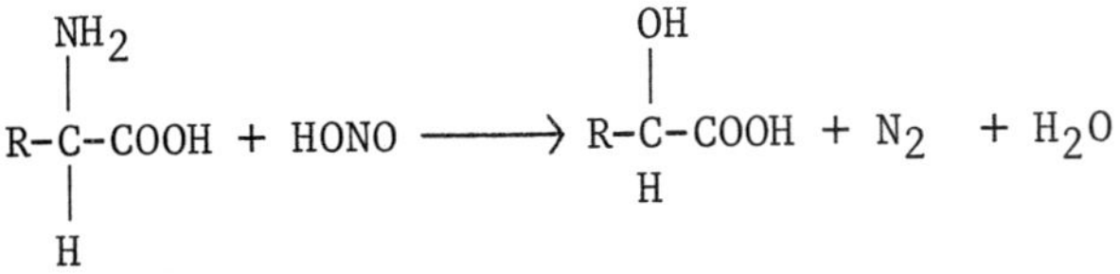

$$R-\overset{\overset{\displaystyle NH_2}{|}}{\underset{\underset{\displaystyle H}{|}}{C}}-COOH \ + \ HONO \ \longrightarrow \ R-\overset{\overset{\displaystyle OH}{|}}{\underset{\underset{\displaystyle H}{|}}{C}}-COOH \ + \ N_2 \ + \ H_2O$$

6. Oxidative Deamination

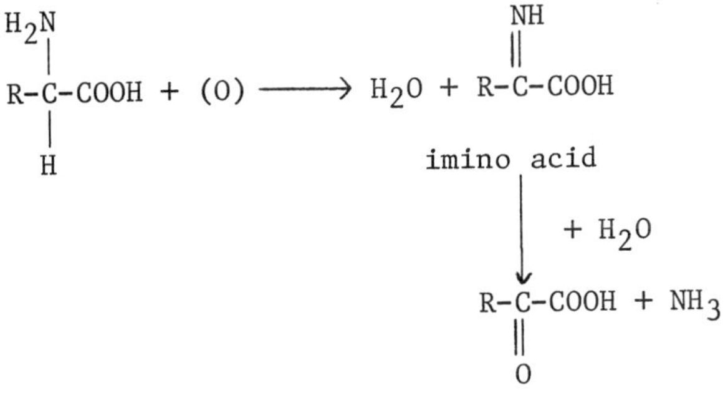

$$R-\overset{\overset{\displaystyle H_2N}{|}}{\underset{\underset{\displaystyle H}{|}}{C}}-COOH \ + \ (O) \ \longrightarrow \ H_2O \ + \ R-\overset{\overset{\displaystyle NH}{\|}}{C}-COOH$$

imino acid

$$\Big\downarrow \ + \ H_2O$$

$$R-\overset{\overset{\displaystyle}{}}{\underset{\underset{\displaystyle O}{\|}}{C}}-COOH \ + \ NH_3$$

α-keto acid

7. Reaction with Formaldehyde (see also page 16)

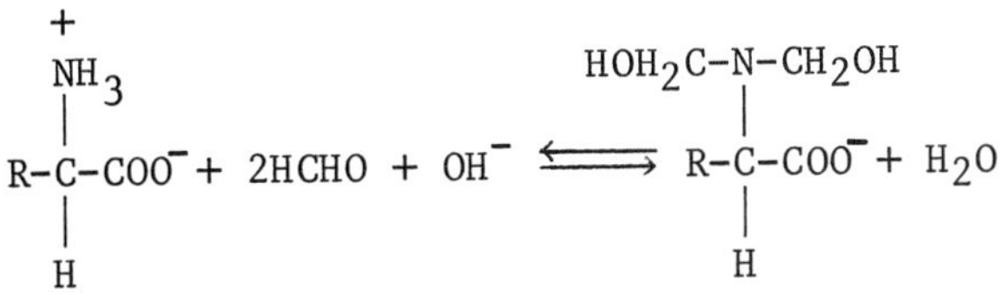

$$R-\overset{\overset{\displaystyle \overset{+}{N}H_3}{|}}{\underset{\underset{\displaystyle H}{|}}{C}}-COO^- \ + \ 2HCHO \ + \ OH^- \ \xleftrightharpoons \ R-\overset{\overset{\displaystyle HOH_2C-N-CH_2OH}{|}}{\underset{\underset{\displaystyle H}{|}}{C}}-COO^- \ + \ H_2O$$

dimethylol amino acid

8. Reaction with aromatic aldehydes in the presence of alkali (Schiff base formation)

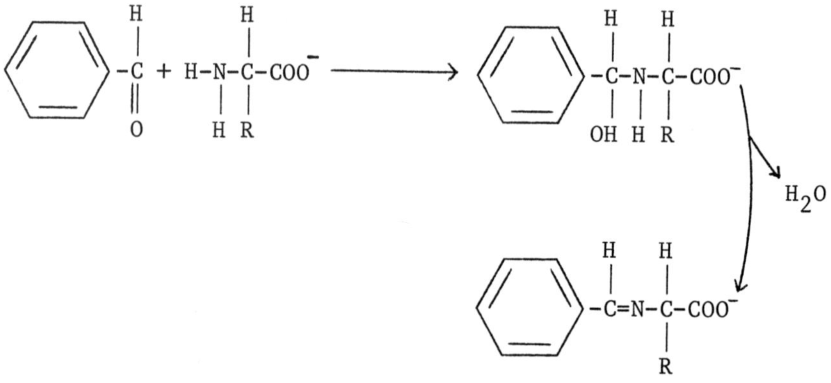

Schiff base

9. Reaction with Ninhydrin. (Specific for free carboxyl groups adjacent to $-NH_2$ or $>NH$ groups and hence, for α-amino and imino acids.)

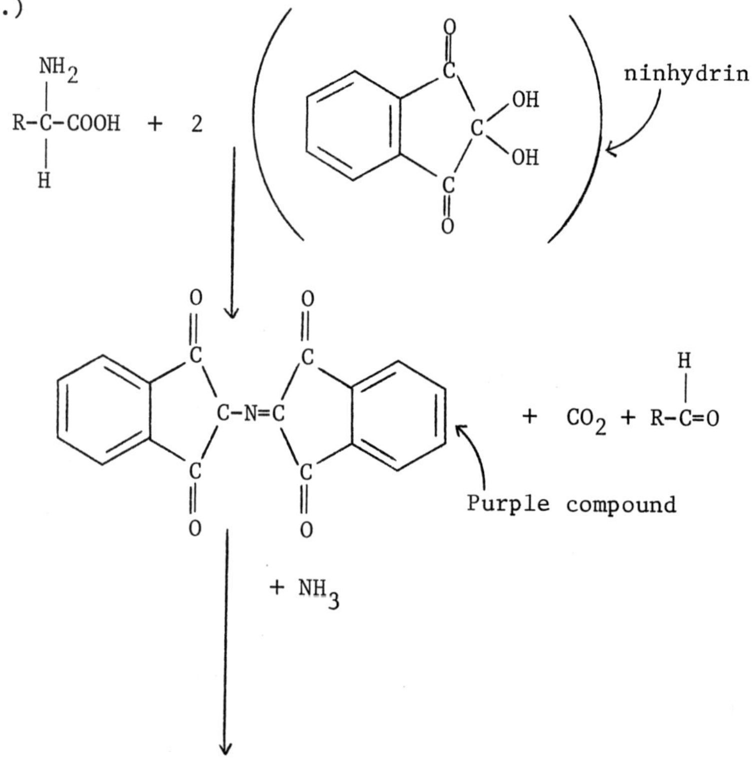

Purple compound

$+ NH_3$

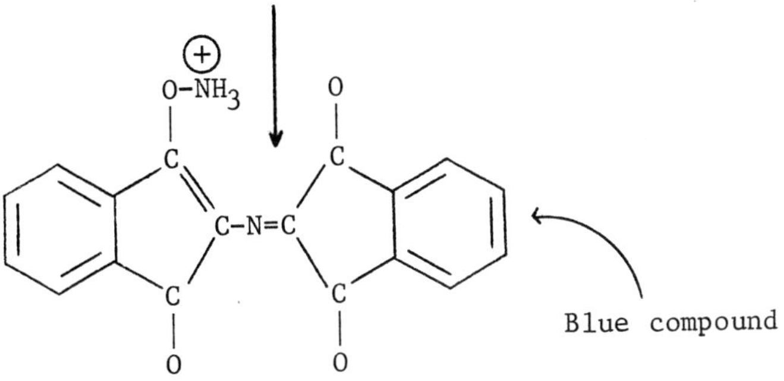

Blue compound

Blue or purple colors are obtained for all amino acids except proline and hydroxyproline, which give a yellow color. This is the reaction used to determine a.a.'s in automatic amino acid analyzers.

10. Reaction with CO_2 to form a carbamino group

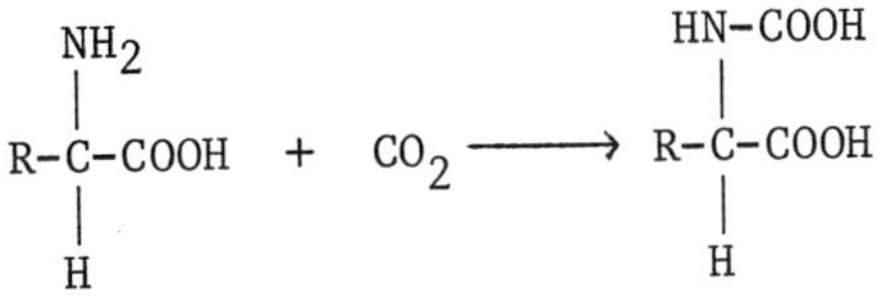

11. Chelation of amino acids with metal ions

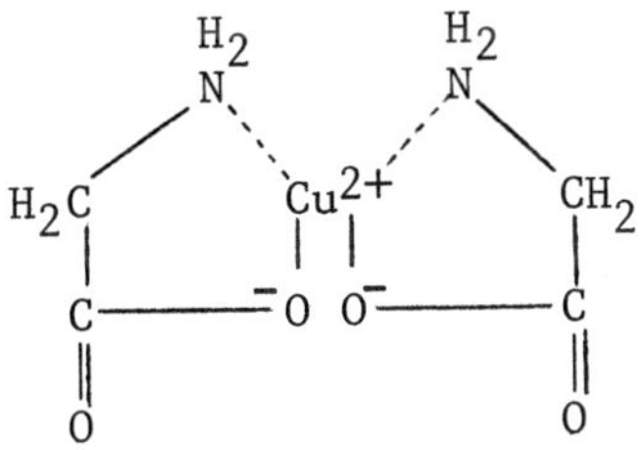

Copper diglycinate

Chelates are nonionic. Amino acids and other chelate formers (e.g., EDTA, penicillamine) form soluble metal complexes. Cu^{2+} chelates of peptide bonds are the basis of the biuret measurement of protein. This is the best general method for protein but it is not as sensitive as other techniques.

12. Peptide Linkage. This is the most common linkage which attaches one amino acid to another and is the result of a dehydration reaction between the α-amino group of one a.a. and the carboxyl group of another (see top of next page).

$$H_2N - \overset{\overset{\displaystyle H}{|}}{\underset{\underset{\displaystyle R_1}{|}}{C}} - COOH \qquad + \qquad H_2N - \overset{\overset{\displaystyle H}{|}}{\underset{\underset{\displaystyle R_2}{|}}{C}} - COOH$$

$$\longrightarrow H_2O$$

$$H_2N - \overset{\overset{\displaystyle H}{|}}{\underset{\underset{\displaystyle R_1}{|}}{C}} - \overset{\overset{\displaystyle O}{\|}}{C} - \overset{\overset{\displaystyle H}{|}}{N} - \overset{\overset{\displaystyle H}{|}}{\underset{\underset{\displaystyle R_2}{|}}{C}} - COOH$$

peptide
bond

Amino Acid Analysis, Sequence Determination and Chemical Synthesis of Proteins

1. In order to understand the structure and function of proteins, it is essential to know

    a. the number and kinds of amino acids present in the protein, and
    b. the order in which the amino acids are connected (called sequence or primary structure).

    This information is needed, for example, in assessing the effect of mutations on the amino acid sequence; understanding the mechanisms of enzyme-catalyzed reactions; synthesis of peptides to obtain specific therapeutic effects. In the last example, the use of synthetic, species-specific peptides may eliminate undesirable hypersensitivity reactions.

2. The first protein sequenced was bovine insulin (M.W. 6,000) in 1955 by Sanger. This molecule is composed of two peptide chains of 21 and 30 amino acids each, linked by two interchain disulfide bonds. This work was significant not only for the methodology which it established,but also because it demonstrated that proteins are characterized by unique primary structures.

3. In 1960, Hirs, Moore, and Stein, and Anfinsen reported the first primary structure of an enzyme. The enzyme was ribonuclease (M.W. 13,700) containing a single peptide chain of 124 amino acid residues with four intrachain disulfide bonds. These investigators

established many of the procedures which are currently used in
sequence analysis. These include the use of ion exchange resins
for the separation of peptides and amino acids, and their
quantitation by the ninhydrin reaction.

4. A general approach to the determination of primary protein structure
is shown below.

    a. Determine amino acid composition of <u>highly purified</u> protein.

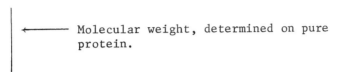

                    Molecular weight, determined on pure
                    protein.

    b. Use composition and molecular weight to determine the number of
residues of each amino acid present per protein molecule to
the nearest whole number.

                    Reduce all disulfide bonds to sulfhydryl
                    groups.

    c. Determine N- and C-terminal amino acids. A unique residue for
each terminus suggests that the native protein contains only
one peptide chain. (If, for example, two different amino acids
are found in equimolar quantities when the pure protein is
treated with carboxypeptidase, this would imply at least two
different chains, present in equimolar amounts, each having a
different C-terminal amino acid. In this case, one might
expect to find two amino terminal residues also.)

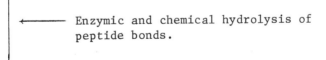

                    Enzymic and chemical hydrolysis of
                    peptide bonds.

    d. <u>Different</u>, preferably overlapping sets of smaller peptides are
obtained next. Each of these is sequenced and, by looking at
the overlapping parts, the sequence of the original protein is
assembled (see example next). Most of the actual sequencing
is done by the Edman reaction, a procedure which results in the
stepwise removal of amino acids from the amino terminus of a
peptide.

Example - overlapping peptide mapping

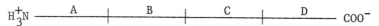

Cleavage of a protein gave four peptides: A, B, C, and D. When
these were separated and sequenced, it was not known whether the
correct order was ABCD or ACBD. A second hydrolysis (by a
different method) was done giving three peptides

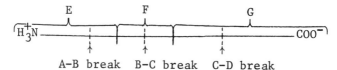

A-B break   B-C break   C-D break

By looking at the sequence of fragments E, F, and G containing,
respectively, the bonds which were cleaved in the first hydrolysis,
it was deduced that the correct order was ABCD. This was the only
way in which the proper sequence could be maintained in going from
A to B, B to C, and C to D.

5. Determination of Amino Acid Composition. The a.a. composition of
a protein is the number of each of the 20 common amino acids
present per mole of the protein. This is obtained by the following
procedure:

   a. The protein is completely hydrolyzed using acid (6 N HCl,
   24 hours or longer at 110°C) into a mixture of its constituent
   amino acids and the resultant hydrolysate is evaporated to
   dryness. The lability of some amino acids (notably tryptophan,
   serine, and threonine) requires alkaline hydrolysis for the
   determination of these compounds (performed on a separate
   sample of the protein).

   b. The hydrolysate is dissolved in a small volume of an acidic
   buffer to obtain the protonated form of the amino acids. It
   is then placed on the top of a cation exchange resin column
   (Dowex50). The resin is an insoluble synthetic polymer
   containing $-SO_3^-$ Na^+ groups. It can bind to a protonated amino
   acid with the release (in this case) of NaCl.

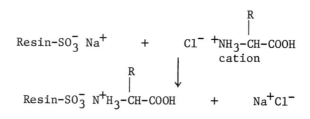

Thus all of the (cationic) amino acids bind to the resin rapidly and stay at the top of the column. The intensity of binding to the resin is dependent upon many factors. These include the number of positive charges on the molecule, the nature and size of the R-group (affecting the adsorption of the amino acids onto resin) and the pK_a of the group involved in the interaction. For example, basic amino acids (lysine, histidine, and arginine) all have more than one positive charge and, hence, bind more tightly than the other amino acids. On the other hand, acidic amino acids are bound less tightly.

c. In addition to the ionic binding to the $-SO_3^-$ groups on the resin, the amino acids are held by hydrophobic and van der Waals interactions. The nature of these bonds helps to determine the resin adsorption characteristics of amino acids having different sidechains. This means that, depending on the R-group, various amino acids bind with different strengths and are released at different times during the elution.

d. The amino acids (cations) from the column are differentially eluted by using a stepwise pH gradient varying between pH 3 and pH 5.

e. As each amino acid appears in the effluent volume, it is reacted with ninhydrin and the color intensity is measured spectrophotometrically to detect and quantitate the amino acid.

f. Using ribonuclease hydrolysate, the amino acids are eluted from the column in the order: Asp, Thr, Ser, Pro, Glu, Gly, Ala, Val, Cys, Met, Ile, Leu, Tyr, Phe, ammonia (produced in the hydrolysis of glutamine and asparagine, each of which have one acid amide $-CONH_2$ group), Lys, His, and Arg.

g. The process of separation and quantitation of amino acid has been automated. In this method, the acidic and neutral amino acids are separated on a "long" column and the basic amino acids on a "short" column. On the short column, the acidic and neutral amino acids emerge unseparated prior to the elution of the basic components. (Recall that the basic amino acids are more tightly bound than the rest of the amino acids.) The analyzer is capable of detecting as little as 10 nanomoles of an amino acid and a complete analysis can be obtained in about four hours.

6. Determination of Sequence

    a. Determination of the terminal residues. The terminal residues
        may be identified by both chemical and enzymatic methods.
        The C-terminal residue can be determined by hydrazinolysis,
        a method which liberates the hydrazide derivatives of all
        amino acids whose carboxyl group is involved in a peptide
        linkage. The C-terminal residue is liberated as the free
        amino acid. The carboxypeptidases are enzymes which digest
        proteins from the carboxyl terminus and are often useful for
        determining this residue. The amino terminal residue is
        readily identified by reaction of the α-amino group with
        reagents such as fluorodinitrobenzene, dansyl chloride or
        phenylisothiocyanate. The derivatized amino acid is isolated
        by hydrolysis and may be characterized by thin layer
        chromatography. Enzymatically, leucine amino peptidase, an
        exopeptidase which attacks proteins from the amino terminus,
        is used in determining the N-terminal amino acid.

    b. Cleavage of Disulfide Bonds. The disulfide bonds of a protein
        may be cleaved oxidatively or reduced and alkylated. Treatment
        of the native protein with performic acid, a powerful
        oxidizing agent, breaks disulfide bonds and converts cysteine
        residues to cysteic acid. Reduction of the disulfide linkage
        by thiols such as β-mercaptoethanol yields reactive sulfhydryl
        groups. These may be stabilized by alkylation with iodoacetate
        or ethyleneimine, yielding the carboxymethyl- or aminoethyl-
        derivatives respectively.

    c. Hydrolysis of Peptide Bonds. Specific hydrolysis of a protein
        is important in terms of reproducibility of an experiment and
        yield of the resulting peptides. This may be achieved both
        chemically and enzymatically. Reagents such as N-bromo-
        succinimide and cyanogen bromide will hydrolyze proteins at
        tryptophan and methionine residues respectively. The most
        specific protease is trypsin, which hydrolyzes the peptide
        linkages following the basic residues lysine and arginine.
        The purification of the products of hydrolysis is often the
        most challenging aspect of sequence determination. Both anion
        and cation exchange resins are used extensively, due to the
        ionic nature of peptides. Preparative paper chromatography and
        electrophoresis are also useful. The purified peptides are
        analyzed for amino acid composition and terminal residues.
        Small peptides may be sequenced directly but large peptides
        must be further hydrolyzed. Proteases such as chymotrypsin,
        pepsin, and papain, which are much less specific than trypsin,
        are often utilized.

d. The sequences of purified peptides and proteins are determined by reaction with phenylisothiocyanate (Edman's reagent). This compound couples with the free alpha amino group of the N-terminal amino acid, giving the phenylthiocarbamyl (PTC) derivative of the peptide or protein. This is then cleaved to a peptide or protein with one less residue, and the thiazolinone of the N-terminal amino acid. This rearranges to the phenylthiohydantoin (PTH) derivative which is extracted from the reaction mixture, converted to a volatile derivative, and identified by gas chromatography. These reactions are illustrated below, using a tripeptide.

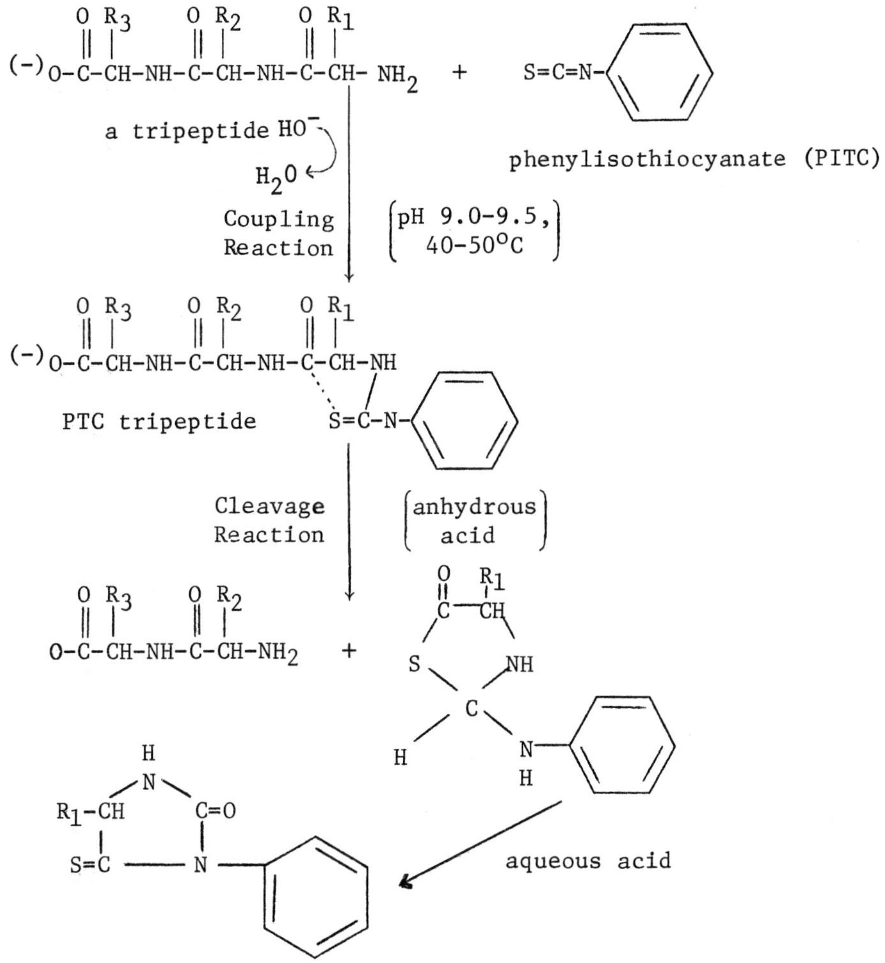

phenylthiohydantoin (PTH) of the amino acid

e. The PTH derivative may also be converted to the dansyl
derivative which is highly fluorescent and can be detected
with greater sensitivity. Alternatively, the amino acid
composition of the remaining peptide may be determined and the
extracted N-terminal residue surmised by the difference from
the original composition (subtractive analysis). The recovered
peptide may be coupled again with the Edman reagent and the
entire process repeated to identify the penultimate residue
of the original peptide. The Edman degradation can be
continued for about ten steps, depending on the properties of
the particular peptide.

f. The utility of the Edman reaction has led to the development
of the protein sequenator. This instrument is programmed to
conduct these reactions continuously, delivering the PTH
derivative at a rate of about one every 90 minutes. These
derivatives are analyzed by gas chromatography. The sequenator
has been used to verify the first 60 residues from the amino
terminus of whale myoglobin. Currently, this instrument is
useful only for proteins and large peptides; small peptides
must be sequenced manually.

g. Mass spectrometry provides another approach to sequence
determination. Although not widely used currently, the high
sensitivity and inherent simplicity of this technique offers
much promise. The method consists of bombarding peptides with
electrons causing fragmentation at the peptide linkages and
yielding a mixture of various pieces of the peptide. The
fragments are separated on the basis of their mass to charge
ratio and the original sequence is deduced by locating the
products of sequential fragmentation. Thus peaks would be
observed for the whole peptide, the peptide minus the terminal
amino acid, the peptide minus the terminal and penultimate
amino acids, etc. The application of this technique to
mixtures of peptides, with on line computer analysis of the
spectrum, offers the potential for simplification of the
entire sequencing procedure.

7. The Chemical Synthesis of Proteins

a. In Merrifield Solid Phase Synthesis, the carboxyl group of the
C-terminal amino acid of the peptide to be synthesized is
covalently linked to a resin.

$$H_2N - C(R_1)(H) - C(=O) - O^- Na^+ \quad + \quad Cl - C(H)(H) - \text{resin} \longrightarrow$$

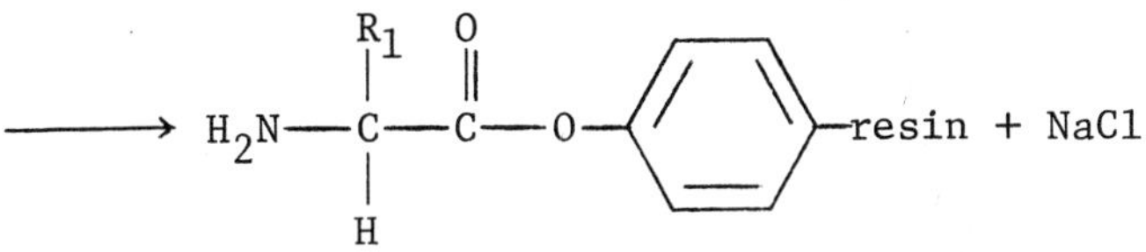

$$\longrightarrow \quad H_2N-\underset{\underset{H}{|}}{\overset{\overset{R_1}{|}}{C}}-\overset{\overset{O}{\|}}{C}-O-\bigcirc-\text{resin} + NaCl$$

b. The amino group of the next amino acid to be added is protected with a t-BOC group and the carboxyl group is then coupled to the amino acid bound to the resin.

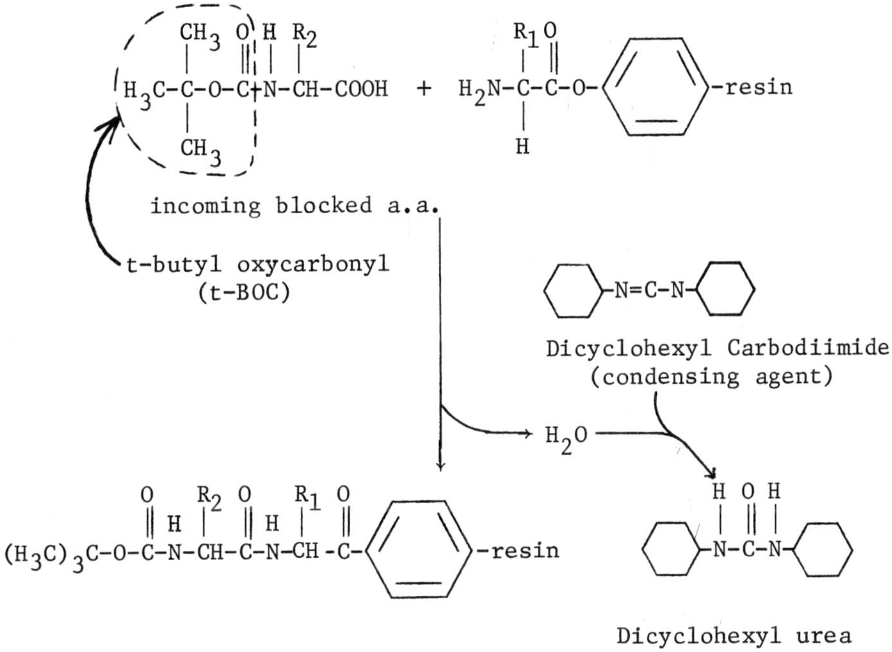

incoming blocked a.a.

t-butyl oxycarbonyl
(t-BOC)

Dicyclohexyl Carbodiimide
(condensing agent)

Dicyclohexyl urea

c. The protecting group (t-BOC) is removed by treatment with acid, which converts it to isobutylene and CO_2, both gaseous products. The free amino group is now ready for the next sequential addition.

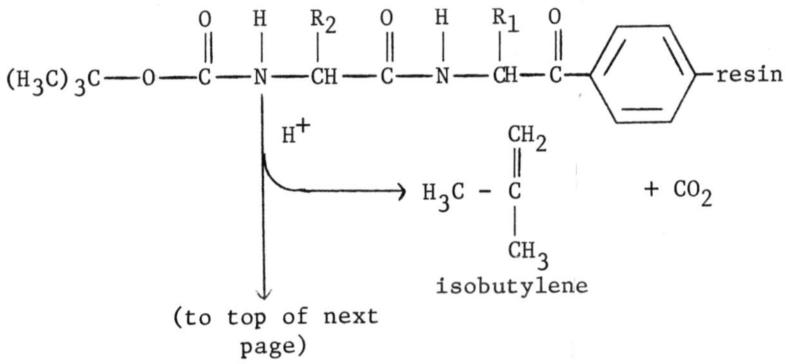

isobutylene

(to top of next
page)

$$\downarrow$$

$$H_2N - \underset{\underset{R_1}{|}}{CH} - \underset{\underset{O}{\|}}{C} - \underset{\underset{H}{|}}{N} - \underset{\underset{R_1}{|}}{C} - \underset{\underset{O}{\|}}{C} - \bigcirc\text{-resin}$$

After the last amino acid has been added, the peptide is cleaved from the resin. The advantages of this method include

1) quantitative yields of products.
2) non-purification of intermediate peptides in the synthesis.
3) susceptibility to automation.
4) ease and rapidity.

Classical methods of peptide synthesis are also of importance. They can use the same reactions given above, but they are done in a homogeneous solution, rather than bound to insoluble resin beads.

Notable achievements in peptide synthesis include ribonuclease (124 amino acids) and insulin (2 chains of 21 and 30 amino acids, respectively). It is noteworthy that both have been synthesized by homogeneous, classical methods as well as by solid phase techniques, although the solid phase synthesis was much more rapid.

The Structure of Proteins

The forces that determine protein structure include

1. Covalent bonding, the sharing of an electron pair by two atoms, one electron (originally) coming from each atom. Bond energies are about 30-100 kcal/mole of bonds. Important examples in protein structure include peptide, disulfide, ester, and amide bonds. They can be between the reactive groups of amino acids in the same chain (intrachain) or of different polypeptide chains (interchain).

2. Coordinate covalent bonding, the sharing of an electron pair by two atoms, with both electrons (originally) coming from the same atom. These are similar to covalent bonds, but with much lower energies (4-5 kcal/mole of bonds). Consequently, they are much more labile (are made and broken much more readily). The electron

pair donor is called a ligand or Lewis base and the acceptor is the central atom (since it frequently can accept more than one pair of electrons) or Lewis acid. These bonds are important in all interactions between transition metals and biological molecules, such as Fe^{+2} in hemoglobin and the cytochromes, and Co^{+3} in vitamin B_{12}.

3. Ionic forces, the coulombic attraction between two groups of opposite charge. The bond energy is about 10-20 kcal/mole of bonds and is strongly dependent upon distance. These bonds are found in bonding between positive residues (α-ammonium, ϵ-ammonium, guanidinium, imidazolium) and negatively charged groups (ionized forms of α-carboxyl, β-carboxyl, γ-carboxyl, phosphate, sulfate).

4. Hydrogen bonding, the sharing of a hydrogen atom between two electronegative atoms which have unbonded electrons. The bond energy is 2-10 kcal/mole of bonds. These bonds are extremely important in water-water interactions and their existence explains many of the unusual properties of water and ice. In proteins, groups having a hydrogen atom which can be shared include $\supset$N-H (peptide nitrogen, imidazole, indole), -OH (serine, threonine, tyrosine, hydroxyproline), $-NH_2$ and $-NH_3^+$ (arginine, lysine, α-amino), and -CONH (carbamino). Groups which can accept the sharing of a hydrogen include $-COO^-$ (aspartate, glutamate, α-carboxylate), -S-S- (disulfide), and $\supset$C=O (in peptide and ester linkages).

5. Van der Waals' attractive forces, also called London forces. These operate between all atoms, ions, and molecules and are due to a fixed dipole in one molecule inducing an oscillating dipole in another molecule by distortion of the charge (electron) cloud. The positive end of a fixed dipole will pull an electron cloud toward it, the negative end will push it away. The strength of these interactions is strongly dependent on distance, varying as $1/r^6$, where r is the interatomic separation. These bonds are particularly important in the non-polar interior structure of proteins, providing attractive forces between non-polar sidechains.

6. Hydrophobic interactions, causing non-polar sidechains to cling together in polar solvents, especially water. The actual energy for this process seems to arise from hydrogen-bonding forces between water molecules, rather than from attractions between the hydrophobic groups. The bond energy for this interaction is not well defined or understood. These forces are also important in lipid-lipid interaction in membranes.

7. <u>Electrostatic repulsion</u>, between charged groups of like charge. These are just the opposite of ionic (attractive) forces above. They depend on distance and upon the charges of the interacting groups according to Coulomb's law: q_1q_2/r^2 , where q_1 and q_2 are the charges and r is the interatomic separation.

8. <u>Van der Waals' repulsive forces</u>, operating between atoms at very short distances. They result from the induction of induced dipoles by the mutual repulsion of electron clouds. Since there is no involvement of a fixed dipole (as there was in Van der Waals' attractive forces), the distance dependence is even greater, $1/r^{12}$ in this case. These forces operate when atoms not actually bonded to each other try to approach more closely than a minimum distance. They are the underlying force in steric hindrance between all atoms.

<u>Types of Protein Structure</u>

There are several levels at which polypeptide structure is considered. The most basic one is the <u>primary structure</u>: the <u>number</u>, <u>kind</u>, and <u>order</u> of the amino acids in the chain. For example:

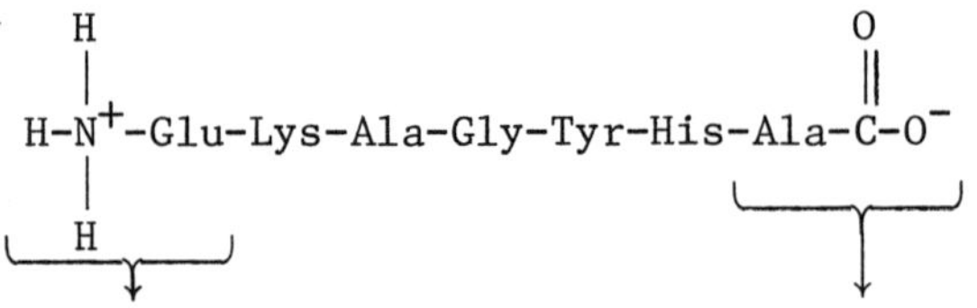

$$\begin{array}{ccc} \text{H} & & \text{O} \\ | & & \| \\ \text{H-N}^+\text{-Glu-Lys-Ala-Gly-Tyr-His-Ala-C-O}^- \\ | & & \\ \text{H} & & \end{array}$$

N-terminal amino acid   C-terminal amino acid

Note that the peptide is written starting with the amino acid having a free α-amino group (N-terminus) and ending with the residue having a free α-carboxyl group (C-terminus). This is a convention, and naming is conducted in the same manner. The proper name for the above peptide is

glutamyllysylalanylglycyltyrosylhistidylalanine.

The C-terminal residue is always given the name of the free amino acid while <u>all</u> other residues have the root of the residue name plus -yl (e.g.,<u>glutamate</u> → <u>glutamyl</u>; <u>tyrosine</u> → <u>tyrosyl</u>).

The residues are linked covalently by <u>peptide bonds</u>. This is a very important linkage and is illustrated below. The bond lengths and angles are average values and will vary somewhat (but not much) depending on the amino acids linked and the molecule of which they are part.

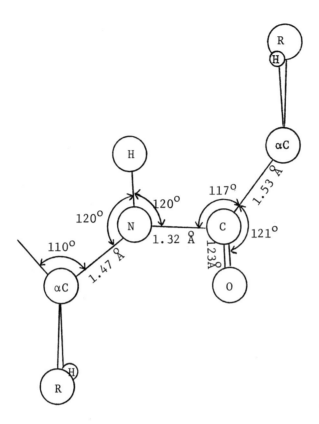

It is also important to notice that the peptide group is in a
planar, trans configuration, with very little rotation or twisting
around the bond linking the α-amino nitrogen of one amino acid and
the carbonyl carbon of the next one (the "peptide" bond).   This
is due to an amido-imido tautomerization, lending partial double-
bond character to the N-C bond.   This is illustrated below, with
the transition state, (II), being probably what actually exists
in nature.

| (I) | (II) | (III) |
| (amido form) | | (imido form) |

The α-amino proton is shared by the nitrogen and oxygen atoms, and the N-C and C-O bonds are both (roughly) "one-and-one-half" bonds (not single, not double). The planarity and rigidity follow from this, since there is no free rotation around any but single bonds. The nitrogen, carbon, and oxygen atoms involved are all partially sp^2, partially sp^3 hybrids. Many features of protein structure (α-helices, pleated sheets, etc.) are possible, due in part to the geometry of the peptide groups in the protein backbone. Based largely on these properties, Pauling, Corey and Branson in 1951 were the first to postulate the existence of helices and pleated sheets in protein molecules.

The folding of parts of polypeptide chains into specific structures held together by hydrogen bonds is referred to as secondary structure. The most common secondary structure types are the left-handed α-helix, the parallel and antiparallel β-pleated sheets, and the random coil. A particular protein might possess only one kind of secondary structure (α-keratin and silk protein consist entirely of α-helix and β-sheet, respectively) or it may have more than one kind (hemoglobin contains both α-helical and random coil regions). Most globular proteins have mixed structures.

The essential features of some secondary structures are shown below.

1. Random Coil

   This term is misleading. The structure is random in the sense that there is no repeating pattern to the way in which each residue of the peptide chain interacts with other residues (i.e., the Nth residue bonded to the N-3 and N+3 residues, as in an α-helix). However, given a particular sequence, there is only one (or at most 2 or 3) ways in which it will coil itself. This conformation will either have the minimum energy or will be in a local energy minimum (see diagram below).

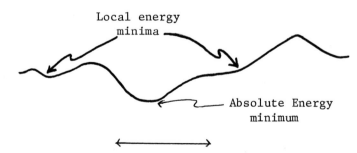

Energy of the conformation (increasing upwards)

Local energy minima

Absolute Energy minimum

changes in molecular conformation move the system along this axis

Since energy must be added to a molecule to make it change conformation (move from a valley over a hill into another valley in the diagram), the molecule can remain trapped in a conformation corresponding to any minimum in the energy map, even though it would prefer to be at the absolute minimum in internal energy.

This concept (of a molecule seeking a preferred, low-energy shape) is the basis for postulating that the primary structure (sequence) of a polypeptide determines the secondary and tertiary structures of the molecule.   There are two principal snags to this hypothesis.

I. If there **is** more than one peptide chain in the molecule (as in insulin), the way in which the chains associate may be determined by factors other than their amino acid sequence.   (This is different than tertiary structure, discussed below, since the insulin chains cannot assume their normal secondary and tertiary structures separately from each other.)

II. If there are more than two cysteine residues in a polypeptide, there is more than one way in which cystine (-S-S-) bridges can form (i.e., cys 1 can form a disulfide bond with cys 2 or cys 3).   Since these bridges are covalent bonds and can be formed under conditions where the weaker forces determining coiling are not operative, these forces will not be powerful enough to break the disulfide bonds and pull the molecule back to the true minimum energy structure.

2. Alpha helix (Left-handed) is shown below. It is intended to illustrate hydrogen bonding and is <u>not</u> drawn to scale.

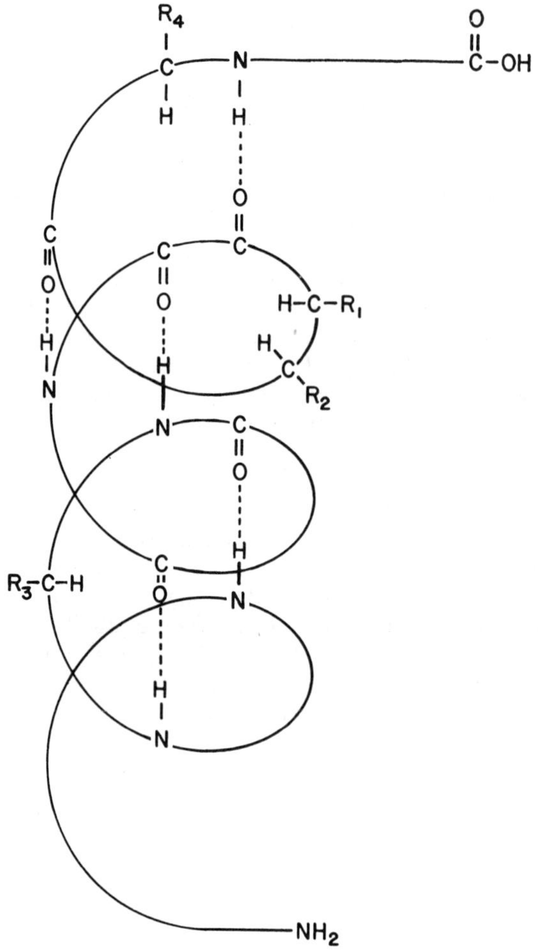

There are 3.6 amino acid residues in each complete turn of the helix (and, hence, 100° of turn per residue). Hydrogen bonds between coils of the helix form the "surface" of the structure (see diagram). If the N-terminus is considered down, an amino group always hydrogen bonds to a carbonyl group above it and since carboxyl groups alternate in the backbone, hydrogen bonds alternate up and down in the helix. The amino proton and carboxyl oxygen of a residue (N) are hydrogen bonded, respectively, to the carbonyl oxygen of residue (N+3) and the amino proton of residue (N-3). This is shown in the uncoiled, schematic diagram on the top of page 47.

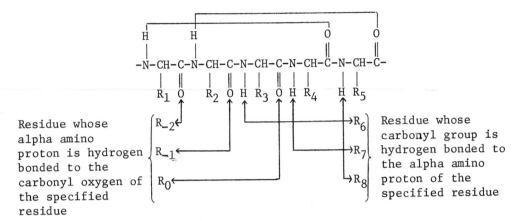

Residue whose alpha amino proton is hydrogen bonded to the carbonyl oxygen of the specified residue

Residue whose carbonyl group is hydrogen bonded to the alpha amino proton of the specified residue

Note also that
a) hydrogen bonds are parallel to the helix axis.
b) the peptide groups are planar and in the trans configuration.
c) the R-groups (sidechains) are roughly perpendicular to the helix axis.

3. The beta (pleated sheet) structure is harder to visualize than is the α-helix. A side (edge-on) view of such a structure would appear as shown below. The hydrogen bonds, between the α-carbonyl oxygens and the α-amino protons, extend into and out of the page in planes perpendicular to the page. One important feature of this structure is that the R-groups (sidechains) are above and below the sheets and nearly perpendicular to them. Amino acids with bulky R-groups tend to not form pleated sheets because their sidechains interfere with each other.

Edge-on View of a Pleated Sheet Structure

The α-carbons always serve as "corners" in the representation. Note, however, that since the α-carbons and the nitrogens are more or less tetrahedral and the carbonyl carbons are trigonal,

some of the atoms shown above to be in the plane of the paper
are actually above or below that plane. The isometric
projection of a pleated sheet (below) will help to clarify
this. Note that the R-groups extend up and down from the
edges of the folds, perpendicular to the edges.

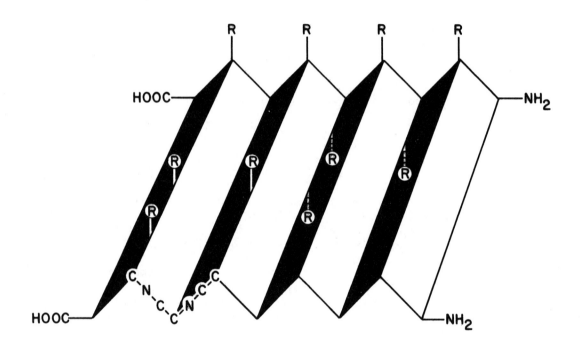

The hydrogen bonds can be arranged two ways, yielding parallel
and antiparallel pleated sheets. The diagrams below illustrate
this. They are top views of the pleated sheets. The hydrogen
bonds (dotted lines) extend into and out of the paper
diagonally. The ones which do not appear to connect to
anything extend to the adjacent chains in the sheet.

```
                              (COOH)   (H₂N)
   R-C                 R-C-H            R-C-H            R-C-H        R-C-H
       \                    \                \                \            \
        N                    N-H· · ·O=C      N-H                N-H·        N-H·
         \                                       \                   \          \
          C    · · ·O=C      · · ·O=C        N-H· · ·        ·O=C      ·O=C        ·
           \                                     \              /         /
            C-R            H-C-R      H-C-R        H-C-R      H-C-R
           /              /          /              /          /
          N    · · ·H-N          ·H-N       C=O· · ·   ·H-N        ·H-N         ·
         /                                     \            \          \
        C                    C=O· · ·H-N       C=O·        C=O·
       /                        \                /           /
      R-C              R-C-H              R-C-H      R-C-H      R-C-H
         \                  \                  \          \          \
          N                  N-H              O=C          N-H·        N-H·
           \                    \                \            \          \
            C    · · ·O=C        ·O=C       N-H· · ·   ·O=C      ·O=C        ·
             \                                   \          /        /
              C-R              H-C-R   H-C-R       H-C-R      H-C-R
             /                /       /             /          /
                            (H₂N)           (COOH)      (H₂N)      (H₂N)

Edge-on
view (same
as above           anti-parallel
diagram)             chains                        parallel chains
```

The edge-on view at the left is what one would see if the
hydrogen-bonded structures were stood on edge. (This amounts
to rotating them 90° around an axis in the plane of and parallel
to the long edge of the paper.) The hydrogen bonds would then
be in the planes indicated by the lines in the edge view
connecting the R-group (α) carbon atoms. In the parallel and
anti-parallel drawings, an R-group extending to the left is
above the sheet and one to the right is below it.

In the beta structures, the hydrogen bonds are usually said to
be interchain rather than intrachain (as they are in the
α-helix). This is completely correct in, for example, silk.
Frequently, however, a single peptide chain will fold back
upon itself and regions of pleated sheet structure are formed
between different parts of the same chain. Such conformations
have actually been identified in crystal structures of
globular proteins.

4. There are other types of secondary structures which are found
    in proteins. One important example is present in collagen.
    Here, the peptide chains are twisted together into a three-
    stranded helix. The resultant "three-stranded rope" is then
    twisted into a superhelix. This is similar, in principle,
    to the double-stranded superhelix found in nuclear DNA.

5. The amino acid sequence (primary structure) strongly influences
    the types of secondary structure which are present in a
    protein. Below is a list of amino acids classified according
    to the sort of secondary structure which they tend to prefer.
    These divisions are not absolute and have considerable overlap
    (glutamic acid residues are found in α-helices; alanine
    residues occur in β-structures).

Table 2. Preferred Secondary Structure of Some Amino Acids

α-helix formers[a]	non-α-helix[b]	β-structure[c]	Random Coil[d]	α-helix breaking[e]
Ala Try Gln	Ser	Gly	Glu ⎫ (−)	Pro
Leu Cys Tyr	Thr		Asp ⎭	Hypro
Phe Met Asn	Ile		Lys ⎫ (+)	
	Val		Arg ⎭	

Notes: [a] Gln, Asn = glutamine , asparagine; these residues readily
        form α-helices in aqueous solution.

[b] Thr, Ile , Val - two substitutions on the β carbon causing
    steric problems in fitting them into an α-helix; (notice that
    most amino acids have a β-carbon which is a $>$CH$_2$ group).
    Ability of ser to form sidechain H-bonds causes it to prefer
    other structures.

[c] Because of its small (-H) sidechain, there is little steric
    hindrance even though the R-groups in a sheet are very close
    together.

[d] Ionic repulsion prevents close approximation of glutamic and
    aspartic sidechains above about pH 4, and of lysyl and
    arginyl sidechains below about pH 9.

[e] Proline and hydroxyproline , because their α-amino group is
    part of a ring, are unable to fit into the rigid geometry
    required by the α-helix. Also, when incorporated into a
    peptide, they lack an α-amino proton for hydrogen bond
    formation. Therefore these amino acids are never part of an
    α-helix. (They do, however, form other helices, as found in
    collagen.)

C. The <u>tertiary structure</u> involves the arrangement and interrelationship of the folded chains (secondary structure) of a protein into a specific shape which is maintained by salt bonds, hydrogen bonds, –S–S– bridges, Van der Waals' forces, and hydrophobic interactions. The hydrophobic "bond" is considered to be a major force in maintaining the shape or conformation of proteins. (The word <u>bond</u> in this context is really a misnomer.) Hydrophobic bonds are interactions between non-polar sidechains of alanine, valine, leucine, isoleucine and phenylalanine within H_2O envelopes.

1. Hydrophobic interactions may involve the sidechains of several molecules or may occur between chains on the same molecule (as shown below). They occur because of the inability of non-polar sidechains to interact with water molecules either ionically or through hydrogen bonds.

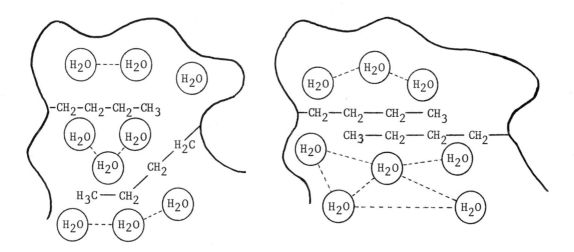

(dashed lines represent
hydrogen bonds)

(I) No hydrophobic interaction between hydrocarbon sidechains; increased non-polar surface area.

(II) Hydrophobic clustering of lipid-like sidechains; water "sees" less lipid surface.

Without going into detail, the general idea is that, when hydrophobic (water-hating; oil or lipid-like) molecules are put into an aqueous solution, they disturb the water structure. To minimize this disturbance (and to return to the lowest possible free energy state), the hydrophobic groups clump together as much as possible so that the surface to volume ratio of the hydrophobic material is minimal. (Two small spheres, each of volume 1 unit, have a greater total surface area than one larger sphere of volume 2 units.) Once the chains are brought close together, Van der Waals' attractive forces can operate to assist in holding them there. The van der Waals' forces are quite weak, however, and exactly where most of the "hydrophobic bond energy" comes from is not clear. The probable source is the free energy made available when water is able to achieve a more stable structure. This can be envisioned in two possible ways.

a. The water which surrounds a hydrophobic group is ice-like (highly ordered) because of the limitations imposed on its movement by the presence of the hydrophobic material. A decrease in hydrophobic surface area causes a decrease in order and, hence, an increase in entropy which helps to lower the free energy of the solution.

b. Hydrophobic regions in an aqueous solution break up hydrogen-bond networks. A decrease in the area of hydrophobic surface permits more hydrogen bonds to form, thus lowering the free energy of the solution.

Final answers about the properties and causes of hydrophobic interactions await more definitive experimental data.

2. A schematic diagram of the tertiary structure of lysozyme is shown below. This structure was determined by single crystal x-ray diffraction techniques. This method is an extremely powerful one for elucidating molecular structures, provided suitable crystals of the material can be prepared. This example is given to show a protein molecule which contains several types of secondary structure. It should be noted that, in general, hydrophobic regions are buried inside the molecule while hydrophilic (water-loving; charged) groups appear on the surface, exposed to the aqueous environment. The arrowheads along the backbone indicate the chain direction, from the amino terminus to the carboxyl terminus.

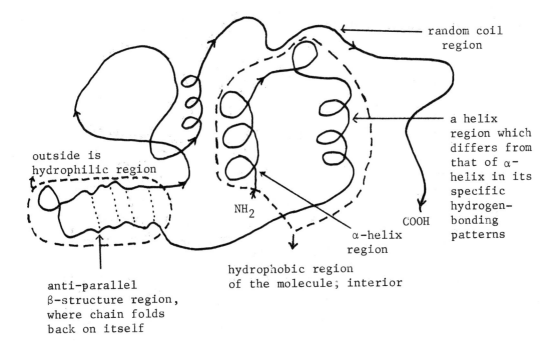

random coil
region

a helix
region which
differs from
that of α-
helix in its
specific
hydrogen-
bonding
patterns

outside is
hydrophilic region

NH₂

COOH

α-helix
region

hydrophobic region
of the molecule; interior

anti-parallel
β-structure region,
where chain folds
back on itself

Quaternary Structure is the association of similar or dissimilar
protein subunits into oligomers (small polymers) or polymers.

1. The subunits are held together by non-covalent forces.
    Consequently, they can be dissociated under relatively mild
    conditions.

2. Many proteins require more than one protein subunit in their
    aggregate structures to perform their biological function
    (hemoglobin, fatty acyl synthetase, actomyosin in muscle, etc.).

3. Enzymes which catalyze the same chemical reaction but which
    have different kinetic properties (ability to bind substrate,
    maximum velocity, etc.) are called isoenzymes or isozymes.
    Important examples include lactic acid dehydrogenase, alkaline
    phosphatase, hexokinase, and hemoglobin (it is somewhat
    incorrect to call hemoglobin an enzyme, though). The isozymes
    of a given enzyme are electrophoretically separable and are
    generally composed of varying proportions of several subunits.

4. The schematic diagram below shows how some of the non-covalent
    forces function in stabilizing quaternary protein structure.

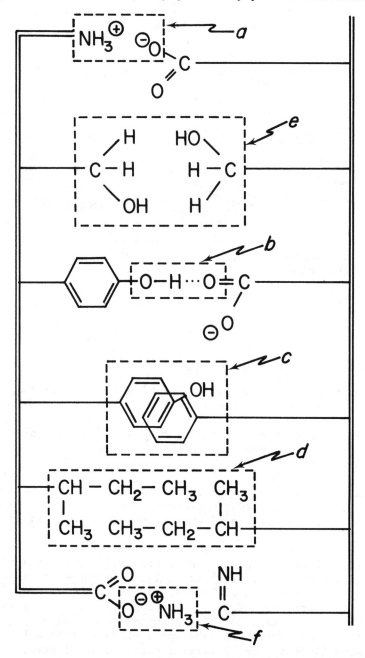

a. Electrostatic attraction between the N-terminal amino group
    and a sidechain carboxyl (from aspartate or glutamate).

b. Hydrogen bond between the phenolic proton of tyrosine and the carbonyl oxygen of a sidechain carboxyl group.

c. Van der Waals' interaction (ring stacking) between the benzene rings of phenylalanine and tyrosine. This is probably also stabilized by hydrophobic forces.

d. Same as (c), between two isoleucyl groups. Here, the hydrophobic forces are probably greater relative to the van der Waals' forces. The residues need not be the same for such interactions.

e. Van der Waals' interaction between two serine residues. Here, the groups are hydrophilic, so that the hydrophobic contribution is minimal.

f. Same as (a), between an arginyl sidechain and the C-terminal carboxyl group.

Denaturation of Proteins

1. Denaturation involves changes in the physical, chemical, and biological properties of a protein molecule.

2. Some of the changes in properties are

    a. Decreased solubility.
    b. Alteration in the internal structure and in the arrangement of peptide chains, which does not involve the breaking of peptide bonds.
    c. Decreased symmetry (e.g., loss of helical structure).
    d. Increased chemical reactivity, particularly of ionizable and sulfhydryl groups.
    e. Increased susceptibility to hydrolysis by proteolytic enzymes.
    f. Decrease or total loss of the original biological activity.

3. The causes of denaturation include

    a. A significant change in the pH of the protein solution.
    b. Temperature changes (particularly high temperature).
    c. Ultraviolet radiation.
    d. Ultrasonic vibration (known as "sonication").
    e. Vigorous shaking or stirring of aqueous solutions which spreads the protein in a thin film over the surfaces of air bubbles (i.e., foaming of the protein solution).
    f. High concentrations of neutral polar compounds such as urea or guanidine. These compounds break hydrogen bonds, allowing new ones to form.

g. Treatment with organic solvents such as ethanol, acetone, etc.
    This sort of denaturation is minimized at low temperatures
    (2^0 to 50^0C).
h. Grinding of proteins which results in mechanical deformation
    and the breakage of peptide chains.

4. The extent and reversibility of denaturation are dependent upon the
    complexity of the protein and the intensity and duration of the
    denaturing treatment. Normally, denaturation is an irreversible
    process, although there are exceptions such as:

    a. The denaturation of hemoglobin with acid and renaturation by
        neutralization under appropriate conditions.
    b. The heat denaturation of pancreatic ribonuclease and
        renaturation on cooling.

    It can generally be stated that denaturation is reversible if no
    disulfide linkages present in the native protein are broken. The
    greatest block to successful denaturation is the proper reformation
    of the disulfide bonds.

denaturation ⟶ pH, temperature   renaturation ⟶ under optimal conditions

specific tertiary
structure of
α-helix, β-structure,
and random coil; all
molecules of a given
protein have the
same shape

chains unfolded and
randomly arranged;
each chain of slightly
different shape

each protein
molecule returns
to its original
shape

5. A knowledge of the denaturability of proteins is very important in
    clinical laboratory work. If the activity of an enzyme is being
    measured to evaluate the health of a patient, a false result can be
    obtained because of changes in the activity brought about by
    denaturation. Care must be taken to avoid this during specimen
    collection, handling, and assay.

Classification of Proteins

Simple Proteins

Fibrous proteins are insoluble in water and resistant to
proteolytic enzyme digestion. They are elongated molecules

which may consist of several coiled peptide chains tightly
linked.

1. Collagens

    a. Proteins of connective tissue.
    b. 30% of the total protein in a mammal is collagen.
    c. Contain large amounts of proline, glycine,
        hydroxyproline and hydroxylysine.
    d. Contain no tryptophan.
    e. Collagens, when boiled in water or dilute acid, yield
        gelatins which are soluble and digestible by enzymes.

2. Elastins are present in tendons, arteries and other elastic
    tissues. They cannot be converted to gelatins by boiling
    in water or dilute acid.

3. Keratins are proteins of hair, nails, etc. They contain
    large amounts of cystine. Human hair is 14% cystine.

Globular Proteins are soluble in water and in salt solutions.
In solution these molecules are spheroids or ellipsoids. All
known enzymes, oxygen-carrying proteins, and protein hormones
are members of this class.

1. Albumins: soluble; coagulable by heat.
2. Globulins: insoluble in water; soluble in dilute neutral
    salt solutions; coagulable by heat.
3. Histones: basic proteins found complexed with nucleic
    acids; contain large amounts of arginine and lysine and
    very few aromatic amino acids.
4. Protamines: basic proteins found complexed with nucleic
    acids, particularly in certain fish sperm; rich in
    arginine.

Conjugated Proteins are complex. They are combined with non-amino
acid substances.

Nucleoproteins: $\left.\begin{matrix}\text{DNA}\\\text{RNA}\end{matrix}\right\}$ + proteins

Mucoproteins (or mucoids): carbohydrate (more than 4%)+ protein

Glycoproteins: small amount of carbohydrates (less than 4%)
+ protein

Lipoproteins: water soluble $\left.\begin{matrix}\\\\\end{matrix}\right\}$ both contain lipid

Proteolipids: insoluble in water   and protein

Others: hemoproteins (protein + heme); other metalloproteins; flavoproteins (protein + flavin); phosphoproteins

Some Properties of Proteins

1. <u>Amphoterism</u> is the ability to behave as an acid or base, depending on the conditions.

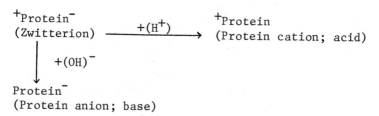

$^+$Protein$^-$
(Zwitterion) $\xrightarrow{+(H^+)}$ $^+$Protein
(Protein cation; acid)

$+(OH)^-$

Protein$^-$
(Protein anion; base)

a. <u>Isoelectric pH</u> (pI). The pH at which a protein does not migrate in an electric field. The protein exists in the zwitterion form with a net charge of 0. There are an equal number of cationic and anionic sites on each molecule.

b. <u>Isoelectric precipitation</u>. Many proteins are easily precipitated when the pH is adjusted to their isoelectric point (i.e., many proteins have their minimum solubility when pH = pI).

c. <u>Electrophoretic mobility</u> is zero at isoelectric pH.

d. Because different proteins have different pI values, electrophoresis at different pH's can be used for purification.

2. <u>Solubility</u> can be changed in several ways.

a. Effect of salt concentration (salting-in and salting-out phenomena). By adding ionic solutes to a protein solution, the affinity of protein molecules for each other compared to the affinity of protein molecules for H_2O may be altered, thus changing protein solubility.

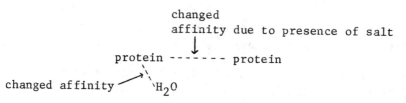

changed
affinity due to presence of salt

protein - - - - - - - protein

changed affinity → H_2O

Salting-in phenomenon. Adding <u>small</u> amounts of ionic solutes
(NaCl, $(NH_4)_2SO_4$, etc.) decreases the protein-protein interaction
(affinity) but increases protein-H_2O interaction, leading to
solubilization of the protein. The protein-protein affinity is
caused by interactions between oppositely charged groups on
the protein molecules.

$$(-\overset{\overset{\textstyle O}{\|}}{C}-O^- \quad \text{and} \quad {}^+H_3N-)$$

and is decreased by salts (X^+ and Y^-) which provide ions that
bind to the charged protein groups and lower their interaction.

$$(-\overset{\overset{\textstyle O}{\|}}{C}-O^-\cdots\cdots X^+ \; Y^-\cdots\cdots H_3^+N)$$

Salting-out phenomenon. Adding <u>large</u> amounts of ionic solutes
results in protein precipitation. The mechanism is not well
understood. It is explained as the possible "dehydration of
active water", which increases the interaction between solute
molecules. It is important in one method of protein separation
and purification.

b. <u>pH</u> (see isoelectric precipitation, page 58).

c. <u>Precipitation with non-polar solvents</u>. Non-polar solvents may
    also cause selective "dehydration of active water" around
    protein molecules and thus increase protein-protein interaction.
    Each protein has a given amount of "active water" associated
    with it. Proteins can be separated from one another by varying
    the concentration of non-polar solvent.

<u>Purification of Proteins</u> is very important in any attempt to describe
exactly the properties of a molecule. Impurities can cause completely
erroneous results. Some methods are

1. Differential solubility
2. Specific precipitation
3. Chromatography
4. Preparative electrophoresis
5. Preparative ultracentrifugation
6. Selective enzyme digestion
7. Others

Estimations of the Amount of Protein Present

1. Kjeldahl Procedure (measurement of total nitrogen present)

$$\text{Organic N} \xrightarrow[\substack{\text{catalyst} \\ \text{(Cu, Se, or Hg)}}]{H_2SO_4} CO_2 + H_2O + NH_4HSO_4$$

$$NH_4HSO_4 + 2NaOH \xrightarrow{\text{distillation}} NH_3\uparrow + SO_4^{=} + 2Na^{+} + 2H_2O$$

The NH_3 produced by distillation is passed into a known amount of
standard acid and at the end of distillation, the acid remaining
is titrated with a standard base. The acid consumed is then a
measure of NH_3 produced which in turn is a measure of nitrogen
present in the protein. This procedure requires much time and
very large amounts of protein. In general, proteins contain about
16% nitrogen. This means that when the weight of nitrogen measured
by the above method (taking into account any non-protein nitrogen)
is multiplied by the factor 6.25 (100/16) one gets the weight of
the total protein present.

2. U.V. absorption at 280 nm, due to tyrosine, phenylalanine, and
tryptophan residues (aromatic groups). Since different proteins
have differing amounts of these residues (some have none of them),
the method has inherent inaccuracies. It is also not very
successful in the presence of nucleic acids and other substances
which absorb near this wavelength.

3. Biuret Method. peptide + $CuSO_4$ + alkali $\longrightarrow$ violet color

This procedure is specific for peptide bonds, but is not very
sensitive. The violet color is produced by the complex shown below.

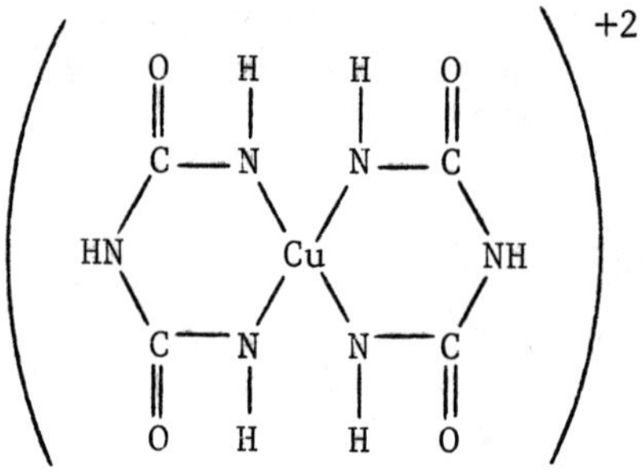

4.   Folin - Ciocalteu Reaction

protein + phosphomolybdo ⟶ blue color
                    tungstic acid

The color is due to a reaction with tyrosine. The procedure is
very sensitive but cannot be used in the presence of other phenols.
Also, proteins contain varying amounts of tyrosine, some even
having none. Thus, the method has some of the same drawbacks that
U.V. absorption has.

3

Enzymes

*with John H. Bloor, M.S.**

1. Almost all of the chemical changes (absorption, digestion, metabolism, locomotion, putrefaction, etc.) that take place in a living organism are speeded up by enzymes (catalysts). Without these catalysts the reactions proceed too slowly for biological systems to function at any significant rate.

2. A <u>catalyst</u> is defined by Ostwald as an agent which affects the velocity of a chemical reaction without appearing in the final products of that reaction. In an enzyme catalyzed reaction (A + B $\rightleftarrows$ C + D), the enzyme influences the reaction velocity of both forward and backward reactions to the same extent. However, the direction in which the reaction proceeds is dependent upon mass-law considerations and the availability of free energy. Enzymes also show the typical features of catalysts such as a) not being consumed in the reaction, b) being needed only in minute quantities, c) reversible and irreversible inhibition, etc. This is discussed much more fully later on, in the section on enzyme kinetics.

3. Enzymes are proteins and are produced by living cells. They possess a high degree of specificity (i.e., they usually catalyze only one type of reaction, frequently acting only on one molecular species) and are classified according to the type of reactions they catalyze.

    a. oxidation-reduction
    b. transfer of groups
    c. hydrolysis of compounds
    d. non-hydrolytic removal of groups
    e. isomerization
    f. joining of two molecules with the breaking of a pyrophosphate bond.

    The following table is a list consisting of examples of enzymes in the above classes. It is by no means complete with respect to classes, subclasses, or examples.

---

* Presently at Pathology Associates Medical Laboratories, Honolulu, Hawaii.

Table 3. Classification of Enzymes[*]

Main Class and Subclasses	Examples

1. Oxidoreductases

    1.1 Acting on the CH-OH group of donors

        1.1.1 with NAD or NADP as acceptor   Alcohol dehydrogenase, lactate dehydrogenase

        1.1.3 with O_2 as acceptor   Glucose oxidase

    1.2 Acting on the aldehyde or keto-group of donors

        1.2.1 with NAD or NADP as acceptor   Glyceraldehyde-3-phosphate dehydrogenase

        1.2.3 with O_2 as acceptor   Xanthine oxidase

    1.3 Acting on the CH-CH group of donors

        1.3.1 with NAD or NADP as acceptor   Dihydrouracil dehydrogenase

        1.3.2 with a cytochrome as an acceptor   Acyl-CoA dehydrogenase

    1.4 Acting on the $CH-NH_2$ group of donors

        1.4.3 with O_2 as an acceptor   Amino acid oxidases

2. Transferases

    2.1 Transferring C_1-groups

        2.1.1 Methyltransferases   Guanidoacetate methyltransferase

        2.1.2 Hydroxymethyltransferases and formyltransferases   Serine hydroxymethyltransferase

        2.1.3 Carboxyltransferase and carbamoyltransferase   Ornithine carbamoyltransferase

* From Pritham, Gordon H.: Anderson's Essentials of Biochemistry, St. Louis, 1968, The C.V. Mosby Co.; compiled from Karlson, P.: Introduction to Modern Biochemistry, ed. 2, New York, 1965, Academic Press, Inc., and Harrow, B., and Mazur, A.: Textbook of Biochemistry, ed. 9, Philadelphia, 1966, W.B. Saunders Co.

Main Class and Subclasses	Examples
2.3 Acyltransferases	Choline acetyltransferase
2.4 Glycosyltransferases	Maltose phosphorylase
2.6 Transferring N-containing groups 2.6.1 Aminotransferases	Transaminases
2.7 Transferring phosphorus-containing groups 2.7.1 Phosphotransferases with an alcohol group as an acceptor	Glucokinase

2.8 Transferring sulfur-containing groups
    2.8.3 CoA-transferases
        2.8.3.1 Acetyl-CoA: propionate CoA-transferase
	Propionate CoA-transferase

3. Hydrolases	

3.1 Cleaving ester linkages
    3.1.1 Carboxylic ester hydrolases
    3.1.3 Phosphoric monoester hydrolases
    3.1.4 Phosphoric diester hydrolases
	Esterases, lipases
	
	Phosphatases
	
	Snake venom phospho-diesterase

3.2 Cleaving glycosides
    3.2.1 Glycoside hydrolases
    3.2.2 N-Glycoside hydrolases
	Amylase, β-Glucosidase, etc.
	Nucleosidases

3.4 Cleaving peptide linkages
    3.4.1 α-Aminopeptide amino acid hydrolases
    3.4.2 α-Carboxypeptide amino acid hydrolases
    3.4.4 Peptidopeptide hydrolases (=endopeptidases)
	Leucine aminopeptidase
	
	Carboxypeptidases
	
	Pepsin, trypsin, chymotrypsin

Main Class and Subclasses	Examples
3.5 Acting on C-N bonds other than peptide bonds	
3.5.1.5 Urea amidohydrolase	Urease
3.6 Acting on acid anhydride bonds	
3.6.1.3 ATP phosphohydrolase	ATPase

4. Lyases

4.1 C-C lyases	
4.1.1 Carboxy lyases	Pyruvate decarboxylase
4.1.2 Aldehyde lyases	Aldolase
4.2 C-O lyases	
4.2.1 Hydrolyases	Fumarate hydratase (=fumarase)
4.3 C-N lyases	Histidine-ammonia lyase (=histidase)

5. Isomerases

5.1 Racemases and epimerases	
5.1.3 Acting on carbohydrates	Ribulose-5-phosphate epimerase
5.2 **Cis**-trans isomerases	Maleylacetoacetate isomerase
5.3 Intramolecular oxidoreductases	
5.3.1 Interconverting aldoses and ketoses	Glucosephosphate isomerase
5.4 Intramolecular transferases	Methylmalonyl-CoA mutase

6. Ligases

6.1 Forming C-O bonds	
6.1.1 Amino acid-RNA ligases	Amino acid-activating enzymes
6.3 Forming C-N bonds	
6.3.1 Acid-ammonia ligases	Glutamine synthetase
6.3.2 Acid-amino acid ligases	Peptide synthetase, glutathione synthetase
6.4 Forming C-C bonds	
6.4.1 Carboxylases	Acetyl-CoA carboxylase

Thermodynamics and Kinetics of Biological Reactions

An important aspect of biochemistry is concerned with the elucidation of chemical interconversions in biological systems. Since these reactions primarily involve the orderly release, storage, and utilization of energy, it is logical that a knowledge of chemical energetics (thermodynamics) is required for their discussion. While studies of metabolism, nutrition, and physiological homeostasis may not directly invoke it, an awareness of thermodynamics contributes to the understanding of these subjects. A number of reactions which are important in living systems result in the transfer of electrons, in addition to transfer of energy. To handle these, some background in electrochemistry is necessary. This can be directly related to more classical thermodynamics.

The actual energy transfers, though, are only part of the story. Thermodynamics treats only initial and final states, with no word about how one group of molecules (the reactants) are converted to another, quite different group (the products). Chemical kinetics is needed for this.

Many of the reactions needed by the body would proceed too slowly to be of use, however, were it not for the catalytic abilities of enzymes. To deal with the ways in which enzymes alter the kinetics of chemical reactions, enzyme kinetics has developed as a separate field. As a prelude, then, to talking about specific metabolic processes, this section discusses the fundamental language of thermodynamics and kinetics and the manner in which they are related to biological systems.

Thermodynamics

1. During the conversion of reactant molecules (A + B) to product molecules (C + D), there is a change in the potential energy (called Gibbs free energy by chemists and symbolized by G or F) of the atoms and molecules involved. This is shown below by plotting the variation in the free energy of the participating molecules as the reaction progresses. The phrase "progress of the reaction" was selected to label the abscissa of this diagram because of its generality. It is intended to include all changes in bond lengths and angles (i.e., in the shapes and atomic interactions) of the reactant molecules as they are smoothly converted to product molecules. "Smoothly" is used to imply reversibility. If both reactants and products are present in a reaction mixture, there will be molecules of reactant being converted to product and "product" molecules becoming "reactant" molecules at any given moment. Thus, the equilibrium state is dynamic, not static.

Figure 5. Potential Energy vs. Progress of Reaction

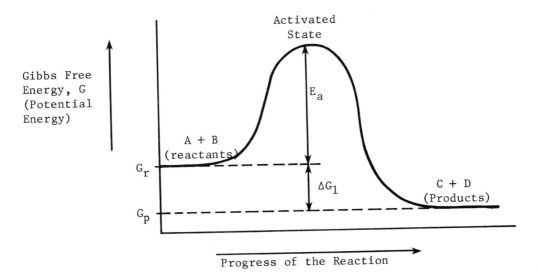

In this diagram E_a = activation energy (free energy of activation; discussed further under kinetics)

G_p = free energy of the products

G_r = free energy of the reactants

$\Delta G_1 = (G_p - G_r) = (G_{C+D} - G_{A+B}) < 0$

(The symbol Δ indicates a change or the difference between two quantities.)

Since ΔG is negative, energy is released in converting reactants to products and the reaction is exergonic. If ΔG is positive, energy is absorbed from the environment and the reaction is endergonic.

Thermodynamics deals with (among other things) the size (magnitude) and sign (+ or -) of ΔG, the difference in free energy between the reactants and products. For a given set of reactants and products at constant pressure (for a given reaction), ΔG depends only on the temperature and the concentrations of the reactants and products. Physically, ΔG is the amount of useful work which can be obtained from a chemical reaction at a specified temperature and set of concentrations.

2. Suppose that one starts with (C + D) and wants to convert them
    to (A + B). This is the reverse of the first reaction.
    ΔG is defined as

$$\Delta G_{-1} = (G_{A+B} - G_{C+D}) = -(G_{C+D} - G_{A+B}) = -\Delta G_1 \text{ and } \Delta G_{-1} > 0.$$

(Note that ΔG_1 is negative so that $-\Delta G_1$ is positive.) The (-1)
in the subscript indicates the reverse of reaction designated
by the subscript (1). The reverse of an exergonic reaction
is always an endergonic reaction. Similarly, for (Y + Z) →
(W + X) (the reverse of the reaction portrayed in Figure 6),

$$\Delta G_{-2} = (G_{W+X} - G_{Y+Z}) = -(G_{Y+Z} - G_{W+X}) = -\Delta G_2 \text{ and } \Delta G_{-2} < 0.$$

Figure 6. Potential Energy vs. Progress of Reaction

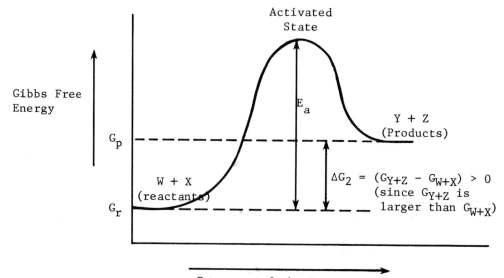

Every reaction is thermodynamically reversible. If, however,
ΔG is very large, so that a great deal of energy must be
supplied for reversal, a reaction may be considered
irreversible for all practical purposes. This means that
under the conditions being used, the back-reaction proceeds to
such a small extent that it need not be considered. (This is
different from saying that a reaction proceeds so slowly that
it can be ignored. That situation will be considered under
kinetics, below.)

3. The dependence of ΔG on temperature and concentration may be written

$$\Delta G = \Delta G^\circ + RT \ \ln \ \frac{(Products)}{(Reactants)}$$

where ΔG° = standard free energy change in calories/mole of
           reactant consumed,
       R = gas constant (1.987 Cal. mole^{-1} deg.^{-1}),
       T = absolute temperature (= $^\circ C$ + 273), Kelvin
     $\ln$ = natural (base e) logarithm,
and (products), (reactants) are molar concentrations (for
solvents and solutes) or pressures in atmospheres (for gases).

Since at equilibrium, $\Delta G = 0$, this equation can be rewritten

$$\Delta G^\circ = -RT \ \ln \ K_{eq}$$

where $K_{eq} = \frac{(Products)}{(Reactants)}$, with the concentrations measured at
equilibrium.

The <u>equilibrium constant</u>, K_{eq}, depends only on the temperature
of the system. This equation provides a way of determining
ΔG° from equilibrium concentration data.

Another commonly used form is

$$\Delta G^\circ = -RT(\ln 10)(\log K_{eq}) = -2.303 \ RT \ \log K_{eq}$$

where log means base 10 logarithm.

The "standard" state for which ΔG° is defined is arbitrary.
By far the most common choice, though, is one in which the
temperature is $25^\circ C$, all gases are present at pressures of
one atmosphere, all liquids and solids are pure (i.e., not
mixtures), and all solutes have unit activity. (Activity is
similar to concentration but it takes into account interactions
of solute molecules with each other and with the solvent.)
Activities are difficult to measure, however, and molar
concentrations are commonly used in their place. Since
concentration and activity are almost equal in dilute solutions
and many biological reactions occur at quite low concentrations,
this approximation is acceptable.

A physical interpretation of ΔG° can be obtained by noting that,
if

$$\frac{(Products)}{(Reactants)} = 1 \text{ then } \Delta G = \Delta G^\circ.$$

ΔG^O is the amount of useful work which can be obtained by conversion of one mole of each of the reactants in their standard states to one mole of each of the products in their standard states.

The standard free energy change for a reaction is also useful for predicting whether a reaction will occur or not <u>under standard conditions</u>. If $\Delta G^O < 0$, then the reaction will occur spontaneously, provided standard conditions prevail; if $\Delta G^O > 0$, the reaction will not occur by itself. Note, however, that it is <u>really ΔG</u> which determines whether or not a reaction will occur under conditions <u>different from the standard state</u>, such as those existing within a cell. Conclusions based on ΔG^O concerning what reactions will and will not occur in the body can be quite erroneous.

4. In biological systems two other conditions are generally assumed.

   a. Since most solutes are present at fairly low concentrations (less than 0.1 M) and water is present at high concentration (55.6 M for pure water), it is assumed that the water concentration is constant. If water enters into a reaction it is incorporated into K_{eq}.

   b. If H^+ or OH^- participates in a reaction, their concentrations influence ΔG^O. As most biological reactions occur in systems buffered to pH $\simeq$ 7, it is useful to define $\Delta G^{O'}$ as being the value of ΔG^O measured at pH = 7 (sometimes other pH's are used). This corresponds to $(H^+) = 10^{-7}$ M. $\Delta G^{O'}$ values can be compared to each other but not to values of ΔG^O. Although this convention is usually followed, if one sees ΔG^O (without the prime) in a biochemical context, it is advisable to check at what pH the measurement was made.

These two conventions are illustrated by the hydrolysis of ethyl acetate at $25^O C$, followed by dissociation of the acetic acid produced.

$$
\underset{CH_3C-O-CH_2CH_3}{\overset{O \atop \parallel}{}} + H_2O \rightleftarrows \underset{CH_3C-O^-}{\overset{O \atop \parallel}{}} + H^+ + HOCH_2CH_3
$$

$$
K_{eq} = \frac{(CH_3COO^-)(H^+)(HOCH_2CH_3)}{(CH_3COOCH_2CH_3)(H_2O)} = 5.8 \times 10^{-6}
$$

ΔG^o = -2.303 RT log K_{eq}

$\quad$ = -(1.987 $\,$ cal $\,$ mole^{-1} deg $^{-1}$)(298°K)(log 5.8 x 10^{-6})

$\quad$ = -592 log 5.8 x 10^{-6}

$\quad$ = +3100 cal $\,$ mole^{-1} = +3.1 kcal $\,$ mole^{-1}

Alternatively, setting (H$^+$) = 10^{-7} M (pH = 7) and (H$_2$O) = 55.5 M,

$$\dot{K}' = K_{eq} \frac{(H_2O)}{(H^+)} = \frac{(CH_3COO^-)(HOCH_2CH_3)}{(CH_3COOCH_2CH_3)}$$

$$= 5.8 \times 10^{-6} \frac{55.5}{1 \times 10^{-7}} = 3.22 \times 10^{+3}$$

and $\Delta G^{o'}$ = -592 log 3.22 x 10^{+3}

$$= -2076 \text{ cal } \text{ mole}^{-1} = -2.08 \text{ kcal } \text{ mole}^{-1}$$

Although under standard conditions the reaction does not occur spontaneously ($\Delta G^o > 0$), if more commonly encountered concentrations of water and hydrogen ion are present, the reaction would proceed by itself.

5.$\quad$a.$\quad$In order to understand the concept of free energy, an example (described by Baldwin) is useful. Consider a self-operative heat engine which functions by taking in an agent (e.g., steam) at temperature T_1 and releasing it at T_2 ($T_1 > T_2$). The heat extracted (designated Q) is converted as completely as possible into useful work. If W represents the maximum useful work available,

$$W = Q \frac{T_1 - T_2}{T_1} = Q - Q \frac{T_2}{T_1}$$

$\quad\quad$where T_1 and T_2 are absolute (Kelvin) temperatures. Unless $T_2 = 0$ or $T_1 = \infty$, the useful work is always less than the total energy supplied, by a factor of Q $(T_2/T_1) = (Q/T_1)T_2$.

$\quad$b.$\quad$The ratio (Q/T_1) is known as the <u>entropy</u> of the system and is designated by S. The amount of energy unavailable for useful work (= T_2S) is that which is lost in the process of energy transfer. It can be thought of as the amount of randomness or disorder introduced into the system during the transfer.

c. For chemical systems, one can define a quantity G by the equation G = H − TS. Here, G (free energy) is analogous to W, the amount of useful work available, H (enthalpy) is analogous to Q, the heat content of the system at constant pressure, S (entropy) is analogous to (Q/T_1), the wasted heat energy, and T is the absolute temperature. (Although it is not shown here, this and the following functions can all be derived with complete rigor from first principles of classical thermodynamics.)

This relationship is more commonly used to describe <u>changes</u> in these quantities. If a system goes from state I (with G = G_I, H = H_I, S = S_I) to state II (G = G_{II}, H = H_{II}, S = S_{II}) at constant temperature, one can write

$$G_{II} - G_I = (H_{II} - H_I) - T (S_{II} - S_I)$$

or, in condensed form,

$$\Delta G = \Delta H - T\ \Delta S$$

where Δ again is read as <u>change</u>.

6. The enthalpy, H, is related to the internal energy of a system by the equation

$$H = E + PV$$

where E = internal energy,
     P = external pressure on the system, and
     V = volume of the system.

In terms of changes between states, this becomes

$$\Delta H = \Delta E + \Delta(PV) = \Delta E + P\Delta V + V\Delta P.$$

At constant pressure, $\Delta P = 0$ and $\Delta H = \Delta E + P\Delta V$. If, in addition, the volume is constant, then $\Delta V = 0$ and $\Delta H = \Delta E$.

In many biological reactions $\Delta V = \Delta P = 0$. Consequently, for a change of state, ΔH, the maximum amount of energy which can be released as heat is equal to ΔE, the change in the internal energy of the system.

ΔH can be measured for a particular material by burning the material in a bomb calorimeter at constant pressure. The heat released is measured as the temperature rise in a large water bath surrounding the combustion chamber. The values of **ΔH** obtained in this manner are indicative of the total energy available in a compound when it is completely oxidized. The factors which usually contribute to ΔH under these conditions are the heat of fusion (melting), the heat of vaporization (boiling), and the heat of combustion (bond making and breaking). The first two are important, for example, if at the standard temperature (25°C) a solid is combusted to a combination of liquids and gases, as in the examples below.

Just as with ΔG, ΔH for a process is equal to -ΔH for the reverse process. Thus: ΔH (liquid → gas) = -ΔH (gas → liquid); ΔH (liquid → solid) = -ΔH (solid → liquid); ΔH (making a bond) = -ΔH (breaking the same bond).

The following reactions were all run at one atmosphere pressure. The subscripts s, l, and g (solid, liquid, and gas, respectively) indicate the phase of the material under one atmosphere pressure at the given temperature. The ΔH° values given are standard heats of formation since all reactants are in their standard (natural) state for the temperature given.

a. Oxidation of one mole of glucose (a carbohydrate) to carbon dioxide and water

$$C_6H_{12}O_{6(s)} + 6O_{2(g)} \rightarrow 6H_2O_{(l)} + 6CO_{2(g)}$$

at 20°C, ΔH° = -673 kcal /mole

b. Oxidation of one mole of palmitic acid (a fatty acid) to carbon dioxide and water

$$C_{16}H_{32}O_{2(s)} + 23O_{2(g)} \rightarrow 16CO_{2(g)} + 16H_2O_{(l)}$$

at 20°C, ΔH° = -2,380 kcal /mole

c. Oxidation of one mole of glycine (an amino acid) to carbon dioxide, water, and nitrogen

$$C_2H_5O_2N_{(s)} + 2\text{-}1/4\ O_{2(g)} \rightarrow 2CO_{2(g)} + 2\text{-}1/2\ H_2O_{(l)} + 1/2\ N_{2(g)}$$

at 25°C, ΔH° = -233 kcal /mole

ΔG and ΔG^o are more useful to biochemists than are ΔH and ΔH^o, because the Gibbs free energy is the _useful_ work (or energy) available from a reaction. Values of ΔH can be misleading when used to discuss the amount of useable energy which a reaction can supply.

7. In living systems, equilibrium is the exception rather than the rule. In fact, life has been defined as the ability to utilize energy from an external source to maintain chemical reactions in a non-equilibrium state. Death corresponds to the attainment of equilibrium by these same reactions.

    In a reaction sequence

$$A \rightarrow B \rightarrow C$$

The reaction $A \rightarrow B$ can be prevented from reaching equilibrium by removing B (converting it to C) faster than it can be made from A. This will be discussed further under kinetics. All reactions in the body are interrelated and the system as a whole is in some sort of balance, although individual reactions usually are not in equilibrium. This condition is known as a steady-state. A change in concentration of any one component (product of one reaction which is used as reactant by another reaction) shifts the concentration of all the other components which are linked to it by a sequence of chemical reactions. This results in the attainment of a new steady-state.

Of interest here is the value of ΔG for a non-equilibrium system. This can be determined from ΔG^o and the actual concentrations of products and reactants in the cell, by use of the equation (given previously)

$$\Delta G = \Delta G^o + 2.303 \ RT \ \log_{10} \frac{\text{(Products)}}{\text{(Reactants)}}$$

If ΔG is negative, then under cellular conditions the reaction proceeds to form products. If ΔG is positive, then the materials called reactants here are what is actually being made in the cell.

Consider the interconversion of 1,3-diphosphoglyceric acid (1,3-DPG) and 2,3-diphosphoglyceric acid (2,3-DPG) in the red blood cell. Suppose that at equilibrium, (1,3-DPG) = 0.1 M and (2,3-DPG) = 0.9 M. Then

$$K_{eq} = \frac{(2,3\text{-DPG})}{(1,3\text{-DPG})} = \frac{0.9}{0.1} = 9.$$

At 37°C (human body temperature)

$$\Delta G^O = -RT \ln K_{eq}$$

$$= -2.303 \times 1.987 \times 310 \log_{10} 9$$

$$= -1353 \text{ calories/mole}$$

Under standard conditions, 1,3-DPG is converted to 2,3-DPG spontaneously.

In the red cell, the <u>actual</u> (non-equilibrium) concentrations are about

$$(1,3\text{-DPG}) = 1 \times 10^{-6} \text{ M}$$

$$(2,3\text{-DPG}) = 4.3 \times 10^{-3} \text{ M}$$

$$\Delta G = \Delta G^O + RT \ln \frac{(2,3\text{-DPG})}{(1,3\text{-DPG})}$$

$$= -1353 + 2.303 \times 1.987 \times 310 \times \log_{10} \frac{4.3 \times 10_{-3}}{1 \times 10^{-6}}$$

$$= -1353 + 5154$$

$$= +3801 \text{ calories/mole}$$

Thus, under conditions existing in the red cell, it would appear that there is no synthesis of 2,3-DPG occurring. Instead 1,3-DPG is being formed from 2,3-DPG. A reaction which is <u>favored</u> on the basis of the relative stability of the reactants and products ($\Delta G^O < 0$) can be <u>reversed</u> by adjusting the concentrations (greatly elevating the concentration of the most stable component) to make $\Delta G > 0$.

In the red cell, the situation is actually more complicated. 2,3-DPG binds strongly to deoxyhemoglobin

$$\frac{(HbDPG)}{(Hb)} = 1.4 \times 10^5 \quad ,$$

thus making much of it unavailable for reconversion to 1,3-DPG. In a completely deoxygenated red cell, the amount of <u>free</u> 2,3-DPG is about 3.8×10^{-5} M.

This gives

$$\Delta G = -1353 + 1419 \log_{10} \left(\frac{3.8 \times 10^{-5}}{1 \times 10^{-6}} \right)$$

$$= +888 \text{ calories/mole}$$

This would still favor 1,3-DPG formation, but the situation is much nearer equilibrium ($\Delta G = 0$) than before. As a result of coupling two reactions

$$1,3\text{-DPG} \overset{\leftarrow}{\rightarrow} 2,3\text{-DPG}$$

$$2,3\text{-DPG} + \text{Hb} \overset{\leftarrow}{\rightarrow} 2,3\text{-DPG} \cdot \text{Hb}$$

(where Hb ≡ hemoglobin),

the overall process is brought much closer to equilibrium. Even this is an oversimplification, though, and the values used are rather inexact. In the red cell there is a dynamic balance among all the components which adjusts the 2,3-DPG level to an optimum.

8.   Oxidation and reduction reactions

Oxidation is the loss of electrons or hydride (H^-) ions (but not hydrogen (H^+) ions) by a molecule, atom, or ion. (The term oxidation was originally applied to a group of reactions in which some material combined with oxygen (O_2). It is now recognized that these reactions belong to the more general class of conversions which result in electron loss by a compound and the name is now used for this larger group. In some, but by no means all such reactions, oxygen is the electron acceptor and one or more atoms of oxygen become part of the product molecule.)

Reduction is the gain of electrons or hydride (H^-) ions by a molecule, atom, or ion.

Note that transfer of one hydride ion results in the transfer of two electrons.

The amount of work required to add or remove the electrons is called the electromotive potential or force (emf) and is designated E or ε. It is measured in volts (joules/coulomb, where a coulomb is a unit of charge, a quantity of electrons).

The standard emf, E^O (or, at pH 7, $E^{O'}$) is the emf measured
when the temperature is $25^O C$ and the materials being oxidized
or reduced are present at concentrations of 1.0 M. In
biological systems, $E^{O'}$ is most commonly used.

A half-reaction is one in which electrons or hydride ions are
written explicitly. Values of E^O and $E^{O'}$ are tabulated with
the half-reactions for which they are measured. For example,

$$NADH \rightarrow NAD^+ + H^- \qquad E^{O'} = -0.32 \text{ volts}$$

and $H_2O \rightarrow 1/2\ O_2 + 2H^+ + 2e^- \qquad E^{O'} = +0.816 \text{ volts}$

are half-reactions since they show electrons (e^-) and hydride
ions (H^-).

Alternatively, these half-reactions can be written

$$NAD^+ + H^- \rightarrow NADH \qquad E^{O'} = +0.32 \text{ volts}$$

$$2e^- + 2H^+ + 1/2\ O_2 \rightarrow H_2O \qquad E^{O'} = -0.816 \text{ volts}$$

Notice that the sign of $E^{O'}$ changes. Changing the direction
of a reaction reverses the sign of the potential change (just
as with ΔG and ΔH).

The oxidation potential is actually a statement of the ease
of removing electrons from a material compared to the ease
of removing electrons from hydrogen in the half-reaction,

$$H_2 \rightarrow 2H^+ + 2e^- \qquad E^O = 0.00 \text{ volts}$$

(Note that, in this reaction, E^O is defined as zero, thus
fixing the scale of E^O value for other reactions.) If it is
easier to remove electrons from something, then $E^{O'}$ will be
negative; if it is harder, $E^{O'}$ is positive. E^O for the hydrogen
half-reaction is used as the zero-point even when talking
about $E^{O'}$ values. Thus,

$$H_2 \rightarrow 2H^+ + 2e^- \qquad E^{O'} = -0.42 \text{ volts}$$

It is easier to remove electrons from hydrogen (and produce H^+)
when $(H^+) = 10^{-7}$ M than when $(H^+) = 1.0$ M. The choice of a
negative sign for the standard emf to mean "easier to remove
electrons" is arbitrary. Although the "negative-easier" choice
is used in this book, the International Union of Pure and
Applied Chemistry has recommended the opposite convention. The
older convention is retained here because it is commonly
encountered in medically and biologically related texts and
literature.

Since free electrons combine exceedingly rapidly with
whatever is at hand, half-reactions never occur by themselves.
There must always be something accepting electrons just as
fast as they are being released. The substance <u>releasing</u>
electrons (or H^-) is the <u>reductant</u> or <u>reducing agent</u> (since
it is oxidized); and the substance <u>accepting</u> electrons is
the <u>oxidant</u> or <u>oxidizing agent</u> (since it is reduced). Two
half-reactions, when combined, give a <u>redox</u> (oxidation-
reduction) <u>reaction</u>. When balanced, such reactions <u>never show</u>
<u>free (uncombined) electrons</u>. For example,

$$H^+ + NADH + 1/2 \; O_2 \rightarrow NAD^+ + H_2O$$

$$E^{O'}_T = -0.32 - (+0.816) = -1.136 \text{ volts}$$

By definition, $E^{O'}_T$, termed the net or total potential or emf,
for this reaction is calculated according to

$$E^{O'}_T = E^{O'} \text{ (reductant)} - E^{O'} \text{ (oxidant).}$$

Under standard conditions (and pH = 7), the reaction will occur
as written if $\underline{E^{O'}_T < 0}$. Otherwise the reaction will proceed
from right to left (the reverse of the way in which it is
written).

Notice also that $E^{O'}$ is the amount of energy (work) <u>per coulomb</u>
rather than the <u>total</u> energy required for oxidation.
Consequently, although the amount of work changes, if the
amount (number of moles) of material transformed is altered,
$E^{O'}$ -- and $E^{O'}_T$ -- are <u>independent</u> of the number of moles oxidized
or reduced.

9. When $E^{O'}_T$ in volts is converted to calories/mole, the result is
    $\Delta G^{O'}$ for the redox reaction being considered. This is
    accomplished by means of the equation

$$G^{O'} = \frac{n \; E^{O'}_T \; F}{4.184} = (23,061) \; n \; E^{O'}_T$$

where F = the faraday (= 96,487 coulombs/gram-equivalent),
      n = number of faradays (gram-equivalents) of electrons
          transferred per mole of material oxidized or reduced,
  $E^{O'}_T$ = $E^{O'}$ (reductant) - E^O (oxidant) in volts, and
  $\Delta G^{O'}$ = the standard free energy change in calories per mole
          of material oxidized or reduced (since there are
          4.184 joules/calories).

For illustration, the reaction

$$H^+ + NADH + 1/2 \; O_2 \to NAD^+ + H_2O \qquad E_T^{O\prime} = -1.136 \text{ volts}$$

is used. For this,

> n = 4 gram-equivalents of electrons per mole O_2 (since the half-reaction $O_2 + 4e^- \to 20^{-2}$ involves four electrons),

> = 2 gram-equivalents of electrons per mole H_2O, NAD^+, or NADH (since the half-reactions for one mole of each of these materials involves only two electrons).

Then $\Delta G^{O\prime} = 23{,}061 \times (-1.136) \times n = -26{,}197n$

> $\Delta G^{O\prime} = -26{,}197 \times 2 = -52{,}394$ calories/mole of H_2O, NADH, or NAD^+

> $= -26{,}197 \times 4 = -104{,}788$ calories/mole of O_2 transformed.

As in all reactions, the actual value of $\Delta G^{O\prime}$ depends on the reactant or product for which it is calculated.

10. The importance of understanding free energy changes and redox reactions can be seen by noting that all life on earth is based on the redox reaction,

$$mCO_2 + mH_2O + \text{energy} \; \underset{\text{animals}}{\overset{\text{plants}}{\rightleftarrows}} \; mO_2 + (CH_2O)_m$$

$E_T^{O\prime} \simeq +1.24$ volts

$\Delta G^{O\prime} \simeq +114{,}300m$ calories/mole of carbohydrate

In animals, carbohydrate (represented by $(CH_2O)_m$) and oxygen are consumed and energy, water, and carbon dioxide are released. The energy is stored, primarily as adenosine triphosphate (ATP), to be used when needed by cellular processes. In plants, energy from the sun interacts with and is absorbed by the chloroplasts of green plants, causing water to be oxidized and carbon dioxide reduced, producing oxygen and carbohydrate.

In order to understand the many steps between carbohydrate and carbon dioxide and water, a knowledge of the energetics involved is fundamental. Oxidation potentials are of particular

use in discussing the initial and terminal steps of the
process: the trapping of light energy by chloroplasts during
photosynthesis, and the formation of water by the reduction
of oxygen via the electron transport chain. After the energy
from carbohydrate oxidation is stored as ATP, further
transfers of it are usually spoken of in terms of free energy
changes.

General Kinetics

Thermodynamics deals with the relative energies and states of
reactants and products, not caring how one travels between these
states. Kinetics, then, is complementary, considering how fast
a reaction occurs and the actual pathway (energetic, structural) --
termed the mechanism -- which it follows. In fact, there is no way,
using kinetics, to decide whether a reaction will or will not occur,
provided an effectively infinite period of time is available. This
will be discussed below in more detail.

1. Reaction Rates

The conversion of $A \xrightarrow{k} B$ (where k is the rate constant,
discussed below) can be represented diagrammatically as the
increase of product, B, with respect to time.

Figure 7.   Time Course of a Chemical Reaction

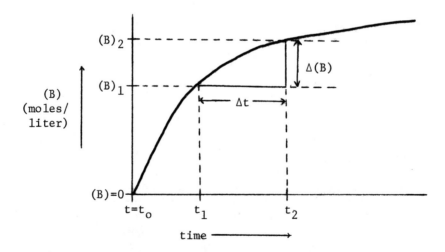

where (B) means the molar concentration of substance B. The average velocity ($\bar{v}$) for the period t_1 to t_2 is the slope,

$$\bar{v} = \frac{\Delta(B)}{\Delta t} = \frac{(B)_2 - (B)_1}{t_2 - t_1}$$

The <u>instantaneous</u> velocity at some time t' ($\neq t_o$) is

$$v = \frac{d(B(t))}{dt} \qquad \text{evaluated at t'}$$

The expression $\dfrac{d(B(t))}{dt}$ is the first derivative of (B(t)) with respect to t.

It is the limit of the slope $\dfrac{\Delta(B)}{\Delta t}$ as $\Delta t \to 0$, or

$$v = \lim_{\Delta t \to 0} \bar{v} = \lim_{\Delta t \to 0} \frac{\Delta(B)}{\Delta t} = \frac{d(B(t))}{dt}$$

Alternatively, one can consider the change in concentration of the reactants, Thus,

$$v = \frac{d(B(t))}{dt} = -\frac{d(A(t))}{dt}$$

since (A), the reactant concentration, decreases at the same rate as (B), the product concentration, increases. The expressions (B(t)) and (A(t)) are used to imply that one has a detailed knowledge (i.e., an analytic function) of the way in which the concentrations of A and B vary as time (and the reaction) progresses. In reality, however, this is seldom true. To avoid this implication and to make a more compact notation, (B) and (A) will be used from now on instead of, respectively, (B(t)) and (A(t)).

Consideration of the chemistry of the situation provides a way around the need for this analytic function. The rate of formation of B should increase if (A), the reactant concentration, increases. It is reasonable, then, to write

$$\frac{d(B)}{dt} = k(A)^n$$

where k is the <u>rate constant</u> (mentioned above), and n = 0,1,2... is the <u>order</u> of the reaction with respect to A. (The order is zero, first, second, etc., with respect to A if n = 0,1,2, etc.)

The rate constant is the proportionality constant between the rate of formation of B and the molar concentration of A. It is characteristic of a particular reaction. The units of k depend on the order of the reaction. For zero-order, they are moles liters^{-1} time^{-1} (time is frequently in seconds). For first-order, they are time^{-1}; second order is liters moles^{-1} time^{-1}, etc. They are whatever is necessary to make d(B)/dt have units of moles liters^{-1} time^{-1}. Notice that, if n = 0, v = k (zero-order reaction). This means that (B) changes in a constant way, <u>independent of the concentration of reactants</u>. This is especially important in enzyme kinetics. A plot of (B) versus t for such a reaction is a <u>straight line</u>.

A somewhat more complicated example is

$$A + B \xrightarrow{\quad k \quad} C + D$$

$$v = \frac{d(C)}{dt} = \frac{d(D)}{dt} = \frac{d(Products)}{dt} = k(A)(B)$$

This reaction is first order with respect to A, first-order with respect to B, and second-order overall. Most reactions are zero-, first-, or second-order.

The concept of reaction order can be related to the number of molecules which must collide simultaneously for the reaction to occur. In a first-order reaction, no collisions are required (since only one reactant molecule is involved). Every molecule having sufficient free energy to surmount the activation barrier will spontaneously convert to products. In a second-order reaction, not only must two molecules have enough free energy, they have to collide with each other for the products to form. A third-order reaction requires the simultaneous meeting of three molecules, a very unlikely event. Because of this, reactions of order higher than second are very seldom encountered in simple chemical conversions.

Higher <u>apparent</u> orders are encountered, however, in some cases where an overall rate equation is written for a process which actually proceeds in several steps, via one or more intermediates. In such situations, the individual steps will seldom if ever involve a third-or higher order process. This is sometimes encountered in enzyme catalyzed reactions. The concept of reaction order for such an overall process is somewhat artificial and of little use.

Zero-order reactions can be accounted for in two ways. If a process is truly zero-order, <u>with respect to all reactants</u> (and catalysts, in the case of enzymes), then either the activation energy is zero or every molecule has sufficient energy to overcome the activation barrier. This kind of reaction is seldom if ever found in homogeneous reactions in gases or solutions.

Alternatively, a reaction can be zero-order <u>with respect to one or more (but not all) of the reactants</u>. This is a very important case, especially in enzyme kinetics and clinical assays utilizing enzymes (see the section on the uses of enzymes in the clinical laboratory). This can be explained by noting that one of the reactants (or a catalyst, such as an enzyme) is in limited supply. Increasing the availability (concentrations) of the other reactants can result in no increase in the velocity beyond that dictated by the limiting reagent. This makes the rate independent of the concentrations of the non-limiting materials and, hence, zero-order with respect to those materials.

As was pointed out in the section on thermodynamics, all reactions are theoretically reversible. The correct way to write $A \rightarrow B$, then, is

$$A \underset{k_{-1}}{\overset{k_1}{\rightleftarrows}} B$$

where k_1 is the rate constant for the conversion of A to B and k_{-1} is the rate constant for the reverse reaction, conversion of B to A. The reaction is kinetically reversible if $k_1 \approx k_{-1}$; it is irreversible for all practical purposes (proceeds in the reverse direction so slowly that it can be ignored) if $k_{-1} \ll k_1$. The rate of a reversible reaction is written

$$v = \frac{d(B)}{dt} = k_1(A) - k_{-1}(B).$$

For $A + B \underset{k_{-1}}{\overset{k_1}{\rightleftarrows}} C + D$

$$v = \frac{d(Products)}{dt} = \frac{d(C)}{dt} = \frac{d(D)}{dt} = k_1(A)(B) - k_{-1}(C)(D).$$

These kinetic schemes (ways of expressing reaction velocities as functions of concentrations and rate constants) can become very complicated, if there are many linked (sequential) reactions and intermediates. They are very important in biological systems, however, since enzymes make reversible many otherwise kinetically irreversible reactions. They will be discussed further in enzyme kinetics.

2. The rate at which a reaction will proceed (measured by k) is directly related to the amount of energy which must be supplied before reactants and products can be interconverted. This activation energy, E_a, comes from the kinetic energy possessed by the reactants. This may be translational and rotational energy (needed if two molecules must collide to react); or vibrational and electronic energy (useful when one molecule rearranges itself or eliminates some atoms to form the product). The larger the activation energy, the slower the reaction rate, and the smaller the rate constant.

E_a is a free energy, so it has both ΔH and ΔS terms (see page 71, 5, above). This means that, not only must the reactants have enough energy (ΔH) to make and break the requisite bonds, but they must also be properly oriented (ΔS) for the products to form. This is useful in thinking about how catalysts function (see **page 86, 5, below**).

Figure 8.   Illustration of Forward and Reverse Activation Energies

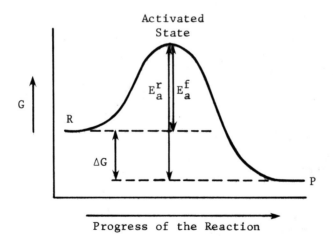

Progress of the Reaction

For this free energy diagram

$$R \underset{k_{-1}}{\overset{k_1}{\rightleftarrows}} P$$

E_a^f = G(activated state) - G(reactants) = activation energy for
$$R \to P$$

E_a^r = G(activated state) - G(products) = activation energy for
$$P \to R$$

$E_a^r = E_a^f - \Delta G$; r and f = reverse and forward, respectively.

Since $\Delta G < 0$ (the reaction as written is exergonic), $E_a^r > E_a^f$.

3. The <u>Arrhenius Equation</u> interrelates rate constants and activation energies

$$k = Ae^{-E_s/RT}$$

where E_s = Arrhenius activation energy ($= E_a + T\Delta S_a$) in
calories/mole;
   R = gas constant = 1.987 calories mole^{-1} degrees K^{-1};
   T = absolute temperature ($^\circ$K);
   A = the Arrhenius pre-exponential factor, a constant;
   k = rate constant; and
   e = 2.71828, the base of the natural logarithms.

From the Arrhenius equation, it is apparent that the larger E_s (or E_A) is, the smaller is the value of k. The basis for kinetic irreversibility (see page 80, 1, above) is the same, then, as that for thermodynamic irreversibility: the more energy a process takes (the more positive is ΔG or E_a), the more difficult and less likely it is to occur.

Although values for A are not well understood, the Arrhenius equation can be used to calculate E_s by rearranging it to

$$\ln k = \left(\frac{-E_s}{R}\right)\left(\frac{1}{T}\right) + \ln A.$$

The slope of a plot of $\ln k$ versus $(1/T)$ equals $(-E_s/R)$, from which E_s can be derived. In order to get E_a, ΔS_a (entropy of activation) is needed.

4. For a reversible reaction, the equilibrium state is one in which the concentrations of reactants and products are such that the forward and reverse rates are equal. This leads to the conclusion that

$$K_{eq} = \frac{k_1}{k_{-1}},$$

which is demonstrated below.

For a reversible reaction, $R \underset{k_{-1}}{\overset{k_1}{\rightleftharpoons}} P$,

the rate of expression is $\frac{d(P)}{dt} = k_1(R) - k_{-1}(P)$.

By definition, the concentrations of products and reactants are constant at equilibrium so that

$$\frac{d(P)}{dt} = 0 = k_1(R) - k_{-1}(P)$$

This can be rearranged to $\frac{k_1}{k_{-1}} = \frac{(P)}{(R)}$

and, since $K_{eq} \equiv \frac{(P)}{(R)}$, it is true that $\frac{k_1}{k_{-1}} = K_{eq}$.

This is another idea which is important in understanding enzyme kinetics. If some reversible step of an enzyme-catalyzed reaction has low enough activation energy, it can be treated as a true equilibrium, simplifying the mathematics of the system.

5. So far, K_{eq} has always been written as a dissociation constant. That is

$$K_{eq} = \frac{(products)}{(reactants)} = K_{diss} = \frac{k_1}{k_{-1}}$$

It is just as correct, though less common, to write it as an association constant.

$$K_{assoc} = \frac{(reactants)}{(products)} = \frac{1}{K_{diss}} = \frac{k_{-1}}{k_1}$$

The more stable the materials termed products are, the larger
is K_{diss} and the smaller is K_{assoc}. This follows from the
thermodynamic arguments presented previously.

6. Catalysts and catalysis

Although ΔG^o and K_{eq} depend solely on temperature, E_a can be
decreased for most reactions by catalysts. Since rate constants
are functions of E_a, catalysts also increase them and,
consequently, reaction rates. Catalysts do not affect ΔG^o or
K_{eq}.

Catalysts are not altered in reactions that they catalyze.
If X is a catalyst for A $\overset{\rightarrow}{\underset{\leftarrow}{}}$ B, then

$$X + A \overset{\rightarrow}{\underset{\leftarrow}{}} X + B$$

Although A and B are interconverted, X is unchanged in the net
reaction. It may and probably does exist, at some intermediate
step, in an altered form, but it always is returned unchanged.
For this reason, a very small (catalytic) amount of X can cause
the reaction of a great deal of A and B.

Since catalysts function by lowering E_a, they speed up forward
and reverse reactions equally. This is essentially saying, in
a different way, that catalysts do not affect K_{eq}. This can
be seen in the potential energy diagram below, for A $\overset{\rightarrow}{\underset{\leftarrow}{}}$ B.

Figure 9.   Effect of a Catalyst on the Activation and Free Energies
               of a Chemical Reaction

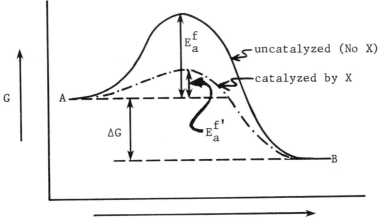

Progress of the Reaction

Catalysts do, however, hasten the <u>attainment</u> of equilibrium. For a given system, ΔG (a function of temperature and concentration) always determines whether or not a reaction will occur.

Since E_a is a free energy change, it is correct to write

$$E_a = \Delta H_a - T\Delta S_a.$$

A catalyst can

i) <u>decrease</u> H_a by changing the electronic distribution in the reactants, making it easier to make or break a bond;

ii) <u>increase</u> S_a by binding and orienting the reactant molecules (decreasing their randomness), making successful collisions more likely.

Both of these effects work to lower E_a. The extent to which each occurs in a particular situation depends on the reaction and the catalyst.

<u>Enzyme Kinetics</u>

Although the thermodynamics of biological systems can be readily handled in terms of classical ideas and methods, the kinetics of bodily processes are more clearly understood when discussed in the language of enzyme kinetics. This is due primarily to the many types of enzymes involved and the myriad ways in which they can function. The principles of enzyme kinetics, however, are no different from those of classical chemical kinetics.

1. Enzymes as catalysts

Most biological reactions would be kinetically irreversible were it not for the class of catalysts known as <u>enzymes</u>. These are proteins which can (usually in association with one or more <u>coenzymes or cofactors</u>) greatly accelerate biochemical reactions. The diagram below shows schematically what an enzyme does.

Figure 10. Effect of an Enzyme on the Activation and Free Energies of a Chemical Reactions

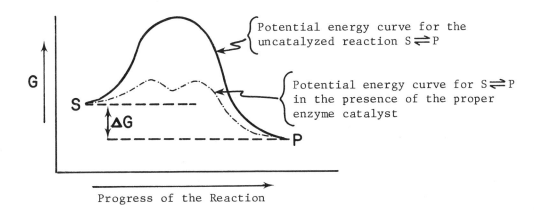

The catalyzed reaction can be written

$$S \underset{k_{-1}}{\overset{k_1}{\rightleftharpoons}} P$$

where R (for reactant) is replaced by S (for substrate). The substances which an enzyme can convert to product are termed substrates. For a freely reversible reaction, the product for one direction is the substrate for the reverse.

The dip (local energy minimum) in the enzyme catalyzed curve indicates the presence of a metastable intermediate (an intermediate having more free energy than either S or P, but which requires some activation energy for conversion to either one). This reaction can be written

$$E + S \underset{k_{-1}}{\overset{k_1}{\rightleftharpoons}} ES \underset{k_{-2}}{\overset{k_2}{\rightleftharpoons}} E + P$$

The intermediate is ES, an enzyme-substrate complex. (Notice that the enzyme (catalyst), E, is regenerated, as it must be.) It is also apparent from the curve and the rate equation that there are four activation energies. The original rate constants and activation energies have no bearing on the rates of the enzyme-catalyzed reaction.

The scheme given above is the simplest one involving intermediates. Some enzyme catalyzed reactions may have no metastable intermediates (as with $X + A \rightleftarrows X + B$, where the activated state represents an unstable intermediate); or they may have many intermediates, with one or more possible kinetic pathways from S to P. The description of such reactions requires a knowledge of the <u>mechanisms</u> by which they occur, and this must be worked out for each reaction and enzyme. Cases are known where one enzyme can catalyze two different (though similar) reactions by apparently different mechanisms (e.g. hydrolysis of amide and ester bonds by chymotrypsin). The general procedure for obtaining this information is to measure rates and rate constants for as many processes as possible, then fit them together into a self-consistent pattern which is chemically plausible. Certain approximations are usually possible which greatly simplify this task.

2. The rate at which an enzyme-catalyzed reaction occurs is influenced by several factors. These include

    a. <u>Temperature</u>. The rates of all chemical reactions are increased by a rise in temperature. Such a rise causes the average kinetic energy and the average velocities of the molecules to increase, resulting in a higher probability of effective (reaction-causing) collisions. Enzymes, however, become denatured and inactivated at high temperatures. Below the denaturation temperature for an enzyme, it is approximately true that the reaction rate will double for every rise in temperature of $10^{\circ}C$. Since the exact ratio of the new (higher temperature) rate to the old rate is termed Q_n where n is the temperature change in degrees centigrade, this is equivalent to saying that $Q_{10} \simeq 2$. Most enzymes have an optimal temperature, that is, one at which they catalyze at a maximum rate. This is usually close to the normal temperature experienced by the organism of which they are a part. In man, most enzymes have an optimum temperature of $37^{\circ}C$.

    b. <u>pH</u>. The activity of most enzymes depends on pH because of the involvement of H^{+} or OH^{-} in the reaction, denaturation at certain pH's, or for other reasons. As with temperature, most enzymes have an optimum pH at which their activity is maximal. It is usually at or near the pH of the fluid in which the enzyme must function. Thus, most enzymes have their highest activity between pH 5 and pH 8 (e.g., pH of human blood is ~7.4). Pepsin, which must operate at the low pH of gastric juice, has maximal activity at pH $\simeq 2$.

c. Concentration of the Enzyme. The reaction rate increases
in direct proportion to the concentration of the enzyme
catalyzing it, suggesting that the enzyme–substrate
interaction obeys the mass–action law. In fact, at a fixed
substrate concentration, the rate of a reaction is
frequently used as a measure of the enzyme concentration
(see uses of enzymes in the clinical laboratory).

d. Concentration of the Substrate. At a fixed enzyme
concentration, the initial velocity (before much substrate
is consumed) increases at first with increasing substrate
concentration. Eventually a maximum is reached and adding
more substrate has no further influence on velocity (v).
This is shown in the diagram below. The curve is known
as a rectangular hyperbola and, in general shape, is
characteristic of all non-allosteric enzymes (allosterism
will be discussed later, under enzyme control).

Figure 11. Effect of Initial Substrate Concentration on the Velocity
of an Enzyme Catalyzed Reaction, where the Concentration
of the Enzyme is Kept Constant

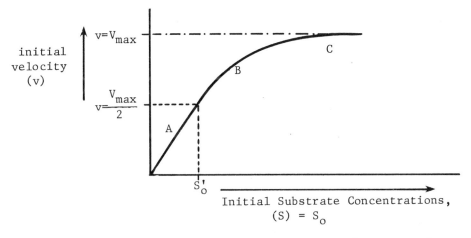

Segment A: linear; first-order with respect to substrate and enzyme
($v = k(E)S_0$); initial substrate concentration is low and
limits the velocity.

Segment B: curved; first-order with respect to enzyme; mixed (first
and zero) order with respect to substrate.

Segment C: almost linear; first-order with respect to enzyme; very
close to zero-order with respect to substrate ($v = k(E)$);
at infinite substrate concentration ($S_0 = \infty$), $v = V_{max}$, a
limiting value for the velocity; as $S_0 \to \infty$, $v \to V_{max}$.

In segment C, effectively all of the enzyme molecules have substrate bound to them. As soon as one reacts and leaves the enzyme, another substrate molecule binds. Increasing (S) cannot speed up the reaction because the enzyme is being fully utilized already. The value of V_{max} (maximum velocity) is an important kinetic parameter, characteristic of a particular enzyme and substrate at specified temperature and pH.

Another important point is the value of S_o which results in the velocity being half-maximal. In the above diagram,

$$v = \frac{V_{max}}{2} \text{ when } S_o = S_o'$$

The value S_o' (in moles/liter) is called the Michaelis constant, symbolized by K_m. It is a ratio of rate constants, the exact form of which (in terms of rate constants) depends on the kinetic scheme for the reaction. Both V_{max} and K_m are discussed below in more detail.

In the body, substrate depletion is an important control mechanism. When a reaction is no longer needed, the supply of substrate for it is reduced or stopped and, as the existing pool runs out, the enzyme catalyzing the reaction ceases to function for lack of reactant.

e. Presence of Inhibitors. The rates of most enzyme-catalyzed reactions can be decreased by a variety of substances, collectively called inhibitors. They can be general in their action (e.g.,urea and other denaturing agents); type specific (acting against phosphorylases, for example); or specific for one enzyme (as are substrate analogues). They are also classified according to the way that they affect the observed kinetics. Inhibitors will be discussed more fully later on.

f. Presence of Allosteric Effectors. A group of enzymes (called allosteric enzymes) exhibits unusual kinetics. Their activity can be increased or decreased by substances termed allosteric effectors. Each allosteric enzyme is influenced by its own set of effectors which can include a wide variety of compounds. The effectors are usually related in some way to the products of the enzyme or to intermediates in some pathway of which the enzyme is part. As the name allosteric (other site) implies, the effectors indirectly influence what happens at the active site by

binding to some part of the enzyme separate from the
substrate binding site. This group of enzymes will be
discussed more thoroughly later on.

It should be noted that paragraphs a-e (above) apply to
some extent to all enzymes, while allosteric effects are
observed with relatively few of them.

3. Michaelis-Menten Treatment

One procedure for handling enzyme kinetics which has found
wide applicability was worked out   by L. Michaelis and
M.L. Menten in 1913. Again, consider the reaction

(I)
$$E + S \underset{k_{-1}}{\overset{k_1}{\rightleftarrows}} ES \xrightarrow{k_2} E + P$$

Here, the release of product is essentially irreversible
and the reverse reaction ($E + P \xrightarrow{k_{-2}} ES$) does not take place
to any appreciable extent. This can also be described by saying
that k_{-2} is much less than k_2, corresponding to a large,
negative ΔG from ES to ($E + P$). Define

   E_o = total concentration of enzyme present (moles/liter)

      = $(E) + (ES)$

   S_o = initial substrate concentration (moles/liter).

Assume that   a) $S_o \gg E_o$ so that as substrate is consumed,
                  the substrate concentration (S) changes very
                  little from its initial value of S_o.
               b) Only initial velocities are measured. This
                  is necessary to insure that, during the period
                  of measurement, (S) = S_o to a good approximation.
               c) (ES) is constant during the period of
                  observation (steady-state condition). This
                  means that
                   i) enough time has elapsed since mixing E
                      and S for (ES) to build up,
                  ii) not enough time has passed for the rate
                      of formation of ES to decrease due to
                      substrate depletion, and
                 iii) $k_2 < k_{-1}$ so that (ES) can build up.
                      If this is not true, ES breaks down as
                      fast as it is formed and a steady-state
                      can never be established.

From this

rate of formation of ES = $k_1(E)(S) = k_1[E_o - (ES)]S_o$

rate of breakdown of ES = $k_{-1}(ES) + k_2(ES) = (k_{-1} + k_2)(ES)$

In steady-state

rate of formation = rate of breakdown so that

$$k_1[E_o - (ES)]S_o = (k_{-1} + k_2)(ES)$$

(II) $$\frac{[E_o - (ES)]S_o}{(ES)} = \frac{k_{-1} + k_2}{k_1} \equiv K_m, \text{ the Michaelis constant}$$
$$\text{(mentioned previously)}$$

Notice that K_m is <u>not</u> an equilibrium constant, although it is a ratio of rate constants, because ES → E + P is irreversible.

Using the definition of K_m, equation (II) can be rearranged to

$$S_o E_o - S_o(ES) = K_m(ES)$$

$$S_o E_o = (ES)(K_m + S_o)$$

(III) $$S_o = \frac{(ES)}{E_o}(K_m + S_o)$$

It is usually difficult to measure (ES) in a reaction mixture. Consequently, equation (III) is not very useful from an experimental point of view. On the other hand, the velocity, v, and the maximum velocity, V_{max}, are readily measured by a variety of methods.

Previously, V_{max} was defined as the value which v approaches as $S_o \to \infty$. At $S_o = \infty$, all enzyme molecules have substrate bound to them so that $(E) = 0$ and $E_o = (ES)$.

Using this and $v \equiv \frac{d(P)}{dt} = k_2(ES)$, one can write $V_{max} = k_2 E_o$.

These expressions can both be solved for k_2, giving

$$k_2 = \frac{v}{(ES)} \text{ and } k_2 = \frac{V_{max}}{E_o} .$$

Combining these, then, give

$$\frac{v}{(ES)} = \frac{V_{max}}{E_o} \text{ or } \frac{v}{V_{max}} = \frac{(ES)}{E_o}$$

Substituting this into equation (III),

$$S_o = \frac{v}{V_{max}} (K_m + S_o)$$

(IV)
$$v = \frac{S_o V_{max}}{K_m + S_o}$$

Equation (IV) is known as the Michaelis-Menten Equation. Regarding this relationship

a) K_m is a constant characteristic of an enzyme and a particular substrate. It is independent of enzyme and substrate concentrations.

b) V_{max} (= $k_2 E_o$) depends on enzyme concentration. For a particular enzyme, it is largely independent of the specific substrate used.

c) K_m and V_{max} may be influenced by pH, temperature, and other factors.

d) A plot of v versus S_o (shown before) is a rectangular hyperbola.

e) If an enzyme binds more than one substrate, the K_m values for the various substrates can be used as a relative measure of the affinity of the enzyme for each substrate (the smaller the value of K_m, the higher is the enzyme's affinity for that substrate).

f) In a metabolic pathway, K_m values for enzymes catalyzing the sequential reactions may indicate the rate-limiting step for the pathway (the highest K_m roughly corresponding to the slowest step).

4. It was stated previously that $K_m = (S_o/2)$, where $(S_o/2)$ is the substrate value for which $v = (V_{max}/2)$. This can be demonstrated by substituting for S_o and v in equation (III).

$$\frac{V_{max}}{2} = \frac{V_{max}(S_o/2)}{K_m + (S_o/2)}$$

$$(K_m + (S_o/2))V_{max} = 2V_{max}(S_o/2)$$

$$K_m + (S_o/2) = 2(S_o/2)$$

and
$$K_m = \frac{S_o}{2} \text{ , when } v = \frac{V_{max}}{2}$$

From equation (IV) it can also be seen that

a. v <u>increases</u> if <u>V_{max} increases</u> at constant S_o and K_m, and

b. v <u>decreases</u> if <u>K_m increases</u> at constant S_o and V_{max}.

5. Straight lines are more amenable to data evaluation than are curves and it is, therefore, convenient to recast equation (IV) in some form which can be plotted as straight lines. There are two such forms commonly used.

a. The <u>Lineweaver–Burke form</u> is arrived at by taking reciprocals of both sides of equation (IV) and rearranging

$$\frac{1}{v} = \frac{K_m + S_o}{S_o V_{max}}$$

(V)
$$\frac{1}{v} = \frac{K_m}{V_{max}} \left(\frac{1}{S_o}\right) + \frac{1}{V_{max}}$$

According to (V), if the system obeys Michaelis–Menten kinetics, a plot of $1/v$ versus $1/S_o$ will be straight line of slope (K_m/V_{max}) and a $(1/v)$ intercept of $(1/V_{max})$. Further, the $(1/S_o)$ intercept, occurring when $(1/v) = 0$, is $(-1/K_m)$. This is seen by setting $(1/v)$ equal to zero.

$$0 = \frac{K_M}{V_{max}} \left(\frac{1}{S_o}\right) + \frac{1}{V_{max}}$$

then,

$$\frac{1}{S_o} = \frac{-(1/V_{max})}{(K_m/V_{max})}$$

$$\frac{1}{S_o} = - \frac{1}{K_m} \quad .$$

This is shown below diagramatically in a <u>Lineweaver–Burke plot</u>.

Figure 12.   Lineweaver-Burke Plot of the Michaelis-Menten Equation

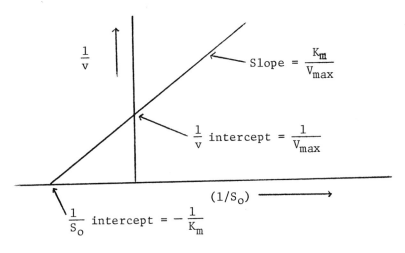

b.   The <u>Eadie-Hofstee form</u> is obtained by multiplying both sides of equation (IV) by $(K_m + S_o)/S_o$ and rearranging

$$v \left(\frac{K_m + S_o}{S_o} \right) = V_{max}$$

$$v \left(\frac{K_m}{S_o} \right) + v = V_{max}$$

(VI) $$v = -K_m \left(\frac{v}{S_o} \right) + V_{max}$$

It can be seen from equation (VI) that a plot of v versus (v/S_o) will have a slope of $-K_m$, a v-intercept of V_{max}, and a (v/S_o)-intercept of (V_{max}/K_m).   The diagram below, illustrating this, is an <u>Eadie-Hofstee plot</u>.

Figure 13.   Eadie-Hofstee Plot of the Michaelis-Menten Equation

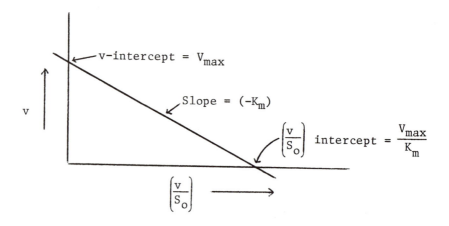

Of these two methods, the Lineweaver-Burke plot is more commonly used. It must be remembered that all of these conclusions are based on the premise that the system under study obeys an equation of the same form as 3(I) (page 93) and that the assumptions (3a, b, c, page 93) are correct. Systems with different reaction equations can be treated in a similar manner, but equations II, III, and IV must be rederived for each case. One example, frequently encountered, involves ES going to EP, an additional intermediate which then breaks down to products.

6. Enzyme inhibition is one of the most important ways in which enzyme activity is both artificially and naturally regulated. It is the manner in which most therapeutic drugs perform their function, usually acting on a specific enzyme. Inhibitor studies have contributed much of the information presently available about enzyme kinetics and mechanisms. Trypsin, a highly active proteolytic enzyme distributed in many body tissues, might rapidly hydrolyze much of the protein in the body were it not for the presence of specific trypsin inhibitors in the plasma and many tissues. A correlation has been demonstrated between a lack of α_1 Trypsin inhibitor and the occurrence of one form of pulmonary emphysema.

    Inhibition can now be considered in terms of the effect that an inhibitor has on the form of equations 3(IV), 3(V), and 3(VI) and, consequently, on their graphs. A general scheme for enzyme inhibition can be drawn.

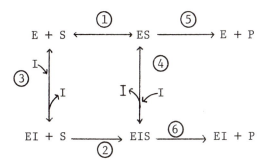

where I represents an inhibitor molecule.

The normal enzyme-substrate reaction is (1) followed by (5).
If reaction (3) is irreversible (EI is formed but cannot be
broken down), the inhibition is termed <u>irreversible</u>; if (3)
is reversible, the inhibition is <u>reversible</u>. Irreversible
inhibition cannot be treated by Michaelis-Menten methods, since
the amount of functional enzyme is not constant. An irreversible
inhibitor is covalently linked to the enzyme, at the active
site or elsewhere.

Reversible inhibition can be divided into several types, each
of which can be further subdivided. The two most important
types are discussed here.

a. <u>Competitive (dead-end) inhibition</u> occurs when inhibitor (I)
    binds to <u>the same site</u> as, and more strongly than, S does.
    The presence of I thus prevents the binding of S,
    inactivating the enzyme until the EI complex dissociates.
    The EI complex is thus a "dead-end" form. It can only go
    back to E + I. This corresponds to saying that reactions
    (2) and (4) (and hence (6)) in the diagram above cannot
    occur. Experimentally, competitive inhibition can always
    be overcome by increasing substrate concentration. The
    more substrate molecules present, the more likely it is
    that S will bind rather than I, and a greater fraction of
    the enzyme will be coupled with S. The degree of inhibition
    depends on the ratio of the concentration of I to the
    concentration of S, and not just on the concentration of I.
    The Lineweaver-Burke and v versus S_0 plots are affected as
    shown below.

Figure 14.   Effect of a Competitive Inhibitor on the Kinetics of an
Enzyme-Catalyzed Reaction

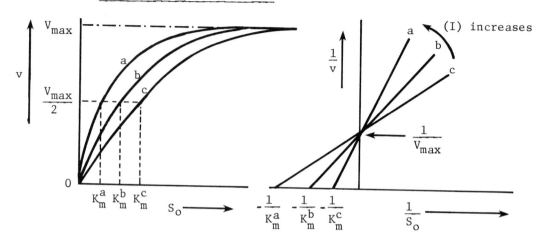

Line a is the curve for no inhibitor; b and c correspond
to increasing amounts of inhibitor; $K_m^a < K_m^b < K_m^c$;
V_{max} = constant.

Note that: i) V_{max} is unchanged (the potential number of
substrate binding sites is not altered) and

ii) K_m for substrate increases (the apparent
ability of the enzyme to bind substrate
decreases, due to some sites being occupied
by (I) with increasing inhibitor concentration).

b. Non-competitive inhibition occurs when the inhibitor binds
to the enzyme at a site partly or completely different from
the active site. This is not the same as allosteric
repression, however, because the system still obeys
Michaelis-Menten kinetics. Non-competitive inhibition
cannot be overcome or decreased by increasing S_o. (Because
of this it is sometimes referred to as pseudo-irreversible
inhibition.) The amount of inhibition is independent of
the ratio (I)/(S); rather, it depends on (I) alone. In
terms of the inhibition scheme given initially, all six
reactions can occur, but reaction (5) is much faster than
(6), the latter being, in some cases, negligibly slow.
The alteration of the Lineweaver-Burke and Michaelis-Menten
plots is seen in the next diagram.

Figure 15. Effect of a Non-Competitive Inhibitor on the Kinetics of an Enzyme-Catalyzed Reaction

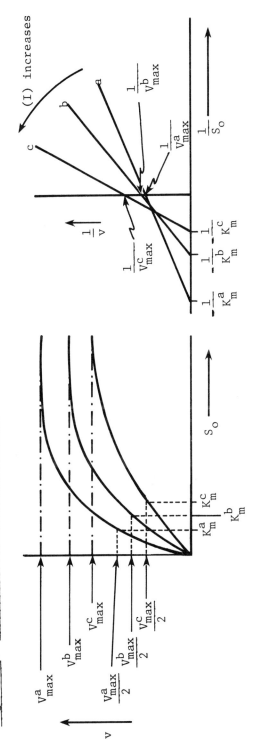

Line a is the curve for no inhibitor; b and c correspond to increasing amounts of inhibitor; $K_m^a < K_m^b < K_m^c$; $V_{max}^a > V_{max}^b > V_{max}^c$.

Notice in the Lineweaver-Burke plot that, although all three lines pass through a common point, the point is to the left of the vertical axis. Since both intercepts change with increasing (I), V_{max} and K_m are both altered by a non-competitive inhibitor. The exact position of the intersection is indicative of which is affected more strongly. If it is near the $(1/v)$ axis, V_{max} changes little; if it is closer to the $(1/S_o)$ axis, K_m is almost unaffected. Although the interaction can fall on the $(1/S_o)$ axis, it is probably not correct to interpret this as meaning that K_m(hence, enzyme affinity for substrate) is unchanged by inhibitor binding. Rather, it is indicative of a more complex system than was first assumed. Such a case (with K_m apparently constant) is termed <u>pure</u> non-competitive inhibition. Since V_{max} is decreased by non-competitive inhibitors, the number of potential substrate binding sites must be decreased.

7. Examples of enzyme inhibition

    a. Irreversible inhibition (covalent linkage of inhibitor to
        enzyme):

        i) Enzymes containing free cysteinyl (sulfhydryl) groups
            can react with alkylating agents.

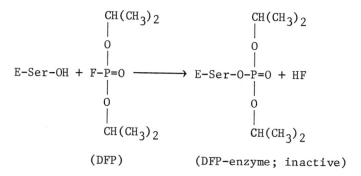

            (iodoacetamide)   (inactive covalent
                                          derivative of E-SH)

        ii) Enzymes having seryl hydroxyl groups can be
            phosphorylated by diisopropylfluorophosphate (DFP).

$$\begin{array}{ccc}
CH(CH_3)_2 & & CH(CH_3)_2 \\
| & & | \\
O & & O \\
| & & | \\
E\text{-}Ser\text{-}OH + F\text{-}P{=}O & \longrightarrow & E\text{-}Ser\text{-}O\text{-}P{=}O + HF \\
| & & | \\
O & & O \\
| & & | \\
CH(CH_3)_2 & & CH(CH_3)_2
\end{array}$$

           (DFP)   (DFP-enzyme; inactive)

A specific example is the DFP-inhibition of acetylcholinesterase, the enzyme which inactivates the
neurotransmitter, acetylcholine, by hydrolysis. (Esterases
and other hydrolases frequently have serine as part of
their active site, so DFP inactivates many enzymes of
this type.) The normal reaction is shown below. The
enzyme is represented by

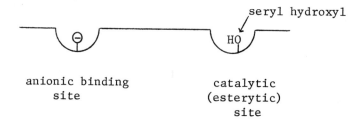

          anionic binding   catalytic
              site   (esterytic)
                                       site

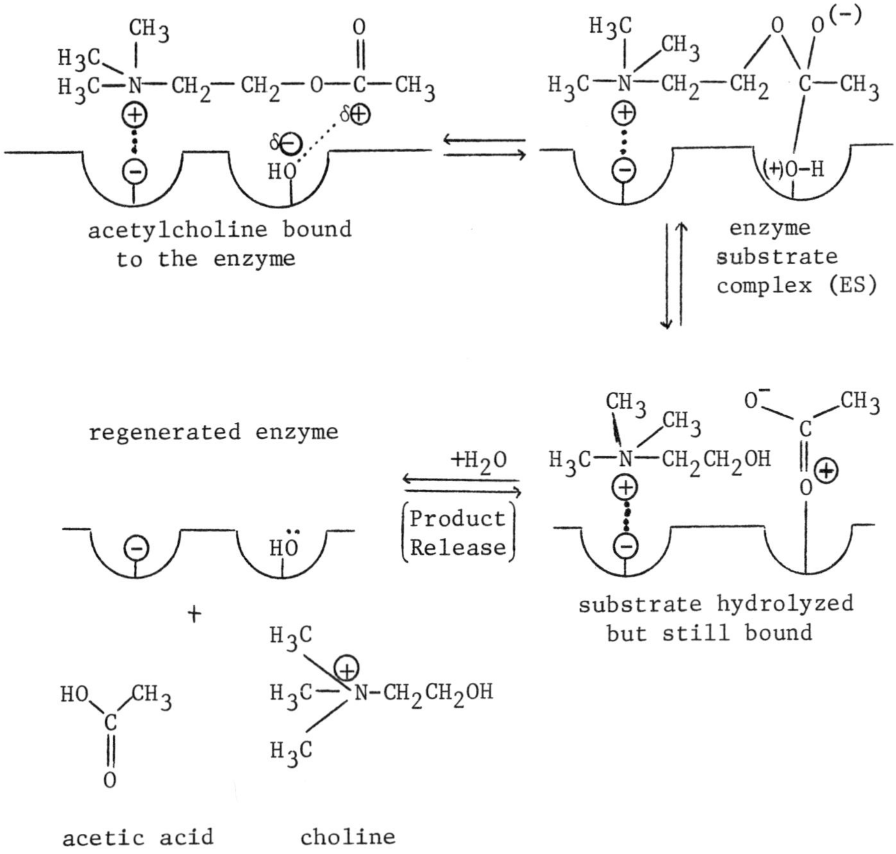

$$CH_3$$
$$H_3C\diagdown \quad |$$
$$H_3C-N-CH_2-CH_2-O-C-CH_3$$
$$\overset{\oplus}{}$$

acetylcholine bound
to the enzyme

$\rightleftharpoons$

enzyme
substrate
complex (ES)

$\overset{+H_2O}{\rightleftharpoons}$ (Product Release)

substrate hydrolyzed
but still bound

regenerated enzyme

$+$

$$HO\diagdown \diagup CH_3$$
$$C$$
$$\|$$
$$O$$

acetic acid

$$H_3C$$
$$H_3C-\overset{\oplus}{\diagup}N-CH_2CH_2OH$$
$$H_3C$$

choline

DFP can also bind to the enzyme, form a complex similar
to ES, and be hydrolyzed. Hydrolysis of the diisopropyl-
phosphate derivative is very much slower than is the
release of acetic acid. This is shown below.

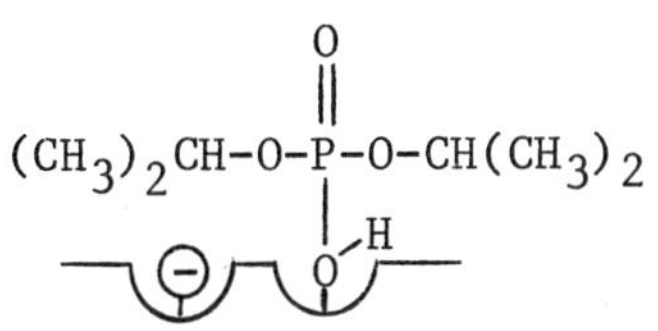

$$O$$
$$\|$$
$$(CH_3)_2CH-O-P-O-CH(CH_3)_2$$

$$H_7C_3-O \quad O-C_3H_7$$
$$O=P-F$$

In this particular case, the acetylcholinesterase can be reactivated by hydroxamic acids, and the general formula is

$$R'-\overset{\displaystyle O}{\overset{\displaystyle \|}{C}}-NH-OH$$

Pralidoxime (pyridine-2-aldoxime methiodide), a drug used clinically to treat DFP poisoning, increases the rate of phosphoryl enzyme hydrolysis by the mechanism below, where

$$R = -CH(CH_3)_2 \quad \text{and}$$

is pralidoxime

DFP-pralidoxyl-enzyme·

regenerated enzyme

Pralidoxime is most active in relieving inhibition of
skeletal muscle acetylcholinesterase. The effects
(increased tracheobronchial and salivary secretion,
bronchoconstriction, central respiratory paralysis and
other CNS disorders) of autonomic system acetylcholinesterase inhibition can be better antagonized
by atropine, which blocks the acetylcholine receptor
sites. Atropine is especially effective in relieving
the excessive secretion and bronchoconstriction.
Notice that it does not relieve the inhibition but
renders ineffective the accumulated acetylcholine.

iii) Metalloenzymes can be inactivated by forming <u>stable</u>
<u>complexes</u> with the metal. For example, cytochrome
oxidase, the terminal enzyme in mitochondrial
electron transport, is inactivated by CN^- (cyanide
ion). Cytochrome oxidase contains both copper and
iron. The normal reactions are

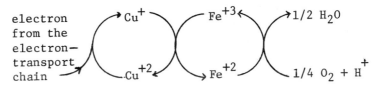

Presumably, CN^- forms a loose complex with Fe^{+2}. Then Fe^{+2} is oxidized to Fe^{+3} which forms a very stable complex with CN^-. This complex cannot be reduced by Cu^+, so the flow of electrons and the uptake of O_2 are stopped. Hydrogen sulfide (H_2S) toxicity operates by the same mechanism.

Cyanide and sulfide toxicity are handled clinically by making use of the ability of methemoglobin (oxidized hemoglobin, $HbFe^{+3}$) to bind CN^- and HS^- as strongly as does cytochrome oxidase. The level of methemoglobin in the erythrocytes is increased by administration of nitrites ($NaNO_2$ solution intravenously; amyl nitrite by inhalation). This causes a shift of the equilibrium.

$$HbFe^{+3} \;+\; CytFe^{+3}(CN^-) \rightleftharpoons HbFe^{+3}(CN^-) \;+\; CytFe^{+3},$$

to the right, releasing the inhibition of cytochrome oxidase. Cyanomethemoglobin is no more toxic than methemoglobin and can be removed by the normal processes which degrade erythrocytes. Excess, uncomplexed cyanide is removed by administration of sodium thiosulfate.

$$CN^- \;+\; Na_2S_2O_3 \; \underset{SCN^- \text{ oxidase}}{\overset{\text{sulfurtransferase}}{\rightleftharpoons}} \; SCN^- \;+\; Na_2SO_3.$$

Thiocyanate (SCN^-) is relatively non-toxic and can be eliminated in the urine. The sulfurtransferases, also known as rhodaneses, are mitochondrial enzymes found in liver, kidney, and other tissues. The reverse reaction, catalyzed by thiocyanate oxidase, is slow. Its action can, however, occasionally cause a return of the toxic manifestations. Cyanide is normally found in man complexed to Co^{+3} in vitamin B_{12} (cyanocobalamin).

b. <u>Competitive inhibition</u> usually or always involves a substrate analogue which can bind in the same place as the substrate but which can only react very slowly or not at all. When used in clinical and biological situations (i.e., with whole cells or tissues), competitive inhibitors are frequently called antagonists or antimetabolites of the substrates with which they compete.

i) The reaction catalyzed by succinate dehydrogenase is
    shown below.

$$\begin{array}{c} \text{COOH} \\ | \\ \text{CH}_2 \\ | \\ \text{CH}_2 \\ | \\ \text{COOH} \end{array} \;+\; \text{FAD} \;\;\underset{\text{dehydrogenase}}{\overset{\text{succinate}}{\rightleftharpoons}}\;\; \begin{array}{c} \text{COOH} \\ | \\ \text{C--H} \\ || \\ \text{H--C} \\ | \\ \text{COOH} \end{array} \;+\; \text{FADH}_2$$

   Succinate   Fumarate

   (where FAD is a coenzyme which serves as a hydride ion
   acceptor). This enzyme is competitively inhibited
   by malonate which acts as an analogue for succinate.

$$\begin{array}{c} \text{COOH} \\ | \\ \text{CH} \\ | \\ \text{COOH}_2 \end{array}$$

                 malonate

ii) Folic acid (pteroylglutamic acid) is a coenzyme used
     in certain biosynthetic reactions. Although it is a
     vitamin in man, some bacteria synthesize it using
     para-aminobenzoic acid (PABA). Sulfonamides are
     chemically and physically similar to PABA and can
     competitively inhibit the synthesis of folic acid,
     thus leading to inhibition of the growth of
     susceptible organisms. The structures are shown below.

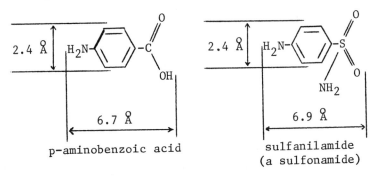

p-aminobenzoic acid   sulfanilamide
                               (a sulfonamide)

iii) Prior to its functioning as a coenzyme, folic acid must undergo two reductions, first to dihydrofolate (FH_2), then to tetrahydrofolate (FH_4). The reductases catalyzing these reactions are competitively inhibited by amethopterin (methotrexate) and aminopterin, making FH_4 unavailable. Since FH_4 is needed for the synthesis of DNA precursors, a deficiency of it does most harm to these cells synthesizing DNA rapidly. Certain types of malignancies, such as the leukemias, exhibit an extremely high rate of cell division and are consequently particularly susceptible to folate antagonists. These reactions and the structures of the compounds are discussed in the section on one-carbon metabolism.

iv) Xanthine oxidase (X.O.) is an enzyme of purine catabolism which catalyzes the oxidation of hypoxanthine to xanthine and xanthine to uric acid. Allopurinol, an analogue of hypoxanthine, competitively inhibits X.O. Its structure along with that of hypoxanthine is shown below.

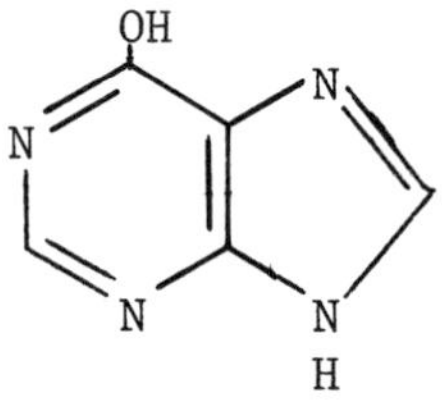

hypoxanthine 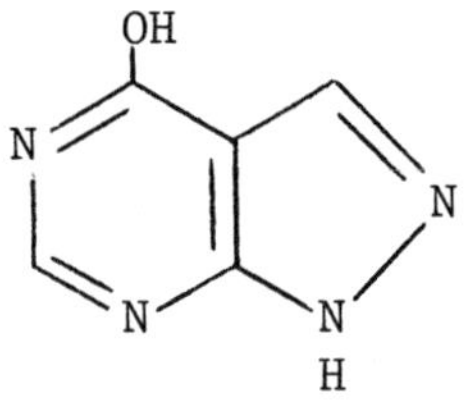   allopurinol

Allopurinol is useful for reducing blood uric acid levels in hyperuricemic conditions such as gout and is further discussed under gout. Xanthine oxidase also provides an interesting example of non-competitive inhibition (see below).

c. <u>Non-competitive inhibition</u>

i) It was mentioned in b(iv), above, that xanthine oxidase also can be non-competitively inhibited. In addition to being a competitive inhibitor, allopurinol functions as a substrate for X.O., about 45-65% of an administered dose being oxidized to oxypurinol (alloxanthine) within a few hours. Oxypurinol bears the same relationship to xanthine that allopurinol does to hypoxanthine.

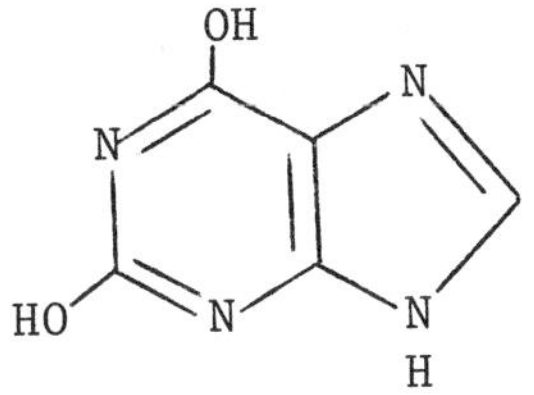

xanthine

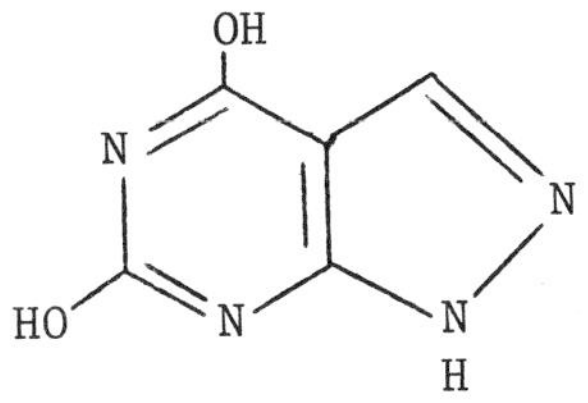

oxypurinol
(alloxanthine)

In the absence of xanthine, allopurinol causes non-competitive inhibition, apparently by combining with a transition state of the enzyme during the oxidation of allopurinol to oxypurinol. Both K_m and V_{max} for allopurinol are less than for xanthine. In the presence of xanthine, allopurinol exhibits only competitive inhibition but oxypurinol behaves non-competitively, probably by the same mechanism as does allopurinol. The K_m of oxypurinol is similar to that for xanthine, but V_{max} is much less and the inhibition cannot be overcome by excess xanthine. Xanthine oxidase activity is also controlled by the reversible reduction of a disulfide bond (see F(ii), below, under regulation of enzyme activity in the body).

ii) In addition to being susceptible to irreversible inhibition, enzymes having cysteinyl sulfhydryl groups can be non-competitively inhibited by heavy metal ions (Ag^+, Pb^{+2}, Hg^{+2}, etc.). The general sort of reaction is shown below with Hg^{+2}.

$$E\text{-}SH + Hg^{+2} \rightleftharpoons E\text{-}S\text{-}Hg^+ + H^+.$$

Heavy metals can also form covalent bonds with $-COO^-$ and histidyl residues.

Allosteric Control of Enzyme Function

1. Allosterism was mentioned previously in sections C.2.f and C.6.b under enzyme kinetics. It is frequently observed in enzymes that catalyze the first step in a pathway which is unique to that pathway (the committed step). In this way, an end product of the pathway quite unlike the initial substrate

can shut off its own synthesis when it has reached an appropriate concentration. This conserves material and shunts the initial substrate into alternate metabolic routes. Well studied examples are found in the interactions of glycolysis, gluconeogenesis, fatty acid synthesis and oxidation, and the TCA cycle. These will be discussed later.

2. An allosteric enzyme which has been extensively studied is aspartate transcarbamylase. It catalyzes the reaction:

$$H_2N-\overset{\overset{O}{\|}}{C}-O-\overset{\overset{O}{\|}}{\underset{\underset{OH}{|}}{P}}-OH \;+\; HOOC-CH_2-\overset{\overset{NH_2}{|}}{CH}-COOH$$

carbamyl   aspartic acid
phosphate

aspartate
transcarbamylase

$$\overset{HOOC}{\underset{O}{\overset{H_2N}{\underset{\overset{|}{\underset{N}{\overset{C}{\diagup}}}}{\diagdown}}}} \quad \overset{CH_2}{\underset{HC\diagdown_{COOH}}{|}} \quad + \; P_i$$
H

carbamyl aspartate

Carbamyl aspartate is eventually converted to the pyrimidine nucleotides which are used in a number of processes including nucleic acid synthesis.

Aspartate transcarbamylase has a molecular weight of 310,000 and is made up of twelve polypeptides. Six peptides, designated α, have a molecular weight of 33,000 each; the other six (β-chains) have a molecular weight of 17,000 each. The α-chains combine into two groups of three, forming type A subunits of formula α_3. The β-chains form three B subunits each of formula β_2. The A subunits bind substrate but have no control sites and are termed catalytic subunits. The B subunits are regulatory subunits with control sites but no catalytic activity. The chain molecular weights do not add up exactly to 310,000 due to difficulties in measurement.

When pyrimidine nucleoside phosphates accumulate in the cell, they bind to some of the control sites and allosterically inhibit the enzyme. CTP is the most effective inhibitor. In

addition, the <u>purine</u> nucleoside phosphates (adenine and guanine
mono-, di-, and triphosphate: AMP, ADP, ATP, GMP, GDP, and
GTP) allosterically <u>activate</u> this enzyme. This is logical
since, in double-stranded DNA, the ratio $(A+G)/(T+C) = 1$. For
most efficient DNA synthesis, then, purine and pyrimidine
nucleotides should be available in equal amounts. Finally,
the substrates themselves act as allosteric activators
(positive allosteric effectors).

The physical change in the enzyme seems to be one of "swelling"
upon binding substrate and releasing CTP (and presumably other
negative effectors) and "contracting" into an inactive (or less
active) state upon binding CTP. The higher activity form
appears to dissociate into free subunits more readily than
does the low activity state. This is summarized in the
diagram below.

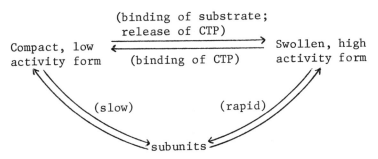

This example illustrates four important general characteristics
of allosteric enzymes and their control.

Allosteric enzymes contain multiple subunits. The subunits
may be identical or different. Each subunit may or may not
be a separate polypeptide chain.

Changes between active and inactive forms are associated
with changes in quaternary structure: the way in which
the subunits interact with each other.

Control sites and catalytic sites are chemically and
spatially separate. They may be on different subunits but
are not necessarily so.

For a particular enzyme, allosteric effectors can include
both substrate molecules and compounds very different from
the substrate. Depending on the enzyme, members of either
class may be positive or negative effectors.

3. The kinetics of allosteric systems are rather different from
    those of enzymes obeying Michaelis-Menten kinetics. For an
    allosteric enzyme, a plot of v versus S_0 is sigmoidal
    (S-shaped) rather than rectangular hyperbolic, as shown below.

<u>Figure 16.</u>   Comparison of Allosteric and Non-Allosteric Kinetics

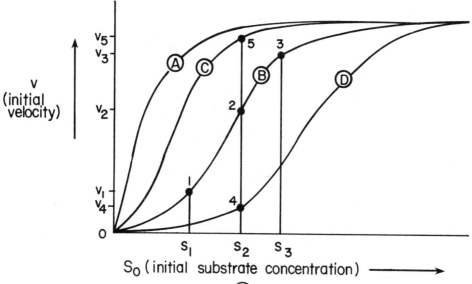

In this diagram   Curve (A) is a rectangular hyperbola,
                   characteristic of Michaelis-Menten kinetics.

                   Curve (B) is a typical sigmoid curve for an
                   allosteric enzyme; the only possible
                   effector present is substrate.

                   Curve (C) shows the effect of adding a
                   positive allosteric effector to the system
                   of Curve (B).

                   Curve (D) illustrates the result of adding
                   an allosteric inhibitor to the system of
                   Curve (B).

Notice that

i)   Curve Ⓑ is sigmoidal even though only substrate is present.
     If an enzyme is allosteric, it will not obey Michaelis-
     Menten kinetics even in the absence of a modifier.

ii) At constant S_0 (for example at $S_0 = S_2$), the activity of
     an allosteric enzyme can be controlled by changing the
     concentrations of positive and negative effectors.

$$no\ effector,\ v = v_2$$
$$positive\ effector,\ v = v_5 > v_2$$
$$negative\ effector,\ v = v_4 < v_2$$

     The important things to consider are the relative
     affinity of the enzyme for the effectors and the ratio of
     their concentrations.

iii) Although an allosteric enzyme is insensitive to changes
     in S_0 at low and high S_0-values, a large change in
     velocity is observed for small changes in S_0 at
     intermediate values (e.g. $v_1 \to v_3$ when $S_1 \to S_3$). The
     rate of a non-allosteric enzyme-catalyzed reaction is
     most sensitive to changes in S_0 at low concentrations
     of substrate.

The reason for the sigmoidal curve can be interpreted
physically. At low S_0, the slope is small because it is
difficult for the reaction to occur. As S_0 increases, the
enzyme is activated in some way, making the reaction proceed
more rapidly, and the slope increases. Eventually, the enzyme
is working at a maximal rate, the slope decreases again, and
further increases in S_0 have no effect on velocity. Although
this appears to be the most common situation, cases are known
where a substrate can actually cause a <u>decrease</u> in enzyme
activity (for example, the binding of NAD^+ to rabbit muscle
D-glyceraldehyde-3-phosphate dehydrogenase). The v versus S_0
curves for such enzymes will not appear as above, although the
enzymes are allosteric. There are many variations on the
basic theme of allosterism.

4. There have been several models devised to explain the anomalous
    (non-Michaelis-Menten) kinetics exhibited by allosteric
    enzymes. The most important ones are those due to Monod, Wyman,
    and Changeux (MWC); Koshland, Nemethy, and Filmer (KNF); and
    Eigen (EIG).

a. Several things are common to all three models.

   i) Allosteric enzymes are assumed to be oligomers (small polymers) of identical subunits. (It is possible, however, to modify the models to include more than one type of subunit.) A subunit may or may not be a single polypeptide chain.

   ii) There are multiple catalytic and regulatory binding sites on the subunits.

   iii) The change from active to inactive forms involves non-covalent forces and results in alteration of the quaternary structure of the enzyme. Changes in the subunit (tertiary) conformation also generally occur.

Specific aspects of the models are given below.

b. MWC

   i) There are two conformations, designated R and T, of an allosteric enzyme. These differ in tertiary and quaternary structure; in their affinities for substrates and for positive and negative allosteric effectors; and in catalytic activity.

   ii) All binding sites in a particular conformation (R or T) are completely identical. This means that the only differences in affinity that can be observed are between R enzyme molecules and T enzyme molecules. On a given enzyme molecule, the binding of one substrate molecule does not affect the ease of binding of a second substrate molecule of the same type.

   iii) In a given enzyme molecule, all subunits are either in the R-form or the T-form. There are no mixtures of R- and T-subunits in the same enzyme molecule.

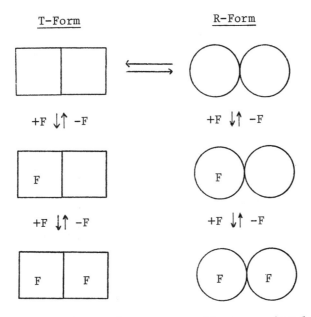

(where F is a substrate or effector molecule)

With nothing bound, the R- and T-forms are in equilibrium, with one form frequently predominating. As soon as a substrate (or effector) molecule binds, the enzyme molecule is "committed" to the form which it had when binding occurred. It cannot change form until all effector, substrate, or product molecules dissociate from it. (For example, the conversion of $\boxed{F}$ to Ⓕ◯ cannot occur directly.) The binding of substrate or effector <u>does not cause</u> the conformational change.

c. KNF

i)   In the absence of substrate or of allosteric effectors, the enzyme exists in only one conformation.

ii) The binding of the substrate or effector to one of the subunits <u>causes</u> that subunit to change conformation (induced-fit hypothesis).

iii) The basis for cooperativity is the difference in interaction between subunits having substrate bound and those without substrate. That is, there is a difference between circle-circle, circle-square, and square-square interactions in the diagram below.

For example, it would be easier for a square to
become a circle if it already had a circle in contact
with it. This example again uses a dimer.

d. EIG   This is a fusion and extension of the MWC and KNF
schemes. The principle features are relaxations of
the restraints of these models.

i) Changes in subunit conformation can occur whether or not
a substrate or effector is bound.

ii) There can be mixtures of subunits in different
conformations in the same enzyme molecule.

This can best be visualized in a diagram.

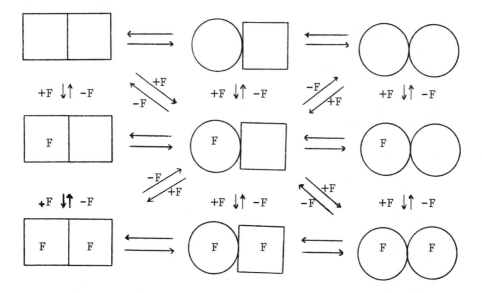

Because of difficulties in drawing, this omits the form
○F , which is in equilibrium with all of the structures
above except for those at the very corners. It is apparent
how the MWC and KNF models can be derived from this model
by elimination of certain forms of the enzyme.

e. The testing of these models on actual systems is just beginning. One major problem is that all three necessarily predict sigmoid curves for allosteric enzymes. Some other measurement must therefore be used to differentiate between the models in a particular system. A comparison of the values predicted by the theories to the actual experimental value of such a property will provide some basis on which to choose between the models. Even then, it will probably be found that each allosteric enzyme falls at some unique point intermediate between the extremes.

Other allosteric systems will be pointed out as they are encountered. Hemoglobin, the classical example, will be discussed later in detail.

1. Isozymes or isoenzymes are physically distinct forms of an enzyme, all of which catalyze the same reaction.

    a. They are electrophoretically separable.

    b. They usually differ in activity (as indicated by K_m and V_{max}, the Michaelis parameters) when the same substrate is used for their assay.

    c. Two or more enzymes are usually termed isozymes only if they occur within a given species.

2. There are several possible molecular differences between a pair of isozymes.

    a. Quaternary structure. Two or more types of subunits can combine in varying proportions. For example, lactate dehydrogenase (LDH) in man is a tetramer composed of different numbers of two kinds of subunits, H and M. The five possible forms (H_4, MH_3, M_2H_2, M_3H and M_4) all can be observed upon the electrophoresis of normal human serum. This will be discussed further under glycolysis.

    b. An enzyme having only one peptide chain (that is, no subunits) may have several forms differing in amino acid sequence. Carbonic anhydrase from human erythrocytes, upon electrophoresis, shows three distinct bands. These show differing activities in catalyzing the reaction.

$$H_2O + CO_2 \rightleftharpoons HCO_3^- + H^+.$$

c. The interaction of an enzyme molecule with various charged
    small molecules can change the electrophoretic mobility and
    kinetics. For example, at least two forms of dihydrofolate
    reductase (discussed under one-carbon metabolism) have been
    observed as separate peaks in starch gel electrophoresis.
    These have been shown to differ only by a molecule of
    NADPH, which imparts a more positive charge to one form.

d. Some reports of apparent isozymic forms may be due to
    changes in the enzyme occurring during its isolation and
    purification. These would have no meaning in terms of the
    intact animal. They could include denaturation,
    proteolysis, and other covalent and non-covalent alterations
    in the native enzyme.

e. Other possible mechanisms have been proposed for isozyme
    formation, including differential folding of one
    polypeptide chain. At the present time, the ones cited
    above (2.a,b,c) appear to be the most common ones.

3. The biological significance of isozymes is generally not known.
    Their existence may reflect differing needs for a particular
    chemical reaction in various tissues. One must, however, be
    careful to consider the actual reaction conditions in the cell
    before deciding whether or not the observed isozymic forms will
    provide the needed variation. Multiple forms of an enzyme may
    also provide a buffer against genetic changes which could
    endanger a vital cellular process. If there is more than one
    gene whose product will catalyze a reaction, then a single
    mutation cannot eliminate that process from the cell's
    repertoire.

4. Clinically, isozymes are increasing in importance. Changes in
    isozyme patterns - the per cent of a total enzyme activity which
    is due to each isozymic form - of several enzymes have been
    correlated with various diseases and some of these are being
    actively used for diagnosis. Some of the methods used to
    distinguish one isozyme from another will be discussed in
    glycolysis, along with LDH.

Regulation of Enzyme Activity in the Body

Several methods of enzyme control in biological systems, where
temperature and pH are quite constant, have already been mentioned.
There are others which are also important. The principal mechanisms
are summarized below. A knowledge of them is necessary both to
understand the influences of foreign materials (drugs, poisons) on

the body and to explain the interdependence of bodily functions.
All of the preceding material on kinetics, allosterism, and
isozymes is ultimately useful in this way.

1. Inhibition. Competitive, non-competitive, irreversible, others;
    substrate analogues are of particular importance; mode of
    action of many drugs.

2. Allosterism. Usually involves multisubunit enzymes; binding
    of a molecule similar to or unlike the substrate to a site
    distinct from the catalytic site; can be positive or negative;
    frequently exhibited by a key control enzyme at the beginning
    of a pathway; examples include the control of hemoglobin
    oxygenation by O_2 and 2,3-diphosphoglycerate (although hemoglobin
    is not truly an enzyme), a)of pyruvate carboxylase by acetyl-
    CoA, b)of phosphofructokinase by AMP, ADP, ATP, and citrate,
    and many others.

3. Isozymes. Multiple isomeric forms of the same enzyme; catalyze
    the same reaction but differ in K_m, V_{max} or both; generally
    separable by electrophoresis; examples include lactic acid
    dehydrogenase, alkaline phosphatase, carbonic anhydrase, and
    many more.

4. Activation of Latent Enzymes

    a. Enzyme is synthesized in an inactive (zymogen, pro-protein,
        proenzyme) form which is activated, frequently by cleavage
        of one or more peptide bonds, before or after secretion.
        Examples include chymotrypsin (chymotrypsinogen), elastase
        (proelastase), pepsin (pepsinogen), and many others.

    b. Enzyme is converted back and forth between active and
        inactive states, depending on the needs of the body,
        usually by a reversible covalent modification such as
        phosphorylation or oxidation and reduction of disulfide
        bonds; mode of control exerted by epinephrine, insulin,
        and other hormones; frequently mediated by cyclic AMP;
        examples include glycogen synthetase and phosphorylase,
        xanthine oxidase (a further control on this enzyme), and
        probably adipose tissue lipase and palmityl-CoA synthetase.

5. Control of mRNA Translation. Similar, in principle, to 4 above;
    less conservative of material but able to respond more rapidly;
    less sensitive than 1, 2, or 4; examples include heme activation
    of globin synthesis in reticulocytes (this and other controls
    of hemoglobin synthesis are discussed under the control of

protein synthesis and under hemoglobin); ACTH activation of desmolase (cholesterol oxidase) synthesis in the adrenal cortex (mechanism 4 (b) is also involved); large increase in protein synthesis observed upon fertilization of an ovum; albumin synthesis may be partly regulated in this manner; see also under albumin synthesis and regulation of protein synthesis.

6. Control of DNA Transcription. Prevention of mRNA synthesis from the DNA template coding for an enzyme; one explanation for inducible and repressible enzyme systems in bacteria and perhaps in higher animals; conserves energy and raw materials; probable mode of action of steroids; the differentiation of different cell types is at least partly controlled in this way; see under regulation of protein synthesis.

Coenzymes

1. Many enzymes, in order to perform their catalytic function, require the presence of small, non-protein molecules. These are termed coenzymes, cofactors or prosthetic groups more or less interchangeably. In this text, these designations will be used with complete interchangeability. The enzyme with bound cofactor is called a holoenzyme, the protein portion being the apoenzyme.

$$apoenzyme \ + \ cofactor \ = \ holoenzyme$$

As implied in the above reaction, cofactor binding is frequently reversible. In some cases, however, it is covalently bound to the apoenzyme. If the linkage is non-covalent, the cofactor is said to be dialyzable (i.e., because the cofactor is much smaller than the apoenzyme, the two can be separated by dialysis across a semipermeable membrane).

Coenzymes also

a. Are generally stable towards heat and other agents which denature proteins.

b. Are frequently derived from vitamins. This is one of the principle functions for vitamins.

c. Are recycled in many cases. In such situations only small, catalytic amounts of them are needed for the conversion of a great deal of reactant to product (see   page 86, 5,

catalysts). Biotin-requiring carboxylations serve as a
good example.

d. Function as cosubstrates. They are chemically altered
during the reaction and must be regenerated by another
enzyme. The distinction between the substrate and the
cofactor in these cases is that the cofactor is returned
to its original form while the substrate usually undergoes
further chemical modifications. Dehydrogenase reactions
provide a good example of this.

2. A list of some important cofactors is given below. It is not
complete but includes the principal ones. Their structure,
synthesis, and function are discussed with the reactions they
catalyze and under the vitamins from which they are derived.

a. Coenzymes involved in hydride (H^-) ion and electron transfer.

i)   Nicotinamide adenine dinucleotide (NAD^+); also known
as diphosphopyridine nucleotide (DPN^+); reduced form
is NADH (DPNH).

ii) Nicotinamide adenine dinucleotide phosphate ($NADP^+$);
also known as triphosphopyridine nucleotide (TPN^+);
reduced form is NADPH (TPNH).

(i and ii, above, both require niacin (nicotinic acid;
nicotinamide) for their synthesis. This is a vitamin in
man, although small amounts are also formed from the
essential amino acid tryptophan.)

iii) Flavin mononucleotide (FMN); reduced form is $FMNH_2$.

iv) Flavin adenine dinucleotide (FAD); reduced form is
$FADH_2$.

(iii and iv, above, are derived from the vitamin riboflavin.
They are not truly flavin nucleotides since the flavin
ring system is attached to ribitol, a sugar alcohol,
rather than ribose.)

v)   Lipoic acid (6,8-dithio-n-octanoic acid); not a vitamin
in man, since it apparently can be synthesized in
sufficient quantity; involved in acetyl-group transfer
during oxidative decarboxylations of amino acids.

vi) Coenzyme Q (CoQ; ubiquinone); a group of closely related
compounds, differing only in the length of a sidechain;
can be synthesized in man from farnesyl pyrophosphate,
an intermediate in cholesterol biosynthesis; structure
is similar to that of vitamin K_1 but no interrelationship
has been established.

b. Coenzymes participating in group transfer reactions.

i) Coenzyme A (CoA, CoASH); -acetyl and other acyl group
transfers; requires the vitamin pantothenic acid for
its synthesis.

ii) Thiamine pyrophosphate (TPP; Cocarboxylase); used for
oxidative and nonoxidative decarboxylations of amino
acids and formation of α-ketols (e.g. for the
transketolase-catalyzed steps of the HMP shunt);
derived from thiamine (vitamin B_1).

iii) Pyridoxal phosphate; cofactor for an unusually wide
variety of reaction types including amino acid
racemization, decarboxylation and transamination and
the elimination of water and hydrogen sulfide; usually
associated with reactions of amino acids; derived from
pyridoxine, pyridoxal, and pyridoxamine (collectively
called vitamin B_6).

iv) Tetrahydrofolic acid (FH_4); carrier of one-carbon
fragments, as formyl, methylene, methenyl, and
formimino groups; derived from folic acid (folacin).

v) Biotin; carboxylation reactions; tightly bound to the
apoenzyme in an amide linkage to the ϵ-amino group of
lysyl residue; biotin is itself a vitamin.

vi) Cobamide coenzyme (5,6-dimethylbenzimidazole cobamide;
5'-dehydroadenosyl cobalamine); contains cobalt bound
in a porphyrin-like carrier ring system; unusual in
that it contains a cobalt-carbon (organometallic)
bond; involved in methyl-transfer reactions; very
small amounts required by the body; derived from
cyanocobalamin (vitamin B_{12}).

vii) Adenosine triphosphate (ATP); can contribute phosphate,
adenosine, and adenosine monophosphate (AMP) for
various purposes; has many other functions; does not

contain any vitamin portion and hence the complete
molecule is synthesized in the body.

viii) Cytidine diphosphate (CDP); carrier of phosphoryl
choline, diacylglycerides, and other molecules during
phospholipid synthesis; CDP does not contain any
vitamin-derived group and can, consequently, be
synthesized entirely in the body.

ix) Uridine diphosphate (UDP); carrier of monosaccharides
and their derivatives in a variety of reactions; see
bilirubin, lactose, galactose and mannose metabolism,
glycogen synthesis, and other pathways; as with ATP
and CDP, this molecule can be completely synthesized
in the human body since it contains no vitamin moiety.

x) Phosphoadenosine phosphosulfate (PAPS; "active sulfate");
sulfate donor in synthesis of sulfur-containing
mucopolysaccharides; sulfate donor in detoxification of
sterols, steroids and other compounds; see metabolism
of the sulfur-containing amino acids; derived from
ATP and an inorganic sulfate, hence, contains no vitamin
portions and can be synthesized completely by man.

xi) S-adenosyl methionine (SAM; "active methionine"); methyl
group donor in biosynthetic reactions; formed from ATP
and the essential amino acid methionine.

c. Metalloenzymes. In addition to cobamide (cobalt) requiring
enzymes, many others require a variety of metals for
activity, maintenance of tertiary and quaternary structure,
of both. If these metals are present complexed to a
porphyrin ring system, the coenzyme is the metalloporphyrin.
A few examples are given below to indicate the diversity
of metals involved.

i) Mg^{+2} is required by most enzymes using ATP. The active
form of ATP is a Mg^{+2}-ATP^{-4} complex.

ii) Ca^{+2} is involved in a wide variety of processes,
notably muscle contraction, blood clotting, nerve
impulse transmission, and cAMP mediated processes.
The exact way in which interaction occurs between
Ca^{+2} and protein is not known.

iii) Fe^{+2}/Fe^{+3} (ferrous/ferric iron); iron is needed in hemoglobin, the cytochrome chain of oxidative phosphorylation, non-heme iron proteins, and other enzyme systems.

iv) Cu^{+1}/Cu^{+2} (cuprous/cupric copper); cytochrome oxidase in mitochondrial electron transport contains iron and copper. Tyrosinase and other oxidases also require copper.

v) Zn^{+2} (zinc); is used by lactic acid dehydrogenase, the alcohol dehydrogenases, carbonic anhydrase and other enzymes.

vi) Mo^{+6} (molybdenum ion); is part of xanthine oxidase (in addition to iron), and of other oxidases and dehydrogenases.

vii) Mn^{+2} (manganous manganese ion); is required by acetyl-CoA carboxylase, deoxyribonuclease (Mg^{+2} can replace Mn^{+2} in this case) and other enzymes.

viii) Other metals may be involved in trace amounts as cofactors for some enzymes. Because of the minute quantity of the metal usually required, studies of this sort are difficult.

The Use of Enzymes in the Clinical Laboratory

1. There are two principal ways in which enzymes are used clinically. They are

   a. determinations of the amount of an enzyme (or enzyme activity) present in a biological sample,

   b. the use of enzymes as specific reagents to measure the concentration of other enzymes of non-enzymic molecules (e.g., substrates) in biological material.

Examples of these are given below and in other sections of this book.

All clinical tests involving enzymes make use of kinetics, regardless of the role that the enzyme plays in the procedure. In the section on enzyme kinetics it was pointed out that the velocity (measured as the change in some property with respect to time) of an enzyme-catalyzed reaction depends on, among other things, the concentrations of enzymes and substrate. In assays in which the enzyme activity is being determined in a biological sample, the substrate is present in great excess and the concentration of the enzyme determines the velocity. When an enzyme is used as a reagent, it is present in excess and the concentration of the compound being assayed (the substrate) is the factor limiting the velocity.

2. a. Again referring to enzyme kinetics, if the initial substrate concentration, S_0, is sufficiently high, the velocity of an enzyme-catalyzed reaction will be independent of S_0 (zero-order with respect to substrate; $v = k(E)$). In addition, if the period of measurement is short enough, little substrate is consumed. The substrate concentration can then be considered constant ($= S_0$) and the reaction will be zero-order during the entire assay. For such a system, if X is the property being measured, a plot of X v.s. time is a straight line as shown in the diagram below.

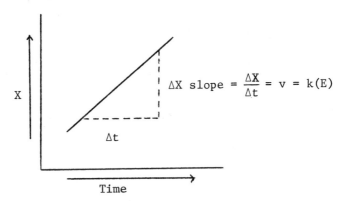

(optical density, oxygen pressure, amount of acid produced, or any other property which can be monitored)

ΔX slope $= \dfrac{\Delta X}{\Delta t} = v = k(E)$

If all other conditions (temperature, pH, S_0, etc.) are constant, the slope (velocity) depends only on (E). After k is determined using a sample having a known enzyme activity, this relationship can be used to quantitate enzyme activities in biological materials.

This is the principle behind <u>fixed incubation time (two-point) assays</u>. In such an assay the property being monitored is measured twice: immediately after mixing the reactants and again after some time interval. The velocity is the change in the property over the time interval. If the X v.s. time curve is linear, this is valid. In some cases, however, the velocity is not constant initially, although it may become so

later (if a lag period occurs, for example). This results in
a false value for the enzyme activity. To avoid this problem,
methods have been developed for continuously monitoring the
rates of some reactions. These are called <u>kinetic assays</u>.
(This nomenclature is unfortunate, since, as was indicated
previously, <u>all</u> enzyme assays are kinetic in the sense that
they all employ enzyme kinetics.) If such an assay is used,
a continuous plot of the property being measured as a function
of time is obtained and an appropriate linear portion of the
curve can be selected. One major cause for uncertainty in the
enzyme assay results can thus be eliminated.

b.   Enzymes can be either "plasma-specific" or "non-plasma-
specific". Plasma specific enzymes are those which are
normally present in the plasma at appreciable levels, perform
their primary function in the plasma, and are not found in
cells to any appreciable extent. Non-plasma-specific enzymes
are normally absent or at low concentration in the plasma,
are primarily intracellular enzymes, and are elevated in plasma
only when a large number of cells are injured or when the rate
of enzyme synthesis is increased to an abnormal level. The
latter group can be further divided into enzymes present in
most cells of the body (e.g.,glycolytic enzymes) and those
found principally in one or two tissues (tissue- or organ-
specific enzymes). In some instances (lactic acid dehydrogenase;
alkaline and acid phosphatases), one of several isozymic forms
can be tissue-specific. If such an enzyme is found to be
elevated in the serum, it is usually indicative of the
destruction of a large number of cells in the tissue or organ
which contains the enzyme. Examples will be discussed below
and in specific sections of this book.

c.   An interesting problem associated with this is the clearing of
non-plasma-specific enzymes from the plasma. Since their
elevation is transient, continuing for only a short time after
cessation of their release from damaged tissue, there must be
some mechanisms by which they are eliminated. Although enzymes
are too large to be filtered by the glomerulus, if they are
first acted upon by serum proteases, they may be partially
removed in this way. They may also be removed to some extent
through the biliary system of the liver. These processes
are not well understood at the present time.

d.   Enzyme-catalyzed reactions are frequently very complicated,
requiring one or more cofactors for their occurrence in
addition to enzyme and substrate. If an enzyme normally present
only within a cell is being measured in a serum sample, one must

be certain that all necessary cofactors and activators are present in sufficient concentration for the expected reaction to occur. Concentrations may differ markedly in going from the interior of a cell to the plasma.

e. In order to understand what an enzyme activity means in terms of normal values, it is necessary to know the units in which it is measured. Thus, alkaline phosphatase activity may be expressed in Bodansky units, King-Armstrong units, or Bessey-Lowry-Brock units, all of which give different numerical values for the normal range. In general, one unit of an enzyme will transform some defined quantity of substrate in a given length of time under specified conditions. To standardize reports of enzyme activity, the Commission on Enzymes of the International Union of Biochemistry has proposed that the unit of enzyme activity be defined as <u>that quantity of enzyme which will catalyze the reaction of one micromole of substrate per minute</u>, and that this unit be called the International Unit (U. or I.U.). They also recommend a standard temperature of 30°C.

f. Under ideal conditions, it is essential to measure both the concentration of the enzyme present and its activity. A low activity may be due either to the absence of any enzyme or to a lack of <u>functional</u> enzyme. The concentration of the enzyme is not commonly measured, however, because of the unavailability of pure human enzyme standards. This is either due to the fact that the enzyme has not yet been isolated in a pure form or that it is not available in sufficient quantities for its use to be practical. Therefore, the common practice is to measure enzyme activity. Herein lies one of the main problems in this field, that is, the lack of an accurate reference method for each enzyme assayed.

g. Glucose-6-phosphate dehydrogenase (G6PD) is a non-plasma specific enzyme found in most cells of the body. A deficiency of this enzyme in the erythrocyte has been implicated as the cause of primaquine sensitivity (see under hexose monophosphate shunt).

Red cell G6PD can be specifically assayed by preparing a red cell hemolysate. The enzyme catalyzes the reaction:

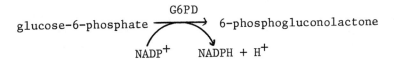

$$\text{glucose-6-phosphate} \xrightarrow{\text{G6PD}} \text{6-phosphogluconolactone}$$
$$\text{NADP}^+ \qquad \text{NADPH} + \text{H}^+$$

Since NADPH absorbs light at 340 nm while NADP$^+$ does not (see Figure 17).

Figure 17. Absorption Spectrum of the Oxidized and Reduced Forms of NADPH. The equivalent spectra for NADH are essentially the same.

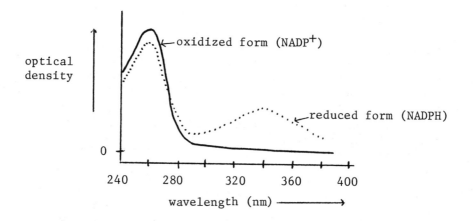

the rate of the reaction can be followed by measuring the change in optical density at this wavelength with time. This assay not only illustrates the measurement of a tissue enzyme, but it is also an example of a pyridine-nucleotide linked assay. The difference in optical density between NADP$^+$ and NADPH (also NAD$^+$ and NADH) is so readily measured, that a number of assays make use of it either directly, as here, or indirectly. In the latter situation, the reaction actually being measured is linked by one or more steps to a reaction in which a pyridine nucleotide is oxidized or reduced. An example of this is the measurement of serum triglyceride, described under plasma lipids.

In practice, the molar absorptivity (absorbance or optical density of a one molar solution one centimeter thick) of NADH is used in the following way:

$$A = a \; b \; c$$

where A = observed absorbance (optical density, O.D.),
    a = molar absorptivity,
    b = pathlength of the spectrophotometer cell in
        centimeters, and
    c = concentration of the solution in moles/liter
        (molarity, M).

This relationship is known as <u>Beer's Law</u>.

For NADH (and NADPH), at 340 nm, a = 6.2 x 10^6.

If b = 1 cm and c = 1 M, then A = 6.2 x 10^6 O.D. units.

If b = 1 cm and c = 1 μM (= 10^{-6} M), then A = 6.2 O.D. units.

If b = 1 cm and c is 1 μmole in 3 ml (= 1/3 μM = 0.3 x 10^{-6} M), then A = 2.1 O.D. units.

In reporting the results of NADH assays, spectrophotometric units are sometimes used.

One spectrophotometric unit of enzyme activity = a change of 2.1 O.D. units,

and International units = spectrophotometric units/2.1.

h. A number of other enzymes are also very important in clinical diagnosis. One example, discussed under glycolysis, is lactic acid dehydrogenase (LDH) and its isozymes. Others include creatine phosphokinase (CPK), alkaline and acid phosphatase, lipase, amylase (discussed in the next section), the trans- aminases (glutamate-oxaloacetate transaminase (GOT; aspartate transaminase) and glutamate-pyruvate transaminase (GPT; alanine transaminase)), and many more.

3. a. If an enzyme is being used as a reagent, the substrate is the limiting quantity, as was pointed out earlier. The amount of substrate consumed or product formed is proportional to the initial substrate concentration, S_o. The initial enzyme concentration, E_o, is set high enough to avoid saturation by any substrate concentrations which are likely to be encountered (i.e.,(E) will never limit the observed velocity).

Unlike the assays discussed so far, the velocity here varies with time since the substrate concentration is not constant. Consequently, a fixed incubation time is used for all standards and unknowns. The average velocity over this time interval, when plotted <u>v.s.</u> S_o, should give a straight line from which unknown values of S_o can be calculated.

b. One example of an assay of this type is the use of glucose oxidase to measure serum glucose levels. This enzyme catalyzes the reaction

$$\beta\text{-D-glucose} + O_2 + H_2O \xrightarrow{\text{glucose oxidase}} H_2O_2 + \text{gluconic acid.}$$

The reaction can be followed by measuring P_{O_2} with a Clark oxygen electrode. If, in addition, peroxidase and ferrocyanide are added, the reactions

$$H_2O_2 \xrightarrow{\text{Peroxidase}} H_2O + \text{Nascent oxygen}$$

$$\text{Nascent oxygen} + \text{ferrocyanide} \longrightarrow \text{ferricyanide}$$

can occur. Since ferrocyanide is colorless while ferricyanide is red, the reaction can now be followed spectrophotometrically. Nascent oxygen is a more reactive form of oxygen than O_2 (molecular oxygen), and is probably atomic oxygen.

c.   The serum level of amylase is elevated in acute pancreatitis. Its assay provides an interesting example of the use of linked reactions and of the use of enzymes as reagents in measuring the activity of another enzyme. The serum to be assayed is added to a reaction mixture containing starch, maltase, and glucose oxidase, resulting in the reactions

$$\text{starch} \xrightarrow{\text{amylase}} \text{maltose} + \text{glucose}$$

$$\text{maltose} \xrightarrow{\text{maltase}} \beta\text{-D-glucose}$$

$$\beta\text{-D-glucose} + O_2 \xrightarrow{\text{glucose oxidase}} \text{gluconic acid} + H_2O_2.$$

The reaction is followed the same way as in the measurement of glucose in the preceding example. As in that example, peroxidase and ferrocyanide can be used for a colorimetric procedure.

4.   There are several other important factors which pertain to all measurements involving enzymes in the clinical laboratory.

a.   It is important that only S_o or (E) vary in a given enzyme assay. All other reaction conditions (temperature, pH, other molecules present) must be constant. The analyst will otherwise be unable to decide whether an observed value is due to the component being measured or the result of other unrelated effects.

b.   Enzymes are very sensitive to temperature and other denaturing agents, and to inhibition by drugs, other foreign substances, and, in some cases, natural, endogenous inhibitors. One must be very careful then in assessing an enzyme activity value, that the observed activity is not altered by effects such as those mentioned above.

4
Carbohydrates

Structures and Classes

Carbohydrates (also called sugars or saccharides) are divided into four major groups.

1. Monosaccharides cannot be hydrolyzed into a simpler form. Their general formula is $C_n(H_2O)_n$. The simplest compound having this formula is

$$CH_2O \ (or \ H-\overset{\overset{\displaystyle O}{\|}}{C}-H)$$

named formaldehyde, but it is not usually termed a carbohydrate. The monosaccharides are called trioses, tetroses, pentoses, hexoses and so on, according to the number of carbon atoms in the molecule. They are further divided into aldo-sugars and keto-sugars. The prefix indicates the presence of an aldehyde or ketone group. The monosaccharides of biological importance in man are

Trioses

D-glyceraldehyde (glycerose; an aldotriose)
D-dihydroxyacetone (a ketotriose)

Tetroses

D-erythrose (an aldotetrose)

Pentoses

D-xylulose (a ketopentose)
D-ribose (an aldopentose)
D-deoxyribose (an aldopentose)
D-xylose (an aldopentose)
D-lyxose (an aldopentose)

Hexoses

D-glucose (an aldohexose)
D-galactose (an aldohexose)
D-mannose (an aldohexose)
D-fructose (a ketohexose)

2. <u>Disaccharides</u> are composed of <u>two</u> of the same or different monosaccharides. The general formula is $C_n(H_2O)_{n-1}$.

    Examples include lactose, maltose, cellobiose (all reducing sugars) and sucrose (a non-reducing sugar).

3. <u>Polysaccharides</u> are composed of many monosaccharide units. For those of biological importance, the general formula is $(C_6(H_2O)_5)_n$. This indicates a polymer of hexoses.

    a. <u>Homopolysaccharides</u> contain only one type of monosaccharide unit. Examples include

        cellulose.......... a polymer of glucose
        glycogen........... a polymer of glucose
        starch............. a polymer of glucose
        inulin............. a polymer of fructose

    b. <u>Heteropolysaccharides</u> contain two or more monosaccharide units. Examples include: agar-agar, vegetable gums, mucilages, and mucin.

    c. <u>Mucopolysaccharides</u> are nitrogen containing polysaccharides such as heparin, chondroitin sulfate, hyaluronic acid and chitin.

4. <u>Miscellaneous Compounds</u> (carbohydrate derivatives): Sialic acid, vitamin C, streptomycin, inositol, glucuronic acid, sorbitol, galactitol, xylitol, and others.

<u>Structural Considerations of Monosaccharides</u>

1. Assignment of absolute configuration, designated as D or L (not to be confused with d and l, which indicate the sign of rotation of the plane polarized light) is based on the position of substituent groups about the penultimate (next to last) carbon. (Refer to the material on the optical activity of the amino acids.)

2. If two compounds differ from one another in configuration around one carbon, they are termed <u>epimers</u>. Galactose and glucose are epimers with respect to carbon 4; mannose and glucose are epimers with respect to carbon 2. The interconversion of glucose and galactose is known as <u>epimerization</u> and the enzyme which catalyzes this reaction is an epimerase. The number of optical isomers of a compound depends on the number of asymmetric carbon atoms (carbons with four <u>different</u> substituent groups). If n = the number of asymmetric carbons, 2^n = the number of optical isomers.

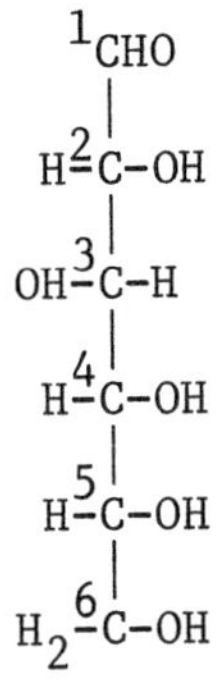

$$^1CHO$$
$$H^2C-OH$$
$$OH^3C-H$$
$$H^4C-OH$$
$$H^5C-OH$$
$$H_2^6C-OH$$

The structure that is written in the adjacent form has carbons, 2, 3, 4, 5 as asymmetric centers. Therefore the number of <u>optical</u> isomers expected is 2^4 = 16 isomers (of which D-glucose, shown here, is <u>one</u> such isomer).

3. The observed optical activity of glucose, however, indicates that there were two forms of D-glucose itself. For this to be so, there must be another asymmetric carbon. This additional asymmetric center is generated at the carbon atom of an aldehyde (C_1 of glucose) or keto (C_2 of fructose) group by reaction of either one of these groups with an alcohol group in the same sugar molecule. This gives rise to an internal <u>hemiacetal</u> (or hemiketal) structure. A hemiacetal or hemiketal is a compound formed by reaction of an aldehyde or ketone with an alcohol.

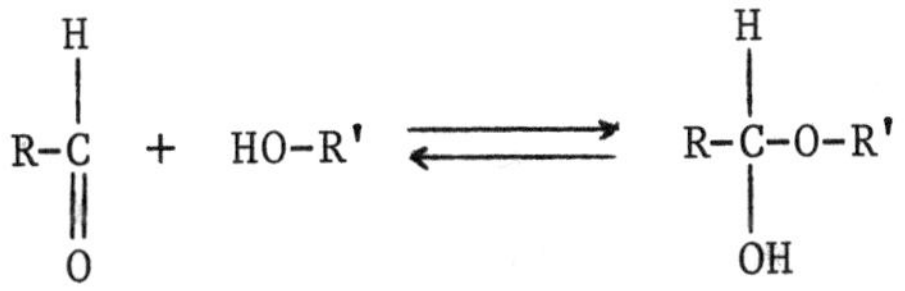

$$
\begin{array}{ccc}
\text{H} & & \text{H} \\
| & & | \\
\text{R-C} + \text{HO-R'} \rightleftharpoons & & \text{R-C-O-R'} \\
\| & & | \\
\text{O} & & \text{OH}
\end{array}
$$

Note that the reactions are <u>reversible</u> so that the reactants and products are readily interconvertible.

In the case of glucose, the functional groups of carbon 1 (aldehyde) and carbon 5 (alcohol) will react to give the cyclic hemiacetal forms shown below.

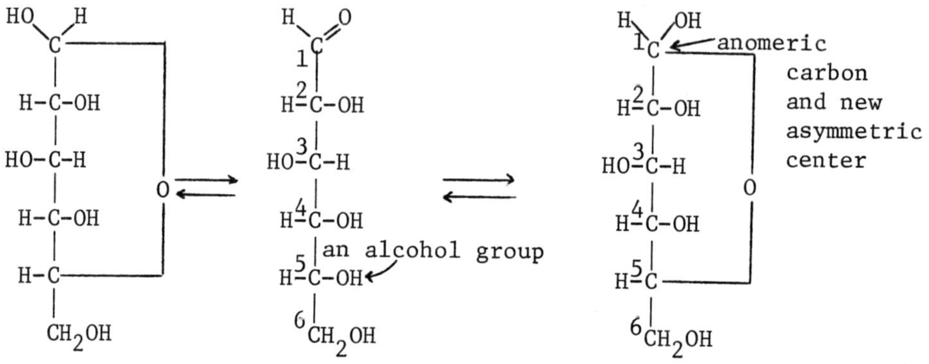

β-D-glucose (a   D-glucose   α-D-glucose (a cyclic
cyclic hemiacetal)   hemiacetal)

The α and β forms of glucose designated in the above structures are
known as <u>anomers</u>. Notice the α and β forms differ only in the
configuration around carbon 1. Also note that the cyclic
hemiacetal forms are in equilibrium with the open=chain structure
in aqueous systems.

Thus the formation of the hemiacetal structure introduces a <u>new</u>
<u>asymmetric center</u> at C_1 (which now has four different groups
attached to it) of D-glucose. Therefore the total number of
isomers of glucose is $2^5 = 32$.

The α and β cyclic forms of D-glucose (known as anomers) have
different optical rotations. These values, however, will not be
of equal magnitude with opposite signs as one would expect from
enantiomorphs (mirror images). This is because the α and β isomers
as a whole are not mirror images of each other. They differ in
configuration about the anomeric carbon (1) but have the same
configuration at other asymmetric carbons (2,3,4,5).

Crystalline glucose is a mixture of the α and β forms (ordinary
shelf crystalline glucose is largely the α-form). When glucose
is dissolved in water, the optical rotation of the solution
gradually changes and attains an equilibrium value. This change
in optical rotation is called <u>mutarotation</u>. It represents the
interconversion of α and β forms to an equilibrium mixture of them.
This can be represented as

                                 equilibrium mixture
              pure α-D-glucose ⟶ of α and β forms ⟵ pure β-D-glucose

optical rotation   +112°   +52.5°   +19°

Mutarotation can occur because of the presence of open-chain glucose (due to the equilibrium in solution) which can cyclize to either the α-or-β form with equal probability.

4. Ring structure representations suggested by Haworth and called Haworth projections are commonly used. They are shown below. The Fischer Projections are named for Emil Fischer who did much work on the structure of carbohydrates.

For glucose and other aldoses

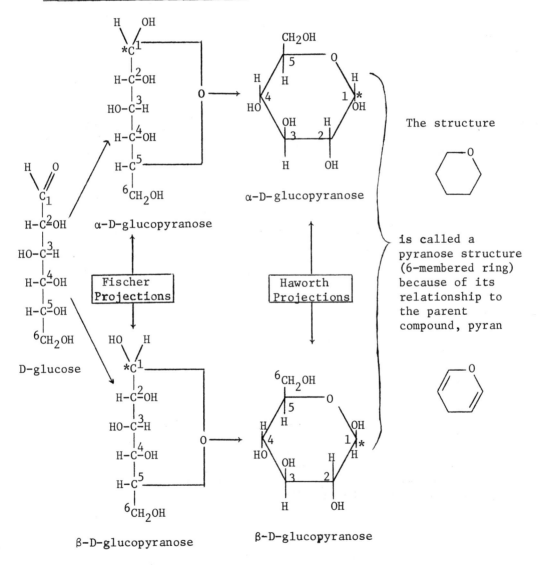

The structure

is called a pyranose structure (6-membered ring) because of its relationship to the parent compound, pyran

For fructose and other ketoses

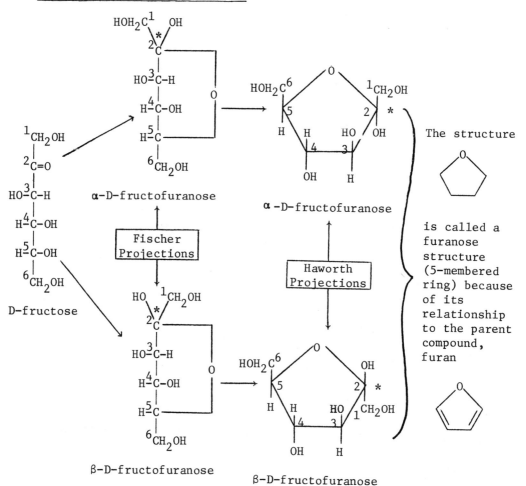

α-D-fructofuranose

β-D-fructofuranose

The structure

is called a
furanose
structure
(5-membered
ring) because
of its
relationship
to the parent
compound,
furan

Note:

1) The anomeric carbon is 1 in glucose. In fructose it is #2.
   They are indicated by asterisks in the drawings above. Notice
   that in an α-sugar the hydroxyl group on the anomeric carbon is
   below the ring; in a β-sugar, it is above. In a D-sugar the
   -CH₂OH containing carbon 6 is above the ring; in an L-sugar, it
   is below. Also, right on the Fischer projections corresponds
   to down on the Haworth projections.

2) It is important to understand the <u>spatial</u> relationship shown by Haworth's ring structures. If the ring is considered to be lying flat on the page, then the groups sticking up from the ring (shown with arrows) are coming up from the page, while the groups below the ring (dotted lines) are below the plane of the paper.

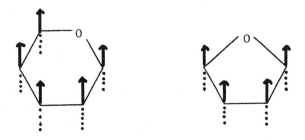

For this reason, if one were to "flip" the ring over, rotating it on an axis as shown, the groups that were above the page end up below the page.

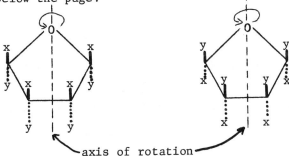

axis of rotation

Thus β-D-fructofuranose can be written

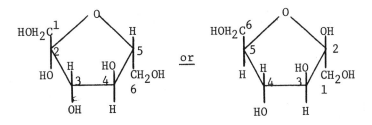

or

Similarly, α-D-glucopyranose can be either

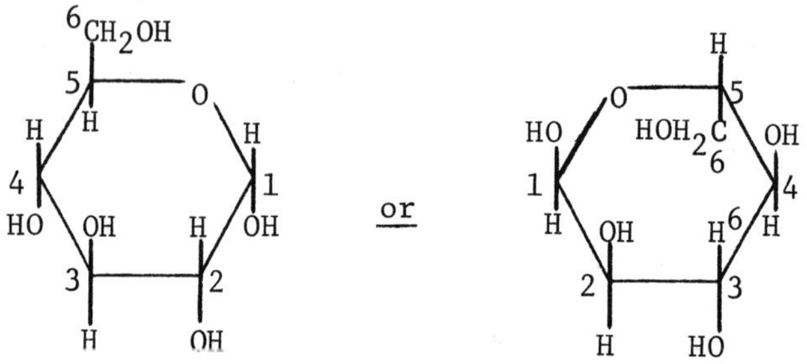

or

3) Although the names glucopyranose, fructofuranose, etc. are
really correct, for convenience the pyranose and furanose will
usually be dropped from the names used from now on. One should
realize that, whenever possible, the ring structures are
what actually exist in solution.

Some Reactions of Monosaccharides

1. Formation of hemiacetals and hemiketals has already been discussed.

2. Glycosides (compounds formed between the <u>hemiacetal</u> or <u>hemiketal</u>
hydroxyl group and an aglycone)

Glycosides = carbohydrate + non-carbohydrate (the aglycone could
           residue   residue   be methyl alcohol,
                      (aglycone)   glycerol, a sterol,
                                  a phenol, etc.)

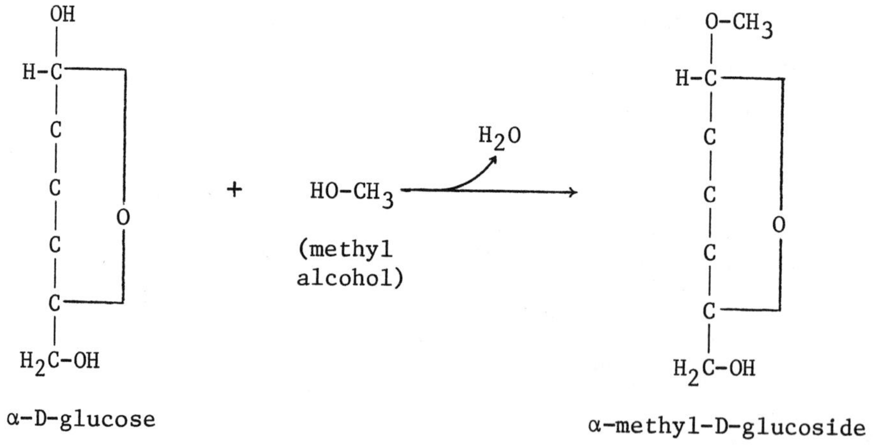

α-D-glucose

+   HO-CH$_3$

(methyl
alcohol)

$\xrightarrow{\text{H}_2\text{O}}$

α-methyl-D-glucoside

When a hemiacetal or hemiketal participates in such a bond, it is called an acetal or ketal (the prefix hemi- is dropped). Such compounds are <u>not</u> in equilibrium with an open-chain form and <u>do not</u> exhibit mutarotation. Glycosides are found in many drugs, spices and constituents of animal tissue. A <u>glucoside</u> is a glycoside formed by glucose.

Examples:

a.

Vanillin-D-Glucoside (Glucovanillin; vanilloside): A β-D-glucoside (natural source of vanilla flavor)

b. A cardiac α-D-glycoside (stimulates cardiac muscle contraction)

aglycone (a sterol)

3. Formation of disaccharides (many of these contain glycosidic bonds).

    Example: Sucrose

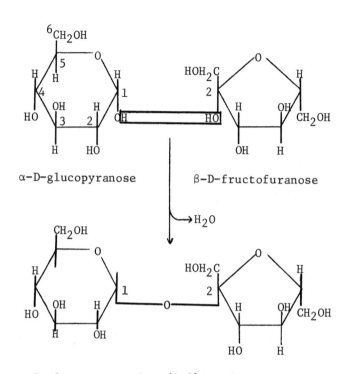

α-D-glucopyranose   β-D-fructofuranose

α-D-glucopyranosyl - (1→2)β-D-fructofuranoside

Note: the D-fructose has been "flipped" over as described
       previously. The anomeric carbons are joined in sucrose.

Sucrose is a <u>non-reducing sugar</u> since both aldehyde and ketone
carbons participate in the glycosidic bond. Disaccharides can
also be named as glycosides, since they are formed by substituting
a monosaccharide for an aglycone. In a glycosidic bond between
two sugars, the arrow indicates the direction of the bond, from
the non-reducing to the reducing (free aldehyde or ketone group)
end of the molecule.

4. Formation of phosphoric acid esters

    Example: Formation of D-glucose-6-phosphate

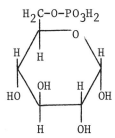

5. Methylation of hydroxyl groups

6. In solutions of strong mineral acids, sugars are usually dehydrated.

    Example:

$$
\begin{array}{c}
\text{H-C=O} \\
|\\
\text{H-C-OH} \\
|\\
\text{H-C-OH} \\
|\\
\text{H-C-OH} \\
|\\
\text{H}_2\text{C-OH}
\end{array}
\quad \xrightarrow[\text{heat}]{\text{H}^+} \quad
\begin{array}{c}
\text{HC=O} \\
|\\
\text{C} \\
\|\\
\text{HC} \\
|\quad\text{O}\\
\text{HC} \\
\|\\
\text{HC}
\end{array}
\quad + \quad 3\ \text{H}_2\text{O}
$$

    Any aldopentose   Furfural

7. Reactions in alkaline solution usually involve enol formation

$$
\begin{array}{c}
\text{O} \\
\|\quad\text{H}\\
\text{-C-C-} \\
\quad\text{H}
\end{array}
\quad \underset{\longleftarrow}{\overset{(\text{OH}^-)}{\longrightarrow}} \quad
\begin{array}{c}
\text{O}^-\quad\text{H}\\
|\quad\quad|\\
\text{-C=C-}
\end{array}
+ \quad \text{H}_2\text{O}
$$

    Ketone or aldehyde   Enolate

8. There are a number of reactions in which sugar molecules react with
    various reagents giving products which absorb light in the visible
    region. Some of these reactions have been used in the determination
    of sugars or of molecules which contain sugar residues. Two such
    examples are the reaction of orcinol with pentoses (used in the
    determination of ribonucleic acid) and the reaction of

diphenylamine with 2-deoxypentoses (used in the determination of deoxyribonucleic acid).

9. Reduction of monosaccharides

Example:

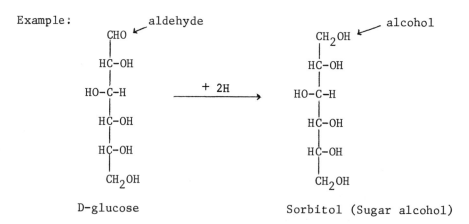

D-glucose   Sorbitol (Sugar alcohol)

Note that, in this reaction, the aldehyde group on carbon 1 has been reduced to an alcohol. This reaction is discussed again on pages 259, 263, and 277-8.

10. Primary oxidation of aldoses

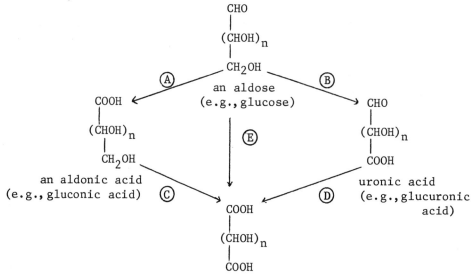

an aldonic acid   uronic acid
(e.g., gluconic acid)   (e.g., glucuronic
                                   acid)

a saccharic or aldaric
acid (e.g., glucosaccharic
or glucaric acid)

Reactions Ⓐ and Ⓓ are the oxidation of the terminal aldehyde group, reactions Ⓑ and Ⓒ are the oxidation of the terminal alcohol group, and reaction Ⓔ is the oxidation of both groups simultaneously.

Other Disaccharides

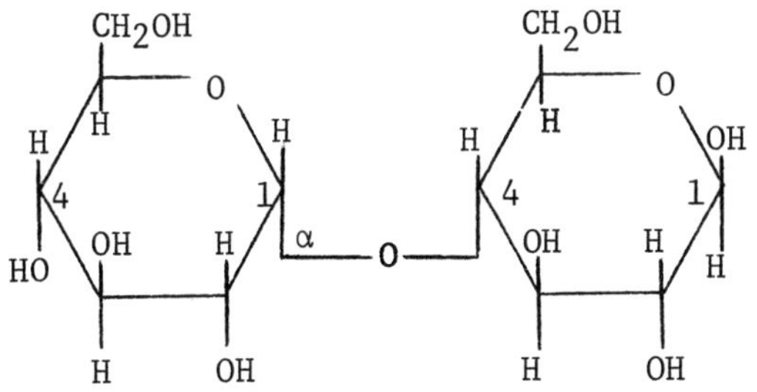

This is a reducing sugar, since at least one of the aldehyde carbons is not involved in a glycosidic bond, and can undergo mutarotation.

β-Maltose: α-D-glucopyranosyl-(1→4)-β-D-glucopyranose

Note: The anomeric carbon of one glucose molecule is joined to a non-anomeric carbon on the other (unlike sucrose). The free anomeric carbon may be in the β-configuration (as shown above), or it may be in the α-configuration, in which case it is called α-maltose (α-D-glucopyranosyl-(1→4)-α-D-glucopyranose).

Lactose

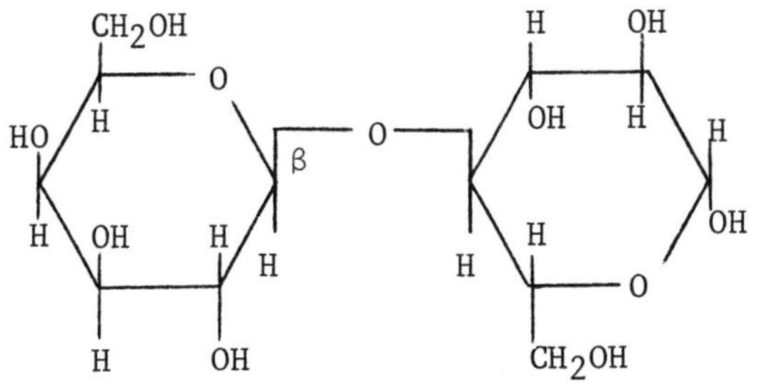

β-D-galactopyranosyl-(1→4)-β-D-glucopyranose.

Note: the second glucose molecule has been flipped over, with the axis of rotation as shown.

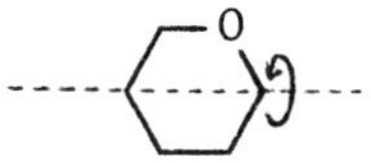

Lactose can also be written

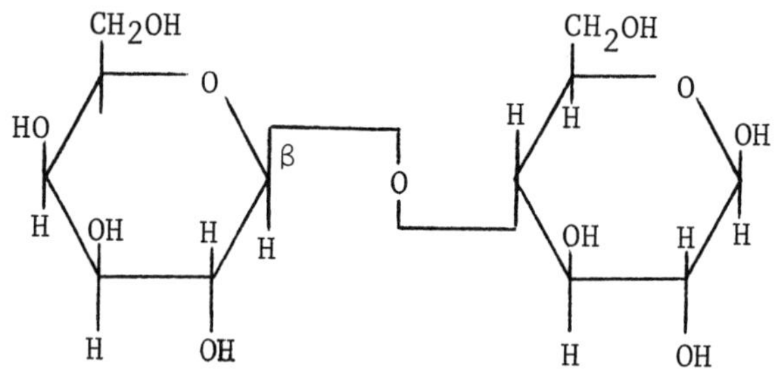

Polysaccharides

Cellulose is the most abundant <u>structural</u> polysaccharide of plants and the most abundant organic substance on earth. It occurs almost exclusively in plants.

Structure   1.   Total hydrolysis yields only D-glucose.
2.   <u>Partial</u> hydrolysis yields the disaccharide cellobiose (β-form): β-D-glucopyranosyl-(1$\rightarrow$4)-β-D-glucopyranose.
3.   Hydrolysis of fully methylated cellulose yields only 2,3,6-tri-0-methyl glucose, showing a lack of branching in the cellulose molecule.

All of this evidence indicates that cellulose is a linear array of D-glucopyranose residues linked by β-(1$\rightarrow$4) glycosidic bonds.

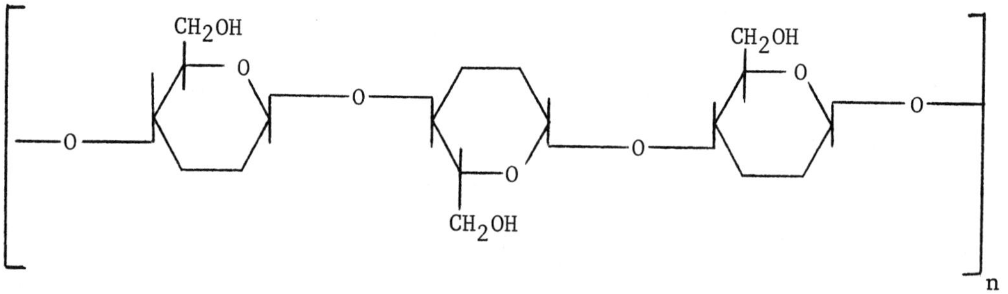

Glucosyl residues are alternately "rotated" 180°, a consequence of the β-glucosyl bond. The above structure can also be written as follows:

(The hydrogen and hydroxyls on the glucose residues have been omitted for clarity.)

Starches

1. Carbohydrate (energy) reservoir in plants.

2. The basic subunit involves the linkage of successive D-glucose molecules by α-(1→4)-glycosidic bonds.

3.

Comparing this structure to that of cellulose and other structural polysaccharides one observes that, in the structural polysaccharides, the monosaccharide units are joined together with β-linkages in contrast to the nutritional carbohydrates (starch, glycogen) which are joined by α-linkages. Mammals do not have the enzymes necessary to digest β-linked compounds. Even cows and horses who live on cellulose do not digest it themselves. Bacteria in their intestinal tracts perform this function.

4. Starch falls into two distinct classes

    a. long unbranched chains (amyloses) having only α(1→4) bonds, and

    b. branched chain polysaccharides (amylopectins) having α(1→6) bonds at the branch-points but α(1→4) linkages everywhere else.

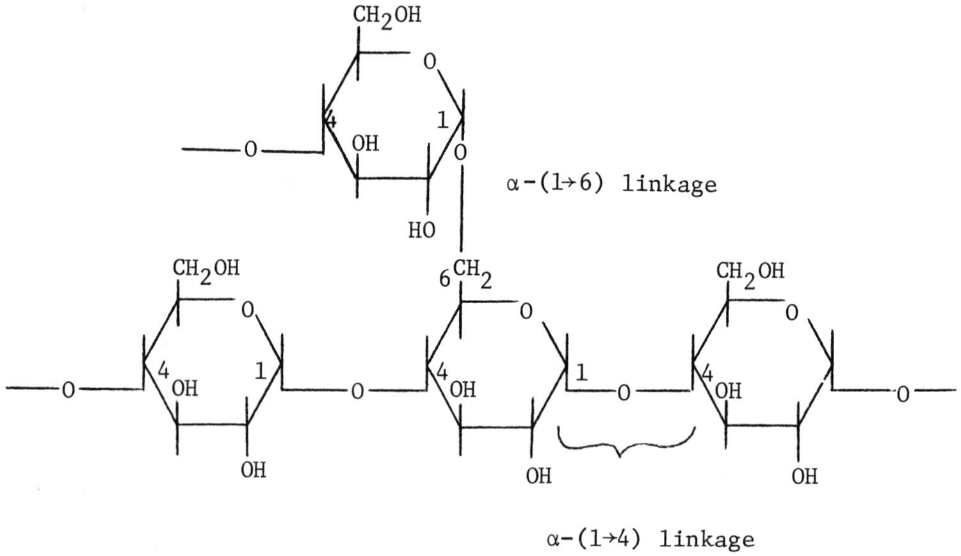

α-(1→6) linkage

α-(1→4) linkage

5. Starches vary widely in their molecular weights and molecular weights of several million are not uncommon.

6. Starches are important in human nutrition. They can be broken down in the alimentary tract to maltose and glucose units and then absorbed. The enzymes that hydrolyze starches are known as amylases. α-amylases (pancreatic and salivary) hydrolyze α(1→4) bonds wherever they occur in polymers of glucose which contain at least three residues. There are also β-amylases which liberate maltose units. They start at the reducing ends of a starch molecule and cleave every other α(1→4) linkage until a branch-point α(1→6) linkage is reached. β-amylases do not hydrolyze β-linkages.

Glycogen (A Polyglucose)

1. Serves as nutritional reservoir in animal tissues.

2. The structure is similar to amylopectin but is more highly branched.

3. Polydisperse (wide range of chain lengths); molecular weights may be as great as 1×10^8.

4. The linear linkages are α(1→4) and the branch-point linkages are α(1→6), as in amylopectin.

Hyaluronic Acid (A Mucopolysaccharide)

1. Present in the connective tissues, synovial fluid, and vitreous fluid; usually found associated with proteins. They may act as lubricants and shock absorbants in the joints.

2. Composed of equimolar quantities of

   a. D-glucuronic acid
   b. N-acetyl-D-glucosamine (an amino sugar)

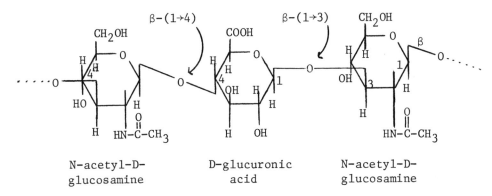

Chondroitin Sulfates

1. They are widely distributed in connecting tissues and serve as a structural material (cartilage, tendons, and bones).

2. There are three chondroitin sulfates: A, B, and C. Their compositions are

   A: (D-glucuronic acid)—Ⓘ—(N-acetyl-D-galactosamine-4-sulfate)
   —ⒾⒾ—(D-glucuronic acid)

   B: (L-iduronic acid)—Ⓘ—(N-**acetyl**-D-galactosamine-4-sulfate)
   —ⒾⒾ—(**L**-iduronic acid)

   C: (D-glucuronic acid)—Ⓘ—(N-acetyl-D-galactosamine-6-sulfate)
   —ⒾⒾ—(D-glucuronic acid)

   Note: Ⓘ indicates a β(1→3) linkage and ⒾⒾ indicates a β(1→4) linkage.

<u>Kerato Sulfate</u> is a polysaccharide present in the cornea and connective tissue (nucleus pulposus and costal cartilage, for example). Its structure is

$$(\beta(1\to4)) \qquad\qquad\qquad\qquad (\beta(1\to3))$$
(D-galactose)----(2-N-acetyl-3-deoxy-D-glucose-6-sulfate)---(D-galactose)

Heparin

1. The exact structure and biological function is <u>not</u> clear. It is a strongly acidic substance with a molecular weight of about 17,000.

2. It occurs in the granules of the circulating basophils and in the granules of the mast cells which are present in many tissues (liver, lungs, walls of arteries). Mast cells also contain histamine. In anaphylaxis (undesirable antigen-antibody reactions), heparin is released from the mast cells leading to a lack of coagulation of blood.

3. It functions as an anticoagulant by preventing the activation of factors VIII and IX and, in association with a plasma factor inhibits the action of thrombin.

4. Heparin causes the liberation of lipoprotein lipase. It presumably functions as a cofactor in the action of this enzyme on triglycerides which are associated with chylomicrons and other lipoproteins. This results in the clearing of the turbidity of a lipemic plasma.

5. It is bound to protein.

6. $\alpha-(1\to4)$ linkages are involved in its structure. A probable unit of heparin is

Sialic Acids

a. Widely distributed in tissues; particularly present in mucins
   (mucoproteins of saliva) and <u>blood group substances</u>.

b. Sialic acids are most often N-acetyl derivatives of neuraminic acid.
   A hydroxyl group may be acetylated in some instances.

c. Neuraminic acid is a 3-deoxy, 5-amino, nine carbon sugar. It is
   the condensation product of pyruvic acid and mannosamine.

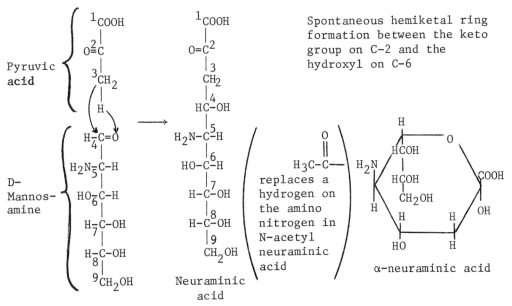

Mucopolysaccharides in Connective Tissue

Connective tissue, in its broadest definition, encompasses a wide
variety of substances including cartilage, synovial fluid, bone, and
dermis. It consists of fibers (composed mostly of the proteins
collagen and elastin), ground substance (acid mucopolysaccharides,
water, and salts), and cells such as fibroblasts and mast cells. The
fibroblasts synthesize collagen and elastin, while heparin is the
principal product of the mast cells. The synthesis of collagen is
discussed in more detail later (see protein synthesis).

The acid mucopolysaccharide composition varies from one connective
tissue to another. Hyaluronic acid is the main one in vitreous humor,
synovial fluid, and other connective tissues in which there is increased
water content. Chondroitin-4-sulfate predominates in cartilage. The

nature of the interaction between the polysaccharide and protein parts
of these mucopolysaccharides is not well understood. In several cases
it appears to be an ether linkage between the 1-hydroxyl group of a
xylose or N-acetylgalactosamine and a seryl hydroxyl group on the
protein.

The synthesis of the mucopolysaccharides is a subject of active
investigation. Chondromucoproteins (those containing chondroitin
sulfate) have been the most thoroughly studied. Their elaboration is
summarized in the following diagram.

Rough endoplasmic reticulum   Smooth endoplasmic reticulum:
(polyribosomes); protein   ─────▶ polysaccharide is elongated;
portion synthesized and   sulfate groups are added.
sugar-protein links formed.   │
                                                 │
                                                 ▼
Completed molecules are   Golgi apparatus: molecules of
secreted by the cell.   ◀───────── chondromucoprotein are
                                     finished and packaged for
                                     secretion.

It is important to notice that, in all of these steps, the enzymes
involved are closely associated with (bound to) membranes. The
degradation of mucopolysaccharides is accomplished by several enzymes.
These include hyaluronidase (degrades hyaluronic acid and the
chondroitin sulfates to oligosaccharides); β-glucuronidase (removes
uronic acid residues); several N-acetylhexosaminidases (remove
N-acetylglucosamine, etc.); at least one sulfatase enzyme (removes
sulfate residues from the sulfated polysaccharides).

A number of mucopolysaccharidoses have been reported. The classic ones
are Hurler's syndrome and the closely related Hunter's syndrome. In
both diseases, dermatan sulfate and heparin sulfate are found in the
tissues and urine. Dermatan sulfate occurs in the fibroblasts and
gangliosides accumulate in the brain. The only known biochemical lesion
is a decrease in β-galactosidase activity in tissues. The biochemical
picture of these and the other mucopolysaccharide disorders is still
quite unclear. It is apparent, however, that there are a number of
distinct diseases and that they are heritable. Several other
disorders, involving glycolipids, are discussed under lipids.

Chemistry of Blood Group Substances (see also the chapter on
Immunochemistry)

1. They are composed of polysaccharides and proteins and are found in
    many places, including erythrocyte membranes, saliva, gastric mucin,
    cystic fluids, and elsewhere.

2. They are water soluble compounds with molecular weights ranging
    from about 200,000 to 2 million.

3. They are immunologically distinct from one another, yet they show
    many similarities in chemical composition. The immunological
    specificity is associated with the type of carbohydrate present at
    the terminal portion of the mucopolysaccharide. Apparently, the
    polypeptide portion plays no role in conferring immunological
    specificity to the molecule.

Digestion of Carbohydrates

1. The major ingested carbohydrates are

    a. starch, a polymer of glucose joined by α-linkages. This
        includes both the branched-chain (amylopectin) and straight-
        chain (amylose) forms;

    b. lactose (galactose-glucose) and sucrose (fructose-glucose)
        which are disaccharides;

    c. the monosaccharides glucose and fructose.

2. Digestion of polysaccharides begins in the mouth with the action of
    salivary amylase (ptyalin). This process is terminated by
    swallowing, since the acid gastric juice inactivates salivary
    amylase activity.

3. The principle digestion of polysaccharides takes place in the
    intestines by the action of pancreatic α-amylase. The products of
    hydrolysis of starch and other disaccharides present in the diet
    are further hydrolyzed, into monosaccharides, by specific
    disaccharidases which are present on the mucosal cell surface.

4. The monosaccharides glucose and galactose are absorbed into the
    epithelial cells by an active transport process which is dependent
    upon the availability of energy and Na^+. It appears that this
    process is mediated by a membrane-bound carrier which transports
    both Na^+ and glucose into the cells simultaneously. Later, Na^+
    is pumped into the extracellular fluid space, a process which may

provide the energy for the active glucose transport. Since K⁺
facilitates the metabolism of glucose to glycogen and lactate, it
indirectly stimulates glucose uptake. It is of interest to point
out that insulin does not play a role in the transport of glucose
from the intestinal lumen into the mucosal cells. Fructose, in
contrast to glucose and galactose, is absorbed more slowly and by
a facilitated diffusion process. The monosaccharides are then
transferred from the mucosal cells to the capillary blood of the
portal circulatory system. Figure 18 is a schematic diagram
illustrating the points discussed above.

Figure 18.   Digestion of Dietary Carbohydrates

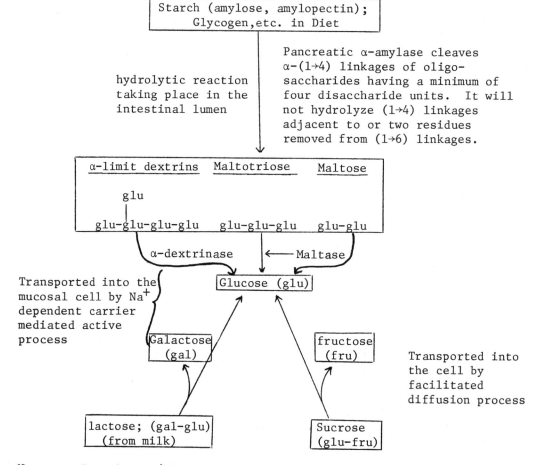

Note: α-Dextrinase (also known as isomaltase), maltase, lactase, and
sucrase are all mucosal cell surface enzymes which act in the
wall of the intestines.

5. A number of inherited deficiencies of disaccharidases have been observed in humans. Such a deficiency causes intolerance to a given disaccharide. The most common type is the lactase deficiency which is found with high frequency (about 70%) in black and oriental populations and to a lesser degree (about 15%) in white populations. The disorder is probably inherited as an autosomal recessive trait. It is discussed further with galactose metabolism.

6. Deficiency of a specific disaccharide leads to gas production due to the action of intestinal bacteria on the unabsorbed disaccharide. This results in abdominal fullness and bloating. In addition, diarrhea occurs due to osmotic effects caused both by the accumulation of the unhydrolyzed disaccharide and its metabolites (monosaccharides and fatty acids). These problems can be overcome by avoiding the intake of the offending disaccharide.

7. Deficiencies of sucrase and α-dextrinase have also been reported. In one defect of glucose and galactose absorption, the carrier-mediated process has been implicated as the probable cause.

Carbohydrate Metabolism

A major function of carbohydrates is to provide energy for body tissues to use for metabolic processes. The energy is released by oxidation of the carbohydrates. The first step in this process is the activation of the carbohydrate by phosphorylation. The carbohydrates directly used by the cells for fuel are the monosaccharides: glucose, fructose, galactose, and mannose. The latter two are converted (epimerized) to glucose before being metabolized. Glucose is the most extensively used form. The fate of pentose sugars is somewhat uncertain except for one important case. D-ribose and D-deoxyribose are used in the synthesis of nucleotides and, hence, nucleic acids.

The remainder of this chapter will deal largely with pathways of carbohydrate metabolism. It is important to keep in mind the overall purpose of these processes and to note that they are not isolated from each other or from other metabolic pathways. Notice which pathways provide alternate methods for accomplishing the same thing under different metabolic conditions. Pay special attention to the reactions that are capable of responding to changes in metabolic conditions. Often these are the rate-controlling steps. It is also important to see how one cycle or process is connected to others. Below is a list of some of the main subdivisions of carbohydrate metabolism. At the end of the chapter is a diagram illustrating some of the important connecting points.

1. Glycolysis - the oxidation of glucose to pyruvate and lactate (also called the Embden-Meyerhof pathway).

2. Citric Acid Cycle - the oxidation of the acetyl-CoA to carbon
    dioxide. The acetyl-CoA may come from carbohydrate, fat, or
    protein metabolism (also called the Krebs Cycle or tricarboxylic
    acid cycle).

3. Electron Transport and Oxidative Phosphorylation - electron
    transport is the final process in the oxidation of fuels (not just
    carbohydrates). Electrons are passed from higher energy molecules
    to molecular oxygen, giving off much free energy. This energy
    may be used to phosphorylate ADP, forming a high-energy molecule
    (ATP) that can be used by the tissues for other metabolic
    processes. Such ATP synthesis is called oxidative phosphorylation.

4. Gluconeogenesis - the formation of glucose from non-carbohydrate
    sources such as amino acids and glycerol.

5. Glycogenesis - the synthesis of glycogen from glucose.

6. Glycogenolysis - the breakdown of glycogen to produce glucose
    (and eventually lactate and pyruvate).

7. Hexose Monophosphate Shunt - an alternative pathway to glycolysis
    and the citric acid cycle. Again, glucose is oxidized to carbon
    dioxide and water. (Also called the direct oxidative pathway,
    the pentose phosphate cycle, the phosphogluconate oxidative pathway).

8. Uronic Acid Pathway - another alternative oxidative pathway for
    glucose. Glucose is converted to glucuronic acid, ascorbic acid
    (in some animals other than man) and pentoses.

9. Fructose Metabolism - the oxidation of fructose to pyruvate. It
    is connected to glucose metabolism.

10. Galactose Metabolism - the conversion of galactose to glucose
    and the synthesis of lactose.

11. Amino Sugar Metabolism - the synthesis of amino sugars from
    glucose and other sugars and the further synthesis of glycoproteins
    and mucopolysaccharides.

Figure 19.   Glycolysis (Embden-Meyerhof Pathway)

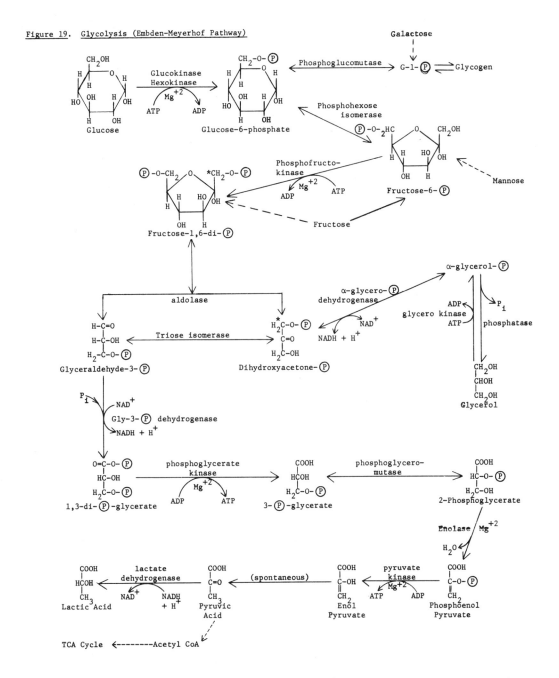

Note: P = phosphate; dotted lines indicate multiple-reaction pathways while solid lines represent single-step
conversions; in fructose-1,6-di P , * indicates the phosphorylated carbon which becomes part of dihydroxyacetone P .

Comments on E-M Pathway

1. All of the enzymes of E-M Pathway are found in the <u>extra mitochondrial soluble fraction of the cell</u> (in the cytoplasm).

2.

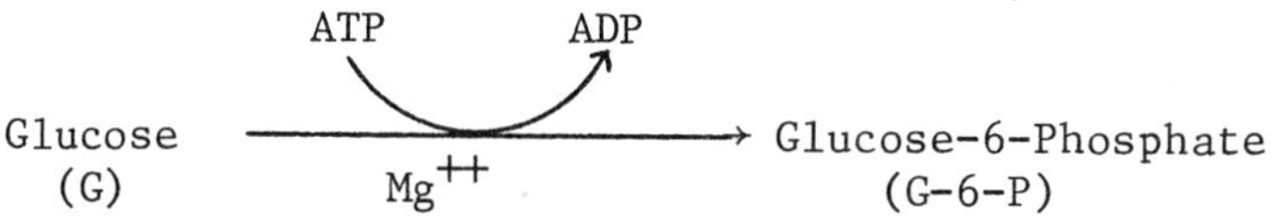

$$\text{Glucose} \xrightarrow[\text{Mg}^{++}]{\text{ATP} \quad \text{ADP}} \text{Glucose-6-Phosphate}$$

Glucose   Glucose-6-Phosphate
(G)   Mg^{++} (G-6-P)

The equilibrium constant favors the forward reaction (K > 300). This is an important step in regulating the rate at which glucose is metabolized. There are two types of enzymes that can catalyze this reaction

  a. <u>Hexokinase</u> is a non-specific enzyme which acts upon glucose, glucosamine, 2-deoxyglucose, fructose and mannose; and glucokinase which is specific for D-glucose.

  b. <u>Glucokinase</u> is a hepatic inducible enzyme and its concentration is affected by changes in the blood glucose levels. Hexokinases of mammalian tissues are present in isoenzymic forms and their activity is inhibited by G-6-P (inhibition may be allosteric). Both enzymes require Mg^{+2} (or Mn^{+2}) for activity.

3. Adenosine Triphosphate (ATP) is an important molecule in the cell. It has the structure

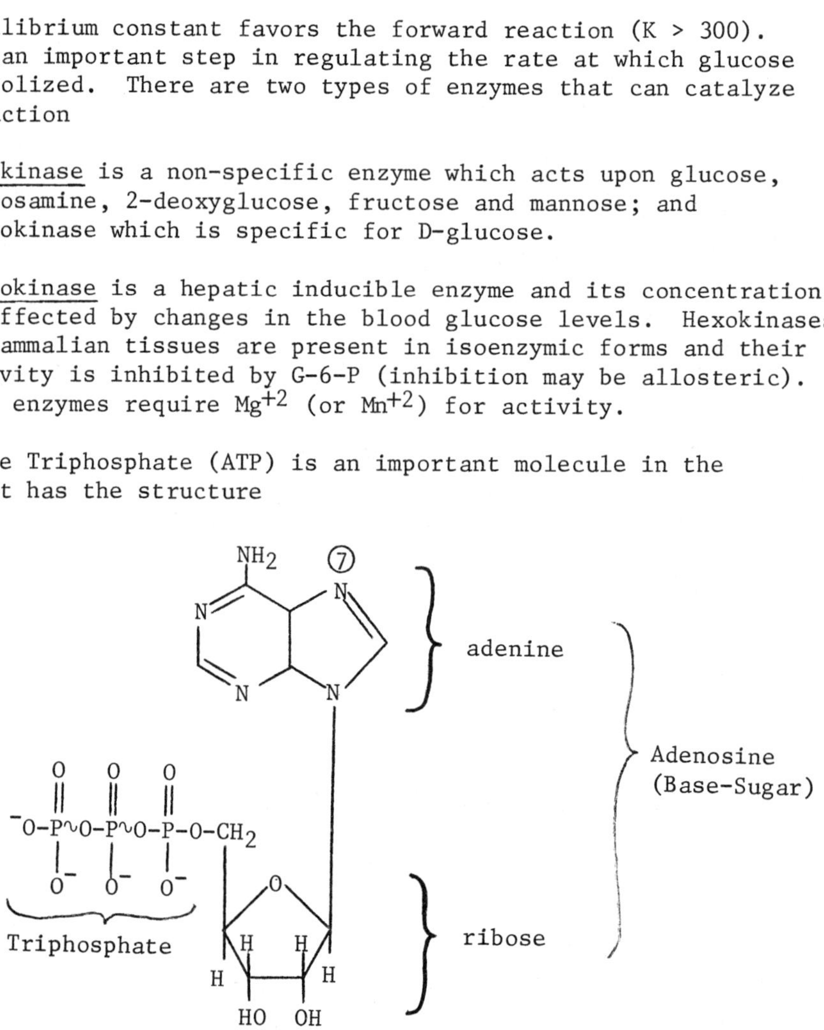

NH$_2$ ⑦

adenine

Adenosine
(Base-Sugar)

$$^-O-P\sim O-P\sim O-P-O-CH_2$$

Triphosphate   ribose

HO OH

where "$\sim$" indicates a "high-energy" bond (one having a large, negative $\Delta G^{o\prime}$ of hydrolysis). (This is a different meaning of high-energy bond than that used by physical chemists.) In the cell, most ATP exists as $(Mg^{+2})(ATP^{-4})$ with the Mg^{+2} chelated to two of the anionic oxygens on the two terminal phosphate groups. In cases where Mn^{+2} replaces the Mg^{+2}, the 7-nitrogen of the adenine ring is also a ligand of the metal ion. Since these complexes are the active forms of ATP, Mg^{+2} (or, in some cases, Mn^{+2}) is essential for practically all reactions involving ATP.

Although ATP has many functions in common with other nucleoside triphosphates (see under nucleic acids), it also serves as the principle "active storage" form of energy within the cell. The principle reaction used in the cell is

$$ATP^{-4} + H_2O \longrightarrow ADP^{-3} + HPO_4^{-2} + H^+ \quad \Delta G^{o\prime} = -7.3 \text{ Kcal/mole}$$

The hydrolysis of the second anhydride bond has the same free energy change

$$ADP^{-3} + H_2O \longrightarrow AMP^{-2} + HPO_4^{-2} + H^+ \quad \Delta G^{o\prime} = -7.3 \text{ Kcal/mole}$$

but it is less frequently used. The reasons for the large, negative $\Delta G^{o\prime}$ of hydrolysis, discussed below in terms of ATP^{-4}, apply equally to ADP^{-3}. They are

a. There are more resonance forms available to ADP^{-3} + there are for ATP^{-4}. These mean that the valence electrons can attain a lower energy state in the hydrolyzed product than in ATP^{-4}. (Note: P_i = inorganic phosphate, HPO_4^{-2} at pH = 7.)

b. ATP^{-4} has 3.8 (since one of the four protons is only about 80% ionized at pH = 7) closely spaced negative charges which repel each other strongly. In the hydrolyzed material ($ADP^{-3} + P_i$), some of the charges have been carried away by the P_i. This not only favors breaking the bond, it also keeps P_i and ADP^{-3} from coming close enough to each other to recombine.

When fuels are metabolized, the energy released is stored by using it to synthesize the high-energy phosphate anhydride bonds in ATP. These bonds are later hydrolyzed as needed in such a way that the free energy of hydrolysis released is coupled to some bodily process which needs energy. One very important property of ATP is that, despite the large negative $\Delta G^{o\prime}$ of hydrolysis, this molecule is stable for long periods of time in aqueous solution. This is due to a large activation energy for the hydrolytic reaction.

4. Glucose-6-Phosphate (G-6-P) is an important junction compound.

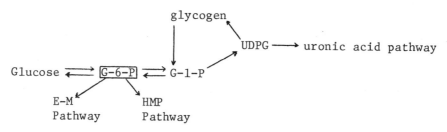

5. Phosphofructokinase

$$\text{Fructose-6-P} \xrightarrow[\text{Mg}^{++}]{\text{ATP} \quad \text{ADP}} \text{Fructose-1,6-diP}$$

a. This reaction is essentially irreversible.

b. It is considered to play a major role in the regulation of the rate of glycolysis. The enzyme is an allosteric one, inhibited by high concentrations of ATP or citrate and activated by AMP or ADP.

c. It is an inducible enzyme.

6. Glyceraldehyde-3-P Dehydrogenase Reaction

$$\text{glyceraldehyde-3-P} \xrightarrow[\text{P}_\text{i}]{\text{NAD}^+ \quad \text{NADH} + \text{H}^+} \text{1,3-diphosphoglycerate}$$

a. Gly-3-P dehydrogenase is an enzyme made up of four identical subunits. Each subunit contains a sulfhydryl (-SH) group which participates in the formation of an intermediate compound with the aldehyde moiety. The enzyme is inhibited by Hg^+ (non-competitively) and iodoacetamide or iodoacetate (irreversibly). Both of these react with the sulfhydryl group present in the active site.

b. The aldehyde group is oxidized by NAD^+ (a co-substrate) which is converted to NADH.

c. The structure of NAD^+ is

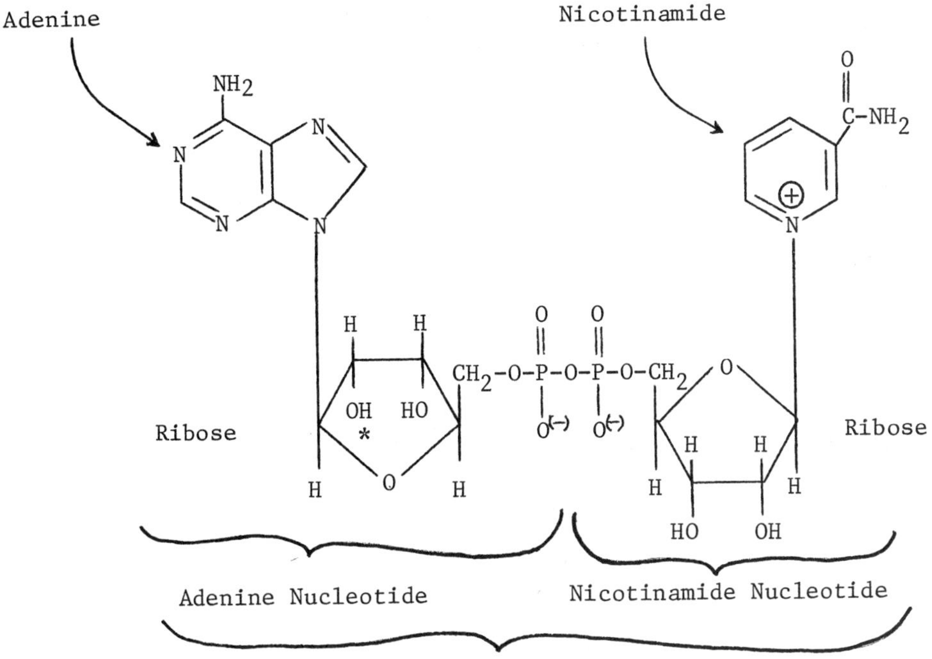

Adenine

Nicotinamide

Adenine Nucleotide   Nicotinamide Nucleotide

Nicotinamide-adenine Dinucleotide

* -H is substituted with $\begin{matrix} O \\ \parallel \\ -P-OH \\ \mid \\ OH \end{matrix}$ in $NADP^+$.

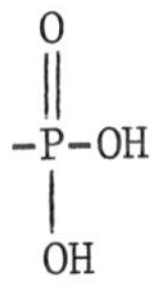

The functional part of NAD^+ and $NADP^+$ is the nicotinamide ring. Nicotinamide is a vitamin required for the synthesis of NAD^+ and $NADP^+$. It can be made to a limited extent from tryptophan (see amino acid metabolism).

d. The mechanism of the glyceraldehyde-3-P dehydrogenation reaction is shown below. The hydride ion transfer is typical of reactions involving NAD^+ and $NADP^+$.

i) Gly-3-(P) binds to the -SH group of glyceraldehyde-3-(P) dehydrogenase.

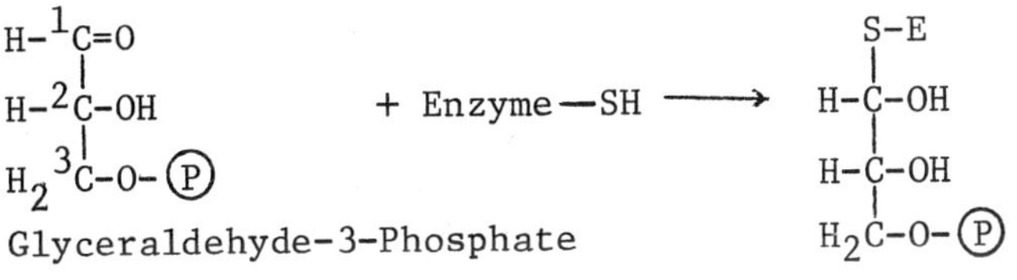

Glyceraldehyde-3-Phosphate

ii) A hydride ion ($H:^-$) is transferred from the 1 carbon of gly-3-P to the para position of the nicotinamide ring.

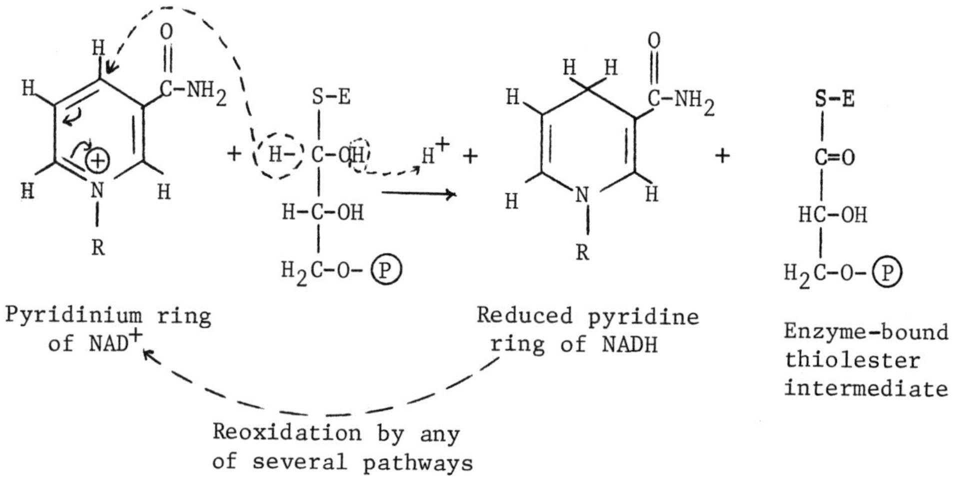

Pyridinium ring
of NAD^+

Reduced pyridine
ring of NADH

Enzyme-bound
thiolester
intermediate

Reoxidation by any
of several pathways

iii) Phosphorylation by inorganic phosphate, release of 1,3-DPG and regeneration of the enzyme.

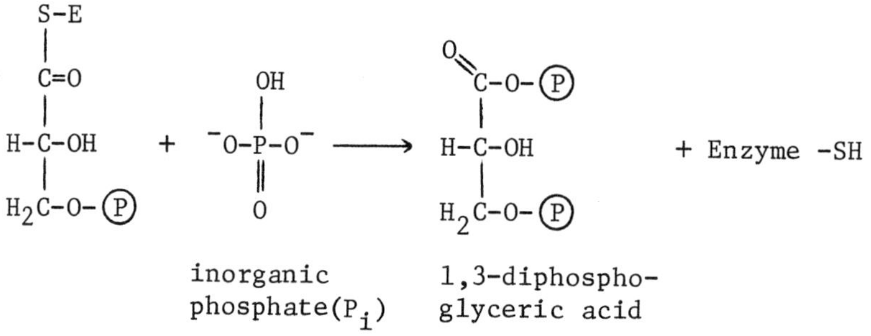

inorganic
phosphate(P_i)

1,3-diphospho-
glyceric acid

7. The reoxidation of reduced NADH can be accomplished by the following reactions catalyzed by the appropriate dehydrogenases.

Pyruvate ⟶ Lactate (discussed below)

Dihydroxyacetone-ⓟ ⟶ α-glycero-ⓟ

Oxaloacetate ⟶ Malate

In aerobic cells, the latter two reactions participate in transporting NADH into mitochondria to be eventually oxidized by oxygen in the electron transport chain (discussed later in the section on mitochondria).

8. Production of 2,3-diphosphoglyceric acid (2,3-DPGA or 2,3-DPG).

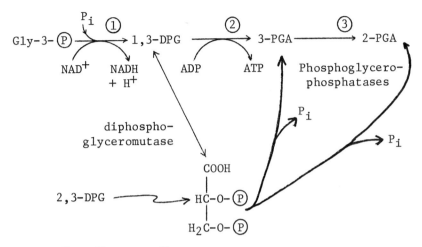

Reactions ①, ②, and ③ are part of the E-M pathway. 2,3-DPG is produced in the human red blood cells and accumulates in these cells. This compound plays an important role in oxygen release from hemoglobin in the tissues by lowering the affinity of hemoglobin for oxygen (this aspect will be discussed in the chapter on hemoglobin).

9. The conversion of 2-PGA to PEP is catalyzed by enolase, requires Mg^{2+}, and is inhibited by fluoride. The inhibition may be due to the formation of a magnesium fluorophosphate complex.

10. Conversion of Pyruvic Acid to Lactic Acid

  a. Under <u>anaerobic</u> conditions, pyruvate is converted to lactic acid by lactate dehydrogenase, LDH. Lactate formation and hence the reoxidation of the NADH formed in the oxidation of glyceraldehyde-3-P allows glycolysis to proceed.

  b. LDH exists as several <u>isozymes</u>. Active LDH (M.W. = 130,000) consists of <u>four subunits</u> of <u>two types</u>: M and H. The letters stand for skeletal <u>m</u>uscle and <u>h</u>eart muscle, the tissues in which the respective types of subunit predominate. This is discussed below. Only the tetrameric molecule possesses catalytic activity. Synthesis of the M and H subunits are controlled by distinct genetic loci.

c. At pH 8.6, isozymes of LDH bear different net charges and hence migrate to different regions in an electric field. This is a characteristic property of isozymes. Its usefulness is discussed below.

d. Different tissues have different types of LDH isozymes suited to perform the function of the tissue in which they are present. Heart muscle, which functions under aerobic conditions, has primarily the LDH isozyme made up of four subunits of H (H_4). The tissues that contain LDH-H_4 isozyme produce very little lactic acid because the pyruvic acid formed in glycolysis is channeled into pathways in which it can be oxidized completely (to CO_2 + H_2O), yielding the greater amount of energy needed to sustain a continuous mechanical performance. This happens because pyruvate inhibits the H-isozymes of LDH. As a result, the enzyme is used for the conversion of lactate to pyruvate in these cells

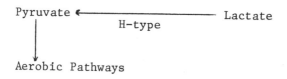

e. On the other hand, in the functioning skeletal muscle, there are times of oxygen deprivation (anaerobic conditions). Under these conditions, muscle (mostly LDH-M_4) isozymes (which are only weakly inhibited by pyruvate) convert large amounts of pyruvic acid to lactic acid at a rapid rate. In the relaxation intervals, the oxygen supply increases over oxygen utilization and lactic acid gets reconverted to pyruvate which is completely oxidized via the TCA cycle and oxidative phosphorylation. In vigorous skeletal muscle activity, however, the large amount of lactic acid produced passes into the blood and is transferred to the liver. There it gets metabolized to other products such as pyruvic acid, glucose, etc. This process, known as the <u>Cori cycle</u> is illustrated in Figure 20.

Figure 20. The Cori Cycle

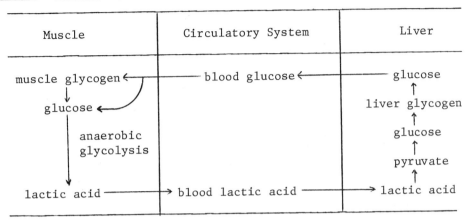

f. Lactic acid dehydrogenase is an intracellular enzyme and the
    isozyme profile is related to the type of metabolic activity
    of the tissue. If there are any changes in the cell
    permeability or if there is cell damage and destruction, the
    LDH isozymes are spilled into the blood. The serum LDH levels
    are then elevated and the serum isozyme pattern is indicative
    of the tissue that is damaged. In myocardial infarction,
    for example, LDH levels are increased (primarily due to H_4
    and H_3M) within about 72 hours after infarction. The high
    levels may be maintained for over a week.

g. LDH Isozyme Profiles in Man

Table 4. LDH Isozyme Profiles in Man

LDH* Isozyme U.S. Nomenclature	International Nomenclature	Subunit Composition	Relative Concentration** (% of total LDH activity) Heart Muscle	Erythro-cytes	Skeletal Muscle	Liver
LD-5	LD-1	H_4	60	42	4	2
LD-4	LD-2	H_3M	33	44	7	6
LD-3	LD-3	H_2M_2	7	10	17	15
LD-2	LD-4	HM_3	Trace ($<$1)	4	16	13
LD-1	LD-5	M_4	Trace ($<$1)	Trace ($<$1)	56	64

(see top of next page for key)

* The International Nomenclature will be used in this text.

** These are average values, taken from several reports in the literature.

From these profiles it is apparent that no tissues are "pure" in having only one isozymic form of LDH. Generally two (sometimes three) forms predominate in a particular tissue.

h. The <u>total serum LDH</u> can be determined by an NAD^+-NADH assay (described earlier). Since the reaction is reversible, either pyruvate or lactate can be used as the substrate depending upon the isozyme being measured. Since the heart isozymes are inhibited by pyruvate, lactate is the preferred substrate for these forms.

i. In addition to total activity, it is important to know the amount of activity due to each isozyme (the isozyme profile). There are several ways to do this.

i) The isozymes are separable by electrophoresis (polyacrylamide gel, cellulose acetate membrane, etc.), giving the qualitative pattern shown below at pH = 8.8, 0.05 ionic strength tris-barbital buffer.

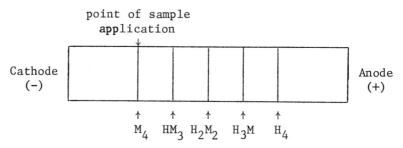

point of sample application

Cathode (-)   Anode (+)

M_4 HM_3 H_2M_2 H_3M H_4

The M-subunits are less positively charged than the H-subunits. As the proportion of M in the tetramer increases, the molecule is less strongly attracted to the anode. When only M is present, the isozyme does not move from the point of application or may even migrate slightly towards the cathode.

ii) The location and amount of LDH activity is determined following the separation, by putting the gel or membrane in contact with agar or a solution containing lactate, NAD^+ and a color reagent consisting of phenazine methosulfate

(PMS) and a tetrazolium dye. Wherever LDH is present, the series of reactions shown below takes place and the gel or membrane is stained blue. In this diagram, 2,3,5-triphenyltetrazolium chloride (TTC) is used as an example, although more sensitive dyes are available. PMS serves as an intermediate electron carrier.

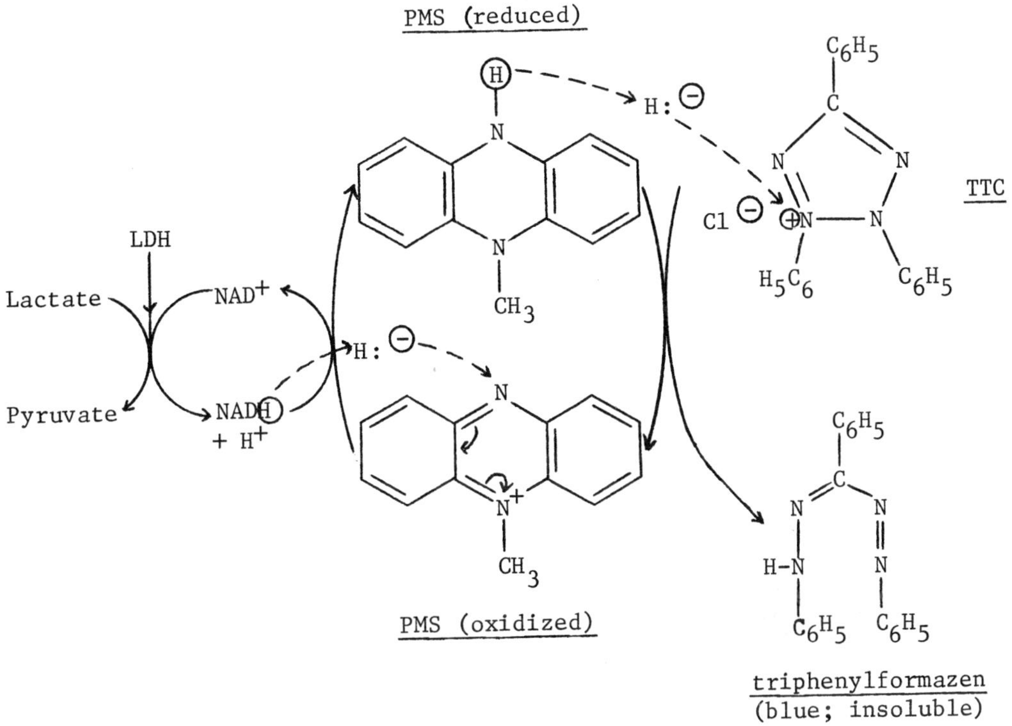

PMS (reduced)

TTC

LDH

Lactate

Pyruvate

PMS (oxidized)

triphenylformazen
(blue; insoluble)

After staining, the gel or membrane can be scanned with an optical densitometer. The areas under the peaks in the resulting graph are proportional to the activities of the different isozymes. This method is highly specific for detecting only LDH activity. It is important to realize (as was pointed out in the section on clinical uses of enzymes) that the activity, and not the amount of enzyme, is being measured. This is because of the differing K_m values for lactate of the five isozymes.

iii) M-subunits are much more susceptible to low temperatures, heat, urea, and other denaturing agents than are H-subunits. Treatment of a sample with denaturing agents destroys M_4 and M_3H activity. Loss of LDH activity upon heating has been used as an index of relative isozyme activities, as shown below.

Sample	Treatment	Source of Activity
1	Untreated serum	LD 1-5
2	Serum, heated to 57°	LD 1-4
3	Serum, heated to 65°	LD-1

Total LDH activity = Sample 1

Activity due to LD-1 = Sample 3

Activity due to LD-5 = Sample 1 - Sample 3

Activity due to LD 2-4 = Sample 2 - Sample 3

Results of this method correlate well with those done by electrophoretic methods.

11. Reversibility of Glycolysis: Although all reactions are technically reversible (see thermodynamics), the following reactions are physiologically irreversible (i.e., irreversible for all practical purposes in the body). These are processes involving large losses of free energy (exergonic reactions). They are catalyzed by the enzymes

a. Hexokinase (glucokinase)
b. Phosphofructokinase
c. Pyruvate Kinase

Alternate routes are available, however, for the synthesis of glucose from lactate, glycerol, and amino acids (gluconeogenesis). Enzymes other than those of glycolysis are used.

Note: The phosphoglycerate kinase reaction is also an exergonic reaction and is not easily reversible, but an alternate route does not exist for this reaction. It is reversed by mass law considerations when 3-(P)-glycerate and ATP accumulate.

12. Lactose is produced by the E-M pathway (glycolysis) in skeletal muscle when the muscle is performing under anoxic conditions and in erythrocytes. In the latter case, the enzymes which oxidize pyruvic acid are absent. This lactic acid is carried by the blood to the liver, where it is converted to glucose (Cori cycle already discussed).

13. The net result of anaerobic glycolysis is

$$\text{Glucose} + 2 \text{ ADP} + 2 \text{ P}_i \quad \rightarrow 2 \text{ lactic acid} + 2 \text{ ATP}$$

14. Pasteur Effect. In many circumstances glycolysis is decreased by the presence of O_2.

aerobic conditions	anaerobic conditions
low glucose consumption	high glucose consumption
low lactic acid production	high lactic acid production

The mechanism is uncertain but the following are possible explanations:

a. The competition between glycolysis and oxidative phosphorylation for ADP and inorganic phosphate.

b. Inactivation of glycolytic enzymes by oxidation of sulfhydryl groups.

c. Inhibition of phosphofructokinase by ATP (citrate accentuates the effect in the presence of ATP). This may be the major effect.

d. Reversal of glycolysis under aerobic conditions (gluconeogenesis). This would cause an apparent decrease in glycolysis when the glucose concentration is being measured.

15. The reciprocal of the Pasteur effect is the Crabtree effect. High concentrations of glucose will inhibit cellular respiration when studied in isolated systems such as ascites tumor cells. The probable mechanism for this effect is more effective competition by glycolysis for inorganic phosphate and NADH. This can lead to a deficiency of materials available to carry on oxidative

phosphorylation.   These interrelationships are shown in
Figure 21.

<u>Figure 21</u>.   <u>Competitive Reactions Which Lead to the Pasteur and</u>
<u>Crabtree Effects</u>

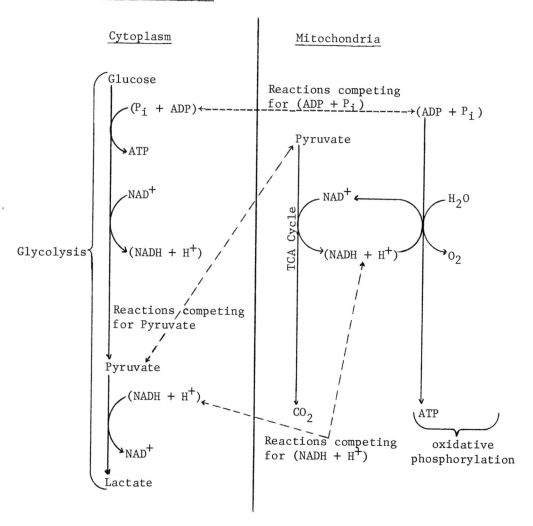

16. Biochemical Lesions of Glycolysis

    a. i) <u>Pyruvate kinase (PK) deficiency</u> in the red cells has been reported. Pyruvate kinase catalyzes the reaction

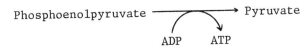

Phosphoenolpyruvate ———————→ Pyruvate

ADP   ATP

       ii) The deficiency (classified as a type of congenital nonspherocytic hemolytic anemia) leads to inadequate production of ATP causing premature destruction of red cells (auto-hemolysis). The lysis is decreased to a small degree by glucose and to a greater extent by ATP, ADP or AMP.

      iii) The lesion appears to be transmitted by an autosomal recessive gene. The enzyme is virtually absent in the red cells obtained from the affected individuals. Heterozygous carriers have approximately half of the normal PK activity (compared to 5-25% in homozygotes) but there is a wide range of values in both groups.

    b. In a number of other cases of hereditary hemolytic anemia specific glycolytic enzymes have been shown to be deficient. These enzymes include hexokinase, glucose phosphate isomerase, phosphofructokinase, triosephosphate isomerase, 2,3-diphosphoglyceromutase, and phosphoglycerate kinase. With the exception of the PK deficiency, discussed above, all such diseases are rare.

Citric Acid Cycle (Tricarboxylic Acid (TCA) Cycle; Krebs Cycle)

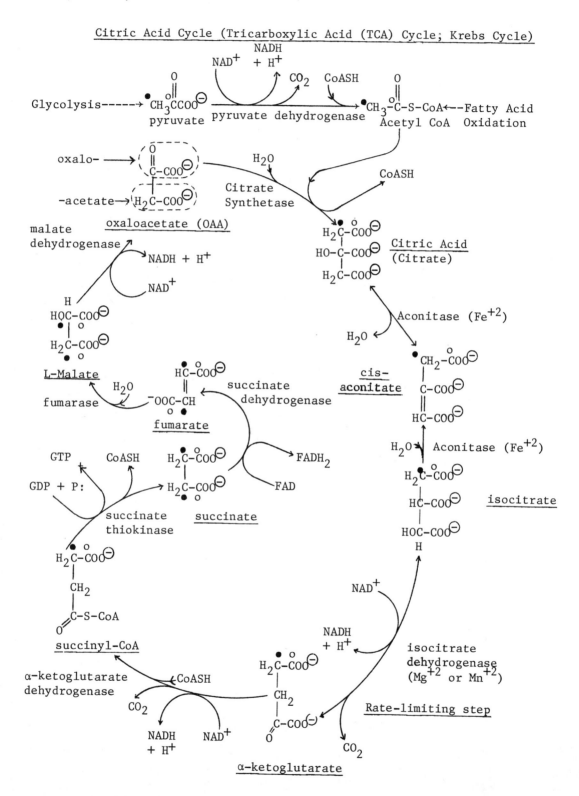

Comments on Citric Acid Cycle. The complete oxidation of glucose which finally produces CO_2 starts with the conversion of pyruvate to acetyl-CoA with concomitant reduction of $NAD^+ \rightarrow NADH + H^+$. In subsequent reactions, FAD and additional NAD^+ is reduced, and CO_2 is formed. Sir Hans Krebs (whose name is frequently associated with the cycle) received the Nobel Prize in 1953 for his work on intermediary metabolism.

1. All of the citric acid cycle enzymes are found in the mitochondrial fraction of the cell and are in close proximity to the enzymes of the respiratory chain, either as components of the cristae membranes or as soluble enzymes of the mitochondrial matrix. The enzymes aconitase, fumarase and malate dehydrogenase are of the latter type, while pyruvic acid oxidodecarboxylase (pyruvate dehydrogenase), ketoglutaric oxidodecarboxylase (α-ketoglutarate dehydrogenase), succinate dehydrogenase, and the others are membrane bound.

2. Conversion of pyruvic acid to acetyl-CoA (active acetate)

    a. The five reactions are catalyzed by three different enzymes. They function as a multi-enzyme complex and are collectively known as pyruvic oxidodecarboxylase. For all practical purposes this is an irreversible reaction. Six coenzymes are required. They are Mg^{+2}, TPP, lipoic acid, CoASH, FAD and NAD^+.

    b. Overall Reaction

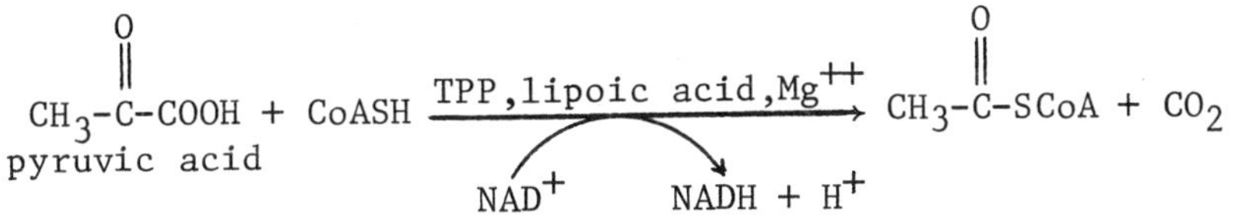

$$CH_3-\overset{\overset{O}{\|}}{C}-COOH + CoASH \xrightarrow[NAD^+ \quad NADH + H^+]{TPP, lipoic\ acid, Mg^{++}} CH_3-\overset{\overset{O}{\|}}{C}-SCoA + CO_2$$

    pyruvic acid

    c. Some details of the reaction

    i)
$$CH_3-\overset{\overset{O}{\|}}{C}-COOH + TPP \xrightarrow{\substack{pyruvic\ acid \\ decarboxylase}} PP-T-\overset{\overset{OH}{|}}{\underset{H}{C}}-CH_3$$

    (an α-keto acid)   (Thiamine Pyrophosphate)   "active" acetaldehyde

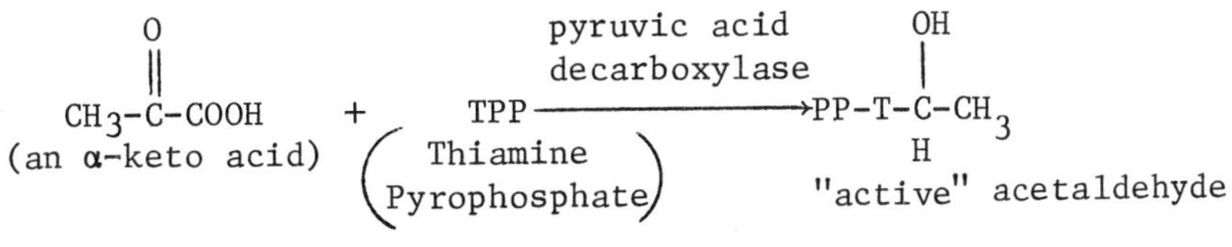

    In the above reaction the two carbon aldehyde groups of pyruvate are transferred to thiamine pyrophosphate (TPP). The details of the reactions are described below. (Note: TPP is the pyrophosphate ester of Vitamin B_1.)

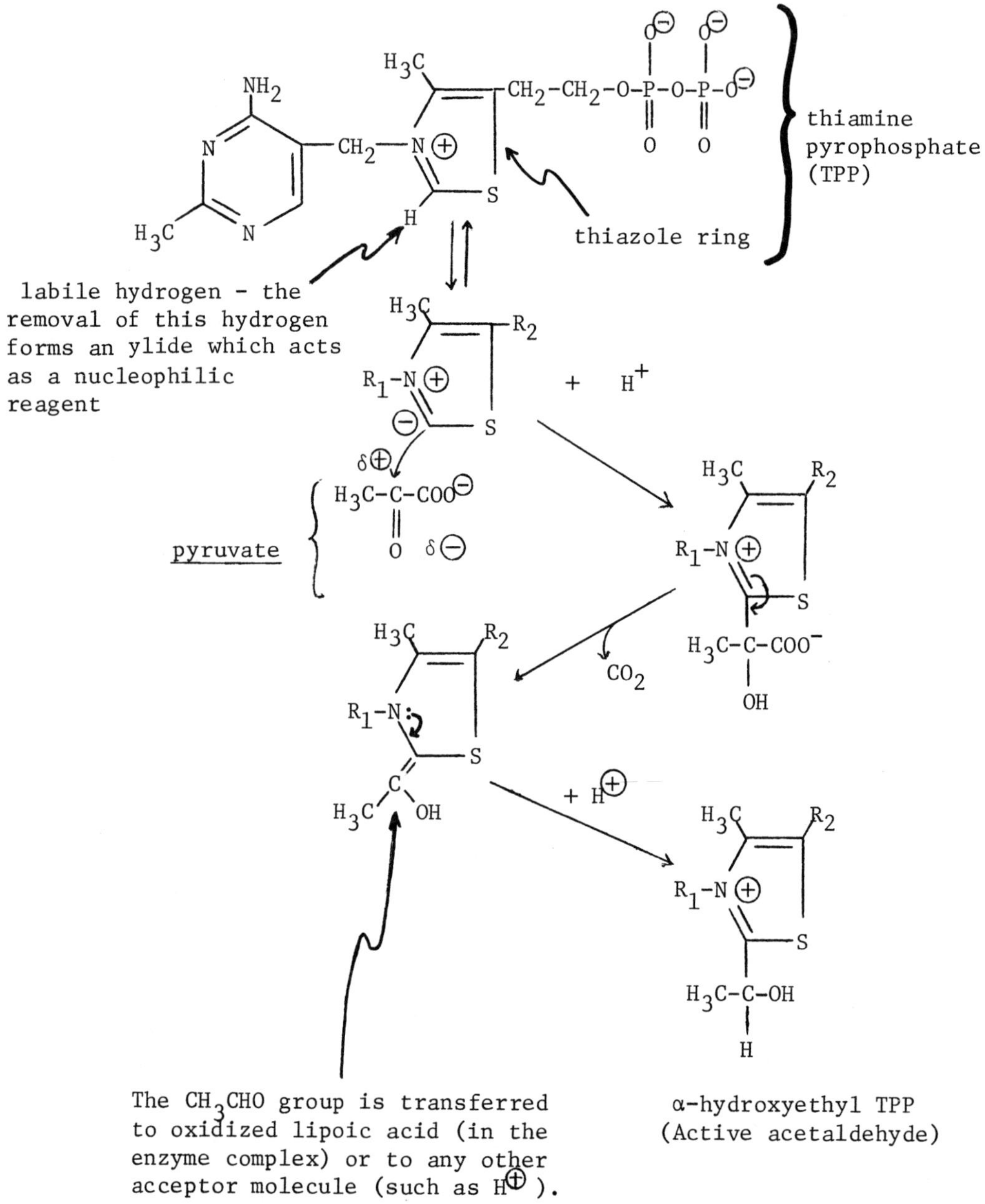

$$
\begin{array}{l}
\text{thiamine} \\
\text{pyrophosphate} \\
\text{(TPP)}
\end{array}
$$

thiazole ring

labile hydrogen - the removal of this hydrogen forms an ylide which acts as a nucleophilic reagent

$+ \ H^+$

pyruvate

$$\text{H}_3\text{C-C-COO}^{\ominus}$$

CO_2

$+ \ H^{\oplus}$

The CH_3CHO group is transferred to oxidized lipoic acid (in the enzyme complex) or to any other acceptor molecule (such as $H^{\oplus}$).

α–hydroxyethyl TPP
(Active acetaldehyde)

Thiamine pyrophosphate thus functions as a coenzyme for the decarboxylation of pyruvic acid, an α-keto acid. This coenzyme also participates (by a similar mechanism) in the decarboxylation reaction of α-ketoglutarate (another α-keto acid), one of the later reactions in the citric acid cycle. TPP also functions as a coenzyme in the transketolase reactions of the HMP-shunt pathway (discussed later). TPP is obtained from thiamine as follows:

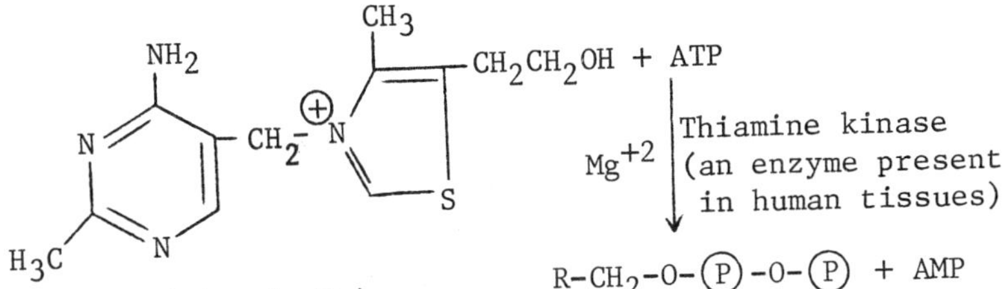

(Vitamin B$_1$)

Thiamine kinase
Mg^{+2} (an enzyme present in human tissues)

R-CH$_2$-O-(P)-O-(P) + AMP
Thiamine Pyrophosphate

ii)

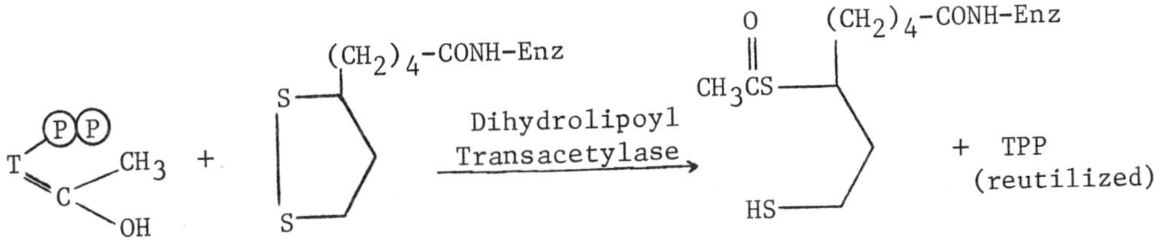

Acetyl TPP   Oxidized lipoyl   S-acetylhydrolipoyl-Enz
                  enzyme

Lipoic acid (considered by some as a vitamin) has the structure

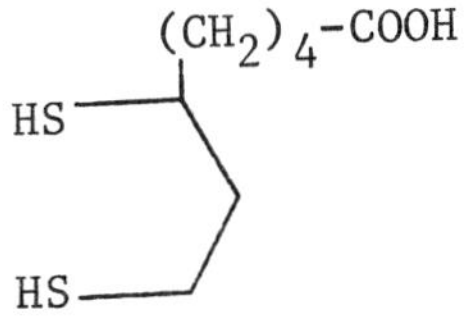

(CH$_2$)$_4$-COOH

6,8-dithio-octanoic acid

It is bound to the protein by a peptide linkage involving the epsilon amino (ε-NH$_2$) group of lysine. It is thought to transfer the reaction product from one active center of the enzyme complex to another by virtue of its ability to move over a relatively large area of the enzyme surface.

iii)

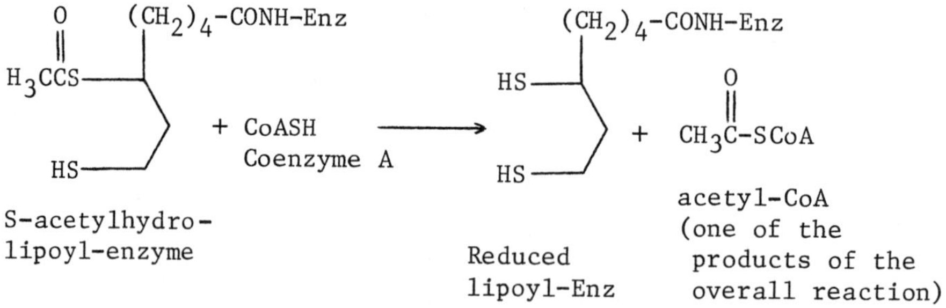

S-acetylhydro-
lipoyl-enzyme

+ CoASH
Coenzyme A

Reduced
lipoyl-Enz

acetyl-CoA
(one of the
products of the
overall reaction)

The structure of Coenzyme A

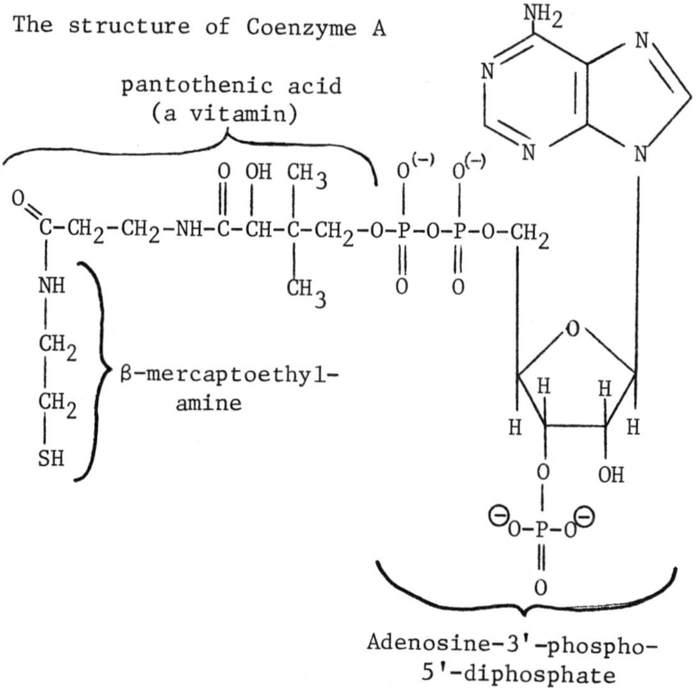

pantothenic acid
(a vitamin)

β-mercaptoethyl-
amine

Adenosine-3'-phospho-
5'-diphosphate

<u>Acetyl-CoA</u>, a key intermediate, is not only obtained by the oxidation of pyruvic acid, but also from the oxidation of fatty acids and amino acids. In the acetyl-CoA molecule, the acetate group is in an "active" state, and hence can easily function in a variety of synthetic reactions. The acetate is attached to the -SH group in a thioester linkage which has a $\Delta G^{o\prime}$ of 7.5 kcal/mole (similar to that of ATP).

iv) In the next two reactions the oxidized-lipoyl enzyme is regenerated from the reduced–lipoyl enzyme.

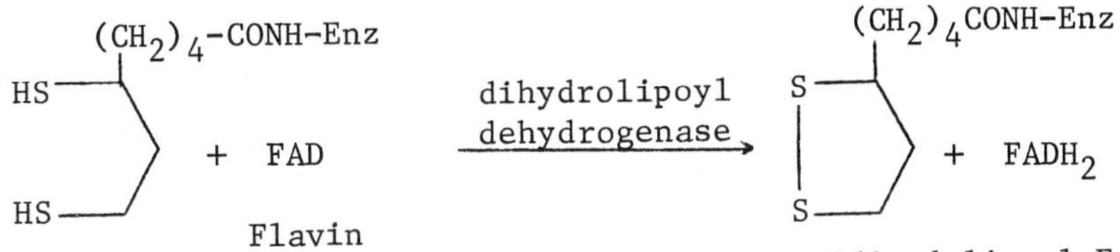

$$(CH_2)_4-CONH-Enz$$

HS—

$+$ FAD

HS—

Reduced   Flavin
lipoyl-Enz   Adenine
              Dinucleotide

dihydrolipoyl
dehydrogenase
$\longrightarrow$

$$(CH_2)_4 CONH-Enz$$

S—

$+$ FADH$_2$

S—

Oxidized lipoyl-Enz
(reutilized)

<u>Flavin adenine dinucleotide</u> is the hydrogen acceptor in the above reaction and has the structure

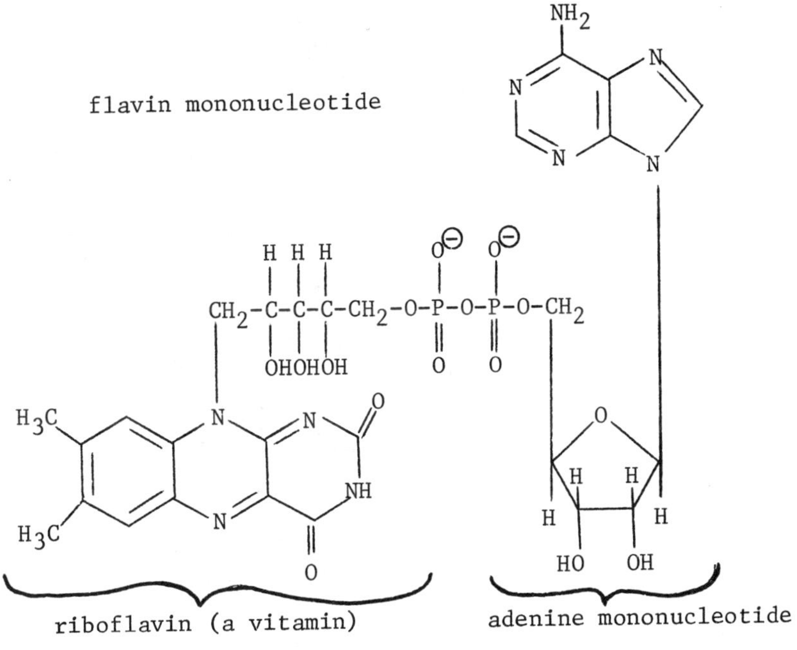

flavin mononucleotide

riboflavin (a vitamin)   adenine mononucleotide

v) In the final reaction, FAD is regenerated with the production of NADH.

$$FADH_2 \ + \ NAD^+ \longrightarrow NADH \ + \ H^+ \ + \ FAD \ (reutilized)$$

d. The pyruvic oxidodecarboxylase complex obtained from E. coli
    consists of three enzymes: decarboxylase (mol wt 183,000),
    transacetylase (mol wt 70,000), and dehydrogenase (mol wt
    112,000) in the ratio of 12:24:6. The complex has also been
    isolated from a variety of other sources, but most of the work
    has been performed using preparations from E. coli. Following
    is a schematic representation of the complex and its reactions.

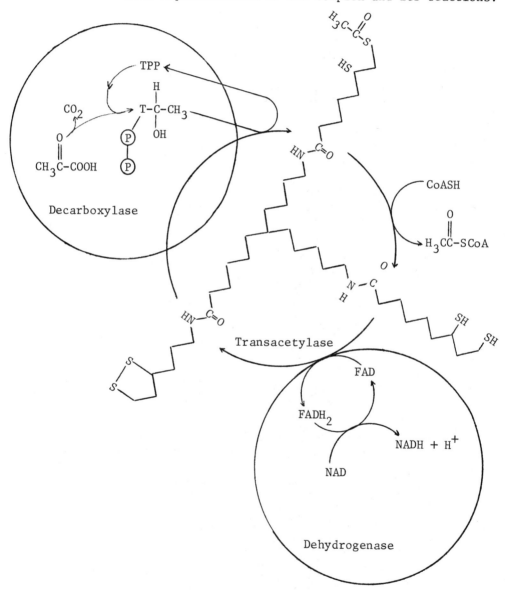

3. Although citric acid is a symmetrical molecule, it reacts in an asymmetric manner with the enzyme aconitase (only one of the two $-CH_2-COOH$ groups on the citric acid molecules is modified).

$$\begin{array}{l} \bullet \quad\ o \\ CH_2- \ COOH \\ | \\ HO-C-COOH \\ | \\ CH_2-COOH \\ \text{citric acid} \end{array}$$

: enzyme acts on the oxaloacetate part of the molecule ($\bullet$ and o indicate atoms derived from acetyl-CoA).

Three-point attachment of the enzyme to the substrate molecule (citric acid) accounts for the asymmetrical activity. Condensation of acetyl-CoA with oxaloacetate is also stereospecific with respect to the keto group of oxaloacetate.

4.

(a) isocitrate $\xrightarrow[\quad Mn^{+2} \quad]{\substack{NADP^+ \quad NADPH + H^+}}$ (oxalosuccinate) $\xrightarrow{CO_2}$ α-ketoglutarate

(b) isocitrate $\xrightarrow[\quad Mg^{+2} \text{ or } Mn^{+2} \quad]{\substack{NAD^+ \quad NADH + H^+ \quad CO_2}}$ α-ketoglutarate

This is the rate-limiting step in the TCA cycle. Two isocitrate dehydrogenases are found in most animal and plant tissues. The cytoplasmic one (which, in most tissues, also occurs in mitochondria but to a much lesser extent) catalyzes reaction i), above. It is non-allosteric, requires $NADP^+$ and Mn^{+2}, and has oxalosuccinate as an enzyme-bound intermediate. The mitochondrial enzyme (catalyzing reaction ii) is allosteric (positive modulators are ADP and NAD^+; negative ones are ATP and NADH), specific for NAD^+, can use either Mg^{+2} or Mn^{+2}, and does not have oxalosuccinate as a free or bound intermediate.

5. Conversion of α-ketoglutarate to succinyl-CoA: this oxidative decarboxylation reaction of an α-keto acid, catalyzed by the α-ketoglutarate dehydrogenase (oxidodecarboxylase) multi-enzyme complex is identical to the conversion of pyruvate to acetyl CoA, with respect to the mechanism and the cofactors required. It is considered to be the irreversible step in the TCA cycle due to the release of CO_2.

6.

$$\text{Succinyl-CoA} \xrightarrow{\text{succinyl thiokinase}} \text{Succinic Acid}$$

GDP (or IDP) + P_i GTP (or ITP)

(GDP = guanosine diphosphate; IDP = inosinediphosphate)

This is the only reaction in the citric acid cycle which directly generates a high-energy phosphate compound. It is called a <u>substrate-level phosphorylation</u> as distinguished from oxidative phosphorylation where oxygen is the terminal electron acceptor. Note that the glycolytic reactions 1,3-DPG → 3PGA and PEP → pyruvic acid also are substrate-level phosphorylation reactions. The GTP formed can be converted to ATP in the following reaction.

$$\text{GTP} + \text{ADP} \xrightleftharpoons{\text{nucleosidediphosphate kinase}} \text{ATP} + \text{GDP}$$

(or NTP   (or NDP
in general)   in general)

7.

$$\text{HOOC-CH}_2\text{-COOH} \xrightarrow[\text{FAD} \quad \text{FADH}_2]{\text{succinic dehydrogenase}} \text{HOOC-C=C-COOH}$$

succinate   fumarate

Note: The product formed is the <u>trans</u> isomer.

This dehydrogenation is the only reaction in the <u>citric acid cycle</u> which involves the direct transfer of H to a flavoprotein <u>without</u> participation of NAD^+. The enzyme contains FAD (structure given previously) and non-heme iron. The oxidation-reduction involves only the flavin ring system. It proceeds as shown below. (R) represents the remainder of the FAD molecule.

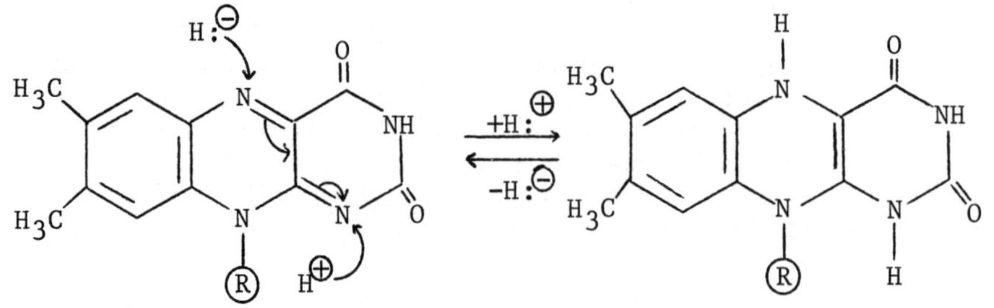

The two-electron reduction of the flavin ring can also occur in two one-electron steps with the electrons coming from some source other than a hydride ion. In such a case, both hydrogens are derived from H^+ in the solution. This sort of mechanism is typical of all redox reactions involving flavin coenzymes.

8. Labeling experiment: $^\bullet CH_3-{}^oC-S-CoA$: C^{14} labeled acetyl-CoA.
    ($\bullet$ and o indicate radioactive methyl and carbonyl carbons, respectively.)

    a. In the first turn of the cycle labeled CO_2 does not appear.

    b. Labeled CO_2 appears in the second and subsequent turns of the cycle.

    c. Succinate is the randomization point in the cycle because it is a symmetrical molecule.

    This labeling is indicated on the diagram of the citric acid cycle. They show that the carbons which are lost as CO_2 on the initial turn of the cycle, are <u>not</u> the carbons which were added on that turn as acetyl-CoA (a, above). Because of the randomization at succinic acid, some label will be found in all carbons of the oxaloacetate formed on the first turn. However, one OAA molecule will <u>not</u> have <u>all four</u> carbon atoms labeled after completion of <u>just one turn</u>.

9. At each turn of the cycle, oxaloacetate is regenerated and can combine with the next acetyl-CoA molecule. As can be seen in Figure 22, though, some citric acid cycle intermediates are utilized in the synthesis of other molecules. This can lead to a deficiency of the intermediates causing a slowdown of oxidation unless oxaloacetate is replenished through other reactions (anaplerotic reactions). One principle reaction in animal tissues which generates oxaloacetate is

$$\text{Pyruvic acid} + CO_2 + ATP \xrightarrow[\text{biotin, } Mg^{+2}]{\substack{\text{Pyruvic acid} \\ \text{carboxylase}}} \text{oxaloacetate} + ADP + P_i$$

Figure 22.   Some of the Interrelationships of Carbohydrate Metabolism
with Other Metabolites and Pathways (for *, see page 181)

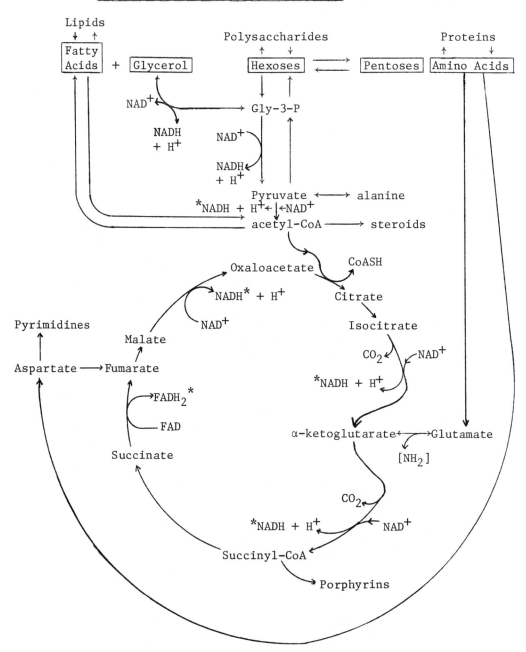

Note that, since one glucose molecule gives rise to two gly-3-P molecules, 10 NADH molecules and 2 FADH$_2$ molecules (twice the number of stars (*) in the diagram) are produced per mole of glucose oxidized. These reduced coenzymes are reoxidized in the electron transport chain. Oxidative phosphorylation, when coupled to electron-transport, produces three moles of ATP per mole of mitochondrial NADH and two moles of ATP per mole of mitochondrial FADH$_2$. This is shown schematically below.

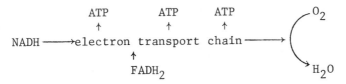

The result is the synthesis of (3 x 10) + (2 x 2) = 34 moles of ATP per mole of glucose oxidized, excluding substrate-level phosphorylation and ATP consumed. The overall balance of the oxidation of one mole of glucose to CO$_2$ + H$_2$O is shown below.

2ATP+glucose ⟶ 2ADP+F-1,6-di(P)   (hexokinase, phosphofructokinase)

F-1,6-di(P) ⟶ 2(gly-3-(P))   (aldolase)

2(gly-3-(P))+2P$_i$ → 2(1,3-DPG)   (gly-3-(P) dehydrogenase)

2ADP+2(1,3-DPG) → 2ATP+2(3-PG)   (phosphoglycerate kinase)

2(3-PG) ⟶ 2PEP   (glycolytic enzymes)

2ADP+2PEP ⟶ 2ATP+2(PA)   (pyruvate kinase)

2GDP+2P$_i$ ⟶ 2GTP   (succinate thiokinase)

2GTP+2ADP ⟶ 2ATP+2GDP   (nucleoside diphosphate kinase)

Glucose+4ADP+4P$_i$ → 2(PA)+4ATP   total of substrate-level conversions

Adding these four ATP's to the 34 obtained via electron transport-coupled oxidative phosphorylation results in <u>38 ATP synthesized per mole of glucose oxidized</u>. There can be one minor variation in this number. When the NADH reducing-power produced by gly-3-P dehydrogenase in the cytoplasm is transferred to the mitochondria, part of it is used to reduce a flavin coenzyme. When this is subsequently oxidized, only two moles of ATP are produced per mole of flavin. A mole of NADH (worth 3 moles of ATP) has thus been changed to a mole of flavin coenzyme (worth only 2 moles of ATP). Taking this into account, only 36 ATP would be produced per mole of glucose. Actually, the correct number is between 36 and 38. This (and the NADH transfers) are discussed more fully in the section on mitochondria.

10. Glyoxalate Cycle

    a. The oxidation of pyruvic acid to acetyl-CoA is an irreversible process. Consequently, the net synthesis of glucose (gluconeogenesis) from acetyl-CoA must come about through bypass reactions. Such reactions are operative in plants and microorganisms but are absent in man and other animals. Therefore, fatty acids, which give acetyl-CoA on oxidation, cannot be converted to glucose in man. The only exceptions to this are the odd-carbon-number fatty acids which give one mole of propionyl-CoA per mole of fatty acid. The propionyl-CoA can be converted to pyruvate or succinate which can subsequently be used to form glucose. The contribution of such routes is very small, though, and is not significant under normal dietary circumstances.

    It is important to realize that net synthesis is being discussed here. If labeled acetyl-CoA is used, some of the label will appear in newly made glucose since the carbons added as acetyl-CoA at one turn of the cycle are not the ones lost as CO_2 on the same turn (pointed out before). However, as will be seen in gluconeogenesis, the only route from acetyl-CoA to glucose is through the two CO_2-releasing reactions. The net effect is that, although two carbons are added as acetyl-CoA, two carbons must be lost as CO_2 prior to the formation of glucose. This inability of humans to convert lipids to carbohydrates presents a major problem in diabetes mellitus. This will be discussed later.

    b. The key enzymes which make up the bypass in plants and microorganisms are isocitrate lyase (also known as isocitritase) and malate synthetase. They catalyze the reactions shown below.

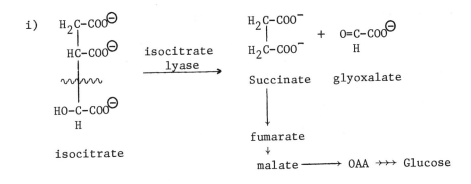

i)

$$H_2C-COO^\ominus$$
$$HC-COO^\ominus$$
$$\xrightarrow{\text{isocitrate lyase}}$$

isocitrate

H_2C-COO^- + $0=C-COO^\ominus$
H_2C-COO^- H

Succinate   glyoxalate

↓

fumarate
↓
malate ⟶ OAA ⇢ Glucose

ii)

$$0=C-COO^- + CH_3CSCoA \xrightarrow{\text{malate synthetase}} HO-CH-COO^-$$
$$H \qquad\qquad\qquad\qquad\qquad\qquad H_2C-COO^-$$

glyoxalate acetyl-CoA   L-Malate

c. The interaction of glyoxalate cycle with the citric acid cycle is shown in the following diagram. Note that the glyoxalate pathway <u>bypasses</u> the CO_2-releasing reactions.

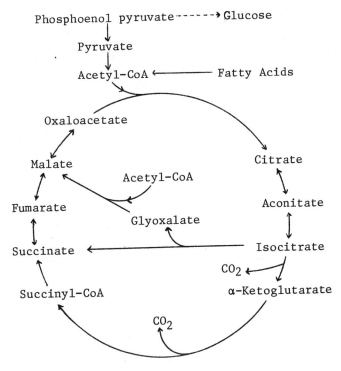

Gluconeogenesis. Reversal of Glycolysis and Formation of Glucose

1. Gluconeogenesis (which means formation of new sugar) is the process
    of formation of glucose from non-carbohydrate sources such as
    amino acids, glycerol and lactic acid. It takes place when the
    energy requirements of the cell are at a minimal level and an energy
    source (ATP) is available.

2. Gluconeogenesis becomes a very important source for supplying
    glucose to various tissues (liver, adipose tissue, brain, etc.)
    when glucose is not otherwise available. The unavailability may
    be due to a number of factors such as starvation.

3. In the human, fatty acids cannot serve as a source of new glucose
    (with the possible exception of the last three carbon atoms of
    odd numbered fatty acids) as was shown in the discussion of the
    glyoxalate cycle.

4. Muscles (under anaerobic conditions) and erythrocytes produce lactic
    acid which is converted to glucose via the gluconeogenic mechanism.
    Glycerol, produced from the hydrolysis of triglycerides in adipose
    tissue, is transported to the liver and can be converted to glucose.
    Glycerol is not metabolized in the adipose tissue due to the lack
    of glycerokinase, the enzyme which catalyzes the conversion of
    glycerol to α-glycerophosphate.

5. The conversion of lactic acid, glycerol, and amino acids into
    glucose and glycogen takes place principally in the liver and
    kidney. A number of hormones play roles in the regulation of these
    reactions. They will be referred to in the appropriate sections.

6. The magnitude of the energy barriers in the following exergonic
    (energy-releasing) reactions prevents their reversal and, hence,
    the reversal of the E-M pathway.

    a. phosphoenol pyruvate + ADP $\xrightarrow{\text{pyruvate kinase}}$ pyruvate + ATP

    $$\Delta G^{o\prime} = -5.7 \text{ kcal/mole of PEP}$$

    b. fructose-6-P + ATP $\xrightarrow{\text{phosphofructokinase}}$ fructose-1,6-DiP + ADP

    $$\Delta G^{o\prime} = -3.4 \text{ kcal/mole F-6-P}$$

c. glucose + ATP $\xrightarrow[\text{hexokinase}]{\text{glucokinase or}}$ glucose-6-P + ADP

$\Delta G^{o\prime} = -3.4$ kcal/mole of glucose

These reactions were mentioned previously under glycolysis.

7. The cell has evolved a series of reactions to circumvent these exergonic reactions and thereby "reverse" glycolysis.

   a. Pyruvate is chosen as a starting point to consider gluconeogenesis since the reaction, pyruvate → PEP, is the first one which must be bypassed. Pyruvate produced in the cytoplasm is first transported into the mitochondria, where it is carboxylated to oxaloacetate. The reaction is

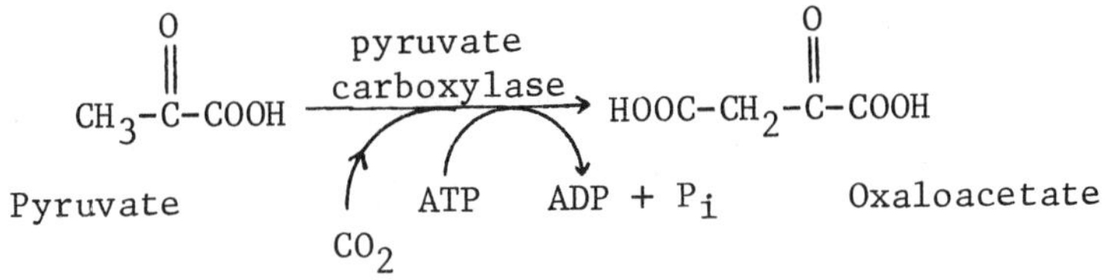

Biotin, a vitamin, is used in the above reaction. It is tightly bound to the carboxylase enzyme by means of the ϵ-NH_2 group of a lysine. It can react with CO_2 forming "active CO_2" in an ATP requiring reaction. The activated CO_2 can then be transferred to various acceptor molecules. Other CO_2 fixation reactions in the human will be dealt with later. Most of them also use biotin as a coenzyme.

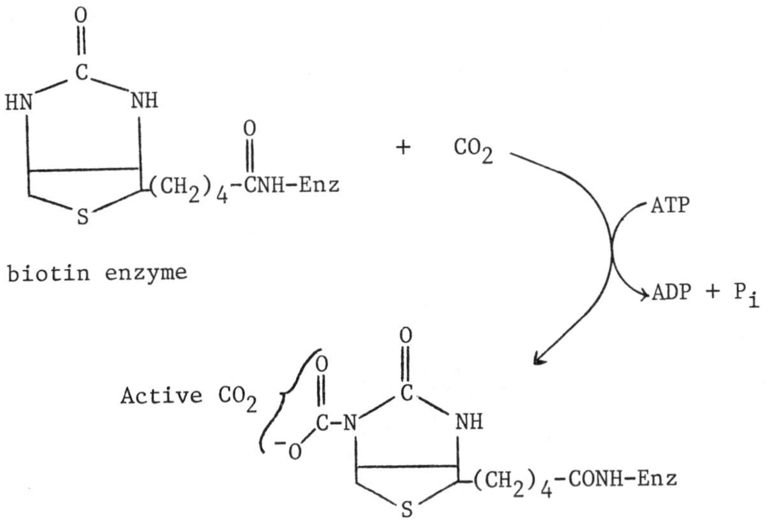

Pyruvate carboxylase is found in the mitochondria. It is an allosteric enzyme, inactive in the absence of acetyl-CoA (acetyl-CoA is a positive modulator). This means that pyruvate can be converted back to glucose only in the presence of high acetyl-CoA levels, a condition that occurs when there is a high level of ATP in the cell.

b. OAA (oxaloacetate) does not readily diffuse out of mitochondria so OAA is converted into <u>malate</u>.

$$\text{Oxaloacetate} + \text{NADH} + \text{H}^+ \xrightarrow[\text{malate dehydrogenase}]{\text{mitochondrial}} \text{NAD}^+ + \text{Malate}$$

Malate diffuses out to the cytoplasm.

c. In the cytoplasm the following reactions occur:

$$\text{Malate} \xrightarrow[\substack{\text{cytoplasmic malate} \\ \text{dehydrogenase}}]{\overset{\text{NAD}^+ \quad \text{NADH} + \text{H}^+}{\curvearrowright}} \text{OAA}$$

$$\text{OAA} + \text{GTP (or ITP)} \xrightarrow[\substack{\text{pyruvate} \\ \text{carboxykinase}}]{\text{phosphoenol-}} \underset{\substack{| \\ \text{CH}_2}}{\overset{\text{COOH} \\ |}{\text{C-O-}\textcircled{P}}} + \text{CO}_2 + \text{GDP (or IDP)}$$

phosphoenol pyruvate (PEP)

8. From PEP to fructose-1,6-di $\textcircled{P}$ the usual glycolytic enzymes are used.

$$\text{PEP} \xrightarrow[\substack{\text{Reversal of} \\ \text{E-M Pathway}}]{} \text{F-1,6-diP}$$

9. The next irreversible step is F-6- P $\xrightarrow[\substack{\\ \text{ATP} \quad \text{ADP}}]{}$ F-1,6-di $\textcircled{P}$.

The reverse reaction is catalyzed by fructose-1,6-diphosphatase and inorganic phosphate is released. The reaction is

$$\text{F-1,6-diP} + \text{H}_2\text{O} \xrightarrow{\substack{\text{fructose-1,6-} \\ \text{diphosphatase}}} \text{F-6-P} + \text{P}_i$$

$$\Delta G^{o\prime} = -4.0 \text{ Kcal/mole of F-1,6-di}\textcircled{P}.$$

This is an allosteric enzyme, present in liver, kidney, and
striated muscle, but absent in adipose tissue. Low concentrations
of AMP and ADP, and high concentrations of ATP and citrate,
favor the phosphatase reaction and inhibit the phosphofructokinase
reaction.

10.

$$\text{F-6-P} \xrightarrow{\substack{\text{Reversal of the}\\ \text{E-M Pathway}}} \text{G-6-P}$$

11. The third irreversible reaction is bypassed by glucose-6-
phosphatase.

$$\text{G-6-P} \xrightarrow{\text{G-6-Phosphatase}} \text{Glucose} + P_i$$

$$\Delta G^{o\prime} = -4.0 \text{ kcal/mole of G-6-P}$$

G-6-phosphatase is present in the intestine, liver, and kidney
but is _not_ found in adipose tissue or muscle tissue. It is
a microsomal enzyme (found in the endoplasmic reticulum) and is
responsible for providing blood glucose. A deficiency of this
enzyme has been implicated as the cause of von Gierke's disease
(discussed under glycogenolysis).

Summary of the Reactions of Gluconeogenesis

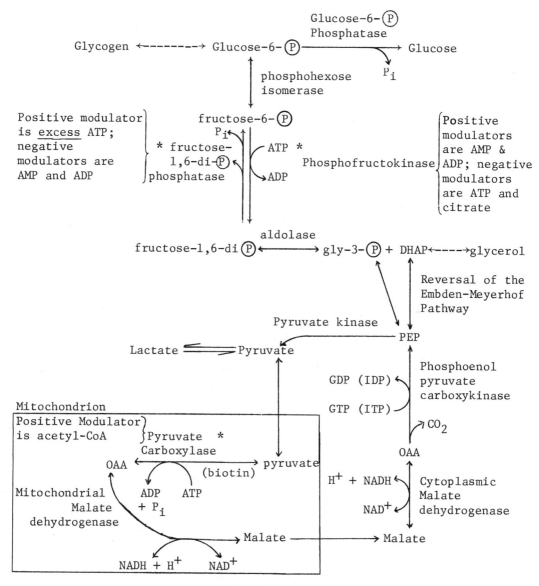

* indicates a control point catalyzed by an allosteric enzyme.

Note that such steps occur at the beginning and the end of the pathways. This avoids wasteful accumulation of intermediates. The dashed lines in this diagram indicate reaction pathways for the interconversion of the indicated materials.

Summary of the Net Effect of Gluconeogenesis

$$2(CH_3\overset{\overset{\text{O}}{\|}}{C}COOH) + 4ATP + 2GTP + 2NADH + 2H^+ + 6H_2O$$

pyruvate

$$\longrightarrow C_6H_{12}O_6 + 2NAD^+ + 4ADP + 2GDP + 6P_i$$

glucose

In contrast, the net effect of glycolysis is

$$C_6H_{12}O_6 + 2ADP + 2P_i + 2NAD^+ \longrightarrow 2CH_3\overset{\overset{\text{O}}{\|}}{C}COOH + 2ATP + 2NADH + 2H^+ + 2H_2O$$

glucose   pyruvate

Biochemical Lesion of Gluconeogenesis

1. A deficiency of hepatic fructose-1,6-diphosphatase has been
    observed in humans. This condition is associated with recurrent
    attacks of lactic acidosis and fasting hypoglycemia. The diagnosis
    of the enzyme deficiency is made by either using a specimen of
    liver obtained by laparotomy or by a more recent method using white
    blood cells. The patients show normal levels of liver glycogen
    and of the other enzymes associated with carbohydrate metabolism.

2. Assay of F-1,6-diphosphatase

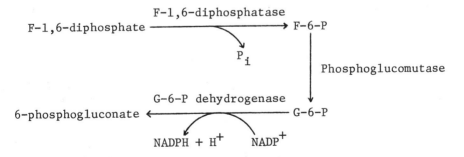

F-1,6-diphosphatase catalyzes the conversion of F-1,6-diP to F-6-P.
The F-6-P produced is converted to G-6-P which is then oxidized to
6-phosphogluconate. The conversion of F-1,6-diP to F-6-P is thus
linked to the reduction of $NADP^+$. This type of assay was discussed
under the use of enzymes in the clinical laboratory.

Allosteric Control of Carbohydrate Metabolism

Although no metabolic pathway or group of pathways ever functions alone within the body, the reactions by means of which the body manipulates its energy supply can be discussed as a separate entity. Similarly, although practically every sort of control mechanism can be found in these pathways, the allosteric effects serve as an illustration of the logic of the control processes in the cell. Some of these are summarized below. Several of the paths have already been discussed. Others will be covered later.

Allosteric Modifiers

	Enzyme	positive	negative
①	fructose-1,6-diphosphatase	excess ATP	ADP, AMP
②	phosphofructokinase	AMP, ADP	ATP, citrate
③	pyruvate carboxylase	acetyl-CoA, ATP	---
④	acetyl-CoA carboxylase	citrate, isocitrate	long-chain acyl-CoA (fatty acyl-CoA)
⑤	citrate synthetase	---	fatty acyl-CoA, ATP
⑥	isocitrate dehydrogenase	ADP, NAD^+	ATP, NADH

The following points should be particularly noted concerning these reactions:

a. The end-products of a pathway usually shut off an enzyme at the beginning of that pathway (ATP shuts off glycolysis and the TCA cycle; fatty acyl-CoA shuts off fatty acid synthesis).

b. Buildup of the products from one pathway can turn on an alternate route of metabolism for a precursor (excess citrate switches on fatty acid synthesis and gluconeogenesis (glucose synthesis) to consume acetyl-CoA and pyruvate).

c. A decrease in ATP as a source of energy for biological reactions and an increase in the products ADP and AMP turns off storage routes (fatty acid synthesis and gluconeogenesis) and activates supply pathways (glycolysis, TCA cycle).

d. The enzymes glycogen synthetase and glycogen phosphorylase (reactions 7 and 8 in Figure 23) are indirectly controlled by allosteric modifiers including cyclic adenosine monophosphate (cAMP). This is discussed in detail under glycogen metabolism.

e. Two compounds which are key intermediates in these pathways are glucose-6-phosphate and acetyl-CoA. They serve as branch or junction points for a number of metabolic routes.

Figure 23.   Allosteric Control Mechanisms of Some of the Pathways
             Involved in Energy Metabolism.

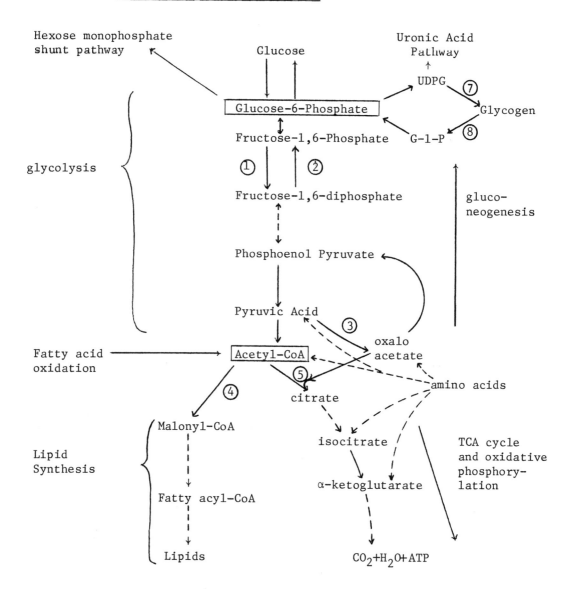

The dotted lines indicate multi-step reaction pathways while solid lines
represent single-step conversions.

Mitochondrial Electron Transport and Energy Capture (Oxidative
Phosphorylation) with Richard J. Guillory*, Ph.D.

1. In this section, we shall discuss the capture and storage (in the
    ATP molecule) of the energy released by the oxidation of reduced
    substrates (NADH and $FADH_2$) produced during the catabolism of
    glucose. Because of its involvement in many energy-requiring
    reactions within the cell, ATP is sometimes referred to as the
    "currency of metabolism".

2. The citric acid cycle functions as a common channel for the
    oxidation of metabolites and the production of reducing equivalents
    from a variety of foods including carbohydrates, proteins, and
    lipids. A reducing equivalent is one mole of electrons in the
    form of one gram-equivalent of a reduced electron carrier. Thus
    one mole of NAD^+ gains two moles (two times Avogadro's number) of
    electrons (as well as one mole of H^+) in being reduced to one mole
    of NADH. One mole of NADH is therefore two reducing equivalents.
    Note that the two electrons and one H^+ actually are transferred
    together, in many cases, as a hydride ion ($H:^-$).

3. The following steps are those in which reduced pyridine nucleotide
    (NADH) and reduced flavin-adenine dinucleotide ($FADH_2$) are formed
    during the breakdown of glucose via the Embden-Meyerhof pathway
    and the citric acid cycle.

    a. NADH-producing reactions

        (i)   Glyceraldehyde-3-Ⓟ ⟶ 1,3-diphosphoglycerate

        (ii)   Pyruvate ⟶ acetyl-CoA

        (iii) Isocitrate ⟶ α-ketoglutarate

        (iv)   α-ketoglutarate ⟶ succinyl-CoA

        (v)   Malate ⟶ oxaloacetate

        The reactions (ii)-(v) take place in the mitochondria and
        reaction (i) takes place in the cytoplasm. Since mitochondrial
        membranes are not directly permeable to NADH, two shuttle pathways
        have been found which transport the reducing equivalents of
        cytoplasmic NADH into the intramitochondrial matrix. These
        reactions are discussed in this section.

* Professor and Chairman, Department of Biochemistry & Biophysics,
University of Hawaii School of Medicine, Honolulu, Hawaii.

b. $FADH_2$-producing reaction

Succinate ⎯⎯⎯⎯⎯→Fumarate

The above reaction, catalyzed by succinate dehydrogenase, is part of the citric acid cycle and operates within the mitochondria. The enzyme appears to have a mitochondrial structural role as well as a catalytic one.

4. Electrons are catalytically removed together with protons, from the reduced substrate (e.g., NADH, $FADH_2$). They are transported over a series of coupled oxidation-reduction reactions to oxygen (O_2), which is the ultimate electron acceptor or electron sink. These coupled redox reactions take place in the mitochondria. The flow starts with the electrons having a relatively high potential energy and ends (at oxygen) with them at a low potential. During this electron flow or transport, a portion of the energy lost by the electrons is conserved by a biological energy transducing mechanism (a mechanism which changes the form of the energy from electrical to chemical). Since the energy is stored by using it to phosphorylate ADP to ATP, the overall coupled process is termed oxidative phosphorylation.

Mitochondria

1. Mitochondria are present in the cytoplasm of almost all aerobic eukaryotic cells. They are frequently found in close proximity to the fuel sources and to the structures which require ATP for maintenance and functional activity,(e.g.,the contractile mechanisms, energy-dependent transport systems, secretory processes, etc.).

2. The number of mitochondria present within a single cell varies from one type of cell to another. By way of an example, a rat liver cell contains about 1,000 mitochondria, while a giant amoeba (Chaos chaos) has some 10,000 mitochondria. In a given cell, the number of mitochondria.may also depend on the cell's stage of development and/or functional activity.

3. The size and shape of mitochondria also varies considerably from one cell type to another. Even within the same cell, mitochondria can undergo rapid changes in volume and shape depending upon the metabolic state of the cell. In general, they are about 1 μ in diameter and 3 to 4 μ in length. Mitochondria have been found to aggregate end to end forming long filamentous structures.

4. Mitochondria consist of about 30% lipid (primarily phospholipids) and 70% protein on a dry weight basis. They also contain specific types of DNA and RNA. The HeLa cell mitochondrial DNA is a circular duplex, made up of a heavy strand and a light strand.

5. Mitochondria undergo self-duplication, with fission taking place following cell division. They have their own protein synthesizing system (details of which are discussed under protein synthesis) but appear to be incapable of synthesizing completely all of the required structural proteins, lipid components, and enzymes. For example, HeLa cell mitochondrial DNA has about 15,000 base pairs corresponding to about 25 genes coding for 25 proteins. Since a mitochondrion has more than 25 different proteins, synthesis of the remaining proteins must require nuclear DNA. In other words, mitochondria are not completely autonomous and are dependent upon other parts of the cell for their synthesis.

6. The protein synthesis system of mitochondria appears to be analogous to that of bacterial systems. This observation, along with the similarity in size and the presence of circular DNA, have been viewed as suggesting that the mitochondria of eucaryotic cells arose from invading bacteria. A probable initial parasitic relationship may have later developed into a beneficial (symbiotic) relationship between the host and the parasite (anaerobic eukaryotic cell and mitochondrion, respectively).

7. Mitochondria consist of two membranes: an outer one which is a smooth, closed system surrounding the inner membrane. These membranes are separated from each other by about 50 to 100 Å. The intermembranal space contains the enzymes responsible for translocation of metabolites, in addition to adenylate kinase and nucleoside diphosphokinase. The inner membrane has many folds directed towards the interior of the mitochondrion. These invaginations, known as cristae, are presumably present to increase the surface area of the inner membrane. This provides more space for the attachment of the multi-enzyme clusters, each about 220 Å long, which carry on electron transport. Attached to these transport complexes are the inner membrane spheres, having a diameter of about 80 Å. These spheres contain ATP-synthesizing enzymes. Under the proper conditions, the spheres also show an ATPase activity, presumably due to a reversal of the normal ATP-synthetic process. It should be pointed out that the inner membrane spheres are not seen in intact mitochondrial preparations. They are observed only in submitochondrial particles prepared by the sonic disruption of mitochondria.

8. The inner membrane surrounds the matrix of the mitochondria and contains

    a. all of the factors essential for oxidative phosphorylation;
    b. the citric acid cycle enzymes;
    c. the fatty acid oxidation enzymes;
    d. the mitochondrial DNA;
    e. ribosomes and other accessories essential for protein synthesis.

9. The membranes, both inner and outer, although performing different functions, have structures similar to that of other biological membranes. Each membrane consists of a middle, non-polar layer (primarily phospholipids), on either side of which are protein layers. Polar regions of the proteins appear to be present on both sides of a single membrane.

10. There is a variable transport of molecules through the mitochondrial membrane. For example, mitochondrial membranes are permeable to pyruvate, while oxaloacetate, ATP and NADH enter with difficulty, if at all. The latter group of molecules is transported only by special carrier mechanisms. There are many specific translocating enzymes (carrier molecules) which participate in this transport. The equimolar exchange of external ADP with internal ATP maintains proper nucleotide pools. A compound isolated from the plant Atractylis gummafera, known as atractyloside, inhibits the exchange of ATP between the inside and outside of mitochondria. Accidental ingestion of this plant has been known to cause death. The active transport mechanism of the mitochondrial membrane are able to accumulate a number of components such as Ca^{++}, phosphate, and other ions, all of which appear to be essential for the proper functioning of the organelle.

    A highly schematic diagram of a mitochondrion is shown in Figure 24.

Electron Transport

1. Electrons are transported in the mitochondria from substrate (NADH, $FADH_2$) to oxygen through intermediate carriers. These carriers are arranged in a definite spatial relationship based upon their individual oxidation-reduction potentials. Electrons flow from a higher (potential) energy level to a lower energy state.

2. At specific locations in the electron-transport chain, part of the energy released by electron transport is conserved in the form of the phosphoric acid anhydride bonds of ATP. This is accomplished by directly coupling the energy-releasing process to the synthesis of ATP from ADP and P_i.

Figure 24. Schematic Diagram of a Mitochondrion

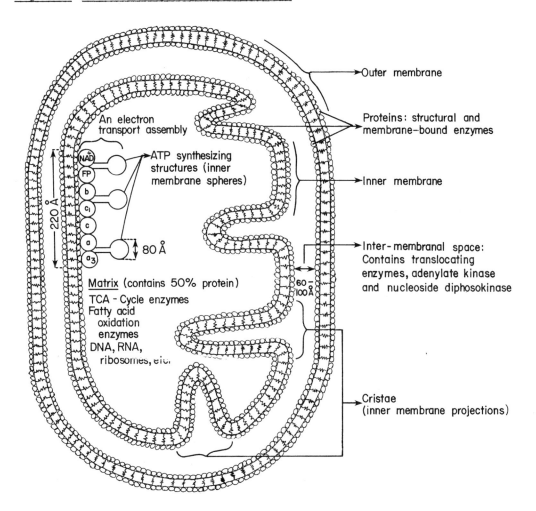

The above diagram is not drawn to scale. It should be pointed out that a liver mitochondrion contains between 15,000 and 30,000 electron-transport assemblies, depending upon the functional activity of the cell.

3. Using NADH as the substrate, in the presence of sufficient amounts of the other required factors, the overall oxidation reaction coupled to phosphorylation can be written:

$$NADH + H^+ + 3ADP + 3P_i + 1/2\ O_2 \longrightarrow NAD^+ + 3ATP + H_2O$$

Before discussing details of the specific reactions and the members of the electron-transport chain, a schematic diagram of the flow of electrons with the relevant energy changes is presented in Figure 25.

Carriers Present Within the Electron-Transport Assembly

1. Nicotinamide Adenine Dinucleotide is the co-substrate which accepts hydride ion (hydrogen with two electrons, $H:^-$) from a suitable donor and is converted to NADH. The structural details of this reaction have been outlined under glycolysis.

2. Flavoproteins. The NADH produced is oxidized by NADH dehydrogenase, an enzyme which contains flavin mononucleotide (FMN) as a coenzyme. The enzyme appears to be associated with non-heme iron (iron bound directly to a protein rather than through a porphyrin group). The role of non-heme iron in the oxidation of NADH by NADH dehydrogenase is not yet completely understood. The structure of FMN was presented in the discussion of the TCA cycle.

3. Coenzyme Q (also known as ubiquinone, a quinone found ubiquitously).

    a. Quinone-coenzymes occur in the lipid fraction of the membrane and are composed of a group having the following basic structure.

(oxidized form)

The predominant coenzyme Q species isolated from mammalian tissues has ten isoprenoid units ($n = 10$, hence CoQ_{10}).

Figure 25. <u>Mitochondrial Electron-Transport System</u>

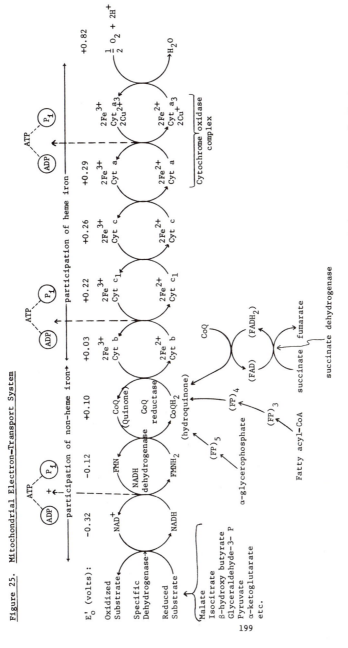

Note: (1) FP = flavoprotein. There are several of these, designated $(FP)_3$, $(FP)_4$, etc. All contain flavin or some derivative of it as a coenzyme.

(2) The relative position of CoQ to cytochrome b is not certain. The E_o' value for CoQ was measured in 95% ethanol, a highly artificial situation.

(3) Most of the E_o' values given are for heart muscle mitochondrial enzymes. There is considerable variation in reported values due to differing methods of preparation and measurement.

b. Oxidation-reduction reaction

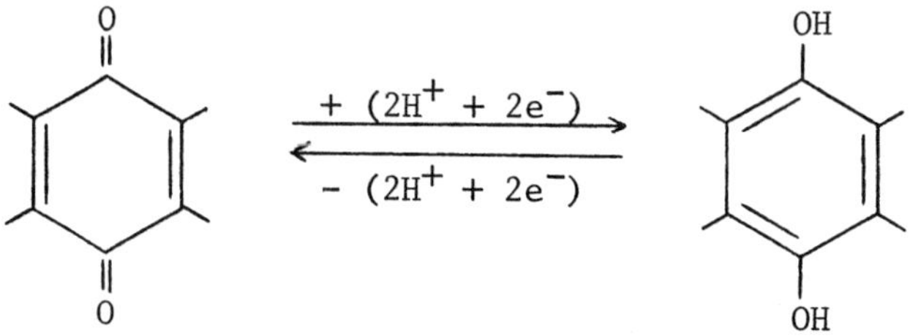

quinone   hydroquinone

c. Vitamin K_1 has a structure similar to that of coenzyme Q. In mycobacteria, which do not contain CoQ, vitamin K has a comparable functional role. In animals, however, CoQ has no known vitamin precursors and is presumably synthesized within the body.

Structure of Vitamin K_1

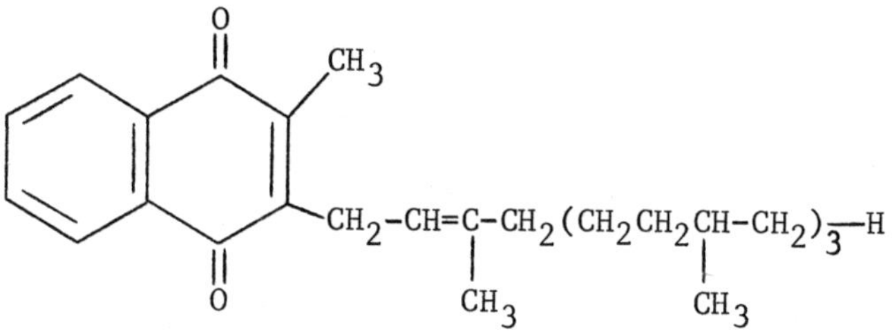

The function of Vitamin K_1 in humans is discussed in the section on vitamins.

4. Cytochromes

a. These are proteins containing iron porphyrins as prosthetic groups (heme-proteins). The iron undergoes repeated oxidations and reductions:

$$Fe^{3+} + e^- \rightleftharpoons Fe^{2+}$$

(Note: This does not happen in hemoglobin, which is also a heme-protein.) Cytochromes can be differentiated by their absorption spectra, as can the oxidation state of any specific cytochrome. The characteristic spectra of the different cytochromes are due to differences in the sidechains of the porphyrin ring.

b. Porphyrin ring systems

i) These are made up of four pyrrole rings joined together by methylene bridges. The parent compound is porphin, shown below.

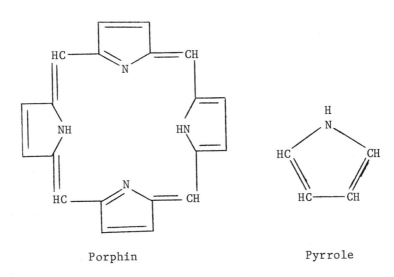

Porphin   Pyrrole

ii) Porphyrins form chelate complexes with metal ions (metalloporphyrins) such as iron. This type of complex can be schematically represented as

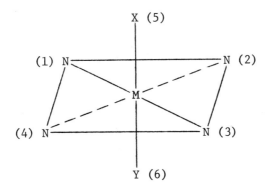

where M is a metal ion (central metal ion) such as Fe^{+2} or Fe^{+3}. Iron can form six coordinate-covalent bonds with any six ligands which have pairs of electrons in non-bonding orbitals (lone pairs of electrons). These complexes have

a shape or symmetry which is octahedral. In the iron-
porphyrin complex, four of the ligands are the four
nitrogens of the porphyrin ring (N(1) to N(4), above).
The other two positions, X and Y, are occupied by H_2O, an
amino acid sidechain or some other ligand. When the
metalloporphyrin is bound to a protein, the fifth and
sixth positions are usually occupied by amino acid sidechains in the protein. A frequently encountered iron-
porphyrin complex is iron-protoporphyrin IX (for porphyrin
names, see the section on porphyrin synthesis). If the
iron is present as Fe^{+2}, this complex is called <u>heme</u> (or
protoheme); if it is Fe^{+3}, the name is changed to <u>hemin</u>.
These designations arose because the complex was first
found in hemoglobin.

c. <u>Cytochrome b</u>

i) This is the first cytochrome of the electron-transport
    sequence. As a prosthetic group it contains a heme (an
    iron-protoporphyrin IX complex). This coenzyme is also
    present in the heme proteins hemoglobin and myoglobin.
    There are significant differences in other respects,
    however, between hemoglobin, myoglobin, and the
    cytochromes. These are presented below in Table 5.
    Synthesis of protoporphyrin IX and other aspects of the
    porphyrins are discussed under the section on hemoglobin.

ii) The prosthetic group of cytochrome b (and hemoglobin
     and myoglobin) is

In cytochrome b, the (5) and (6) positions of the iron are
occupied by protein ligands and are therefore inaccessible
for binding with external ligands such as O_2, CO, CN^-, etc.

iii) Cytochrome b accepts electrons from reduced coenzyme Q ($CoQH_2$) and passes them on to cytochromes of higher (more positive) oxidation-reduction potential.

iv) Cytochrome b is tightly bound to the lipoprotein membrane.

v) The table below provides a comparison of some of the properties of the cytochromes with those of hemoglobin and myoglobin.

Table 5. Comparison of Different Types of Metalloporphyrin-Containing Proteins

	Cytochromes*	Hemoglobin & Myoglobin
5 position occupied by	R-groups of specific amino acid residues	Imidazole group of a histidine residue
6 position occupied by	R-groups of specific a.a. residues; unavailable for binding with O_2, CO, CN, etc.	H_2O (in deoxy Hb or deoxy myoglobin) or oxygen (O_2) or CO, etc. (in oxyhemo-globin or carbon monoxy-hemoglobin and myoglobin)
Valence State of Iron	$Fe^{2+} \rightleftarrows Fe^{3+} + e^-$ (oxidation changes back and forth during the normal functioning of the cytochromes)	The normal valence state of Fe^{2+} is unchanged when O_2 is gained or lost. When the reaction $Fe^{2+} \longrightarrow Fe^{3+}$ takes place via an oxidizing agent (e.g.,ferricyanide), methemoglobin is produced methemoglobin cannot function as an oxygen carrier

* Cytochrome oxidase differs from the other cytochromes in a number of ways (see later).

d. Cytochrome c

    i) Prosthetic Group

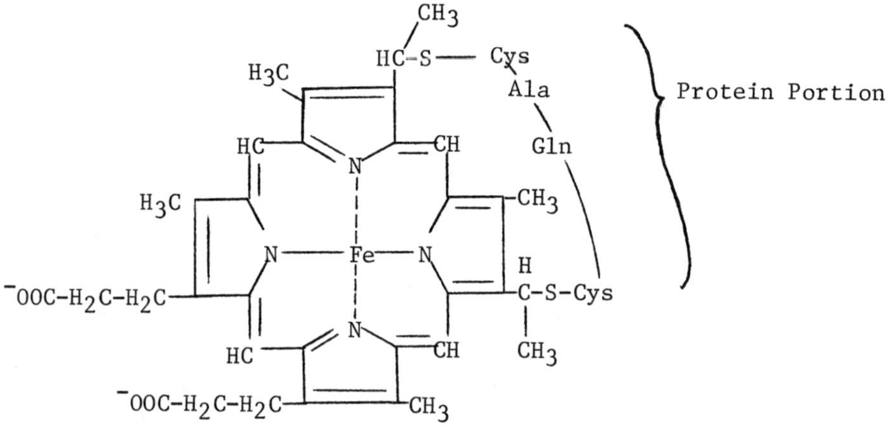

    Note the linkage of the heme group to the protein through cysteine residues.

    ii) This is the central member of the electron-carrier system and has an intermediate redox potential. It links the oxidation of cytochrome b to the cytochrome oxidase complex. The molecule is relatively loosely bound to the respiratory chain system and is easily extracted in a fairly pure form.

e. Prosthetic Group of Cytochrome a

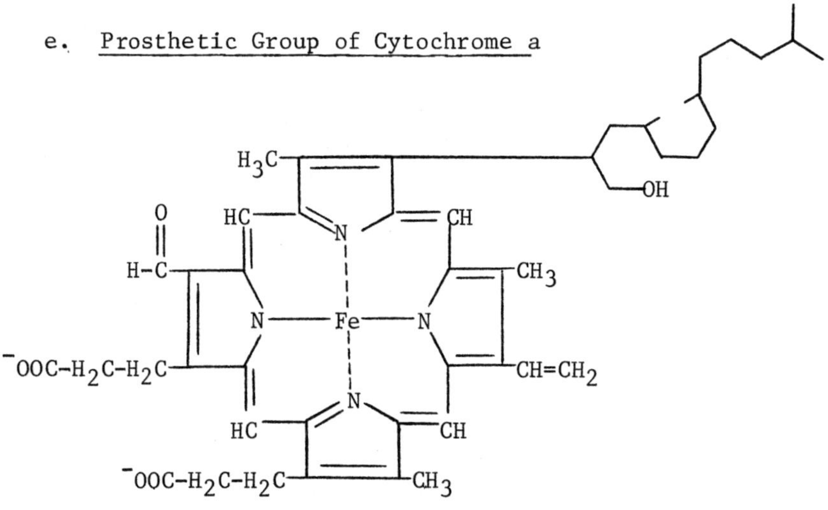

f. Cytochrome a/a$_3$ (collectively termed cytochrome oxidase)

    i) This is the terminal member of the electron-transport assembly. It reoxidizes cytochrome c and reduces oxygen and has a high (positive) redox potential.

    ii) The oxidase contains copper in addition to an iron-porphyrin system. The copper also undergoes an oxidation-reduction reaction ($Cu^{2+} + e^- \rightleftharpoons Cu^+$) during electron transport.

    iii) The oxidase is inhibited by CN^-, H_2S, N_3^- (azide), and CO. The details on CN^- binding have already been discussed in the section on irreversible inhibitors of enzymes.

g. It should be pointed out that, with the possible exception of NAD^+ and cytochrome c, the electron-transport chain members are intimately associated with the lipids of the membranes. It is likely that electron transport and energy conservation take place in the hydrophobic environment of the membrane.

h. Other forms of mitochondrial-type electron-transport systems occur in microsomes (pieces and vesicles of endoplasmic reticulum and other intracellular membranous structures formed during cellular fractionation). Two examples are given below.

    i) In rat liver microsomes, there is a pathway which uses oxygen for the ω-oxidation fatty acids (oxidation at the carbon farthest removed from the carboxyl group). Three enzymes have been shown to be part of this system: a unique cytochrome (known as B420 or P450), an NADPH-cytochrome P450 reductase, and a heat stable lipid factor. The numbers 420 and 450 attached to these cytochromes indicate the wavelengths of maximum absorption of the cytochrome and its carbonmonoxy derivative, respectively. The system shows broad specificity, being able to hydroxylate cyclohexane, fatty acids, n-alkanes, and a variety of drugs and other compounds having O- and N-methyl groups.

    ii) Other NADPH and O_2-requiring hydroxylases are present in microsomes from the adrenal gland. Some of these are involved in the oxidation of endogenous and exogenous aromatic compounds and of steroids. Details of some of the microsomal oxidase systems are given in appropriate sections.

Experiments Used in the Establishment of the Sequence of the Electron-Transport Assembly

1. Spectroscopic evidence has been used for

    a. identification of the presence of suspected or postulated carriers in various subcellular fractions;

    b. determination of the amount of the carrier present;

    c. studies on the kinetics of reduction and reoxidation reactions.

2. Fragmentation and reconstitution of submitochondrial particles

    By means of mechanical disruption and detergent treatment, four different complexes have been obtained from mitochondria. The functions of these complexes and their interrelationships are shown in the following diagram.

Figure 26.   Composition, Interrelationship, and Function of Protein Complexes Obtained from Mitochondrial Fractionation

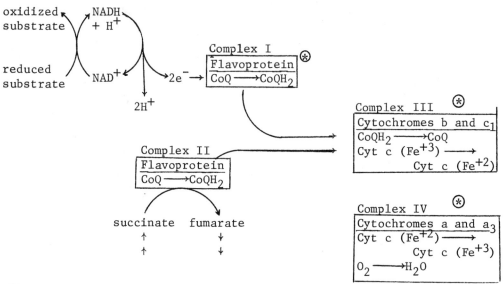

⊛ At these points ATP is synthesized by the reaction ADP + P_i ⟶ ATP.

3. <u>Inhibitors</u> that block the passage of electrons at selected places have been used in understanding the nature of the carriers, the routes of the flow of electrons and the sites of ATP synthesis.

<u>Figure 27</u>.   <u>Inhibitors of Electron Transport and the Specific Transfers Which They Inhibit</u>

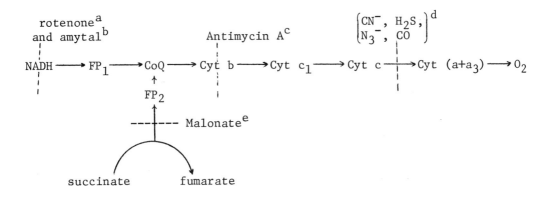

a. Rotenone is obtained from the roots of the tropical plants <u>Derris elliptica</u> and <u>Lonchocarpus nicou</u>.

b. Amytal is a barbiturate.

c. Antimycin A is an antibiotic isolated from the bacteria <u>Streptomyces</u>.

d. All of these molecules form stable complexes with Fe^{+3} preventing its reduction to Fe^{+2}. This prevents further e^- transfer to O_2 and was discussed earlier in greater detail (see enzyme inhibition).

e. Malonate is an analogue of succinate. It competitively inhibits succinic acid dehydrogenase (see TCA cycle).

4. a. <u>Standard redox-potential data</u> is useful since electrons flow from compounds having a higher $E^{o\prime}$ (more negative value) to compounds with a lower $E^{o\prime}$ (more positive value). A knowledge of $E^{o\prime}$ values for the members of the electron-transport chain permits the placing of the individual carriers in the proper order for concerted electron transport.

b. Some of the $E^{O'}$ and $\Delta G^{O'}$ (= 23,061 $nE^{O'}_T$; see p. 78) values of the electron-transport system components are shown below.

Figure 28. Free Energy Changes Associated with Electron Transport

$$|\longleftarrow E^{O'}_T = -1.14 \text{ volts}; \Delta G^{O'} = -52.6 \text{ kcal/mole NADH} \longrightarrow|$$

$$NADH \rightarrow FP_1(FMN) \rightarrow CoQ \rightarrow Cyt\ b \rightarrow Cyt\ c_1 \rightarrow Cyt\ c \rightarrow Cyt\ (a+a_3) \rightarrow O_2$$

	Complex I	Complex III	Complex IV
$E^{O'}$ (volts):	-0.32 -0.12	$+0.03$ $+0.22$	$+0.29$ $+0.82$
$E^{O'}_T$ (volts):	-0.20	-0.19	-0.53
$\Delta G^{O'}$ (kcal/mole):	-9.2	-8.8	-24.4

Recall that these complexes are also the sites of ATP synthesis (see Figure 26).

c. The formation of ATP can be described by the reaction

$$ADP + P_i \longrightarrow ATP + H_2O \qquad \Delta G^{O'} = +7.3 \text{ Kcal/mole ATP}$$

The $\Delta G^{O'}$ values of each of the three complexes in Figure 28 are sufficiently negative (i.e., enough free energy is released) to drive ATP synthesis. This can be seen by noting that the $\Delta G^{O'}$ values for the three complexes are all less than -7.3 kcal. Experimentally, it has been found that three molecules of ATP are synthesized for each pair of electrons passed from NADH to oxygen (for each atom of oxygen reduced). The ratio of moles of ATP synthesized to gram atoms of oxygen reduced is called the P/O ratio (or "P to O ratio"). For oxidative phosphorylation under normal conditions, P/O = 3.

d. Since 52.6 kcal/mole of NADH are released in the overall process, it is thermodynamically possible to synthesize

$$\frac{52.6 \text{ kcal/mole NADH}}{7.3 \text{ kcal/mole ATP}} = 7 \text{ moles ATP/mole NADH}$$

Under standard conditions, at pH = 7, then, oxidative
phosphorylation is ∿43% (= 3/7 x 100%) efficient. Under actual
cellular conditions, however, ΔG for ATP formation is probably
about 10-12 kcal/mole. If the values of $\Delta G^{o\prime}$ for the electron
carriers are similar to their actual intracellular (ΔG) values,
then the efficiency will be better than 43%. The difficulty
of discussing intracellular processes in terms of isolated
systems (i.e., values of $\Delta G^{o\prime}$ rather than ΔG) is once again
apparent. The energy not used to synthesize ATP appears as
heat and is important for maintaining the constant body
temperature of warm-blooded animals.

3. The P/O ratio for a given substrate is dependent upon the energy
    level at which the compound transfers its electrons to the
    electron-transport chain. P/O ratios are given below for some
    of the electron-transport substrates together with the oxidized
    products.

	P:O Ratio
Pyruvate ⟶ Acetyl-CoA	3
Isocitrate ⟶ α-Ketoglutarate	3
α-Ketoglutarate ⟶ Succinate	4←(includes 1 substrate level phosphorylation)
Succinate ⟶ Fumarate	2
Malate ⟶ Oxaloacetate	3

Regulation of Coupled Electron Transport

1. Oxidation and phosphorylation are tightly coupled. In other words,
    when respiration occurs there must be concomitant phosphorylation
    of ADP. Certain substances, such as 2,4-dinitrophenol, can
    uncouple the two processes and are known as uncoupling agents.
    Uncouplers act by dissociating the oxidative process from
    phosphorylation, permitting electron transfer and oxygen consumption
    without the formation of ATP. Respiration, as measured by O_2
    uptake, can occur in the absence of ADP or P_i. Such agents have
    been very useful in elucidating certain aspects of oxidative
    phosphorylation.

2. Normally, in a coupled system, once all of the ADP or P_i have been
    used up, very little respiration takes place. Even if NADH is
    present, it is not reoxidized. If an uncoupler is added to such a
    system, O_2 uptake resumes and NAD^+ is produced. In the normal
    cell, the amount of ADP present is dependent on the expenditure
    of ATP.

3. There are four other states (types, conditions) of respiratory
    control in addition to the (ADP + P_i) supply. These are
    described in the table below, as originally defined by Chance and
    Williams.

Control State	Conditions which are satisfactory to maintain normal respiration	Condition which controls or limits respiratory rate
1	O_2, respiratory chain	substrate and ADP concentration
2	O_2, ADP, respiratory chain	substrate concentration
3	O_2, ADP, substrate	functional capacity of the respiratory chain
4	O_2, substrate, respiratory chain	ADP concentration
5	ADP, substrate, respiratory chain	O_2 concentration

In this table, substrate refers to NADH, $FMNH_2$, $FADH_2$, or some source
of these reduced coenzymes.

Most resting cells are controlled by the availability of ADP (state 4).
The ADP-ATP cycle is shown in the next diagram.

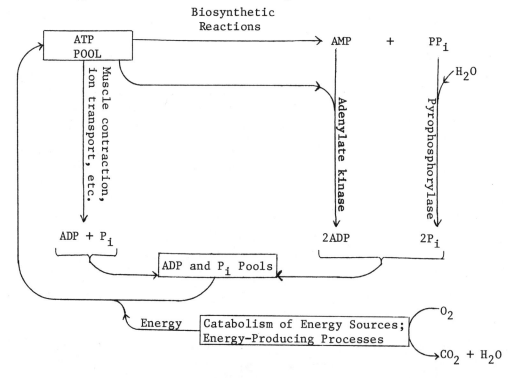

Possible Mechanisms for the Conservation of Energy During Electron Transport

1. Chemical coupling (biochemical coupling)
2. Chemiosmotic coupling
3. Conformational coupling (mechanochemical coupling)

1. Chemical Coupling

    a. This hypothesis postulates that energy is conserved by the formation of high-energy intermediates as electrons are passed from one carrier to the next. The energy is ultimately stored as a phosphoric acid anhydride bond in ATP.

    b. After many years of intensive investigation, no high energy intermediates have been isolated.

    c. Oxidative phosphorylation occurs only in the presence of reasonably intact membrane structures and hence structural organization of components must play a major role. If high energy intermediates do exist, it is possible that they are tightly bound to a membrane. This might explain why they have not yet been isolated.

    d. The name "chemical coupling" is unfortunate since any possible mechanisms involve chemical reactions to a great extent. It is generally used to imply purely chemical as opposed to one involving another sort of energy transfer at some stage.

2. Chemiosmotic Coupling (largely developed by P. Mitchell)

    a. It is assumed (and experimentally shown) that the mitochondrial membrane is impermeable to protons (H^+) and hydroxyl ions (OH^-).

    b. During electron transport in intact mitochondria, protons are released to the outside of the mitochondria. This results in the establishment of a proton gradient across the membrane, with a high concentration of H^+ (low pH) outside the mitochondria and a low concentration of H^+ (high pH) inside. The energy released during electron transport is stored by using it to maintain this gradient.   Gradients of other cations may also be involved.

    c. A vectorial (anisotropic) ATPase is located in the membrane. It is oriented so that, to hydrolyze ATP, it must draw H^+ from the inside of the mitochondria and OH^- from the outside.

Because of the concentrations which are actually present, the reverse reaction, $ADP + P_i \longrightarrow ATP + H^+ + OH^-$ occurs. The H^+ is released to the inside of the membrane, the OH^- goes to the outside. The ATP synthesis is actually driven, then, by the rapid reaction of H^+ with the OH^- inside and of the OH^- with the H^+ outside. The energy for the formation of the phosphoric acid anhydride bond is supplied by the large, negative value of ΔG which results from water formation under the existing H^+ and OH^- concentrations.

d.   Figure 29 is a schematic drawing of a possible arrangement of the enzymes involved. Any spatial association of phosphorylation with the individual components of the electron-transport chain (complexes I, III, and IV for example) has been ignored. FMN and CoQ are actually associated with enzymes and the oxidation of NADH involves a specific dehydrogenase. These enzymes are not explicitly shown.

3.   Conformational Coupling

a.   Mitochondria undergo changes in the size and shape of their cristae during active respiration stimulated by ADP.

b.   These conformational changes may be the driving force for the formation of ATP. At least conceptually, this can be seen as a reversal of the process occurring in muscle which is hydrolyzing ATP to perform work. It is important to realize that none of the above three mechanisms by itself completely explains oxidative phosphorylation. Quite possibly the true mechanism will be represented by some combination of them all.

Entry of Cytoplasmic NADH into Mitochondria

The mitochondrial membrane is impermeable to NADH, NAD^+, NADPH, and $NADP^+$. The NADH and NADPH formed in the cytoplasm (e.g., by glycolysis and the HMP shunt) have their reducing power (i.e., their electrons) transferred from the extramitochondrial to the intramitochondrial space by two shuttle mechanisms. These are

a.   the glycerolphosphate shuttle--unidirectional; occurs in liver cells and insect flight muscle, and

b.   the malate shuttle--can function in either direction since all reactions are reversible; occurs in liver and other cells . They are shown diagrammatically in Figures 30 and 31.

Figure 29. Possible Mechanism for the Coupling of Electron Transport
to Oxidative Phosphorylation by Chemiosmosis

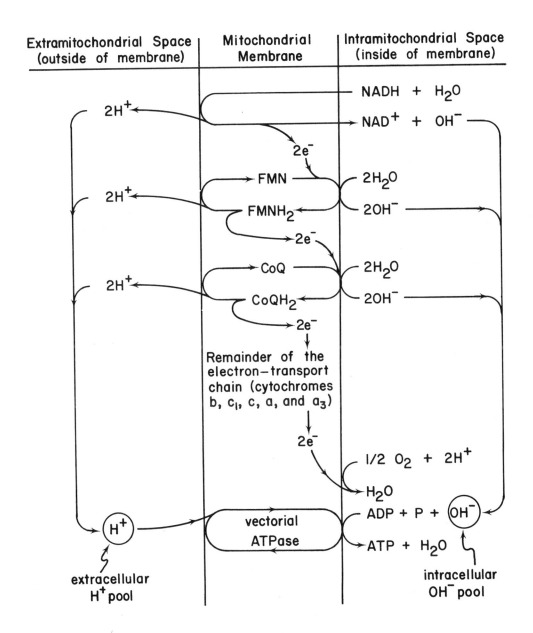

<u>Figure 30.</u>   <u>Glycerolphosphate Shuttle for Transporting Cytoplasmic</u>
<u>Reducing Equivalents into the Mitochondria</u>

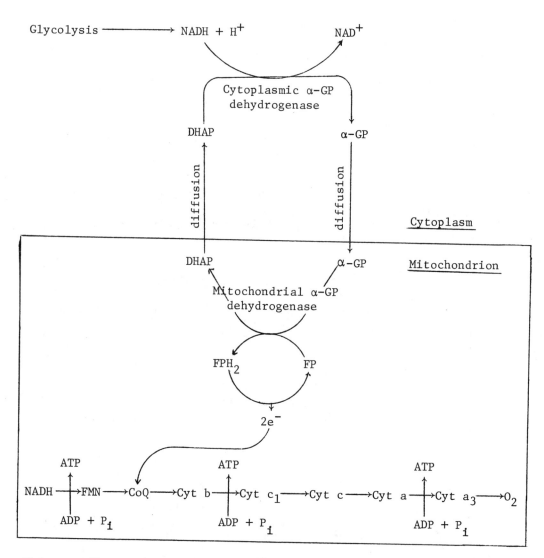

Note: α-GP = α-glycerophosphate; DHAP = dihydroxyacetonephosphate.

a. Since the intramitochondrial α-GP dehydrogenase is linked to a flavoprotein, the electrons are put into the electron-transport chain at the CoQ level rather than at the NADH dehydrogenase level. One phosphorylation site is thus bypassed and only two ATP's are synthesized for every two electrons. An NADH molecule put through this pathway produces only two ATP's, whereas an NADH formed within the mitochondrion (by, say, the TCA cycle) or one transported by the malate shuttle (see Figure 31) produces three ATP's. (This was mentioned previously in discussing the energy output from glycolysis and the TCA cycle.)

In the presence of oxygen, the intramitochondrial supply of reducing substances is depleted and cytoplasmic NADH is needed as a source of electrons. Under these conditions, α-GP dehydrogenase has a higher affinity for NADH than does lactic acid dehydrogenase (LDH) and the α-GP shuttle operates.

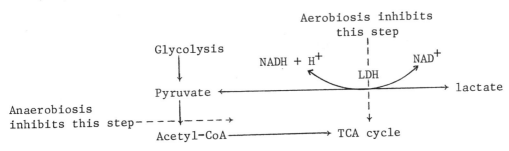

In the above diagram, the presence or absence of oxygen causes the indicated effects by several mechanisms other than the differential affinity of these particular enzymes for NADH.

Most cancer cells accumulate lactate during respiration (i.e., in the presence of oxygen). This could be due to a lack of α-GP dehydrogenase or a lack of control over glucose transport. Such cells show alterations in the structure and function of cell-surface membranes relative to normal cells. These changes include a loss of contact inhibition and changes in the membrane glycoprotein. Similar alterations could also occur in the mitochondrial membranes.

Figure 31.   Malate Shuttle for the Transport of Cytoplasmic Reducing
Equivalents into the Mitochondria

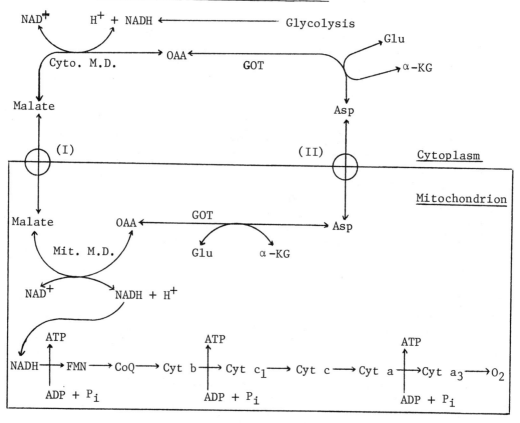

Note: OAA = oxaloacetic acid
         Asp = aspartic acid
         Glu = glutamic acid
        α-KG = α-ketoglutarate
         GOT = glutamate-oxaloacetate transaminase
Cyto. M.D. = cytoplasmic malate dehydrogenase
Mit. M.D. = mitochondrial malate dehydrogenase
       (I) = a membrane-bound malate-succinate carrier
      (II) = a membrane-bound aspartate-specific carrier.

b. When cytoplasmic NADH is oxidized via this pathway, 3 moles of ATP are produced per mole of NADH. It is not clear which pathway (this or the glycerolphosphate shuttle) operates in the normal cell, but it appears more likely that it is this one.

In adipose cells obtained from obese patients, the activity of both cytoplasmic and mitochondrial α-GP dehydrogenase is lower than that of lean control individuals. This implies that the increased lipogenesis which caused the obesity was due to the availability of greater amounts of α-GP (used in triglyceride synthesis) and to an increased efficiency in the use of metabolites (malate shuttle pathway) by the obese patients.

Water of Oxidation (Metabolic Water)

1. Water that is produced as a metabolite in the electron-transport system is known as water of oxidation. It contributes to the total body-water needs. In humans living in a temperate climate, the average intake and internal production of water is about 2600 ml per day. The water sources with their approximate contributions are indicated below.

Source	Amount (ml/day)
Water and other beverages	1200
Water present in solid foods	1100
Water of oxidation	300
	2600 ml/day

To maintain a balance within the body, water is lost via the lungs, skin (perspiration), feces, and urine.

2. In animals that live under conditions of limited external water supply (e.g., desert animals), oxidative metabolism becomes the principal source of water. These animals lose very minimal amounts of water through perspiration, urination, and defecation. In addition, they produce more metabolic water by increasing lipid oxidation and decreasing glycolysis.

3. The table below indicates the amount of water produced and the energy liberated in the oxidation of carbohydrates, proteins, and lipids. These values were obtained by burning the respective materials in a bomb calorimeter.

Substance	Water of Oxidation (gm H_2O/gm material)	Energy Released (kcal/gm material)
Carbohydrate	0.55	4.2
Protein	0.41	5.6
Lipid	1.07	9.3

Of the three, lipids have the highest yield/gm of both energy and water. This is because they have a higher ratio of (carbon + hydrogen) to oxygen than do either proteins or carbohydrates. Consider, for example, stearic acid ($C_{18}H_{36}O_2$) and glucose ($C_6H_{12}O_6$).

For stearic acid $(C + H)/O = 54/2 = 27$

For glucose $(C + H)/O = 18/6 = 3$

Because stearic acid contains less oxygen than glucose, more oxygen must be supplied by respiration. This means during the oxidation of stearic acid, a greater number of electrons pass down the electron-transport chain, more ATP molecules are synthesized, and more new water is formed by the reduction of oxygen than during the analogous catabolism of glucose.

Glycogenesis (Glycogen Synthesis)

1. The synthesis of glycogen (the principle storage form of carbohydrate in mammals) occurs in every tissue of the body. It is most active in liver and muscle.

   Structure of glycogen:

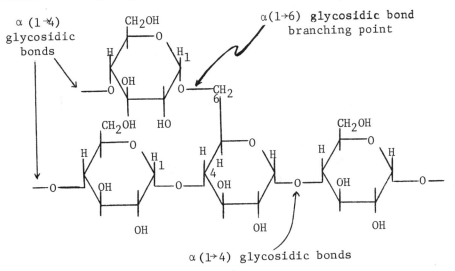

   The monomer units which polymerize to form glycogen are glucose molecules. Glycogen is a polyglucose, as are starch and cellulose.

2. The first two steps of the biosynthesis are

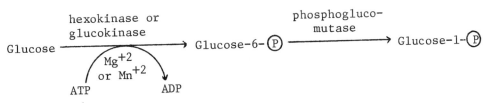

3. The third step involves the formation of a nucleoside diphosphate sugar derivative, uridine diphosphate glucose (UDPG).

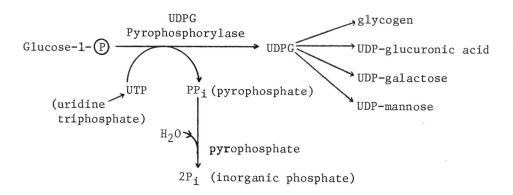

UDPG is an important intermediate in several pathways besides glycogen synthesis. It participates in the formation of UDP-glucuronic acid (as part of the uronic acid pathway) and is involved in the interconversion of other sugars. For example, the epimerization of galactose and mannose to glucose require UDP-galactose and UDP-mannose as intermediates.

Structure of UDPG

This reaction is one example of the functioning of nucleoside phosphates as carriers of small molecules during the synthesis of larger ones. This and other roles of nucleoside phosphates are discussed under nucleic acids and in other appropriate sections.

4.   The glucosyl group of UDPG is transferred to the terminal glucose residue at the non-reducing end of an amylase chain (primer) to form an $\alpha-(1\rightarrow4)$ glycosidic linkage. The primer may be a polyglucose or an oligoglucosaccharide (small glucose polymer) of at least four residues.

$$\text{UDP-glucose} + (\text{glucose})_n \xrightarrow{\quad\text{Glycogen}\quad\text{Synthetase}\quad} \text{UDP} + (\text{glucose})_{n+1}$$

Primer

Amylose [$\alpha(1\rightarrow4)$ polyglucose] chain

Part of the free energy of glucose phosphorylation is conserved in the $\alpha(1\rightarrow4)$ glycosidic bond.

5.   Glycogen synthetase cannot make the $\alpha(1\rightarrow6)$ bonds that occur at the branch points of glycogen. For this, another enzyme, amylo $(1\rightarrow4, 1\rightarrow6)$ transglucosylase (branching enzyme) is needed. After the synthetase has added about eight to twelve glucose units to the non-reducing end of an amylose chain, the transglucosylase transfers six or seven of these to the sixth carbon of the glucose residue which was about the 12th residue from the end. The new molecule now has two non-reducing ends and, hence, two growing points. This is shown on the next page.

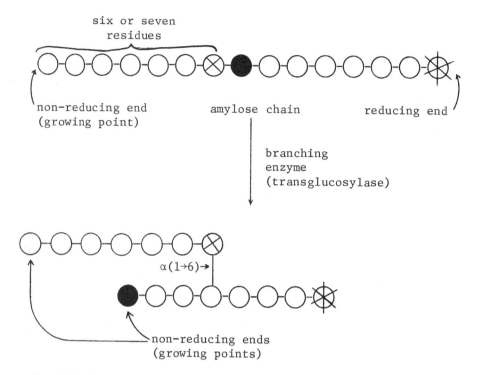

In this diagram, each circle represents a glucose residue.

The $\alpha(1\rightarrow6)$ bonds occur about every 8-12 glucose units. Since the sidechains can be branched in the same way as the main chain, the result is a very highly branched molecule.

Glycogen Synthetase (Glycogen Synthase; UDP-α-D-glucose-glycogen-glucosyltransferase)

1. Glycogen synthetase is a key enzyme and control point in glycogenesis. It exists in two forms: (1) a phosphorylated form, called glycogen synthetase-D because it has little or no activity by itself but is greatly stimulated by glucose-6-phosphate (and thus is dependent on G-6-P) and (2) a dephosphorylated form, called glycogen synthetase-I (because it is independent of G-6-P and is active in its absence). The phosphate group is attached to a serine residue by an ester linkage. The activation by G-6-P appears to be of the allosteric type. Glycogen synthetase-D activity is prevented by the presence of UDP. The two forms of glycogen synthetase may be interconverted by the appropriate enzymes. One of these enzymes, glycogen synthetase kinase (GSK),

also exists in an active and inactive form. (Kinase is a general
name for any enzyme which catalyzes the phosphorylation of
another compound.) The inactive form is transformed to the
active form in the presence of the positive modulator, cyclic AMP,
which is formed from ATP by the enzyme, adenyl cyclase. This
enzyme, present in the plasma membrane of many cells, is stimulated
by epinephrine. Insulin depresses the level of cyclic AMP in some
tissues. These relationships are shown in the Figure 32.

Cyclic AMP may activate GSK by increasing the affinity of the
enzyme for Mg^{+2}. It is known that   1) in the absence of Mg^{+2},
the maximum stimulation caused by cyclic AMP is only a 20% increase
over that in the absence of the cyclic nucleotide; and 2) at high
Mg^{+2} concentration where the kinase is already maximally stimulated,
cyclic AMP cannot further increase the activity.

Note: In the glycogen synthesis system, phosphorylation converts
the glycogen synthetase enzyme from an active form to a less active
form and dephosphorylation transforms the enzyme back to the
active form. In the glycogen breakdown (glycogen phosphorylase)
system, the opposite effect occurs. Phosphorylation of a key
enzyme activates it, while dephosphorylation inactivates it. All
of these systems are controlled by cyclic AMP levels. These are
discussed below.

Note that the effect of epinephrine is to make glycogen synthesis
dependent upon G-6-P. Thus, in a fight-or-flight situation where
energy demands are high and G-6-P levels are consequently low
(since G-6-P has been used up for energy production), epinephrine
release turns off the glycogen synthesis while promoting glycogen
breakdown. Note also that glycogen synthetase phosphatase is
inhibited by glycogen (negative feedback).

Comments on Cyclic AMP

1. The formation of cyclic-3',5'-adenylic acid (cyclic AMP, cAMP) and
    its action is implicated in the functioning of many hormones.
    Some hormones exercise their influence, at least in part, by
    causing the formation of cAMP in their target tissues. Cyclic
    AMP thus acts like an intracellular hormone or "second messenger",
    the first messenger being the hormone or other stimulus which
    invokes it. Ca^{++} ion is often required for cyclic AMP action.
    Binding of a hormone at specific sites on target tissue cell
    membranes causes an increase in cell-wall permeability to Ca^{++}.
    After the hormone is removed from the binding site or inactivated,
    the membrane again becomes impermeable to Ca^{++}. An ATP-requiring
    Ca^{++} pump returns the intracellular Ca^{++} concentration to a low level.

Figure 32.   Regulation of Glycogen Synthetase Activity Within the Cell

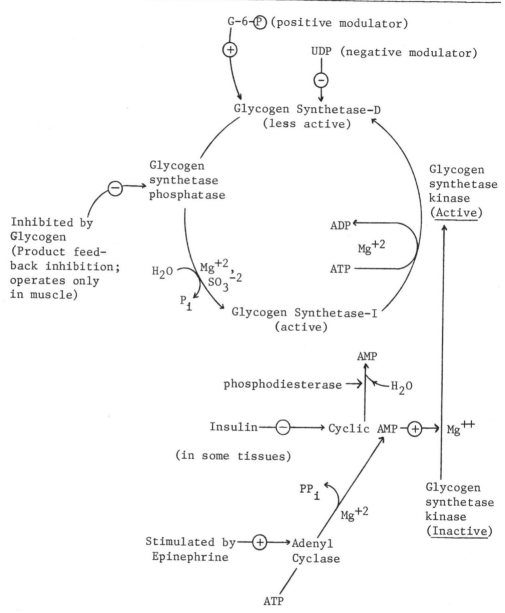

In this diagram, a ⊕ or ⊖ means that a substance increases or decreases respectively the level or activity of the compound or enzyme indicated.

Adenyl cyclase, the enzyme which catalyzes formation of cyclic AMP, is present in the plasma membrane of many cells. It is especially abundant in liver cells. The structure of cAMP is shown below.

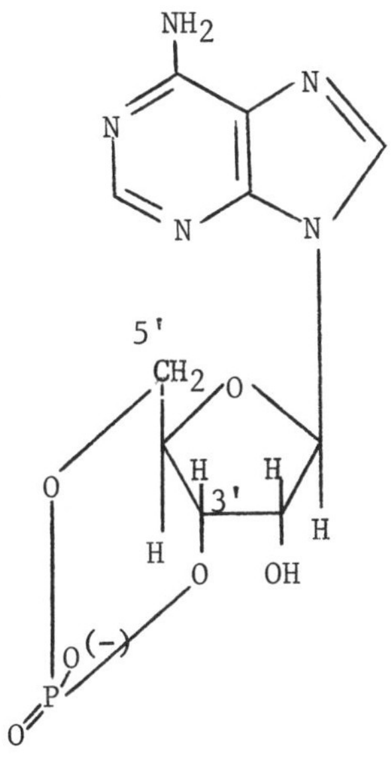

2. Figure 33 illustrates the synthesis and breakdown of cAMP and indicates the general way in which it has been shown to exert its influence on cellular processes.

3. A number of hormones can alter the levels of cAMP, bringing about diverse responses. These responses include substrate mobilization, hormone release, changes in the permeability of the cell membrane, sensory and neural excitation, and melanocyte dispersion. The specificity of the hormone action on a target cell depends upon the presence of unique hormone binding sites (receptor sites) on the cell membrane, as well as the enzymic makeup of the cell. For example, in liver cells both epinephrine and glucagon can increase the level of cAMP, but only epinephrine can bring about such a response in a muscle cell. This is explained on the basis of the absence of receptor sites for glucagon on the muscle cell membrane. In general, it appears that the binding of a hormone molecule to a receptor site sets into action a chain of events which, in one way or another, releases a constraint on the appropriate intracellular systems.

Figure 33.   Mechanism by Which Cyclic AMP Mediates the Effect of Hormones on Intracellular Processes

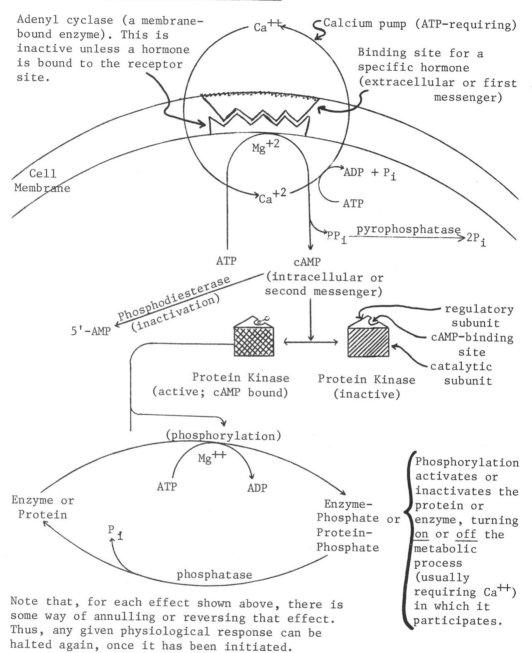

Adenyl cyclase (a membrane-bound enzyme). This is inactive unless a hormone is bound to the receptor site.

Calcium pump (ATP-requiring)

Binding site for a specific hormone (extracellular or first messenger)

Ca^{++}

Cell Membrane

Mg^{+2}

$ADP + P_i$

Ca^{+2}

ATP

PP_i —pyrophosphatase→ $2P_i$

ATP

cAMP (intracellular or second messenger)

5'-AMP ← Phosphodiesterase (inactivation)

regulatory subunit
cAMP-binding site
catalytic subunit

Protein Kinase (active; cAMP bound)

Protein Kinase (inactive)

(phosphorylation)

Mg^{++}

ATP   ADP

Enzyme or Protein

P_i

Enzyme-Phosphate or Protein-Phosphate

phosphatase

Phosphorylation activates or inactivates the protein or enzyme, turning on or off the metabolic process (usually requiring Ca^{++}) in which it participates.

Note that, for each effect shown above, there is some way of annulling or reversing that effect. Thus, any given physiological response can be halted again, once it has been initiated.

4. Until recently, the concept of receptor sites was largely
    hypothetical. Now, however, receptors have been isolated from
    cell membranes and have been shown to bind appropriate hormones
    in appreciable quantities at physiological concentrations. Some
    examples of hormones bound and tissues from which the receptors
    were isolated (in parenthesis) are: ACTH (adrenal cortex);
    vasopressin (kidney); insulin (liver, adipose tissue, lymphocytes,
    fibroblasts); epinephrine (erythrocyte, spleen capsule); and
    glucagon (liver). Norepinephrine has been shown to bind to
    β-adrenergic receptors isolated from cardiac tissue. This
    binding does not activate the adenyl cyclase known to be present,
    however, unless phosphatidyl inositol (known to be present in
    myocardial cell membranes but removed during receptor preparation)
    is added. Diseases associated with hyper- or hyposensitivity to
    specific hormones may be the result of an increase or decrease
    (respectively) in the number of active hormone receptors in the
    affected tissues. In the genetically obese laboratory mouse, a
    marked decrease in insulin-binding sites in both liver and
    adipose tissue has been demonstrated. Other diseases which are
    likely candidates for similar explanations include
    pseudohypoparathyroidism and vasopressin-resistant diabetes
    insipidus. Catecholamine hypersensitivity appears to occur in
    the aging process, certain cases of hypertension, hyperthyroidism,
    and following denervation. These may eventually be related to an
    increase in catecholamine receptor sites.

5. In addition to the control of cAMP levels through the modulation
    of adenyl cyclase activity, the concentration of cAMP may also
    be controlled by variations in the activity of the phosphodiesterase.
    Recall that this enzyme degrades cAMP to 5'-AMP by hydrolysis of
    the 3'-ester bond. The methyl xanthines (caffeine, theophylline,
    and theobromine) are known to inhibit the diesterase. It has
    also been observed in certain tissues (liver, adipose) that
    insulin depresses the level of cAMP; and that in many tissues,
    prostaglandins are antagonists to the action of cAMP although they
    do not necessarily alter the cAMP concentration. The mechanisms
    by which insulin and prostaglandins function are still unclear.
    Prostaglandins have, in fact, been shown to stimulate cAMP formation
    in intact fat pads while inhibiting it in isolated fat cells.

    Besides the central role which cyclic AMP plays in many cellular
    functions of higher organisms, it is also found in unicellular
    organisms (e.g., E. coli and cellular slime molds). In these
    organisms (and in a wide variety of species studied, up to and
    including man), an increase in cAMP synthesis is a general response
    to a decrease in glucose supply (i.e., to hunger). In the slime
    molds, if the glucose level remains low for a sufficient time, the

cAMP eventually causes aggregation of the free-living amoeboid form into a multicellular, sporulating form, better able to survive the food deprivation.

6. In addition to the well-documented role of cyclic AMP, other cyclic nucleotides (particularly cyclic 3',5'-guanylic acid, cGMP) have been implicated in intracellular control processes.

7. In order to understand the action of cyclic AMP, a convenient, sensitive method must be available for measurement of cAMP concentrations. Several have been developed and will be discussed here.

   a. The original method, used by Sutherland (who received the Nobel Prize for his work on cyclic AMP) and his coworkers, makes use of the activation of phosphorylase kinase by cAMP (see the section on glycogenolysis). The rate of phosphorylation of substrate by the kinase is proportional to the cAMP concentration in the sample. Other enzymatic methods have also been developed.

   b. The dilution of a known amount of a radioactive compound by unlabeled molecules of the same compound present in a sample is the basis for both radioimmunoassays (RIA; see also the measurement of insulin levels) and competitive protein-binding assays (CPB assays). Both types of assays are known for their high specificity (measure only the desired compound) and sensitivity (can, for example, measure 1-2 nanomoles or less of cAMP per gram of tissue), and their relative ease of performance. The methods differ only with respect to the source of the protein used to measure the dilution. The general scheme for both of these is shown in Figure 34. Several radioactive atoms have been used as labels. Frequently, ^{125}I, ^{131}I, and ^{3}H are the ones employed. In both methods, a standard curve is prepared and used to calculate the unknown concentrations. Since these are essentially isotope dilution techniques, the concentration of unknown is inversely proportional to the amount of radioactivity bound in the final separation.

      i) The use of radioimmunoassays began in the late 1950's when Berson and Yalow published a method for insulin. Since then, assays have appeared to measure a wide variety of peptide hormones and other, non-protein substances. In the preparation of the antibody needed for the RIA, a small molecule such as cAMP is not antigenic by itself. In order to induce antibody formation in the animal, the small

Figure 34. Flow Chart for Radioimmunoassay and Competitive Protein-
           Binding Assay

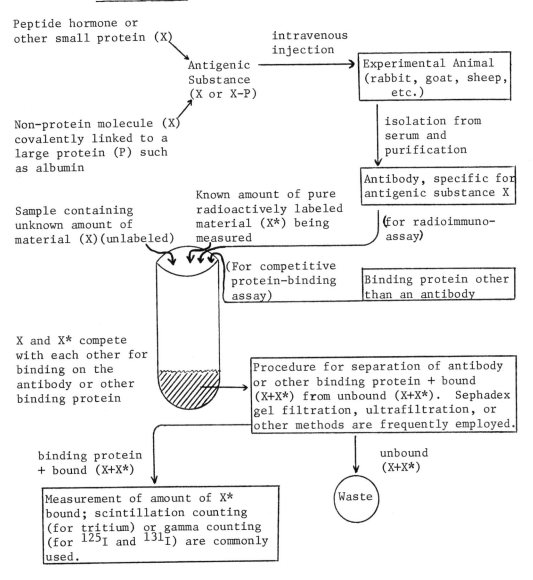

Note: X is the hapten, the part of the antigen molecule for which
       the antibody is specific.

molecule (termed the hapten, the part of an antigen
molecule for which an antibody is specific) must be coupled
to a much larger protein molecule such as albumin. The
nature of the immune response is discussed more fully in
the chapter on immunochemistry.

In the "double-antibody" modification of RIA, a second
antibody is prepared using the first antibody (the one
against X, the substance being measured) as an antigen.
When the two antibodies are employed together, a more
complete separation of free and bound (X+X*) is affected.

ii) Competitive protein-binding assays make use of the ability
of certain proteins and enzymes to bind selectively to
one type of small molecule. For example, one of the
protein kinases involved in the control of glycogen
metabolism is activated by the binding of cAMP. Other
cyclic nucleotides and related compounds have essentially
no effect on the kinase activity. An assay has been
developed based on this selective binding which uses
protein kinase isolated from muscle. The procedure is
essentially identical to the radioimmunoassay for cAMP,
except that the kinase replaces the antibody to cyclic
AMP. Another similar assay has been published using an
uncharacterized cAMP-binding protein fraction from beef
adrenal cortex. The use of biological molecules to
selectively bind other molecules is also the basis of
the purification technique known as <u>affinity
chromatography</u>.

iii) One interesting and potentially important source of
specific binding proteins is hormone receptor sites
isolated from cell membranes. The application of these
proteins to the assay of extracellular (first messenger)
hormones is equivalent to the use of, say, protein
kinase (one of the intracellular receptors for cAMP, an
intracellular hormone or second messenger) for cAMP
assays. The name <u>radioreceptor assays</u> has been coined
for this group of methods. If the receptor proteins can
be highly purified, they should prove to be unusually
specific for the appropriate hormone. To date, clinically
applicable assays of this sort have been published for
ACTH, human chorionic gonadotrophin (HCS), and prolactin.

<u>Glycogenolysis</u> (glycogen breakdown) is catalyzed by the following
enzymes:

A. <u>Glycogen phosphorylase</u>: forms glucose-1-Ⓟ from glycogen by
adding an inorganic phosphate across the α(1→4) linkages;
cannot hydrolyze the three α(1→4) bonds adjacent to an α(1→6)
branch-point.

B. <u>Oligo-α(1→4)→α(1→4) glucan transferase</u>: transfers the three
glucose residues attached to a branch-point residue to the
non-reducing end of the main chain, thus exposing the branch-
point.

C. <u>Debranching enzyme</u> (amylo-(1→6)glucosidase): hydrolyzes the
α(1→6) glucosyl linkages present at branch-points, releasing
one molecule of free glucose per branch-point.

1. These enzymes work in a cyclic manner to completely degrade
glycogen to glucose and glucose-1-Ⓟ. The sequence is A, B, C, A,
B, C, etc. The glucose-1-Ⓟ undergoes the transformations shown
below prior to entering into other metabolic reactions.

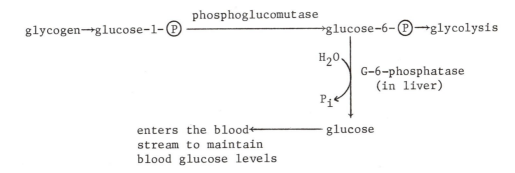

Note that the action of debranching enzyme causes the release of
some free glucose. Thus, it is possible for a small (but not
significant) rise in blood sugar to occur even in the absence of
glucose-6-phosphatase. This happens in von Gierke's disease,
discussed later in this section. The rate-limiting step of
glycogenolysis is the formation of glucose-1-Ⓟ by the action of
glycogen phosphorylase. This enzyme is a key control point and is
discussed later in detail. The phosphorylase found in the liver
differs significantly from that found in muscle.

2. The cyclic functioning of enzymes A, B, and C in glycogen
breakdown is shown in Figure 35. Each circle in the chains
represents a glucose residue. The solid circles are four residues
removed from α(1→6) branch-points and hence are the beginning of

Figure 35.   Action of the Glycogen-Breakdown Enzymes on a Typical (Although Small)
Glycogen Molecule

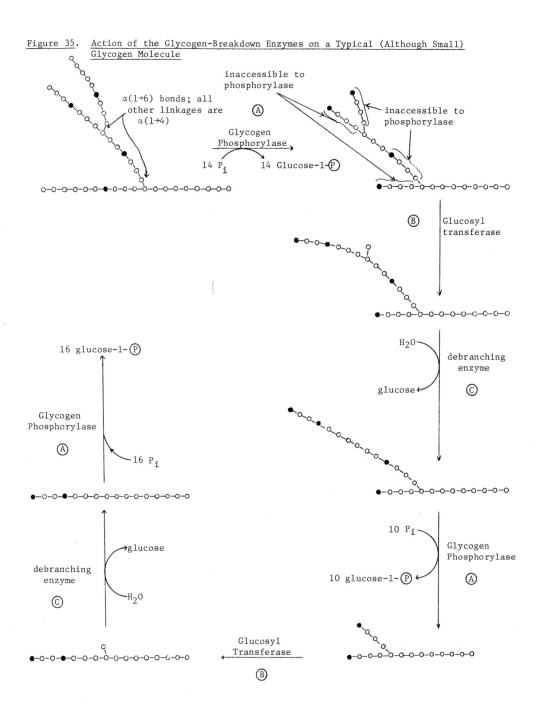

stretches inaccessible to the phosphorylase without prior
transferase action. The overall process shown can be summarized
by the equation

$$(\text{glucose})_{42} + 2\ H_2O + 40\ P_i \longrightarrow 2\ \text{glucose} + 40\ \text{glucose-1-}\textcircled{P}\ .$$

The polymers of glucose which are left at each step, following the
action of the three enzymes are called <u>limit dextrins</u>. A dextrin
is any polymeric product of the incomplete hydrolysis of a
polysaccharide. Since these molecules are degraded as completely
as the specific enzymes are capable, they are limit **dextrins**.
See limit dextrinosis (Cori's disease, type III glycogen storage
disorder) later in this section.

3. Glycogen breakdown is initiated in both liver and muscle by
    glycogen phosphorylase. This enzyme specifically catalyzes the
    phosphorolysis of the glucopyranosyl-$\alpha(1{\rightarrow}4)$-glucopyranose linkages
    in glycogen, releasing glucose-1-$\textcircled{P}$. The phosphorylases from liver
    and muscle are immunologically different, but they are similar in
    a number of other ways.

    Both appear to be allosteric enzymes.

    Pyridoxal phosphate is required as a cofactor for both enzymes.
    In the case of the muscle phosphorylase, there is one
    molecule bound per subunit. The function of the pyridoxal
    phosphate is not known.

    Both enzymes undergo an interconversion between a
    phosphorylated, active form and a dephosphorylated, inactive
    form. The phosphates are attached to seryl or threonyl
    residues by ester bonds, as in the synthetase enzyme.

    Epinephrine causes the activation of both the phosphorylases.
    Its effect is mediated by cyclic AMP.

Of the two enzymes, the muscle phosphorylase has been much more
thoroughly studied.

a. Figure 36 indicates the way in which the activity of muscle
    glycogen phosphorylase is controlled. In this diagram, PCMB =
    p-chloromercuricbenzoate. This is a reagent which reacts with
    sulfhydryl groups, thereby inactivating them (see the section
    on enzyme inhibition by heavy metals). Although it appears that
    at sufficiently high (saturating) 5'-AMP levels, the activities
    of phosphorylases a and b are similar, at the actual substrate
    and 5'-AMP levels found within the cell, control of phosphorylase

Figure 36.   Regulation of Muscle Glycogen Phosphorylase

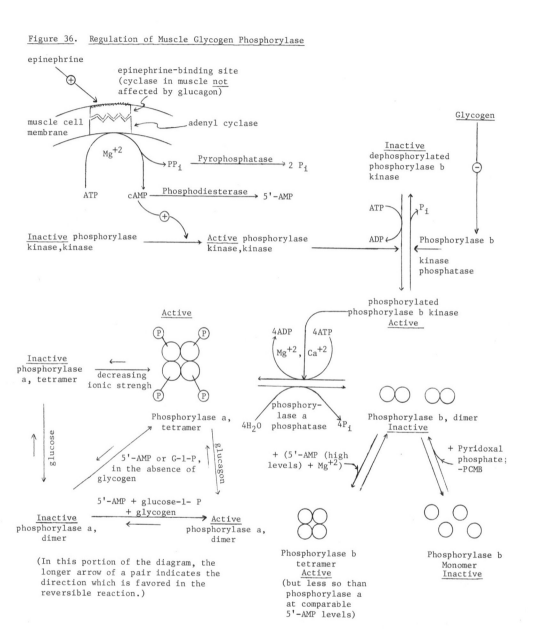

activity results from the a $\rightleftarrows$ b conversion. In addition, in at least some species, the phosphorylase a dimer is more active than the tetramer.

b. As was pointed out above, the mechanism of the liver glycogen phosphorylase system has been studied less thoroughly than that of muscle. Figure 37 illustrates the general pattern, which is quite similar to that in muscle. Particular points to note about the liver system are

   i) The depressive effect of insulin on glycogen breakdown.

  ii) The stimulatory effect of glucagon as well as epinephrine.

 iii) The absence of a phosphorylase kinase,kinase. The cAMP activates the dephosphophosphorylase kinase directly.

  iv) The liver dephosphophosphorylase (analogous to muscle phosphorylase b) is not activated by 5'-AMP.

   v) Although it has not been conclusively demonstrated, liver phosphorylase is probably a dimer in both the phospho- and dephospho-forms.

4. a. Although the control of glycogen synthesis and breakdown appears very complicated, at least part of the difficulty is that some or perhaps all of the effects (allosteric modulators, dissociation of subunits, etc.) observed _in vitro_ probably play a minor role _in vivo_. For example, although it appears that at sufficiently high (saturating) levels of 5'-AMP the activities of muscle phosphorylases a and b are similar, at the actual substrate and 5'-AMP concentration found within the cell, this does not occur. In the body, the important controls seem to be

   i) <u>cAMP</u> stimulates glycogenolysis and inhibits glycogenesis.

  ii) <u>Glycogen</u> accumulation enhances the rate of its own breakdown and inhibits further synthesis.

 iii) Adjustments in the ratios of (phosphorylase a/phosphorylase b) and (synthetase D/synthetase I) are the real sources of control.

b. In muscle, the conversion of phosphorylase b to phosphorylase a may be the mechanism by which glycogenolysis is coupled to muscle contraction. The basis for this lies in the activation of the muscle phosphorylase kinase by Ca^{+2} and the

Figure 37.   Regulation of Liver Glycogen Phosphorylase

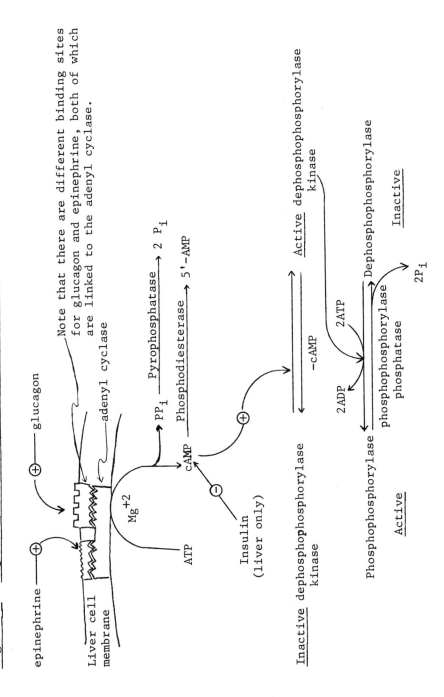

involvement of Ca^{+2} in the a $\rightleftarrows$ b interconversion. At present, this is largely speculation.

5. Liver and kidney have glucose-6-phosphatase which catalyzes the last reaction in glycogenolysis.

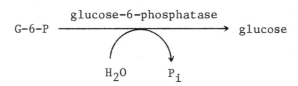

$$G-6-P \xrightarrow{\text{glucose-6-phosphatase}} glucose$$

$$H_2O \qquad P_i$$

Muscle does not have this enzyme. Thus, glucose can be added to the blood by the former two tissues, especially liver, but muscle glycogen <u>does not</u> participate in the maintenance of blood glucose levels. Muscle glycogen is used primarily to supply the energy needs of the muscle.

<u>Glycogen Storage Disorders</u>

These are a group of inherited diseases characterized by the deposition of abnormally large amounts of glycogen in the tissues. These are presented in Tables 6 and 7.

Table 6. Biochemical Lesions Associated with the Glycogen Storage Disorders

Type	Name	Decrease or Deficiency of	Glycogen Structure
I	von Gierke's disease (hepato-renal glycogenosis)	Glucose-6-phosphatase in the cells of liver and of renal convoluted tubules	Normal
II	Pompe's disease (generalized glycogenosis)	Lysosomal α-1,4-glucosidase (acid maltase)	Normal
III	Cori's disease (limit dextrinosis)	Amylo-1,6-glucosidase (debrancher enzyme)	Abnormal; outer chains missing or very short; increased number of branched points
IV	Andersen's disease (brancher deficiency; amylopectinosis)	$(1,4\rightarrow1,6)$-transglucosylase (brancher enzyme)	Abnormal; very long inner and outer unbranched-chains
V	McArdle's Syndrome	Muscle glycogen phosphorylase (myophosphorylase)	Normal
VI	Hers' disease	Liver glycogen phosphorylase (hepatophosphorylase)	Normal
VII	---	Muscle phosphofructokinase	Normal
VIII	---	Hepatic dephosphophosphorylase kinase	Normal

Table 7. Clinical Characteristics of the Glycogen Storage Disorders

Cori-Type	Name	Comments
I	von Gierke's	Hypoglycemia; lack of glycogenolysis under the stimulus of epinephrine or glucagon; ketosis, hyperlipemia; hepatomegaly; autosomal recessive.
II	Pompe's	How this deficiency leads to glycogen storage is not well understood. In some cases, the heart is the main organ involved; in others, the nervous system is severely affected; autosomal recessive.
III	Cori's	Hypoglycemia; diminished hyperglycemic response to epinephrine or glucagon; normal hyperglycemic response to fructose or galactose; autosomal recessive.
IV	Andersen's	Rare, or difficult to recognize; liver cirrhosis and storage of abnormal glycogen; diminished hyperglycemic response to epinephrine; abnormal liver function; autosomal recessive.
V	McArdle's	High muscle glycogen content (2.5 to 4.1% versus 0.2 to 0.9% normal); fall in blood lactate and pyruvate after exercise (normal is sharp rise); normal hyperglycemic response to epinephrine (thus normal hepatic enzyme); autosomal recessive.
VI	Hers'	Not as serious as G-6-phosphatase deficiency; liver cannot make glucose from glycogen, but can make it from pyruvate; mild hypoglycemia and ketosis; hepatomegaly; probably more than one disease (Type VIII, below, was originally part of this group).
VII	---	Shows properties similar to Type V; autosomal recessive.
VIII	---	Hypoglycemia; enhanced liver glycogen content; hepatomegaly; X-linked trait.

Figure 38.   Recapitulation of Glycogenesis and Glycogenolysis

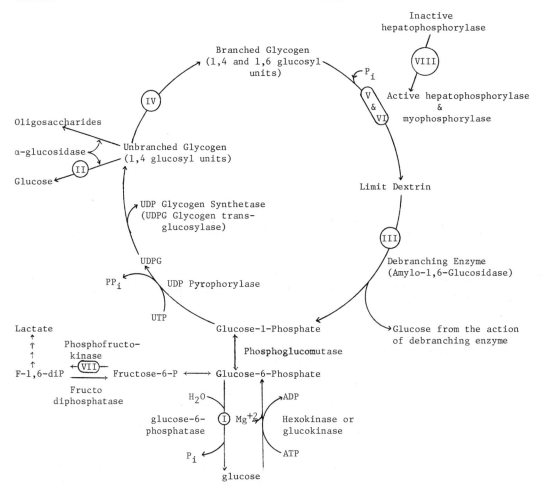

The numbers indicate the Cori-type of the glycogen storage disorder.

Glycogen Storage Disorders

Cori-Type	Name	Deficiency
I	von Gierke's	Glucose-6-phosphatase
II	Pompe's	α-1,4-glucosidase (probably works as shown on diagram)
III	Cori's	Amylo-1,6-glucosidase (debrancher enzyme)
IV	Andersen's	(1,4→1,6)-transglucosylase (brancher enzyme)
V	McArdle's	Myophosphorylase
VI	Hers'	Hepatophosphorylase (+ perhaps others)
VII	---	Phosphofructokinase
VIII	---	Hepatic dephosphophosphorylase kinase

In addition, in the presence of glycolytic enzymes,

$$\text{Gly-3-}\textcircled{P} \xrightleftharpoons{\substack{\text{triosephosphate} \\ \text{isomerase}}} \text{DHAP}$$

$$\text{Gly-3-}\textcircled{P} + \text{DHAP} \xrightleftharpoons{\text{aldolase}} \text{F-1,6-di}\textcircled{P} \xrightarrow{\text{F-6-diphosphatase}} \text{F-6-}\textcircled{P} + \text{P}_i$$

$$\text{F-6-}\textcircled{P} \xrightleftharpoons{\text{hexosephosphate isomerase}} \text{G-6-}\textcircled{P}$$

To obtain two molecules of Gly-3-$\textcircled{P}$ for the synthesis of F-6-$\textcircled{P}$ consider six molecules of glucose:

$$6 \text{ glucose} \longrightarrow 4 \text{ F-6-}\textcircled{P} + 2 \text{ Gly-3-}\textcircled{P} \;\; \left.\begin{array}{c} \\ \searrow \text{ F-6-}\textcircled{P} \end{array}\right\} 6\text{G-6-}\textcircled{P}$$

Doubling the entire pathway, then, one arrives at

$$6 \text{ glucose} + 6 \text{ H}_2\text{O} + 12 \text{ NADP}^+ + 6 \text{ ATP}$$

$$\longrightarrow 5 \text{ G-6-}\textcircled{P} + 6 \text{ CO}_2 + 12 \text{ NADPH} + 12 \text{ H}^+ + 6 \text{ ADP} + \text{P}_i.$$

The <u>net</u> conversion is the oxidation of one mole of glucose to CO_2 and reduced NADP$^+$, with the hydrolysis of one mole of ATP.

$$C_6H_{12}O_6 + 6 \text{ H}_2\text{O} + \text{ATP} + 12 \text{ NADP}^+$$

$$\longrightarrow 6 \text{ CO}_2 + \text{ADP} + \text{P}_i + 12 \text{ NADPH} + 12\text{H}^+$$

The glucose-6-$\textcircled{P}$ formed can reenter the glycolytic pathway.

Comments on the HMP Shunt

1. The reaction of this pathway **is** catalyzed by enzymes found in the extramitochondrial cytoplasm of animal cells. It is therefore termed a soluble (as opposed to membrane-bound) enzyme system.

Figure 39. Hexose Monophosphate (HMP) Shunt (also known as the pentose pathway, direct oxidative pathway, and phosphogluconate oxidative pathway)

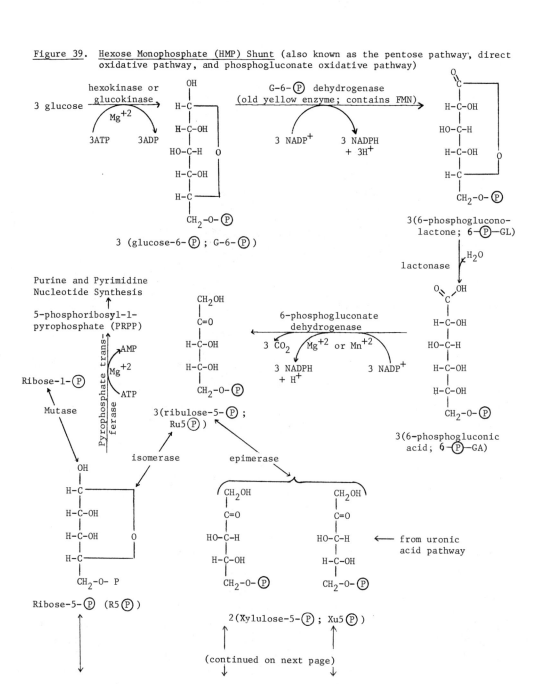

(continued on next page)

Figure 39. Hexose Monophosphate (HMP) Shunt (continued)

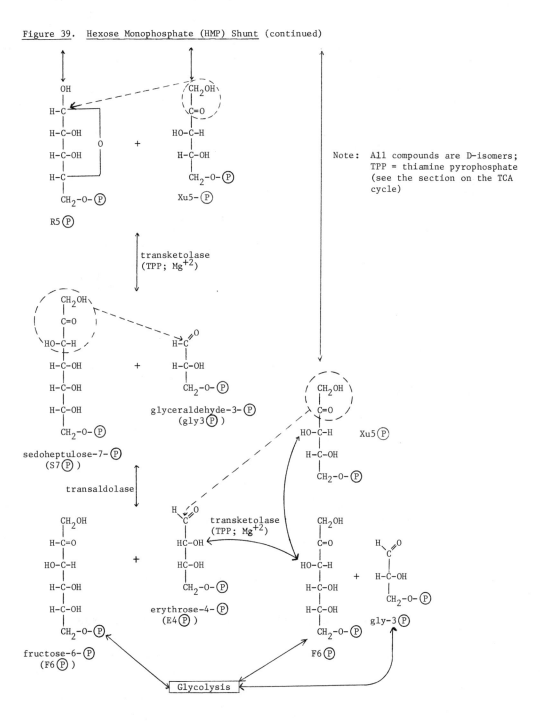

Note: All compounds are D-isomers;
TPP = thiamine pyrophosphate
(see the section on the TCA
cycle)

2. The pentose shunt has been demonstrated in liver, lactating mammary tissue, adipose tissue, leukocytes, testes, and the adrenal cortex.

3. Due to the absence of glucose-6-Ⓟ dehydrogenase, the complete pathway does not function in striated muscle. The 5-phosphoribosyl-1-pyrophosphate (PRPP), which is needed for nucleotide synthesis in the muscle cells, originates from fructose-6-Ⓟ and glyceraldehyde-3-Ⓟ formed in glycolysis. These compounds are converted to ribose-5-Ⓟ by the action of the two transketolases, the transaldolase, and the epimerase and isomerase which are part of the HMP shunt (see nucleotide synthesis).

4. Concerning specific reactions in this sequence

    a. The conversion of 6-phosphogluconic acid to ribulose-5-Ⓟ is an oxidative decarboxylation analagous to those catalyzed by the malic enzyme (malate + NADP$^+$ → pyruvate + CO_2 + NADPH + H$^+$; this is another important source of NADPH) and isocitrate dehydrogenase (see TCA cycle). The reaction presumably involves 6-phospho-3-ketogluconic acid as an enzyme-bound intermediate. The free compound has never been isolated, however.

$$
\begin{array}{l}
\text{COOH} \\
| \\
\text{H-C-OH} \\
| \\
\text{C=O} \\
| \\
\text{H-C-OH} \\
| \\
\text{H-C-OH} \\
| \\
\text{CH}_2\text{-O-}Ⓟ
\end{array}
\qquad \text{6-phospho-3-ketogluconic acid}
$$

    b. The two reactions catalyzed by transketolases are similar to the formation of acetyl-CoA by the oxidative decarboxylation of pyruvic acid. Thiamine pyrophosphate and magnesium are required for all three conversions.

5. This pathway requires only one ATP molecule for each glucose molecule oxidized, and it is independent of the TCA cycle components. The shunt is energetically comparable to glycolysis and the TCA cycle if the NADPH is transhydrogenated to NADH.

$$\text{NAD}^+ + \text{NADPH} \underset{\text{transhydrogenase}}{\rightleftharpoons} \text{NADH} + \text{NADP}^+$$

The NADH can be oxidized via the electron-transport pathway.   For each glucose oxidized, 36 ATP can be produced in this way.   The net result is the formation of 36 - 1 = 35 ATP formed per glucose oxidized in the HMP shunt.   This is slightly less than the 36 or 38 ATP formed during glycolysis and the TCA cycle.

6.   The use of the HMP shunt for glucose oxidation to provide ATP is probably of very minor importance.   The NADPH reducing power must be transferred to the intramitochondrial space, which can be inefficient, depending on the shuttle used (see the section on mitochondria).   The real importance of this pathway is more likely.

   a.   Provision of a mechanism for the formation of pentoses needed in the synthesis of nucleotides.

   b.   A route for the interconversion of pentoses, and of pentoses and hexoses.

   c.   A source for NADPH which serves as a reducing agent in the synthesis of fatty acids, steroids, glutathione (GSH; see G-6-$\textcircled{P}$ dehydrogenase deficiency), etc.   This reducing power is made available to the extramitochondrial part of the cell, where most anabolic (synthetic) processes take place.

7.   It appears that the shunt pathway is stimulated in any tissue which requires NADPH.   Thus, many tissues in which this pathway operates specialize in synthesizing fatty acid or steroids, both NADPH-requiring activities.

Biochemical Lesions of Carbohydrate Metabolism and the Red Blood Cells

1.   Circulating erythrocytes (red blood cells, RBC's) are biconcave discs having no nuclei.   They are made in the bone marrow and have an average circulatory life expectancy of about 120 days.   Their primary functions are the delivery of oxygen to the tissues and participation in the transport of CO_2 to the lungs.

2.   Figure 40 shows the development of mature erythrocytes from the primitive nucleated cells.   The pattern of the intracellular changes which eventually provide the metabolism required for the specialized function of the red blood cell is indicated.   Also included in the diagram is the development of the other cells such as white cells, whose metabolism are discussed later in this section.

Figure 40.   Differentiation of Stem Cells and Characteristics of the Developing Red Blood Cells

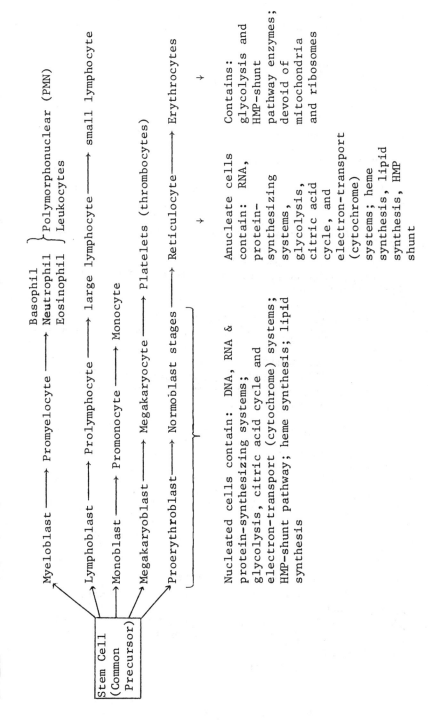

The mature erythrocytes possess   (a) glycolytic pathway to
provide energy and 2,3-diphosphoglycerate (DPG), a key intermediate
which plays a role in oxygen delivery to the tissues (discussed
under hemoglobin); and (b) HMP-shunt pathway to yield NADPH
which helps maintain the -SH groups of sulfhydryl-containing
proteins in a reduced state. Such proteins function as enzymes
(glyceraldehyde-3- Ⓟ dehydrogenase) and as parts of the membrane
system in the erythrocytes.

3.   The normal production and release of erythrocytes (normal
     erythropoiesis) is under the control of a glycoprotein hormone
     (erythropoietin) which stimulates conversion of some stem cells to
     proerythroblasts in anemic or hypoxic situations. The mechanism
     of action of the hormone is presumably through the stimulation of
     mRNA and protein-synthesizing system (discussed later).
     Erythropoietin is produced in the plasma by the action of a
     factor (erythropoietic factor) coming from either kidney or liver
     on a plasma globulin. The synthesis of the factors is increased
     by a variety of conditions and agents such as hypoxia, androgens,
     cobalt salts, etc.

4.   Drugs can cause a decrease in the number of circulating erythrocytes
     either through impaired production (marrow aplasia caused by
     chloramphenicol) or increased destruction of red cells (hemolysis
     caused by a variety of chemical compounds). This occurs
     particularly in individuals with deficiencies of one or more of
     the enzymes associated with carbohydrate metabolism. Following
     are some examples of such instances.

Glucose-6-Phosphate Dehydrogenase Deficiency

1.   When certain drugs are administered to susceptible persons,an
     acute hemolytic anemia results. Some of these drugs are

     a.   8-aminoquinoline-antimalarials, such as primaquine and
          pamaquine (plasmoquine)

$$\text{HN-CH-CH}_2\text{-CH}_2\text{-CH}_2\text{-N} \Big\langle \begin{array}{c} R_1 \\ R_2 \end{array}$$

with CH$_3$ on the CH, H$_3$CO on the quinoline ring, N in ring.

Primaquine: $R_1 = R_2 = -H$

Pamaquine: $R_1 = R_2 = -CH_3CH_2$

b.   sulfonamides

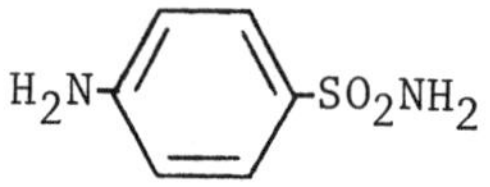

$$H_2N-\!\!\!\langle\bigcirc\rangle\!\!\!-SO_2NH_2$$

sulfanilamide

c.   acetanilide

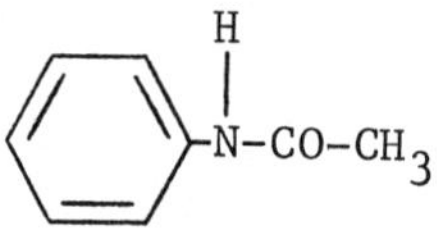

$$\langle\bigcirc\rangle\!\!-\!\!\overset{\overset{H}{|}}{N}\!\!-CO-CH_3$$

d.   phenacetin.

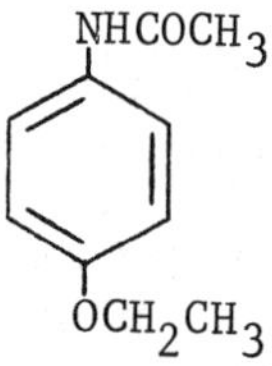

$$\overset{NHCOCH_3}{\underset{OCH_2CH_3}{\langle\bigcirc\rangle}}$$

e.   furadantin and others

f.   ingestion of <u>vicia fava</u> beans by Caucasians of Mediterranean
origin (Favism).

2.   If primaquine is administered daily to sensitive persons an acute
hemolytic crisis develops, beginning on the second or third day.
From 30 to 50% of the patient's RBC's are destroyed and the
patient becomes jaundiced with dark, often black urine. Older
cells (63-76 days old) are destroyed while younger cells (8-21 days)
are not affected. If the patient survives the crisis, recovery
takes place as new cells become preponderant, despite continued
administration of primaquine.

3.   Susceptibility to drug-induced hemolytic disease may be due to a
hereditary deficiency of G-6-(P) dehydrogenase activity in the
erythrocyte, as explained next . Normally the following sequence
of reactions occurs.

a. G-6-Ⓟ dehydrogenase catalyzes the reaction

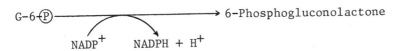

G-6-Ⓟ ⟶ 6-Phosphogluconolactone

NADP$^+$ NADPH + H$^+$

b. The NADPH formed is used in the conversion of oxidized glutathione (GSSG) to reduced glutathione (GSH).

$$\text{GSSG + NADPH + H}^+ \xrightarrow{\text{GSH reductase}} \text{2 GSH + NADP}^+$$

oxidized glutathione   reduced glutathione

(GSSG is actually a dimer of GSH, the molecules being linked by a disulfide bond). GSH plays a very important role in red cell metabolism, by "protecting" the −SH groups of different substances, such as hemoglobin, catalase, and the lipoprotein of cell membranes. Since GSH is more readily oxidized than the protein sulfhydryls, it serves as a source of oxidizing power to reduce disulfide bonds which may be deleterious to normal cellular function. Glutathione also helps by destroying oxidizing agents which may be present in the cell. In the red cells, the concentration of GSH is 70 mg per 100 cc of red cells. A deficiency of GSH can give rise to hemolysis. It is not yet known exactly how drugs like primaquine affect the G-6-Ⓟ dehydrogenase-deficient red cells to cause hemolysis. Some of these interrelationships are shown in Figure 41.

For an explanation of the numbers in Figure 41, see page 252.
Abbreviations

G-6-Ⓟ = glucose-6-phosphate

Gly-3-Ⓟ = glyceraldehyde-3-phosphate

DPG(1,3;2,3) = diphosphoglycerate

3-Ⓟ GA = 3-phosphoglyceric acid

6-Ⓟ GL = 6-phosphogluconolactone

6-Ⓟ GLD, G-6-Ⓟ D, Gly -3-Ⓟ D = the respective dehydrogenases

GSH; GSSG = reduced and oxidized glutathione, respectively.

Figure 41.   Interrelationships of Carbohydrate Metabolism in the Erythrocyte

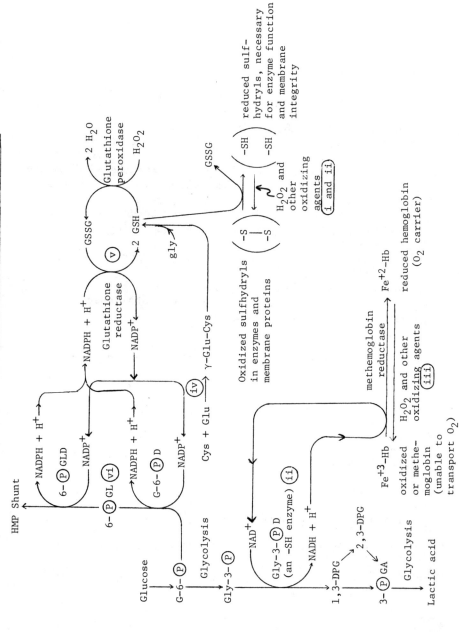

For a key to the numbers (i-vi), see page 252.   The abbreviations used in this figure are explained on page 249.

4. Structure and biosynthesis of glutathione

i) Structure

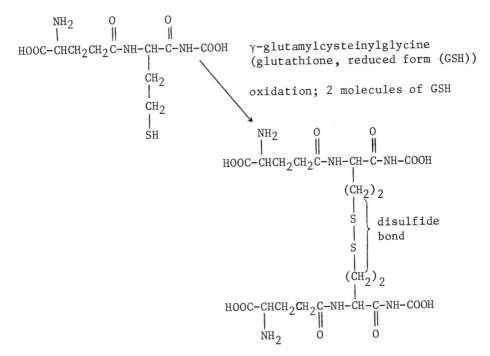

γ-glutamylcysteinylglycine
(glutathione, reduced form (GSH))

oxidation; 2 molecules of GSH

ii) Synthesis

It is important to realize that the synthesis of this tripeptide takes place in the <u>absence</u> of the normal protein-synthesizing systems (involving mRNA, tRNA, ribosomes and appropriate enzyme systems, discussed later). Glutathione can consequently be formed even in the anucleate, mature erythrocyte.

L-Cys + L-Glu $\xrightarrow[\text{Mg}^{+2}]{\text{γ-GC Synthetase}}$ γ-Glu-Cys

ATP   ADP + P$_i$

ATP

Gly   Mg^{+2};
        GSH Synthetase

ADP + P$_i$

γ-glutamylcysteinylglycine
(γ-Glu-Cys-Gly; reduced glutathione)

5. There are several factors that may contribute to the hemolysis (the numbers below refer to Figure 41). These include

    i) Damage to the cell membrane when sulfhydryl groups are oxidized (a process reversed by GSH).

    ii) Interruption of normal pathways of metabolism inside the cell (Gly-3-$\textcircled{P}$ dehydrogenase, for example, is an important enzyme that requires GSH).

    iii) Degradation of hemoglobin, and the shifting of the equilibrium between oxyhemoglobin and methemoglobin (which can be influenced by a lack of NADH or by the enzyme glutathione peroxidase that requires GSH to degrade H_2O_2).

    iv) Biochemical lesions in the synthesis of GSH.

    v) Glutathione reductase is a flavin enzyme. Deficiency of the vitamin riboflavin results in lowering of the reductase activity.

    vi) Other deficiencies in the production of NADPH.

The above factors are all interrelated. For example, a lack of Gly-3-$\textcircled{P}$ dehydrogenase would cause both an interruption of glycolysis and a decrease in NADH. The lack of NADH may result in an increase in methemoglobin due to a decrease in the reaction

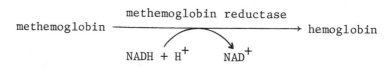

which requires NADH. In addition, the oxyhemoglobin-deoxyhemoglobin equilibrium would be affected, with an increase in the level of oxyhemoglobin at a given oxygen pressure. This would occur because of a deficiency in 2,3-DPG, which lowers the oxygen affinity of hemoglobin by an allosteric mechanism. It is formed from 1,3-DPG, the product of Gly-3-$\textcircled{P}$ dehydrogenase.

$$\text{Gly-3-}\textcircled{P} \longrightarrow \text{1,3-DPG} \longrightarrow \text{2,3-DPG}$$

Both of these topics are discussed further in the section on hemoglobin. Notice that the general pattern in the diagram is one in which harmful changes are brought about by H_2O_2 and other oxidizing agents, both exogenous (e.g., nitrites) and endogenous.

These are prevented or reversed by reducing equivalents derived from carbohydrate oxidation. The reductions are mediated by pyridine nucleotides, glutathione, and specific reductase and dehydrogenase enzymes.

6. Several different types of glucose-6-(P) dehydrogenase deficiencies are known: Negro, non-Negro, Northern Italian, Chicago, Oklahoma and Seattle.

7. G-6-(P) dehydrogenase deficiency can be detected by an NADP-linked assay

    Primaquine sensitivity can be detected in three ways.

    a. Quantitative determination of G-6-(P) dehydrogenase in red cells.

    b. GSH determination after aerobic incubation of cells with acetyl phenylhydrazine (which oxidizes GSH to GSSG).

    c. Test for the reduction of methylene blue by the methemoglobin formed upon the addition of $NaNO_2$ to the red cells (qualitative procedure). In G-6-(P) dehydrogenase deficiency, the methemoglobin fails to be reduced and its brown color can be recognized visually.

8. G-6-(P) dehydrogenase deficiency is inherited as a sex-linked trait and is partially dominant.

    $\overline{X}Y$ = full expression (where $\overline{}$ indicates an X-chromosome carrying the deficiency)

    $\overline{X}\,\overline{X}$ = full expression

    $\overline{X}X$ = heterozygote (not affected; such persons have two populations of red cells; a. cells with normal enzyme activity and b. cells with G-6-PD deficiency).

9. The geographic distribution of G-6-(P) dehydrogenase deficiency gene could be due to the fact that the condition may confer some protection against malaria. Malarial parasites (Falciparum malaria) require GSH and hence, the HMP-shunt pathway for growth. The G-6-(P) dehydrogenase heterozygotes have enough GSH to carry on metabolism but not enough for the parasites, so the parasites cannot survive. This is similar to the protection conferred by the sickle cell-trait (see hemoglobin diseases).

10. It has been reported that the red blood cells obtained from selenium deficient rats show reduced activity of glutathione peroxidase. Apparently the enzyme contains high levels of selenium. Note, however, that selenium in higher concentration is toxic to most organisms.

Carbohydrate Metabolism in Polymorphonuclear (PMN) Leukocytes

1. An important function of leukocytes is phagocytosis - a process which involves the engulfing and destroying of particulate material (e.g., bacteria). During phagocytosis there is a significant increase in oxygen consumption, glycogenolysis, oxidation of glucose through the HMP-shunt pathway, and glycolysis (indicated by excess lactate production).

2. A suggested mechanism for the bactericidal action of PMN leukocytes is the production of hydrogen peroxide (H_2O_2) which performs the intracellular killing of bacteria. The H_2O_2 is produced as follows:

   a. NADPH synthesis in the first two steps of the HMP-shunt pathway.

   b. $NADPH + H^+ + O_2 \xrightarrow{\text{NADPH oxidase}} NADP^+ + H_2O_2$

3. Some of the evidence for the postulated role of H_2O_2 in the intracellular killing of bacteria by PMN leukocytes is given below.

   a. The H_2O_2 concentration increases during phagocytosis along with an increase in HMP-shunt pathway activity and oxygen consumption.

   b. Increased NADPH oxidase activity is observed during phagocytosis.

   c. Intracellular killing is impaired when oxygen is excluded (anaerobic conditions). This result is not obtained with all bacteria studied, however, suggesting the presence of preformed H_2O_2 and/or the action of other bactericidal agents.

   d. Enhanced killing in the presence of added (exogenous) H_2O_2.

   e. Impairment of killing activity by the addition of catalase. This enzyme breaks down H_2O_2 very rapidly by catalyzing the reaction.

   $$2\ H_2O_2 \xrightarrow{\text{catalase}} 2\ H_2O + O_2$$

4. It is important to point out that the action of H_2O_2 in the intracellular killing process does not eliminate the presence of other antibacterial factors such as myeloperoxidase, the lysosomal enzyme system, etc. This is discussed further in the chapter on immunochemistry.

5. In a metabolic abnormality known as <u>chronic granulomatous disease</u> (CGD) (discussed further in the chapter on immunochemistry), the patient's leukocytes fail to produce hydrogen peroxide. The leukocytes are capable of engulfing the bacteria but are unable to perform the killing process. These patients are subject to recurrent suppurative infections. The biochemical lesions are related to <u>HMP-shunt pathway</u> activity as well as to <u>NADPH oxidase</u> <u>systems</u>. CGD is transmitted as an X-linked recessive trait. Other diseases with deficiencies in bactericidal activity have been observed. In some of these disorders, lesions of glutathione peroxidase and myeloperoxidase have been reported. Note: Both peroxidases and catalase have H_2O_2 as a substrate and water as a product. Catalase uses another molecule of H_2O_2 as the reducing agent while peroxidases use a number of other compounds (such as GSH) as a source of electrons. Examples of the reactions catalyzed are

$$2 \ H_2O_2 \xrightarrow{\text{catalase}} 2 \ H_2O + O_2$$

$$2 \ GSH + H_2O_2 \xrightarrow{\substack{\text{glutathione} \\ \text{peroxidase}}} G\text{-}S\text{-}S\text{-}G + 2 \ H_2O.$$

Figure 42.   Uronic Acid Pathway (Glucuronate Pathway)

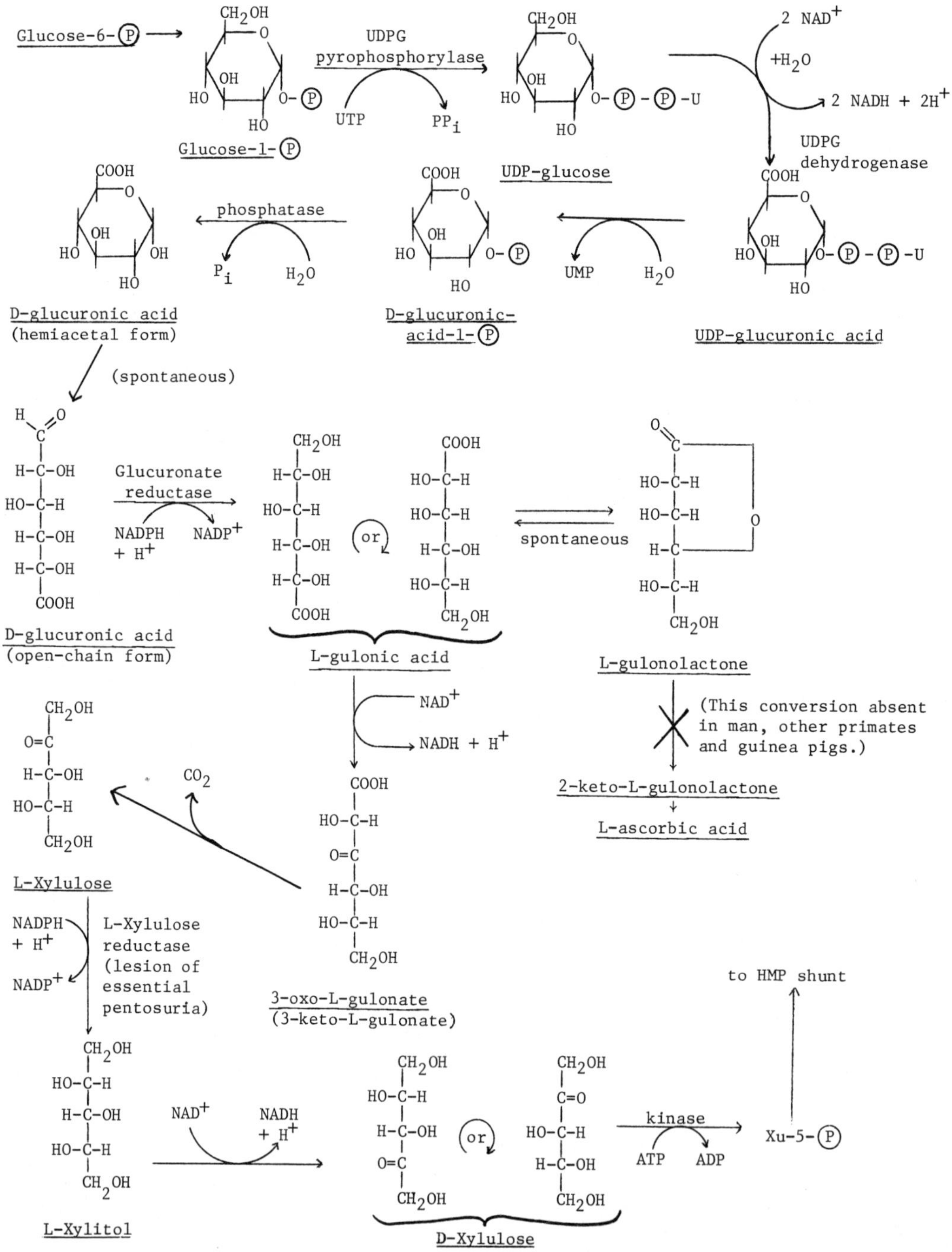

Glucose-6-(P) →

CH₂OH

Glucose-1-(P)

UDPG
pyrophosphorylase

UTP   PPᵢ

UDP-glucose

2 NAD⁺
+H₂O

2 NADH + 2H⁺

UDPG
dehydrogenase

COOH

O-(P)-(P)-U

UDP-glucuronic acid

phosphatase

Pᵢ   H₂O

COOH

O-(P)

D-glucuronic-
acid-1-(P)

UMP   H₂O

COOH

D-glucuronic acid
(hemiacetal form)

(spontaneous)

D-glucuronic acid
(open-chain form)

Glucuronate
reductase

NADPH   NADP⁺
+ H⁺

(or)

L-gulonic acid

spontaneous

L-gulonolactone

(This conversion absent
in man, other primates
and guinea pigs.)

2-keto-L-gulonolactone
↓
L-ascorbic acid

NAD⁺

NADH + H⁺

3-oxo-L-gulonate
(3-keto-L-gulonate)

CO₂

L-Xylulose

NADPH
+ H⁺

NADP⁺

L-Xylulose
reductase
(lesion of
essential
pentosuria)

L-Xylitol

NAD⁺   NADH
+ H⁺

(or)

D-Xylulose

kinase

ATP   ADP

Xu-5-(P)

to HMP shunt

Comments on the Uronic Acid Pathway

1. These reactions occur in the tissues of animals and higher plants. In man and other primates, and guinea pigs, the enzyme is absent which converts L-gulonolactone to 2-keto-L-gulonate. The lack of this pathway makes ascorbic acid a vitamin for these species.

2. A known biochemical lesion of this pathway is the lack of L-xylulose dehydrogenase. Since this enzyme catalyzes the conversion of L-xylulose to xylitol, its absence causes an accumulation of L-xylulose, which appears in the urine. The disease is known as essential pentosuria (see below).

3. It is of interest to point out that the two oxidation steps require NAD$^+$, whereas the two reductions require NADPH. It is a normal occurrence that whenever a reducing process (anabolic reaction) is required, NADPH is the preferred cofactor.

4. A key intermediate that is formed in the pathway is UDP-glucuronic acid. Its functions are as follows:

    a. UDP-glucuronic acid is "active" glucuronic acid. It participates in the incorporation of glucuronic acid into chondroitin sulfate and other polysaccharides.

    b. Glucuronic acid conjugation with steroids, certain drugs, bilirubin, etc. is important for purposes of detoxification.

        Example:

        Bilirubin + 2 UDPG $\xrightarrow{\text{glucuronyl transferase}}$ Bilirubin diglucuronide + 2 UDP

        (The reactions with bilirubin will be discussed later.) The reaction of the conjugating compounds with glucuronic acid are of two chemical types: glucosidic (ether) and ester.

Glucosidic (Ether) Type

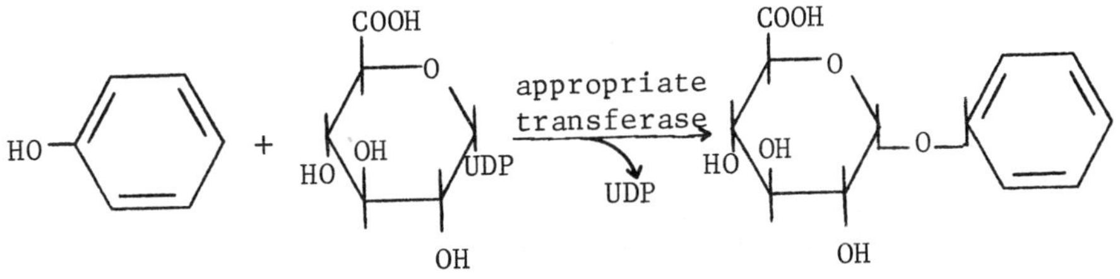

Phenol   UDP-β-glucuronate   phenylglucuronide

Ester Type (as in bilirubin conjugates)

Benzoic Acid

β-glucuronic acid
monobenzoate

Pentosurias

This is a group of diseases characterized by the urinary elevation of
one or more of the pentoses. Normally present in the urine are
L-xylulose (up to 60 mg/24 hrs), D-ribose (up to 15 mg/24 hrs),and
D-ribulose (traces).

1. **Essential Pentosuria** is characterized by excretion of excessive
    amounts of the pentose L-xylulose (1-4 gm/24 hrs in the urine).

    a. No apparent disturbance in physiological function has been
        demonstrated (i.e.,the condition is innocuous).

    b. It is often mistakenly diagnosed as diabetes mellitus (due to
        the presence of a reducing substance in the urine).

    c. It is a recessively inherited biochemical disorder occurring
        in persons of all age groups.

    d. The biochemical lesion is a deficiency of L-xylulose
        dehydrogenase.

    e. Diagnostic measures are indicated below. One might suspect
        this disease if

        i) reducing substances are present in the urine without the
            other symptoms of diabetes mellitus,

        ii) negative results are found when glucose is tested for by
            any of the enzyme methods specific for glucose.

    f. More specific methods include

        i) Reduction of Benedict's reagent at low temperature:
            L-xylulose is a strong reducing agent in contrast to glucose

and most other urinary sugars. Benedict's reagent is reduced by L-xylulose at 55°C in 10 minutes or at room temperature in 3 hours. (Caution: Fructose will also reduce this reagent at low temperature.)

   ii)   Paper chromatography: Using as solvent n-butanol, ethanol, and water (50:10:40), L-xylulose has an $R_f = 0.26$ which exceeds all other commonly observed urinary sugars. It also gives a red color when treated with orcinol-trichloroacetic acid reagent. This is probably the most convenient method.

   iii)   Determination of the melting point of the phenylosazone derivative of the sugar.

2.   Other types of pentosuria include alimentary pentosuria (L-arabinose and L-xylose are found in the urine following ingestion of large quantities of fruit) and ribosuria (elevation of urinary D-ribose levels associated with muscular dystrophy).

Fructose Metabolism

1.   Fructose (a ketohexose) is widely distributed in plants and is an important source of dietary carbohydrate, accounting for 1/6 to 1/3 of the total carbohydrate intake. It is found as free fructose and, linked to glucose, occurs as sucrose. Inulin, a polymer of fructose, is present in chicory and sweet potato. Since inulin is relatively resistant to hydrolysis in the intestine, this is an unimportant source of fructose.

2.   Fructose is present in significant quantities in prostate and seminal fluid. It is presumably synthesized in the prostate gland by way of the following reactions:

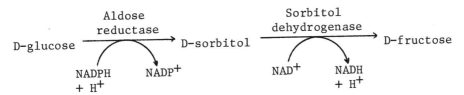

For other aspects of this conversion, see galactose metabolism and diabetes.

3.   For the detection of fructose, glucose is first removed with the enzyme glucose oxidase. Fructose is then determined either by chromatographic or colorimetric methods.

4. Fructose can enter the Embden-Meyerhof Pathway by several routes. Notice in particular that fructokinase forms the 1-Ⓟ, unlike hexokinase and glucokinase.

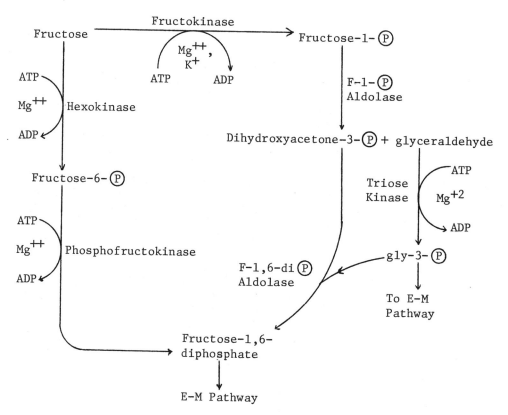

5. In essential fructosuria

    a. Fructose is found in the urine, due to a lack of the enzyme fructokinase.

    b. The condition is harmless and asymptomatic.

    c. Glucose and galactose metabolism are normal.

    d. Insulin has no effect on fructose levels.

    e. The disorder is inherited as an autosomal recessive trait.

6. Hereditary fructose intolerance

    a. Occurs in infants during the weaning period. Symptoms appear
        only after ingestion of fructose. Vomiting occurs shortly
        after fructose-containing foods are eaten.

    b. Occurrence in children gives rise to jaundice, vomiting,
        hyperbilirubinemia, albuminuria, aminoaciduria, hepatomegaly
        (liver enlargement), and failure to thrive.

    c. A strong aversion for fruits and sweets is characteristic.

    d. Fructose is found in the urine only after fructose ingestion.
        High blood fructose levels are reached during oral fructose
        tolerance tests.

    e. The rise in fructose level is accompanied by a fall of blood
        glucose to severely hypoglycemic levels. Glucose and galactose
        tolerance tests are normal, however.

    f. The primary enzyme defect is a deficiency of liver F-1-(P)
        aldolase, which catalyzes the reaction

        F-1-(P) ──────────→ D-glyceraldehyde + DHAP

        Although this enzyme is virtually absent in the disorder,
        F-1,6-di(P) aldolase activity is normal.

    g. Fructose-induced hypoglycemia is probably caused by a block
        in the release of glucose from the liver resulting in ATP
        depletion.

Galactose Metabolism

1. The principle dietary source of galactose is the disaccharide lactose
    (β-D-galactopyranosyl-(1$\rightarrow$4)-β-D-glucopyranose).

    This molecule is hydrolyzed to galactose and glucose by the enzyme
    lactase, found in the intestinal microvillae.

    lactose ──lactase──→ galactose + glucose
              ↗
            H_2O

Although most children possess this enzyme, it frequently disappears as they mature if they belong to certain susceptible racial groups. For example, Orientals and African and American Negroes frequently display lactose intolerance whereas, among Northern European and American whites, intolerance to the disaccharide is uncommon, even in adults. This is perhaps due to the fact that milk is a major component in the diet of adults in the latter groups. Further evidence that intolerance is related to dietary habits has been obtained by a comparison of neighboring tribal groups in Nigeria. It has been found that those with a pastoral lifestyle are generally (greater than 75%) lactose-tolerant as adults, while groups which are agrarian show a high frequency (70-98%) of intolerance. The symptoms of lactose intolerance (watery, explosive diarrhea, cramps, belching, and flatulence) result from undigested lactose passing into the large intestine. Once there, it increases the osmolarity of the feces and hence the water content. In addition, it is fermented by bacterial action, producing organic acids and carbon dioxide.

2.   After the release of galactose from lactose, the following reactions provide for its entry into glycolysis.

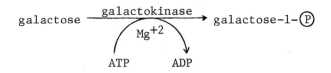

galactose-1-Ⓟ + UDP-glucose $\xrightarrow[\text{(block in galactosemia)}]{\substack{\text{Galactose-1- Ⓟ}\\ \text{uridyl}\\ \text{transferase}}}$ UDP-galactose + glucose-1-Ⓟ

UDP-galactose $\underset{}{\overset{\substack{\text{UDP-Galactose-}\\ \text{4-Epimerase}}}{\longleftrightarrow}}$ UDP-glucose $\longrightarrow$ glycogen $\longleftrightarrow$ glycolysis

3.   Galactosemia is the inability to metabolize galactose.

   a.   It is a rare congenital disease of infants inherited as an autosomal recessive trait.

b. The biochemical lesion is a lack of galactose-1-Ⓟ uridyl transferase. This results in the accumulation of galactose-1-Ⓟ.

c. Clinical manifestations include nutritional failure (infants lose weight), hepatosplenomegaly, mental retardation, jaundice, and the development of cataracts. Since the lens tissue is freely permeable to galactose but relatively impermeable to galactitol (the polyalcohol derived from this sugar), the last symptom may be due to the formation of galactitol by the reaction below.

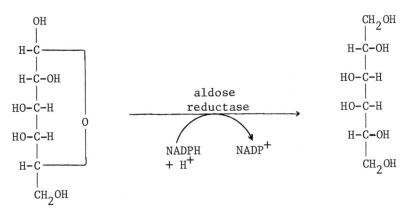

D-galactose   D-galactitol

(See also fructose metabolism and diabetes). The accumulation of the alcohol causes an increase in the osmolarity and water uptake of the aqueous humor, resulting in swelling of the lens. Myopia and tissue damage occur, leading ultimately to the opacity associated with cataracts. It is not clear whether the other symptoms also result from galactitol accumulation, from galactose-1-Ⓟ accumulation, or from some other aspect of the impaired galactose metabolism.

d. Laboratory findings in galactosemia include galactosuria, amino aciduria albuminuria, ketonuria, and impaired galactose tolerance (abnormal galactose tolerance test).

e. The enzyme UDP-galactose pyrophosphorylase has been identified in human liver. It catalyzes the reaction

Galactose-1-Ⓟ + UTP $\longrightarrow$ UDP-galactose + PP$_i$

Its activity is probably less than 1% of that of the transferase enzyme, however, so its role in galactose metabolism appears to be minor.

f. A good diagnostic test for this disorder is the measurement of galactose-1-(P) uridyl transferase enzyme activity in red cells. If the diagnosis is made early and milk (lactose) and other galactose sources are withdrawn from the diet, the symptoms and findings recede and normal development appears to ensue. Even when early postnatal diagnosis is made, however, there may be some prenatal damage due to exposure of the fetus to galactose ingested by the mother.

4. Galactokinase Deficiency. Hypergalactosemia and galactosuria have been observed in a few individuals due to the deficiency of galactokinase, the enzyme which catalyzes the reaction

$$\text{Galactose} + \text{ATP} \longrightarrow \text{Galactose-1-} \textcircled{P} + \text{ADP}$$

Unlike galactosemic patients, the individuals with galactokinase deficiency do not develop hepatic and renal problems. Instead, they showed the formation of cataracts at an early age (less than one year after birth) possibly due to the mechanism discussed under galactosemia. Increased incidence of cataracts was also observed in heterozygous individuals, who also exhibited a mild form of galactose intolerance.

5. Galactose is necessary for the synthesis of a number of biomolecules. These include lactose (in the lactating mammary gland; discussed below), glycolipids, cerebrosides, chondromucoids, and mucoproteins. A galactosemic individual is able to synthesize these compounds, even on a galactose-free diet, because of the reversibility of the UDP-galactose-4-epimerase reaction. The UDP-glucose needed for this pathway is readily synthesized from glucose-1-(P) and UTP in the reaction catalyzed by UDP-glucose-pyrophosphorylase.

6. Lactose (milk sugar) is synthesized by the following reaction

$$\text{UDP-galactose} + \text{glucose} \xrightarrow[\text{(+}\alpha\text{-lactalbumin)}]{\substack{\text{galactosyl} \\ \text{transferase}}} \text{lactose} + \text{UDP}$$

Galactosyl transferase (also called lactose synthetase) requires alpha-lactalbumin, a protein found in milk, as a cofactor. In the absence of α-lactalbumin, galactosyl transferase catalyzes the reaction

$$\text{UDP-galactose} + \text{N-acetylglucosamine} \xrightarrow{\substack{\text{galactosyl} \\ \text{transferase}}} \text{N-acetyllactosamine} + \text{UDP.}$$

Alpha-lactalbumin synthesis is initiated late in pregnancy by a
decrease in the progesterone level.

Metabolism of Amino Sugars

1. These compounds are important components in many complex
    polysaccharides. They are frequently found in glyco- and
    mucoproteins and lipopolysaccharides. The structures of two
    representative compounds are given below.

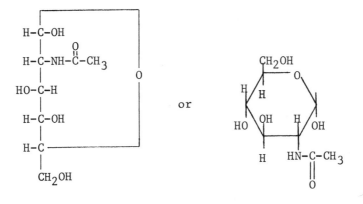

D-glucosamine-6-(P)

N-acetylglucosamine

2. Some of the pathways which involve these amino sugars are shown
    in Figure 43.

Figure 43.   Metabolism of Amino Sugars

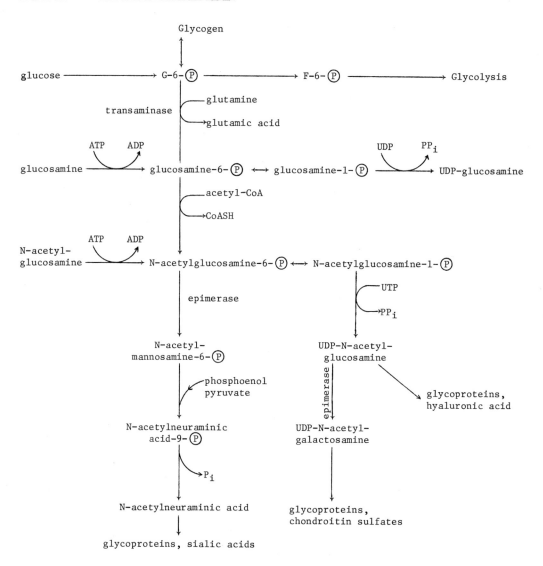

Blood Glucose

1. At any given time, the blood glucose level is dependent upon the
    amount of glucose <u>entering</u> the blood from dietary sources,
    glycogenolysis and gluconeogenesis and the amount that is being
    <u>removed</u> by oxidative and biosynthetic processes. These are
    illustrated in the following diagram.

Figure 44.   <u>Factors Which Affect Blood Glucose Levels</u>

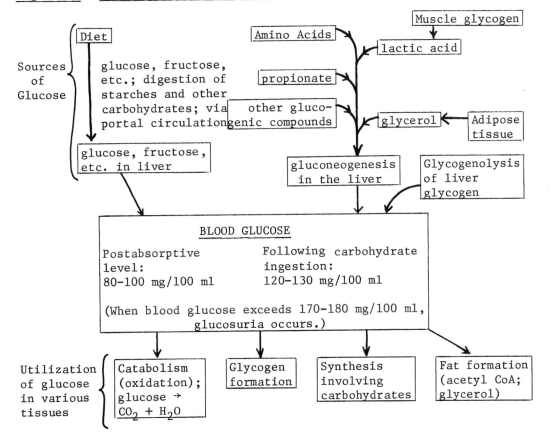

2. The addition of glucose to the blood stream through hepatic
    glycogenolysis can make only a minor contribution to the blood
    glucose level. This is due to the limited supply of glycogen.
    A liver weighing about 1500 grams can provide only about 45-50 grams
    (3-4%) of glucose. This is probably most useful for short periods
    of time in emergency situations. Note that epinephrine and
    glucagon stimulate hepatic glycogenolysis (previously discussed
    under glycogen). In conditions where there is sustained lack of
    glucose, however, (e.g., starvation) gluconeogenesis plays a
    principal role in providing blood glucose. The muscle glycogen
    cannot furnish glucose to be added to the blood stream. This is
    due to the absence in muscle tissue of the enzyme glucose-6-
    phosphatase, which catalyzes the cleavage of G-6-Ⓟ to free glucose
    and phosphate. The muscle also lacks some of the transaminases
    needed to convert amino acids to keto acids which then can be used
    for gluconeogenesis.

3. In the resting body, the brain consumes about two-thirds of the
    blood glucose. Most of the other one-third is used by the
    erythrocytes and skeletal muscle. The brain <u>must</u> have an energy
    source or behavioral changes, confusion, and coma result. In
    prolonged starvation, the body meets this need in two stages.
    Initially, muscle protein is broken down and the amino acids are
    used to synthesize glucose (gluconeogenesis). In particular,
    alanine is released from the muscle protein and carried in the
    blood to the liver and kidney where gluconeogenesis occurs. This
    is known as the <u>alanine cycle</u> and is similar to the Cori cycle for
    lactate. Both pathways simply recycle a fixed amount of glucose.
    The steps involved are indicated in Figure 45.

Figure 45.   The Alanine Cycle

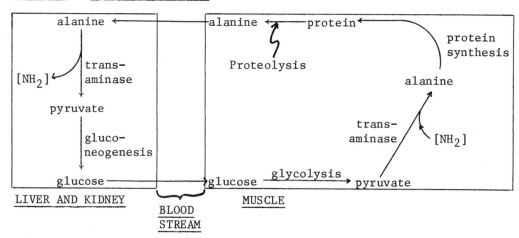

This is a heavy drain on body protein, however, and if the glucose supply remains low, the brain adapts to the use of ketone bodies (acetoacetic acid, acetone, and β-hydroxybutyric acid) as a source of energy. The most important of these three are acetoacetic acid and β-hydroxybutyric acid. The synthesis of these compounds from acetyl-CoA is discussed in the section on ketosis in lipid metabolism. Normally the body produces only small amounts of the ketone bodies. In the absence of glycolysis, acetyl-CoA is primarily derived from fats, however, so ketone body synthesis directly relieves the need to catabolize muscle protein for energy, drawing upon the adipose tissue instead. It has been shown that the brain possesses the enzymes necessary for utilizing this new energy source. The mechanism by which the body adapts to ketone body synthesis is not yet known, but it is probable that hormones (insulin: lower than normal during starvation; glucagon: higher than normal, relative to insulin, during starvation) are involved.

4. A number of hormones influence the blood glucose levels. Briefly, epinephrine, glucagon, ACTH, cortisol, and growth hormone tend to raise blood glucose levels (hyperglycemic substances) while insulin lowers the levels (hypoglycemic hormone). These are discussed more fully in the section on diabetes.

Relations of the Endocrine Glands to Carbohydrate Metabolism

1. The major control of carbohydrate metabolism is affected through hormones excreted by tissues, especially the pancreas, anterior pituitary (adenohypophysis), and adrenal cortex.

2. Each endocrine organ generally secretes more than one hormone and these often exert apparently opposing effects.

3. The adenohypophysis produces not only hormones having a direct effect, but also, through tropic hormones, regulates the other endocrine organs (see Chapter 13).

```
                        hormones → direct effect on carbohydrate metabolism
Adenohypophysis
                        tropic hormones
                                   ↘
                                    other endocrine organs
                                              ↘
                                               hormone
                                                   ↘
                                                    effect on carbohydrate metabolism
```

Some authors use the term trophic (nourishing; functioning in nutrition) rather than tropic (bringing about a change) for those hormones which regulate the levels of other hormones. Despite their rather different roots, the terms have become interchangeable in this context. Because of the interrelations of pathways (fat, protein, carbohydrate), it is difficult to distinguish between primary and secondary hormonal effects.

<u>Methods Used in the Study of Hormonal Regulation of Carbohydrate Metabolism</u>

1. Carbohydrate metabolism is primarily under the control of insulin (produced by pancreatic β-cells) which promotes the metabolism of glucose and the transport of glucose into cells; and glucagon (produced by pancreatic α-cells) which mobilizes glucose in the liver for export into the blood. Hormones from the adenohypophysis and adrenal cortex generally oppose the action of insulin.

2. In experimental animals diabetes may be produced by

    a. pancreatectomy

    b. destruction of the β-cells (which are part of the islets of Langerhans) with drugs like alloxan

    Studies involving such artificial diabetic states have contributed to the understanding of the naturally occurring condition.

3. The removal of one or more of the endocrine organs controlling carbohydrate metabolism causes the remaining organs to take over the control.

For example: <u>After</u>	<u>Carbohydrate Metabolism is under Control of</u>
pancreatectomy	adrenal + hypophyseal hormones
adrenalectomy hypophysectomy	pancreatic hormones

After	Carbohydrate Metabolism is under Control of
adrenalectomy pancreatectomy	hypophyseal hormones
hypophysectomy pancreatectomy	adrenal hormones (animal is extremely sensitive to the hypoglycemic action of insulin)
hypophysectomy pancreatectomy adrenalectomy	all hormonal control of carbohydrate metabolism is abolished

Synthesis and Secretion of Insulin

1. Evidence has been presented to show that insulin is synthesized as a single polypeptide chain by the ribosomes of the rough endoplasmic reticulum in the classical protein-synthesizing system. The peptide chain then folds on itself making the proper -S-S- bridges to yield proinsulin. Proinsulin is activated by removal of the C-peptide, containing 33 amino acid residues. This is indicated in Figure 46. Notice the presence of three disulfide bonds (two interchain, one intrachain) in both insulin and proinsulin. The amino acid sequence is due to Sanger and his coworkers. For his central role in the work, Sanger was awarded a Nobel Prize in 1957.

2. Proinsulin is put into storage granules (β-granules) in the Golgi apparatus. The transformation of proinsulin to insulin, involving proteolysis, takes place presumably during the granule formation and maturation.

Proinsulin (Molecular Weight about 9,000) $\xrightarrow{\text{Trypsin and/or converting enzyme}}$ Insulin (Molecular Weight about 6,000) + C-peptide (Molecular Weight about 3,000)

Insulin (and some _unactivated_ proinsulin) stored in the granules is secreted under the influence of a suitable stimulus (see below). The granules move toward the cell's plasma membrane, the granule membranes fuse with the plasma membrane, and the contents of the granules are emptied into the blood stream. This process is called exocytosis or emiocytosis and involves no loss of cytoplasm from the cell. It is probably mediated by cAMP. Proinsulin has significantly less biological activity than insulin as measured using several different assay procedures (see measurement of insulin levels).

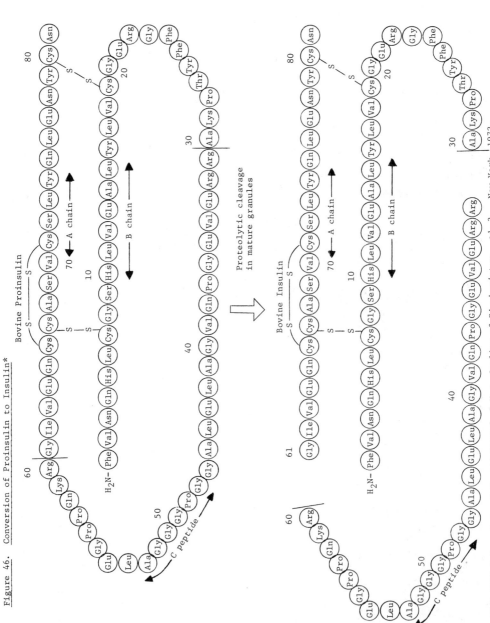

Figure 46.   Conversion of Proinsulin to Insulin*

* Modified from Conn, E.C. and Stumpf, P.K.: Outline of Biochemistry, ed. 3 , New York, 1972, John Wiley and Sons, Inc., Reproduced with permission.

It is of interest to point out that the synthesis of proinsulin as a precursor to insulin probably occurs to insure the proper folding of the peptide and correct formation of the disulfide bonds. As was pointed out earlier, if there are more than two cysteine residues in a polypeptide (and, in this case, more than one chain, in the active form), the disulfide bonds can form in more than one way, leading to one or more types of incorrect (inactive) molecules. Even in the proinsulin, it is not clear whether the proper disulfide links form in response to a preferred conformation or if there is some more complicated mechanism which specifies the proper alignment. Isolated A and B chains, prepared by chemical synthesis or by reduction of the disulfide bonds in purified native insulin, when subjected to oxidizing conditions, recombine in several inactive forms despite their having the correct amino acid sequence.

3. Insulins isolated from different species show minor variations in amino acid composition and sequence. Although this does not significantly alter their biological activity, prolonged administration of insulin leads to the production of anti-insulin antibodies if the donor and recipient are of different species. These antibodies <u>do</u> interfere with the effectiveness of the insulin. In diabetics who still retain some β-cell function, the antibodies have been shown to react with the patient's own proinsulin and insulin, thereby further reducing the amount of available hormone.

4. The fate of the C-peptide, which is biologically inactive, is unclear. It is secreted by the pancreas in a 1:1 ratio with insulin and is present in the circulation. Recently, radioimmunoassay measurements (see measurement of cAMP and insulin) of C-peptide immunoreactivity (i.e., amount of material which reacts with antibodies against C-peptide) have been used to evaluate pancreatic β-cell function in patients with insulin-treated diabetes. Direct determination of insulin levels is hampered by the presence of antibodies synthesized by the patient in response to the exogenous insulin used in the treatment. Both free C-peptide and proinsulin (which still contains the C-peptide region) are measured in these assays. The contribution of each of these materials to the total immunoreactivity can be determined by performing a preliminary separation by gel filtration. The results of this study are discussed in the section on diabetes.

5. The predominant physiologic stimulator of insulin secretion is glucose. The glucose level sensing device of the β-cells is not well understood. It has been suggested that this process is mediated through the formation of cyclic AMP. Note, however, that insulin <u>reduces</u> the cyclic AMP level in some other cells (e.g., adipose tissue cells) where insulin-mediated processes take place.

Other agents which stimulate the insulin release are the amino
acids (leu, arg, lys, and phe), glucagon (a hyperglycemic hormone
produced by the α-cells of pancreas which functions by increasing
hepatic glycogenolysis; it also has a direct effect on the β-cells);
secretin and pancreozymin (hormones which activate the exocrine
function of the pancreas); corticotropin; growth hormone; sulfonyl
urea compounds such as tolbutamide, used in the treatment of
maturity onset diabetes; theophylline (an inhibitor of
phosphodiesterase, the enzyme which inactivates cAMP); and
stimulation of the right vagal nerve. Examples of <u>inhibitors</u> of
insulin secretion include: α-adrenergic stimulating agents
(epinephrine and norepinephrine; epinephrine also causes hepatic
and muscle glycogenolysis); and β-adrenergic blocking agents
(thiazide diuretics, 2-deoxyglucose).

<u>Metabolic Effects of Insulin</u>

1. Insulin is required for the entry of glucose into muscle (skeletal,
   cardiac, and smooth), adipose tissue, leukocytes, the mammary gland,
   and the pituitary. Tissues which are freely permeable to glucose
   include brain, intestinal mucosa, lens, retina, nerve, kidney,
   erythrocytes, blood vessels, and islet cells. Although the liver
   is not dependent on insulin for glucose uptake, the hormone still
   influences this organ's glucose metabolism by stimulating the liver
   glucokinase activity.

2. Insulin increases glycogen formation and oxidation of carbohydrates
   in muscle and liver. It suppresses the principal gluconeogenic
   enzymes: pyruvate carboxylase, PEP carboxykinase, fructose-1,6-
   diphosphatase and G-6-phosphatase, thus reducing the formation of
   sugar. It enhances the utilization of glucose by inducing
   glucokinase (in the liver), phosphofructokinase, and pyruvate
   kinase. Insulin also stimulates the HMP-shunt pathway and
   lipogenesis (discussed under lipid metabolism) in the adipose
   tissue.

3. Insulin has an anabolic effect and affects growth. This is due to

   a. its amino acid sparing action (glucose can be used for energy
      rather than amino acids),

   b. enhanced transportation of amino acids into cells and their
      incorporation into proteins. Growth is a complex process and
      is determined by a number of genetic, nutritional and hormonal
      growth hormone (thyroxine, insulin and androgens) factors.

4. It is important to note that insulin has none of these effects in experimental cell-free extracts. This (and other evidence) suggests that insulin exerts its primary effect at the cell membrane.

Measurement of Insulin Levels

1. A number of bioassays have been developed for the measurement of insulin. These are based on the occurrence or magnitude of a physiological response evoked by administration of an insulin-containing sample to an intact animal or a piece of isolated tissue. Such methods include

    a. hypoglycemic response in rabbits;

    b. mouse convulsive assay;

    c. glucose uptake by rat epididymal fat pad;

    d. glucose uptake by isolated rat diaphragm.

    Bioassay procedures are frequently criticized for being inaccurate, insensitive, nonspecific, and time consuming. Few studies have been published, however, which specifically compare their results to those of radioimmunoassays and other new techniques.

2. The measurement of insulin by radioimmunoassay (RIA) was published in the late 1950's by Berson and Yallow. It was the first example of the method and because of its high sensitivity and specificity, it made possible a wide range of new studies and measurements. As of 1973, at least 53 different substances could be measured by kits which are available commercially. Other methods for additional materials are currently being used in research. A discussion of this method and the related competitive protein-binding technique are presented in the section on cyclic AMP.

    One difficulty in the use of immunologic methods for the determination of insulin and other peptides is the partial species-specificity in their amino acid sequence. Antibodies produced in response to insulin of one species may bind insulin from another species to a lesser extent. Thus, if porcine insulin is used to prepare antibodies which are then employed to measure human insulin levels, care must be taken to standardize the assay against human insulin.

    Another problem encountered specifically with peptide hormone RIA's is the purification and labeling of the radioactive hormone. Hormones are only transiently present in biological fluids and then usually at low concentrations; and they are not readily synthesized by chemical means. These problems are frequently of lesser magnitude with other molecules.

In molecules isolated from living systems, radioactive labeling must usually be done after purification. One method which has found wide application is the iodination of tyrosyl residues in the molecule with ^{131}I (or ^{125}I) in the following reaction:

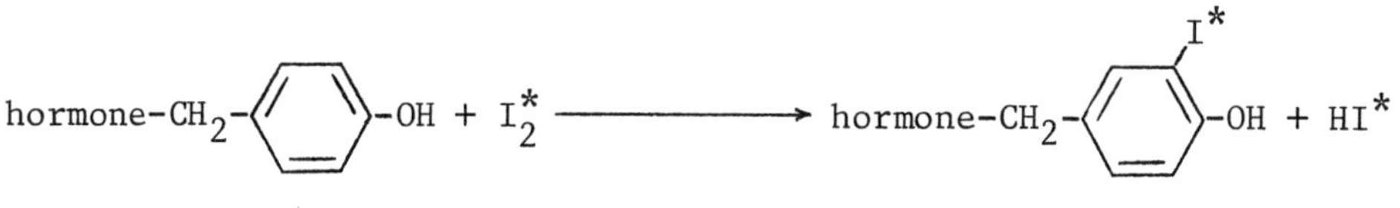

$$\text{hormone-CH}_2\text{-}\langle\bigcirc\rangle\text{-OH} + I_2^* \longrightarrow \text{hormone-CH}_2\text{-}\langle\bigcirc\rangle\text{-OH} + HI^*$$

where * indicates radioactivity. This procedure depends on the presence of a tyrosyl residue in the peptide (phenylalanine will not work). In some non-peptide RIA's, tyrosyl groups are coupled to the molecule which the assay is intended to measure, to provide a site for labeling.

In the insulin assay, the presence of proinsulin complicates matters (as it does in the C-peptide radioimmunoassay). It cross-reacts with the insulin antibodies and is also a contaminant in the insulin used to prepare the antibodies. The assay may also fail to measure all of the physiologically active insulin in the body fluids.

The Diabetic State

1. Although the term <u>diabetes</u> is defined as a deficiency condition indicated by the discharge of an abnormal quantity of urine (polyuria), it is usually used to mean diabetes mellitus. The latter is a generalized disorder of protein, carbohydrate, and lipid metabolism, characterized by a number of clinical findings. Diabetes mellitus is a very complex disease whose cause is not well understood. The discussion presented below is intended to cover only the most important features.

    a. The blood sugar increases to high levels (hyperglycemia) and glucose is found in the urine (glucosuria). The hyperglycemia and glucosuria persist during fasting.

    b. Liver glycogen falls to very low levels. Muscle glycogen is decreased but much less as compared to liver glycogen.

    c. Ingested and endogenous glucose is not taken into muscle and fat cells but is instead excreted to a great extent in the urine.

d. The respiratory quotient, which is an indication of the type
    of fuel being metabolized, falls to around 0.71 (the value
    for fat oxidation) and is not raised by ingestion of glucose.

e. The rate of tissue-protein breakdown is markedly accelerated
    (increased urinary nitrogen excretion occurs).

f. The injection of glucose into normal individuals results in
    a rise in blood pyruvate and lactate. This does not happen
    in a diabetic patient.

g. In the diabetic, large quantities of ketone bodies
    (acetoacetic acid, β-hydroxybutyric acid, and acetone) are
    produced due to increased fatty acid metabolism and decreased
    capacity to oxidize ketone bodies in the muscles. This can
    lead to diabetic coma.

    Ketone bodies ⟶ severe acidosis ⟶ coma ⟶ death

h. Excretion of large amounts of glucose and ketone bodies results
    in water and salt loss leading to dehydration and severe thirst.
    An abnormally large amount of urine is excreted (polyuria).

i. There is an increased rate of mobilization of triglycerides
    and unesterified fatty acids from adipose tissue into the
    blood.

2. If the diabetes is caused by a lack of insulin, and if the patient
    is not "insulin-resistant", an injection of insulin promptly
    corrects all the foregoing metabolic disturbances. An overdose of
    insulin is very dangerous, producing insulin shock due to
    hypoglycemia. This condition superficially resembles a diabetic
    coma. In the insulin-resistant form of diabetes, even excessive
    amounts of insulin will not produce insulin shock.

3. A normal adult secretes about 2 mg (45 units) of insulin per day.
    A diabetic patient being treated with insulin receives about
    60 to 70 units per day.

4. As was indicated under the metabolic effects of insulin, a number
    of tissues do not require the hormone for the entry of glucose into
    the cells. Consequently, in the hyperglycemia associated with
    diabetes, the intracellular glucose of these tissues attains a
    level similar to that of the blood. This condition is unaffected
    by insulin unless total blood glucose can be reduced. Within
    these cells, the excess glucose is reduced to sorbitol by aldose
    reductase (see also galactosemia).

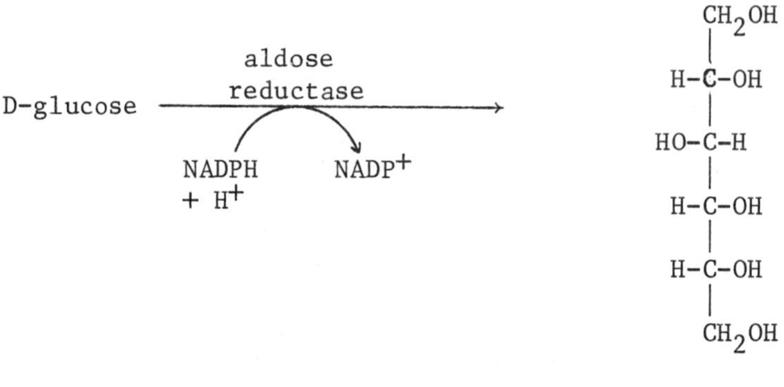

D-glucose $\xrightarrow[\substack{NADPH \\ + H^+}]{\substack{aldose \\ reductase}} $ NADP$^+$

$$
\begin{array}{c}
CH_2OH \\
| \\
H-C-OH \\
| \\
HO-C-H \\
| \\
H-C-OH \\
| \\
H-C-OH \\
| \\
CH_2OH
\end{array}
$$

Sorbitol

Part of the sorbitol is oxidized to fructose by sorbitol dehydrogenase.

Sorbitol $\xrightarrow[\substack{NAD^+ \\ \ }]{\substack{Sorbitol \\ dehydrogenase}}$ NADH + H$^+$ → D-fructose

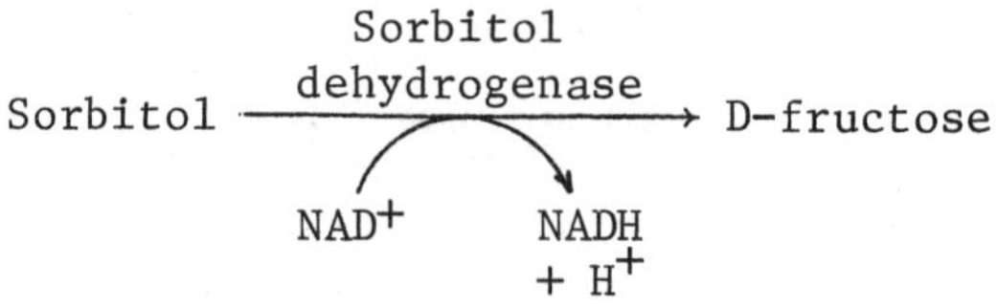

Although the cell membranes are freely permeable to glucose, fructose and sorbitol do not readily pass through and they are retained within the cells. Even if the hyperglycemia is ultimately controlled by insulin, any sorbitol and fructose which may have accumulated remain until they are removed via other metabolic pathways.

Large amounts of sorbitol and fructose within the cell cause hypertonicity and water retention. Many of the physiological and pathological alterations associated with diabetes can be related to these effects. For example:

a. Sorbitol accumulation in the lens may result in cataract formation (see also galactose metabolism). This is supported by studies of cataract formation in diabetic rats and rats on high-galactose diets.

b. Sorbitol elevation in the peripheral nerves is associated with peripheral neuropathy.

c. In erythrocytes, sorbitol will displace 2,3-DPG, reducing the oxygen-carrying ability of the blood (see hemoglobin).

d. Other changes include vascular problems leading to
    atherosclerosis, nephropathy, and retinopathy.

A probable mechanism by which sugar alcohols (sorbitol, galactitol)
bring about these abnormalities is indicated below.

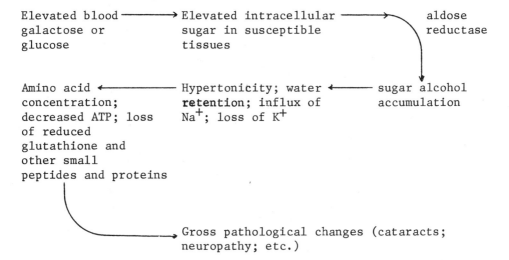

Elevated blood ⟶ Elevated intracellular ⟶ aldose
galactose or   sugar in susceptible   reductase
glucose   tissues

Amino acid ⟵ Hypertonicity; water ⟵ sugar alcohol
concentration;   **retention**; influx of   accumulation
decreased ATP; loss Na^+; loss of K^+
of reduced
glutathione and
other small
peptides and proteins

⟶ Gross pathological changes (cataracts;
  neuropathy; etc.)

Aldose reductase, the key enzyme in this process, has been found
in the affected tissue cells at high concentration.

5. Basement Membrane Changes in Diabetes

One of the clinical manifestations of diabetes which is poorly
understood is the change in the basement membrane (BM), leading to
its thickening in tissues such as the small blood vessels
(microangiopathy) and the renal glomerulus. The BM is made up of
a fibrillar network of proteins and its functions are to support
cells and to act as a selective filter (particularly in the case
of the glomerulus).

The BM-protein is a fibrous glycoprotein closely related to
collagen. It is distinguished from collagen by the presence of
mannose, hexosamine, sialic acids, and fucose, in addition to the
glucose and galactose residues found in collagen. BM-protein also
has a large number of half-cystine residues (few of which occur
in collagen) and a greater number of hydroxylysine residues than
collagen.

The biosynthesis of BM-protein is similar to that of collagen, requiring many post-ribosomal modifications. These include hydroxylation reactions and the addition of sugar residues (see collagen synthesis). In alloxan diabetic rats, increases have been measured in the amount of lysine which is hydroxylated, the number of glucosylgalactose units present (attached to the δ-hydroxyl group of lysine), and the activity of the glycosyltransferases, needed for the formation of this type of bond. The presence of these extra, bulky sidechains could disrupt the packing of the protein fibers in the membranes, resulting in the observed thickening and increased porosity. If insulin is administered in the early stages of diabetes, the changes in the membranes are reversed. Prolongation of the disease progressively decreases the reversibility, suggesting the value of early diagnosis and treatment.

Although the biochemical lesions in diabetes are still not clear, this work indicates that they do not directly involve coding for the enzymes responsible for the membrane changes (i.e., lysinehydroxylase, glycosyltransferases, etc.). Rather, the observed BM-protein alterations probably result from the decreased insulin supply, the increased intracellular glucose concentration (in insulin-independent tissues such as the kidney and the blood vessels), or both. This is analogous to the proposed mechanism for cataract formation (via the sorbitol pathway) in diabetes mellitus and galactosemia.

6. <u>Clinical diabetes</u> in young persons is frequently due to a lack of insulin (juvenile-onset diabetes). This lack of insulin may be relative rather than absolute (i.e., there may be some insulin produced but not enough). The production of abnormal insulin (mutated amino acid sequence; failure to activate proinsulin) has the same effect as a deficiency in the quantity of insulin secreted. These types of diabetes can usually be treated by administration of an adequate insulin supply.

7. Maturity-onset diabetes, appearing in adults, is usually insulin-resistant. The insulin supply is adequate and even elevated, due to overstimulation of the pancreas by the high blood glucose. In some patients, there may be a delay in the pancreatic response to hyperglycemia followed by an oversecretion of insulin. Other possible causes include

   a. The inability to use or metabolize glucose in the tissues,

   b. The presence of an insulin antagonist (e.g., an insulin antibody or an insulinase),

c. The inability of normally insulin-sensitive tissues to respond to the hormone.

8. Diabetes is found in association with several rare inherited diseases such as Weiner's, Klinefelter's, Refsum's and Turner's syndromes; Friedreich's ataxia; and others. It is not yet clear whether gout and diabetes are significantly associated. The malfunction of other endocrine glands which influence carbohydrate metabolism may also produce diabetes. Diabetes insipidus is a failure of the kidneys to respond to or a deficiency of, antidiuretic hormone. This results in polyuria, polydipsia (excessive thirst), and persistant hypotonicity of the urine.

9. Morphological changes in the pancreas which are associated with diabetes include

   a. Primary pathology changes of the pancreas.

   b. Reduction in granulation of the β-cells (this does not necessarily imply a diminution in insulin production).

   c. Anatomic changes in the islets of Langerhans (seen in autopsy cases).

   d. In many cases there are no morphological changes. Furthermore, in an individual whose entire pancreas has been removed (as in carcinoma) the syndrome of diabetes is not as severe as that of many patients with spontaneous diabetes. A severe diabetic needs 80-150 units of insulin per day, whereas patients with pancreatectomy need only about 40 units per day. It appears that the requirement for insulin is greater in the diabetic than in normal persons and that the pancreas of the diabetic is unable to meet this increased requirement.

10. Although it is clear that diabetes mellitus is a familial disorder and that a predisposition for the disease is at least partly inherited, the mode of genetic transmission is far from clear.

    a. A single recessive (autosomal) defect with incomplete penetrance (absence of phenotypic expression in some homozygous individuals) is still the explanation preferred by some geneticists. While able to account for many of the observed characteristics of the inheritance pattern, there are, however, several aspects which it cannot explain. Even if this should turn out to be basically correct, it is apparent that the defect confers only a predisposition for the disease. A variety of other factors (age of the subject, diet, exercise, sex, infections, etc.) strongly influence the ultimate manifestation of the diabetic syndrome.

b. Recent evidence tends to favor a multifactorial inheritance pattern. Several specific genes must be present for the characteristic phenotype to appear. Environmental conditioning quite probably plays an important role here also.

c. At the present,there is no reason to exclude a combination of both mechanisms. Several distinct genotypes may result in the same or highly similar phenotypes.

11. Biochemical defects associated with diabetes include

a. The capacity of the liver to deposit and hold glucose as glycogen is practically lost in the diabetic. The rate of glycogenolysis is also enhanced.

b. The glucokinase reaction is severely impaired.

c. There is a decreased rate of glucose transport into certain tissue cells (particularly muscle and adipose tissue).

d. There is an increase of G-6-phosphatase activity in the liver.

$$\text{G-6-}\textcircled{P} \xrightarrow{\text{Glucose-6-Phosphatase}} \text{glucose} + P_i$$

e. Glucose formation from amino acids is increased (gluconeogenesis).

12. Insulin antagonists, mentioned previously, can cause ineffectiveness of insulin.

a. Insulin antibodies, produced as a result of insulin therapy, are found in the β and γ globulin fractions of plasma. Antibody production is a complexity and must be considered in the treatment of diabetes.

b. Another type of insulin antagonist is distinguishable from antibodies in that it is present without prior administration of exogenous insulin. It occurs in the β lipoprotein fraction of serum.

c. Physiological antagonists include growth hormone, thyroxine, and cortisol.

13. Breakdown of Insulin

   a. Insulin released into the portal blood system through the
      pancreatic vein reaches the liver before it enters the
      systemic blood circulation to exert its effects on the
      insulin-dependent tissues such as muscle and adipose. In
      the liver, in addition to producing its physiologic effect,
      insulin is degraded by reductive cleavage.

   b. The enzyme system that catalyzes the reductive cleavage of
      the interchain disulfide bonds is shown below.

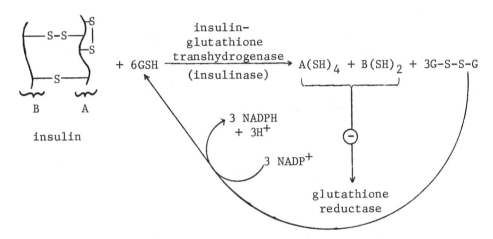

      Insulin inactivation is self-limiting because the G-S-S-G
      reductase is inhibited by the reduced insulin chains so that
      no more GSH (which is required for the reduction of insulin)
      is produced.

   c. The A and B chains are further degraded by proteolytic enzymes
      (peptidases).

14. Stimulation of Insulin Secretion

   a. Sulfonylureas are drugs used in maturity-onset diabetes.

$$H_3C-\!\!\bigcirc\!\!-SO_2-NH-\overset{\overset{O}{\|}}{C}-\overset{H}{N}-C_4H_9$$

      Tolbutamide

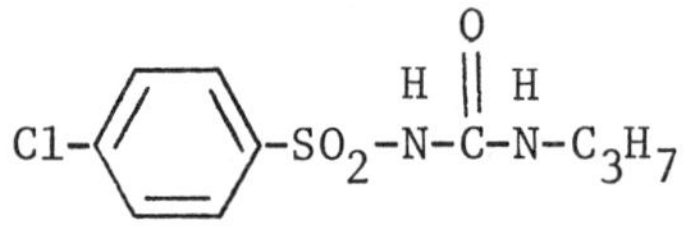

$$Cl-\langle\bigcirc\rangle-SO_2-N-\overset{\overset{\displaystyle O}{\|}}{\underset{}{C}}-N-C_3H_7$$

with H above each N.

Chlorpropamide

They presumably act by increasing the secretion of insulin by β cells. The efficiency depends upon the presence of functional islet tissue. These drugs have no effect on pancreatectomized animals.

b. Biguanidines act by increasing the uptake of glucose by cells. Following their administration,there is an increase in anaerobic glycolysis with production of lactate. These compounds do not produce hypoglycemia in the normal individuals and do not stimulate the β cells. Example: Phenethylbiguanidine (Phenformin)

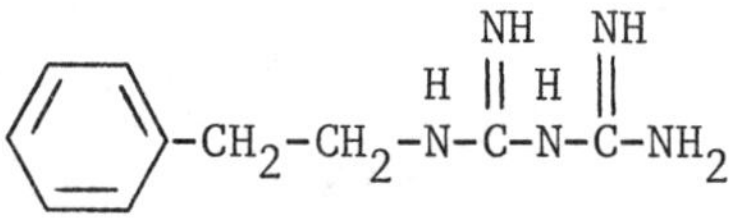

$$\langle\bigcirc\rangle-CH_2-CH_2-N-\overset{\overset{\displaystyle NH}{\|}}{\underset{}{C}}-N-\overset{\overset{\displaystyle NH}{\|}}{\underset{}{C}}-NH_2$$

with H above each N.

Biguanidines have been shown to inhibit electron transport in the mitochondria. A side-effect of this compound is lactic acidosis.

Summary of Action by Hormones on Carbohydrate Metabolism
(Notice that all of these hormones are hyperglycemic, except for insulin.)

1. Insulin

a. Plays a central role in the removal of glucose from blood, lipid synthesis, (in adipose tissue and liver), glycogen formation (principally in muscle),and glucose oxidation (in all insulin-dependent tissues).

b. Is secreted into the blood as a direct response to hyperglycemia.

c. Has an immediate effect in increasing the glucose uptake by certain tissues; produces hypoglycemia.

d. Is produced by the β cells of the pancreas.

2. Glucagon (29 amino acid residues)

    a. Release is stimulated by hypoglycemia.

    b. Accelerates glycogenolysis in the liver.

    c. Enhances gluconeogenesis from amino acids and lactate.

    d. Is produced by the α cells of the pancreas.

3. Anterior Pituitary: GH (growth hormone) and ACTH (adrenocorticotropin)

    a. Tend to elevate blood glucose levels.

    b. GH decreases glucose uptake by the tissues.

    c. GH → adipose tissue → releases free FA → inhibits glucose utilization.

    d. Chronic administration of GH leads to diabetes (by producing chronic hyperglycemia which causes β cell exhaustion).

    e. ACTH stimulates the adrenal cortex.

4. Adrenal Cortex: glucocorticoids (hyperglycemic agent)

    a. Increase gluconeogenesis by the following mechanism:

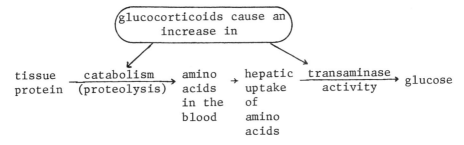

    b. Inhibit utilization of glucose in extrahepatic tissues.

5. Adrenal Medulla: epinephrine stimulates glycogenolysis in liver and muscle. It also inhibits glycogenesis.

6. Thyroid hormone (thyroxine, triiodothyronine)

    a. Has diabetogenic action.

    b. Fasting blood sugar is elevated in hyperthyroid and decreased in hypothyroid patients.

    c. Effects on carbohydrate metabolism are not understood.

Carbohydrate Metabolism and Mechanical Functions in Biological Systems

1. Movement in biological systems is coupled to carbohydrate metabolism via ATP. During muscle contraction and other mechanical processes, ATP is hydrolyzed to ADP + P_i and the energy originally stored during oxidation is used to do useful work.

2. The immediate sources of ATP for rapid bursts of muscular contractions are threefold.

   a. During times of muscular relaxation, ATP is used by the enzyme creatine phosphokinase (CPK) to phosphorylate creatine (a substituted guanidine; see glycine metabolism). Since the nitrogen-phosphorous bond is unstable (has a high free energy of hydrolysis; $\Delta G^{o'}$ = -10.3 kcal/mole), it can be used to phosphorylate ADP, so the reaction (shown below) is freely reversible.

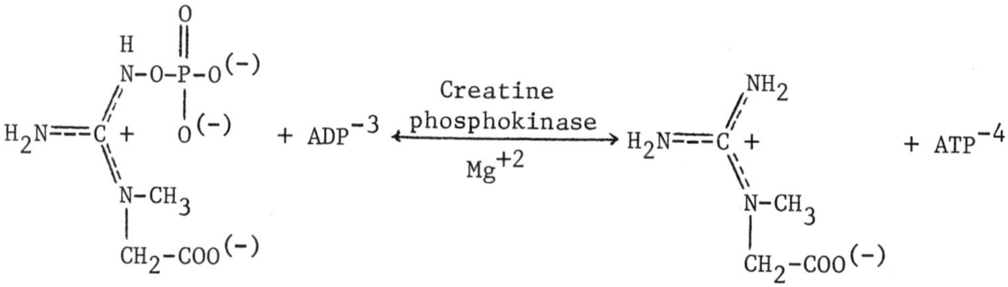

Creatine phosphate   Creatine

   Creatine phosphate is thus another storage form of high-energy phosphate bonds, especially in muscle tissue.

   b. Adenylate kinase (myokinase) reuses ADP to make ATP.

$$2 \text{ ADP} \xrightarrow{\text{adenylate kinase}} \text{ATP} + \text{AMP}$$

   c. ATP synthesized in glycolysis is also a source of energy. Initially, in the working muscle, oxygen stored in oxymyoglobin is available to maintain aerobic glycolysis. When this supply is used up, anaerobic glycolysis begins with the production of lactic acid. The lactate diffuses from the muscle into the blood, where it is carried to the liver and either further oxidized or used for gluconeogenesis. Glucose formed in this way can return to the muscle, via the circulation. This is known as the Cori cycle (see glycolysis).

3. Muscles which are capable of long-term or sustained activity depend
    upon the oxidation of free fatty acids, ketone bodies, and pyruvate
    (either from glucose or lactate) under aerobic conditions to produce
    ATP. Examples of this are the cardiac and diaphragm muscles.

4. Of the three known systems in plants and animals which transform
    chemical energy (ATP) into mechanical energy, the best-studied of
    these is striated muscle.

    a. Muscle contains actin and myosin. <u>In vivo</u>, and under appropriate
        conditions <u>in vitro</u>, these are non-covalently combined as
        actomyosin, the contractile protein in muscle.

        i)   Actin (F-actin) is a polymer, probably helical, of subunits
             (G-actin molecules). The subunits have a molecular weight
             of about 47,000 daltons. The active form is F-actin,
             which is able to stimulate the ATPase activity in myosin.

        ii) Myosin has properties of both fibrous and globular
             proteins and a molecular weight of about 5×10^5 daltons.
             It appears to have the structure shown schematically below.

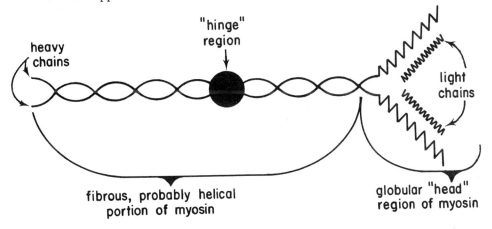

             Notice that (1) there are four peptide chains, two light
             and two heavy; (2) the globular part of the molecule
             contains the ATPase activity and, presumably, is the
             region which functions during muscle contraction; and
             (3) myosin is susceptible to hydrolysis by proteolytic
             enzymes at the "hinge" point. The function of this
             region is hypothetical only.

        iii) Other proteins such as troponin A and B, tropomyosin,
             Z-band protein, and actinin α and β have been isolated
             from muscle but are not well purified or characterized.

b. Myosin and actin are associated in the manner indicated below.

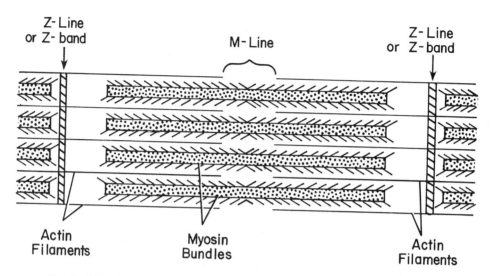

Z-Line
or Z-band

M-Line

Z-Line
or Z-band

Actin
Filaments

Myosin
Bundles

Actin
Filaments

The helical F-actin filaments are linked by protein at the
Z-band. The myosin bundles are made of myosin molecules with
the fibrous portions parallel to each other and the globular
regions, containing the ATPase, extending from the bundles
like arms. Contraction is thought to occur by a sliding of
actin and myosin molecules along one another, propelled by
a "pulling" motion of the myosin "arms". During contraction
the M-line becomes narrower and the Z-lines move close to
each other. In this sliding-filament hypothesis, proposed
by Huxley, the fibers are assumed to always maintain a
constant length. It should be emphasized that details of
this model are still hypothetical, supported mostly by indirect
evidence. The nature of the interaction between the myosin
and actin is not known. The way in which the chemical energy
of ATP is used to move the protein molecules is also unclear.

c. Ca^{+2} appears to be intimately involved in the activity of
the myosin ATPase during muscle contraction. In the resting
muscle, Ca^{+2} is stored in certain regions of the endoplasmic
(sarcoplasmic) reticulum. The electrical discharge which
triggers muscle contraction causes Ca^{+2} release and the Ca^{+2},
in turn, stimulates the ATPase and muscle contraction. During
relaxation, the Ca^{+2} is reaccumulated in the storage sites and
the ATP is regenerated. Recently, it has been postulated that
Ca^{+2} may perform these functions by interacting with troponin
(mentioned above). Together, they may cause the myosin arms
to contact the actin filaments, thereby activating the ATPase.
Actin is known to stimulate the myosin ATPase _in vitro_. The
reaccumulation of Ca^{+2} in the endoplasmic reticulum may be
defective in some types of muscular dystrophy.

d. Two major groups of inherited diseases of muscle are the
    periodic paralyses and the muscular dystrophies. Both have
    multiple etiologies and, in general, the biochemical lesions
    have not yet been defined.

5. <u>Microfilaments</u> are a second energy-transducing system within the
    cell. They are associated particularly with cytokinesis (cleavage
    of the cytoplasm during cell division) in animal cells and
    movement of non-ciliated cells. (Cytokinesis in plant cells seems
    to involve microtubules only. These are discussed below.) They
    have been shown to contain an actin-like protein but nothing
    similar to myosin. Although they have not been chemically
    well-characterized, they may represent a primitive actomyosin-muscle-
    like system. The disruption of microfilamentous contractile
    systems has been postulated as the mechanism by which cytochalasins
    halt a number of cellular processes (see below).

6. While the two previous systems have been filamentous and best
    adapted to pulling, <u>microtubules</u> are found where a pushing force
    is needed. Processes which appear to involve microtubules include
    the motion of sperm tails and other flagella, ciliary function,
    and spindle formation and nuclear division during mitosis.

    a. The arrangement of microtubules in cilia and flagella is the
        "9 + 2" pattern (9 pairs of tubules around the rim with two
        tubules in the center), illustrated below in a schematic
        cross-section of a cilium.

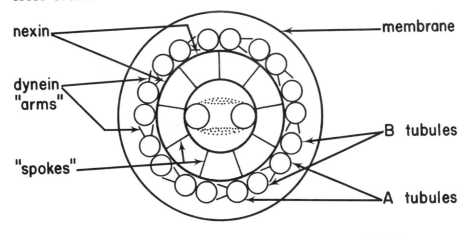

The basic molecular components of this structure which have
been identified are:

i) <u>tubulin</u> - a globular protein which stacks upon itself to
    form the walls of the tubules; it appears analogous to
    actin in muscles; sometimes called flagellin in certain
    flagella.

      ii)   <u>dynein</u> - this protein has the only ATPase activity in the flagellum; extends from the walls of the A tubules; probably equivalent to the myosin "head" region in muscle.

      iii)   the <u>spokes</u> and the <u>nexin band</u> are composed of proteins and can be hydrolyzed by proteases. Their function appears structural, maintaining the tubule arrangement during ciliary and flagellar movement.

b.   The A and B tubules of a pair are slightly different, both chemically and morphologically. Theoretical and experimental studies favor a "sliding-tubule" mechanism, similar, perhaps to the sliding-filament mechanism proposed for muscle. In this model, the dynein arms transduce the energy released by ATP hydrolysis into motion by attaching to the adjacent B tubule, bending (thereby sliding one tubule relative to the other), releasing, and returning to their original position. It has been shown that ATP is hydrolyzed in this process.

c.   Several compounds, the classical one being colchicine, apparently disrupt microtubular movement. Colchicine is noted for halting mitosis in metaphase, with destruction of the spindle structure which is probably microtubular in nature. Colchicine is used in the treatment of acute gout and the mechanism of its action is discussed under gout.

7.   The <u>cytochalasins</u> are an important class of compounds which produce a number of interesting effects when applied to cells. They are proving to be very useful tools in cytological research.

a.   The first cytochalasins discovered were A and B, shown below.

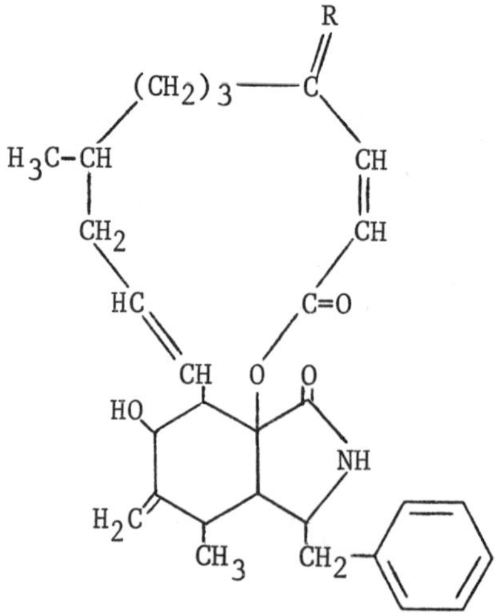

R = 0 in cytochalasin A

R = H and OH (one bond to the carbon from each) in cytochalasin B

b. General effects which these compounds have on cells include

    i) interference with cytoplasmic cleavage (cytokinesis);

    ii) inhibition of cell movement;

    iii) cell enucleation (extrusion of the nucleus from a cell).

c. Specific processes inhibited by cytochalasins include:

    i) clot retraction (platelets);

    ii) phagocytosis;

    iii) platelet aggregation;

    iv) cardiac muscle contraction;

    v) thyroid secretion and growth hormone release.

d. On the basis of limited evidence, some authors claim that the
    cytochalasins function strictly or primarily by disrupting
    microfilaments and, from this, infer that all of the above
    processes involve microfilaments. This remains to be proven.
    Elucidation of the mechanism of action of the cytochalasins
    will go along with an increased knowledge of the mechanisms
    for subcellular movement.

Oxidation of Alcohols

1. Ethyl Alcohol

    a. Ethyl alcohol, which is rapidly absorbed from the stomach,
        small intestine, and colon, is oxidized in the liver to acetyl-
        CoA.

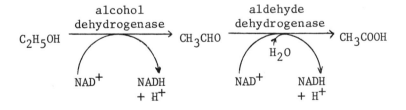

$$CH_3COOH \xrightarrow[\substack{CoASH \quad ATP \qquad AMP \\ + \\ PP_i}]{\text{Activating enzyme}} CH_3-\overset{\overset{\displaystyle O}{\|}}{C}-SCoA + H_2O$$

Alcohol dehydrogenase is a zinc-containing enzyme present in the soluble fraction of the cell. It possesses free sulfhydryl groups which are essential for enzyme activity. It is not clear whether the acetyldehyde produced gets converted to acetyl-CoA via the production of acetic acid (as shown in the preceding diagram) or by a direct reaction. The rate of oxidation of alcohol is independent of the blood alcohol concentration.

b. It has been reported that alcohol is also oxidized by hepatic microsomes which are part of the smooth endoplasmic reticulum of the cell. Other drugs such as barbiturates are metabolized here. The microsomal system consists of NADPH, microsomal hemoprotein P-450, etc. (see the section on microsomes, under mitochondria). The observation that alcohol potentiates the effect of barbiturates is explained partly on the basis that both the compounds are competing for the same metabolic degradation process.

c. Disulfiram is a compound used in the treatment of chronic alcoholism. It inhibits the aldehyde dehydrogenase by competing with NAD^+ for binding sites on the enzyme molecule. This leads to accumulation of acetaldehyde giving rise to such undesirable effects as nausea, vomiting, thirst, sweating, headache, etc.

Structure

$$\underset{\substack{H_5C_2}}{\overset{\substack{H_5C_2}}{}}N-\underset{\|}{\overset{}{C}}-S-S-\underset{\|}{\overset{}{C}}-N\overset{\substack{C_2H_5}}{\underset{\substack{C_2H_5}}{}}$$
$$\qquad\qquad\quad S\qquad S$$

Disulfiram

d. The acetyl-CoA produced can either be oxidized in the citric acid cycle or converted to cholesterol, fatty acids or other lipid components of the cell. It is known that alcohol can enhance the synthesis of lipids in the liver leading to fatty liver. This results from two things.

i) The NADH generated is used to convert dihydroxyacetone-phosphate to α-glycerophosphate (needed for triglyceride synthesis);

ii) Acetyl-CoA gets converted to acyl-CoA (activated fatty acids); α-glycerolphosphate and acyl-CoA give rise to triglycerides (discussed under lipid metabolism).

2. Methyl Alcohol Degradation

a. Methyl alcohol is usually introduced into the body accidentally and is metabolized in the liver as follows:

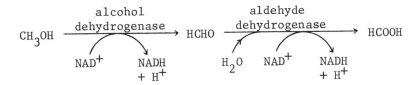

The initial oxidation produces formaldehyde which damages the retinal cells, leading to blindness. The second oxidation step generates formic acid which produces acidosis.

b. The initial dehydrogenase which acts on methyl alcohol to produce formaldehyde is the same enzyme that oxidizes the ethyl alcohol to acetaldehyde. This observation has been used in the treatment of methanol toxicity. Ethanol is administered to ameliorate the toxic effects of methanol. The ethanol competes with the methanol for enzyme and NAD^+, thereby reducing the rate of formaldehyde synthesis to the point where the aldehyde dehydrogenase can remove it rapidly enough to prevent any buildup. The acidosis is corrected by administering bicarbonates, etc.

Measurement of Parameters of Clinical Importance in Carbohydrate Metabolism

1. Glucose is measured by both non-enzymatic methods (using the reducing ability of glucose) and an enzymatic method (glucose oxidase). The latter procedure is highly specific for glucose.

    a. Non-enzymatic methods

        i) Alkaline ferricyanide method

$$Fe(CN)_6^{-3} \xrightarrow[\text{heat}]{\text{glucose, OH}^-} Fe(CN)_6^{-4}$$

Ferricyanide	Ferrocyanide
(yellow)	(colorless)

Disappearance of the yellow color is measured at 420 nm. Note that the iron atom is reduced from +3 to +2.

        ii) Copper reduction method

$$Cu^{+2} \xrightarrow[\text{heat}]{\text{glucose, OH}^-} CuOH$$

$$2\ CuOH \longrightarrow Cu_2O + 2\ H_2O$$

$$\text{Phosphomolybdic acid (colorless)} + Cu_2O \longrightarrow \text{lower oxides of molybdenum (blue)} + Cu^{+2}$$

The final blue color is measured spectrophotometrically at 490 (or 620 or 660) nm.

        iii) O-toluidine methods

$$\text{O-toluidine} \xrightarrow[\text{heat}]{\text{glucose,HOAc}} \text{glycosylamine} \longleftrightarrow \text{Schiff's base derivative}$$

O-toluidine gives a final green color, measurable at 630 nm. Other aromatic amines give different colors. Glucose, galactose, and mannose all give identical reactions. Pentoses give an orange color, absorbing around 480 nm.

b. <u>Enzymatic methods</u>

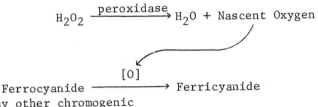

$$glucose + O_2 + H_2O \xrightarrow{\text{glucose oxidase}} gluconic\ acid + H_2O_2$$

$$H_2O_2 \xrightarrow{\text{peroxidase}} H_2O + Nascent\ Oxygen$$

$$\text{Ferrocyanide} \xrightarrow{\text{[O]}} \text{Ferricyanide}$$
(or any other chromogenic
    oxygen acceptor)

Glucose oxidase acts only upon β-D-glucose. In an aqueous
solution of glucose, 35% of the sugar is in the α-form,
65% in the β-form. The conversion of α-to β-form takes place
by mass action as the β-form is consumed in the glucose
oxidase reaction. This process is speeded up by the enzyme
mutarotase, found as a contaminant in commercial preparations
of glucose oxidase. The glucose oxidase reaction is discussed
further in the section on the uses of enzymes in the clinical
laboratory.

2. <u>Glucose Tolerance Tests</u>

A tolerance test is useful in assessing the functional capacity
of the pancreas to secrete insulin for the utilization of
administered glucose. In essence, the method consists of giving
a known amount of glucose to a fasting patient and measuring
glucose in the blood at periodic intervals. In normal persons,
the glucose level rises initially (160 mg/100 ml plasma) and then
drops to or below fasting (80-100 mg/100 ml) levels. In a
diabetic, the glucose peak reaches a higher level and takes a
longer time to return to the fasting level. Galactose and fructose
tolerance tests are also used to assess the metabolism of these
sugars.

3. <u>Insulin Assay</u>. Already discussed under insulin.

4. <u>Amylase</u>

$$\text{starch} \xrightarrow{\text{amylase}} \text{maltose} + \text{glucose}$$

α-Amylases produce short-chain polysaccharides (dextrins) in addition to the above products. β-amylases yield mostly maltose as an end product. This assay is detailed under the uses of enzymes in the clinical laboratory. Blood levels of α-amylase are increased in acute pancreatitis and pancreatic carcinoma (as are serum lipase levels). They are lowered in conditions involving pancreatic insufficiency.

5. Lactic Acid, Pyruvic Acid, and Lactic Acid Dehydrogenase (LDH)

These can be measured by using appropriate variations of NAD^+-NADH oxidation-reduction reactions which are followed spectrophotometrically at 340 nm (see uses of enzymes in the clinical laboratory). These parameters are useful in evaluating a variety of conditions such as liver disease, congestive heart failure, muscular dystrophy, thiamine deficiency, neoplastic diseases, diabetes mellitus, and others. LDH is discussed extensively in the section on glycolysis. Note: Lactate accumulation can give rise to metabolic acidosis.

6. Creatine Phosphokinase (CPK)

   a.

   $$\text{Creatine phosphate} + \text{ADP} \xrightarrow{\text{CPK}} \text{Creatine} + \text{ATP}$$

   $$\text{ATP} + \text{glucose} \xrightarrow{\text{hexokinase}} \text{ADP} + \text{glucose-6-}\textcircled{P}$$

   $$\text{glucose-6-}\textcircled{P} \xrightarrow[\substack{NADP^+ \quad NADPH \\ + H^+}]{\text{G-6-}\textcircled{P}\ \text{dehydrogenase}} \text{6-phosphogluconate}$$

   This is another example of a pyridine-nucleotide linked assay (see the uses of enzymes in the clinical laboratory). Under appropriate conditions, the increase in absorbance at 340 nm is proportional to CPK activity.

   b. The CPK level in the blood increases due to a myocardial infarction as early as six hours after the infarct. Muscle trauma (such as an intramuscular injection), unrelated to a myocardial infarct, also gives rise to an elevated serum CPK.

5
Nucleic Acids

1. Nucleic Acids. There are two types of nucleic acids: ribonucleic acids (RNA) and deoxyribonucleic acids (DNA). The "deoxy" refers to the absence of an -OH in the 2-position of the ribose molecules. They are called acids because, due to the presence of the phosphate groups, they behave chemically as acids. The diagram below illustrates their components.

Constituents of Nucleic Acids

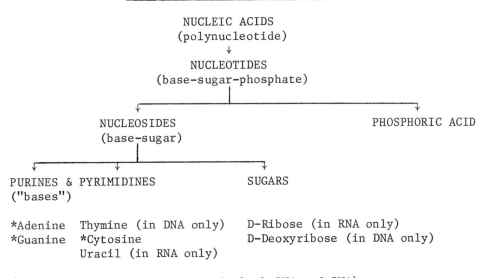

(* means that the base occurs in both DNA and RNA)

2. Mononucleotides (or just nucleotides) are structural units of nucleic acids. They consist of

   Purine or pyrimidine (nitrogenous base)

   D-ribose or 2-deoxy-D-ribose (sugar)

   Phosphoric acid

   a. The nitrogenous bases found in most nucleotides are indicated next. The numbering systems are those used by Chemical Abstracts.

297

Purines

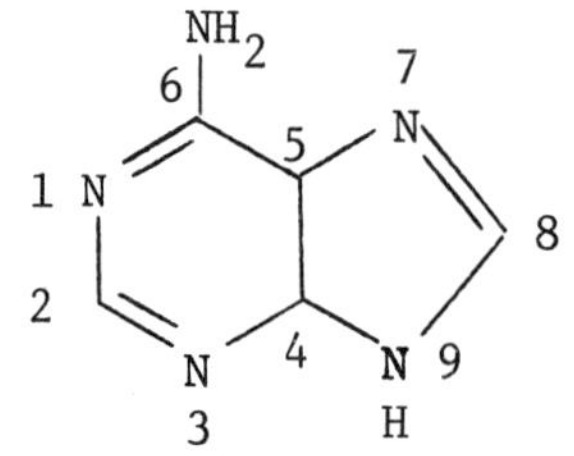

NH$_2$
6
7
5
1 N
N
8
2
N   4   N 9
3   H

O
N
HN
N
H$_2$N   N   N
H

Adenine: 6-aminopurine 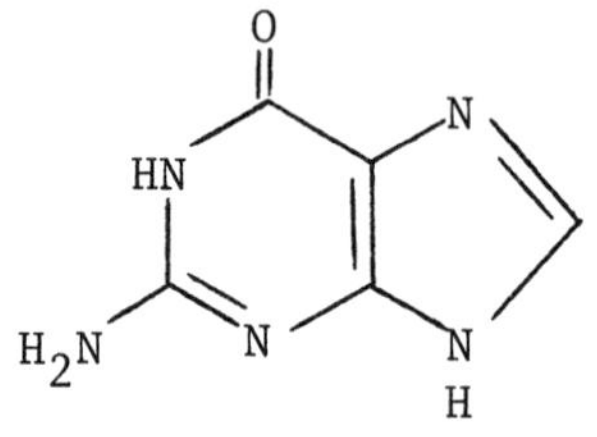   Guanine: 2-amino-6-oxypurine

Pyrimidines

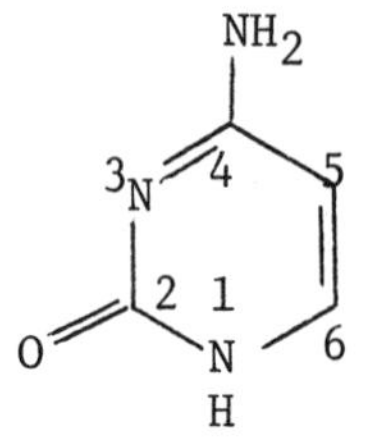

NH$_2$
3 N   4   5
2   1
O   N   6
H

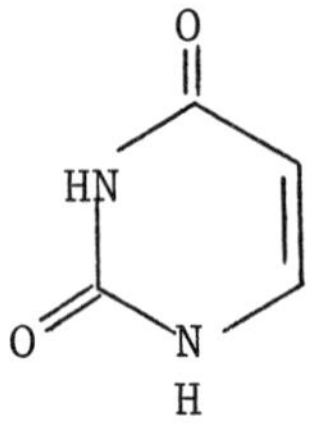

O
HN
O   N
H

O
HN
CH$_3$
O   N
H

Cytosine: 2 oxy-   Uracil: 2,4-   Thymine: 5-methyl-
            4 amino-   dioxy-   2,4-dioxy-
            pyrimidine   pyrimidine   pyrimidine

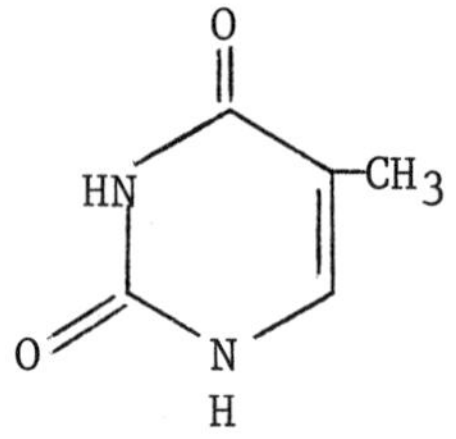

Other bases, particularly methylated derivatives of the
five indicated above, are found in some instances.
For example, many (probably all) transfer RNA (tRNA)
molecules contain inosine (a purine), pseudouridine
(a pyrimidine), and others (see under RNA).

b. In oxypurines and oxypyrimidines, an equilibrium exists between
keto and enol forms. This interconversion, known as
tautomerization, involves the exchange of two protons between
different positions on the molecule. This is illustrated
below for uracil. This property is very important in nucleic
acid structure and the genetic code, since the ability of the
keto and enol forms to hydrogen bond differs. The tautomers
favored in the equilibrium are indicated on the above
page and this form will be used throughout the book. The
constant presence of at least small amounts of the other form
may be a contributory factor in the rate of spontaneous
mutation.

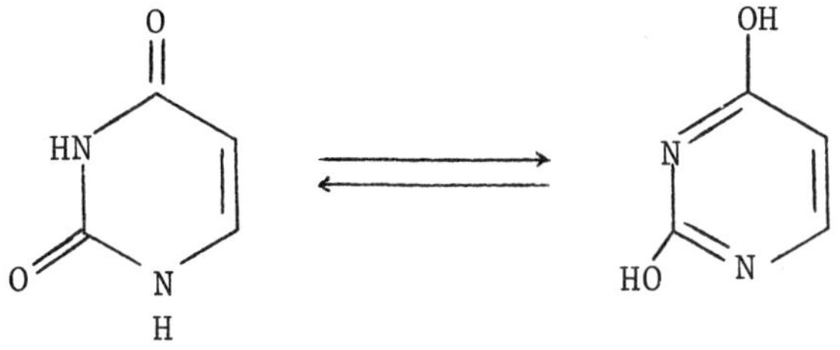

Lactam or keto form   Lactim or enol form
    of Uracil   of Uracil

c. Nucleosides are β-N-glycosides of pyrimidine and purine bases.
    Examples are shown below.

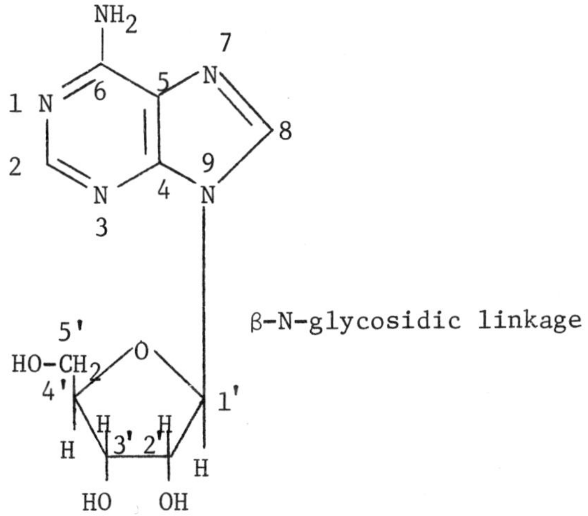

β-N-glycosidic linkage

This compound is <u>adenosine</u> or <u>adenine ribonucleoside</u>
(9-β-D-ribofuranosyladenine). If the -OH on the 2'-position
of the ribose is replaced by -H, the corresponding
2'-deoxynucleoside (in this case 2'-deoxyadenosine) is formed.
Note that it is the N-9 position on the purine ring which is
involved in the N-glycosidic bond. In pyrimidine nucleosides,
the N-1 nitrogen forms the linkage.

d. <u>Nucleotides</u> are formed from nucleosides by addition of a
    phosphate group. The phosphate is attached in an ester
    linkage, usually to the 5'-hydroxyl but sometimes to the
    3'-hydroxyl of the ribose. If the position (5' or 3') is not
    specified, the 5'-ester can generally be assumed. For example,
    adenine-5'-nucleotide is more usually known as adenylic acid
    (since the phosphate group is acidic) or adenosine

monophosphate (AMP). If additional (one or two) phosphate
groups are attached to the first phosphate (by anhydride
linkages), the compounds formed are called adenosine diphosphate
(ADP) and adenosine triphosphate (ATP). Notice that
nucleoside di- and triphosphates are exactly the same as
nucleotide mono- and diphosphates. The word nucleotide already
indicates the presence of one phosphate group. If a deoxy
sugar is present, the term deoxy is used as a prefix with
the nucleotide name. When bases other than adenine are
involved, the corresponding names (guanosine, guanylic acid;
cytosine, cytidylic acid; uridine, urydylic acid; thymidine,
thymidylic acid) are used. A summary of these names is given
with purine and pyrimidine nucleotide biosynthesis.

Functions of Nucleotides

1. By means of the ATP $\overset{\rightarrow}{\leftarrow}$ ADP + P$_i$ interconversion, ATP serves as the
    <u>primary carrier of chemical energy</u> in the cell. (Other nucleoside
    triphosphates (NTP's) are also involved, but to a much smaller
    extent.) The energy can be used for mechanical function (as in
    the ATPase of muscle) and to drive chemical reactions (as in the
    synthesis of glutathione). For reactions of this sort,
    only the terminal phosphate anhydride is usually hydrolyzed,
    giving ADP and P$_i$.

2. Different parts of NTP molecules are used as <u>carriers and
    activators</u> of a variety of groups in biosynthetic reactions.
    Several examples are given below.

    a. Adenosine Triphosphate

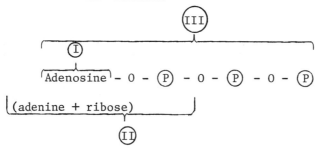

    I . adenosine, as <u>S</u>-adenosylmethionine, carries "active
         methyl" groups for biological methylations (see
         metabolism of the sulfur-containing amino acids).

    II . AMP, as part of 3'-phosphoadenosine-5'-phosphosulfate
          (PAPS; see metabolism of sulfur-containing amino acids)
          provides "active sulfate" for synthesis of the

chondroitin sulfates, etc. AMP is also the carrier of
amino acids during the loading (activation) of the
tRNA's used in protein synthesis.

III . ADP is part of coenzyme A. In this and other molecules
containing NDP's (see below), even though ATP is the
immediate source of adenosine, usually only one
phosphate comes along with it. The other phosphate was
already part of the molecule to which the adenosine
monophosphate is being added. For example:

4'-phosphopantetheine + ATP $\longrightarrow$ dephospho-CoA + PP_i.

This also illustrates the release of inorganic
pyrophosphate (PP_i), another feature of many reactions
in which a NMP is incorporated into some other molecule.
The pyrophosphate is rapidly hydrolyzed to inorganic
phosphate by pyrophosphatase.

$$PP_i + H_2O \xrightarrow{\text{pyrophosphatase}} 2\ P_i$$

About 7.3 kcal are released per mole of PP_i hydrolyzed.
This helps to make the resynthesis of ATP by the
reversal of these reactions quite difficult and the
desired product (here, dephospho-CoA) much more stable.

b. Uridine triphosphate is used to synthesize UDP-glucose,
UDP-galactose, UDP-glucuronic acid, and other UDP-sugars.
These compounds are used in the biosynthesis of polysaccharides
and for other reactions. See for example glycogen synthesis,
galactose metabolism, the uronic acid pathway, amino sugar
metabolism, and bilirubin (hemoglobin) metabolism.

c. Cytidine triphosphate serves as a source of CMP for CDP-choline
and CDP-diglyceride in lipid and phospholipid biosynthesis.

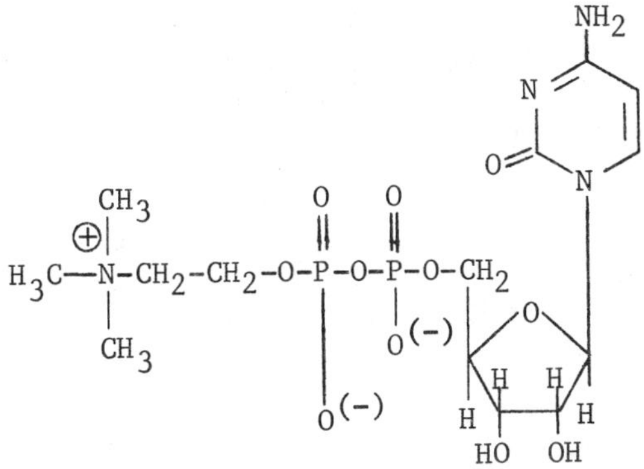

Cytidine
diphosphate
choline
(CDP-choline)

3. NTP's, especially ATP, <u>provide inorganic phosphate and pyrophosphate for biosynthesis.</u>

    a. Phosphorylation reactions include those catalyzed by hexokinase, glucokinase, fructokinase, the protein kinases, and others. These are discussed in a number of places. If X is a molecule which can be phosphorylated, a general reaction is

$$X + NTP \xrightarrow{\text{kinase}} X-P + ADP$$

    b. Pyrophosphate transfers (pyrophosphorylations) are less common. One important example is the formation of 1-phosphoribosyl-5-pyrophosphate (PRPP), the form of ribose used in nucleotide synthesis. The reaction is

$$Ribose-1-P + ATP \longrightarrow PRPP + AMP$$

    This is discussed further under purine and pyrimidine synthesis.

4. A number of <u>coenzymes</u> involved in electron transfer reactions contain nucleotides. These include NAD^+, $NADP^+$, and FAD. FMN's (and part of FAD) are not true nucleotides since they contain D-ribitol, a sugar alcohol, in place of D-ribose. The nitrogenous base in FMN (and part of FAD) is 6,7-dimethylisoalloxazine (flavin), which is not one of the bases found in the nucleic acid nucleotides. Coenzyme A, mentioned above, is another nucleotide coenzyme.

5. ATP, CTP, GTP, and UTP are <u>precursors for the nucleotides in RNA.</u> The deoxynucleotides (dATP, dCTP, dGTP, and dTTP) probably function similarly in DNA synthesis. Some evidence exists, however, that the deoxynucleotide triphosphates are not the <u>immediate</u> precursors of DNA (i.e., they are not substrates for DNA polymerase), at least in <u>E. coli</u>. There may also be two or more intracellular pools of nucleotides which provide material for different processes. In tRNA, the "unusual" nucleotides (those containing bases other than A, C, G, or U) are usually if not always the result of modifications performed on A, C, G, or U <u>after</u> these "usual" nucleotides are already part of the polynucleotide. This is equivalent to postribosomal modification in protein synthesis.

6. The most recently described function of nucleotides is their role as <u>intracellular hormones or "second messengers".</u> This is best typified by cyclic AMP, discussed previously. Other cyclic nucleotides may perform a similar function.

Structure and Properties of DNA

1. A portion of a polynucleotide chain is shown below.

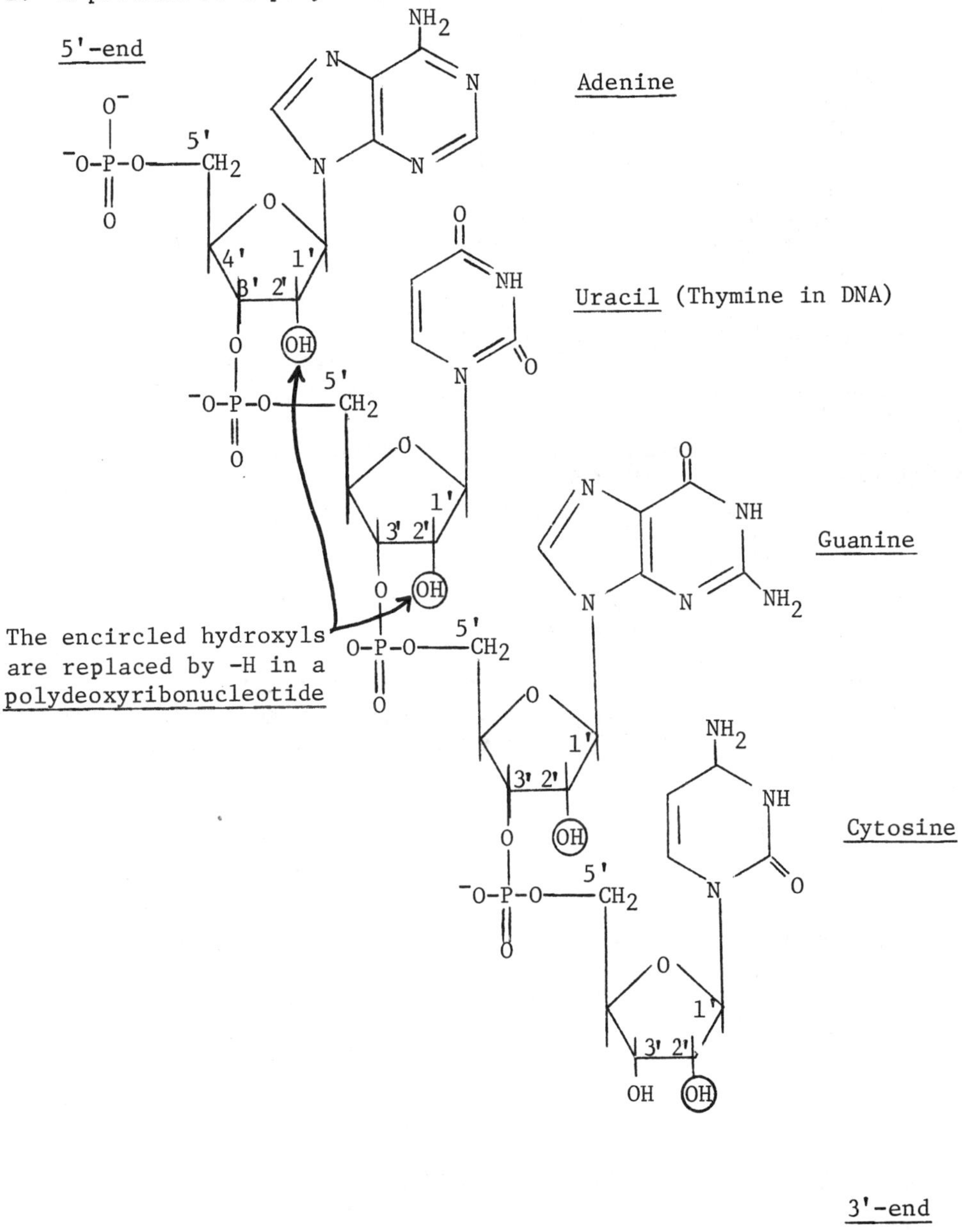

5'-end

Adenine

Uracil (Thymine in DNA)

Guanine

The encircled hydroxyls are replaced by -H in a polydeoxyribonucleotide

Cytosine

3'-end

Regarding this structure, notice that

a. In both DNA and RNA the nucleotides are connected by
    3',5'-phosphodiester bonds.

b. There should be a free hydroxyl group at the 3'-end of the
    polymer of which this sequence is a part, since 5'-triphosphates
    are thought to be the precursors.

c. The presence of a 5'-triphosphate group has never been
    demonstrated in DNA synthesized in vivo or in vitro despite
    the apparent role of the 5'-triphosphates as precursors.

d. The only differences between DNA and RNA are (1) the 2'-OH
    groups in the latter and (2) thymine in DNA v.s. uracil in RNA.

2. Strong evidence supports the theory that DNA is the repository of
    genetics information within the cell. DNA is present primarily
    in the nucleus or nuclear zone. Much smaller amounts are found
    in other locations within the cell, as indicated in Table 8. One
    of the most significant differences between eucaryotic cells
    (those of "higher" plants and animals) and procaryotic cells (small,
    primitive cells) is the way in which their DNA exists. A
    comparison of these two cell types is made in Table 8. Because
    of the simplicity of the procaryotes, much research in enzymology
    and molecular biology has been done with them. Results of such
    work frequently do not apply to mammals (eucaryotes). Great care
    must be taken to avoid drawing conclusions about the functioning of
    higher organisms from data collected in procaryotic systems.

3. Watson and Crick (1953) postulated a double-helical structure for
    DNA based on the following facts which were then available to them.

    a. Chemical Analysis (due largely to I. Chargaff and his coworkers)

        i)   Base equivalence - the number of adenine bases equals
             the number of thymine bases; similarly, the number of
             guanine bases equals the number of cytosine bases (i.e.,
             A/T = 1.0 and G/C = 1.0).

        ii) The sum of the purine nucleotides = the sum of the
             pyrimidine nucleotides (A + G = T + C + MC; MC = 5-
             hydroxymethyl-cytosine, found in a few DNA's in place of
             some or all of the cytosine).

        iii) 6-amino-containing bases = the 6-keto-containing bases
             (A + C + MC = G + T).

Table 8. Comparison of Eucaryotic and Procaryotic Cells

Procaryotic Cells   Eucaryotic Cells

(Eubacteria (including   (All higher plants and animals,
Escherichia coli and   including man)
Bacillus subtilis); the
spirochetes and rickettsiae;
and others)

  I) Small cells   I) Cells are 1000–10,000 times
                                                larger

 II) Cytoplasm is structureless   II) In addition to the outer cell-
     except for storage granules,   membrane, many structured,
     ribosomes, and the nuclear   membrane–bounded organelles occur
     zone; only membrane present   within the cytoplasm. These
     is the cell membrane,   include mitochondria,
     surrounding the entire cell   chloroplasts (in photosynthetic
                                                cells), Golgi bodies, rough and
                                                smooth endoplasmic reticulum,
                                                peroxisomes, lysosomes, and
                                                the nucleus

III) Ribosomes have a   III) Larger ribosomes (sediment at
     sedimentation rate (in the   80S); subunits are 40S and 60S
     ultracentrifuge) of 70S;   (varies somewhat between
     composed of two subunits   plants and animals)
     (30S and 50S)(see under RNA)

 IV) Most DNA exists as a single   IV) DNA is distributed among multiple
     macromolecule (chromosome);   chromosomes (8 in Drosophila
     it is thought that no protein   (fruit fly); 46 in man; 78 in
     is associated with the DNA   chickens); one or more molecules
     (although this is not well   of DNA per chromosome; DNA
     established; in B. subtilis   associated with large quantities
     there is some evidence for   of basic proteins (protamines
     DNA association with a basic   and histones; DNA + protein =
     protein); small amounts of   chromatin); nucleus surrounded
     cytoplasmic DNA occur and   by membrane; nucleolus (site of
     are called plasmids or   ribosomal synthesis) associated
     episomes   with nucleus; some DNA
                                                (0.1–0.2% of total cellular DNA)
                                                found in mitochondria and
                                                chloroplasts; very small amounts
                                                of DNA (satellite DNA) found
                                                free in the cytoplasm

iv) The base ratio (A + T)/(G + C + MC), known as the
     dissymmetry ratio, is a characteristic ratio for a given
     species and shows species variation.

b. <u>X-ray diffraction analysis</u> of DNA fibers by Franklin and
    Wilkins showed a regularity in the physical structure of
    the DNA molecule. (Wilkins shared the Nobel Prize in medicine
    and physiology with Watson and Crick in 1962.)

c. The development, by Cochran, Crick and Vand (1952), of the
    theory of diffraction of helical molecules and the
    postulated (Pauling, Corey, and Branson, 1951) α-helical
    structure for some polypeptides   (Pauling received a Nobel
    Prize for his work on protein structure) were important also.

4. Details of their proposed structure are given below.

a. The basic structure is two chains wound around a common
    (hypothetical) axis, with the ribose-phosphate backbone on the
    outside (away from the center) and the bases pointing in,
    towards the axis. The diameter of the helical cylinder is 20 Å.

b. The bases are planar and in the keto tautomeric form. The
    planes of the bases are all perpendicular to the central axis
    and the distance from one base to the next is 3.4 Å. This
    results in the stacking of the bases, one upon the other,
    not unlike a pile of plates (albeit a twisted one, since the
    stack must follow the curve of the helix). There are ten
    nucleotides in each complete turn of the helix. The
    van der Waal's and hydrophobic interactions resulting from
    the base stacking provide at least part of the energy needed
    to stabilize the structure.

c. The two chains are complementary and are of opposite polarity
    to each other. This is shown schematically below. In the
    drawing, the four bases are represented by their initial
    letters (A, C, T, G) and the ribose-phosphate backbone is
    indicated by the dashes between the letters; the dotted
    lines between bases represent hydrogen bonding.

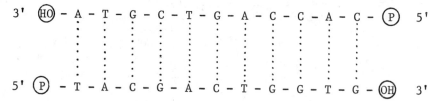

Note the opposite polarity of the strands (one runs 3' → 5'
while the adjacent (complementary) one runs 5' → 3') and the
complementarity of the bases (A always opposite T; G always
opposite C). Notice in each case, a purine is paired
with a pyrimidine.

d. The complementarity indicated above is necessary to explain
the observed base ratios (chemical composition). The origin
of the specificity (A-T and G-C) lies in the ability of these
pairs to form the most stable hydrogen-bonded dimers. The
hydrogen-bonding patterns thought to be significant are
indicated below. As in the case of the α-helix and peptide-
group geometry (mentioned before),knowledge of this pattern
and the base geometries contributed to the prediction of the
double helix.

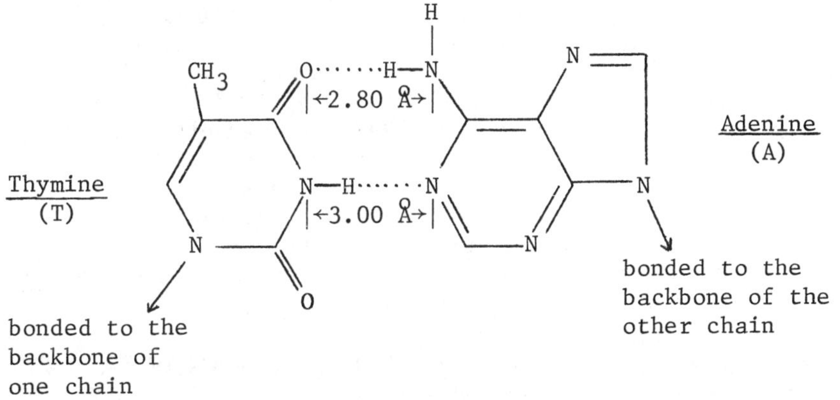

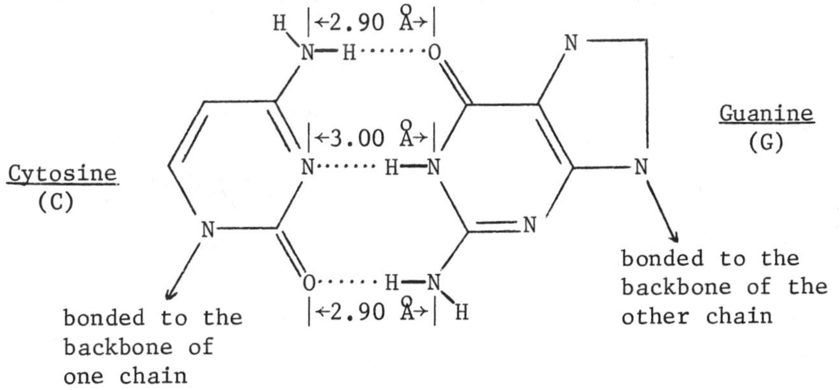

Thymine
(T)

bonded to the
backbone of
one chain

Adenine
(A)

bonded to the
backbone of the
other chain

Cytosine
(C)

bonded to the
backbone of
one chain

Guanine
(G)

bonded to the
backbone of the
other chain

Other hydrogen-bonding patterns (notably those proposed by
Hoogsteen) can also be drawn, but they are probably of lesser
significance (however, see under tRNA).

5. The two strands and the specificity of the base pairing in this structure account for accurate replication of the DNA molecule during mitosis.

6. The DNA duplexes undergo unwinding (denaturation) upon heating, changes in pH, and other denaturing influences (see protein denaturation). Upon denaturation, the optical absorbance at 260 nm increases due to the unstacking of the bases. This change in optical density is a very useful method for measuring the degree of denaturation of a DNA solution under various circumstances. Other changes (decreased viscosity, increase in optical rotation, increase in buoyant density) also occur upon denaturation.

7. Experiments in bacterial transformation (transfer of genetic information from one bacterial cell to another bacterial cell; discovered by Griffith and used extensively by Avery and others), viral transduction (virus-mediated transfer of parts of the bacterial genome from one cell to another), other viral techniques (notably the Hershey-Chase experiments) have helped greatly to establish the role of DNA as the genetic material. Through these experiments, the genetic map (location of the genes relative to one another in the DNA)· of E. coli has been determined. It has been shown that the single chromosome (DNA molecule) in E. coli is a closed loop (circular). Certain other DNA's, isolated from several sources (some viruses, mitochondria, other bacteria), appear also to be circular.

8. It has been observed, in general, that the amount of DNA per cell increases with the increase in complexity of a cell's function. This is indicated in Table 9.

Table 9.   The DNA Content of Some Selected Cells

	DNA in picograms per cell or per virion
Bacteriophage T	0.00024
Bacterium (E. coli)	0.009
Fungi (Neurospora crassa)	0.017
Higher Plant (tobacco)	2.5
Sponge (Tube sponge; diploid)	0.12
Bird (chicken erythrocytes)	2.5
Reptile (alligator)	4.98
Fish (carp)	3.0
Amphibia (Xenopus laevis)	8.4
Mammal (human)	6.0

9. In the genomes of eucaryotic cells (but most probably not in viruses
    and bacteria), a given nucleotide sequence may be repeated many
    times (gene amplification). In other words, many extra copies of
    the same gene are present. This observation is widespread and may
    be true in all higher organisms. It may help to explain why
    amphibians, for example, have more DNA per cell than do mammals.

    The occurrence of multiple copies of some genes was concluded
    largely through the use of DNA-DNA hybridization experiments. In
    doing these, one first isolates DNA from the cells to be tested,
    denatures it to obtain single-stranded DNA, and immobilizes it by
    bonding it to some material (e.g., resin beads) which can be used
    to pack a chromatography column. Labeled (usually tritiated) DNA
    is then prepared by growing more of the cells on a medium
    containing radioactive thymidine. This DNA is similarly isolated
    and denatured, then passed over the beads containing the bound
    "reference" DNA. Renaturation (annealing) of the DNA is a second-
    order process, depending on the frequency with which two genes
    capable of pairing meet each other. Consequently, the more
    copies there are present of one gene, the greater is the probability
    that a particular meeting (collision) will be successful (result
    in a stable, base-paired hybrid). The amount of pairing which
    occurs in a given length of time is therefore directly related to
    the number of copies of each gene present. Alternatively, one
    can say that the rate of hybridization is proportional to the
    number of copies of the same DNA sequence which are present.

    Nucleic acid hybridizations (DNA-DNA, DNA-RNA and RNA-RNA) are
    also used to measure the degree of homology between the genomes
    of different species and hence to evaluate how closely related
    two species are (molecular taxonomy), to measure the number of
    copies of RNA made from a particular gene (DNA molecule), and
    for other purposes. These methods generally measure the maximum
    amount of DNA which can be hybridized, rather than the rate of
    hybridization. Such experiments are easier with procaryotic
    nucleic acids than with eucaryotic ones, due to the much greater
    complexity of the genetic materials in the latter cells. Some
    very interesting data have resulted from these experiments, despite
    the many pitfalls in interpretation of the results.

10. There are at least three reasons why multiple copies of one gene
    might be necessary.

    a. The need for synthesis of large quantities of one type of
        protein or nucleic acid. Thus, many copies of the genes for
        collagen, keratin, and structural proteins of membranes would
        be useful. That this actually occurs has been demonstrated
        in the case of ribosomal RNA.

b. The occurrence, in some proteins, of repeating homologous amino acid sequences. Examples are the immunoglobulins and hemoglobin. In at least the latter case, however, gene amplification has been shown to <u>not</u> occur. Rather, <u>many copies of the mRNA</u> for the hemoglobin peptides are transcribed from a few (2-3) copies of the appropriate gene.

c. A mutation in one copy, leading to the production of an inactive enzyme, will not completely eliminate that enzyme from the cell if there is another good gene for it somewhere else on the DNA. This is also cited as a reason for the occurrence of isozymes (see under proteins and enzymes).

11. Among higher organisms, all of the somatic diploid cells of a given species contain the same amount of DNA. This is independent of the metabolic state of the cell (not modified by diet or environmental circumstances), unlike the RNA content of the same cells. This indicates again the permanence of the DNA complement, a necessary feature for anything which must store and transmit information.

12. The base compositions of DNA specimens vary from one species to another, but the complementarity (A=T; G=C) is present regardless of species.

Replication of DNA

1. Double-stranded DNA undergoes semiconservative replication. This means that the two complementary antiparallel chains separate and each serves as a template for the synthesis of its complement, so that the two daughter double-stranded molecules are identical to the parent, each containing one strand from the parent and one "new" strand. This was postulated by Watson and Crick and experimentally verified by Meselson and Stahl. Although the details are far from clear, it appears that the replication proceeds as the unwinding is occurring, the new synthesis taking place just behind the unwinding-point and moving along with it.

2. There is still a great deal of controversy surrounding the replicative mechanism. Despite the discovery of at least three DNA-dependent-DNA-polymerases, it is not clear which enzyme actually performs the polymerization <u>in vivo</u>. It is also not known for certain exactly what the immediate precursors to the deoxynucleotides in the DNA are.

3. Kornberg and his coworkers were the first to isolate a DNA-dependent-DNA polymerase [(DNA polymerase I), Kornberg enzyme; from <u>E. coli</u>]. For this work Kornberg received the Nobel Prize.

As a polymerase, this enzyme requires

a. The four deoxynucleoside triphosphates (dATP, dTTP, dGTP, dCTP) and Mg^{+2}. All four must be present for synthesis to proceed.

b. A preformed piece of single-stranded DNA (a primer or template). The enzyme does not bind to intact, circular double-stranded DNA, but a nicked or partially damaged double strand can function as a template.

Under these conditions, the enzyme reads the primer strand, starting at the 3' end, and synthesizes a new strand complementary to the primer. The growing point of the new strand is at its 3' terminus. Total synthesis of biologically active DNA of the ØX 174 bacteriophage has been achieved using Kornberg's DNA polymerase and a DNA ligase (connects the 5' end to the 3' end of a strand of DNA, making a circle). This work was published by Goulian, Kornberg and Sinsheimer.

4. The reactions involved in the functioning of DNA polymerase are shown below.

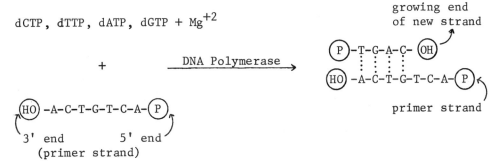

In this reaction, one pyrophosphate (PP_i) is released for each nucleotide added to the polymer. The polymerization is made more irreversible by the action of a pyrophosphatase, which catalyzes the highly exergonic reaction.

$$PP_i + H_2O \xrightarrow{\text{pyrophosphatase}} 2 P_i$$

This was described previously in the section on functions of the nucleotides. The structural relationship of the growing chain to the nucleotides being added is shown in the following diagram.

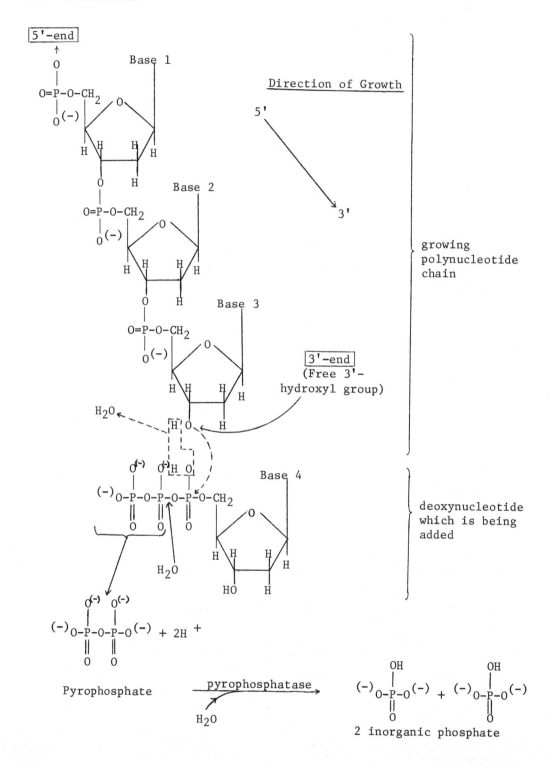

Direction of Growth

5'

3'

growing
polynucleotide
chain

3'-end
(Free 3'-
hydroxyl group)

Base 4

deoxynucleotide
which is being
added

Pyrophosphate

$\xrightarrow{\text{pyrophosphatase}}$

H_2O

2 inorganic phosphate

5. The Kornberg enzyme possesses two types of exonuclease activity in
    addition to the polymerase activity described above. It can
    remove, by hydrolysis, nucleotide residues from either the
    3'-(3' → 5' exonuclease) or 5'-(5' → 3' exonuclease) end of a
    single strand of DNA. This ability may aid in the elimination
    of mismatched nucleotide pairs so that DNA synthesis proceeds
    correctly; or it may be needed by the enzyme in performing DNA
    repair, described below.

6. The accurate transmission of genetic information requires great
    stability of the structure of DNA in order to guarantee precise
    replication. To assist in this, cells have developed certain
    error-correcting mechanisms. One common type of damage which
    most cells can reverse is thymine-dimer formation. Adjacent
    thymine residues in the DNA molecule, upon exposure to ultraviolet
    light, undergo the reaction shown below:

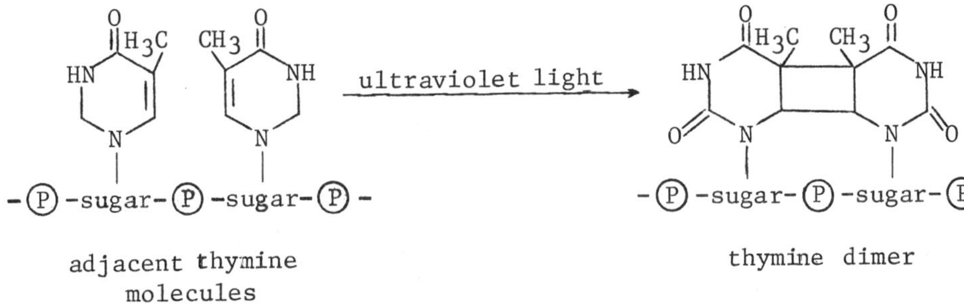

adjacent thymine   thymine dimer
     molecules

These dimers form "lumps" in the DNA and prevent replication. Two
mechanisms are known for dimer removal.

a. Photo-reactivation is just the reverse of the above reaction.
    There is thought to be an enzyme which binds selectively to
    thymine dimers. Upon exposure to visible light the dimers
    are cleaved and the photoreactivating enzyme is released.

b) Thymine dimers can also be reactivated in the dark in a more
    complex process probably involving excision of the dimer and
    resynthesis of a short stretch of DNA. An endonuclease
    (enzyme which hydrolyzes nucleic acids at points other than
    the ends) specific for regions containing thymine dimers has
    been found in Micrococcus luteus. This enzyme makes a nick
    (single cut) in the defective strand and exposes a free 3'-OH
    for a polymerase to act on. The resynthesis of the defective
    region (complementary to the undamaged strand) and final
    excision of the dimer-region are probably performed by DNA

polymerase I. Evidence in support of this includes the discovery of a strain of E. coli mutants which are more sensitive to ultraviolet light than is the wild type and which lack the DNA polymerase I enzyme; and the fact that the requirements for thymine-dimer repair synthesis are the same as those for DNA polymerase I synthesis of DNA. For these and other reasons, Kornberg's enzyme is probably not directly involved in DNA replication in vivo. The use of the complementarity of the two strands of DNA to direct the repair process may indicate an evolutionary reason for the development of double-stranded DNA.

   c. In humans, xeroderma pigmentosum is an autosomal recessive disease in which the mechanism of repair synthesis is defective in the skin fibroblasts. Persons so afflicted are more likely to develop skin cancer and show an abnormal sensitivity to exposure to sunlight. The defect appears to be a lack of the endonuclease needed to recognize the thymine dimers and make the initial nick.

7. The biggest setback to the theory that DNA polymerase I was the replicative enzyme came from the discovery of E. coli mutants capable of mitosis yet lacking in this enzyme. Since then, two other DNA polymerases have been found in these mutants. Both appear to be membrane-bound enzymes (which may explain why they were not isolated earlier). They have been designated DNA polymerase II and DNA polymerase III. It appears that polymerase III may be the in vivo replicative enzyme.

8. DNA replication has been much more thoroughly studied in viruses and procaryotic cells than in eucaryotic ones, as is the case with most processes involving nucleic acids. Much work has been devoted to explaining functions which are largely unique to these "simple" systems, such as replication of circular chromosomes. Bacterial and viral studies are also simplified by the apparent absence of the histones and other basic proteins present in eucaryotic chromosomes. It still remains to be seen just how many of the findings in such systems really apply to man and the other mammals. There is no doubt, however, that much valuable insight and many novel techniques have arisen from the work on primitive genomes.

Types and Properties of RNA

1. RNA (ribonucleic acid) is the second type of nucleotide polymer in the cell. The principal differences between DNA and RNA are the presence of an -OH group in the 2' position of the ribose in RNA, and the occurrence of uracil in place of thymine as one of the four bases in RNA.

There are four major types of RNA in the cell.

a. Ribosomal RNA (rRNA)

b. Transfer RNA (tRNA) also called soluble or sRNA

c. Messenger RNA (mRNA)

> found in all cells; function discussed below

d. Nuclear RNA (nRNA) - found only in eucaryotes; function not known but may be involved in the control of transcription (see under regulation of enzyme synthesis); not well characterized

2. Properties of E. coli RNA's

Type	Sedimentation Coefficient	Approximate Molecular Weight (daltons)	No. of Nucleotide Residues	% of Total Cellular RNA
ribosomal RNA (3 species)	5S	35,000	~120	82
	16S	550,000	~1,500	
	23S	1.1×10^6	~3,000	
transfer RNA (60 species)	4S	23,000 to 30,000	75-90	16
messenger RNA (many species)	6S-25S	25,000 to 1×10^6	75-3000	~2

3. In bacterial cells, all RNA is found in the cytoplasm but in the mammalian cells the RNA is distributed among the various organelles. For example, in a liver cell the nucleus contains 11% of the total cellular RNA, the mitochondria has 15%, the ribosomes have 50%, and 24% is free in the cytoplasm. The most stable RNA's are tRNA and rRNA. The least stable is mRNA.

4. RNA has been shown to be the genetic material in most (if not all) plant viruses (such as the wound tumor and tobacco mosaic viruses); some bacterial viruses or bacteriophages (the Qβ and R[17] viruses of E. coli, for example); and some animal viruses (poliomyelitis and type A influenza; leukemia and solid tumor viruses of mice and chickens; others). Some of these viruses have been shown to possess an RNA-dependent-DNA-polymerase (an enzyme which can form DNA polymers by reading an RNA template; reverse of DNA-dependent-RNA-polymerase) and to require active DNA replication for infection. The presence of this polymerase in some human leukemic cells has been interpreted as suggestive evidence for a viral etiology of some human cancers.

5. Ribosomal RNA (rRNA) is found associated with protein in ribosomes.
    These are roughly spherical bodies found in most cells. Their
    role in protein synthesis is discussed later. There are three
    principal types of ribosomes: procaryotic ribosomes, eucaryotic
    cytoribosomes (found in the cytoplasm and bound to the rough
    endoplasmic reticulum), and mitoribosomes (present in the
    mitochondria and chloroplasts of eucaryotic cells).

    a. The most thoroughly studied ribosomes are those of E. coli.
        They are composed of two subunits, whose composition is
        given in Figure 47.

Figure 47. Molecular Composition of Ribosomes in E. coli

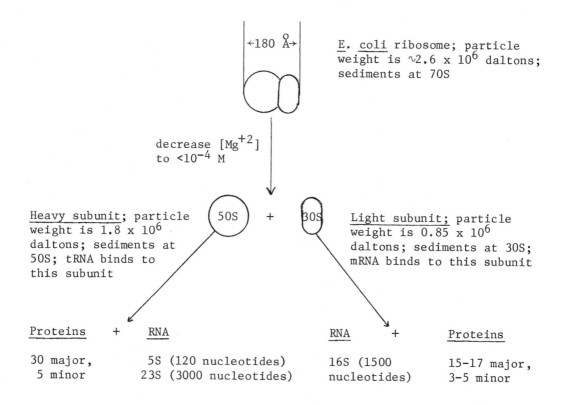

$\leftarrow 180 \overset{o}{A} \rightarrow$

E. coli ribosome; particle weight is ~2.6 x 10^6 daltons; sediments at 70S

decrease [Mg^{+2}] to <10^{-4} M

50S + 30S

Heavy subunit; particle weight is 1.8 x 10^6 daltons; sediments at 50S; tRNA binds to this subunit

Light subunit; particle weight is 0.85 x 10^6 daltons; sediments at 30S; mRNA binds to this subunit

Proteins	+	RNA	RNA	+	Proteins
30 major, 5 minor		5S (120 nucleotides) 23S (3000 nucleotides)	16S (1500 nucleotides)		15-17 major, 3-5 minor

    b. Eucaryotic cytoribosomes, while basically similar from species
        to species, show much more heterogeneity than do procaryotic
        ribosomes. A comparison of the two types is presented below.
        Considering the other major differences between eucaryotes and
        procaryotes, the similarities in functions and properties of
        the ribosomes are probably more important than the differences.

Procaryotic Ribosomes	Eucaryotic Cytoribosomes
60-65% RNA	About 55% RNA
35-40% protein	About 45% protein
Particle weight is about 2.6×10^6 daltons; diameter is 180 Å; sedimentation coefficient of 70S	Somewhat larger, with a particle weight of about 3.6×10^6 daltons and a sedimentation coefficient of 77-80S
Two subunits of sedimentation coefficients 30S and 50S	Two subunits of sedimentation coefficients 38-45S and 58-60S (particle weights of 1.2×10^6 and 2.4×10^6 daltons, respectively)
Large subunit has two rRNA molecules of size 5S and 23S; small subunit has one rRNA of size 16S	Large subunit has 2 or 3 rRNA molecules; small subunit has one rRNA molecule; size of rRNA is somewhat species-dependent
Genes (cistrons) for rRNA synthesis are part of the one bacterial chromosome	rRNA is synthesized on DNA contained in the nucleolar organizer; this becomes part of the nucleolus during mitosis
Ribosomes occur in the cytoplasm either free or associated with mRNA (polysomes)	Ribosomes occur free in the cytoplasm or bound to the rough endoplasmic reticulum; when associated with mRNA, they are called polysomes

c. Mitoribosomes are found in the chloroplasts and mitochondria
    present in the cytoplasm of eucaryotic cells. These organelles
    also have their own supply of DNA which is used for the
    synthesis of some (but by no means all) of the proteins and
    enzymes in these structures. Although these ribosomes resemble
    procaryotic ribosomes with respect to inhibitor sensitivity,
    details of peptide-chain initiation, and other properties,
    they are distinct from them in size and RNA composition. In
    some ways they appear to be intermediate between the
    cytoribosomes and the bacterial ribosomes. For this and other
    reasons, it has been suggested that mitochondria and
    chloroplasts originated as bacterial invaders of eucaryotic
    cells. Eventually they lost the ability to exist outside of
    their hosts and a symbiotic relationship developed. There is

little or no experimental data to substantiate this
other than circumstantial evidence.

d. The proteins associated with the ribosomes are basic in nature
and it appears that the major ones are present only once in
a particular ribosome. Although their exact purposes are not
known, they probably function at least partly as structural
proteins to maintain proper conformation of the RNA for
protein synthesis. Other postulated functions for the
ribosomal proteins include

    i) Control of mRNA translation (e.g., selection of the proper
    mRNA molecules).

    ii) The fulfillment of some of the enzymatic functions needed
    for protein synthesis, such as movement of the ribosome
    along the message, transfer of the amino acids from the
    tRNA to the growing chain (peptidyl transferase), GTPase
    activity, etc.

Few experimental data are available to either confirm or deny
these roles.

6. <u>Transfer RNA</u> (tRNA; also called soluble RNA or sRNA)

a. These RNA molecules are carriers of specific amino acids
during protein synthesis on the ribosomes.

b. Each amino acid has at least one specific corresponding tRNA.
Some have multiple tRNA's (leucine and serine each have
5 corresponding tRNA's in <u>E. coli</u>).

c. RNA contains rather large numbers of minor or "unusual" bases
which are usually methylated forms of the four "usual" bases.
(Any base other than A, G, C, or U is considered "unusual"
in RNA.)

d. Examples of these "unusual" nucleotides are

Pseudo-Uridylic Acid

Ribothymidylic Acid
(5-methyl-uridylic acid)

Other "unusual" bases found in tRNA include 4-thiouracil, N^6-isopentyladenine, 2'-O-methyladenosine (a nucleoside), hypoxanthine, 5-methylcytosine, and several more. These bases are formed by specific enzymes (e.g., tRNA methylases) following polymerization of the tRNA on the DNA template.

e. The structure in Figure 48 is a schematic representation of yeast alanyl tRNA. The general features are quite similar to those of the other tRNA's whose structures have been determined. (These include phenylalanyl, tyrosyl, and two species of seryl tRNA, all from yeast.)

Figure 48. Schematic Structure of Alanine Transfer RNA from Yeast

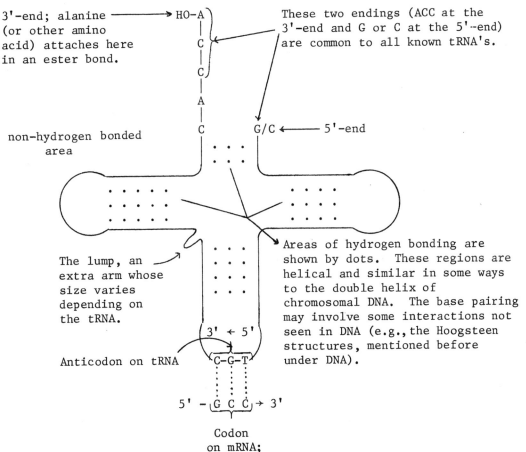

3'-end; alanine ──────→ HO-A (or other amino acid) attaches here in an ester bond.

These two endings (ACC at the 3'-end and G or C at the 5'-end) are common to all known tRNA's.

G/C ←──── 5'-end

non-hydrogen bonded area

The lump, an extra arm whose size varies depending on the tRNA.

Anticodon on tRNA   C-G-T   3' ← 5'

Areas of hydrogen bonding are shown by dots. These regions are helical and similar in some ways to the double helix of chromosomal DNA. The base pairing may involve some interactions not seen in DNA (e.g., the Hoogsteen structures, mentioned before under DNA).

5' - G C C → 3'

Codon on mRNA; mRNA is read in the 5' → 3' direction

Note that the codon on the mRNA (5' → 3') is oriented in the
opposite direction relative to the anticodon on the tRNA
(3' → 5'). This is analogous to the relationship between the
two strands of double-stranded DNA. The fact that the third
letter of the anticodon (here I, for inosine) need not always
be the base which is usually complementary to the third letter
of the codon (here C) is explained by the wobble hypothesis.
Basically this states that due to a certain flexibility or
"wobble" in the recognition of the anticodon bases by those
of the codon, the same tRNA can be recognized by codons which
differ by only the last letter. This is not necessarily true
of all codons or all tRNA molecules, as is indicated by the
occurrence of five seryl tRNA's in E. coli for six codons;
and by the fact that, while asparagine is coded for by AAU and
AAC, the codons AAA and AAG specify lysine (see under genetic
code). In the first case, the six serine codons fall into
two groups.

    i) UCU, UCC, UCA, UCG and

    ii) AGU, AGC.

If the wobble hypothesis was always used, these two groups would
require two seryl tRNA's at most (since the codons are all
either UC- or AG- where the "-" indicates more than one letter).
In the instance of asparagine and lysine, if the last letter
of the codon were totally unimportant, then there would be
frequent errors observed where Asp and Lys were put in the
wrong places in proteins. Moreover, if the third base of a
codon really did not matter, then the three-letter code would be
reduced to an effective two-letter code. For more on this,
refer to the section on the genetic code.

The unusual nucleosides found in yeast alanyl tRNA are
pseudouridine, inosine, dehydrouridine, ribothymidine,
methyl guanosine, dimethyl guanosine, and methyl inosine. They
occur largely (but not exclusively) in or adjacent to the
non-hydrogen-bonded loops and the lump.

f. The general reaction of tRNA with an amino acid is shown below.
    The tRNA, amino acid, and the aminoacyl-tRNA synthetase enzyme
    must all match for the loading of the tRNA to take place.

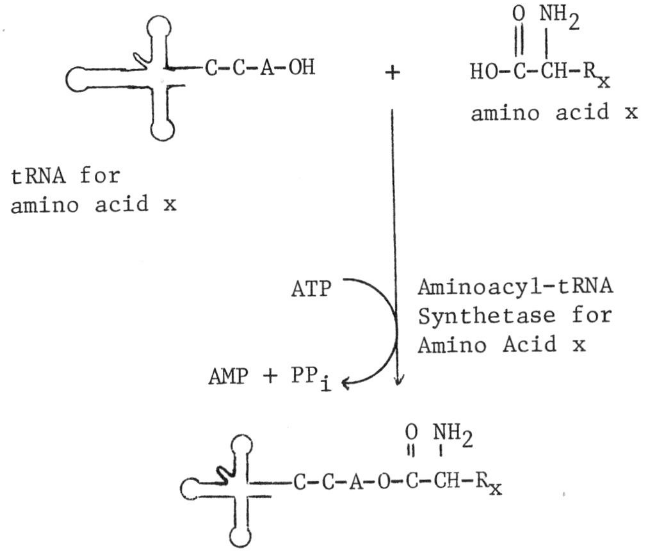

$$\text{tRNA for amino acid x} \quad + \quad \text{HO-C-CH-R}_x \quad \text{(amino acid x)}$$

ATP ⟶
AMP + PP$_i$ ⟵

Aminoacyl-tRNA
Synthetase for
Amino Acid x

Aminoacyl-tRNA$_x$
(loaded or charged tRNA)

The reaction appears to involve an enzyme-bound intermediate
of the form Aminoacyl$_x$ ∿ AMP. Of the twenty common amino acids,
all are attached to tRNA by this reaction except glutamine.
It appears that glu-tRNA (glutamyl-tRNA) is transamidated with
glutamine or asparagine which transfers their amide nitrogen
to the γ-carboxyl of glu to form gln-tRNA (glutamine-tRNA)
(see also protein biosynthesis, below).

g. The first tRNA sequenced was yeast tRNA^{ala} (R.W. Holley and
    coworkers, 1965; Holley received a Nobel Prize for this work
    in 1968). As of 1969,the complete nucleotide sequences of
    twenty tRNA species had been worked out. Work is also
    progressing on the elucidation of the three-dimensional
    structure of tRNA molecules by x-ray diffraction. These
    results indicate a high degree of similarity in the structures
    of many if not all of the different tRNA species.

h. A tRNA molecule must be able to uniquely recognize

    i)   the proper aminoacyl-tRNA synthetase;

    ii) the binding site for tRNA on the large ribosomal subunit;

    iii) the codon on the mRNA.

It has been clearly established that the codon recognition site
is the anticodon base triplet. The ribosomal binding site, which
should be similar or identical on all tRNA's, has not yet been
isolated. The sequence $-G-T-\psi-C-G-$, found in most tRNA's, may
be involved in this interaction. The aminoacyl-tRNA synthetase
recognition site is still unknown, but the anticodon is clearly
not directly involved.

7. <u>Messenger RNA</u> (mRNA) is the ribonucleic acid most similar to DNA.
It is the most unstable (rapidly degraded) RNA with a half-life
varying from 4 to 6 seconds (<u>E. coli</u>) to 8 to 12 hours in rat
liver cells. The half-life is quite dependent on the generation
time of the cell and on the number of proteins which must be made
from each mRNA molecule. mRNA is the intermediate in the
information transfer DNA → RNA → Protein.

a. The synthesis of mRNA (and of the other RNA molecules in the
cell) is shown below. Note the similarity of these reactions
to those catalyzed by DNA polymerase I.

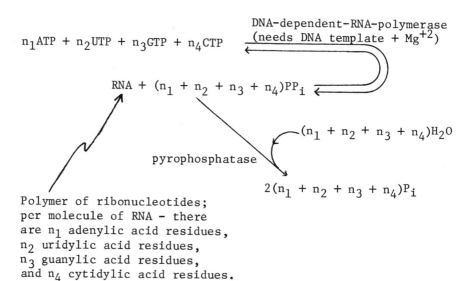

n_1ATP + n_2UTP + n_3GTP + n_4CTP

DNA-dependent-RNA-polymerase
(needs DNA template + Mg^{+2})

RNA + $(n_1 + n_2 + n_3 + n_4)$PP$_i$

$(n_1 + n_2 + n_3 + n_4)$H$_2$O

pyrophosphatase

$2(n_1 + n_2 + n_3 + n_4)$P$_i$

Polymer of ribonucleotides;
per molecule of RNA - there
are n_1 adenylic acid residues,
n_2 uridylic acid residues,
n_3 guanylic acid residues,
and n_4 cytidylic acid residues.

b. Either Mg^{+2} or a 4:1 mixture of $(Mg^{+2} + Mn^{+2})$ is required for
the reaction. If Mn^{+2} alone is used, some deoxyribonucleotides
are incorporated.

c. All four ribonucleoside-5'-triphosphates are required.

d. The enzyme adds mononucleotide units to the 3'-hydroxyl end of the RNA chain so that the direction of RNA synthesis is 5' → 3'. This is the same as the direction in which the DNA is read.

e. The reaction can be reversed by increasing the concentration of pyrophosphate. In the cell a pyrophosphatase (PP_i → 2 P_i) ensures that the reaction proceeds toward the synthesis of RNA.

f. The polymerase enzyme is found in bacterial, plant and animal cells. In the higher organism, it is found in the nuclei and in the mitochondria. Regardless of the source, the same reaction is catalyzed with the same chemical requirements.

g. Unlike DNA polymerase I, RNA polymerase has a complex structure containing five subunits. One of the subunits, designated sigma (or σ), is necessary in order for the RNA polymerase to recognize the DNA start signals for RNA polymerization. There appear to be many different sigmas, each perhaps capable of recognizing different initiator regions on the DNA. It is thought that when viruses invade cells, they provide their own unique sigmas to permit transcription of their genomes by the cells' RNA polymerase molecules.

h. Double-stranded DNA is most active as a template; single stranded DNA shows less activity.

i. The RNA formed has a base composition complementary to that of the template DNA. The frequencies of occurrence of the bases A, T, G, and C in the DNA template are equal, respectively, to the frequencies of occurrence of U, A, C, and G in the RNA formed.

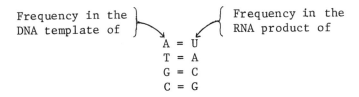

Frequency in the DNA template of $\Bigg\}$ $\Bigg\{$ Frequency in the RNA product of

A = U
T = A
G = C
C = G

j. By using highly purified E. coli enzyme in the presence of double-stranded DNA as a template, it has been shown that only one of the two strands is transcribed. The mechanism is not understood. The strand being transcribed can switch, depending on the region of the DNA.

k. In E. coli the other RNA's (transfer-RNA and ribosomal-RNA) are also made by the enzyme using DNA as template.

1. The product of the RNA polymerase <u>does</u> carry a 5'-triphosphate
    group, unlike the product of Kornberg's enzyme (DNA polymerase
    I). Most RNA chains found <u>in vivo</u> start with either pppA or
    pppG at the 5' end.

m. The antibiotic actinomycin D, inhibits RNA synthesis by binding
    to the DNA, probably adjacent to guanine residues. This
    prevents transcription by the RNA polymerase. Rifampin (also
    rifampicin, rifamycin B, and other closely related compounds
    of the rifamycin family) block RNA synthesis by binding to
    the polymerase and preventing chain initiation. They are
    specific for polymerases of procaryotic origin and do not
    affect those of nuclear, eucaryotic origin. The rifamycins are
    isolated from <u>Streptomyces mediterranei</u> and are currently used
    in the treatment of tuberculosis. Rifampicin is administered
    orally. The structures of actinomycin D and rifampin are shown
    below.

Structure of Rifampin $(C_{43}H_{58}N_4O_{23})$

(Rifamycin B and rifampicin differ in the substituents in the
1 and 2 positions of the Ⓐ ring.)

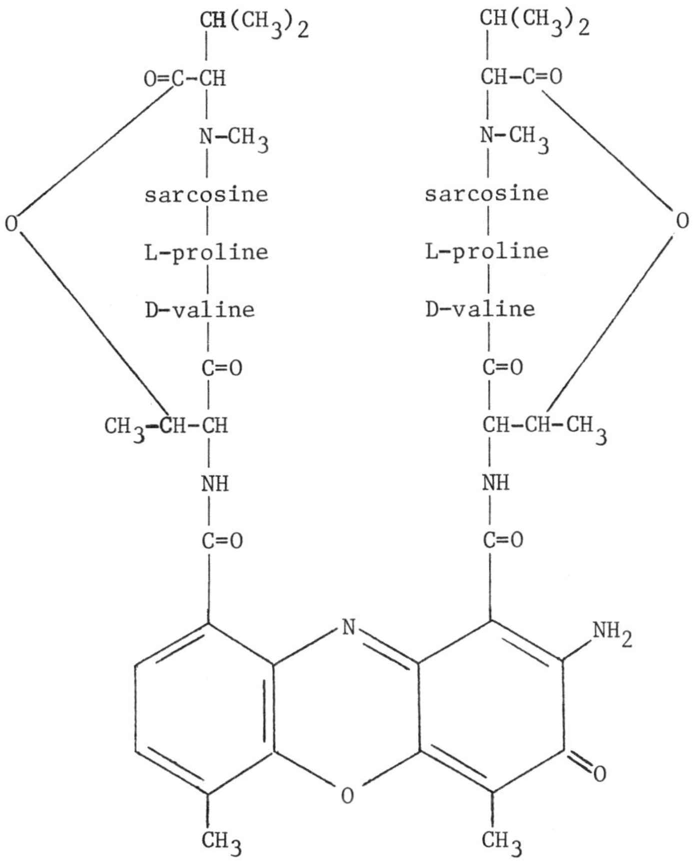

CH(CH$_3$)$_2$ CH(CH$_3$)$_2$

O=C–CH   CH–C=O

N–CH$_3$ N–CH$_3$

sarcosine   sarcosine

L-proline   L-proline

D-valine   D-valine

C=O   C=O

CH$_3$–CH–CH   CH–CH–CH$_3$

NH   NH

C=O   C=O

Structure of Actinomycin D

Nucleases

1. These are enzymes which catalyze the hydrolysis of nucleic acids.
    They have varying degrees of specificity and have proved quite
    useful in the determination of sequence and other properties of
    these compounds. The use of their specificity is the basis of the
    nearest-neighbor analysis, used to prove complementarity of
    various nucleic acid strands.

2. In addition to the obvious difference in specificity between
    ribonucleases and deoxyribonucleases, there are phosphodiesterases
    which are non-specific, hydrolyzing both RNA and DNA. There is
    also the distinction between exonucleases (requiring a free

3' or 5' end to start) and <u>endonucleases</u> (severing polynucleotides
internally; do not require a free 3'- or 5'-hydroxyl group).
Some of the known nucleases are indicated in Table 10, with the
characteristics of their reactions. A polynucleotide chain is
shown schematically below. Notice that, by successive application
of the nucleases described in Table 10, a wide variety of highly
specific products can be obtained.

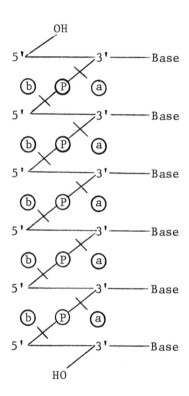

The Genetic Code

1. The sequence of the purines and pyrimidines of mRNA (originally
    transcribed from DNA) provides the information for the sequence
    of the amino acids of proteins (primary structure).

2. The information present in DNA is transcribed into the form of
    RNA. mRNA is formed from the DNA template, catalyzed by RNA-
    polymerase (Jacob and Monod, 1961).

Table 10.   Sources and Specificities of Some Nucleases

a.   Exonucleases

	Substrate	Cleavage Site and Products
i) Snake venom phosphodiesterase	DNA or RNA	All ⓐ linkages, starting at the end with a free 3'-OH group and working toward the 5'-end; releases nucleoside-5'-phosphates.
ii) Bovine spleen phosphodiesterase	DNA or RNA	All ⓑ linkages, starting at the end with a free 5'-OH group and working toward the 3'-end; releases nucleoside-3'-phosphates.

b.   Endonucleases

	Substrate	Cleavage Site and Products
i) Bovine pancreatic deoxyribonuclease (DNAse I)	DNA	Cleaves all ⓐ linkages, but prefers those between purine and pyrimidine bases; releases two smaller polynucleotides, one with a free 3'-OH group, the other with a 5'-phosphate group.
ii) Deoxyribonuclease II (DNAse II); from mammalian spleen and thymus, some bacteria)	DNA	Randomly cleaves all ⓑ linkages; releases two smaller polynucleotides, one with a free 5'-OH group, the other with a 3'-phosphate group.
iii) Pancreatic ribonuclease	RNA	Cleaves all ⓑ linkages in which the phosphate is also bound to the 3'-OH of a pyrimidine nucleoside; releases two smaller polynucleotides, one with a free 5'-OH group, the other with a pyrimidine-3'-phosphate at the 3'-terminus.

iv) Taka-diastase (isolated from the Mold Aspergillus oryzae)

	Substrate	Cleavage Site and Products
(I) Ribonuclease T_1	RNA	Same as pancreatic ribonuclease except that guanine is required in place of a pyrimidine; cleaves ⓑ linkages.
(II) Ribonuclease T_2	RNA	Same as ribonuclease T_1, except that adenine must be on the 3' side, rather than guanine; cleaves ⓑ linkages.

3. The genetic code for a particular amino acid consists of a <u>triplet</u> of <u>non-overlapping</u> bases which act as a code word (called a <u>codon</u>). For example:

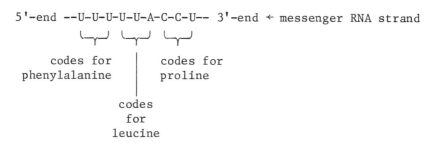

5'-end --U-U-U-U-U-A-C-C-U-- 3'-end ← messenger RNA strand

        codes for   codes for
     phenylalanine   proline

              codes
              for
            leucine

4. Elucidation of the Genetic Code:

   a. If the code words were each made up of two nucleotides and since there are only four bases, there would be only 16 (= 4^2 possible dinucleotide combinations) codons which is not enough to code for the 20 amino acids. There are, however, 64 (= 4^3) trinucleotide combinations, which is more than enough to code for twenty amino acids. This suggested that the code must be a <u>triplet</u> code, with each amino acid specified by a group of <u>three nucleotide bases</u>.

   b. The existence of a triplet code was proved experimentally by using synthetic systems and analyzing the peptide synthesized; and by binding studies of aminoacyl-tRNA's to mRNA triplets.

5. There may be more than one codon for a given amino acid. For example: <u>GUU</u>, <u>GUC</u>, and <u>GUA</u> all code for valine. This characteristic is referred to as the <u>degeneracy</u> of the code.

6. Note that the first two nucleotides in each of these triplets are the same in all codons for a given amino acid, but that the third nucleotide is variable. This suggests that the third base is less important in specifying the incorporation of an amino acid. The wobble hypothesis (see under tRNA) is partly based on this observation.

7. Of the 64 possible triplets, 61 have been shown to code for amino acids. The 3 triplets which do not code for any amino acid are chain-terminating triplets (also called nonsense triplets) and they signal completion of a polypeptide chain. They are UAA, UAG, and UGA. The codons UAG and UAA are also called amber and ochre codons, respectively.

8. There may be initiating codons as well. In E. coli for example,
    the chain-initiating codon is AUG which codes for N-formylmethionyl-
    tRNA (fMet-tRNA).

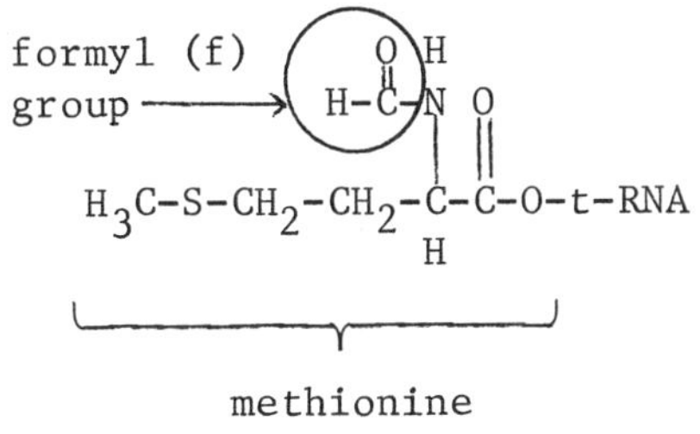

$$\text{formyl (f) group} \longrightarrow$$

H$_3$C-S-CH$_2$-CH$_2$-C-C-O-t-RNA

methionine

One of the valine codons (GUG) can also bind fMet tRNA (bearing
anticodon UAC). This codon may also be involved in chain
initiation.

9. AUG also codes for methionine, raising the question of how one
    codon can code for two amino acids. After the synthesis of a
    peptide chain is terminated, the formyl group and, in some cases,
    the methionine are cleaved from the N-terminus of the polypeptide.

10. In E. coli,there are two types of tRNA which can have methionine
    attached to them. Only one of these can be recognized by the
    enzyme which attaches a formyl group to the Met residue. They
    are therefore designated tRNA^{Met} and tRNA^{fMet}. Both of these
    can bind to the AUG codon but only tRNA^{fMet} can be recognized
    by the valine codon GUG. In eucaryotic cells, which use Met
    rather than fMet as an initiating amino acid, there are also two
    distinct tRNA^{Met} species. One of these can be formylated
    by the enzyme from E. coli, the other cannot. While the
    significance of all this is not yet completely clear, it
    suggests that there is something more than the genetic code which
    is used for the recognition of the tRNA needed for peptide-
    chain initiation.

11. The genetic code is universal, the same code words being used to
    designate the same amino acids in all organisms (see also under
    loading of tRNA in the section on protein synthesis).

12. The genetic code is shown in Table 11. It is important to
    realize that these codons are the bases found in the mRNA and
    that the sequence in the DNA is complementary to this. The
    relationship of DNA, mRNA, and protein sequence is shown below.
    Notice also that the direction of reading of both DNA and RNA is
    important.

Table 11.   The Genetic Code

First Position (5' end)	Second Position				Third Position (3' end)
	U	C	A	G	
U	Phe	Ser	Tyr	Cys	U
	Phe	Ser	Tyr	Cys	C
	Leu	Ser	Term[a]	Term[a]	A
	Leu	Ser	Term[a]	Trp	G
C	Leu	Pro	His	Arg	U
	Leu	Pro	His	Arg	C
	Leu	Pro	Gln	Arg	A
	Leu	Pro	Gln	Arg	G
A	Ile	Thr	Asn	Ser	U
	Ile	Thr	Asn	Ser	C
	Ile	Thr	Lys	Arg	A
	Met[b]	Thr	Lys	Arg	G
G	Val	Ala	Asp	Gly	U
	Val	Ala	Asp	Gly	C
	Val	Ala	Glu	Gly	A
	Val[b]	Ala	Glu	Gly	G

[a] Chain-terminating codons

[b] Chain-initiating codons

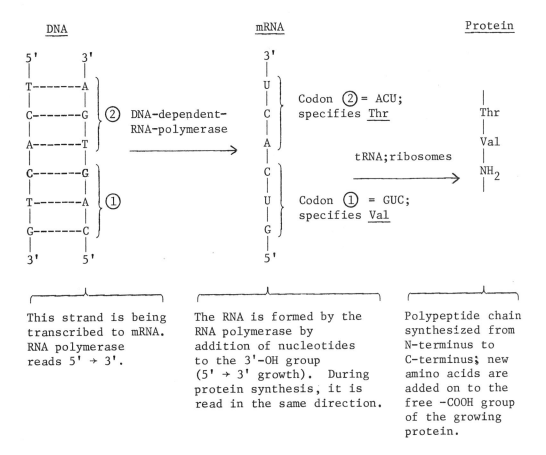

DNA	mRNA	Protein

This strand is being transcribed to mRNA. RNA polymerase reads 5' → 3'.

The RNA is formed by the RNA polymerase by addition of nucleotides to the 3'-OH group (5' → 3' growth). During protein synthesis, it is read in the same direction.

Polypeptide chain synthesized from N-terminus to C-terminus; new amino acids are added on to the free -COOH group of the growing protein.

Mutations

Mutations have been known for a long time in terms of their phenotypic or outward expression. In some cases, variation in a specific protein has been implicated as the cause of the modified phenotype. In others, such as an alteration in eye color or in height, the new phenotype appears to involve a more complex change in the molecular makeup. Diseases which can be shown to be inherited are called "inborn errors of metabolism", a term coined by Garrod in 1909. Now that DNA has been established as the genetic material, research has been concentrated on relating heritable characteristics to changes in the chromosomes and in DNA.

A major breakthrough in this work was the discovery of the genetic code (see above). Changes in the amino acid sequence of a protein could be related to a specific alteration in the composition of the DNA. Certain general types of mutations (changes in the base sequence of the DNA) were then recognized and a start could be

made toward attributing them to particular chemical events. The
mechanisms of action of some of the known mutagens (things which
increase the number of mutations over that found normally) were
elucidated in terms of the chemistry of DNA. These basic types
and the way in which mutagens may cause them are shown below with
examples. The changes indicated are based on the wild-type DNA
sequence ("normal" or reference sequence; the one found in the
unmutated genome) below.

```
Wild-type
DNA sequence:   -G-C- A -C-
                 . .  .  .
                 . .  .  .
                -C-G- T -G-
                      ↑
```

The examples will indicate changes (mutations)
involving this base pair. Note that the
pair must be considered even though the protein
is specified by just one strand, because the
DNA in the chromosomes is double stranded.

1. Transition. One purine-pyrimidine base pair is replaced by another.
    Example: A··T is replaced by G··C .

```
    -G-C- A -C-                          -G-C- G -C-
     . .  .  .                            . .  .  .
     . .  .  .        ⟶                   . .  .  .
    -C-G- T -G-                          -C-G- C -G-
```

These mutations can occur spontaneously. One possible mechanism
involves tautomerization (discussed below) of the adenine to the
enol form during DNA replication. In this form, A can pair with
C, resulting in insertion of a cytosine in the strand being
synthesized in place of thymine. In the next replicative cycle,
the C will specify a G in its complementary strand and one of
the resultant cells will now have a G-C pair in place of the
original A-T pair. These are called copy-errors and require DNA
replication for their appearance.

Mutations of this type may also be induced by 5-bromouracil (5BU)
which resembles thymine and which may be incorporated into one
DNA strand in place of thymine during replication. Since 5BU pairs
with G (unlike T, which pairs with A), T and A are replaced with
C and G. Similarly, 2-amino-purine may be read as either A or G.
Nitrous acid (HNO_2) deaminates A, forming hypoxanthine which pairs
with C.

2. <u>Transversion</u>. A purine-pyrimidine base pair is replaced by a pyrimidine-purine pair.

Example: $\boxed{A \cdot \cdot T}$ is replaced by $\boxed{T \cdot \cdot A}$.

```
    -G-C- A -C-                      -G-C- T -C-
        .  .  .              →           .  .  .
        .  .  .                          .  .  .
    -C-G- T -G-                      -C-G- A -G-
```

This type of mutation is commonly observed in spontaneous mutations in some species. About half of the mutations of the α-and β-chains of hemoglobin are transversions. The mechanism is unknown and no agents that specifically cause this type of mutation have been found.

3. <u>Insertion</u>. The insertion of one or more extra nucleotides

```
    -G-C- A -C-                      -G-C- A - G -C-
        .  .  .          →               .  .  .  .
        .  .  .                          .  .  .  .
    -C-G- T -G-                      -C-G- T - C -G-
                                                ↑
                                         (new base pair)
```

This is one type of frame-shift mutation (see below). It may be caused by acridine orange and proflavin . These molecules can become inserted between two DNA bases (a process called intercalation). They spread the bases farther apart than normal and cause addition of an extra base in the complementary strand during replication. Insertion of <u>more</u> than one base probably occurs during lysogeny by a virus.

Acridine

4. <u>Deletion</u>. The deletion of one or more nucleotides

```
    -G-C- A -C-                      -G-C-C-
        .  .  .          →               .  .  .            A,T   deleted bases
        .  .  .                          .  .  .             ↑
    -C-G- T -G-                      -C-G-G-
```

This is also a frame-shift mutation. It may be caused by hydrolysis (and then loss) of a base due to high temperature or pH change, by covalent cross-linking agents, or by alkylating or deaminating agents. The latter three cause the altering of bases so that pairing cannot occur.

5. Transition and transversion mutations affect at most one codon in the DNA. Due to the existence of chain initiation and termination codons and the degeneracy of the code, a variety of results can occur. It is important to remember that a mutation occurs in the DNA whereas the codon change actually is manifest in the mRNA. The box around one letter of the codon indicates the letter affected.

Table 12.   Changes in Amino Acid Sequence Brought About by Mutation

Mutation	mRNA Change	Old Codon	New Codon	Change in Protein
G → A (transition)	A → G	CU A̲	CU G̲	Leu → Leu (no change)
		U A̲ G	U G̲ G	term → Trp (no termination; next gene read)
		A̲ UG	G̲ UG	Met → Val (may result in loss of initiation site or incorrect amino acid in protein)
A → T (transversion)	U → A	U U̲ A	U A̲ A	Leu → term (premature chain termination)
		AG U̲	AG A̲	Ser → Arg (if Ser is active site, as in many enzymes, loss of activity occurs)
		U̲ UA	A̲ UA	Leu → Ile (probably no change in structure)

Mutations which produce a new chain termination codon are called nonsense mutations while those which replace one amino acid codon by another amino acid codon are called missense mutations. These are all hypothetical examples. Some actual mutations are shown in the section on abnormal hemoglobins.

6. Insertion and deletion mutations result in a shift of the reading frame of the DNA molecule (frame-shift mutations). In some organisms, the majority of the spontaneous mutations observed appear to be of these types.

    A frame-shift mutation affects all of the information in the strand from the point of mutation up to the next initiation or control site. Consider the loss of -G-C-C-A- from the sequence shown below.

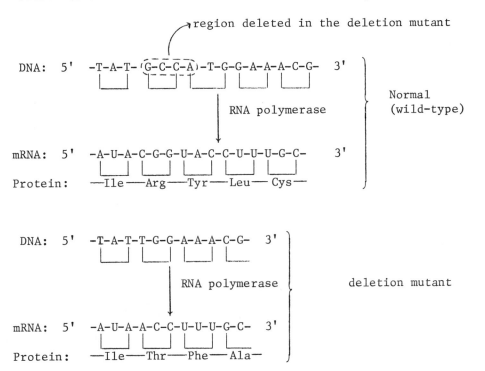

The first amino acid (Ile) is unaffected, since it precedes the deleted region. The succeeding amino acids are all altered and the chain is shortened by one amino acid. (Notice that GC- is enough information to specify alanine, since all four codons (GCI, GCC, GCA, GCG) are for the same amino acid.) If, by chance, the deletion or addition starts between codons and deletes or adds a number of bases which is a multiple of three, then there is no shift in the reading frame. A net loss or gain of codons is all that occurs.

The actual deletion and insertion events only affect one strand of the DNA, initially. During replication, a new strand is made, complementary to the affected sequence. This new, double-stranded DNA is shorter in both strands.

7. Transition mutations resulting from copy-errors (above) cannot occur unless DNA replication is in process. Other mutations can occur in the absence of replication, but require replication for their expression. One can also imagine a mutagenic event which affects a gene necessary for normal cell function. This could both occur and influence the cell in the absence of any DNA replication. If such a mutation affected a process vital to the cell's survival, the cell would probably die prior to mitosis and the mutation would not be transmitted.

8. Errors (mutations) can occur in regions of the DNA which code for rRNA, tRNA, nRNA, and sequences directly involved in the control of transcription (discussed later). These mutations may have a more drastic effect than a change in the sequence of one protein. Loss of control of transcription can result in over- or under-production of one or more gene products. Defects in rRNA or tRNA can completely upset protein synthesis. Regions of the DNA which are used for the control of transcription are called control genes, as opposed to structural genes which specify functional polypeptides.

9. Mutagens usually only affect one strand at a time of a double-stranded DNA. If a mutation occurs in an organism, in most cases half of the progeny of the organism will be normal, the other half mutant. This is further evidence in support of a semi-conservative mechanism for DNA replication (see above).

10. Mutations are frequently deleterious and even lethal to an organism, but they are not necessarily so. They provide the basis for genetic variations which is necessary for evolution to occur. Scientifically, they have been extremely valuable as a tool for gaining knowledge about the mechanism of inheritance and cell function.

Protein Biosynthesis

The overall process is one in which mRNA is read by the ribosomes, one codon at a time. The proper tRNA binds at each codon and adds its amino acid to the growing chain.

There are five major stages.

1. Activation of the Amino Acids
2. Initiation of the Polypeptide Chain
3. The Elongation Process
4. Termination of the Polypeptide Chain
5. Post mRNA-Ribosomal Modifications of Proteins

Although steps 1 and 5 have been studied extensively in both eucaryotic and procaryotic systems, steps 2, 3, and 4 are much more clearly worked out for procaryotes. Steps 1-4 are discussed below in terms of the mechanism elucidated in E. coli and other procaryotes. As more data accumulate on eucaryotic protein synthesis, it appears to be similar in many ways to the process described here.

1. Activation of the Amino Acids. This stage requires ATP, amino acids, tRNA, aminoacyl-tRNA synthetases, and Mg^{++}. It occurs in two stages:

aminoacyl-adenylic acid-enzyme complex ("activated" amino acid)

Aminoacyl tRNA ("loaded" tRNA)

Note: In this and the other diagrams in this section, tRNA is
represented schematically by C -C-A-OH , where

-C-C-A-OH is the nucleotide sequence and the free 3'-OH
group found in all known tRNA molecules. It should be
understood that this drawing actually represents a
nucleotide polymer, containing 75-90 nucleotide bases.
Another representation might therefore be

$$(G/C) \text{---} (N)_{75-90} \text{---} C \text{---} C \text{---} A \text{---} OH.$$

a. The aminoacyl-tRNA synthetases are highly specific for both
the amino acid and the corresponding tRNA. The enzyme possesses
three sites for binding: (i) the specific amino acid,
(ii) the specific tRNA, and (iii) ATP.

b. In the eucaryotic cells, mitochondria contain their own tRNA
and aminoacyl synthetases which are different from those
found in non-mitochondrial systems.

c. The mRNA-ribosome complex does not recognize the aminoacyl
portion of the loaded tRNA. The specificity of the tRNA-mRNA
interaction is entirely due to the relationship between the
tRNA anticodon and the mRNA codon. This was demonstrated by
the following experiment.

Cysteine, attached to cysteinyl-specific tRNA was converted
to alanine in the reaction

$$\text{(Cys-specific tRNA)} - O - \overset{\overset{\displaystyle O}{\|}}{C} - CH - CH_2 - SH$$

$$\underset{\text{cysteine}}{NH_2}$$

hydrogenation in
the presence of H_2
Raney Nickel $\rightarrow H_2S$

$$\text{(Cys-specific tRNA)} - O - \overset{\overset{\displaystyle O}{\|}}{C} - CH - CH_3$$

$$\underset{\text{alanine}}{NH_2}$$

When this tRNA, having the anticodon for cysteine but carrying
the amino acid alanine, was used in a protein synthesizing
system, alanine was incorporated into the polypeptide product
where cysteine should have been.

d. The anticodon region (recognition site) of the tRNA consists
    of a triplet of bases which aligns with the codon on the mRNA
    molecule (see previously under the structure of tRNA). The
    matching between the anticodon region of the tRNA and the codon
    region of the mRNA presumably occurs by standard base-pairing
    (A with U and G with C). It appears that the requirements for
    base pairing in the positions of the first two nucleotides
    (starting from the 3' end of the anticodon) is more stringent
    than in the third nucleotide position (at the 5' end of the
    anticodon) (see the wobble hypothesis, under the structure
    of tRNA).

e. The anticodons can contain unusual bases as do other parts of
    tRNA. For this reason, and because the base-pairing of the
    5' end of the anticodon with the 3' end of the codon is not
    as rigidly controlled, the anticodons are often different from
    what would be predicted by base pairing. This is illustrated
    in Table 13.

f. Although the genetic code is universal (the code word for a
    particular amino acid is the same in all organisms), the tRNA
    of one organism may not function in another organism.
    Presumably this is due to the presence of a different nucleotide
    sequence (at regions other than the anticodon) in some of the
    tRNA's of different species. These nucleotide sequences
    recognize the activating enzymes of the particular species.

g. There may be more than one tRNA specific for the same amino
    acid (just as there may be more than once codon for an amino
    acid). However, there need not be a separate activating enzyme
    for each of the several tRNA's for one amino acid (since the
    tRNA binds to the enzyme and the mRNA template at different
    sites, the enzyme-binding sites may be identical even though
    the mRNA template-binding sites are different).

2. Initiation of the Polypeptide Chain. Requirements are the
    initiating aminoacyl-tRNA (formyl Met-tRNA in bacteria), mRNA, GTP,
    initiating factors (F_1, F_2, F_3), 30S ribosomal subunit and 50S
    ribosomal subunit. In eucaryotic systems, it appears that a special
    Met-tRNA is used for chain initiation. This tRNA is similar in some
    respects to the bacterial fMet-tRNA and it cannot be used to
    insert a Met residue into a growing peptide chain except at the
    N-terminus.

Table 13. Anticodon Assignments for Some Transfer RNA's

Amino Acid	mRNA Codon	Anticodon expected in tRNA on the basis of standard base pairing	Literature Reported Anti-codons in tRNA
	(5' → 3')	(3' → 5')	(3' → 5')
Tyrosine	UAC UAU	AUG AUA	AψG
Phenylalanine	UUC UUU	AAG AAA	AAGm
Alanine	GCA GCG GCC GCU	CGU CGC CGG CGA	CGI
Methionine (fMet)	AUG	UAC	UAC (for tRNA^{fMet})
Isoleucine	AUU AUC AUA	UAA UAG UAU	UAI
Valine	GUU GUC GUA GUG	CAA CAG CAU CAC	CAI

Note: The methionine anticodon is the one predicted by standard base pairing; ψ = pseudouridine; Gm = 2'-O-methyl guanosine; all anticodons are from yeast tRNA molecules except that for fMet.

Probable Sequence of Events. (Note that the shape of the ribosomal subunits in the diagram below is highly schematic):

a. The initiating aminoacyl-tRNA (in bacteria) is formed from methionine tRNA by formylation of the α-amino nitrogen (for structure, see under tRNA) in the reaction:

Met-tRNA + N^{10}-formyl tetrahydrofolate

$$\longrightarrow \text{fMet-tRNA + tetrahydrofolate}$$

b. The inactive ribosome (70S) composed of two subunits (30S, 50S) is activated by initiation factors

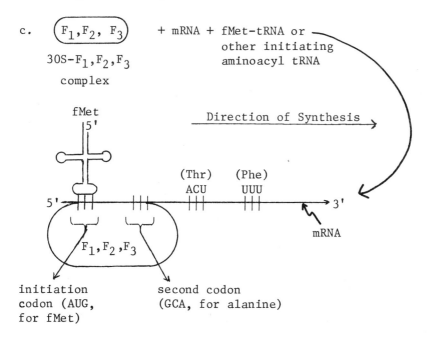

In mammalian systems, initiation factors M_1, M_2, and M_3 have been isolated. These appear to function in a manner similar to F_1, F_2, and F_3 in E. coli.

c. (F_1, F_2, F_3) + mRNA + fMet-tRNA or other initiating aminoacyl tRNA

30S-F_1,F_2,F_3

complex

Initiation complex, containing mRNA; 30S ribosomal subunit; initiation factors F_1, F_2, and F_3; and fMet-tRNA.

d.   Initiation complex + 50S ribosomal subunit

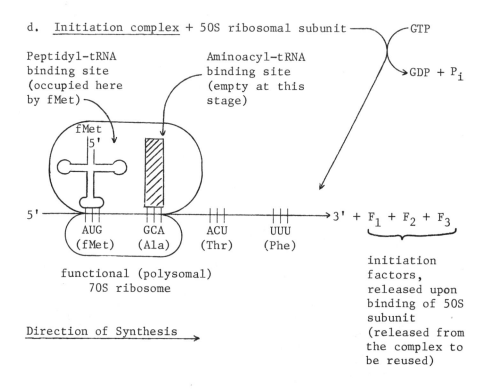

functional (polysomal)
70S ribosome

Direction of Synthesis ⟶

3.   The elongation process consists of three steps.

Codon-directed binding of the proper aminoacyl-tRNA to the aminoacyl binding site (A) of the 50S subunit.

Transfer of the growing peptide chain to the α-amino group of the incoming amino acid (peptidyl transfer).

Translocation of the tRNA bearing the growing peptide to the peptidyl site (P) and release of the discharged (empty) tRNA located there. This permits repetition of step (a) and addition of the next amino acid residue.

a.   The codon-directed binding of the aminoacyl-tRNA to the A-site requires GTP and factor T (a cytoplasmic protein composed of two parts: Tu and Ts), as well as the proper loaded tRNA (Ala-tRNA in this case).

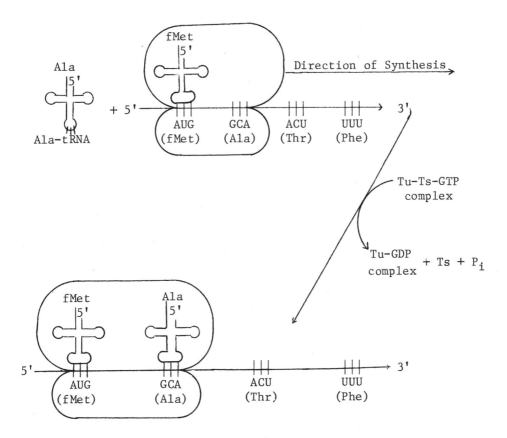

Note that the direction of synthesis is from 5' to 3' on the mRNA. In the diagrams above, the codons for the amino acids are indicated (AUG for N-formyl methionine, for example). The mRNA and tRNA are aligned so that the codon is read from 5' → 3' by an anticodon which is positioned in the 3' → 5' direction.

b. Formation of the peptide bond requires peptidyl transferase, an enzyme which is part of the 50S ribosome. This step requires no ATP or GTP. The energy needed to form the peptide bond probably comes from hydrolysis of the ester linkage between the peptide chain and the tRNA.

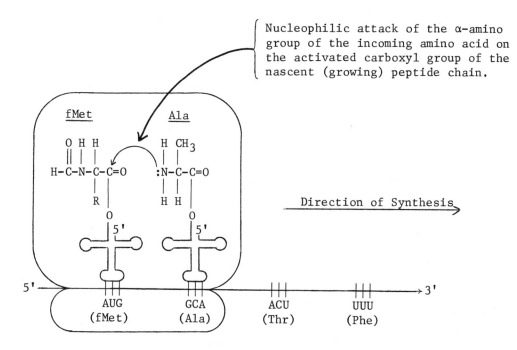

Nucleophilic attack of the α-amino group of the incoming amino acid on the activated carboxyl group of the nascent (growing) peptide chain.

peptidyl transferase

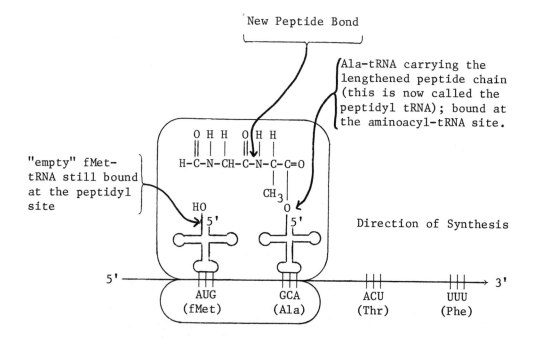

New Peptide Bond

Ala-tRNA carrying the lengthened peptide chain (this is now called the peptidyl tRNA); bound at the aminoacyl-tRNA site.

"empty" fMet-tRNA still bound at the peptidyl site

c. <u>Translocation Reaction</u>. Peptidyl-tRNA is physically shifted
from the aminoacyl site to the peptidyl site, displacing
the empty tRNA from the peptidyl site. Energy for the
conformational change in the ribosome is provided by the
hydrolysis of GTP to GDP + P_i

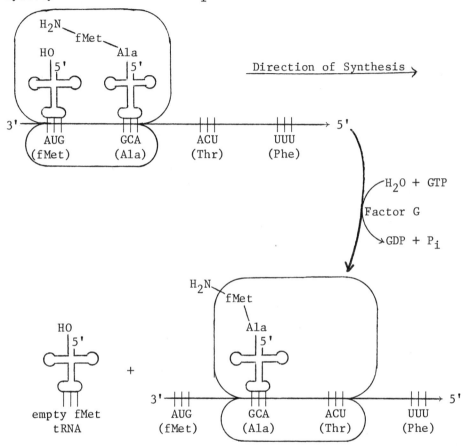

In procaryotic cells, factor G, a protein factor, is also
needed for the translocation reaction. In eucaryotic cells,
a factor known as T_2 is required which performs a function quite
similar to that of factor G. One important <u>difference</u> between
these factors is that T_2 can be inactivated by diphtheria toxin
in the presence of NAD^+. Under these conditions the following
reaction occurs.

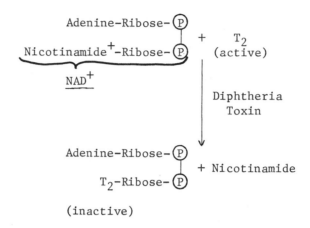

(inactive)

The T_2 is bound in a covalent link to the ribose originally bearing the nicotinamide. Diphtheria toxin does not inactivate factor G, the procaryotic elongation factor.

d. These three reactions are cyclically repeated until one of the termination codons is reached. Although it is indicated here that two GTP molecules are hydrolyzed (one when the aminoacyl-tRNA binds to the ribosome, the other during translocation), this is not known for certain. Studies intended specifically to measure the GTP:amino acid ratio have produced results of 1:1. The value of 2:1 is obtained when (as here) the synthesis is considered as a series of isolated steps, rather than a continuous, synchronous process.

e. Ribosomes undergo complex changes in shape during each bond-making cycle. It has been shown that the mRNA is bound to the 30S subunit at a specific site which is about 30 nucleotide units long. Although it is not completely clear just what role the rRNA plays in this binding, it has been suggested that Mg^{+2} ions form bridges between the rRNA and mRNA bases. Whatever the exact mechanism is, there is definitely a portion of the mRNA molecule which cannot be hydrolyzed by RNAse as long as ribosomes are bound to it.

4. Termination of the Polypeptide Chain

a. Termination is signaled by a termination codon in the mRNA after the last amino acid has been added.

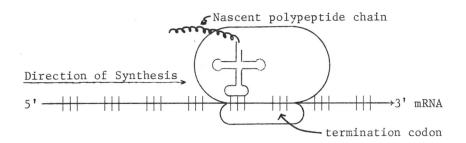

To terminate the synthesis, two interactions have to be ended.
 i) the tRNA (anticodon)-mRNA (codon) interaction, and
ii) the tRNA-polypeptide interaction.

b. The termination codon promotes the activity of the protein
    release factors S and R_1 or R_2 (R_1 functions with the UAA and
    UAG terminator codons; R_2 translates UAA and UGA). They have
    been isolated from the soluble portion of the cell but perform
    their function bound to the ribosome. The completed
    polypeptide chain is first released from the mRNA-ribosome-
    peptidyl tRNA complex (possibly by the action of the peptidyl
    transferase enzyme), followed by the empty tRNA and the
    ribosome. GTP is not required for this process.

$$mRNA\text{-}ribosome\text{-}peptidyl\ tRNA\ complex$$

$$H_2O \rightarrow \bigg| \begin{array}{l} \text{hydrolytic release} \\ \text{of polypeptide} \end{array}$$

$$mRNA\text{-}ribosome\text{-}tRNA + completed\ polypeptide\ complex$$
$$\downarrow$$
$$mRNA + ribosomal\text{-}tRNA\ complex$$
$$\downarrow$$
$$70S\ ribosomes + tRNA$$
$$\downarrow$$
$$F_1, F_2, F_3$$
$$\downarrow$$

30S + 50S subunits,
ready for next round
of protein synthesis

c. Since the direction of growth of the polypeptide chains is

$$NH_2\text{-terminal end} \longrightarrow COOH\text{-terminal end,}$$

the C-terminus is the one attached to the tRNA. Consequently,
final release of the polypeptide involves hydrolysis of an
ester bond.

5. Post-Ribosomal Modifications of Proteins

    a. After their synthesis on mRNA-ribosome complexes, most
        proteins are modified by hydroxylation (collagen);
        phosphorylation (glycogen phosphorylase, ribosomal proteins,
        etc.); acetylation (histones); methylation (histones,
        muscle proteins, cytochrome c, brain protein); addition of
        various carbohydrate groups (glycoproteins); or removal of a
        peptide (activation of insulin, proteolytic enzymes).

    b. The exact biological significance of these post-ribosomal
        modifications is not known, but some of the possibilities
        include

        i)   protection against proteolytic enzymes (e.g.,in the
              case of collagen and elastin);

        ii)   self-assembly of single proteins into an active,
              multisubunit form (as in the formation of the collagen
              superhelix from collagen monomers; discussed later);

        iii) initiation or termination of a biological function
              (methylation and acetylation of histones; activation
              of insulin and proteolytic enzymes; phosphorylation and
              dephosphorylation of certain enzymes in glycogen
              metabolism);

        iv)   identification of proteins intended for export from the
              cell (e.g.,the glycoproteins); and

        v)   introduction of molecular homogeneity following synthesis
              on the mRNA ribosome complex (see the immunoglobulins in
              the chapter on immunochemistry).

    c. Methylation predominantly occurs on the sidechains of the
        amino acids lysine and arginine, and on free carboxyl groups.
        Phosphorylation usually occurs at serine hydroxyls.

    d. Extensive hydroxylation of preformed protein occurs in collagen
        (discussed later), primarily at lysyl and prolyl residues.

6. A comparison of some of the "factors" required by eucaryotes and
    procaryotes for various stages of protein synthesis is given in
    Table 14. Some authors use different symbols to designate these
    molecules. It should be recognized that these processes are still
    far from completely understood. It is quite possible that further
    factors will be isolated and that the precise roles of the known
    factors will be clarified.

Table 14.   Factors Required for the Ribosomal Synthesis of Proteins

	Process	Factors Required in Procaryotes	Factors Required in Eucaryotes
a.	Chain Initiation	fMet-tRNA; F_1, F_2, F_3; GTP	Met-tRNA; M1, M2, M3; GTP
b.	Chain Elongation		
	(i) aminoacyl-tRNA binding	T(= Tu + Ts); GTP	T1; GTP
	(ii) translocation	G; GTP	T2; GTP
c.	Chain Termination	termination codon; R_1 or R_2; S	termination codon; release factor (R); GTP

Inhibitors of Protein Synthesis

1. Puromycin has a structure very similar to that of the <u>terminal adenosine of aminoacyl-tRNA</u>, so that it is able to enter the aminoacyl (A) site on the ribosome. Another part of the puromycin molecule resembles the amino acids tyrosine and phenylalanine. Peptidyl transferase will catalyze bond formation between the previously synthesized polypeptide and the puromycin. Because this bond is not easily broken and because the puromycin is not really attached at the A site, the peptidyl-puromycin is released. Thus puromycin affects peptide chain elongation, causing premature termination of synthesis. The released chains are shorter than they should be and have a puromycin molecule covalently attached to the C-terminus (see Figure 49).

2. <u>Chloramphenicol</u> probably interferes with the peptide-forming steps by binding to the 50S subunit. It inhibits protein synthesis by the 70S ribosomes in procaryotic cells and in the mitochondria of eucaryotic cells,but not by the 80S ribosomes of eucaryotic cells. In susceptible organisms, chloramphenicol prevents chain elongation beyond the first peptide bond.

Figure 49. Schematic Diagram Illustrating the Action of Puromycin in Its Inhibition of Protein Biosynthesis

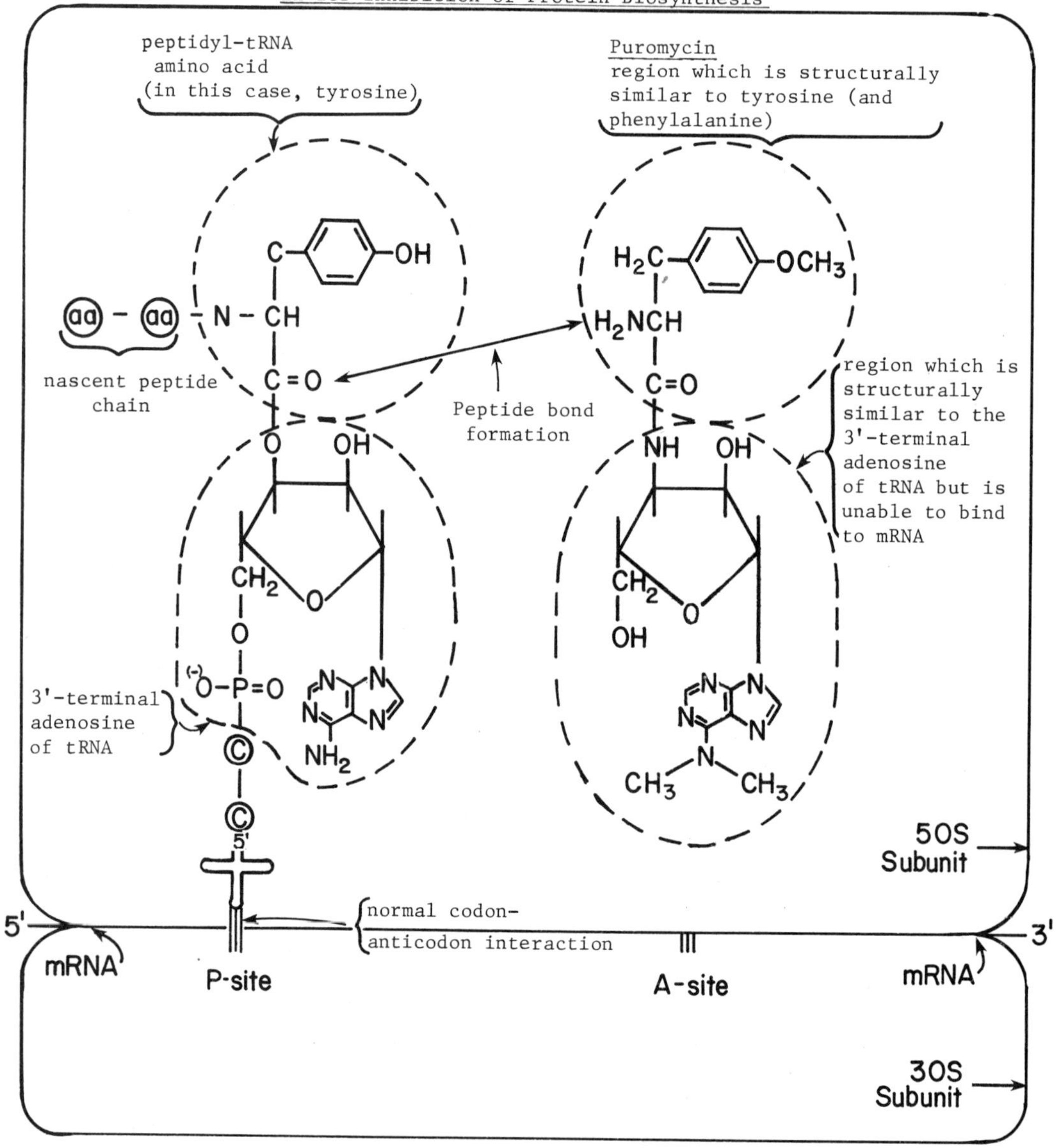

Cycloheximide (also called actidione) inhibits protein synthesis in 80S ribosomes of eucaryotic cells, but not the 70S ribosomes of procaryotes. The results which it produces are similar to those induced by chloramphenicol.

3. Streptomycin (one of the aminoglycoside antibiotics) binds to the 30S ribosomal subunit. It inhibits protein synthesis and also causes misreading of the genetic code, probably by altering the conformation of the 30S subunit so that aminoacyl-tRNA's are less firmly bound to codons and are thus less specific. This and related compounds are important antimicrobials.

4. Agents which inhibit DNA transcription and RNA synthesis thereby
    also prevent protein synthesis. Actinomycin D binds to DNA and
    prevents transcription by blocking movement of the RNA polymerase.
    Rifampicin (and other members of the rifamycin family of
    antibiotics) binds directly to the DNA-dependent-RNA-polymerase
    and inhibits transcription in this manner. The rifamycins affect
    procaryotic polymerase but not the polymerase found in eucaryotic
    cells. Rifamycin B occurs naturally and rifampicin is a semi-
    synthetic derivative of it. The structures of actinomycin D and
    rifampin (a typical rifamycin) are given in the section on
    messenger RNA.

Polyribosomes

1. Clusters of ribosomes attached to one another by a strand of mRNA
    are known as polyribosomes (polysomes; ergosomes). In electron
    microscopic studies, the connecting fiber can actually be seen.

2. Polysomes result from the reading of one mRNA molecule by more than
    one ribosome at a time. One of the four peptide subunits of
    hemoglobin contains about 150 amino acids. The corresponding mRNA
    needed to synthesize this peptide has 3 x 150 = 450 nucleotide
    bases. Since each base occupies about 3.4 Å of the mRNA, 450 bases
    is about 1500 Å long. An 80S reticulocyte ribosome is about 220 Å
    in diameter and it has been found that polysomal ribosomes are
    spaced about 50-100 Å apart. Based on these values, there should
    be 5-6 ribosomes in each hemoglobin polysome. This is illustrated
    in Figure 50.

Figure 50.   Polysome for Hemoglobin Subunit Synthesis Containing Five
             Ribosomes Per Message

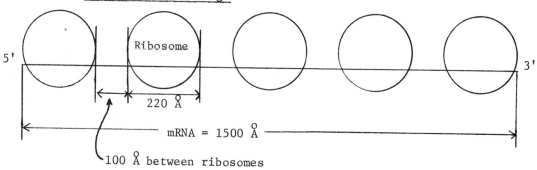

If there were only 50 Å between ribosomes, there would be just about
enough room for a sixth ribosome on the message. The drawing is on
a scale of 100 Å to the centimeter.

3. Polysomes of many sizes have been found. In E. coli they have been found with up to 40 ribosomes per mRNA molecule. From liver, polysomes containing up to 20 ribosomes have been reported.

4. Such synthesis increases the efficiency of utilization of mRNA, since several polypeptide chains can be made simultaneously from one message. Prior to the discovery of polysomes, measurements of the amount of mRNA in a cell gave values which were too low to explain the observed rate of protein synthesis.

5. The direction of translation is 5' → 3' and it is not reversible. mRNA is always read from the 5' → 3' end.

Protein Synthesis in Mitochondria

1. Protein synthesis does take place in mitochondria.

2. Mitochondria possess circular DNA and have a DNA-directed-RNA-polymerase as well as specific forms of tRNA and activating enzymes.

3. Mitochondrial ribosomes are roughly 70S, similar to those of bacteria (procaryotic cells; see under rRNA). Thus, eucaryotic mitochondria are affected by drugs like chloramphenicol.

4. Experiments strongly indicate that the cytochromes and other mitochondrial enzymes are specified by nuclear chromosomes and are synthesized extramitochondrially. Mitochondrial DNA may code for only a few membrane-located proteins. Similarly, chloroplast DNA may also code for some membrane-bound proteins but information contained in the nuclear DNA is necessary for the formation of functional chloroplasts.

Protein Synthesis and Secretion in Eucaryotic Cells

1. As was mentioned previously, procaryotic ribosomes and polysomes are found free in the cytoplasm. In eucaryotic cells, proteins which are to be used within the cell are also synthesized by ribosomes (polysomes) present in the cytoplasm, unattached to any of the other subcellular organelles. An example of this type of synthesis is the formation of hemoglobin in the reticulocytes.

2. Some cells synthesize a large amount of protein which is intended for export (excretion). This protein is made by ribosomes which are bound in rows to the outer surface of parts of the endoplasmic reticulum (ER). The rough endoplasmic reticulum appears rough in the electron microscope because of the ribosomes present as bumps on the surface. It is thought that the large (60S) ribosomal subunit contains the point of attachment to the endoplasmic

reticulum and that the growing peptide extends through the 60S subunit into the cisternae of the reticulum. Thus, the completed peptide is released into the cisternae for transport.

3. Examples of this include the pancreatic exocrine cells (which synthesize proteolytic enzymes for secretion into the intestinal lumen) and the liver (which makes albumin and other serum proteins). Based on autoradiography and electron microscopy, the protein formed by the polysomes of the rough endoplasmic reticulum travels through the smooth endoplasmic reticulum to the Golgi apparatus. There it is packaged as storage granules (zymogen granules, in the case of the pancreas) and kept in the cell until the proper release stimulus is received. This transport (through rough ER to smooth ER to Golgi apparatus) requires K^+ ion and is inhibited by ouabain.

Recapitulation of Some Chemical Reactions Which Occur in Protein Biosynthesis

1. Activation of the amino acid by aminoacyl-tRNA synthetase. E, below, is the synthetase enzyme.

mixed acid anhydride, between an amino acid and a phosphate
(a.a.-AMP- Ⓔ)

2. Loading of the tRNA by transfer of the activated amino acid to the 3'-OH group of the tRNA. This reaction is also catalyzed by the synthetase enzyme, E. ⓒ and Ⓟ represent, respectively, cytosine and phosphate.

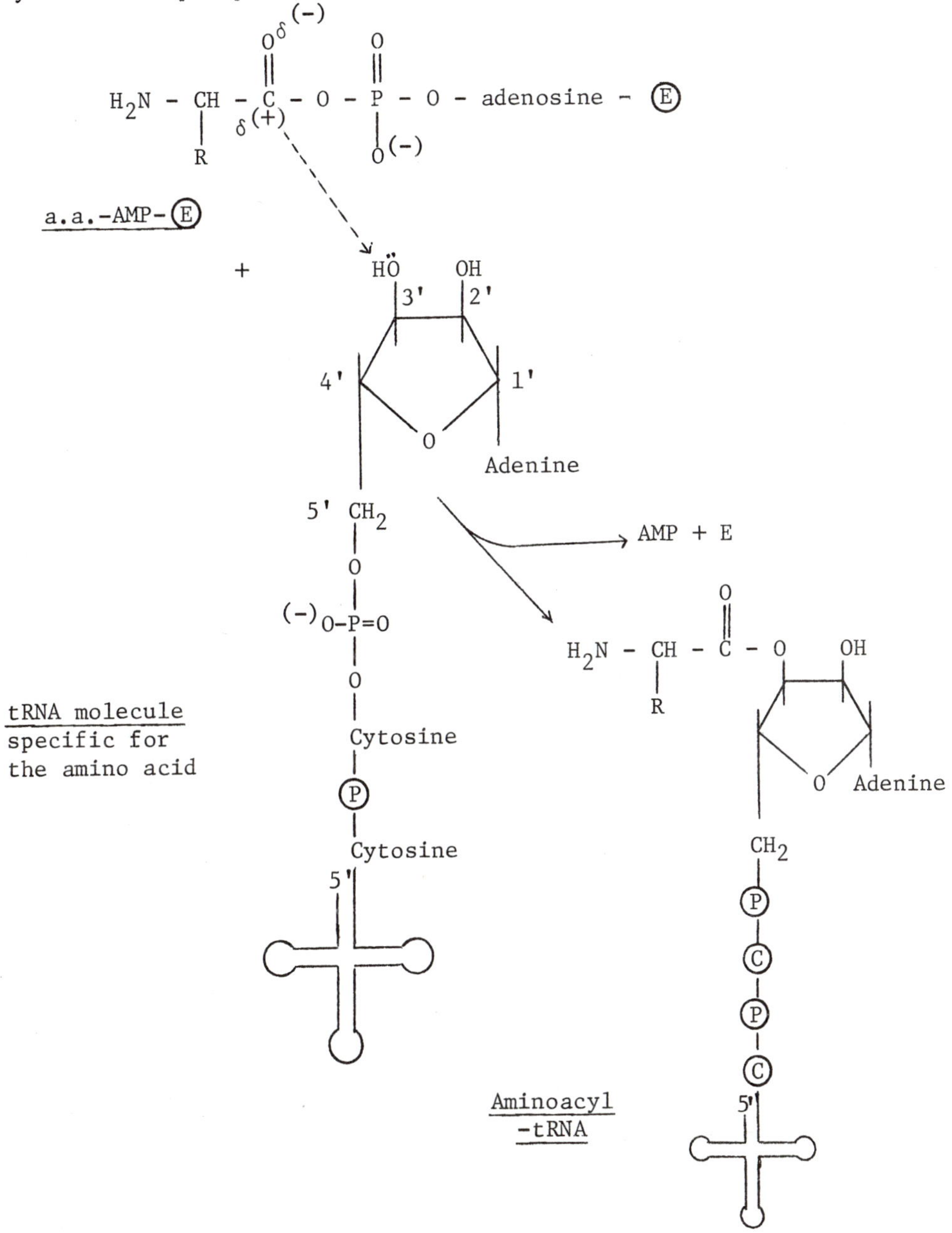

$$H_2N - CH - \overset{\underset{\delta(+)}{\|}}{\underset{R}{C}} - O - \overset{\underset{O(-)}{\|}}{P} - O - adenosine - Ⓔ$$

$\overset{(-)}{O^\delta}$

a.a.-AMP-Ⓔ

$+$

HŌ   OH
|3' |2'

4'   1'

O

Adenine

5' CH$_2$

O

$(-)_{O-P=O}$

O

Cytosine

Ⓟ

Cytosine

5'

tRNA molecule
specific for
the amino acid

$\longrightarrow$ AMP + E

$$H_2N - CH - \overset{\overset{O}{\|}}{C} - O \qquad OH$$

R

O Adenine

CH$_2$

Ⓟ

ⓒ

Ⓟ

ⓒ

5'

Aminoacyl
-tRNA

3. Transfer of the growing peptide chain from the tRNA bound to the
    peptidyl site on the ribosome to the α-amino group of the aminoacyl
    tRNA bound at the aminoacyl site. The reaction is catalyzed by
    peptidyl transferase.

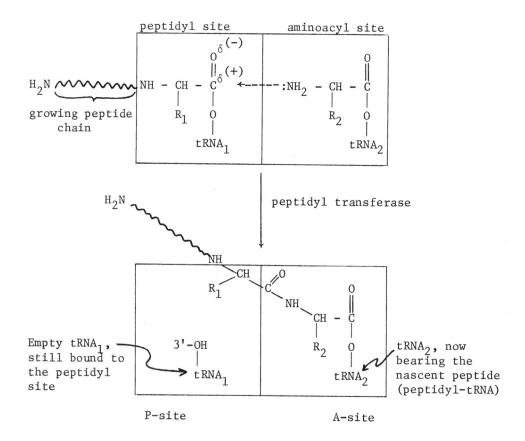

Some Selected Examples of Protein Synthesis in Mammalian Systems

1. Comments on Collagen Biosynthesis

    a. Collagen is the principal protein of the connective tissue.
        Its major function is to provide strength and maintain the
        structural integrity of tissues and organs. In its final form
        (tropocollagen), in the absence of any denaturation or
        degradation, it is insoluble.

Figure 51. **An Overall Schematic Diagram of Protein Biosynthesis in Bacteria (E. coli)**

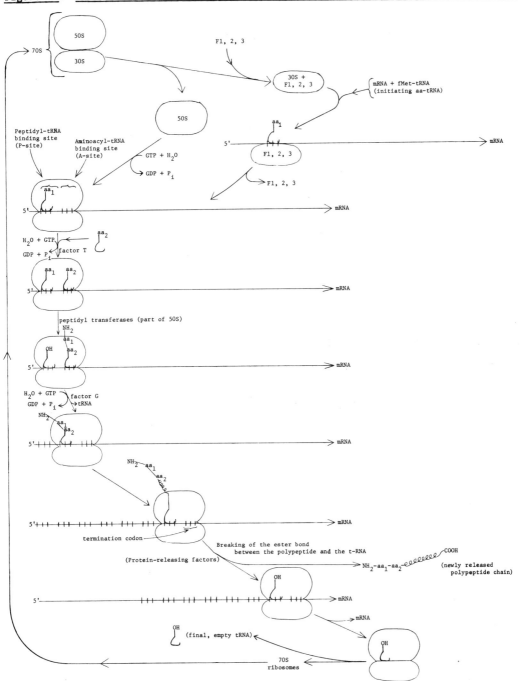

b. Collagen is made up of three polypeptide chains. Two of the
    chains are usually similar to one another and are designated
    α1-chains. The third chain - called an α2-chain - usually
    differs from the other two. The three chains are wound around
    each other to give a rigid, helical structure. The collagen
    helix is <u>not</u> an α-helix. In the α-helix there is only one
    peptide chain and all of the hydrogen bonds are <u>intrachain</u>
    (between different parts of the same chain). In tropocollagen
    (the finished form of collagen), hydrogen bonds are interchain,
    linking the three chains to each other.

c. The peptide chains frequently contain a repeating tripeptide
    unit of glycine with two other amino acids: gly - aa_1 - aa_2.
    Proline and 4-hydroxyproline are most commonly found at
    positions aa_1 and aa_2 respectively. Glycine's R group
    (hydrogen, the smallest of the atoms) and the fixed ring
    structures of proline and hydroxyproline account for the
    structural features of tropocollagen (already discussed,
    page 50). Lysine and δ-hydroxylysine are also found at
    positions aa_1 and aa_2.

    Hydroxyproline and hydroxylysine are almost unique to collagen
    and neither of them is coded for by the DNA. They are formed
    by the action of two apparently different oxygenases
    (hydroxylases) on the protocollagen molecules.

d. Collagen contains small amounts of tyrosine and very little
    tryptophan and cysteine. The absence of cysteine in most
    collagens eliminates the possibility of formation of disulfide
    bonds cross-linking the peptide chains. Covalent cross-linking
    in collagen takes place almost entirely through aldehyde
    and amino groups.

    The aldehyde groups are formed by the action of lysyloxidase
    on the ε-amino groups of lysine and hydroxylysine. The general
    reaction is

$$R-CH_2-NH_2 \xrightarrow{\text{lysyl oxidase}} R-\overset{\overset{\displaystyle O}{\|}}{C}H$$

The specific products are

$$\text{lysine} \xrightarrow{\text{lysyl oxidase}} \text{α-amino adipic acid δ-semialdehyde (AL)}$$

$$\text{δ-hydroxylysine} \xrightarrow{\text{lysyl oxidase}} \text{δ-hydroxy α-amino adipic acid}$$
$$\text{-δ-semialdehyde (HAL)}$$

The cross-links can form by either of two basic reactions:

i) Schiff base formation

$$R_1-\overset{\overset{\text{H}}{|}}{C}=O \; + \; H_2N-CH_2-R_2 \; \underset{\longleftarrow}{\longrightarrow} \; R_1-\overset{\overset{\text{H}}{|}}{C}=N-CH_2-R_2 \; + \; H_2O$$

The residues involved can be any combination of δ-semialdehyde and ϵ-amino group. This is not a very stable linkage. R_1 and R_2 represent two separate collagen polypeptides.

ii) Aldol condensation

$$R_1-CH_2-\overset{\overset{\text{H}}{|}}{C}=O \; + \; \underset{\underset{\text{H}}{|}}{\overset{|}{\underset{C=O}{CH_2-R_2}}} \; \longrightarrow R_1-CH_2-\overset{\overset{\text{H}}{|}}{\underset{\underset{OH}{|}}{C}}-CH-R_2$$

$$\searrow H_2O$$

$$R_1-CH_2-\overset{\overset{\text{H}}{|}}{C}=\underset{\underset{\underset{\text{H}}{|}}{\overset{|}{C=O}}}{C}-R_2$$

The reaction shown is between two AL residues, although two HAL residues or an HAL and an AL residue could undergo an aldol condensation also. This cross-linkage, particularly the product formed after dehydration, is very stable.

The overall process of cross-linking (oxidation and coupling) is called <u>maturation</u> of the protein. The final cross-linked product is insoluble under most conditions unless degradation and denaturation occur. Cross-links similar to these have been found in elastin, another important connective tissue protein found predominantly in elastic tissues (e.g.,walls of the large blood vessels).

Other cross-links are also known in both of these proteins especially the derivatives of desmosine. Most of them appear to be more complex products of the same two reaction types described above.

e. A deficiency in amine oxidase and hence fewer cross-linkages in the collagen molecules can lead to increased destruction of collagen. This situation may be operative in <u>Marfan's Syndrome</u>. Defective cross-linking has also been observed in <u>homocystinuria</u> due to excessive amounts of <u>homocysteine</u> present in the connective tissues. The homocysteine reacts with the aldehyde groups, making them unavailable to form cross-links. In the disease <u>Scleroderma</u>, there is an excessive amount of collagen deposited in the tissue spaces and the tissues. This is due either to a deficiency in the regulatory mechanism of collagen synthesis or defective breakdown process by collagenases. Compounds such as penicillamine (a drug also used in Wilson's disease, an error in copper metabolism) that prevent cross-links by combining with aldehyde groups, can reduce the excessive accumulation of collagen by increasing its catabolism.

f. Steps involved in collagen biosynthesis

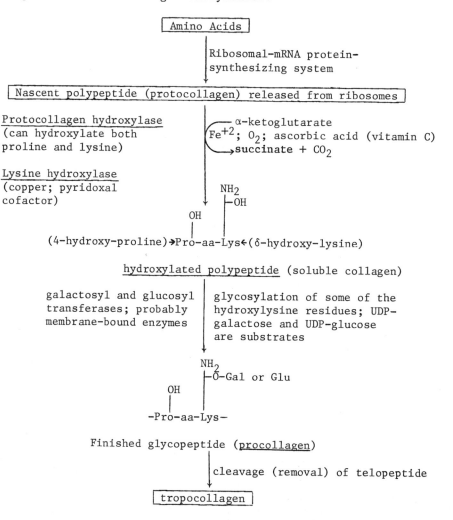

i)   Note that one of the hydroxylation reactions requires the concomitant conversion of α-ketoglutarate to succinate and carbon dioxide. This conversion appears to be different from the decarboxylation reaction of α-KG that takes place in the mitochondria which involves CoA, thiamine pyrophosphate, lipoic acid, etc. The other lysine-specific hydroxylase has not yet been completely characterized, but it does appear different from the better known protocollagen hydroxylase.

ii) The determining factors that establish the extent of hydroxylation and glycosylation of the polypeptide are not understood. Certainly those are methods by which a cell can introduce heterogeneity into peptides with identical amino acid sequence made on the mRNA-ribosome complex.

iii) It should be pointed out that types of collagen differ from one species to another and between tissues in the same species. The differences **include chain type, amino** acid composition, and carbohydrate content.

iv) The distinction between procollagen and tropocollagen is that the peptide chains in the former are longer than in tropocollagen. The purpose of this extra peptide (the telopeptide) is not completely clear. It is possibly involved in the self-assembly of the collagen monomers into the three-stranded helix present in tropocollagen. Following this assembly, the telopeptide is no longer needed and is removed by a protease.

g. The rate of collagen synthesis is stimulated by growth hormone at the level of the mRNA-ribosome mediated protein synthesis. It is inhibited at the hydroxylation step by ascorbic acid (vitamin C) deficiency. At least in vitro, however, the vitamin C requirement of protocollagen hydroxylase can be partically replaced by reduced pteridines.

2. Comments on Albumin Synthesis and Edema

a. Albumin, the principal serum protein, has a molecular weight of about 65,000 daltons and contains about 575 amino acids. In humans, the normal blood plasma range is 3.5-5.0 gm of albumin/100 ml. Albumin is also found in the extravascular spaces, the lymph, and in other biological fluids including amniotic fluid, bile, gastric juice, sweat, tears, etc. In kidney diseases it is found in the urine and it is a major component of edema fluid (see later, under edema).

b. Plasma albumin has two primary functions.

    i) Maintenance of osmotic pressure (see later, under edema).

    ii) Transport of a variety of substances which are bound
non-covalently to albumin. These include metals and
other ions, bilirubin, amino acids, fatty acids, enzymes,
hormones, drugs, and others.

c. Albumin is synthesized by the liver cells (hepatocytes) and
is the principal protein made in the liver. No other sites
of synthesis are known at present. In a 70 kg human, the
liver synthesizes about 12-14 gm of albumin per day. The
half-life of an albumin molecule in man is about 20 days.
The albumin synthetic system is the classical mRNA-ribosome
mediated process discussed earlier in detail. The mRNA for
albumin (575 amino acids) is composed of (575 x 3) + 3 nucleotides
for initiation + 3 nucleotides for termination = 1731 nucleotides.
Allowing about 90 nucleotides (310 $\overset{\circ}{A}$) from the center of one
ribosome to the center of the next, the polysomes should contain
about 19-20 ribosomes per albumin message. This has been found
experimentally by the use of sucrose gradient centrifugation.
Since albumin is an <u>export</u> protein, it travels to the Golgi
apparatus where it is packaged preparatory to secretion. The
exact biochemical transport mechanism is not known but it
requires K^+ and is inhibited by ouabain (see previously under
proteins for export). Albumin which leaks through the blood
vessel walls into the extravascular spaces is returned to the
blood via the lymphatic system.

d. Control of albumin synthesis occurs at both the transcriptional
(DNA → RNA) and translational (RNA → Protein) levels.

    i) Adequate nutrition (protein-nitrogen intake) is basic to
the regulation of all protein synthesis. Although protein
deficiency and overall malnutrition (starvation) differ
somewhat in their molecular effects, they both
result in a decreased rate of protein synthesis due to
alterations in RNA metabolism. Refeeding of starved animals
results in an immediate, threefold increase in the amount
of albumin present in the newly synthesized protein, implying
that the albumin mRNA is more stable (has a longer half-
life) than other mRNA.

    ii) Although all of the amino acids must be present to support
protein synthesis, albumin synthesis seems especially
sensitive to tryptophan. This amino acid appears to

stimulate mRNA synthesis and to increase the activity
of the RNA polymerase which makes rRNA. Tryptophan may
also increase the stability of the ribosome-mRNA-ER complex
on which albumin is synthesized. It is far from clear,
however, why this specific amino acid plays such an
important role in this regulation.

iii) Albumin synthesis is stimulated <u>in vivo</u> by both cortisone
and the thyroid hormones (thyroxine and triiodothyronine)
perhaps synergistically. Cortisone is known to increase
the synthesis of hepatic rRNA, tRNA, and mRNA and it may
promote ribosomal binding to the endoplasmic reticulum.
Thyroid hormones have a similar effect, stimulating mRNA
and rRNA synthesis as well as increasing the binding of
rRNA to the ER and perhaps stimulating the ribosomes.
Testosterone causes the same changes as do the thyroid
hormones. Increased albumin synthesis has been noted in
Cushing's Syndrome (involving hyperactivity of the
adrenal cortex). In hyperthyroidism the rates of both
albumin synthesis and breakdown are increased. In
hypothyroid individuals, these processes are decreased
but, more important, edema can result in this state
from the accumulation of albumin in extravascular spaces.
It should be noted that despite the increase in albumin
synthesis associated with cortisol administration, the
hormone has been considered antianabolic because it
causes an overall negative nitrogen balance.

iv) Insulin may also be needed for maximal albumin synthesis,
although hypoalbuminemia and decreased albumin synthesis
are not usually associated with diabetes mellitus. Growth
hormone enhances albumin synthesis by stimulating rRNA
polymerase activity.

v) <u>In vitro</u> studies have shown that albumin synthesis in
hepatic tissue is influenced by an osmotic regulatory
mechanism. This is a significant observation in view of
albumin's role in maintaining colloidal osmotic pressure.
Albumin synthesis is <u>increased</u> in blood <u>hypotonia</u> and
<u>decreased</u> in <u>hypertonic</u> situations.

e. Albumin levels in disease states

i) Despite earlier studies, it now appears that although serum
albumin levels may decrease in cirrhosis, this
observation has no direct bearing on total liver function,
prognosis, or serum albumin synthesis.

Analbuminemia is a rare, probably inherited lack of serum albumin. The symptoms associated with it are surprisingly mild, considering the protein's central role in osmotic regulation. This is at least partly explained by an increase in serum levels of cholesterol, phospholipid, gamma globulins, fibrinogen, and transferrin. These replace albumin to some extent in maintaining the blood osmolarity. As yet, no specific metabolic defect has been demonstrated in patients having this disorder.

ii) Variations in albumin synthesis may be related in a more complex fashion to cirrhosis and gastrointestinal disease. In the latter case, and in nephrosis, an increased rate of destruction contributes to the hypoalbuminemia. In patients suffering from severe and extensive burns, there is extensive loss of serum, hence water and albumin. This loss is too great to be met by increased synthesis.

f. Edema means swelling and is characterized by an increase in the fluid in intercellular and extravascular spaces. This can be visibly observed in the subcutaneous tissue. The basic mechanisms are related to alterations in the permeability of the blood vessels and to changes in the osmotic and hydrostatic pressures of the blood and extravascular fluids. Figure 52 illustrates Starling's hypothesis and shows the interrelationships of these factors in a normal capillary.

From Figure 52 it is apparent that a shift in blood pressure can cause a change in the direction of filtration (water movement). An increase or decrease in the other pressures involved may result in a similar reversal. The blood and tissue osmotic pressures are the pressure which would be measured across a semipermeable membrane with water on one side and blood or extravascular fluid, respectively, on the other side. The arrows indicate the direction of net water flow under the influence of the individual pressures.

g. Based on the foregoing discussion, four types of edema can be differentiated by their etiologies.

i) If the serum protein concentration (especially the albumin concentration) is decreased, the blood osmotic pressure also decreases. When that pressure falls below its normal minimum (about 23 torr), the net filtration pressure (and the filtration rate) increases at the arterial end of the capillary and the absorption rate decreases at the venous end. The net result is a shifting of water into the

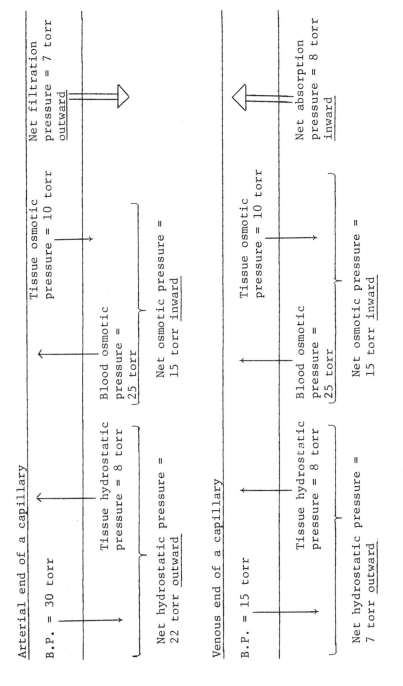

Figure 52. Pressure Interrelationships Between Intravascular and Extravascular Fluids

Note: B.P. = blood pressure; torr = mm of Hg, a unit of pressure; inward refers to from the tissue into the capillary lumen; outward refers to into the tissue from the capillary lumen.

extravascular space. This type of edema can occur in
severe albuminuria (which may be associated with lipoid
nephrosis, the nephrotic stage of glomerulonephritis, and
amyloid nephrosis). It can also result from the decreased
albumin synthesis present in protein starvation and
malnutrition.

ii) A decrease in the blood osmotic pressure and an increase
in the tissue osmotic pressure occurs if the capillary
endothelium is damaged sufficiently to permit passage
of protein into the interstitial fluid. This amounts to
simultaneously decreasing serum osmolarity and increasing
the extravascular fluid osmolarity. Once again, the
filtration rate (arterial capillary) increases and the
absorption rate (venous capillary) decreases. This
mechanism is involved in the production of blisters, hives,
and in other localized inflammation. Toxins, trauma,
and histamine-like substances function similarly (see
further under immunochemistry).

iii) Increased venous blood pressure results in decreased net
absorption pressure and the retention of water in the
interstitial fluid. This can be caused by the presence
of a tourniquet and supposedly is part of the mechanism
of edema formation in congestive heart failure. It is
associated with varicose veins (permanently distended,
tortuous veins, occurring especially in the lower
extremities). Increased venous pressure occurs in
cardiac decompensation (inability of the heart to
maintain adequate circulation), but the exact mechanism
is not clear. One possibility is that failure of the
heart results in renal retention of sodium and, hence,
retention of water, with subsequent edema.

iv) As was mentioned before, protein (especially albumin) lost
to the interstitial fluid is returned to the blood via
the lymphatic system. Obstruction of the lymphatic vessels
thus results in an increase in extravascular osmolarity
and net filtration rate and a decrease in reabsorption.
This situation occurs in elephantiasis, a disease
characterized by hypertrophy of the skin and subcutaneous
tissue, especially of the legs and genitals. It is
caused by obstruction of the lymphatics by filarial worms.
In myxedema, swelling (edema) is caused by the presence
of large quantities of mucoproteins in the lymph vessels.
It is associated with hypothyroidism. The lymph vessels
seem to be metabolically sluggish and unable to remove

interstitial protein as rapidly as it enters. Since it
is the mucoproteins which accumulate in the edema fluid,
there may be a selectivity in the protein removal process.

h. Other aspects of edema, including electrolyte (Na^+, K^+) levels
and hormone interactions,will be discussed in the chapter on
water metabolism and electrolyte balance.

3. The structure and synthesis of the immunoglobulins are discussed
in the chapter on immunochemistry.

4. The synthesis and characteristics of hemoglobin are covered in
detail in the chapter on hemoglobin. See also the next section
on the control of protein synthesis.

Regulation of Protein Synthesis and Other Control Mechanisms

1. It is important that cells produce the proper substances
(qualitative control) in the proper amounts (quantitative control)
at the proper times (temporal control). Energy, raw materials,
and space would be wasted if there **was** no way to control production.
Cellular development and differentiation are based on the cells'
ability to control their molecular makeup and function. There are
three major levels of control within cells:

a. <u>Regulation of enzyme activity</u> has been discussed previously
in the chapter on proteins and enzymes. One of the most
important examples of this type of control is <u>allosterism</u>.
It is exemplified by the reaction below.

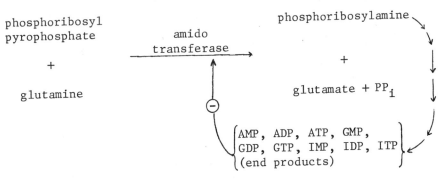

phosphoribosyl
pyrophosphate

+

glutamine

amido
transferase

phosphoribosylamine

+

glutamate + PP_i

AMP, ADP, ATP, GMP,
GDP, GTP, IMP, IDP, ITP
(end products)

Glutamine phosphoribosyl pyrophosphate amidotransferase
catalyzes the first step unique to purine biosynthesis (see
under nucleotide biosynthesis). The end-products of this
pathway are the purine nucleotides, all of which are quite
unlike the substrates or products of the amidotransferase-

catalyzed reaction. They interact allosterically with the
enzyme, decreasing its activity. Thus, accumulation of the
products of a pathway causes a decrease in the activity of
that pathway. This is an illustration of one mechanism of
<u>feedback or end-product inhibition</u>.

b. <u>Regulation of mRNA translation</u> may be a factor in the biosynthesis
of hemoglobin and other proteins. With respect to hemoglobin,
this mechanism will be discussed later in this section. At
least in some cells, the presence of unusual tRNA's may be a
control mechanism. The occurrence of these tRNA's is influenced
by factors exterior to the cell. This type of control
apparently exists in many plants. Cortisone and the thyroid
hormones appear to regulate albumin synthesis at least partly
at this level.

The ability of ribosomes to initiate protein synthesis may
also be modulated by reversible phosphorylation of the
subunits. As was pointed out earlier, ribosomes must dissociate
into subunits before they are able to bind to an initiation
codon. The phosphorylation may inhibit this dissociation.

c. <u>Regulation of gene expression (transcription)</u> is the "highest"
level of control. It permits conservation both of protein-
synthetic materials and of the RNA precursors. Induction and
repression in bacteria; differential expression of the genome
during development of higher animals; and control of the
synthesis of hemoglobin, albumin, and perhaps other molecules,
operate at least partially by this mechanism.

The first and third types of control (a and c) are recognized to be
of major importance in most cells. Using these two levels, cells
can function optimally. The second level is less well studied but
it appears to be important in at least some situations.

Coarse control───────────────⟶ regulation of gene expression

Intermediate control───────────⟶ regulation of mRNA translation
                                (protein synthesis)

Fine control─────────────────⟶ allosteric control of enzymes

Examples of the control of translation and transcription are
discussed below.

2. A further example of the modulation of translation via changes in ribosomal aggregation has been described (see also b, page 368). In bacterial systems, an initiation factor (F_3, see under protein synthesis) is necessary to prevent association of the 30S and 50S subunits and thereby to permit initiation of protein synthesis. An entirely analogous protein factor (tentatively designated IF-3rr) has been demonstrated to be necessary for globin mRNA initiation in rabbit reticulocytes. IF-3rr appears to <u>prevent</u> subunit aggregation following release from the polysomes, rather than actively causing dissociation of 80S ribosomes. Heme (Fe^{+2}-protoporphyrin III complex, see under hemoglobin) is involved in some way, being responsible either for IF-3rr synthesis or activity. In the absence of IF-3rr, all protein synthesis would cease in the reticulocyte. Thus, this factor may represent one step in a coordinated mechanism for regulation of both heme and globin synthesis during erythroid development. This larger role is still only circumstantial. These findings are especially significant because they support the generality of the protein synthetic process originally described in procaryotes.

3. In eucaryotes, all RNA, except that made in the mitochondria, is synthesized within the nuclear membrane (which includes the nucleolus). Recent work has indicated that variable release of this RNA to the cytoplasm and endoplasmic reticulum may be a potential source of control over gene expression. Stretches of polyadenylic acid (poly A) have been found associated with some of the nuclear RNA and there is increasing evidence that the mRNA's of eucaryotic cells are terminated by regions of poly A, probably at the 3'-OH end. This poly A may be involved in selection of the mRNA to be released from the nucleus or in the ability of mRNA to be used for translation. This whole process is still very new and requires much more investigation.

4. Regulation of gene expression is important not only in the cell's economy, but also plays a major role in cell differentiation. For example, the development of a neuron and a pancreatic acinar cell from the same zygote are both controlled by the same DNA. The way in which this DNA is used (how much and what part of it is used for RNA transcription and when this use occurs) must be controlled for the differential development of cells.

5. <u>Enzyme Induction</u>

    Enzymes may be classified according to the conditions under which they are present in a cell. There are two basic types of enzymes.

a. <u>Constitutive enzymes</u> are formed at constant rates and in
    constant amounts. Their presence in a cell is not related to
    the presence or absence of their substrates. They are
    considered to be part of the permanent enzymatic makeup of the
    cell. The enzymes of the glycolytic pathway are examples of
    this group.

b. <u>Inducible enzymes</u> (adaptive enzymes) are always present in at
    least trace amounts, but their concentrations vary in
    proportion to the concentrations of their substrates. The
    classical example is β-galactosidase, which is present in small
    amounts in wild type <u>E. coli</u> and whose concentration increases
    in the presence of lactose.

    lactose
    (galactosyl-4-β-D $\xrightarrow{\text{Induction of} \atop \text{β-galactosidase}}$ glucose + galactose
    glucose

    In this example, the inducer (lactose; however see later, under
    induction mechanisms and the lac operon) is a substrate for the
    enzyme induced. This is the usual case, but compounds similar
    to the substrate may also be inducers and there are substrates
    which are not inducers. The enzyme β-galactosidase is coded
    for by part of the lac (for lactose) operon. The accepted
    mechanism for this process is discussed later in this section.

c. A given enzyme may be constitutive in one organism or in one
    strain of an organism, inducible in another, and absent in a
    third .   In addition, the magnitude of the response to an
    inducer varies from organism to organism and from strain to
    strain, and is genetically determined. Mutations are known
    in bacteria which cause inducible enzymes to become constitutive.

6. <u>Enzyme Repression</u>

If a bacterial strain capable of synthesizing a particular amino
acid is placed in a culture medium containing that amino acid,
synthesis of the amino acid ceases, due to repression. The amino
acid acts as a corepressor of the apparatus (operon) that produces
the enzymes for its synthesis. This is another type of feedback
system very similar to the end-product inhibition of an enzyme.
For example, if <u>E. coli</u> cells are provided with sources of carbon
and NH_4^+, they will synthesize all 20 amino acids. If histidine
is added to the culture medium, the key enzyme for histidine
synthesis (a dehydrogenase) is no longer produced, but enzymes for
the synthesis of other amino acids are still produced.

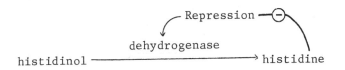

$$\text{histidinol} \xrightarrow{\quad \text{dehydrogenase} \quad} \text{histidine}$$

7. In 1961 and 1962, Jacob and Monod published a genetic mechanism for
    the induction and repression of enzyme systems in bacteria. They
    postulated that in addition to the structural genes which direct
    the synthesis of specific enzymes and other proteins using a mRNA
    template (previously described), there is an operator gene which
    precedes and controls the structural genes. If the operator gene
    is repressed, the structural genes cannot function. The sequence
    of DNA, composed of an operator gene and one or more <u>structural
    genes</u>, is called an <u>operon</u>. It is transcribed as one continuous
    piece of mRNA (a polycistronic messenger). For example, when the
    lac operon is induced, β-galactoside permease and β-thiogalactoside
    transacetylase are synthesized in addition to β-galactosidase. All
    of these enzymes are needed for metabolism of the galactose
    released from lactose.

    There is also a <u>regulator gene</u> on the chromosome, not necessarily
    adjacent to the operon. This gene codes for the mRNA for a
    repressor protein. In a repressible enzyme system (such as the
    histidine synthesis operon), the repressor protein is termed an
    <u>aporepressor</u>, since it is unable to cause repression without the
    binding of the corepressor. The inducer and corepressor substances
    probably cause a conformational change in the repressor **or**
    aporepressor.

    A <u>promoter region</u> is also believed to be present. In the case of
    the lac operon, it precedes the operator and structural gene
    regions. Its role appears to be to provide the actual site for
    initiation of transcription. It is not shown in the succeeding
    diagrams because it was not part of the original model proposed
    by Jacob and Monod.

    When a repressor is bound to the operator gene, the structural
    genes cannot be transcribed and the operon is said to be repressed.
    When the repressor itself is inactivated, the operator gene is
    active and the operon is said to be derepressed.

a. The general pattern of the model is illustrated here.

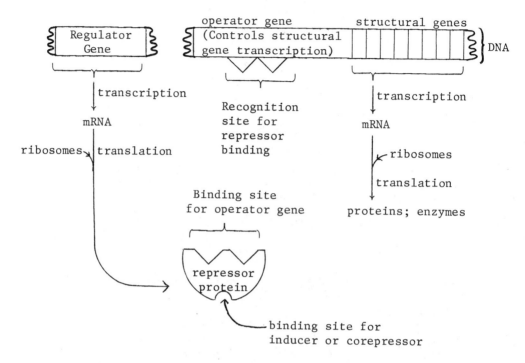

b. In an <u>inducible enzyme system,</u> the repressor is active unless
    <u>inactivated by an inducer</u> such as lactose.

i) <u>Repressed state</u>: the repressor protein is bound to the
    operator gene, blocking transcription of the structural
    genes and thereby preventing synthesis of the proteins
    coded for therein.

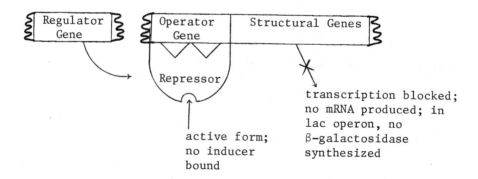

ii) Induced (derepressed state): inducer binds to the repressor
making the repressor unable to bind to the operator gene.
This permits transcription of the operator and structural
genes and synthesis of the coded proteins occurs.

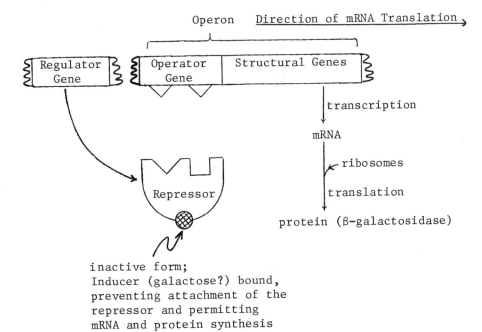

inactive form;
Inducer (galactose?) bound,
preventing attachment of the
repressor and permitting
mRNA and protein synthesis

The above mechanism is self-regulating. Since the enzyme
induced metabolizes the inducer, the inducer will
gradually be removed, the repressor substance will be
free, and the system will go back to the repressed state.

c. Enzyme Repression (another type of end-product repression)

In this case, the repressor is inactive unless activated by a
corepressor.

i) Derepressed state: the operator gene is unblocked since
the repressor, synthesized from the regulator gene, is
inactive. The structural genes are transcribed and synthesis
of the coded protein takes place.

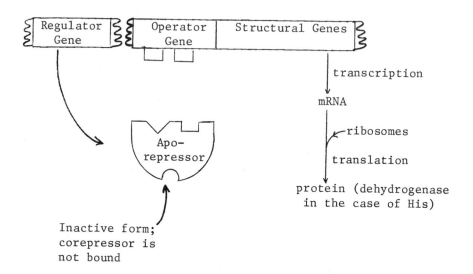

Inactive form;
corepressor is
not bound

ii) <u>Repressed state</u>: the corepressor combines with the inactive repressor to form an active complex, capable of binding to the operator gene and blocking structural gene transcription.

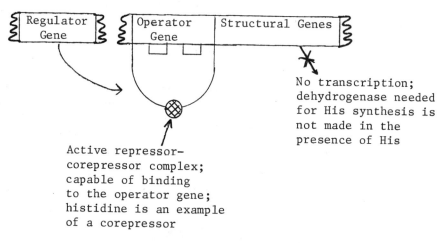

Active repressor-
corepressor complex;
capable of binding
to the operator gene;
histidine is an example
of a corepressor

d. The repression of a group of enzymes by a single repressor is called <u>coordinate repression</u>. The induction of a group of enzymes by a single inducer is called <u>coordinate induction</u>. In the lac and gal operons (responsible for lactose and galactose metabolism, respectively), there are three structural genes each. In the Trp and His operons (tryptophan and histidine synthesis) there are five and fifteen structural genes, respectively.

e. The lac operon is the most thoroughly studied example of an inducible enzyme system. The lac repressor has been isolated and shown to be a protein composed of four subunits and having a molecular weight of about 1.5×10^5 daltons. Its isolation was quite a remarkable feat since it is estimated that there are only about 10 molecules of the repressor per E. coli cell (about 10^{-8} M). Constitutive mutants (ones in which lac operon transcription is maximal in the absence of any inducer) have been found with errors in the regulatory and operator genes. In the regulator mutant, the repressor is unable to recognize the operator binding site; in the operator mutant, normal repressor is present but the operator is abnormal. It is interesting that all known operator mutants are deletions. Regulator mutants are also known in which the inducer substance cannot bind to the repressor molecule and the lac operon is not transcribed, even in the presence of an inducer. These are called super-repressed strains. A curious fact is that the true intracellular inducer of the lac operon is known not to be lactose. Rather, certain galactosides appear to function in this role.

f. As has been pointed out before, the experimental work on which the preceding discussion is based was obtained exclusively in bacterial systems. It is an example of how far our knowledge of procaryotic control mechanisms has advanced but it is not clear to what degree similar thinking can be applied to eucaryotic cells. Some indication of this is given in the following sections on control in higher organisms.

8. The control of enzymes at the transcriptional and translational levels in higher, eucaryotic organisms is less well understood. Several inducible liver enzymes (e.g., glucokinase) have been found. Steroids are known to penetrate into the cell nucleus and bind to the DNA, as are certain antibiotics, carcinogens, and other molecules. This may cause changes in the DNA transcription, both qualitatively and quantitatively. The role of steroids is discussed further under steroid hormones.

a. As mentioned previously, selective derepression of the nuclear DNA is probably the basis for cellular differentiation during development. The concept of genetic totipotence (possession of the entire genome by all diploid cells of higher organisms and, under proper conditions, the expression of these genes).

is supported by many observations and experiments. Histones
(a class of basic proteins associated with DNA; see under
proteins and enzymes) may function as permanent (or long-term)
repressors within differentiated cells. Histones can be
displaced from DNA by polyanions such as pieces of RNA and
phospho- and lipoproteins and they can be phosphorylated by
protein kinases which are activated by cyclic AMP. Phosphorylation of the histones decreases their affinity for DNA. This
effect provides a further mechanism for hormonal regulation
of protein synthesis. Enzymatic methylation of certain
bases in DNA presents still another mechanism **for** permanent
gene repression in differentiation.

b. One important characteristic of malignant cells is their
dedifferentiation, with the concomitant derepression of large
parts of their genomes. This might be due to a lack of
production of (functional) repressors or production of
antibodies to the repressors (assuming that the repressors are
proteins). The antibodies could bind to the repressor, thereby
inactivating it. Another approach to carcinogenesis is
based on the observation that histones can be displaced from
DNA by nuclear RNA (which is a polyanion as are all
polynucleotides). Perhaps viral RNA (or DNA) can similarly
bind to nuclear DNA, remove the histones, and cause derepression.
At the present, this is largely either speculative or based
only on circumstantial evidence.

c. When RNA binds to double-helical DNA, histones are displaced
and the duplex appears to unwind. Since single-stranded DNA's
(or single-stranded regions within duplex DNA) are required for
transcription, this suggests an additional mechanism for
transcriptional control.

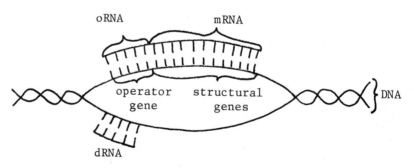

For transcription to occur, derepressor RNA (dRNA) must bind
to the DNA strand complementary to the one to be transcribed.
This causes local unwinding of the DNA duplex. In the next
step, during transcription, the entire operon (operator gene +
structural genes) is transcribed. After release of this large
RNA, it is cleaved by an endonuclease into operator RNA (oRNA,
complementary to the operator gene) and messenger RNA (mRNA,
complementary to the structural genes). Suppose (as shown
above) that the dRNA binds to the DNA region which is
complementary to the operator gene. This provides a self-
regulation of the transcription. The oRNA formed can bind to
the dRNA, giving a double-stranded RNA. The dRNA-oRNA duplex
is unable to bind to the DNA and initiate transcription.
Removal of the dRNA by complexation with oRNA results in
renaturation of the single-stranded region and cessation of
its transcription. Although there are some suggestive
experimental data, this system has yet to be actually
demonstrated either in vivo or in vitro.

d. Although the repressor-inducer model of Jacob and Monod has
    been substantiated in bacterial systems, evidence for it in
    eucaryotic systems is weaker. The clearest demonstration has
    come from the studies of Granick and his coworkers on the
    control of heme synthesis in mammalian reticulocytes. Heme is
    the prosthetic group (cofactor) of hemoglobin, myoglobin, and
    some of the cytochromes. It consists of a porphyrin ring
    (protoporphyrin III; see under porphyrin synthesis) and an atom
    of iron. Mention has already been made (see the section on
    enzyme regulation) of the control of translation by heme and
    initiation factor 3rr (IF-3rr). It appears that heme also
    functions as a corepressor for transcription of the gene for
    δ-aminolevulinic acid synthetase (ALA synthetase). This enzyme
    catalyzes the first unique step in the synthetic pathway for
    protoporphyrin III. This is another example of end-product
    inhibition of the committed step in a synthetic pathway. Just
    as in procaryotes, one can summarize this by the equations

heme + aporepressor $\longrightarrow$ repressor

repressor + operator gene $\longrightarrow$ inactivation of structural
                                           gene transcription

In addition to its effect on transcription and translation, heme
is a negative allosteric effector for ALA synthetase. The
synthesis and role of heme and related compounds is presented
in the sections on hemoglobin and the porphyrins.

e. Repressors and aporepressors studied in bacterial systems seem to be proteins rather than RNA. In eucaryotes, as indicated above, protein and RNA may be involved in transcriptional regulation as repressor and aporepressor molecules. Currently, however, even the existence of operons (or their equivalent) in eucaryotic cells has not been established with any certainty. For reasons such as this, all of the preceding discussion of regulation in eucaryotes should be regarded as a tentative explanation of certain experimental data.

Purine and Pyrimidine Nucleotide Biosynthesis

Before we proceed with the synthesis, it is essential to be familiar with (1) the names of the nucleotides and their relationship to nucleosides and bases; (2) one-carbon transfer reactions which are utilized in the nucleotide synthesis; and (3) synthesis of 5-phosphoribosyl-1-pyrophosphate (PRPP), a compound utilized in the de novo and salvage pathways of nucleotide synthesis.

Nucleotide Nomenclature

Following is a list of nucleotides whose synthesis will be discussed in this chapter. Also given in the list are the corresponding base and nucleosides. Note that in the deoxyribonucleotides, the sugar present is D-deoxyribose rather than D-ribose. This is not specifically shown in the diagram. The deoxyribose is indicated with a "d", as for example dAMP. The only exception to this is thymidylic acid, abbreviated TMP even though it contains deoxyribose.

Base	Nucleoside (Base+sugar)	Nucleotide (Base+sugar+phosphate)
Purines		
Adenine	Adenosine	Adenylic acid (AMP, adenosine monophosphate)
Guanine	Guanosine	Guanylic acid (GMP, guanosine monophosphate)
Hypoxanthine	Inosine	Inosinic acid (IMP, inosine monophosphate)
Pyrimidines		
Uracil	Uridine	Uridylic acid (UMP, uridine monophosphate)
Cytosine	Cytidine	Cytidylic acid (CMP, cytidine monophosphate)
Thymine	*Thymidine	*Thymidylic acid (TMP, thymidine monophosphate)

*Note: In these molecules, the usual sugar present is deoxyribose.

One-Carbon Metabolism

1. One-carbon compounds such as formate and formaldehyde are utilized
    in the biosynthesis of the purines, the pyrimidine thymine, and
    the amino acids, serine and methionine. The addition of a one-
    carbon compound is accomplished by a group of carriers known as
    the folic acid coenzymes. (S-adenosyl methionine, discussed with
    the sulfur-containing amino acids, donates methyl groups for
    biosynthetic reactions. These are known as biological methylation
    reactions and constitute a group of one-carbon transfers distinct
    from those involving folate derivatives.) As their name implies,
    these carriers are derived from folic acid (a vitamin, in man).
    A vitamin is an "accessory food factor", a compound which must be
    present in the diet for normal metabolism. A vitamin cannot be
    synthesized by the organism for which it is a vitamin.

2. Structure of folic acid (F)

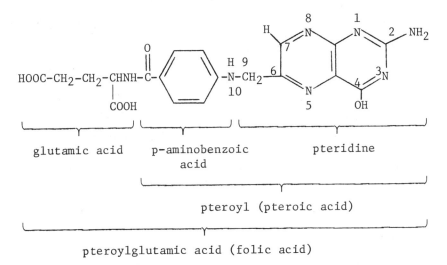

pteroylglutamic acid (folic acid)

3. Folic acid (shown above) is not active as a coenzyme. It must go
    through two reductions, first to dihydrofolic acid (FH_2), and
    then to tetrahydrofolic acid (FH_4) before it may act as a carrier.

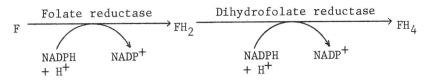

It is not completely clear whether one enzyme catalyzes both reductions or, as indicated above, there are two separate enzymes.

4. Tetrahydrofolate has the structure

where R is the remainder of the molecule, identical to the equivalent portion of folic acid. Thus, FH_4 is 5, 6, 7, 8-tetrahydrofolic acid. The partially reduced dihydrofolate (FH_2) found <u>in vivo</u> appears to be 7,8-dihydrofolate. The other possible isomers (5,6- and 6,7-) have never been unequivocally demonstrated as products of either the chemical or enzymatic reduction.

5. Dihydrofolic acid reductase is very strongly competitively inhibited by methotrexate (amethopterin) which is an analogue of FH_2 and thus is an important "folic acid antagonist". Folic acid antagonists are used to halt the growth of certain fast-growing cancer cells (especially the leukemias). They function by preventing the formation of FH_4. This slows down the synthesis of nucleic acids, which requires FH_4. The sulfonamides (sulfa drugs) interfere with folic acid activity in a different way. They are very similar in structure to a p-aminobenzoic acid (PABA) which is used in the biosynthesis of folic acid. These analogues of PABA interfere with folic acid synthesis in microorganisms which have this synthetic capacity. Since man cannot synthesize folic acid (it is provided in the diet as a vitamin), sulfonamides do not interfere with human metabolism.

<u>Methotrexate</u> (amethopterin) (in aminopterin, another antineoplastic agent, the circled methyl group is replaced by a hydrogen).

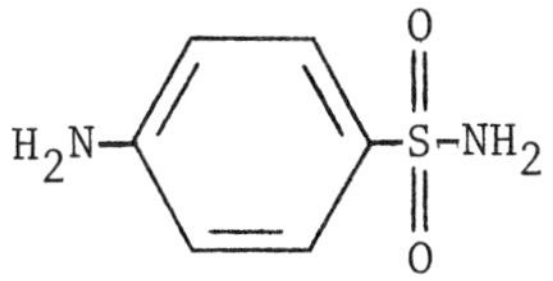

$$H_2N-\text{(benzene ring)}-\overset{\displaystyle O}{\underset{\displaystyle O}{\overset{\|}{\underset{\|}{S}}}}-NH_2$$

<u>Sulfanilamide</u> (one of the sulfonamides)

6. There are several active species within the folic acid group.
    Different species function as one-carbon carriers in different
    metabolic processes. However, in all biosynthetic reactions for
    which a folic acid derivative provides a one-carbon fragment, the
    carbon is carried in a covalent linkage to one or both of the
    nitrogen atoms at the 5 and 10 positions of the pteroic acid
    portion of tetrahydrofolate. The five known forms of carriers are
    shown below. The portions of the structure which are not drawn
    are identical to the corresponding portion of FH_4.

a. N^5, N^{10}-methylene FH_4

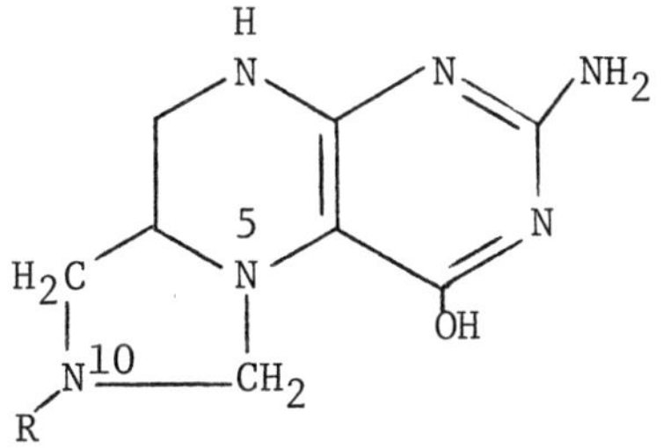

b. N^5, N^{10}-methenyl FH_4

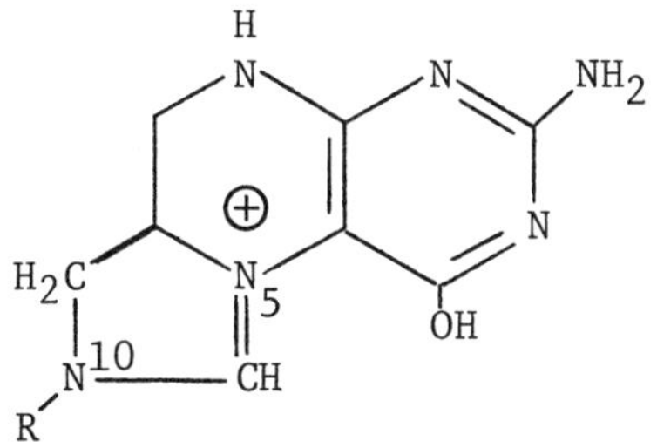

c. N^{10}-formyl FH_4

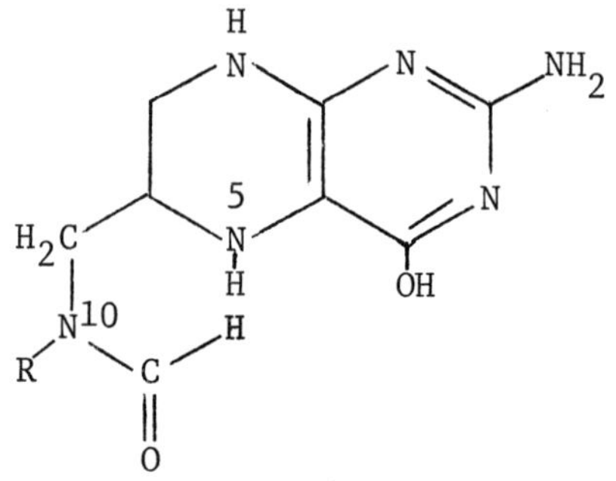

d. N^5-formyl FH_4 (folinic acid)

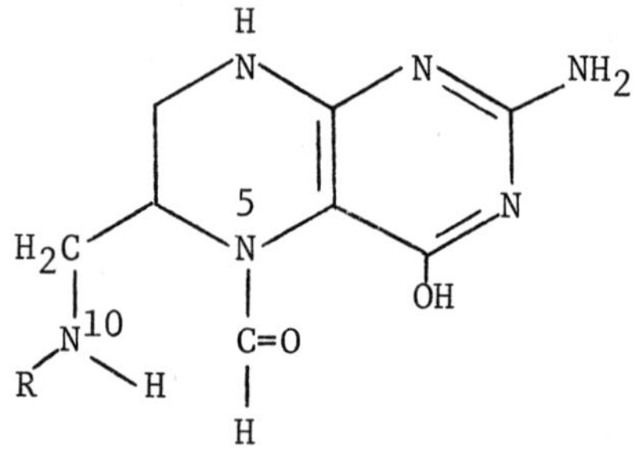

e. N^5-formimino FH_4

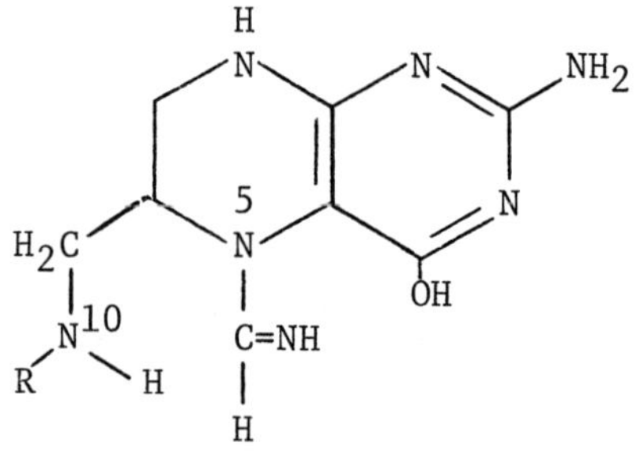

7. The reactions which produce some of the active species of the folate coenzymes are shown here. These interconversions are summarized in Figure 53.

a. Formation of N^5,N^{10}-methylene FH_4 (a <u>formaldehyde</u> derivative of FH_4) may occur in several ways.

i)
$$FH_4 + H\text{-}\overset{\overset{\displaystyle O}{\|}}{C}\text{-}H \xrightarrow{\text{(non-enzymatic)}} N^5,N^{10}\text{-methylene } FH_4$$

   formaldehyde

ii)
$$FH_4 + \underset{\underset{\displaystyle CH_2\text{-}OH}{|}}{NH_2\text{-}CH\text{-}COOH} \xrightarrow[\substack{\text{(pyridoxal} \\ \text{phosphate)}}]{\substack{\text{serine} \\ \text{hydroxymethylase}}} N^5,N^{10}\text{-methylene } FH_4$$

$$+$$

$$NH_2\text{-}CH_2\text{-}COOH$$

   serine

   glycine

iii)
$$N^5,N^{10}\text{-methenyl } FH_4 \xrightarrow[\substack{NADPH \\ +\ H^+}]{\substack{N^5,N^{10}\text{-methylene-}FH_4 \\ \text{dehydrogenase} \\ \\ NADP^+}} N^5,N^{10}\text{-methylene } FH_4$$

The last two reactions (above) are the most important sources of this form of the coenzyme <u>in vivo</u>.

b. Formation of N^5,N^{10}-methenyl FH_4 (a <u>formate</u> derivative of FH_4) may also occur in several ways.

i)

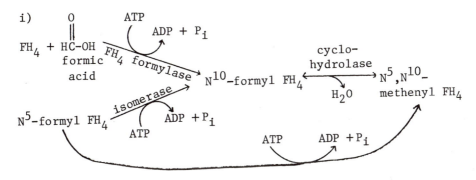

Figure 53. Some Interrelationships of One-Carbon Transfer Reactions Involving Folate Derived Carriers

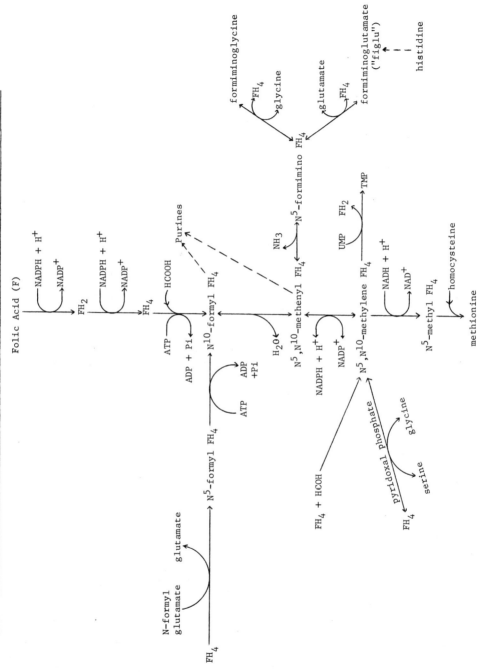

Note: Dashed arrows indicate multiple-step reaction pathways while solid arrows represent direct, single-step reactions.

The two reactions above also show synthetic routes for N^{10}-formyl FH_4. Note that N^5-formyl FH_4 can be converted to N^5,N^{10}-methenyl FH_4 either directly or <u>via</u> N^{10}-formyl FH_4.

ii) Reversal of the tetrahydrofolate dehydrogenase reaction (a, iii, above).

iii) The action of cyclodeaminase on N^5-formimino FH_4 releases NH_3 and N^5,N^{10}-methenyl FH_4.

c. Formation of N^5-formyl FH_4 may occur from N-formyl glutamate and FH_4.

$$\text{N-formyl glutamate} + FH_4 \longrightarrow N^5\text{-formyl } FH_4 + \text{glutamate}$$

d. The N^5 position of FH_4 can also carry a formimino group. The formimino group of N^5-formimino FH_4 can be transferred to other compounds, producing formimino derivatives which are needed in biosynthesis. For example, formimino glycine, a precursor of the purines, can be formed by the reversal of reaction (i) below. The two principal synthetic routes for this FH_4 compound are

i)

$$\underset{\overset{|}{\text{HC=NH}}}{\text{HN-CH}_2\text{-COOH}} + FH_4 \xrightarrow[\text{glycine}]{\text{transferase}} N^5\text{-formimino } FH_4$$

formimino glycine

ii)

$$\underset{\overset{|}{\text{NH}} \; \overset{|}{\text{HC=NH}}}{\text{HOOC-CH}_2\text{-CH}_2\text{-CH-COO}^{(-)}} + FH_4 \xrightarrow[\text{glutamate}]{\text{transferase}} N^5\text{-formimino } FH_4$$

formiminoglutamate

The second reaction is involved in histidine metabolism. If a loading dose of histidine is given to a patient deficient in folic acid, urinary excretion of formiminoglutamic acid ("figlu") is increased. This is known as the <u>figlu excretion test</u> and is useful in the diagnosis of megaloblastic anemias.

8. The formyl group present on the different species of
    tetrahydrofolic acid serves as a one-carbon source in several
    important reactions. Some examples are: the synthesis of the
    purine nucleus using N^5,N^{10}-methenyl FH_4 and N^{10}-formyl FH_4 (see
    next section); the synthesis of N-formyl methionine-tRNA (discussed
    earlier) using N^{10}-formyl FH_4; and the formation of serine from
    glycine, also using N^{10}-formyl FH_4. In mammalian systems,
    N^5,N^{10}-methylene can be reduced to N^5-methyl FH_4 which can transfer
    a methyl group to homocysteine, forming methionine (see under
    sulfur-containing amino acids). N^5,N^{10}-methylene FH_4, in a
    reaction catalyzed by thymidylate synthetase (see later in this
    chapter under pyrimidine nucleotide biosynthesis), supplies a
    methyl group for thymidylic acid synthesis. Hydroxymethyl FH_4
    may participate as an intermediate in some of these reactions.
    These roles of FH_4 derivatives are indicated in Figure 53.

Role of 5-Phosphoribosyl-1-Pyrophosphate (PRPP)

1. Functions of PRPP

    PRPP is a key intermediate and functions in several different ways
    in nucleotide synthesis.

    i)   It is required for _de novo_ synthesis of purine and pyrimidine
         nucleotides.

    ii) _Salvage pathways_ of purine nucleotides take preformed purines
         (derived from dietary sources and nucleic acid catabolism)
         and convert them to the respective nucleotides using PRPP to
         attach a ribose-1-phosphate to the base.

$$\text{(P)-Ribose-(P)-(P)} \longrightarrow \text{Base-Ribose-(P)} + PP_i$$
$$\text{(PRPP)} \qquad\qquad \text{(nucleotide)}$$

2. Synthesis of PRPP

(R-5-P)   (PRPP)

Note: This reaction is catalyzed by a kinase which transfers the
terminal pyrophosphate group of ATP to R-5-P. This is
unusual, since kinases generally transfer only phosphate
groups.

3.   <u>Origin of Ribose-5-P</u>. R-5-P which is utilized in the above synthesis
of PRPP can come from three possible pathways.

a.   The hexose monophosphate shunt (phosphogluconate pathway) is
probably the major source of R-5-P in tissues (liver, bone
marrow) in which glucose can be oxidized aerobically.

b.   Uronic acid pathway. Xylulose-5-P produced in this pathway
is converted to R-5-P as follows:

$$\text{D-Xylulose-5-P} \xrightarrow{\text{Epimerase}} \text{D-Ribulose-5-P} \xrightarrow{\text{Isomerase}} \text{D-Ribose-5-P}$$

The contribution made by this pathway to the supply of R-5-P
may be a minor one. The epimerase and isomerase enzymes are
the same ones which participate in the HMP shunt.

c.   Ribose-5-phosphate can also be produced in tissues (such as
skeletal muscle) in which glucose is metabolized principally
by anaerobic glycolysis. This is shown in the following
diagram, starting from fructose-6-phosphate (F-6-P).

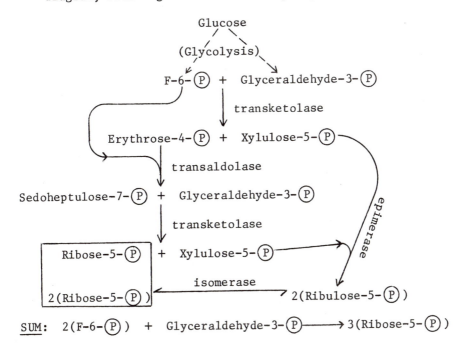

SUM: 2(F-6-P) + Glyceraldehyde-3-P ⟶ 3(Ribose-5-P)

Notice that all of the enzymes in this pathway are also part of the HMP shunt. In the muscle, however, complete operation of the HMP pathway does not occur, due to the absence of glucose-6-phosphate dehydrogenase. These interconversions can occur under either aerobic or anaerobic conditions but they are probably more important in the absence of oxygen.

Purine Nucleotide Biosynthesis

In the diagrams below, the portion of the molecule enclosed by a dashed line was added in the preceding reaction.

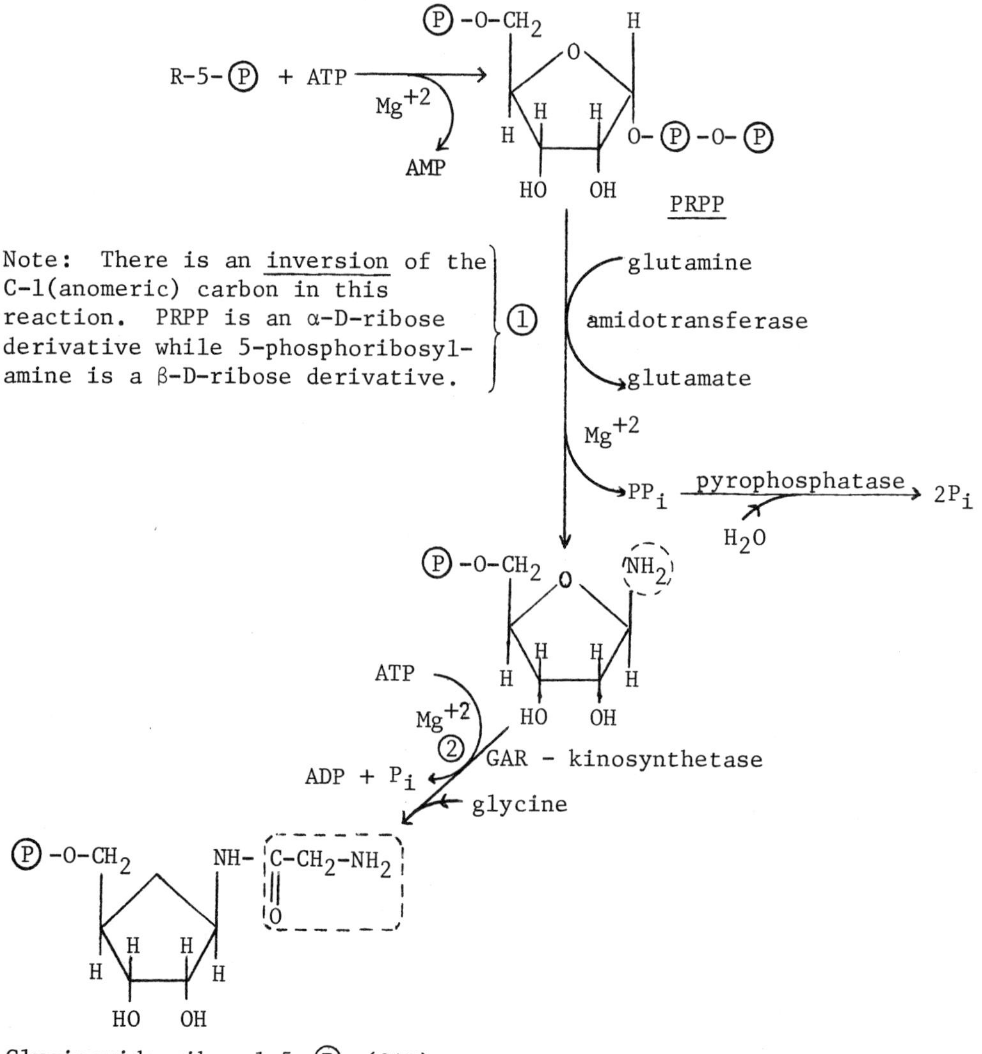

PRPP

Note: There is an <u>inversion</u> of the C-1(anomeric) carbon in this reaction. PRPP is an α-D-ribose derivative while 5-phosphoribosyl-amine is a β-D-ribose derivative.

① glutamine → amidotransferase → glutamate

Mg^{+2}

PP_i → pyrophosphatase → $2P_i$

H_2O

② GAR – kinosynthetase

ATP, Mg^{+2}, ADP + P_i, glycine

Glycinamide ribosyl-5-P (GAR)

GAR can be written in a different form for the sake of convenience. Note that the ribosyl phosphate group is now represented by "R-5-P".

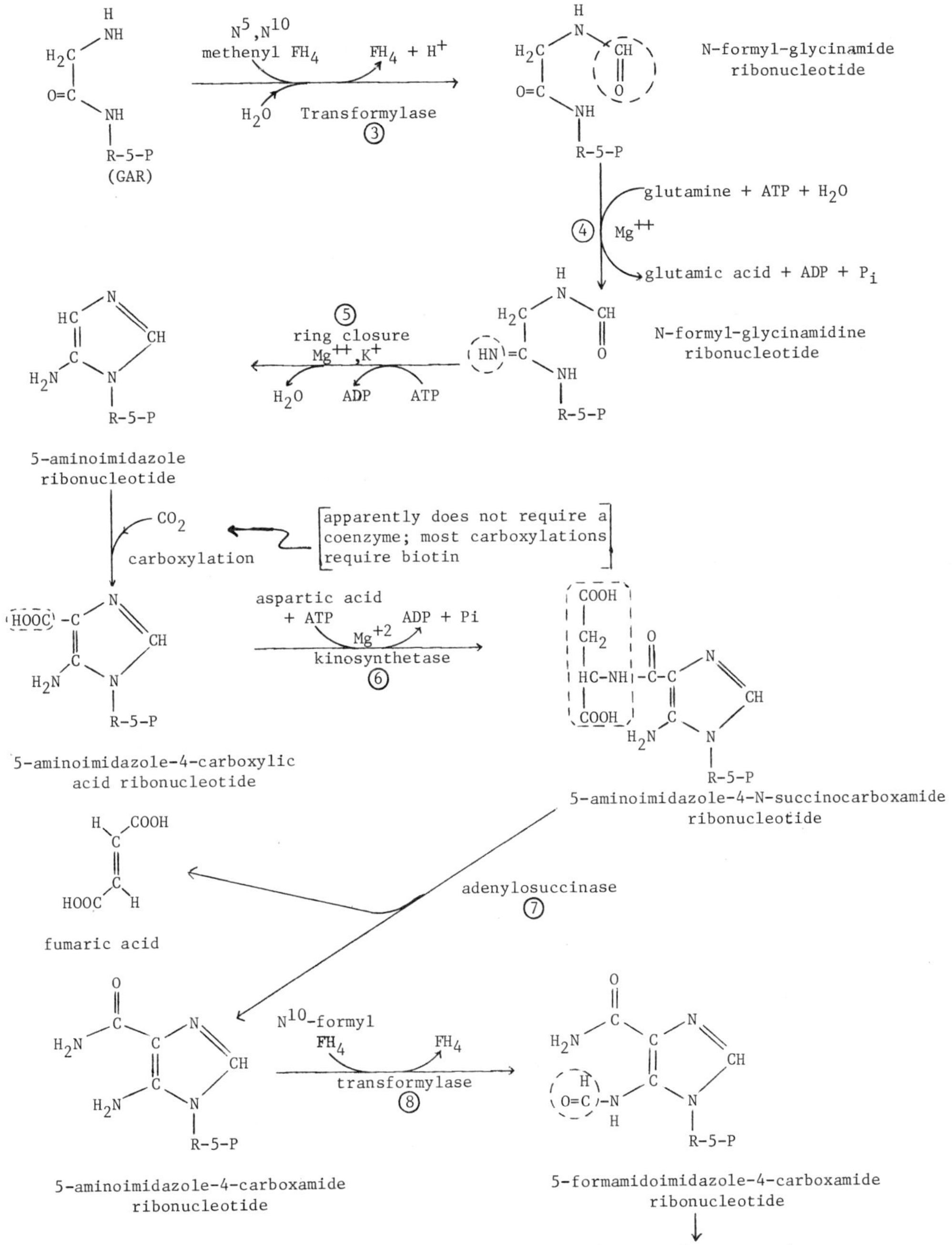
R-5-P
(GAR)
N^5,N^{10}
methenyl FH_4
$FH_4 + H^+$
H_2O
Transformylase
③
N-formyl-glycinamide
ribonucleotide
glutamine + ATP + H_2O
④
Mg^{++}
glutamic acid + ADP + P_i
N-formyl-glycinamidine
ribonucleotide
⑤
ring closure
Mg^{++}, K^+
H_2O
ADP
ATP
5-aminoimidazole
ribonucleotide
CO_2
carboxylation
apparently does not require a coenzyme; most carboxylations require biotin
5-aminoimidazole-4-carboxylic
acid ribonucleotide
aspartic acid
+ ATP
ADP + Pi
Mg^{+2}
kinosynthetase
⑥
5-aminoimidazole-4-N-succinocarboxamide
ribonucleotide
fumaric acid
adenylosuccinase
⑦
N^{10}-formyl
FH_4
FH_4
transformylase
⑧
5-aminoimidazole-4-carboxamide
ribonucleotide
5-formamidoimidazole-4-carboxamide
ribonucleotide

(to top of next page)

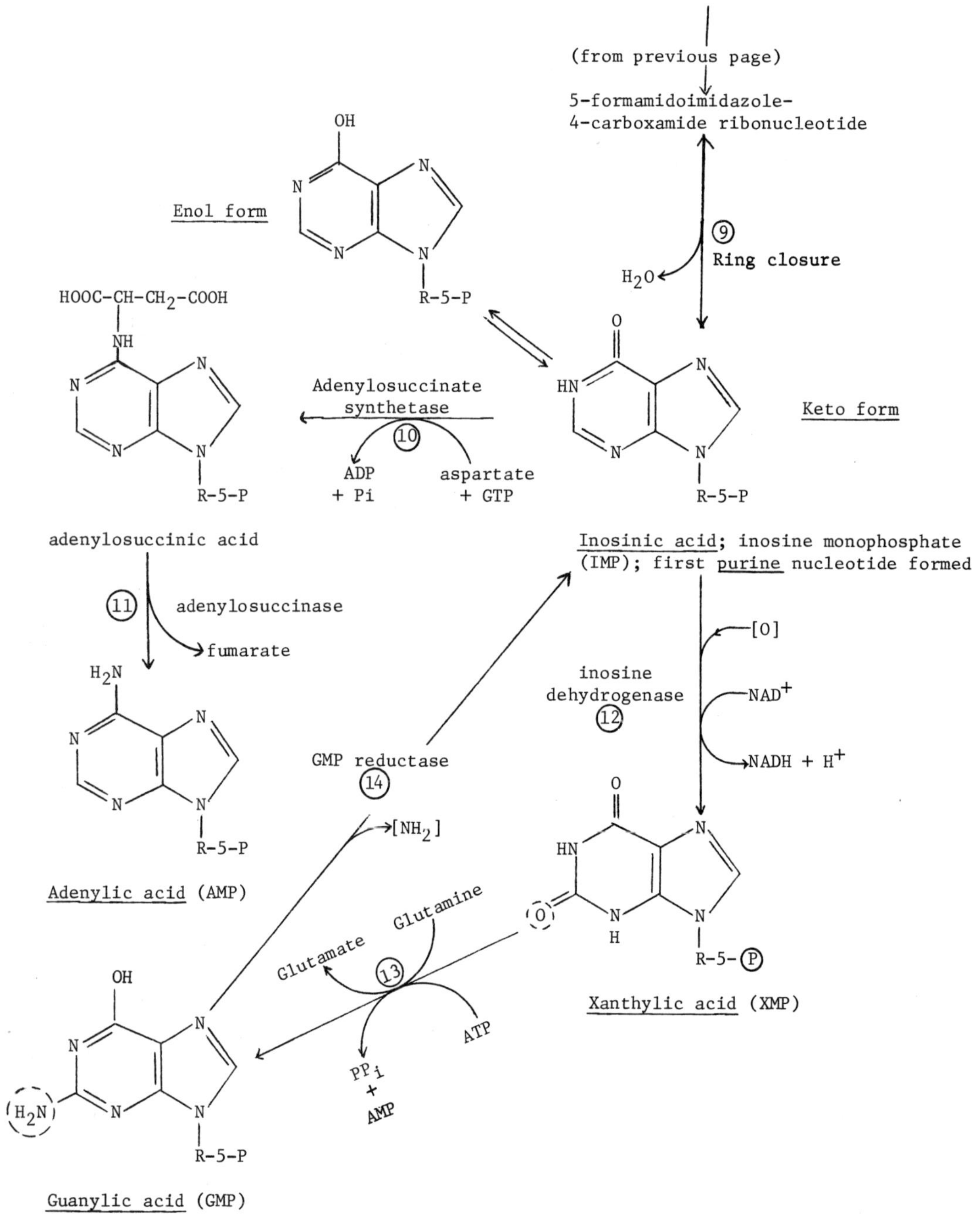

(from previous page)

5-formamidoimidazole-
4-carboxamide ribonucleotide

OH

Enol form

HOOC–CH–CH$_2$–COOH
          |
          NH

R–5–P

$\textcircled{9}$ Ring closure

H_2O

O

HN

Keto form

R–5–P

Adenylosuccinate
synthetase
$\textcircled{10}$

ADP   aspartate
+ Pi   + GTP

adenylosuccinic acid

Inosinic acid; inosine monophosphate
(IMP); first **purine** nucleotide formed

$\textcircled{11}$ adenylosuccinase

fumarate

inosine
dehydrogenase
$\textcircled{12}$

[O]

NAD$^+$

NADH + H$^+$

H_2N

R–5–P

GMP reductase
$\textcircled{14}$

[NH$_2$]

Adenylic acid (AMP)

O

HN

N
H

R–5–$\textcircled{P}$

Xanthylic acid (XMP)

OH

Glutamate   Glutamine   (O)

$\textcircled{13}$

PP$_i$
+
AMP

ATP

(H$_2$N)

R–5–P

Guanylic acid (GMP)

Other reactions of importance in the synthesis of purine nucleotides are the <u>salvage pathways,</u> summarized below. The significance of these is discussed later, following the inhibition of purine synthesis.

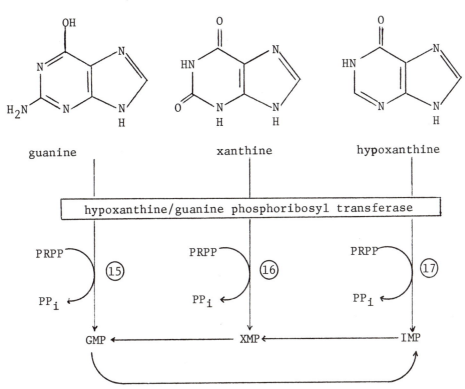

guanine   xanthine   hypoxanthine

hypoxanthine/guanine phosphoribosyl transferase

PRPP   PRPP   PRPP

⑮   ⑯   ⑰

PP$_i$ PP$_i$ PP$_i$

GMP ←   ← XMP ←   ─ IMP

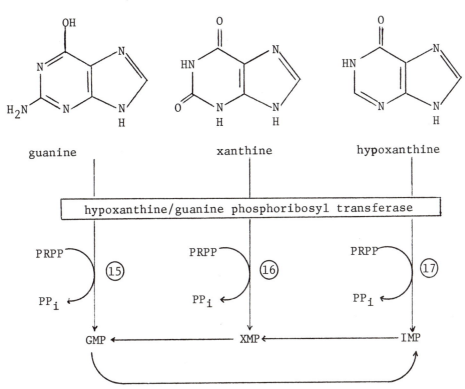

NH$_2$

adenine
phosphoribosyl transferase

⑰   AMP

PRPP   PP$_i$

Adenine

Comments on Purine Biosynthesis

1. Sources of atoms in the purine molecule

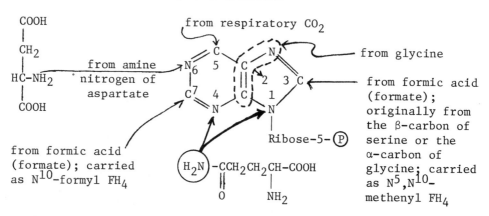

The numbers in this diagram indicate the order of addition of the atoms in the purine nucleus. Note that the three atoms of glycine are added as a group and that glycine is the only amino acid which provides both carbon and nitrogen atoms to the purine ring. (In pyrimidines, aspartate plays an equivalent role.)

2. In purine biosynthesis, the glycosidic bond is formed when the first atom of the purine ring is incorporated. This is in contrast to pyrimidine biosynthesis, where the pyrimidine ring is completely formed prior to addition of ribose-5- P .

3. In E. coli cells, there are several control mechanisms present to regulate purine synthesis.

    a.

Amidotransferase is a multivalent (many binding sites) regulatory enzyme. Feedback inhibition is by ATP, ADP, or AMP, or by GTP, GDP, GMP; each group of nucleotides acting at a separate control site on the enzyme. IMP, IDP, and ITP also inhibit this reaction (see the section on enzyme regulation for more information about this reaction).

b.

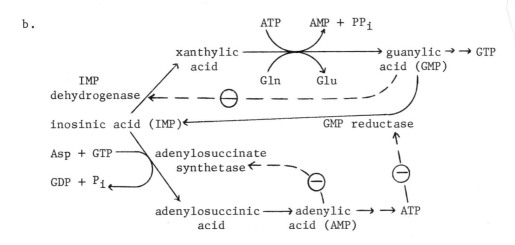

i) ATP is required as a cofactor for GTP synthesis and, reciprocally, GTP is needed to form ATP. This regulates the synthesis of one nucleotide relative to the other.

ii) End-product inhibition also occurs in both pathways above. Adenylosuccinate synthetase is inhibited by AMP and IMP dehydrogenase is inhibited by GMP.

iii) GMP reductase, which converts GMP to IMP, is inhibited by ATP. Thus, an adequate supply of adenine nucleotides prevents their further synthesis at the expense of guanine nucleotides.

c. There is also evidence that guanosine and adenosine suppress the synthesis of some of the enzymes which catalyze steps in purine nucleotide biosynthesis.

d. Some interrelationships and control points in purine biosynthesis are summarized in the next diagram. Heavy lines indicate reactions while dashed lines show control pathways.

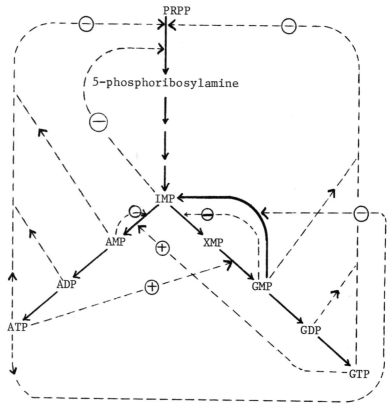

4. Inhibitors of purine nucleotide biosynthesis include

    a. <u>Glutamine Analogues</u>. Azaserine, an antibiotic which has been
        isolated from a species of <u>Streptomyces</u>, is a structural
        analogue of glutamine.

$$N{\equiv}N{=}CH-\overset{\overset{\displaystyle O}{\|}}{C}-O-CH_2-\overset{|}{\underset{\displaystyle NH_2}{C}}H-COOH$$

Azaserine NH_2

Azaserine inhibits steps
1, 4, and 13 of purine
synthesis (see diagram of
purine synthetic pathway).

$$H_2N-\overset{\overset{\displaystyle O}{\|}}{C}-CH_2-CH_2-\overset{|}{\underset{\displaystyle NH_2}{C}}H-COOH$$

Glutamine NH_2

b. Folic acid analogues such as methotrexate (amethopterin) and
    aminopterin block the production of functional folic acid
    derivatives by inhibiting dihydrofolate reductase (FH_2-X→FH_4).
    Step 8 of purine synthesis (see purine synthesis diagram) is
    particularly inhibited. The structures of these compounds are
    given under one-carbon metabolism. Other folic acid analogues
    can also be made by varying other parts of the basic FH_4
    structure. The most potent analogue appears to be methotrexate
    (MTX), in which FH_4 has been given 4-amino and 10-methyl
    substituents. MTX functions as an antagonist without need for
    any preliminary metabolic transformation. Resistance to MTX
    can develop for several reasons, including an increase in
    FH_2-reductase levels and modification of the form of the
    reductase. MTX also affects both the humoral and cellular
    immune responses (see under immunochemistry).

c. Purine Analogues

    i) 6-mercaptopurine

        (I) Mechanism of action. 6-MP is a structural analogue
             of hypoxanthine and can be converted to a nucleotide
             by one of the salvage enzymes.

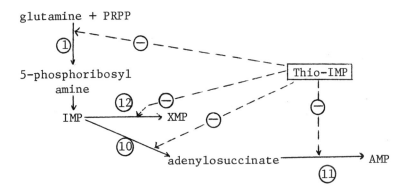

             SH

             6-MP

             hypoxanthine/guanine
               phosphoribosyl
                 transferase
             ─────────────────────→ Thio-IMP (6MP-R-P)
                                        (Active inhibitor)
             PRPP PP_i

             This reaction is similar to salvage pathway reactions
             15, 16, and 17. Thio-IMP prevents the production of
             both AMP and GMP by inhibiting the following reactions:

glutamine + PRPP

①

5-phosphoribosyl
     amine

IMP ⑫ → XMP

⑩

Thio-IMP

adenylosuccinate ────────→ AMP
            ⑪

Although unmetabolized 6-MP, thio-ITP, deoxythio-ITP, and thio-IDP are biologically active in several ways (which may account for some of the side-effects of 6-MP treatment), thio-IMP is the most active metabolic form of 6-MP.

(II)   6-MP is used in the treatment of acute leukemias. Eventually, however, 6-MP-resistant tumor cells do become preponderant. The probable mechanisms of resistance include development of altered specificity or a lack of phosphoribosyl transferases so that thio-IMP (the active inhibitor) is not formed. In support of this mechanism, the resistant cells do respond to 6-methylmercaptopurine ribonucleoside, which is converted to the corresponding nucleotide. Other means for 6-MP-resistance may be alterations in the permeability of the cell and increased rate of destruction of 6-MP.

(III)   6-MP is partially metabolized to 6-thiouric acid (which lacks antitumor activity) by xanthine oxidase (see under proteins and enzymes; purine catabolism), a process similar to the production of uric acid from purines by the same enzyme. Allopurinol is a xanthine oxidase inhibitor used in the treatment of gout (discussed later). It potentiates the action of 6-MP by preventing its conversion to 6-thiouric acid. The potentiation effect of allopurinol is one of the properties that is taken into consideration in 6-MP treatment. 6-MP is also degraded by methylation of the sulfhydryl group and subsequent oxidation of the methyl group. In order to prevent this type of degradation, S-substituted derivatives of 6-MP have been synthesized such as azathioprine.

ii) Azathioprine (imuran) is a derivative of 6-MP and functions as an antiproliferative agent. The compound has its principal use as an immunosuppressive agent. Azathioprine presumably releases 6-MP in the body by reacting with sulfhydryl compounds such as GSH.

iii) 6-Thioguanine is similar in its action to 6-MP. Again,
the most active form is 6-thio-GMP, the ribonucleotide
of the drug.

6-Thioguanine

d. <u>Sulfonamides</u> prevent formation of functional folic acid in
bacteria. They are structural analogues of PABA (see also
under one-carbon metabolism).

a sulfonamide; if R = -H,
the compound is sulfonilamide

5. The primary pathway for purine nucleotide biosynthesis is the one
presented on pages 388-390. There are, however, other secondary
reactions by which purine nucleotides are formed. These are called
"salvage pathways" because they involve the synthesis of purine
nucleotides from free purines and purine nucleosides which are
"salvaged" from dietary sources and from tissue breakdown, as
shown below. These reactions and the structures involved were
given previously on page 391, following the <u>de novo</u> purine
biosynthetic pathway.

a.

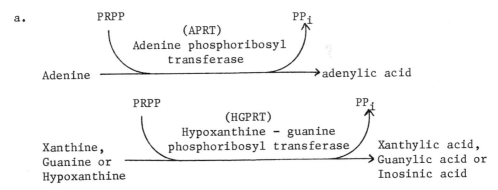

The Lesch-Nyhan Syndrome is an X-linked recessive neurological disorder associated with an overproduction of uric acid (see also gout; hyperuricemia), mental retardation, and a tendency for self-mutilation. In red cells, skin fibroblasts, and brain and liver cells of patients having this disease, an almost complete lack of hypoxanthine/guanine phosphoribosyl transferase activity has been demonstrated. The absence of this salvage enzyme is presumably the reason why 6-MP and imuran are ineffective in controlling purine synthesis in such patients. The connection between the observed symptoms and the lack of the transferase is not clear.

b. Purine nucleosides may be formed from free purines and ribose-1-phosphate

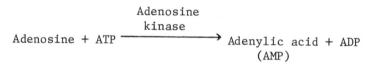

$$\text{free purine} + \text{R-1-P} \xrightarrow{\substack{\text{purine}\\ \text{nucleoside}\\ \text{phosphorylase}}} \text{nucleoside} + P_i$$

The phosphorylase may also be involved in degradative reactions (see purine catabolism, below). Nucleosides may then be converted to nucleotides by nucleoside kinases as in the example here.

$$\text{Adenosine} + \text{ATP} \xrightarrow{\substack{\text{Adenosine}\\ \text{kinase}}} \begin{array}{c}\text{Adenylic acid} + \text{ADP}\\ \text{(AMP)}\end{array}$$

6. Guanylic acid (GMP) and adenylic acid (AMP) are converted to GTP and ATP respectively in several ways.

a. A group of nucleoside monophosphokinases, of which adenylate kinase is one, catalyze the formation of nucleoside diphosphates from nucleoside monophosphates using ATP. If Z is the purine group, then the reaction is as follows:

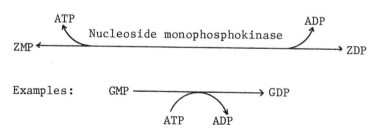

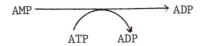

b. Nucleoside diphosphokinases catalyze the formation of a
    nucleoside triphosphate from a nucleoside diphosphate.

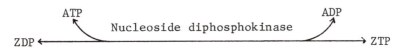

Examples:

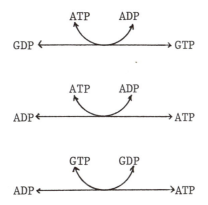

The reactions above are used in the formation of pyrimidine
nucleoside triphosphates as well.

7. Formation of deoxyribonucleotides will be discussed later in this
    chapter. Purine and pyrimidine deoxynucleotides will be covered
    together.

8. Purine Catabolism

    a. The purine nucleotide products of RNA and DNA catabolism are
        the 3'- and 5'-phosphates of adenosine, guanosine,
        deoxyadenosine and deoxyguanosine. The phosphates are removed
        (as P_i) from all of these except adenosine-5'-(P) by appropriate
        phosphatase enzymes. Adenosine-5'-(P) is catabolized as
        shown next.

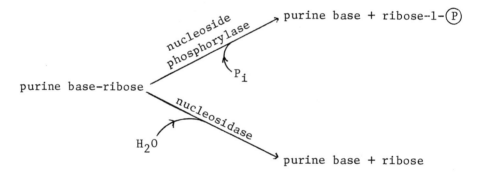

adenosine-5'-(P)   inosine-5'-(P)

The inosine-5'-(P) is converted to inosine and P_i by a
phosphatase.

b.   The purine nucleosides produced by the above reactions next have
the N-glycosidic linkage broken. This can be done by addition
of P_i (catalyzed by a nucleoside phosphorylase) or of H_2O
(catalyzed by a nucleosidase). The general reactions are

purine base-ribose
  nucleoside phosphorylase → purine base + ribose-1-(P) P_i
  nucleosidase → purine base + ribose H_2O

The analogous reactions occur for the deoxyribonucleosides,
producing deoxyribose and deoxyribose-1-(P). The sugars are
metabolized further or excreted. Note that the nucleoside
phosphorylases may also be involved in synthetic (salvage)
pathways .

c.   The purine bases (adenine, guanine, and hypoxanthine) are
further catabolized in the reactions below.

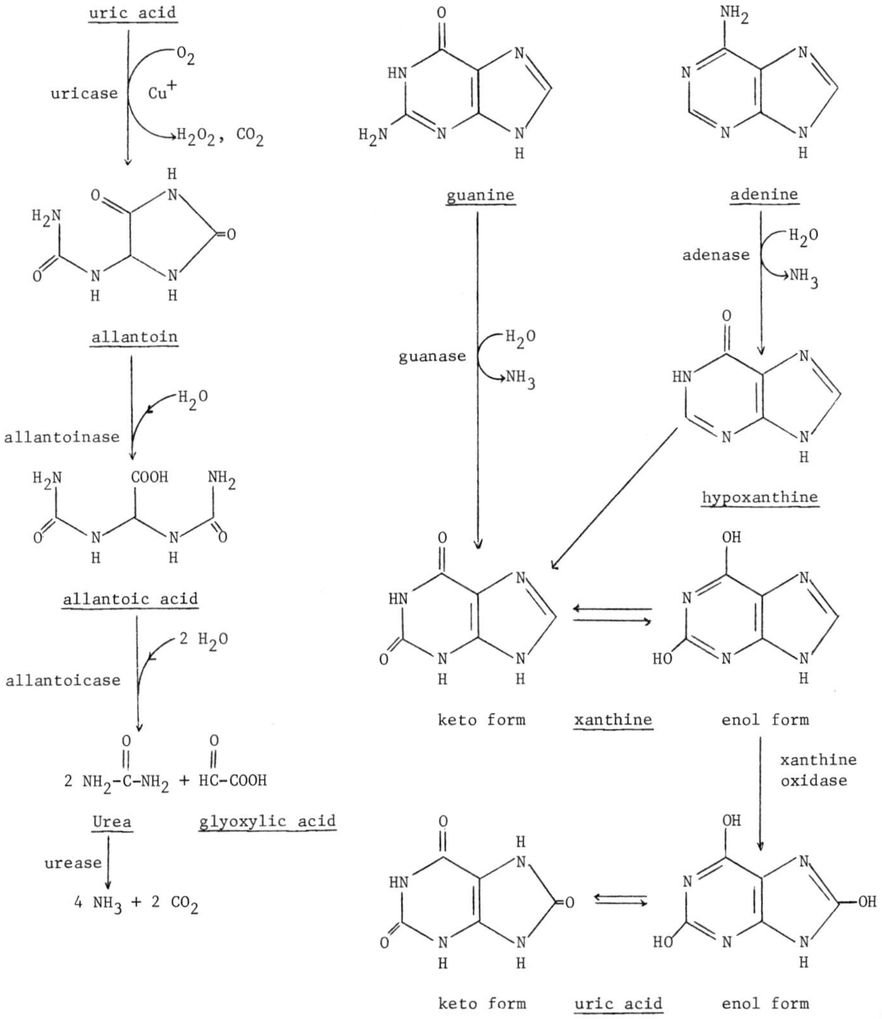

uric acid

O_2

uricase Cu^+

H_2O_2, CO_2

allantoin

$-H_2O$

allantoinase

H_2N COOH NH_2

allantoic acid

2 H_2O

allantoicase

$2 \; NH_2-\overset{\overset{\textstyle O}{\|}}{C}-NH_2 \; + \; HC-COOH$

Urea   glyoxylic acid

urease

$4 \; NH_3 + 2 \; CO_2$

guanine

guanase $-H_2O$ NH_3

keto form   xanthine   enol form

keto form   uric acid   enol form

adenine

adenase H_2O NH_3

hypoxanthine

xanthine oxidase

In man, this is the end-product of purine catabolism. The further reactions (page 401) occur in other animals (see page 402).

d. <u>Xanthine oxidase (X.O.)</u> is an important and interesting enzyme. There are two FAD's, two molybdenum atoms, and eight nonheme iron atoms in each molecule of X.O. This enzyme is discussed further in the section on enzyme regulation (proteins and enzymes chapter) and in the next section, on gout.

9. <u>Final Excretory Products of Purine Metabolism in Some Animals</u>

Animal Group (examples)	Product Excreted
1. some marine invertebrates, crustaceans	ammonia
2. most fish , amphibia (<u>ureotelic</u> animals)	urea
3. some teleost fish	allantoic acid
4. most mammals other than primates; turtles; insects	allantoin
5. primates (including man); dalmatian dogs; birds; some reptiles (<u>uricotelic</u> animals)	uric acid

<u>Gout</u>

1. Characteristics of gout

a. Gout is characterized by elevated levels of urates (salts of uric acid) in the blood (hyperuricemia) and uric acid in the urine (uremia). Virtually all patients with gout (even those who are asymptomatic) have hyperuricemia (urate levels above 6.0 mg/100 ml of plasma). Significant differences in plasma urate levels have been noted between different racial groups. For example, among the Maoris (Polynesians native to New Zealand), the mean serum urate level in the adult male population is greater than 7 mg/100 ml. Although there is a higher incidence of gout in the adult male Maori population, hyperuricemia is not always sufficient, in itself, to cause gout (see below, under primary gout). The disease may remain asymptomatic or it may become clinically manifest with the characteristic attacks of gouty arthritis, the formation of uric acid stones in the kidney, or both.

b. Many of the clinical symptoms of gout arise from two things.

i) The <u>relative insolubility</u> of uric acid (the end-product of purine metabolism in humans) in aqueous solutions (sodium urate is more soluble than the acid itself);

ii) the absence, in man, of the enzyme <u>uricase</u>, which degrades
uric acid to the more-soluble compound allantoin
(mentioned above).

Thus, the accumulation of uric acid and urates above normal
levels results in precipitation and the formation of sodium
urate crystals in the joints, kidney, etc. This leads to the
typical gout syndrome by the mechanism discussed below.

c. <u>Primary gout</u> is a genetically determined disorder of purine
metabolism, seen predominantly in men. The occurrence of gout
in women is uncommon and when it does occur it is usually
found in postmenopausal individuals. There may be a metabolic
or hormonal sexual factor involved. Men normally have a blood
urate concentration which is about 1 mg/100 ml higher than
women, a difference which disappears after menopause. The
different hormonal makeup of men and women may directly
influence the onset of gouty attacks. In the proposed mechanism
for episodes of gouty inflammation, small amounts of sodium
urate initially precipitate and are phagocytized by polymorphonuclear (PMN) leucocytes. After engulfment, the urate
crystals interact with the membranes of the intracellular
lysosomes and with the PMN leucocyte cell membrane. This
interaction occurs due to the ability of urate to hydrogen
bond with groups in the membrane, and it leads to destruction
of the membrane and release of the cytoplasmic and lysosomal
components into the surrounding fluid. The lysosomal enzymes
(proteases, lipases, etc.) proceed to indiscriminately destroy
the tissue and inflammation results (see the section on
immunochemistry). It is reported that 17-β-estradiol (present
in premenopausal women but largely absent in men and
postmenopausal women) confers a greater stability to the
lysosomal membranes than does testosterone.

Primary gout may be due to an overproduction or an underexcretion of uric acid, or, more likely, to a combination of
both. Frequently, siblings and other close relatives of
afflicted individuals have high levels of uric acid but do not
develop gout, indicating that hyperuricemia is not the only
factor involved. Primary <u>renal</u> gout is due to an underexcretion
caused by an enzymatic deficiency that affects the uric acid
transport. Primary <u>metabolic</u> gout is due to an overproduction
of uric acid and its precursors (purines).

d. <u>Secondary gout</u> is an acquired form of the disease, in which
gouty arthritis develops as a complication of hyperuricemia
caused by another disorder (such as leukemia, chronic
nephritis, polycythemia, etc.). This type of hyperuricemia is

usually due to high levels of uric acid caused by abnormally rapid turnover of nucleic acids. Notice that here hyperuricemia is sufficient to produce the clinical symptoms of gout.

e. Gout is very rare in children and adolescents. When it does occur, it may represent a unique form of secondary gout. In fact, it is uncommon before the thirties, the greatest incidence of initial attacks being seen in men in their forties or fifties. Such patients have probably had hyperuricemia for many years but did not experience acute attacks of gout until later in life.

f. In the usual development of the (untreated) disease, acute attacks follow the initial appearance with increasing frequency and usually with increasing severity. There is often a build-up of chalky deposits of sodium monourate crystals, either in nodules (called tophi; singular tophus) or diffusely in cartilage and in tissues around the joints. The arthritic attacks of gout are caused by the deposition of large amounts of urate crystals in the joints, probably in the synovial fluid leukocytes. This chronic gouty arthritis may be grotesquely deforming, totally disabling, and excruciatingly painful. The syndrome has been known since the early Greek and Roman physicians. The modern history of gout dates from 1683 when T. Sydenham, himself a sufferer from the disease, chronicled its symptoms in great detail.

2. Drugs used to treat gout

a. Colchicine (see also the section on microtubules) is used to treat the pain and inflammation of an attack of acute gouty arthritis. Other anti-inflammatory drugs such as adrenocorticotropin, phenylbutazone, corticoids and oxyphenbutazone are also helpful.

Colchicine

The mechanism by which colchicine relieves acute gouty attacks is obscure. Microtubules are involved in leukocytic motion and it is colchicine which apparently disrupts such motile systems. Although this is a possible mode of action, it remains to be demonstrated that this occurs in vivo.

b. Allopurinol, which is structurally similar to hypoxanthine, inhibits the enzyme xanthine oxidase and thus reduces the formation of xanthine and of uric acid. Xanthine oxidase slowly oxidizes the hypoxanthine to alloxanthine (oxypurinol) which is also a xanthine oxidase inhibitor. These interactions are shown diagrammatically below and are discussed further in the section on enzyme inhibition.

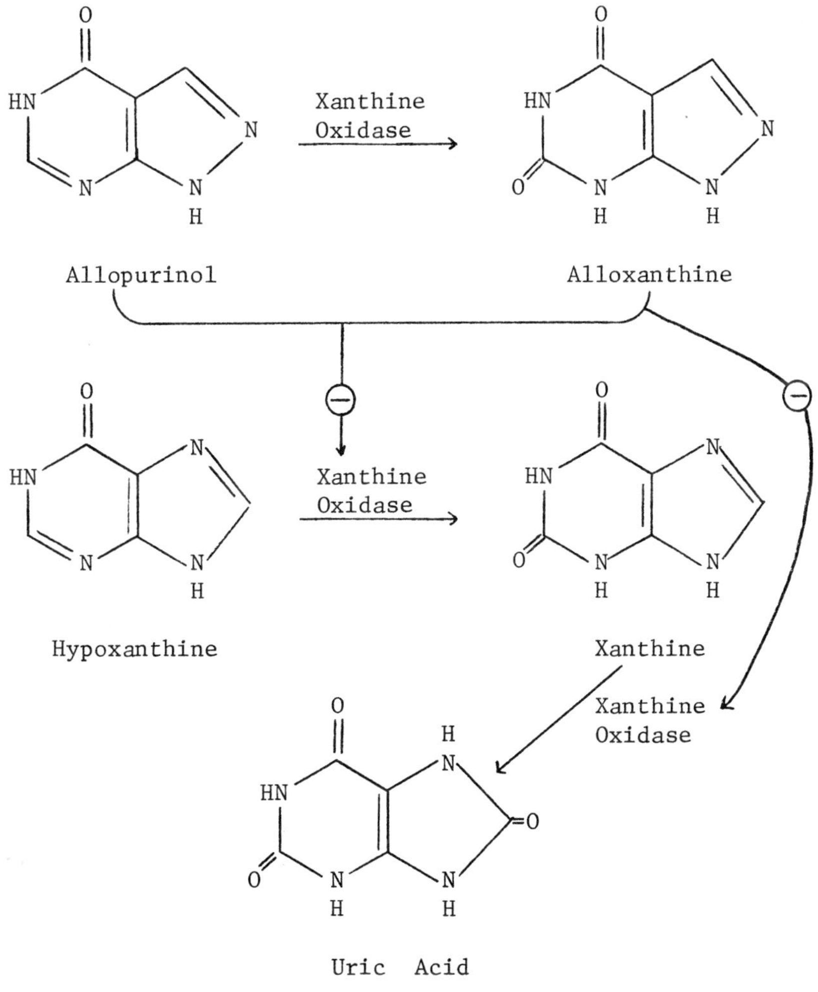

Allopurinol

Xanthine Oxidase

Alloxanthine

Hypoxanthine

Xanthine Oxidase

Xanthine

Xanthine Oxidase

Uric Acid

Figure 54.   A Proposed Mechanism for the Pathogenesis of Acute Gouty Arthritis*

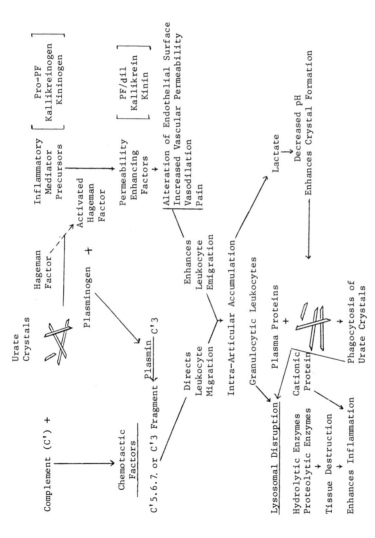

Pro-PF = Propermeability factor; PF/dil = permeability factor/diluted serum. Some of the factors indicated above (complement system, kallikrein, Hageman factor, leukocytic function) are discussed in the chapter on immunochemistry. Note that the urate crystals appear to function in two primary ways: (1) by directly and indirectly activating the complement systems; and (2) through activation of the Hageman factor. Both routes lead to the same ultimate effect.

* Reproduced with permission from "Hageman Factor and Gouty Arthritis", R.W. Kellermeyer, Arthritis and Rheumatism 11, 453, 1968).

Allopurinol also gets converted to allopurinol ribonucleotide. This decreases uric acid production by decreasing overall purine synthesis through depletion of PRPP, a substrate used in the initial step of this pathway. Additionally, the allopurinol ribonucleotide allosterically inhibits PRPP-amidotransferase, the first enzyme needed for <u>de novo</u> purine biosynthesis, by a feedback mechanism. These effects are illustrated below.

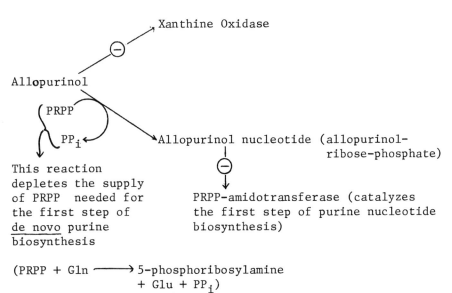

(PRPP + Gln ⟶ 5-phosphoribosylamine
+ Glu + PP_i)

3. <u>Hyperuricemia</u>, which can lead to gout, is produced by a variety of causes and only in a few instances have the biochemical lesions been delineated. The presence of high levels of uric acid in a majority of the individuals who have gout as a clinical disorder is not explained by an established defect in a biochemical process. Lesions that may result in the overproduction of uric acid are listed below.

a. <u>Reactions leading to an overproduction of PRPP</u>

   i) Glutathione reductase variant. The enzyme glutathione reductase catalyzes the following reaction (see also under erythrocyte metabolism):

$$GSSG + NADPH + H^+ \longrightarrow 2 \ GSH + NADP^+$$

The NADPH required in this reaction is produced in the hexose monophosphate shunt pathway (already discussed, see page 245). In a few patients with gout, it has been observed that there is an increased activity of glutathione

reductase. This puts a drain on the supply of NADPH and,
in a cascade process, stimulates the HMP shunt, increases
R-5-P production (R-5-P is a product of the HMP shunt), and
provides more PRPP for purine nucleotide biosynthesis.
This leads to an overproduction of purine nucleotides and
eventually to hyperuricemia. These interrelationships can
be diagrammatically shown as below.

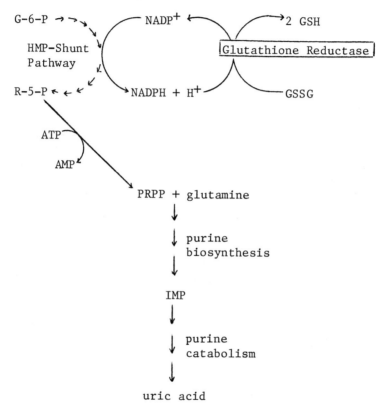

b. <u>Reactions leading to the overproduction of glutamine</u>. Glutamine
is produced as follows:

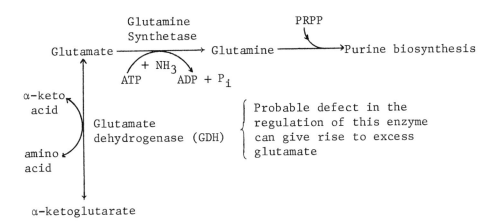

(The details of these reactions will be discussed under amino acid metabolism)

In the above relationships, the GDH-catalyzed reaction is favored in the direction of synthesis of α-ketoglutarate from glutamic acid. Elevated plasma glutamic acid levels have been observed in gouty patients, suggesting the possibility of a defect in the control of GDH activity. It is presumed that the increased amount of glutamic acid is diverted to glutamine synthesis and purine production.

c. Abnormalities in the function of glutamine-PRPP-amidotransferase. This enzyme, an allosteric protein, catalyzes one of the rate controlling steps in purine biosynthesis. Hence, changes in the level of this enzyme or defects in its control can lead to enhanced production of uric acid.

d. Deficiency of Hypoxanthine Guanine Phosphoribosyl-Transferase (HGPRT). This enzyme participates in the salvage (reutilization) pathway of nucleotide formation.

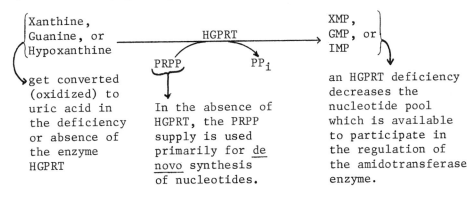

As is indicated above, in the discussion of salvage pathways, an HGPRT deficiency or absence can lead to elevated levels of uric acid. There are two categories of HGPRT enzyme defect which have been observed.

  i)   Virtually complete deficiency of the enzyme occurs in the
       Lesch-Nyhan Syndrome. This is an X-linked disorder,
       affecting only males. It is characterized by excessive
       production of uric acid leading to gouty arthritis, urate
       overexcretion, and the formation of renal stones.
       Neurological disorders such as mental retardation, a
       tendency towards self-mutilation, spasticity, and
       choreoathetosis are also apparent to some degree in all
       cases. The relationship between the neurological symptoms
       and the deficiency of the salvage enzyme is not clear
       at present.

  ii) A partial deficiency of the enzyme has been observed in
       some adult hyperuricemic individuals. However, extensive
       epidemiological studies correlating the enzyme activity
       with uric acid levels in patients with gout is not
       available at this time.

e. Excessive production of organic acids leading to elevated
    levels of uric acid. Lactate, acetoacetate, and β-hydroxy-
    butyrate (the latter two are known as ketone bodies) compete
    with uric acid for secretion in the kidneys. Thus, elevated
    levels of these organic acids reduce the uric acid excretion
    and can increase its levels in the body fluids. Excess lactic
    acid production (lactic acidosis) can occur in G-6-phosphatase
    deficiency and alcohol ingestion. Excess ketone bodies are
    formed in untreated diabetes mellitus, starvation,
    G-6-phosphatase deficiency, etc. Additionally, since alcohol
    oxidation produces NADH and acetyl-CoA, glycolytic activity is
    decreased and glucose is metabolized to R-5-P in the HMP shunt
    under conditions of excessive ethanol ingestion.

    Although untreated diabetes mellitus, starvation, and G-6-P
    phosphatase deficiency all cause an elevation of the ketone
    bodies in the blood, in diabetes, the hyperglycemia actually can
    increase the glomerular filtration rate and thereby improve uric
    acid elimination. As was pointed out in the discussion of
    diabetes, its relationship to gout is not yet clearly understood.

f. A summary of the known biochemical defects which can result in
    hyperuricemia is presented in Figure 55. For more details,
    refer to the preceding material.

Figure 55. A Composite Diagram Depicting the Biochemical Lesions that May Lead to Hyperuricemia

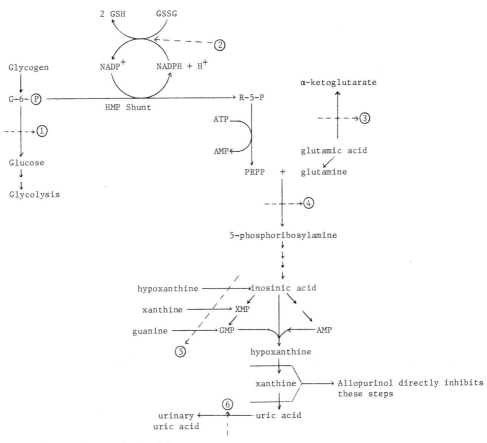

The numbers refer to the key below

Number	Enzyme	Comments
1	glucose-6-phosphatase deficiency	leads to excess PRPP production
2	glutathione reductase overactivity	leads to drain on NADPH and consequent overproduction of PRPP
3	glutamate dehydrogenase deficiency	oversupply of glutamine for purine biosynthesis
4	PRPP-amidotransferase overactivity (loss of control)	increased rate of purine biosynthesis
5	hypoxanthine-guanine phosphoribosyl transferase decrease or absence	degradation of nucleotide bases to uric acid rather than salvage; enzyme is completely absent in Lesch-Nyhan Syndrome patients
6	overproduction of other organic acids	decreased secretion of uric acid due to competition with other organic acids in the kidneys

Biosynthesis of Pyrimidine Nucleotides

1. The first step in the pathway is the formation of carbamyl phosphate

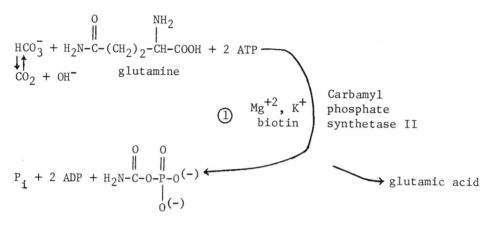

carbamyl phosphate

2. Formation of the pyrimidine ring

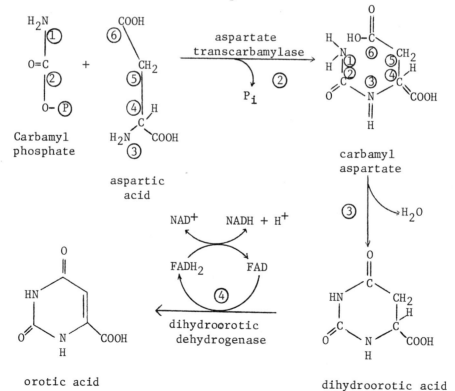

3. Formation of UMP

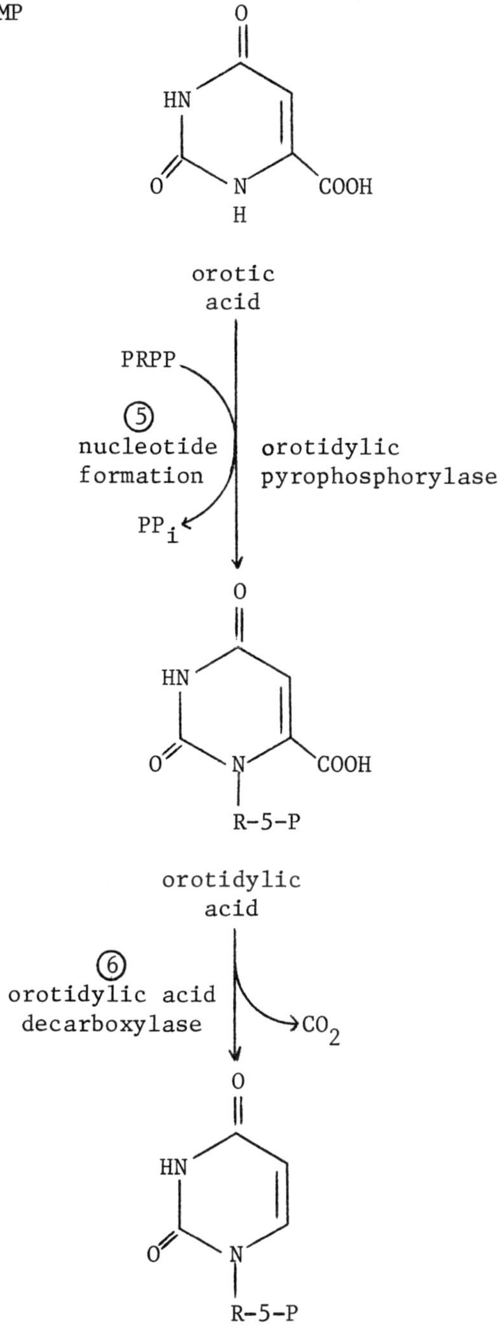
orotic acid
PRPP
⑤ nucleotide formation
orotidylic pyrophosphorylase
PP_i
R-5-P
orotidylic acid
⑥ orotidylic acid decarboxylase
CO_2
R-5-P

uridylic acid (UMP)

4. Conversion of UMP to CTP, dUMP, TMP, TTP, dCMP, dCTP, and
    5-hydroxymethyldeoxycytidylic acid

Uridine-5'-(P) (UMP)

UMP kinase, Mg^{+2}, ATP → ADP

UDP

Ribonucleotide diphosphate reductase

NADPH + H^+ → NADP$^+$ + H_2O

deoxyuridine-5'-pyrophosphate (dUDP)

phosphatase, P_i

dUMP

nucleoside diphosphate kinase, Mg^{+2}, ATP → ADP

UTP

N^5,N^{10}-methylene FH$_4$, Mg^{+2} → FH$_2$

thymidylate synthetase

CTP synthetase

Gln → Glu

ATP + H_2O → ADP + P_i

thymidylic acid (TMP)
deoxyribose-5-(P)

N^5,N^{10}-methylene FH$_4$ → FH$_2$

thymidylate synthetase

2ATP kinases → 2ADP

TTP

cytidine triphosphate (CTP)
Ribose-5-(P)-(P)-(P)

dUMP

AMP → ADP

NH_3, H_2O

dCMP aminohydrolase

CDP

ribonucleoside diphosphate reductase
Mg^{+2}; ATP, NADPH + H^+ → NADP$^+$ + H_2O

dCDP

phosphatase → P_i

dCMP

dHMP synthetase

N^5,N^{10}-methylene-FH$_4$ → FH$_4$

ATP → ADP

dCTP

5-hydroxymethyldeoxycytidylic acid (dHMP)
deoxyribose-5-(P)

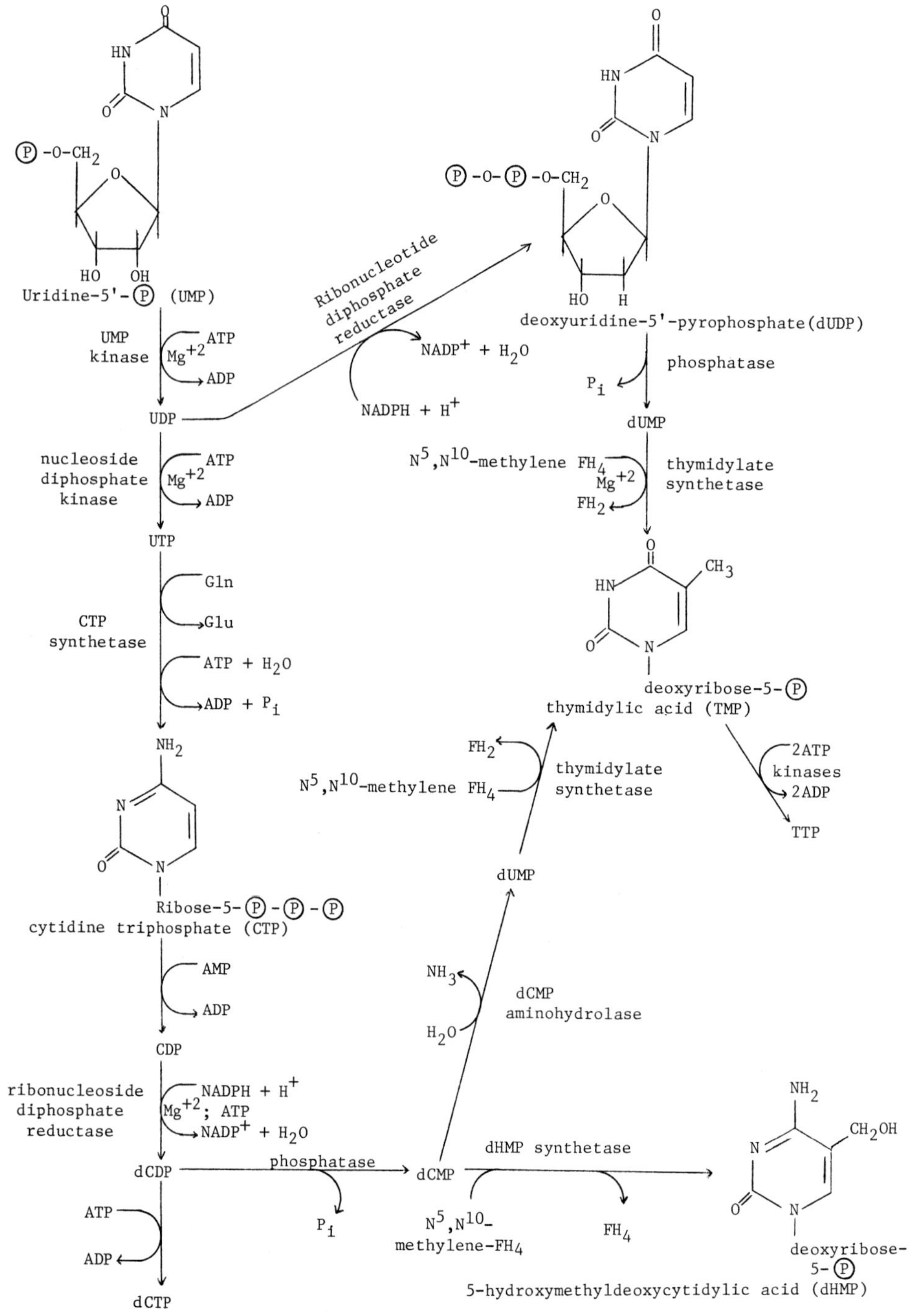

5. The nucleoside and deoxynucleoside triphosphates, needed for nucleic
    acid synthesis, are formed from the mono- and diphosphates by
    kinase enzymes in reactions entirely analogous to those shown for
    the purine nucleotides.

Comments on Pyrimidine Biosynthesis

1. The elements of the pyrimidine ring come from glutamine, CO_2 and
    aspartate.

2. In pyrimidine synthesis, the base is first formed and then
    nucleotide formation takes place with the addition of the sugar and
    phosphoric acid group. This is in contrast to purine biosynthesis,
    where a nucleotide is formed in the first step of the synthesis
    well before either ring closure occurs.

    There are several points of similarity between purine and pyrimidine
    synthesis.

    a. Both require carbon dioxide and the amide nitrogen of glutamine.

    b. Both use an amino acid as a "nucleus".

        i) In purine biosynthesis, glycine supplies two carbons and
            a nitrogen.

        ii) In pyrimidine biosynthesis, aspartic acid provides three
             carbons and a nitrogen.

3. a. In both pyrimidine biosynthesis and the urea cycle (discussed
        under amino acid metabolism), carbamyl phosphate is the
        metabolically active source of carbon dioxide and ammonia
        nitrogen. In pyrimidine biosynthesis, carbamyl phosphate reacts
        with aspartate. In the urea cycle, it reacts with ornithine
        (aspartate enters into the cycle at a later stage) to give
        citrulline.

b. Within the eucaryotic cell there are two separate pools of carbamyl phosphate and two separate enzymes, designated carbamyl phosphate synthetase (CPS) (I) and (II). CPS (I) provides carbamyl phosphate for the urea cycle. It is strictly a mitochondrial enzyme, requires N-acetylglutamine as a positive allosteric activator, and can use either NH_3 or the amide nitrogen of glutamine as a nitrogen source. CPS (II) is found in the cytoplasm and maintains the carbamyl phosphate pool used in pyrimidine nucleotide biosynthesis. Its activity is not influenced by N-acetylglutamine but, unlike CPS (I), it is inhibited by glutamine analogues such as D-carbamyl-L-serine. CPS (II) can only use the amide nitrogen of glutamine as a nitrogen source.

c. The presence of these two physically separated enzymes, each with its own properties, probably reflects the need for the cell to be able to control pyrimidine biosynthesis and amino acid catabolism independently of each other, despite the fact that both pathways require carbamyl phosphate. E. coli is unusual in this respect, having apparently only one carbamyl phosphate synthetase to supply both pathways. This may be reflected in a difference in the control steps in procaryotes and eucaryotes (see below).

d. Although there is an absolute biotin requirement in the CPS reaction, a biotin deficiency is difficult to produce, presumably due to biotin synthesis by the intestinal bacteria. Biotin can, however, be inactivated by <u>avidin</u>, a protein (molecular weight about 45,000 daltons) found in raw egg white. Avidin is readily inactivated by heating. The nature of its interaction with biotin is not known.

4. The conversion of dihydroorotic acid to orotic acid (reaction ④) is catalyzed by a metalloflavoprotein dehydrogenase. This enzyme contains FAD as a prosthetic group. Following reduction of the protein-bound flavin to $FADH_2$, the complex is reoxidized by NAD^+, forming $NADH + H^+$.

5. a. Thymidylate synthetase, the enzyme which catalyzes formation of TMP from dUMP, has been best studied in E. coli. The enzyme isolated from this bacterium is similar to that from L. casei, in that both use N^5,N^{10}-methylene FH_4 as a methyl source <u>and</u> as a reducing agent (note that FH_2 is released in the reaction, rather than FH_4).

b. One substrate for thymidylate synthetase, dUMP, comes either
    from dUDP (by action of a phosphatase) or from dCMP (in a
    reaction catalyzed by dCMP aminohydrolase). dCMP aminohydrolase
    is activated by dCTP and inhibited by dTTP, presumably by an
    allosteric mechanism. In regenerating rat liver and other
    rapidly growing tissues, the aminohydrolase is very active while
    in normal tissue it is almost undetectable. This, and its
    inhibition by dTTP, suggest that it may be a key control point
    in DNA synthesis, along with thymidine kinase (discussed
    below under salvage pathways).

6. In a group of viruses known as the T-even bacteriophages,
    5-hydroxymethyldeoxycytidylic acid (5-HMP) replaces thymine in the
    DNA. Note that it differs from thymine only by the presence of a
    hydroxyl group on the methyl substituent at the 5-position of the
    cytidine ring.

7. Regulation of pyrimidine biosynthesis

    a. Control of pyrimidine biosynthesis has been much more thoroughly
        studied in bacteria than in eucaryotes. In bacteria, several
        important examples of negative allosteric feedback control have
        been elucidated. For example, aspartate transcarbamylase
        (reaction ② in the synthesic pathway) and orotidylic acid
        decarboxylase (reaction ⑥) are both inhibited by UMP. This
        is <u>negative feedback</u>, since a product (UMP) controls its own
        production. Aspartate transcarbamylase is a particularly well-
        studied example of an allosteric enzyme which has both control
        and catalytic capabilities. Substrates and <u>purine</u> nucleotide
        act as positive modulators of this enzyme while <u>pyrimidine</u>
        nucleotides (products of the pathway) are negative effectors.

    b. In mammalian systems, there is evidence that carbamyl phosphate
        synthetase (II) is the site of feedback inhibition by the
        pyrimidine nucleotides. Because there are <u>two</u> CPS enzymes in
        mammals, the formation of pyrimidine-channeled carbamyl
        phosphate is the first unique (committed) step in mammalian
        pyrimidine biosynthesis. If this was the control point in
        bacteria which have only one CPS (see above), both pyrimidine
        biosynthesis <u>and</u> the urea cycle would have to be regulated
        simultaneously, which is not necessarily desirable. The
        committed step (first unique reaction) in <u>bacterial</u> pyrimidine
        biosynthesis is catalyzed by aspartate transcarbamylase. As
        described above, this is the feedback inhibition point of this
        pathway in bacteria. The logic of these processes, to conserve
        material and energy at the earliest feasible step, is very
        important to recognize.

c. Also in mammals, dihydroorotic dehydrogenase (reaction ④) is inhibited by several purines and pyrimidines; and orotidylic acid decarboxylase (reaction ⑥) is inhibited by UMP as it is in E. coli.

8. a. While salvage pathways for purines and purine nucleosides are well established, reutilization of pyrimidines by similar pathways has not been described in any detail. That pyrimidine salvage does occur is evident in the treatment of orotic aciduria. This disease (discussed below) results from a defect in the pyrimidine-synthetic pathway. Many of its symptoms can be relieved by treatment with uridine which must be utilized via a salvage route.

b. The phosphorylation of nucleosides may also be regarded as a salvage pathway. An important example is the reaction catalyzed by thymidine kinase.

$$\text{deoxythymidine} + \text{ATP} \longrightarrow \text{thymidylic acid} + \text{ADP}$$

This enzyme has very low activity in normal rat livers. In partially hepatectomized animals, however, its activity is greatly elevated. There is evidence that it may be one of the rate-controlling enzymes in DNA synthesis. It is inhibited by dTTP (a feedback-process) and by dCTP.

9. Certain pyrimidine derivatives serve as powerful antimetabolites:

a. The metabolism of 5-Fluorouracil (5FU) is shown below.

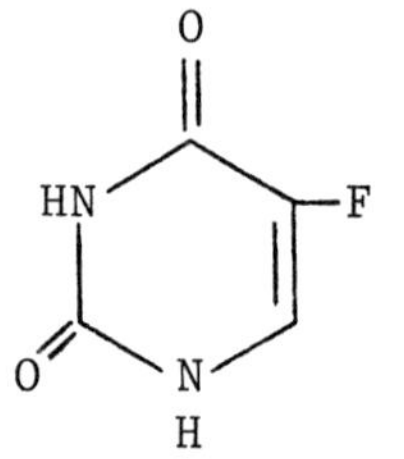

(i) 5-FU → 5-FU-Ribonucleoside
      ↓
   5-FU-Ribonucleotide
      ↓
FUdR → 5'-fluoro-2'-deoxyridine-5'-Ⓟ
       (F-dUMP)
     (inhibits thymidylate synthetase)

5-Fluorouracil

(ii) F-dUMP inhibits (both competitively and non-competitively) the conversion of dUMP to TMP. Note that, in the above reaction, the active, selectively toxic agent is F-dUMP which can also be directly derived from fluorodeoxyuridine (5-fluorouracil deoxyribose; FUdR) by the action of deoxyuridine kinase and ATP.

   (iii)   Both compounds (5FU and FUdR) are used in the treatment
           of carcinomas of the ovary, breast, colon, etc.

b. 5-Iododeoxyuridine (IRdR). This compound has no significant
    action on thymidylate synthetase but the phosphorylated
    derivative of IUdR can be incorporated into DNA in the place
    of thymidylic acid and thus interfere with the growth of the
    cell. IUdR is used in viral infections such as vaccinia and
    herpes.

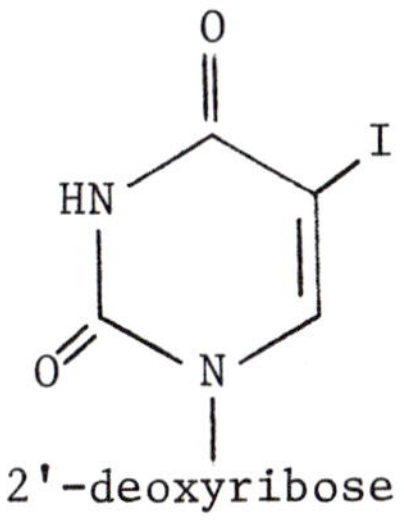

5-iododeoxyuridine

2'-deoxyribose

c. 1-β-D-arabinofuranosyl cytosine (cytosine arabinoside,
    cytarabine) blocks the formation of dCMP and inhibits
    biosynthesis of DNA. It is suggested that cytarabine inhibits
    the conversion of ribonucleotides to deoxyribonucleotides.
    The drug is used to treat acute lymphocytic and acute
    myelocytic leukemias.

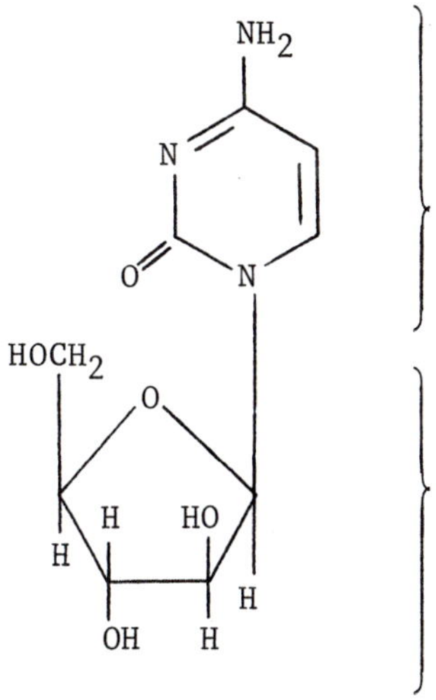

cytidine

arabinose: note the
similarity of this sugar
to 2'-deoxyribose

d. Folic acid analogues like aminopterin, methotrexate, etc. competitively inhibit the dehydrogenases which catalyze the following reactions:

$$F \longrightarrow FH_2 \quad \text{and} \quad FH_2 \longrightarrow FH_4$$

Methotrexate is used to treat coriocarcinoma and childhood leukemia. These folate antagonists are discussed in more detail under one-carbon metabolism and gout.

10. Orotic aciduria is an inherited metabolic disease caused by biochemical lesions in pyrimidine metabolism. In the affected child, it is characterized by an increase in the excretion of orotic acid due to a defect in pyrimidine biosynthesis. In the type I disease, both orotidylic acid pyrophosphorylase (reaction 5) and orotidylic acid decarboxylase (reaction 6) are present but decreased markedly in activity. In the type II disorder, pyrophosphorylase activity is normal or elevated, but the decarboxylase is almost entirely absent. The single patient observed so far to have the type II disease had both orotic aciduria and orotidinuria. Patients also display hypochromic erythrocytes and megaloblastic marrow which are unrelieved by iron, pyridoxine, B_{12}, folic acid or ascorbic acid. Leukopenia is also present. Treatment with uridine (2 to 4 gm/day) results in an excellent reticulocyte response, rise in hemoglobin, an increase in growth, etc. It is important to emphasize that these individuals are dependent upon exogenous pyrimidine supply in order to maintain normal growth and development. This is identical to the need for vitamins and essential amino and fatty acids in all humans. Antagonists to orotidylic acid decarboxylase (6-azauridine and allopurinol) will produce transient orotic aciduria.

11. Since folate and vitamin B_{12} are involved in thymidylic acid synthesis, their deficiencies can also cause megaloblastic anemia.

12. Pyrimidine Catabolism

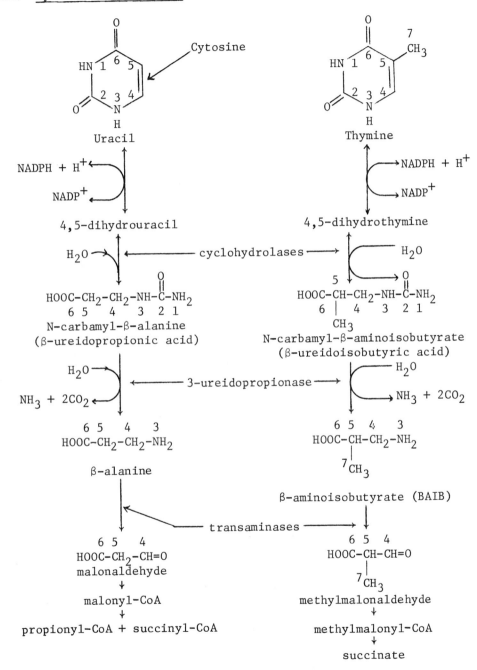

High levels of BAIB are found in urine when excessive tissue destruction takes place. Also, in the Oriental population, familial high-excretors often have been reported, the significance of which is not understood.

Biosynthesis of Deoxyribonucleotides

1. The pathways for TTP synthesis have been discussed and the formation of dCTP has been mentioned previously. Both can be found in the section on pyrimidine nucleotide biosynthesis.

2. In E. coli, the nucleoside diphosphates are the substrates for the reductase; in Lactobacillus leichmanii the triphosphates function in this capacity and, in addition, a cobamide coenzyme is required. Although much work remains to be done, the mammalian system seems most similar to that occurring in E. coli. The enzyme system shown below is from E. coli.

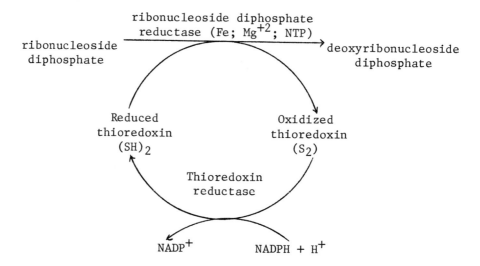

Different nucleoside triphosphates (NTP's) are required for the reduction of the various ribonucleoside diphosphates, implying that they are needed as allosteric effectors rather than as an energy source. For example, ATP must combine with the reductase for it to catalyze the conversion of CDP to dCDP.

3. Control of deoxyribonucleotide biosynthesis appears to be complex.
Ribonucleoside diphosphate reductase (RDR) is inhibited by
hydroxyurea and similar compounds, apparently through their
interactions with the non-heme iron present in RDR. Other effects,
probably all allosteric, are summarized below. Further information
on some of these conversions is given under pyrimidine biosynthesis.

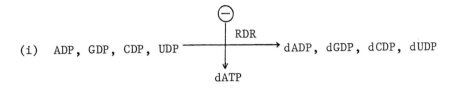

(i) ADP, GDP, CDP, UDP ———→ dADP, dGDP, dCDP, dUDP

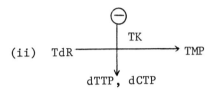

(ii) TdR ———→ TMP

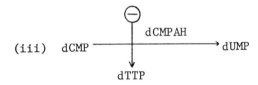

(iii) dCMP ———→ dUMP

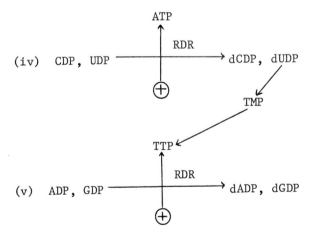

(iv) CDP, UDP ———→ dCDP, dUDP

(v) ADP, GDP ———→ dADP, dGDP

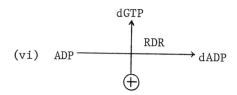

(vi)   ADP

The enzyme abbreviations used in these reactions are: RDR = ribonucleotide diphosphate reductase; TK = thymidine kinase; and dCMPAH = dCMP amino hydrolase. Regarding these reactions, notice that (i-iii) all involve feedback inhibition and tend to prevent the accumulation of unneeded nucleoside triphosphate in the cell. Reactions (iv-vi) serve to maintain a balance between purines and pyrimidines and between the two members of each group. It should be remembered, however, that these control properties were all worked out using <u>in vitro</u> systems. The extent to which they apply <u>in vivo</u> remains to be demonstrated. The controls on reactions (ii) and (iii) have been demonstrated in mammalian systems while those in reactions (iv-vi) have been shown only in <u>E. coli</u>. (The work on the synthesis and control of deoxynucleosides in <u>E. coli</u> has been largely worked out by P. Reichard and his collaborators in Stockholm.) In reaction (i), dATP inhibits all four reductions, whereas it has been demonstrated only for GDP → dGDP in mammals and for CDP → dCDP in birds. Thus, once again, caution should be exercised in asserting that regulation of these pathways as described here functions in mammals.

6
Amino Acid Metabolism

General Scheme of Amino Acid Metabolism

1. Generalized picture of protein metabolism

   Sources (a-c)   Utilization (d-f)

   a. Dietary protein is hydrolyzed   d. Synthesis of body
      to amino acids and absorbed   proteins. Examples:
      from the intestine   structural proteins,
                                              plasma proteins,
            (through portal   enzymes, milk
             circulation)   proteins, hormones

   b. Breakdown of   e. Synthesis of
      tissue protein ──→   necessary non-protein
                        Amino Acids   nitrogen compounds:
                        in the   hormones, choline,
                        Blood   creatine, purines &
   c. Synthesis   pyrimidines,
      (predominantly   coenzymes,
       in the liver)   glutathione, melanin

                    f. Energy Production

   converted to urea
   which is excreted   Amino Acids
   in the urine
                    NH$_3$ ←   → glucose
                                α-ketoacids
        used in
     transaminations   α-ketoacids
                                              → acetyl-CoA; ketone bodies

                    fatty acid
                    metabolism

2. Cellular proteins and their constituent amino acids are in a
   dynamic equilibrium. Degradation and synthesis (to replace the
   existing molecules) are in constant operation.

3. The turnover rate (quantity of a substance synthesized or degraded
   per unit of time) of different proteins in different tissues varies
   enormously. Example:

425

<u>Proteins</u>	<u>Half-life Times in Days</u>
From blood, liver, other internal organs	2.5-10
muscle protein	180
collagen	greater than 1000

4. Some mechanisms that contribute to protein turnover are tissue replacement, breakdown and synthesis of individual tissue proteins, and utilization and replacement of proteins such as digestive enzymes, hormones, serum proteins, etc.

5. Amino acids derived from protein turnover may make a significant contribution toward the energy requirements of the body.

6. Nitrogen Balance

    a. When nitrogen intake (the amount of nitrogen in the diet) equals nitrogen losses (the amount of nitrogen excreted as urine or feces), the body is said to be in <u>nitrogen balance</u>. In such a situation, anabolism = catabolism. This is the condition in normal adults.

    b. When nitrogen intake <u>exceeds</u> nitrogen losses, then the body is retaining nitrogen as tissue protein. This is called <u>positive nitrogen balance</u> and anabolism exceeds catabolism. This condition is characteristic of growing children, patients recovering from emaciating illnesses, and pregnant women.

    c. <u>Kwashiorkor</u> (which means displaced children in the Bantu language) is a disease caused by malnutrition (specifically, by a prolonged insufficient intake of necessary proteins) in children. Some of its characteristics are lack of appropriate cellular development (because of failure to synthesize normal amounts of protein); edema; diarrhea (may be with fatty stools); atrophy of the pancreas (which requires a high rate of protein synthesis) and the intestinal mucosa; gray and scaly skin (due to lack of melanin) which may develop ulcerating patches, etc. <u>Marasmus</u> is another disease associated with malnutrition. It is due, however, to generalized starvation rather than an insufficiency specifically of protein.

7. Amino acids are used as a source of energy. They enter the tricarboxylic acid (TCA) cycle for ultimate oxidation. The relationship of amino acid metabolism to the TCA cycle is shown in Figure 56.

Figure 56. Interrelationships of Amino Acid Metabolism to the TCA Cycle and Energy Production

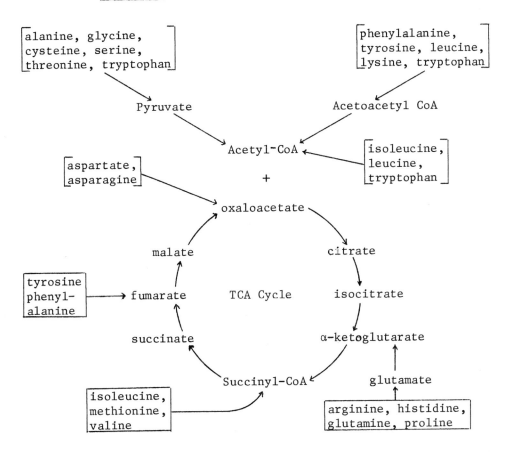

8. Amino acids may be classified according to the types of products that are formed when the amino acids are metabolized. Some amino acids give rise only to acetoacetic acid and other ketone bodies and are called ketogenic amino acids. Others give rise only to glucose or glycogen (or compounds which can be converted to these compounds) and are called glucogenic amino acids. Some amino acids are both glucogenic and ketogenic.

Classification	Amino Acid
Ketogenic only	leucine
Ketogenic and Glucogenic	isoleucine, lysine, phenylalanine, tyrosine, tryptophan, threonine
Glucogenic Only	all other amino acids

9. Essential and Non-Essential Amino Acids

   a. The body can synthesize some amino acids. These are known as non-essential amino acids (because it is not essential that they be available in the diet for normal growth and development). These amino acids are synthesized by the transamination of keto acids which have the appropriate carbon skeletons. Transamination is discussed in detail later in this chapter. Note that the terms essential and non-essential have no meaning with respect to the relative importance of an amino acid in metabolism. They are strictly dietary terms.

   b. Essential amino acids are those which the body cannot synthesize and which must thus be obtained from an external source, the diet. Adequate amounts of each of these essential amino acids are required to maintain proper nitrogen balance. The essential L-amino acids for man are tryptophan, phenylalanine, lysine, threonine, valine, methionine, leucine, isoleucine, and histidine.

10. Amino acids and their derivatives have been shown to influence nerve impulse transmission, both in the central and peripheral nervous systems. The post synaptic effect of these materials may involve cyclic AMP and, hence, adenyl cyclase.

   a. In the brain, compounds of the formula $HOOC-CH_2-(CH_2)_n-NH_2$ where n = 0-4 exert an inhibitory influence on synaptic transmission. One member of this group is γ-aminobutyric acid (GABA), obtained by α-decarboxylation of glutamic acid (see proline and phenylalanine metabolism). Taurine (see metabolism of sulfur-containing amino acids) and alanine have similar depressive effects. Glutamic and aspartic acids, on the other hand, display potent excitatory effects on central neurons. Note that the inhibitors are neutral amino acids while the stimulatory substances are acidic.

b. Glycine appears to have an inhibitory effect on neural
    transmission within the spinal cord. On the other hand,
    acetylcholine, which is metabolically related to glycine,
    is strongly implicated as a neurotransmitter in the central
    nervous system.

c. Aromatic amines, derived from phenylalanine, tyrosine,
    histidine, and tryptophan, have for some time been considered
    as neurotransmitter substances. The most thoroughly studied
    of these materials are the catecholamines, epinephrine and
    norepinephrine (see phenylalanine); serotonin (see tryptophan);
    and histamine (from histidine).

d. Since the 1930's, a group of polypeptides, designated
    "substance P", have been known for their ability to cause
    smooth muscle contraction and vasodilation. Although this
    appears to be truly a peptide effect and not simply due to
    small-molecule impurities, much work remains to be done
    in this area.

11. Protein Synthesis and Storage

$$\text{amino acids} \underset{\text{catabolism}}{\overset{\text{anabolism}}{\rightleftharpoons}} \text{proteins}$$

The mechanism of protein synthesis has already been discussed.
Each tissue of the body has its own characteristic proteins and
most of these proteins are synthesized within their own tissue
cells. However, some tissues also make proteins for export.
For example:

1. The liver synthesizes fibrinogen, albumin, and most of the
    plasma globulins.

2. The lymphocytes of the lymphoid tissue synthesize plasma
    γ-globulins.

3. Protein hormones are synthesized by various organs.

4. Trypsin, chymotrypsin, lipase, and amylase (all digestive
    enzymes) are made in the pancreas.

The body apparently does not store free amino acids and has little
capacity even for the storage of proteins (unlike carbohydrates and
fats). The minimal reserves that are available are proteins that
are incorporated into the structure of the liver (and perhaps
muscle).

Effects of Some Hormones on Protein Metabolism

1. Thyroxine (produced by the thyroid gland) affects the metabolism
    of carbohydrates, proteins and fats. The overall metabolic rate
    of the entire body reflects changes in the amount of thyroxine
    available.

    a. Severe deficiency ⟶ low metabolic rate characterized by

        i) failure to grow, imbecility
        ii) decrease in protein synthesis

    b. Excessive production ⟶ high metabolic rate characterized by

        i)   increased rate of tissue breakdown
        ii) marked emaciation, nervousness
        iii) increased rate of oxidation of food; more amino acids
             are destroyed in oxidation, and thus fewer go into
             protein synthesis

2. Growth hormone (produced by the anterior pituitary) fosters protein
    anabolism, possibly by facilitating amino acid uptake by the cells.
    Underproduction leads to dwarfism and overproduction leads to
    gigantism and acromegaly.

3. Insulin promotes increased utilization of glucose and hence
    increased production of ATP as a source for peptide-bond formation.
    This indirectly affects protein metabolism because there follows
    an increased incorporation of amino acids into proteins, and
    probably increased rate of translation of messenger RNA.

4. The male sex hormones (androgens) stimulate protein synthesis.

5. The hormones of the adrenal cortex promote breakdown of proteins
    and amino acids.

6. Epinephrine lowers the plasma levels of free amino acids, probably
    by facilitating the uptake of amino acids by the cells.

Protein Digestion

1. a. Digestion in the stomach is initiated by food intake which
        produces nervous, chemical, and mechanical stimuli. These
        induce the release of gastrin (a peptide hormone having
        seventeen amino acid residues) by specialized endocrine cells
        (G cells) beneath the surface of the pyloric antral mucosa of
        the stomach. The hormone is stored in granules within these

cells. Gastrin appears to consist of approximately equimolar quantities of two 17-residue peptides called gastrin I and II. The amino acid sequences of these are identical, the only difference being a -SO$_3$H group on the tyrosine in the twelfth position of gastrin II. The sulfate is absent in gastrin I. Another interesting feature in the structure of the gastrins is the occurrence of a pyroglutamyl residue at their C-termini. This is an internal, cyclic lactam of glutamic acid, found also to occur in thyrotropin-releasing-hormone (TRH).

b. The nervous stimulation is associated with the act of ingestion, whereas the chemical and mechanical effects result from actual contact of the food with surface mucosal cells in the antral region. Small molecules (amino acids, choline, lower aliphatic alcohols) appear to be the most active gastrin-releasing factors. A further chemical influence is the inhibition of gastrin release by low pH. Significant quantities of gastrin are not secreted unless the pH of the antrum is near neutrality, as it is when buffered by the presence of food.

2. Gastrin is released into the blood stream where it is carried to its target cells through which it affects several functions of the stomach.

    a. It stimulates motility of the stomach, bringing about an intensification of the churning action which breaks up food and prepares it for intestinal digestion.

    b. It causes the secretion of pepsin, an endopeptidase present in the gastric juice, which hydrolyzes proteins into polypeptides and amino acids. Although the enzyme preferentially hydrolyzes peptide bonds in which tyrosine and phenylalanine participate, its action is not highly specific. Pepsinogen, the zymogen form of pepsin, is secreted by the chief cells of the gastric mucosa. Pepsinogen is converted to the active enzyme by H$^+$ at pH 1.0 to 1.5 (pH of normal gastric content) and also by autocatalysis of pepsin itself. A milk-curdling enzyme known as <u>rennin</u>, also produced by the chief cells (of the fundic glands), assists in the proteolysis of milk proteins by pepsin. The clotting of milk by rennin slows down its passage through the alimentary tract and thus provides time for the action of pepsin in the stomach.

    c. Hydrochloric acid is produced by the parietal cells of the gastric mucosa under the influence of gastrin. As indicated above, this is important because pepsinogen is activated at very low pH's. The hydrochloric acid is produced by the action of carbonic anhydrase in the parietal cells.

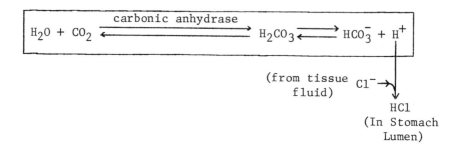

Chloride ions which are present in the gastric juice are derived from the tissue fluids and transported across the gastric epithelial cells into the stomach lumen presumably by a carrier-mediated ATP-dependent process. Acetazolamine, a carbonic anhydrase inhibitor, prevents HCl production. In normal individuals, the gastric content of free HCl is anywhere between 0 and 40 mEq/l. A variety of chemicals stimulate the HCl production, including caffeine, alcohol and histamine. In gastric analysis, histamine is used to test the ability of the parietal cells to produce HCl.

d. Another component of gastric juice is mucus which consists of at least two glycoproteins. Mucus not only coats the stomach lining, protecting it against hydrochloric acid, but it is also essential in the absorption of vitamin B_{12} (see vitamin metabolism).

3. From the stomach, the chyme (partially digested stomach contents) moves into the upper part or bulb of the duodenum. Entry of the acid chyme stimulates the release of bulbogastrone from the mucosal cells of this region. The target cells of this hormone are the G-cells in the pyloric antrum, where it inhibits release of further gastrin. This, and the drop in pH associated with the departure of food from the stomach, shut down the digestive processes in the stomach. The chemical nature of bulbogastrone has not yet been elucidated.

a. As the chyme moves further into the duodenum and into the jejunum, other hormones perform various functions. Enterogastrone is released, primarily from the jejunum, under the stimulation of fat which is in an absorbable form. It acts in much the same way as bulbogastrone, inhibiting gastrin release and, hence, the spectrum of effects caused by gastrin.

Although the name enterogastrone was coined in 1930, the chemical nature of this molecule has so far defied efforts to unravel it. One possible candidate is a substance called gastric inhibitory polypeptide (GIP), having a molecular weight of 5105 and containing 43 amino acid residues. There is a strong sequential homology between GIP, secretin, and glucagon. The principal thing which remains to be demonstrated is whether GIP is released under the same physiological stimuli as enterogastrone.

b. Cholecystokinin-pancreozymin (CCK-PZ) and secretin are also released from cells in the jejunum and lower duodenum.

   i) CCK-PZ release is stimulated by the presence of fats and protein digests (especially long-chain fatty acids and essential amino acids) in the duodenal lumen. It causes contraction of the gallbladder (hence, release of bile needed for fat absorption) and stimulates the exocrine pancreatic cells to secrete pancreatic juice rich in enzymes (amylase, lipase, proteases). It was originally thought that two separate hormones caused these effects, hence the double name for one hormone. CCK-PZ is a polypeptide having 33 amino acid residues. The C-terminal sequence of five amino acids is identical to the same regions of the gastrins.

   ii) Secretin release is stimulated by the low pH of the chyme as it comes from the duodenal bulb. This hormone influences the pancreas to secrete a watery pancreatic fluid, rich in bicarbonate, into the small intestine. The alkalinity of this secretion neutralizes the acid chyme, bringing it to a pH of about 7-8, suitable for the activity of the pancreatic enzymes induced by CCK-PZ. Secretin appears to be controlled by a feedback mechanism (as is gastrin). The increased pH which results from the bicarbonate inhibits secretin release. Porcine secretin is a polypeptide of 27 amino acid residues. It shows a great deal of sequence homology with porcine glucagon.

4. The foregoing material is summarized in Table 15. It should be pointed out that the action of these hormones is not as singular as it is indicated to be in this table and in the preceding discussion. For example, although CCK-PZ has a strong effect on the gallbladder and the pancreatic enzyme secretion, it can also, especially in large doses, bring about gastric and intestinal motility and secretion of bicarbonate (from both the pancreas and the liver), pepsin, gastric acid, and even insulin. Similarly, while gastrin has a group of primary functions, it can also bring about all of the effects which CCK-PZ induces. There are other examples of multiple functions of the G.I. hormones, some of them overlapping each other as do these.

Table 15.   Actions and Properties of Some Gastrointestinal Hormones

Hormone	Produced by	In response to stimulation from	Acts on (target organ)	Primary effects Produced
I. Gastrin	Cells in the pyloric antral region of the stomach	Ingestion of food; antral distension; small-molecular components of diet	Stomach	Secretion of pepsin and HCl; gastric motility increased
II. Bulbogastrone	Mucosal cells of the duodenal bulb	Acid chyme entering the duodenum	Gastrin-secreting cells of the stomach	Decreased gastrin secretion
III. Enterogastrone	Cells in the jejunum and lower duodenum	Absorbable fat entering these regions	Gastrin-secreting cells of the stomach	Decreased gastrin secretion
IV. Cholecystokinin-pancreozymin (CCK-PZ)	Cells in the jejunum and lower duodenum	Fats and partially digested proteins entering these regions	Gallbladder and exocrine pancreas	Contraction of the bladder and release of bile; secretion of pancreatic enzymes
V. Secretin	Cells in the jejunum and lower duodenum	Low pH of the chyme	Pancreas	Secretion of bicarbonate
VI. Intestinal Glucagon	Cells of the walls of the stomach and small intestine	Ingestion of glucose	Pancreas	Insulin secreted

5. Intestinal glucagon (enteroglucagon; glucagon-like immunoreactive factor), mentioned in Table 15, appears to have properties in common with both pancreatic glucagon and secretin. For example, it stimulates the release of insulin and pancreatic bicarbonate. Intestinal glucagon, however, is <u>induced</u> by ingested glucose, whereas pancreatic glucagon is <u>suppressed</u> by elevated blood glucose levels.

6. Although the six compounds discussed above are the best-characterized of the gastrointestinal (G.I.) hormones, a number of other substances have also been implicated in the regulation of gastric function. Some of these have been chemically identified, but parts of the physiological data needed to fully characterize them are missing. Members of this group include

    i) enterocrinin (released by upper small intestine by the presence of chyme; increases the secretion of intestinal digestive juice);

    ii) hepatocrinin (released by upper small intestine by the presence of chyme; increases hepatic bile secretion);

    iii) motilin (polypeptide containing 22 amino acid residues obtained as by-product in secretin isolation; stimulates gastric motility in both fundic and antral pouches; release stimulus not known);

    iv) vasoactive intestinal peptide (VIP) (polypeptide of 28 residues isolated from porcine-upper-intestinal wall; smooth muscle relaxant which thereby increases splanchnic and peripheral blood flow; resembles secretin and CCK-PZ in this respect and in structure; release stimulus and true physiological role unknown).

7. a. Certain proteolytic enzymes are secreted as inactive enzyme precursors called pro-enzymes or zymogens (discussed previously). These are then converted to the active form by hydrolysis of one or more peptide bonds. The pancreatic pro-enzymes listed below are activated as indicated.

$$\text{Trypsinogen} \xrightarrow{\text{trypsin or enterokinase}} \text{Trypsin}$$

$$\text{Chymotrypsinogen} \xrightarrow{\text{trypsin}} \text{Chymotrypsin}$$

$$\text{Procarboxypeptidases A \& B} \xrightarrow{\text{trypsin}} \text{Carboxypeptidases A \& B}$$

$$\text{Proelastase} \xrightarrow{\text{Trypsin}} \text{Elastase}$$

This system, in which digestive enzymes exist as inactive
precursors until they are at a site where their activity is
desirable, has obvious value. Trypsin, chymotrypsin and
elastase are endopeptidases. Carboxypeptidases (and amino
peptidases, mentioned below) are exopeptidases (see Table 16).

b. Several peptide-like substances which inhibit trypsin activity
(trypsin inhibitors) have been reported. One type of trypsin
inhibitor is present with the trypsinogen in the zymogen
granules. The presumed function of this inhibitor is to
protect the pancreatic tissue from the proteolytic action of
trypsin. A trypsin inhibitor (or proteinase inhibitor) is
present in the α_1-globulin fraction of plasma of humans. This
protein, called α_1-trypsin inhibitor, inhibits the activity
of a variety of proteinases such as elastase, collagenase
leukocytic proteases, trypsin, etc. In one type of pulmonary
emphysema, a deficiency of the α_1-trypsin inhibitor has been
observed. Based upon this observation, it has been hypothesized
that the lack of antitrypsin factor leads to destructive action
of the proteolytic enzymes of the lung tissue. A hereditary
deficiency of trypsinogen in an infant has been reported.

c. Among the enzymes produced by the small intestine itself are
several which act on proteins.

  i) Aminopeptidases: cleave terminal amino acids from the
      N-terminal **end** of peptides.

 ii) Dipeptidases: hydrolyze dipeptides into **individual** amino
      acids; require cobalt or manganese ions for activity.

iii) Tripeptidases: hydrolyze tripeptides, releasing a free
      amino acid and a dipeptide.

d. The preferred sites of cleavage (specificities) of some of the
proteolytic digestive enzymes are indicated in Table 16.

e. In cystic fibrosis (CF), an inherited childhood disease for
which a specific biochemical lesion is not known, blockage of
the pancreatic ducts by fibrous tissue and mucus occurs in
about 85% of the patients. This causes a decrease or cessation
in the supply of pancreatic digestive enzymes (proteases,
lipase, amylase) to the G.I. tract, resulting in steatorrhea
and azotorrhea (fatty and nitrogenous stools, respectively).

Table 16. Specificities of Some of the Gastrointestinal Proteolytic Enzymes

Peptide bond
cleaved

$$H_2N ----- NH - CH - \overset{\overset{\displaystyle O}{\parallel}}{C} + NH - CH - C ----- C\overset{\displaystyle O}{\underset{\displaystyle OH}{\diagup}}$$
$$R_1 \qquad\qquad R_2$$

Enzyme	Prefers to cleave peptide bonds in which
a. Trypsin	R_1 = Arg or Lys; R_2 = any amino acid residue
b. Chymotrypsin	R_1 = Aromatic amino acid (Phe, Tyr, Trp); R_2 = any amino acid residue
c. Carboxypeptidase A	R_1 = any amino acid residue; R_2 = any C-terminal residue except Arg, Lys, or Pro
d. Carboxypeptidase B	R_1 = any amino acid residue; R_2 = Arg or Lys at the C-terminus of a polypeptide
e. Elastase	R_1 = Neutral (uncharged) residues; R_2 = any amino acid residue
f. Leucine Aminopeptidase	R_1 = Most N-terminal residues of a polypeptide chain; R_2 = any residue except Pro
g. Pepsin	R_1 = Trp, Phe, Tyr, Met, Leu; R_2 = any amino acid residue

Dipeptidases

$$H_2N - CH - \overset{\overset{\displaystyle O}{\parallel}}{C} + NH - CH - C\overset{\displaystyle O}{\underset{\displaystyle OH}{\diagup}}$$
$$R_1 \qquad\qquad R_2$$

Peptide bond
cleaved

Tripeptidases

$$H_2N - CH - \overset{\overset{\displaystyle O}{\parallel}}{C} + NH - CH - \overset{\overset{\displaystyle O}{\parallel}}{C} + NH - CH - C\overset{\displaystyle O}{\underset{\displaystyle OH}{\diagup}}$$
$$R_1 \qquad\qquad R_2 \qquad\qquad R_3$$

R_1, R_2, R_3 = any amino acid residue is cleaved by a tripeptidase.

Choice of bond cleaved depends on specific enzyme. In a given molecule, only one bond

Note: Of the first 7 enzymes, a, b, e, and g are endopeptidases while c, d, and f are exopeptidases; trypsin is the most specific endopeptidase; pepsin is the least specific endopeptidase.

Electrolyte imbalance (due to increased salinity of the sweat as measured in the sweat test; see the chapter on electrolyte balance) also usually occurs in this disorder. Chronic bronchiolar obstruction (due to mucus deposits) and infection of the lungs are another part of the classic CF syndrome.

 f. Pancreatitis also can result in blockage of the pancreatic secretions. Some of the enzymes accumulate and spill over into the blood stream. Measurement of serum lipase and amylase are useful in the diagnosis of this disorder (see the section on uses of enzymes in the clinical laboratory).

8. Proteins are completely hydrolyzed to their constituent amino acids by the action of the great variety of enzymes described above. These free amino acids are absorbed into the portal blood stream and carried into the general circulation by way of the liver. The absorption of amino acids is accomplished by means of active transport mechanisms. L-isomers are absorbed at a much higher rate than D-isomers.

9. There is evidence that, in some instances, partially digested proteins (small peptides) may be absorbed. These situations may lead to abnormal immunological phenomena if the body makes antibodies to these molecules.

10. Very little is known about intracellular proteolysis. Presumably it takes place at a high rate in the liver cells.

Some General Reactions of Amino Acids

1. Synthesis and Interconversion of Amino Acids

As has been mentioned, essential amino acids must be obtained preformed in the diet. The other amino acids may be synthesized by the conversion of certain α-keto acids to amino acids (the α-keto group is replaced by an amino group). Note that this synthesis by conversion of α-keto acids occurs in several of the reactions discussed below.

2. Deamination of Amino Acids

Removal of the α-amino group of amino acids is the first step in their catabolism. There are two general categories of deamination: oxidative and non-oxidative.

 a. Oxidative deamination is catalyzed by amino acid oxidases, of which there are two major types: L-amino acid oxidase, and D-amino acid oxidase.

i) L-amino acid oxidase

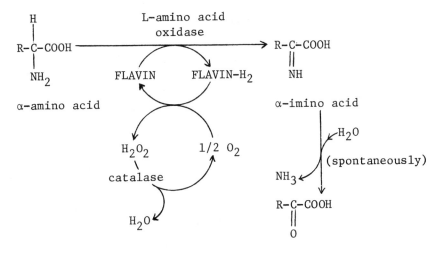

α-keto acid

L-amino acid oxidase occurs in the liver and kidney (it is
not widely distributed) and is a flavoprotein containing
either FMN or FAD as a coenzyme. The enzyme does not attack
glycine, dicarboxylic amino acids, or β-hydroxy amino acids.
The activity of this enzyme is very low and the metabolism
of L-amino acids by this pathway is not of major importance
for the above reasons.

ii) High levels of D-amino acid oxidase (also called glycine
oxidase) activity is found in the liver and kidney. This
enzyme contains FAD and deaminates many amino acids,
including glycine (L-amino acid oxidase is inactive towards
glycine). The oxidation of the glycine is shown below.
The reactions of the other amino acids (D-isomers) are
entirely analogous to this.

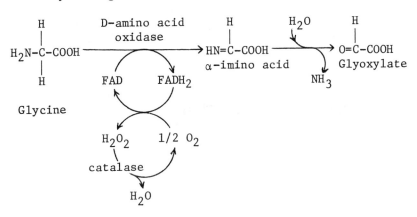

D-amino oxidases are found in <u>peroxisomes</u>, also known as
microbodies. Peroxisomes are single membrane sacs which
bud off from the smooth endoplasmic reticulum. They
contain enzymes which are responsible for the production
of H_2O_2 such as D-amino acid oxidase, L-α-hydroxy acid
oxidase, citrate dehydrogenase and presumably L-amino acid
oxidase. They also contain catalase and peroxidase which
<u>destroy</u> hydrogen peroxide. In the white cells, the killing
of bacteria involves the hydrolytic activity of the
lysosomes and the production of H_2O_2 by the peroxisomes
(see under red cell metabolism).

iii) D-amino acid oxidase is very important because it
catalyzes the conversion of a D-amino acid (which is
normally not used by the body) to an α-keto acid which
no longer has an asymmetrical α-carbon. The keto acid
may be reaminated to form an L-amino acid which can then
be used by the body. Because this conversion from D- to
L-amino acids is possible, the body can use D-amino acids
in the diet as a source of essential amino acids.

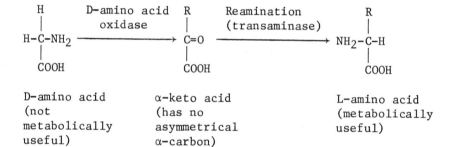

D-amino acid α-keto acid   L-amino acid
(not   (has no   (metabolically
metabolically   asymmetrical   useful)
useful) α-carbon)

iv) The major pathway for the deamination of most L-amino acids
seems to involve L-glutamate dehydrogenase (GDH) rather
than L-amino acid oxidase. This enzyme is widely
distributed in the tissues and is localized intracellularly
in the mitochondria. It has a very high activity. Bovine
GDH is a regulatory enzyme made up of five enzymatically
active monomeric units. It has a molecular weight of about
2×10^6 daltons.

GTP

                              I   II
Polymeric form ⇌ Monomeric ⇌ Monomeric
   of GDH   Form A   Form B

ADP

Equilibrium	Conditions for Change	Substrate Specific
I	Dilution (low protein concentration) promotes the reaction to the right, yielding monomeric units.	Monomer A has primarily GDH activity.
II	Influenced by the regulatory molecules ADP, GTP, NAD$^+$, NADH, steroids, and thyroxine.	Monomer B has primarily alanine dehydrogenase activity.

v) The pathway involving GDH consists of transamination followed by deamination. The amino group from other amino acids is transferred (by transamination) to α-ketoglutarate, forming glutamic acid which is then deaminated with GDH. The overall reaction may be termed <u>transdeamination</u> because it involves transamination and deamination.

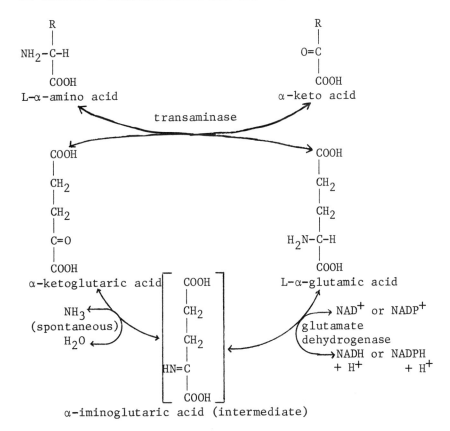

vi) Note that all of the above reactions are reversible. This pathway is important in the synthesis of the non-essential amino acids. The α-keto acids formed can be aminated with the appropriate transaminase.

b. <u>Non-oxidative deamination</u> is accomplished by several specific enzymes, some of which are shown here.

i) Amino acid dehydrases

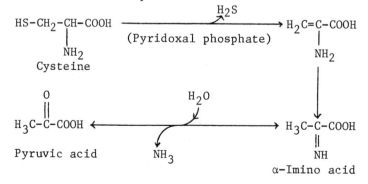

Example:

ii) Amino acid desulfhydrases

3. Reamination

In these reactions, NH_3 is assimilated. Following are some examples.

a. Amination of α-keto acids to give the respective α-amino acids (this has already been mentioned as a mode of synthesis of amino acids).

b. Synthesis of carbamyl phosphate (used in pyrimidine and urea biosynthesis).

c. Glutamine and asparagine synthesis.

4. Transamination of Amino Acids

This combines both deamination and amination reactions. Transaminations are reversible and catalyzed by transaminase enzymes, (also called aminotransferases) which are intracellular enzymes. These enzymes are found in practically all tissues but are especially numerous in heart, brain, kidney, testicle, and liver. The general process of transamination is shown below.

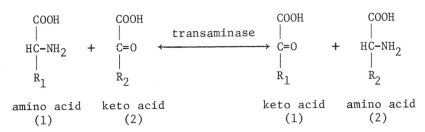

amino acid   keto acid   keto acid   amino acid
   (1)   (2)   (1)   (2)

All naturally occurring amino acids participate in transaminase reactions.

Examples:

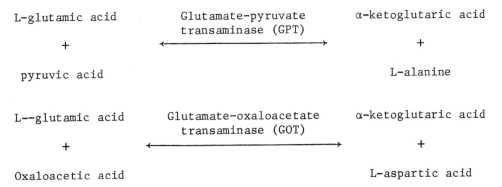

The most common general type of reaction found in animals, plants, and microorganisms is transamination with α-ketoglutaric acid.

L-amino acid   α-keto acid

+   ⟵——————————⟶   +

α-ketoglutaric acid   L-glutamic acid

5. All transaminase reactions appear to have the same mechanism and use the same coenzyme (pyridoxal phosphate, a derivative of vitamin B_6 which is important in many reactions involving α-amino acids). Pyridoxal phosphate serves as a carrier of amino groups or amino acids. The coenzyme is reversibly transformed from its free aldehyde form, pyridoxal phosphate, to its aminated form, pyridoxamine phosphate.

a. Role of pyridoxal phosphate in the conversion of an α-amino acid to an α-keto acid

R = $-H_2C-O-PO_3H_2$

b. Conversion of an α-keto acid to an α-amino acid

α-keto acid 2 + pyridoxamine phosphate $\rightleftharpoons$ (reversal of the above reactions) α-amino acid 2 + pyridoxal phosphate

The two reactions above (a and b) are coupled as shown below.

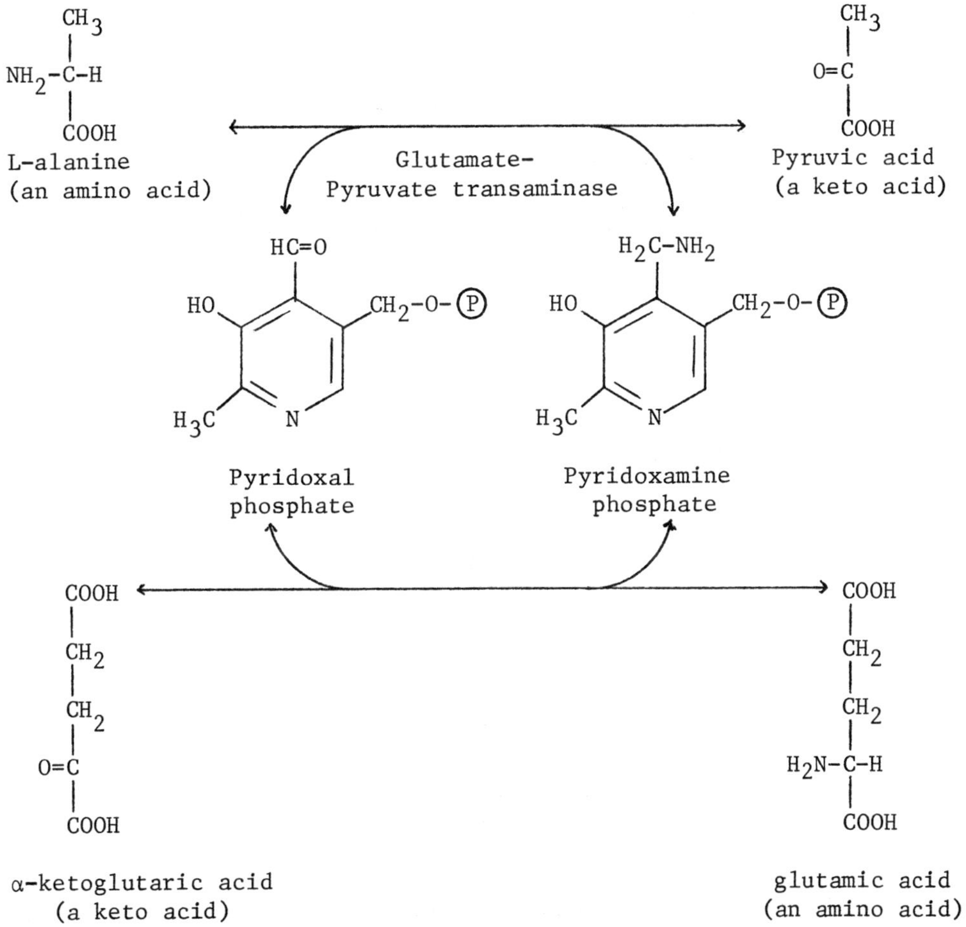
L-alanine
(an amino acid)
Glutamate-
Pyruvate transaminase
Pyruvic acid
(a keto acid)
HC=O
HO
CH2-O-P
H3C
N
H2C-NH2
Pyridoxal
phosphate
Pyridoxamine
phosphate
α-ketoglutaric acid
(a keto acid)
glutamic acid
(an amino acid)

6. Physiological Significance of Transaminases

    a. Two transaminases of particular clinical importance are glutamic oxaloacetic transaminase (GOT) and glutamic pyruvic transaminase (GPT). The reactions which these catalyze were shown above.

    b. Because these are intracellular enzymes, the serum levels of GOT and GPT are normally very low. Any significant tissue breakdown gives rise to high serum transaminase levels. For example, in myocardial infarction, there is an increase in the serum level of GOT (heart muscle contains high concentrations of this enzyme). There are alterations in serum GOT and GPT levels in some of the liver diseases (infectious hepatitis, infectious mononucleosis). One also finds high serum levels of these enzymes in conditions where there is damage to skeletal muscle. When these enzymes are measured in the serum, the values obtained are referred to as SGOT and SGPT activities (where the S refers to serum).

Table 17.   Tissue Distribution of Transaminases in Man*

	GOT	GPT
Heart	156	7.1
Liver	142	44
Skeletal Muscle	99	4.8
Kidney	91	19
Pancreas	28	2
Spleen	14	1.2
Lung	10	0.7
Serum	0.02	0.016

Values expressed in terms of units $\times 10^{-4}$/gm wet tissue homogenate.

----

* From Wróblewski, F. and La Due, J., Proc. Soc. Exp. Biol. Med., 91, 569 (1956).

c. Assay for Transaminases

One of the best methods of measuring transaminase activity
involves coupling the GOT and GPT reaction with the oxidation
of NADH.

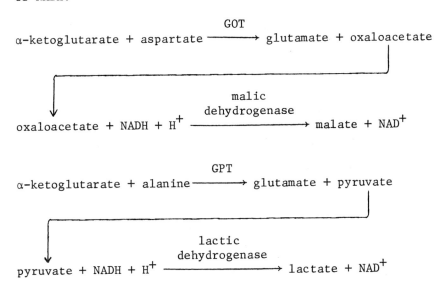

$$\alpha\text{-ketoglutarate} + \text{aspartate} \xrightarrow{\text{GOT}} \text{glutamate} + \text{oxaloacetate}$$

$$\text{oxaloacetate} + \text{NADH} + \text{H}^+ \xrightarrow[\text{dehydrogenase}]{\text{malic}} \text{malate} + \text{NAD}^+$$

$$\alpha\text{-ketoglutarate} + \text{alanine} \xrightarrow{\text{GPT}} \text{glutamate} + \text{pyruvate}$$

$$\text{pyruvate} + \text{NADH} + \text{H}^+ \xrightarrow[\text{dehydrogenase}]{\text{lactic}} \text{lactate} + \text{NAD}^+$$

The oxidation of NADH is measured with a spectrophotometer in
terms of the reduction in absorbance at 340 nm. This method
was discussed in detail in the section on the uses of enzymes
in the clinical laboratory. It is important to point out that
serum from clotted blood should be used for SGOT and SGPT
assays and for blood enzyme measurements such as LDH, CPK, etc.
The materials (EDTA, oxalate, fluoride, heparin) used to
prevent clotting in order to collect plasma can interfere with
enzyme activity.

d. Normal causes for variations in transaminase activity include

    i) Dietary factors: increased (or decreased) pyridoxine intake
      is followed by increased (or decreased) transaminase levels.

   ii) Physical activity: rigorous physical activity produces
      high SGOT resulting from skeletal muscle breakdown.

  iii) Sex differences: insignificant.

   iv) Pregnancy: slight increase in both SGOT and SGPT levels.

    v) Age differences: neonatal SGOT levels are normally very high.

e. Conditions which result in hyper serum transaminase levels

    i) Myocardial infarction. Degree and duration of serum GOT elevation depend upon the magnitude of infarct but SGOT can, at its peak, be as much as ten times the normal value. GOT is high within the first 12 hours following the attack, reaching a maximum at 24-48 hours and returning to normal in 4-7 days (post infarction). In myocardial infarction, the increase of SGOT is greater than the increase in SGPT. Serum GOT levels are widely used to confirm the diagnosis of myocardial infarction.

    ii) Infectious hepatitis (viral). The elevation of GPT is generally more than that of GOT and the elevation may persist for many months after the onset. During the recovery, there is no correlation of symptoms with transaminase levels. The continued high transaminase levels during recovery from hepatitis indicate continuing hepatocellular necrosis.

    iii) Infectious mononucleosis. Alteration in serum GPT and GOT levels appear to parallel the subjective manifestations. Elevations of GPT are more striking than GOT, probably indicating hepatic involvement.

    iv) Renal infarction. High levels of GOT are seen in renal arterial infarction.

    v) Severe burns. The skin contains a relatively low concentration of transaminase, but extensive burns may liberate quite a large quantity of these enzymes.

    vi) Trauma. Transaminase elevations usually here are more transient than those stemming from myocardial infarction, but the diagnosis of myocardial infarction in accident victims and surgical patients is complicated by these elevations.

    vii) Poliomyelitis. In acute paralytic poliomyelitis there is a rise in transaminase activity in cerebrospinal fluid, probably a result of nervous tissue damage. Poliomyelitis during the stage of viremia sometimes produces myocarditis and subclinical liver damage, and thus an increase in serum transaminase levels. These complications probably do not cause the elevated cerebrospinal fluid levels, however.

7. Transamidation

The amide nitrogen of glutamine may serve as a source of amino groups. This is illustrated by a step in the synthesis of glucosamine, in which the amide nitrogen of glutamine is transferred to the keto group of fructose-6-phosphate, a reaction catalyzed by a transamidase.

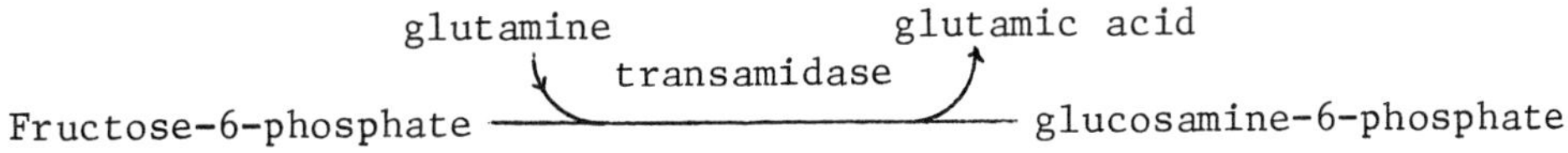

```
                    glutamine              glutamic acid
                         ⌄    transamidase   ⌃
Fructose-6-phosphate ────────────────────────── glucosamine-6-phosphate
```

8. Decarboxylation

Decarboxylation of amino acids to produce amines is catalyzed by pyridoxal-dependent bacterial decarboxylases. Toxic amines (ptomaines) are produced in this way during putrefaction of food protein. Examples of this sort of reaction are shown below.

a.
$$H_2N-(CH_2)_4-\overset{\overset{\displaystyle NH_2}{|}}{CH}-COOH \longrightarrow H_2N-(CH_2)_5-NH_2$$

lysine CO_2 cadaverine

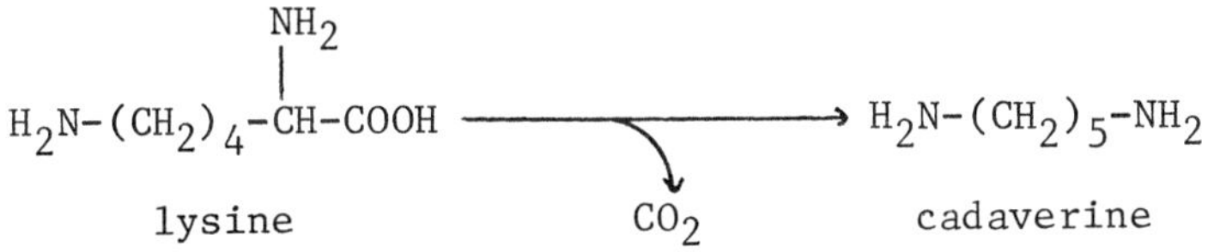

b.
$$H_2N-(CH_2)_3-\overset{\overset{\displaystyle NH_2}{|}}{CH}-COOH \longrightarrow H_2N-(CH_2)_3-NH_2$$

ornithine CO_2 putrescine

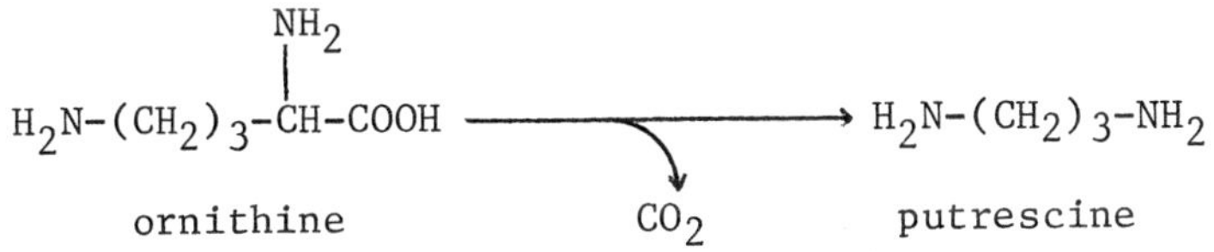

c.
$$H_2N-\overset{\overset{\displaystyle }{\underset{\underset{\displaystyle NH}{||}}{C}}}{}-NH-(CH_2)_3-\overset{\overset{\displaystyle NH_2}{|}}{CH}-COOH \longrightarrow H_2N-\overset{}{\underset{\underset{\displaystyle NH}{||}}{C}}-NH-(CH_2)_4-NH_2$$

arginine CO_2 agmatine

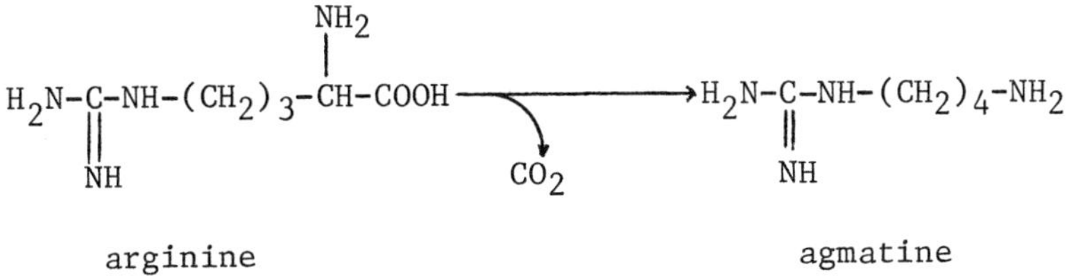

d. The formation of tyramine and histamine from tyrosine and histidine respectively, is shown along with the metabolism of these amino acids.

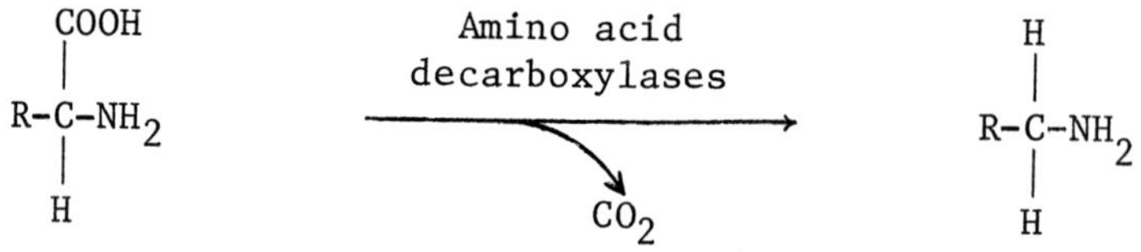

$$\underset{\text{amino acid}}{\overset{\overset{\displaystyle COOH}{|}}{\underset{\underset{\displaystyle H}{|}}{R-C-NH_2}}} \xrightarrow{\makebox[3cm]{}} \underset{\text{ptomaine}}{\overset{\overset{\displaystyle H}{|}}{\underset{\underset{\displaystyle H}{|}}{R-C-NH_2}}}$$

Amino acid decarboxylases

CO_2

Metabolism of Ammonia and Urea Formation

1. Ammonia (NH_3) from amino acids may be used in the amination of keto acids and in the formation of amino acid amides (glutamine and asparagine). It may also be excreted as an ammonium salt, particularly in metabolic acidosis. The majority of the NH_3 in the body, however, is converted to urea and excreted. The <u>principal end-product of protein metabolism in man is urea.</u>

2. Normally only faint traces of NH_3 are present in blood. This is because NH_3 is removed rapidly by the following reactions:

    a. Amination of keto acids

    $$\alpha\text{-Ketoglutarate} \longrightarrow \text{L-Glutamate}$$

    b. Amidation of glutamic acid to form glutamine

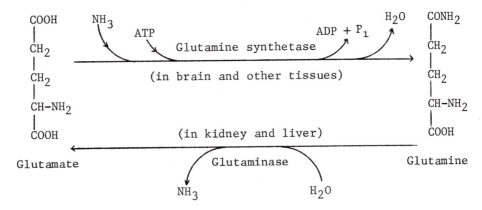

    Glutamine functions in the temporary storage and transport of NH_3, maintaining the NH_3 concentration below toxic levels in tissues. This is particularly important in the brain which is extremely sensitive to NH_3. Glutamine also functions as a donor of amide nitrogen in the synthesis of pyrimidines and purines and of the amino sugars found in the structural polysaccharides. The glutaminase reaction produces NH_3 in the liver and kidney. In the liver, NH_3 is converted to urea. In the kidney, as ammonium ions, it plays a role in the maintenance of acid-base balance.

    c. By far the most important route for the detoxification of NH_3 is the urea cycle (<u>Krebs-Henseleit Cycle</u>) which operates in the liver. It is discussed below in detail.

3. Considerable quantities of NH_3 are produced in the gastrointestinal tract by the action of bacterial enzymes (e.g., urease) on nitrogenous products. This NH_3 is normally absorbed from the intestine and caused by the portal blood to the liver where it is converted to urea.

4. Toxicity of Ammonia

    a. In liver diseases, where there is loss of functional liver tissue, NH_3 levels may rise to high levels. Even minute quantities of NH_3 are toxic to the central nervous system and may produce tremor, slurred speech, blurred vision, and, in severe cases, coma and death. The mechanism of NH_3 toxicity is not clear.

    b. NH_3 levels may also be elevated due to the presence of biochemical lesions in urea biosynthesis (mentioned later in this section).

    c. To reduce the ammonia levels the following treatments have been used:

        i) Administration of glutamate or of ornithine (a urea cycle intermediate) to promote the rate of removal of ammonia.

        ii) Reduction of protein intake.

        iii) Administration of antibiotics to kill the flora of the intestines so that they do not produce ammonia by acting on the nitrogenous products.

Urea Formation (Urea Cycle; Krebs-Henseleit Cycle)

1. Formation of carbamyl phosphate, the first step in urea synthesis, involves CO_2, NH_3, and ATP. N-acetylglutamic acid (AGA) and biotin act as activators. The reaction occurs in two steps.

    a.
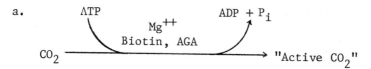

b. "Active CO_2"

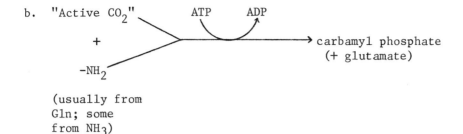

(usually from
Gln; some
from NH_3)

c. These can be summarized as

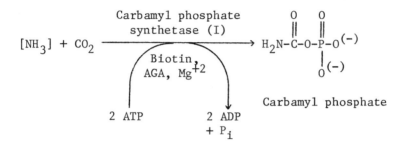

2. a. Carbamyl phosphate synthetase (I) (CPS (I)), the enzyme which
        catalyzes carbamyl phosphate formation to supply the urea cycle,
        is a mitochondrial enzyme. It requires AGA and biotin for
        activity and can use either ammonia or the amide nitrogen of
        glutamine as a nitrogen source. As was discussed under
        pyrimidine biosynthesis, there are two different CP synthetases
        (I and II) which supply the urea cycle and pyrmidine
        biosynthesis respectively. CPS (II) is a cytoplasmic enzyme,
        differing in certain ways from CPS (I).

    b. AGA is an <u>allosteric</u> effector for CPS (I). It acts at a
        non-catalytic site and maintains the enzyme in an active
        configuration. Note that the urea cycle shares a number of
        reactions with the TCA cycle, a pathway which also operates
        within the mitochondria.

3. The remaining reactions of the urea cycle are summarized in
    Figure 57. Note that this pathway is also the route for the
    biosynthesis of <u>arginine,</u> a non-essential amino acid.

4. Biochemical Lesions of the Urea Cycle

    a. Carbamyl phosphate synthetase deficiency (congenital hyperammonemia,
        type I; one known patient): hyperammonemia; mild hyperglycinemia
        and hyperglycinuria; hyperammonemia which is responsive to
        decreased protein intake; increased plasma glutamine level.

Figure 57.   Reactions and Structures of the Urea Synthetic Cycle

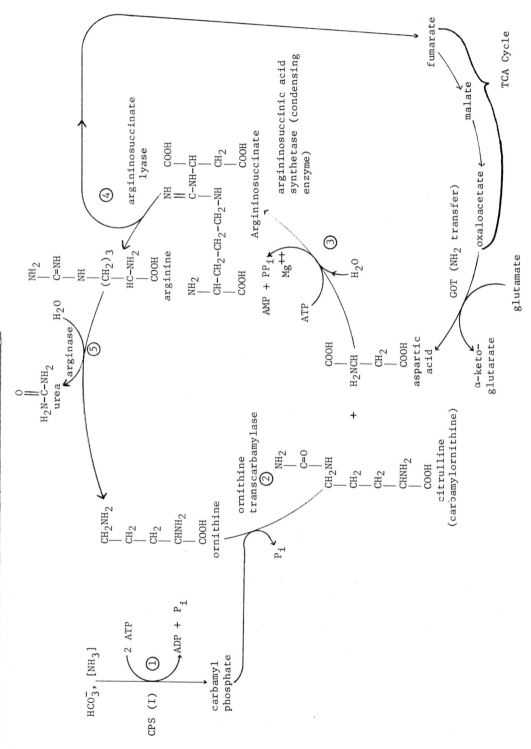

b. Ornithine transcarbamylase deficiency (congenital hyper-
    ammonemia, type II; eight known patients): ammonia
    intoxication; generalized aminoaciduria; neonatal lethargy;
    responsive to low protein diets.

c. Argininosuccinate synthetase deficiency (citrullinemia;
    citrullinuria; two known patients): severe vomiting; coma,
    due to ammonia intoxication; hyperammonemia; responsive to
    low protein diets.

d. Argininosuccinate lyase deficiency (argininosuccinicaciduria;
    twenty-two known patients): accompanied by citrullinuria;
    mental retardation; ammonia intoxication; friable, tufted hair
    and rough skin; responsive to low protein diets.

e. Arginase deficiency (hyperargininemia; two known patients):
    hyperammonemia; mental retardation; epileptic seizures;
    responsive to low protein diets.

Metabolism of Some Individual Amino Acids

Amino Acid Biosynthesis

Mammalian tissues are capable of synthesizing more than half of the
amino acids needed by the body. These amino acids, known as non-essential
amino acids, are made from carbon skeletons derived from lipid and
carbohydrate metabolites or from transformations of essential amino
acids. The nitrogen of the amino groups is obtained from either NH_4^+
ions or from the $-NH_2$ groups of other amino acids. The amino groups
are transferred by means of transaminase reactions. Following is a
list of some non-essential amino acids with their precursors indicated
in parentheses.

    Glutamic acid (α-ketoglutaric acid)
    Aspartic acid (oxaloacetic acid)
    Serine (3-phosphoglyceric acid)
    Glycine (Serine)
    Tyrosine (phenylalanine)
    Proline (glutamic acid → Δ^1 - pyrroline-5-carboxylic acid)
    Hydroxyproline (proline)
    Alanine (pyruvic acid)
    Cysteine & cystine (sulfur atom comes from methionine and carbon
                        atoms from serine)

Some details of the above reactions as well as catabolism of some
individual amino acids are discussed below in the pathways for the
synthesis of the non-essential amino acids.

Glycine Metabolism

1. Metabolic Pathways

    a. Glycine may be formed from serine (using tetrahydrofolate), from glyoxylate (by transamination), and from choline.

    b. Glycine breakdown may involve conversion to serine; to glyoxylate (which is further broken down to form oxalate or formate); or by decarboxylation to N^5,N^{10}-methylene FH_4.

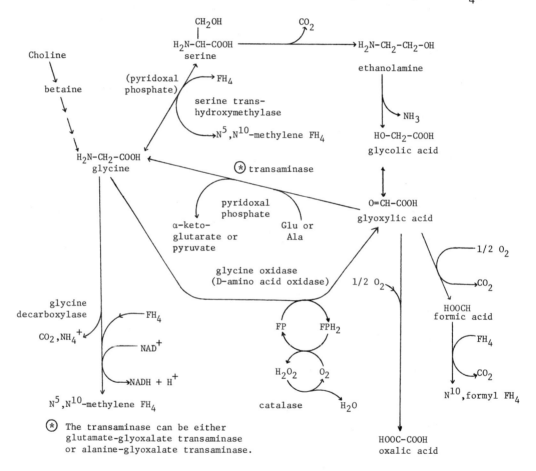

2. Hyperoxaluria

    a. Biochemical lesions.

        i) impairment of oxidation of glyoxylic acid to formic acid.

        ii) inactive glutamic-glyoxylic transaminase or alanine-glyoxylic transaminase.

        If there is impairment of conversion of glyoxylate to formate and to glycine, excess glyoxylate is oxidized to oxalate.

    b. The disease is characterized by high urinary excretion of oxalic acid (unrelated to dietary intake of oxalic acid).

    c. Symptoms include nephrocalcinosis and recurrent urinary tract infection.

    d. Death can occur due to renal failure or hypertension.

3. Glycinuria (rare disease)

    a. The syndrome is characterized by high excretion of glycine in the urine and a tendency to form oxalate renal stones.

    b. The disease is probably due to a defect in renal tubular transport of glycine (decreased reabsorption of glycine).

4. Hyperglycinemia has been observed as a secondary feature in a number of diseases including carbamylphosphate synthetase deficiency (mentioned above), methylmalonic acidemia, and isovaleric acidemia. There are also cases which appear to be due to specific errors in glycine metabolism.

    a. Ketotic hyperglycinemia may be due to a defect in the enzyme system which oxidizes glycine or, more likely, in the biosynthesis of a cofactor for this system.

    b. Non-ketotic hyperglycinemia appears to be caused by a defect in the glycine decarboxylase reaction in which glycine is converted to CO_2, NH_3, and N^5,N^{10}-methylene tetrahydrofolate.

5. <u>Some Special Functions of Glycine</u>

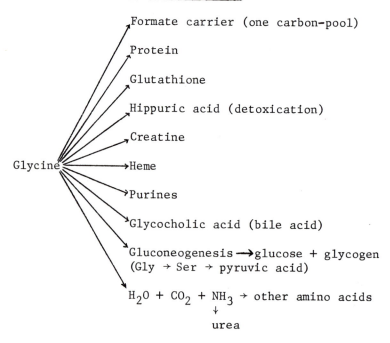

Glycine
- Formate carrier (one carbon-pool)
- Protein
- Glutathione
- Hippuric acid (detoxication)
- Creatine
- Heme
- Purines
- Glycocholic acid (bile acid)
- Gluconeogenesis $\longrightarrow$ glucose + glycogen
  (Gly $\rightarrow$ Ser $\rightarrow$ pyruvic acid)
- $H_2O + CO_2 + NH_3 \rightarrow$ other amino acids
  $\downarrow$
  urea

Metabolism of Creatine and Creatinine

1. Synthesis of creatine, creatine phosphate, and creatinine.

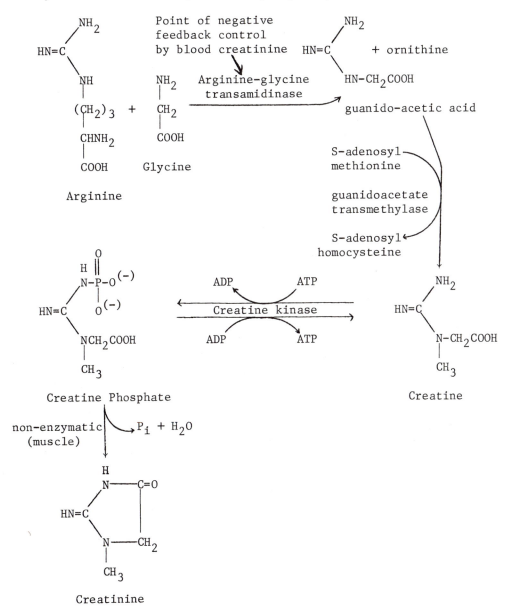

2. Creatine phosphate (also known as phosphocreatine) is a reservoir of high energy phosphate which can readily convert ADP to ATP in muscle and other tissues (see page 286).

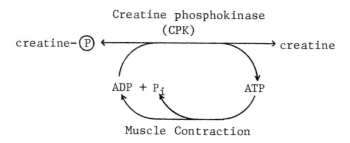

Note: CPK is also known as ATP-creatine phosphotransferase.

3. Striated muscle, heart muscle, testes, liver, and kidney have high concentrations of creatine. Small quantities are found in the brain and still smaller quantities are found in blood.

4. The rate of creatine synthesis (in the liver) is apparently controlled by blood creatine levels which can alter the arginine-glycine transamidinase enzyme activity. The control mechanism is recognized as one of the feedback type.

5. Creatine is often present as creatine phosphate. There is a constant but slow conversion of creatine phosphate to creatinine which has no useful function and is eliminated in the urine.

6. Creatinine excretion is dependent upon skeletal muscle mass and is constant in a healthy individual. The creatinine coefficient is defined as: mgs of (creatinine + creatine) excreted per kilogram of body weight per day. Normal values are 24-26 for men and 20-22 for women. Measurement of the creatinine coefficient is used as one of the kidney function tests.

7. Creatinuria is characterized by excessive excretion of creatine in urine. It may be present in the following conditions:

    a. process of growth
    b. fevers, starvation, diabetes mellitus
    c. extensive tissue destruction
    d. muscular dystrophy
    e. hyperthyroidism

8. Other diseases which are associated with changes in creatine and creatinine metabolism are renal disease, uremia, hypothyroidism, myocardial infarction, gonadal dysfunction, hyperpituitarism, nervous system disorders, and leukemia.

9. Plasma CPK levels are elevated in myocardial infarction as well as in any muscle trauma (see page 286).

Serine Metabolism

1. Serine is <u>not</u> an essential amino acid in man. It is synthesized from 3-phosphoglycerate which is formed from glucose as an intermediate in glycolysis.

2. The pathway of serine synthesis is as follows:

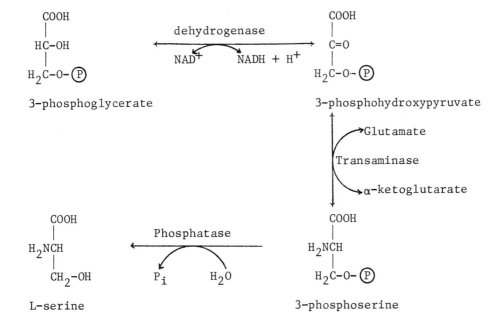

3-phosphoglycerate

3-phosphohydroxypyruvate

L-serine

3-phosphoserine

3. Some of the other reactions in which serine participates are: serine ↔ glycine interconversions, synthesis of cephalins, formation of ethanolamine and choline and the synthesis of cysteine.

Interconversion of Proline, Hydroxyproline, and Glutamic Acid

1. All of these amino acids are <u>non-essential</u> (dietarily dispensable).

2. The pathways of interconversion are shown below.

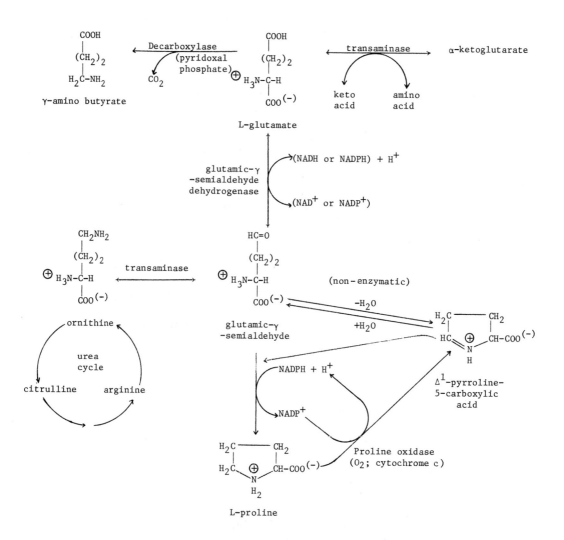

3. Proline is converted to 4-hydroxyproline in a reaction catalyzed by a hydroxylase and requiring molecular oxygen. It takes place predominantly as a post ribosomal modification of proline residues in collagen polypeptides (see collagen synthesis, pages 358-361).

4. GABA is produced by decarboxylation of glutamate. Brain tissue is rich in GABA and its exact role is not understood, although it is implicated in the transmission of nerve impulses (see pages 428-429).

5. Hydroxyproline undergoes catabolism principally in the liver to glyoxylate and pyruvate. The first step in this process is the oxidation of hydroxyproline to Δ^1-pyrroline-3-hydroxy-5-carboxylic acid, catalyzed by hydroxyproline oxidase. Two patients have been reported with hydroxyprolinuria and hydroxyprolinemia, apparently due to a deficiency in this oxidase. Proline metabolism in these individuals appears to be normal, suggesting that there are separate oxidases for proline and hydroxyproline. Both patients were mentally retarded.

6. Hyperprolinemia has been described in a number of patients. Prolinuria, hydroxyprolinuria, and glycinuria generally accompany it. There appear to be two types, due to different enzyme defects.

    a. Type I: proline oxidase is deficient.

    b. Type II: glutamic-γ-semialdehyde dehydrogenase (also known as Δ^1-pyrroline-5-carboxylic acid dehydrogenase) is deficient. In this type, Δ^1-pyrroline-5-carboxylic acid also accumulates in body fluids.

    There is no definite clinical pattern presented by either type of defect other than the aminoaciduria. Attempts have been made to link renal disorders with these diseases but the correlation is poor. Although proline metabolism (by way of GABA) has been implicated in proper neural function, there is no clear relationship between mental abnormalities and faulty proline metabolism.

Histidine Metabolism

1. Histidine is not a required amino acid for man and is synthesized in the liver. The pathway for the synthesis has not yet been completely delineated.

2. The decarboxylated product of histidine is histamine which is produced in the mast cells and various other types of cells found in different parts of the body (see page 429). Some of the principal actions of histamine are: dilatation of arterioles;

changes in capillary permeability (resulting in a fall in blood pressure and the leakage of fluids); constriction of smooth muscles, particularly those of the bronchioles (an effect antagonized by epinephrine); increased secretion of acid gastric juice (hence, histamine is used in the assessment of gastric function); and others. The main pathway for degradation of histamine consists of formation of 1-methyl histamine which is then converted to methyl imidazole acetic acid by an amine oxidase enzyme.

3. Histidine catabolism

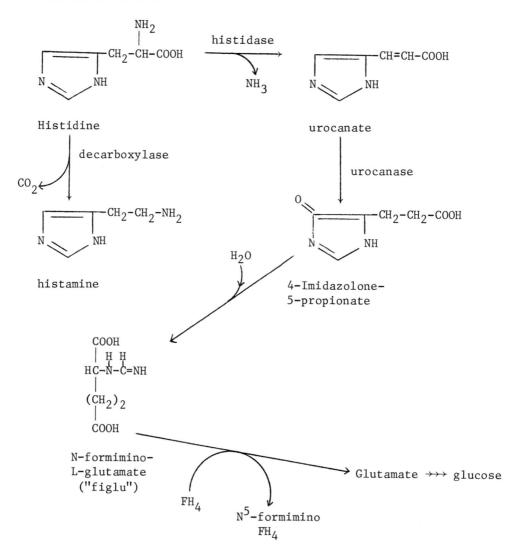

Histidine

urocanate

histamine

4-Imidazolone-
5-propionate

N-formimino-
L-glutamate
("figlu")

Glutamate $\rightarrowtail$ glucose

4. A folic acid deficiency can lead to a lack of utilization of formiminoglutamate ("figlu") leading to its accumulation, with large amounts of it being excreted in the urine (see also one-carbon metabolism, pages 384-385).

Catabolism of Branched-Chain Amino Acids

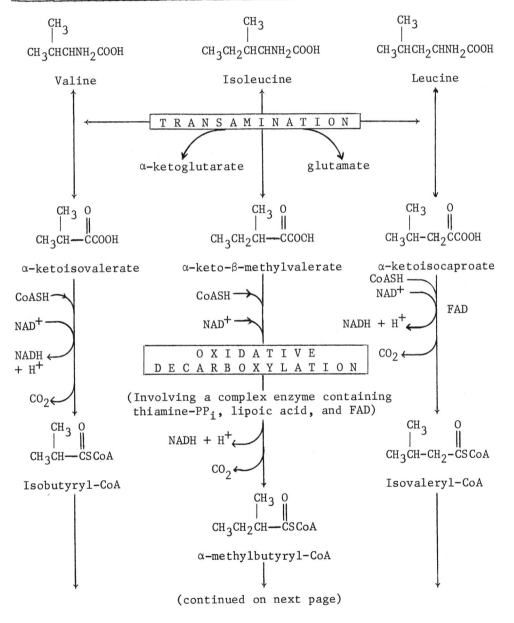

(continued on next page)

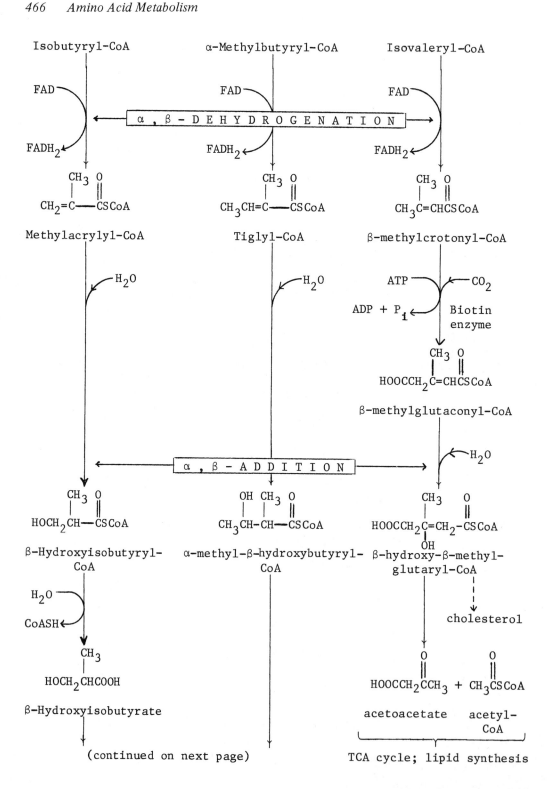

Isobutyryl-CoA   α-Methylbutyryl-CoA   Isovaleryl-CoA

α , β - D E H Y D R O G E N A T I O N

CH₂=C——CSCoA   CH₃CH=C——CSCoA   CH₃C=CHCSCoA

Methylacrylyl-CoA   Tiglyl-CoA   β-methylcrotonyl-CoA

HOOCCH₂C=CHCSCoA

β-methylglutaconyl-CoA

α , β - A D D I T I O N

HOCH₂CH——CSCoA   CH₃CH-CH——CSCoA   HOOCCH₂C=CH₂-CSCoA

β-Hydroxyisobutyryl-   α-methyl-β-hydroxybutyryl-   β-hydroxy-β-methyl-
CoA   CoA   glutaryl-CoA

cholesterol

HOCH₂CHCOOH

β-Hydroxyisobutyrate

HOOCCH₂CCH₃ + CH₃CSCoA

acetoacetate   acetyl-
CoA

(continued on next page)   TCA cycle; lipid synthesis

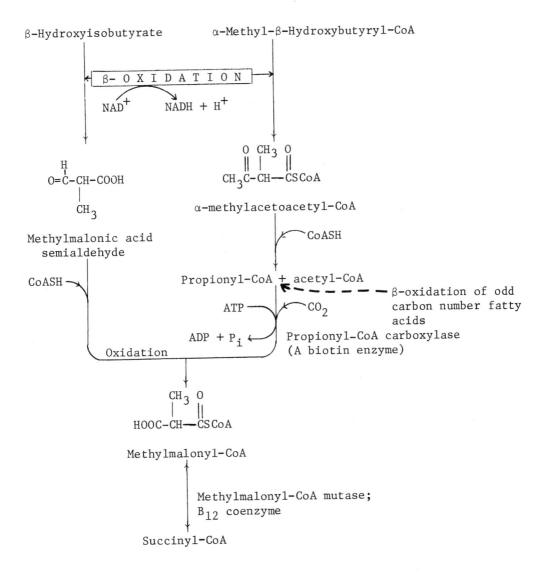

Comments on the Catabolism of Branched-Chain Amino Acids

1. L-valine, L-isoleucine, and L-leucine are all essential amino acids
    for man although their respective keto acids can replace them in
    the diet.

2. All of the branched-chain amino acids transaminate with α-keto-
    glutarate in reversible reactions. One enzyme probably catalyzes
    all of the three transaminase reactions in the heart muscle but
    in the liver there are three different enzymes, one for each
    reaction.

3. The oxidative decarboxylations are irreversible reactions and are
    catalyzed by an enzyme complex which contains NAD$^+$, FAD, thiamine
    PP, and lipoic acid as cofactors. In some children, non-functional
    decarboxylase prevents further catabolism of all three of the
    α-keto acids. These branched-chain keto acid derivatives
    accumulate in the urine. This inborn error of metabolism is
    known as <u>maple syrup urine disease</u> because the characteristic odor
    of the urine resembles that of maple syrup. The disease involves
    severe impairment of the central nervous system, mental retardation,
    and often death.

4. A biochemical lesion in the metabolism of methylmalonic acid,
    which is produced from valine and isoleucine, has been noted. (Note:
    methylmalonic acid can also be obtained from other metabolic
    sources such as methionine, threonine, etc.) This hereditary condition
    is characterized by the presence of large amounts of methymalonic
    acid in the urine, metabolic acidosis, etc.

5. A hereditary deficiency of α-ketoisovaleryl CoA-dehydrogenase has
    also been reported. This enzyme catalyzes the conversion of
    α-ketoisovalerate (formed from valine) to isobutyryl-CoA. Its
    absence causes α-ketoisovaleric acid levels to rise in the
    blood and urine.

6. The pathways of metabolism of these amino acids have similarities
    with carbohydrate and fatty acid metabolism. Examples: the
    oxidative decarboxylation step is similar to pyruvic acid and
    α-ketoglutarate oxidations; the α, β-dehydrogenation and α, β-addition
    are similar to steps in fatty acid oxidation.

7. Note that in the pathway for leucine oxidation, β-hydroxy-β-
    methylglutaryl-CoA is formed. This is an intermediate in the
    biosynthesis of cholesterol. Also, leucine yields acetoacetate
    and acetyl-CoA which are both ketogenic and hence it is the only
    purely ketogenic amino acid. On the other hand, isoleucine is
    both ketogenic and glucogenic, and valine is only glucogenic.

Summary of Overall Catabolism

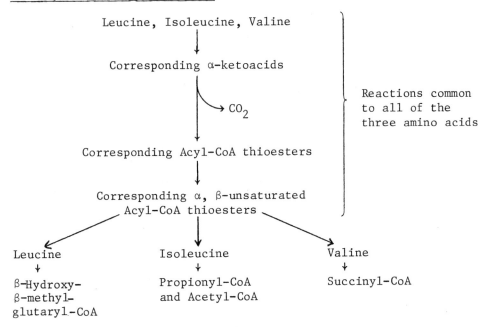

Leucine, Isoleucine, Valine

↓

Corresponding α-ketoacids

→ CO_2

↓

Corresponding Acyl-CoA thioesters

↓

Corresponding α, β-unsaturated Acyl-CoA thioesters

Reactions common to all of the three amino acids

Leucine	Isoleucine	Valine
β-Hydroxy-β-methyl-glutaryl-CoA	Propionyl-CoA and Acetyl-CoA	Succinyl-CoA

Metabolism of the Sulfur-Containing Amino Acids

Methionine and cysteine are the principal sources of the body's organic sulfur. Methionine is an essential amino acid (unless adequate homocysteine and a source of methyl groups are available), but cysteine is not essential since it can be synthesized from methionine. Cystine, another sulfur-containing amino acid, is synthesized from cysteine.

Metabolism of Methionine

1. In the body, methionine primarily is

   a. utilized in protein synthesis,

   b. the principal methyl donor (methyl groups are transferred from methionine to other compounds for detoxication processes, biosynthesis, etc.),

   c. a precursor of cysteine.

2. Formation of S-adenosylmethionine (active methionine), the methyl-donating form of methionine

    a.

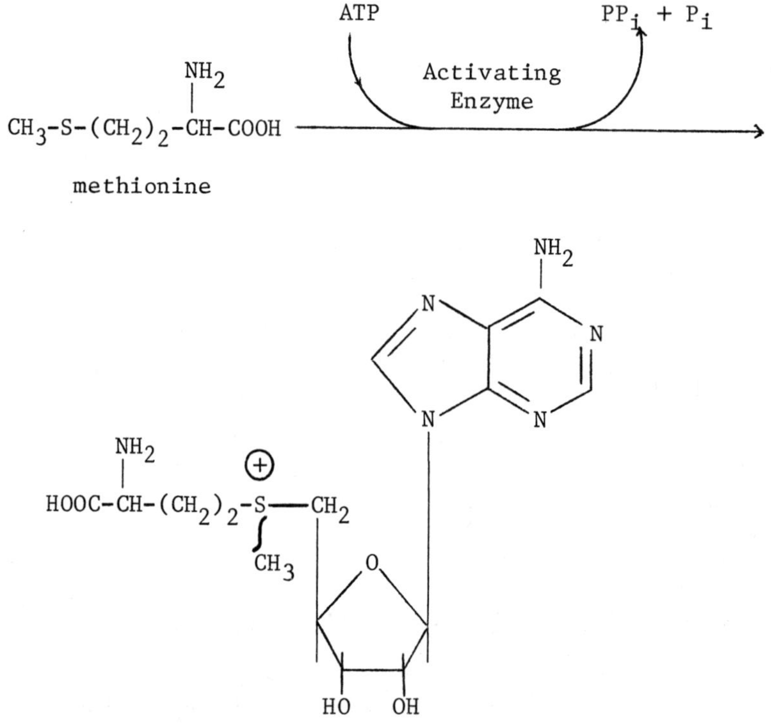

methionine

S-adenosylmethionine (SAM)

    b. S-adenosylmethionine is a sulfonium compound. The carbon-sulfur bond is a "high-energy" bond, having a free energy of hydrolysis of 8 kcal/mole. This energy is available for biosynthetic reactions (transmethylation).

A generalized transmethylation reaction is shown below.

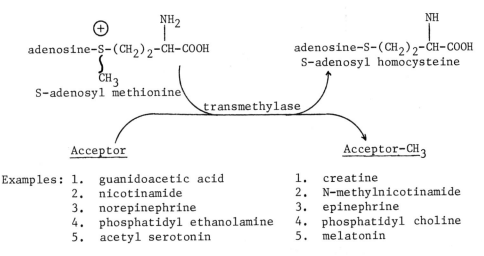

Examples:
	Acceptor		Acceptor-CH₃
1.	guanidoacetic acid	1.	creatine
2.	nicotinamide	2.	N-methylnicotinamide
3.	norepinephrine	3.	epinephrine
4.	phosphatidyl ethanolamine	4.	phosphatidyl choline
5.	acetyl serotonin	5.	melatonin

3. Some Interrelationships between Methionine and Other Amino Acids

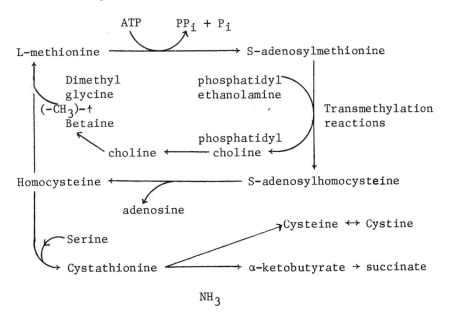

4. Formation of Cystathionine and Cysteine

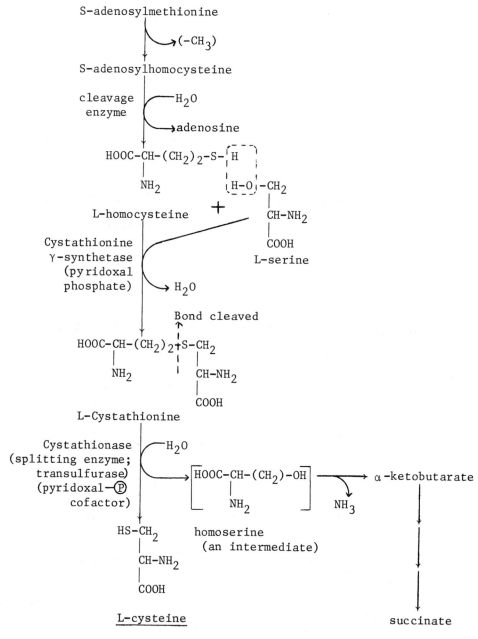

Note: In the above synthesis of cysteine, only the sulfur atom is provided by methionine. The carbon chain and the amino group are obtained from serine.

5. Catabolism of Cystine & Cysteine

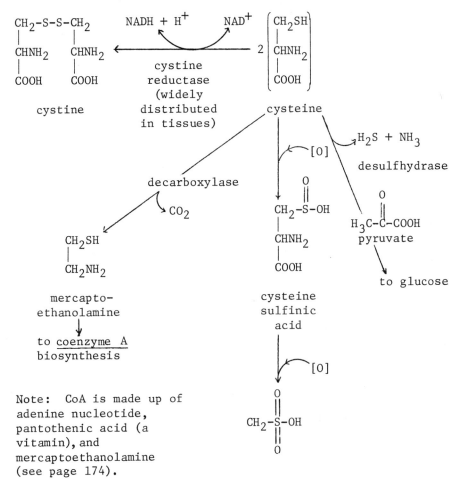

Note: CoA is made up of
adenine nucleotide,
pantothenic acid (a
vitamin), and
mercaptoethanolamine
(see page 174).

Note: Taurine forms conjugates with bile acids to produce
taurocholic acids (see pages 660-661).

6.   Formation of "Active Sulfate"

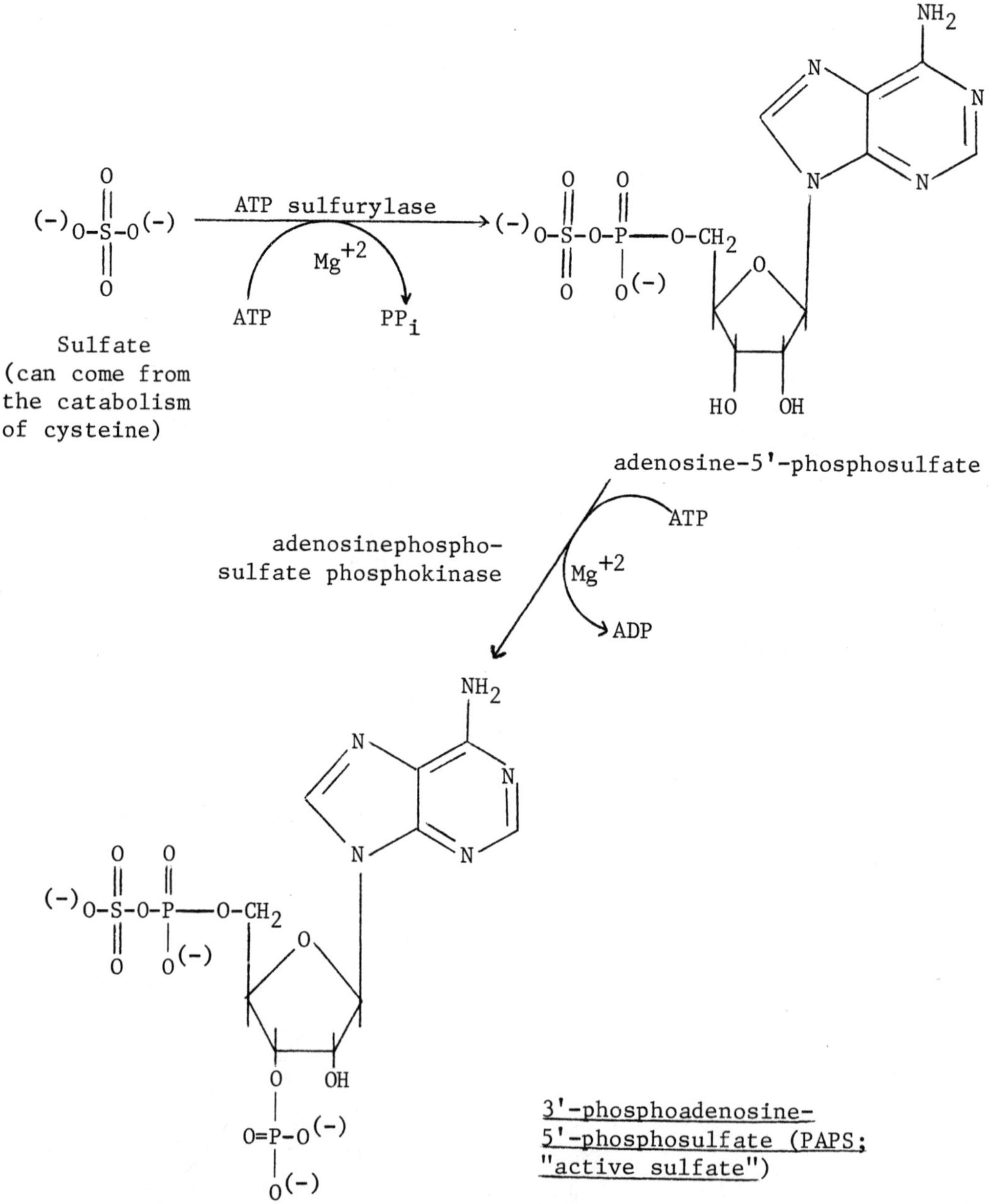

Sulfate
(can come from
the catabolism
of cysteine)

ATP sulfurylase

Mg^{+2}

ATP   PP$_i$

adenosine-5'-phosphosulfate

adenosinephospho-
sulfate phosphokinase

Mg^{+2}

ATP

ADP

3'-phosphoadenosine-
5'-phosphosulfate (PAPS;
"active sulfate")

PAPS is the donor of the sulfate moiety which occurs as an ester
of galactose in compounds such as the chondroitin sulfates and
other sulfate-containing mucopolysaccharides. Phenols, steroids,
etc. also react with PAPS giving the respective sulfate derivatives
which are then eliminated in the urine. This is one of the
detoxication mechanisms which takes place in the liver. These
reactions are summarized as follows:

$$\text{PAPS} + \text{R-OH} \quad \xrightarrow{\text{transferring enzymes}} \quad \text{R-O-}\overset{\displaystyle O}{\underset{\displaystyle O}{\overset{\|}{\underset{\|}{S}}}}\text{-OH}$$

     (phenols

      steroids

      etc.)   sulfate ester

Abnormalities Involving Sulfur-Containing Amino Acids

1. Cystinuria

   a. Excessive excretion of cystine in urine occurs in this
      inherited disease.

   b. There are also defects in renal reabsorptive mechanisms for
      lysine, arginine, and ornithine.

2. Cystinosis

   a. Cystine crystals are deposited in many tissues and organs.
      This disease is also inherited.

   b. General aminoacidura is also present.

   c. Other severe renal impairments occur, and the patients usually
      die at an early age, apparently from renal failure.

3. Homocystinuria

   a. This inherited disease is characterized by abnormally large
      quantities of homocystine in the urine and excessive
      methionine excretion.

   b. Convulsions and mental retardation are common in this disorder.

   c. Decreased activity of **hepatic cystathionine synthetase** is the
      cause.

   d. Supplying excess cysteine in the diet from an early age may be
      beneficial in preventing some of the damage.

4. The Fanconi Syndrome

    a. This is a genetically determined abnormality involving renal
        tubular defects.

    b.* There is a generalized decrease in the capacity to reabsorb all
        amino acids, glucose, calcium, phosphate, Na^+, K^+, uric acid,
        H_2O, etc.

    c. The survival of the patient depends on the severity of the
        disease.

    d. Clinical findings: rickets; acidosis (in severe cases); and
        dehydration.

    e. In children with Fanconi's Syndrome, cystinosis is almost
        always present.

    f. Microdissection of the nephrons of infants and adults who have
        died of the Fanconi Syndrome has demonstrated the presence in
        nephrons of a very thin, apparently non-functioning initial
        segment of the proximal convoluted tubule. The remaining
        portion of the proximal tubules seems to be normal. Whether
        this anatomical abnormality can account for all reabsorptive
        problems remains to be proven. The relationship of renal
        defects to cystinosis is not understood.

    g. Treatment: replacement of lost electrolytes and restoration
        of bone by high vitamin D and calcium intake; administration
        of alkaline salts to correct the acidosis. There is no known
        treatment for cystinosis. If untreated, the severe condition
        (involving acidosis and dehydration) is usually fatal.

Metabolism of Phenylalanine and Tyrosine

1. Phenylalanine is an essential amino acid. If adequate amounts of
    phenylalanine are available, tyrosine is not essential; otherwise it
    is. These two amino acids are involved in the synthesis of a
    variety of important compounds, including thyroxine, melanin,
    epinephrine and norepinephrine.

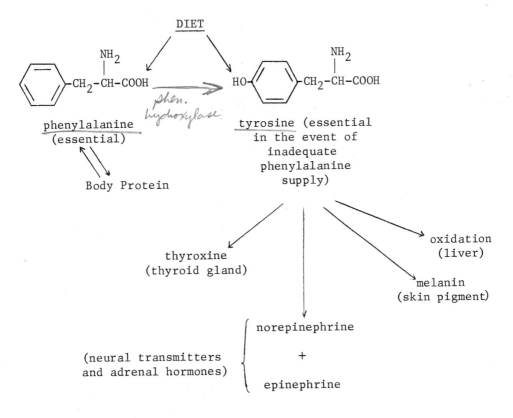

A number of diseases are related to the lesions that are present in the metabolism of phenylalanine and tyrosine. These are phenylketonuria (PKU), alcaptonuria, tyrosinosis, albinism, certain defects in thyroid hormone production, and defects in the production of norepinephrine and epinephrine.

2. Oxidation of Phenylalanine (normal catabolic route)

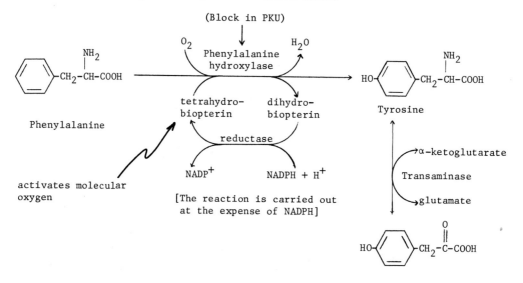

(Block in PKU)

O$_2$ Phenylalanine   H$_2$O
hydroxylase

Phenylalanine   CH$_2$-CH-COOH (NH$_2$)

tetrahydro-biopterin   dihydro-biopterin

reductase

activates molecular oxygen

NADP$^+$ NADPH + H$^+$

[The reaction is carried out at the expense of NADPH]

HO- CH$_2$-CH-COOH (NH$_2$)   Tyrosine

α-ketoglutarate

Transaminase

glutamate

HO- CH$_2$-C-COOH (O)

p-hydroxyphenylpyruvic acid

This reaction consists of hydroxylation, shift of the side chain, and decarboxylation

2 O$_2$

p-hydroxyphenylpyruvic acid oxidase (block in tyrosinosis); ascorbic acid, Cu^{++}

CO$_2$

HO- -OH
CH$_2$-COOH

Homogentisic acid

O$_2$

Homogentisic acid oxidase (block in alcaptonuria)

glucose

COOH   +   CH$_3$
CH   C=O
HC   CH$_2$
COOH   COOH

Fumaric acid   Acetoacetic acid

HOOC, H
C
C   CH$_2$ C   CH$_2$ COOH
H   O   O

Fumarylacetoacetic acid

isomerase (GSH needed as a cofactor)

H-C COOH
H-C
C   CH$_2$ C   CH$_2$ COOH
O   O

Maleylacetoacetic Acid

3.   Structures of the Biopterins

5,6,7,8-tetrahydrobiopterin
(quinonoid form)

7,8-dihydrobiopterin
(quinonoid form)

The reductase in mammalian liver which reduces dihydrobiopterin to tetrahydrobiopterin uses NADPH as a cofactor and is indistinguishable from dihydrofolate reductase (see one-carbon metabolism). Recall that folic acid contains a pteridine ring system, as do the biopterin cofactors. Note also that there can be a tautomerization involving exchange of a proton between the 3-nitrogen and the 4-keto oxygen.

Phenylketonuria

1. This inherited disease is due to the absence in the liver of the enzyme, phenylalanine hydroxylase,which catalyzes the conversion of phenylalanine to tyrosine. This enzyme occurs attached to the membrane of the endoplasmic reticulum. In vitro, the enzyme has been shown to require lysophosphatidylcholine (a component of this membrane) for optimal activity. That part of the dietary intake of L-phenylalanine that would normally be converted to tyrosine accumulates in the body (as phenylalanine or one of its metabolites)

and is responsible for a number of pathological changes in body
tissues. The name arises from the appearance of phenylketones
(as shown below) in the urine.

2. Various metabolites that accumulate in PKU

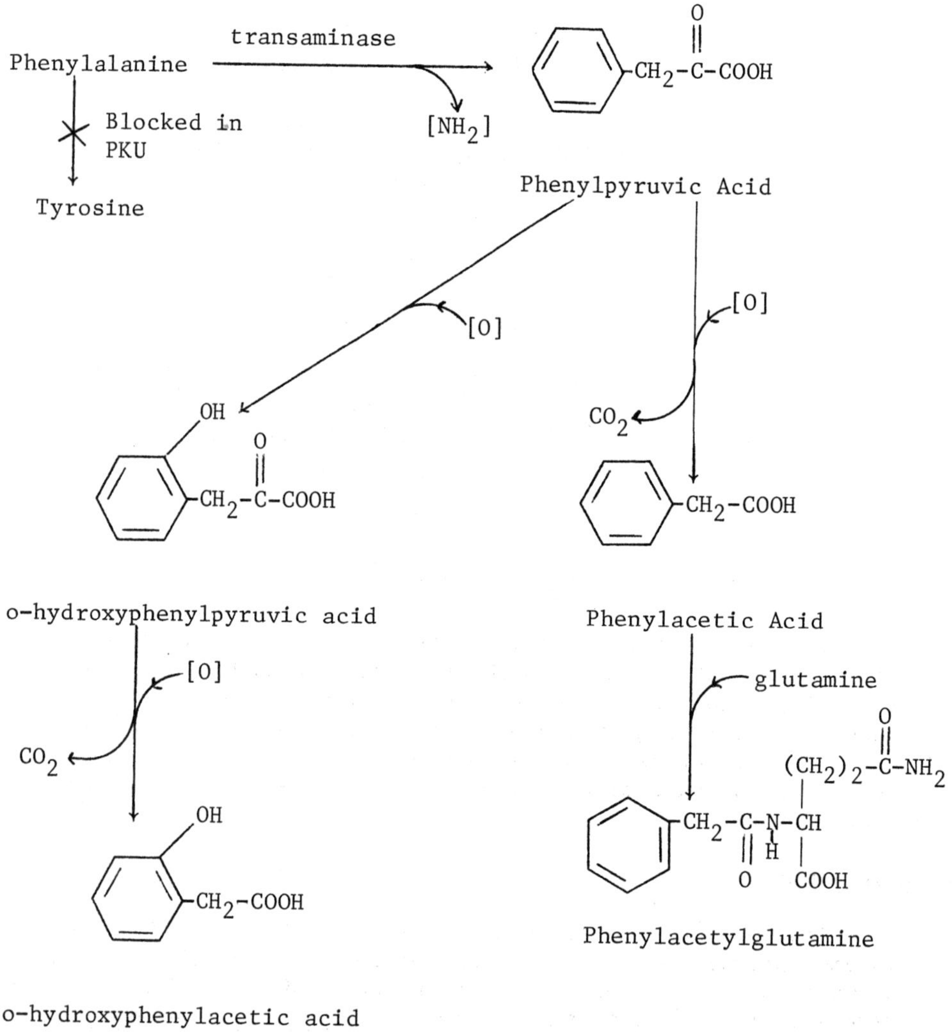

3. The disease is inherited as an autosomal recessive characteristic.
The frequency of occurrence in the population is 1:25,000 in North
European countries.

4. Pathological changes of skin and brain in PKU are due to the accumulation of phenylalanine. The changes can be prevented if a diet very low in phenylalanine is provided very early in life. If this is not done, severe, irreversible damage soon occurs. Small amounts of phenylalanine must be provided, however, for it is an essential amino acid required for the formation of body proteins.

5. A biochemical basis for the severe mental impairment and for some of the other defects has not been discovered. One possibility, involving tryptophan metabolism, is discussed below. Another factor may be that phenylpyruvic acid and phenylacetic acid inhibit glutamic acid decarboxylase. This enzyme catalyzes the reaction.

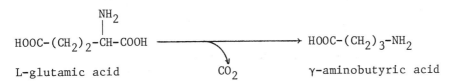

$$\underset{\text{L-glutamic acid}}{HOOC-(CH_2)_2-\overset{\overset{\displaystyle NH_2}{\displaystyle |}}{C}H-COOH} \xrightarrow[\quad CO_2 \quad]{} \underset{\gamma\text{-aminobutyric acid}}{HOOC-(CH_2)_3-NH_2}$$

The product, γ-aminobutyric acid, may be involved in synaptic transmission within the central nervous system (see also hyperprolinemia).

6. Defects in skin and hair pigmentation (light skin and blond hair) may be caused by excess phenylalanine acting as an antimetabolite to tyrosine (which is a precursor of melanin).

7. High phenylalanine levels may disturb the transport of amino acids into cells. It has been postulated that the variation in the overt pattern of PKU may reflect differences in the efficiency of this transport in the presence of elevated phenylalanine concentrations.

8. Apparently in all infants, phenylalanine hydroxylase in the liver matures postnatally and only when it fails to mature (during the biochemical differentiation of liver) as in PKU, does brain damage occur. Therefore, the promptness and consistency of treatment with the proper dietary regimen usually determines the magnitude of irreversible intellectual impairment. The low phenylalanine dietary regimen can be eased at about the age of six. This is apparently due to the fact that the completion of brain differentiation takes place by that age. An enigmatic aspect of this problem is that in some cases, restricted phenylalanine intake does not correct the defect and a few of the untreated PKU patients have normal IQ.

9. In PKU, defective myelination occurs in the brain, there is an increased incidence of epileptic seizures, and abnormal EEG's are common.

10. A defect in tryptophan metabolism also appears to be involved in PKU, as evidenced by the occurrence of abnormal indole derivatives in the urine and low levels of serotonin (a product of tryptophan metabolism) in the blood and brain.   There are several points where phenylalanine and tryptophan metabolism are connected.

   a. The enzyme, 5-hydroxytryptophan decarboxylase which catalyzes the conversion of 5-hydroxytryptophan to serotonin is inhibited in vitro by some of the metabolites of phenylalanine.

   b. Phenylalanine hydroxylase is identical to the liver enzyme that catalyzes the hydroxylation of tryptophan to 5-hydroxy-tryptophan, a precursor of serotonin. Also, in vitro, phenylalanine is found to inhibit the hydroxylation of trypto-phan.

   There has been some speculation that the mental defects related to PKU may be caused by the decreased production of serotonin. The details of tryptophan metabolism are discussed later in this chapter.

11. As was suggested above, there are a number of clinical patterns associated with prolonged and transient phenylalanine elevation and the presence of phenylketones in the urine. These may be genetic variants of "classical" PKU or may involve additional or related enzyme defects.

Methods of Assay for Phenylalanine and Its Metabolites

1. Automated Method

$$Phe + Ninhydrin \xrightarrow[\text{alkaline Cu-tartrate buffer}]{\text{L-leucyl-L-alanine; heat;}} \text{fluorescent complex}$$

The Leu-Ala dipeptide enhances the fluorescence of the Phe-ninhydrin complex.

2. Ferric chloride ($FeCl_3$) methods are based on the reaction

$$Urine + FeCl_3 \text{ (soln.)} \longrightarrow \text{green or blue-green color}$$

This reaction is the basis of the "diaper test" for PKU in newborn infants. It is not sensitive enough, however, to detect the disease until several days after birth, when phenylalanine levels have risen to over 50 mg/100 ml of urine. Also, as was pointed out above, phenylalanine hydroxylase activity matures postnatally and it is the failure to mature which causes PKU. A number of substances interfere with this test, yielding confusing or erroneous results. The reagents have now been incorporated into a dried preparation used to coat test strips which can be employed in place of the ferric chloride solution.

3. Microbiological method. A sensitive assay using a bacterial-inhibition method has been developed by Guthrie. The test is based on the fact that phenylalanine can overcome the inhibitory effect of thionylphenylalanine on the growth of <u>Bacillus subtilis</u>.

4. Measurement of plasma phenylalanine is the most specific diagnostic test for PKU.

Tyrosinosis

1. This is a very rare disease in which p-hydroxyphenylpyruvic acid is excreted.

2. The biochemical lesion is presumed to be p-hydroxyphenylpyruvic acid oxidase or tyrosine transaminase.

3. Another disorder involving tyrosine, in which there is enlargement of liver and spleen and multiple renal tubular defects, has been termed tyrosinemia. It appears to be due to a lack of p-OH phenyl pyruvic acid oxidase.

Alcaptonuria

1. This is a rare, metabolic, hereditary disease in which there is an accumulation and elimination in urine of homogentisic acid. The urine darkens upon exposure to air (due to the oxidation of homogentisic acid).

2. The biochemical lesion is a lack of homogentisic acid oxidase.

3. Homogentisic acid is a good reducing agent and causes positive reactions in reducing tests for glucose which may lead to an erroneous diagnosis of diabetes or renal glycosuria.

4. The clinical features include pigmentation of cartilage and other connective tissues due to the deposition of oxidized homogentisic acid. This generalized pigmentation is called ochronosis. Patients nearly always develop arthritis in later years. The exact relationship between the pigment deposition and arthritis is not yet understood. Alcaptonuria is not known to be associated with any important clinical manifestations except for this arthritis, which may be severe. However, there are indications that other complications may be associated with the disease. This is currently an area of study.

Biosynthesis of Melanin

Tyrosine $\xrightarrow[\text{[O]}]{\text{tyrosinase}}$ 3,4-Dihydroxyphenylalanine (DOPA)

Tyrosine

3,4-Dihydroxyphenylalanine (DOPA)

$\xrightarrow[\text{H}_2\text{O}]{\text{[O]} \quad \text{tyrosinase}}$

DOPA Quinone

$\xleftarrow{\text{[O]}}$

Dopachrome (hallochrome)

Note: Zn^{+2} is present in melanocytes at a high concentration.

$\xrightarrow[\text{CO}_2]{\text{Zn}^{++}}$

5,6-Dihydroxyindole $\xrightarrow[\text{H}_2\text{O}]{\text{[O]}}$ Indole − 5,6-Quinone

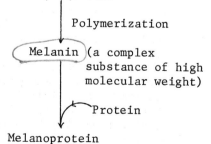

Indole − 5,6-Quinone

↓ Polymerization

Melanin (a complex substance of high molecular weight)

↓ Protein

Melanoprotein

Comments on the Biosynthesis of Melanin

1. Melanin is a polymer of indole nuclei. Its exact structure is not known. Melanin is produced by melanocytes which are normally present in the skin (in hair bulbs and in the dermis), the mucous membranes, the eye, and the nervous system. It is a highly insoluble compound which presents problems in its study.

2. The varying degrees of coloration of skin (white-brown-black) are due to the amount of melanin synthesized in the melanocytes rather than the number of melanocytes. Melanin formation is apparently under both hormonal and neural regulation.

   a. The adenohypophysis secretes melanocyte-stimulating hormone (MSH) which affects skin coloration in fish, reptiles, amphibians, etc., but whose role in mammals is uncertain. It is a polypeptide hormone.

   b. Certain hormonal disturbances are accompanied by changes in pigmentation. For example, in Addison's disease, in which there is an underproduction of cortisol (a hormone produced by the adrenal glands), increased pigmentation of the skin is a frequent clinical finding. Another disease is panhypopituitarism in which there is a deficiency in the production of all of the pituitary hormones. It is characterized by skin pallor. The exact causes of these pigmentation changes are not yet known. A decrease (or increase) in MSH production may be involved or the MSH-like activity of ACTH may account for some of the pigmentation changes.

   c. ACTH has a small amount of MSH-like activity. (ACTH and MSH have overlapping amino acid sequences in their structure.)

   d. Melatonin (N-acetyl-5-methoxy tryptamine) is a hormone present in the pineal body and peripheral nerves. A metabolite of tryptophan, it has been shown to reduce the deposition of melanin in frog melanocytes. In mammals, its synthesis appears to be sensitive to light and it has some inhibitory effect on the ovary. All of this is suggestive of involvement in cyclic or rhythmic processes.

3. A lack of melanin production in the melanocytes gives rise to several hereditary diseases collectively called albinism.

   a. Among the mechanisms that have been investigated as possible causes of albinism are

(i)   deficiency or lack of tyrosinase,

(ii)   lack of melanin polymerization,

(iii) lack of synthesis of the protein matrix of the melanin
       granule,

(iv)   lack of availability of the substrate tyrosine,

(v)   presence of inhibitors of tyrosinase.

Of these, the basic defect appears to be (i).

b. Clinical features of albinism involving the skin include
increased susceptibility to various types of carcinoma (melanin's
major function is as a screen which gives protection from solar
radiation). When the eyes are involved, photophobia, subnormal
visual acuity, strabismus nystagmus (involuntary rapid movement
of the eyeball; vertical, horizontal, or rotatory), absence
of some of the anatomical components of the eye (such as the
sphincter muscle) are some of the aspects observed. The
inheritance pattern of albinism varies from type to type. For
example, universal albinism (characterized by absence of
melanin in the hair bulb, retinal pigment epithelium, skin and
ureal tract) is autosomal recessive; Piebald albinism
(characterized by isolated and scattered patches or absence of
melanin in the skin and hair) is autosomal dominant; and ocular
albinism (characterized by absence of melanin in the retinal
pigment epithelium) is a sex-linked recessive trait.

Epinephrine (Adrenalin) and Norepinephrine (Noradrenalin)

1. These compounds, termed catecholamines, are found in the granulated
vesicles of the adrenal medulla (chromaffin granules) and the
sympathetic nerve endings (peripheral and central nervous systems).
The adrenal medulla, derived from the sympathetic autonomous nervous
system, contains about 80% epinephrine and 20% norepinephrine. The
sympathetic nerve endings, on the other hand, have very little
epinephrine, the principal hormone being norepinephrine.

2. Biosynthesis

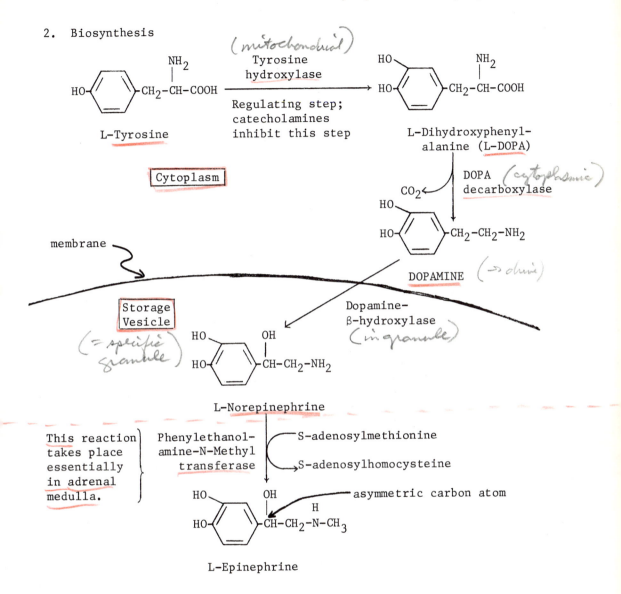

L-Tyrosine

Tyrosine hydroxylase (mitochondrial)

Regulating step; catecholamines inhibit this step

L-Dihydroxyphenyl-alanine (L-DOPA)

Cytoplasm

DOPA decarboxylase (cytoplasmic)

DOPAMINE (→ drive)

membrane

Storage Vesicle (= specific granule)

Dopamine-β-hydroxylase (in granule)

L-Norepinephrine

This reaction takes place essentially in adrenal medulla.

Phenylethanol-amine-N-Methyl transferase

S-adenosylmethionine
→ S-adenosylhomocysteine

asymmetric carbon atom

L-Epinephrine

Biosynthetically the L-isomers of the above compounds are formed and L-DOPA has been shown to be 15 times more effective than D-DOPA.

3. The catecholamines are released from the granules along with other
    materials (such as ATP and enzymes) by an exocytotic process
    involving calcium (see under cyclic AMP). This is brought about
    by a chemical stimulation followed by membrane depolarization.

4. Dopamine, an intermediate in the synthetic pathway, has been found
    to be present in those areas of brain involved in coordinating
    motor activity. It is presumed to function as a neurotransmitter
    in dopaminergic neurons. L-DOPA, the immediate precursor of
    dopamine, is used therapeutically to treat a neurological disorder
    known as paralysis agitans (Parkinson's disease), possibly to
    stimulate dopaminergic fibers. L-DOPA is used, instead of dopamine,
    because the latter does not cross the blood-brain barrier.

5. Regulation of Synthesis

    a. Catecholamine levels regulate tyrosine hydroxylase activity,
        higher concentrations inhibiting the conversion of tyrosine
        to DOPA.

    b. It has been observed that prolonged stimulation of the
        sympathetic system gives rise to elevated levels of tyrosine
        hydroxylase and dopamine-β-hydroxylase in the norepinephrine-
        synthetic system in the nerve endings. Similarly, in epinephrine-
        synthesizing pathways, elevated levels of phenylethanolamine-
        N-methyl transferase have been noted in the adrenal medulla.

    c. Adrenocorticotropic hormone (ACTH) influences the levels of
        the enzymes involved in the biosynthesis of catecholamines.
        Hypophysectomized animals show very low levels of these
        enzymes but, following ACTH administration, the levels are
        restored to normal.

    d. In these (hypophysectomized) animals, dexamethasone (a cortico-
        steroid) restores only the activity of phenylethanolamine-N-
        methyl transferase and not the activity of the other enzymes.

6. Some inhibitors of catecholamine biosynthesis and their storage

    a. α-methyl-p-tyrosine inhibits the tyrosine hydroxylase catalyzed
        reaction - Tyrosine → DOPA - and hence, the formation of
        norepinephrine. This compound is useful in treating the
        hypertension resulting from pheochromocytoma. Pheochromocytoma
        is a tumor of the adrenal medullary chromaffin cells.

    b. Disulfiram (for structure, see page 292) inhibits the dopamine-
        β-hydroxylase reaction.

c. The compounds, reserpine and guanethidine, lower the level of
norepinephrine by preventing the storage of catecholamines in
the vesicles.

d. Tyramine and other related amines displace the noradrenaline
from the storage vesicles due to their structural similarities.
The displaced norepinephrine enters the synaptic cleft and
brings about its physiological action.

7. Functions of Epinephrine and Norepinephrine

a. Norepinephrine is a neurotransmitter (neurohormone) of the
sympathetic nervous system. Epinephrine is the corresponding
adrenal medullary hormone.

b. Norepinephrine is a generalized vasoconstrictor, while
epinephrine's vasoconstriction is restricted to the arterioles
of skin, mucous membrane, and splanchnic viscera. Both
stimulate the heart, although norepinephrine does so to a
lesser degree.

c. Epinephrine brings about the relaxation of smooth muscles such
as those of the stomach, intestine, bronchioles, and urinary
bladder, while contracting the sphincters of the stomach and
bladder. The relaxing action of epinephrine on the bronchiolar
smooth muscle has been used clinically in the relief of
asthmatic episodes.

d. Epinephrine enhances the BMR (basal metabolic rate);
glycogenolysis in liver and muscle (mediated by cAMP and
involving reactions shown on pages 234 and 236); and release
of free fatty acids from adipose tissues (see pages 680 and
682), thus increasing the fatty acids of blood. Norepinephrine
brings about a similar metabolic effect, but to a lesser extent.

8. Catabolism of Epinephrine and Norepinephrine

  a. Following is the catabolic pathway for epinephrine and norepinephrine.

3-methoxy-4-hydroxymandelic acid
(also called vanillylmandelic acid, VMA)

COMT = catechol-O-methyltransferase

MAO = monoamine oxidase

b. All of the intermediates, including epinephrine and
    norepinephrine, can form conjugates with glucuronic acid and
    sulfate, in addition to methylated glucuronides and sulfates.

c. The <u>major</u> metabolites that appear in urine (as glucuronides and
    sulfates) are VMA, metanephrine, and normetanephrine.

d. Monoamine oxidase (MAO) is a mitochondrial enzyme which shows
    a broad specificity. It can be inhibited by hydrazides
    $$\begin{array}{cc} H & H \\ | & | \end{array}$$
    (R-N-N-R), which are used clinically as antidepressants.

e. Catechol-O-methyltransferase (COMT) also has a broad
    specificity and transfers methyl groups from S-adenosylmethionine
    (see page 471) to a variety of catecholamine intermediates.
    It is a cytoplasmic, sulfhydryl enzyme which requires Mg^{++} for
    its activity.

f. Adrenal medullary catecholamines are catabolized (inactivated)
    in the liver by COMT and MAO. Neuronal mitochondrial MAO,
    as well as the MAO and COMT of the effector cells, inactivates
    excess norepinephrine.

Comments on the Biosynthesis of Thyroxine (T_4) and Triiodothyronine) (T_3) (Figure 58)

1. The thyroid gland weighs about 30 g in an adult and is made up of
    many spherical follicles which are well vascularized with blood and
    lymphatic vessels. The gland's total activity is relegated to
    endocrine function. Each follicle is surrounded by a single layer
    of low, cuboidal epithelial cells. The follicles contain the thyroid
    hormones, T_3 (triiodothyronine) and T_4 (thyroxine), bound to
    thyroglobulin (TG) which, in histological studies, appears as a
    colloidal substance.

2. In addition to providing T_3 and T_4, the thyroid gland is also the
    source of calcitonin, a polypeptide hormone involved in calcium
    homeostasis (see calcium metabolism, page 698).

Figure 58.   <u>Biosynthesis of the Thyroid Hormones (Thyroxine and Triiodothyronine)</u>

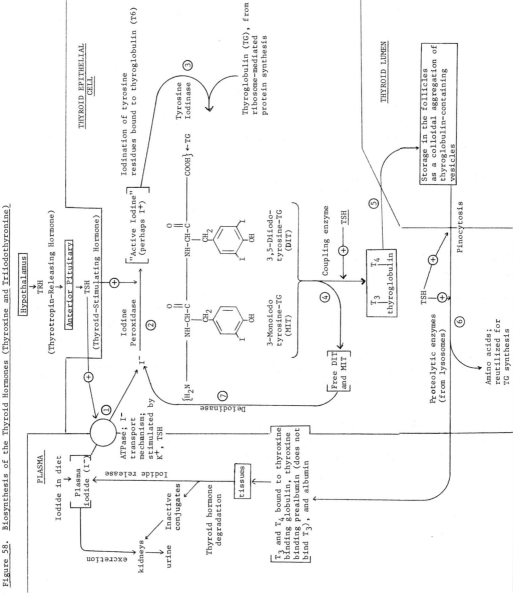

3. The structure of a follicle is shown in Figure 59.

Figure 59. Diagrammatic Representation of a Thyroid Follicle

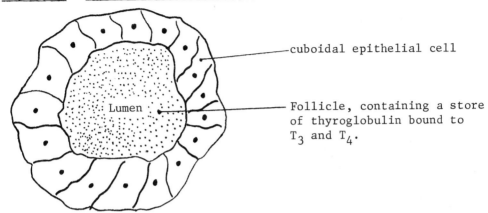

cuboidal epithelial cell

Follicle, containing a store
of thyroglobulin bound to
T_3 and T_4.

Lumen

Note: In a hypothyroid state (e.g., that produced by I^- deficiency),
the follicles are enlarged and the epithelial cells are flattened
(squamous type). In a hyperthyroid state (caused, for example,
by excess TSH secretion),the follicles are small, contain little
colloidal substance, and are linked with columnar cells.

4. The pathway of the synthesis of T_3 and T_4 is shown in Figure 58.
The structures of these hormones are on page 495. Some of the
details of these steps are

a. The normal intake of iodine as iodide (about 100-200 mg/day) is
absorbed from the small intestine into the portal blood
circulatory system. About 1/3 of this is taken up from the
plasma by the thyroid gland, using an iodide pumping mechanism
which works against a concentration gradient. The remaining
iodide (about 2/3) is eliminated in the urine. The trapping or
concentrating of iodide by the thyroid cells is an energy-
dependent process and respiratory inhibitors such as cyanide
and azide, or uncouplers of oxidative phosphorylation such as
dinitrophenol, inhibit the process. Transport of I^- (reaction 1,
in Figure 58) is competitively inhibited by anions of
thiocyanate (SCN^-) and perchlorate (ClO_4^-). Thyroid-
stimulating hormone, also known as thyrotropin, released from
the anterior pituitary (adenohypophysis), promotes iodide
uptake by the gland. TSH is a glycoprotein which also mediates
several other reactions in the biosynthesis and release of T_3
and T_4. The action of TSH may be brought about through the
hormone's ability to increase the level of cAMP. TSH also
stimulates cell proliferation in the thyroid. If T_3 and T_4

levels do not rise in response to TSH, abnormal enlargement
of the thyroid (goiter) occurs.

b. Within the thyroid cell the iodide is oxidized to an active
form of iodine (reaction 2), a process catalyzed by a peroxidase.

$$H_2O_2 + 2\ I^- + 2\ H^+ \xrightarrow[\text{TSH}]{\substack{\text{Iodine} \\ \text{Peroxidase}}} 2\ \text{"active I"} + 2\ H_2O$$

This reaction is inhibited by many HS-containing compounds
such as: 2-thiouracil, thiourea, sulfaguanidine, propyl-
thiouracil, 2-mercaptoimidazole, 5-vinyl-2-thiooxazolidone
(present in yellow turnips), and allylthiourea (present in
mustard). These compounds are known as goitrogens or
antithyroid compounds. Not only do they inhibit the peroxidase-
catalyzed reaction, but they also inhibit the iodination of
tyrosine residues and the coupling reaction of MIT and DIT to
produce T_3 and T_4. A probable mechanism of action for these
compounds is their binding with the free-radical intermediates
produced in the reactions. These compounds thus inhibit the
formation of T_3 and T_4 and, consequently, diminish their output
from the thyroid, resulting in low plasma levels of the hormones.
This leads to enhanced release of TSH by the pituitary gland
(in response to low plasma concentrations of T_3 and T_4)
resulting in a compensatory hypertrophy of the thyroid gland.
This type of enlargement of the gland, in the absence of any
inflammation or malignancy, is known as goiter.

c. Iodination of tyrosine residues present on the thyroglobulin
molecule. Thyroglobulin is a glycoprotein (molecular weight
660,000 daltons) containing about 140 tyrosine residues. It is
synthesized on the ribosomes in the classical protein-synthesizing
pathway and transferred to small vesicles where the tyrosine
gets iodinated (reaction 3). The vesicles are subsequently
stored in the follicles (reaction 5) until their release is
stimulated by TSH. Of the 140 tyrosines in one TG molecule,
about 10 are iodinated to MIT and 5-10 are iodinated to DIT,
even when iodine supplies are adequate.

```
    "active I"                          monoiodotyrosine-TG (MIT)

        +          ─────────────────>            +

    Tyrosine residues                   diiodotyrosine-TG (DIT)
    of Thyroglobulin
```

d. The coupling reaction of DIT with DIT and DIT with MIT, giving rise to T_4 and T_3 (respectively), involves condensation of two substituted phenols to give a corresponding diphenyl ether and a three-carbon compound (possibly serine or alanine) (reaction 4). This coupling process takes place <u>on</u> the thyroglobulin molecule and both the hormones and the three-carbon residue remain attached to the TG molecule until the hormones are required by the body. Of the relatively few MIT and DIT residues in each TG, only about 2-4 of them are used to form T_4. Factors which control the coupling and iodination are probably steric in nature, relating to the location of the tyrosine in the protein.

e. When there is a need for T_3 and T_4, the thyroglobulin reenters the thyroid cells by pinocytosis (a TSH- mediated step). There it undergoes proteolysis catalyzed by lysosomal proteases (reaction 6). T_3 and T_4 are released into the blood where they are transported, bound to specific carrier proteins and to albumins.

f. In the coupling reaction and proteolysis of TG, small amounts of MIT and DIT are released. The iodine contained in these molecules is recovered by the action of deiodinase and made available to the iodide pool of the cell (reaction 7). This reutilization process is quantitatively significant, in that the deficiency of the enzyme, deiodinase, can lead to hypothyroidism due to inadequate iodine supply.

5. Structure and Functions of T_3 and T_4

a. <u>Structures.</u> T_3 and T_4 are derivatives of thyronine.

Thyroxine (T_4) is 3,5,3',5'-<u>tetra</u>iodothyronine.

Triiodothyronine (T_3) is 3,5,3'-<u>tri</u>iodothyronine.

b. T_3 and T_4 are carried in the blood bound **non-covalently to** several proteins. Thyroxine-binding globulin (TBG), thyroxine binding **pre-albumin** (TBPA), and albumin (to a small extent) carry T_4, while T_3 is transported by TBG and albumin but not by TBPA. The ratio of free to bound hormone in the **blood stream** actually determines the extent of the tissue response. T_3 is five times more active than T_4 and it has been suggested that T_3 is the active species <u>in vivo</u>. The half-life of T_4 is 10 days and that of T_3 is 1.5-3 days. The PBI (protein-bound iodine) value is a measure of the amount of $T_3 + T_4$ carried by the serum proteins.

c. T_3 and T_4 have many biological actions. The best-known one is their ability to stimulate oxygen consumption, thereby causing a rise in the overall metabolic rate. They are essential for normal growth and have profound effects on nearly every aspect of metabolism. For example, they enhance free fatty acid release from adipose tissue; exert a direct effect on protein synthesizing systems; influence energy-producing reactions by the swelling of mitochondria and the uncoupling of oxidative phosphorylation; etc. Efforts are still being made to understand the primary function of thyroxine.

6. There are a number of diseases associated with defects in thyroid metabolism.

a. <u>Hypothyroidism</u>, a lack of thyroid hormones, may be caused by many factors. Some of the possible causes are iodine deficiency, administration of goitrogenic substances (already discussed), defects in the enzymes involved in hormone synthesis, autoimmune thyroiditis (antibodies are formed against the body's own tissue, in this case, the thyroid, etc.).

b. <u>Hyperthyroidism</u>, an excess of thyroid hormone, may also be caused by many factors including: a defect in hormonal feedback, adenoma, abnormal thyroid hormone, etc. Drugs that are used in the treatment of hyperthyroidism are inhibitors of synthesis of T_3 and T_4, already mentioned.

c. Abnormalities in iodothyronine binding to serum proteins have also been observed.

7. Summary of the formation of T_3 and T_4

a. Active uptake and concentration of iodide in the thyroid gland.

b. Conversion of I^- to the active elemental form of iodine.

c. Iodination of some of the tyrosine residues of thyroglobulin.

d. Coupling of DIT and MIT residues to form T_3 and T_4.

e. Storage of thyroglobulin in the lumen of the thyroid follicle.

f. Reentry of thyroglobulin by pinocytosis and cleavage with the help of lysosomal proteolytic action to release T_3, T_4 and iodotyrosines.

g. Recovery of iodide from iodotyrosines and transport of T_3 and T_4 in the blood complexed with specific binding proteins.

8. Regulation of T_3 and T_4 production is depicted in Figure 60.

Figure 60. Regulatory Mechanisms Involved in Thyroid Hormone Control

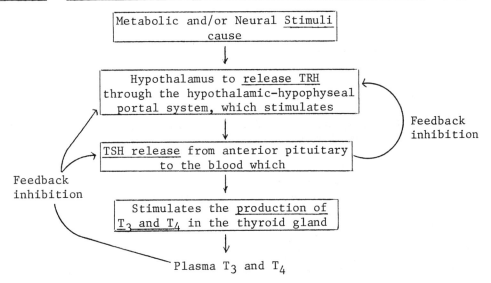

a. The arrows in the figure indicate the self-regulatory feedback mechanism which controls the secretion of the glands. The plasma levels of T_3 and T_4 regulate TSH secretion. When T_3 and T_4 levels decrease, TSH secretion goes up and vice versa. T_3 and T_4 also regulate the TRH secretion in an analogous manner. TRH and TSH levels show a similar reciprocal relationship to each other.

b. TRH (thyrotropin-releasing hormone) has been isolated and purified from porcine and ovine sources. It is a tripeptide with the structure pyroglu-his-proline (NH_2).

Pyroglu ———— His ————Pro-NH$_2$

Pyroglutamic acid is an internal, cyclic lactam of glutamic acid. It is also found in gastrin, one of the G.I. hormones. Pro-NH$_2$ represents the amide of proline, the C-terminal residue of the tripeptide. TRH is released in response to low plasma levels of T$_3$ and T$_4$ and its action on the anterior pituitary leads to TSH release (as well as the release of prolactin, discussed under the chapter on hormones). This release appears to be mediated through cAMP-requiring reactions. In vitro , TSH release can be stimulated by cAMP and by theophylline (which inhibits cAMP breakdown).

9. Inactivation of T$_3$ and T$_4$. These hormones are inactivated by a variety of chemical reactions taking place in the liver, although the liver appears to not be essential for these reactions to occur. Some of these processes are: deiodination; conjugation of the phenolic group with glucuronic acid; oxidative deamination and decarboxylation of the alanine sidechain; O-methylation; and formation of sulfate esters.

10. Laboratory evaluation of thyroid function with respect to T$_3$ and T$_4$. The diverse physiological effects of T$_3$ and T$_4$ makes it essential to perform several different assays in the assessment of their function in order to reflect different aspects of metabolism.

   a. The basal metabolic rate (BMR) measures oxygen consumption. It is, however, altered by other factors in addition to T$_3$ and T$_4$.

   b. Protein-bound iodine (PBI): PBI values are approximately proportional to circulating T$_3$ and T$_4$. This is because the circulating hormones are associated with carrier proteins (already discussed). When the proteins are precipitated, the hormones are carried down along with them. The measurement of protein-precipitable iodine is used frequently in the evaluation of thyroid function. The normal adult range of PBI = 3.2-7.2 µg/100 ml of plasma.

c. Measurement of T_4 by competitive protein-binding method. This method (discussed in detail under cAMP) consists of incubating plasma samples with human thyroxine-binding globulin which is saturated with radioactive T_4. Plasma T_4 (non-radioactive) will displace from the globulin an amount of radioactive T_4 which is proportional to the T_4 concentration in the plasma. The freed, radioactive T_4 is separated from the bound T_4 by using a Sephadex column and the radioactivity is measured in the eluate. By running appropriate standard curves, the determination of unknown T_4 levels in the plasma can be accomplished.

d. Uptake of radioiodine. Anatomic localization of a functioning thyroid gland can be accomplished with the aid of I^{125} or I^{131}, due to the fact the tissue actively and rapidly accumulates iodine. Following ingestion of radioactive iodine, scanning the thyroid area with a radiation detector will reflect the metabolic state of the gland. The higher the **thyroid activity, the more rapidly the iodine will be accumulated there and the greater the radioactivity will be after a given** period of time, Another isotope, ^{99m}Tc (technetium; usually administered as a pertechnate-TcO_4^--salt), is also accumulated by the iodine-trapping mechanism of the thyroid. Consequently, it can be used in place of radioiodine in a similar scanning procedure.

Metabolism of Tryptophan

Comments on Tryptophan Metabolism

1. Tryptophan is an essential amino acid and is required for the synthesis of body proteins. It undergoes reversible deamination, producing indolepyruvic acid, which accounts for the fact that L-tryptophan can be replaced with either the keto acid (indole pyruvic acid) or D-tryptophan (D-tryptophan may be converted to indole pyruvic acid which in turn may be converted to L-tryptophan).

2. Tryptophan is involved in the synthesis of several important compounds.

a. Nicotinic acid (amide) is a vitamin, required in the synthesis
    of NAD$^+$ and NADP$^+$, the cofactors of many dehydrogenase reactions.
    It has been found in humans that 60 mg of tryptophan gives rise
    to 1 mg of nicotamide.

i) <u>Synthesis of NAD$^+$</u> (and some related reactions)

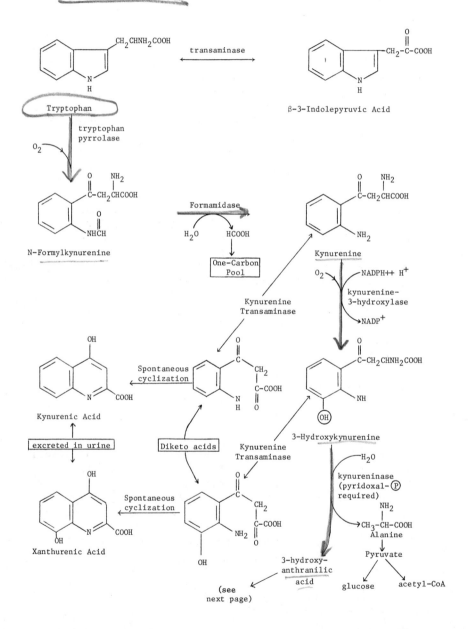

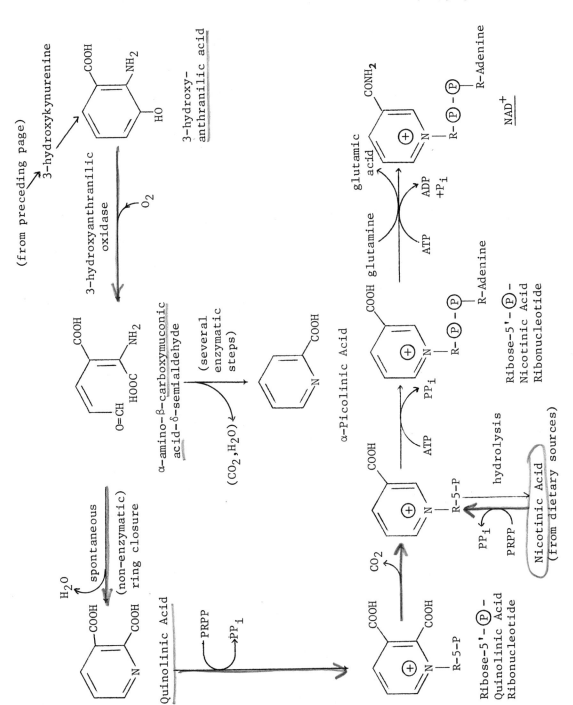

ii) In the synthesis of nicotinic acid from tryptophan, the
first reaction is tryptophan → N-formylkynurenine. This
reaction is catalyzed by tryptophan pyrrolase, an
inducible iron-porphyrin enzyme found in the liver.
Molecular oxygen is incorporated into the product in this
reaction.

iii) Deficiency of vitamin B_6 (pyridoxine) results in a
deficiency in the production of nicotinic acid. This
happens because kynureninase which catalyzes the
reaction:

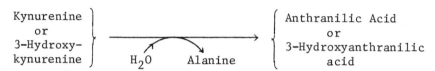

is dependent upon B_6-PO_4 and is very sensitive to its
deficiency. In the absence of this vitamin, the above
reaction will not take place and kynurenine and
3-hydroxykynurenine are converted to the corresponding
diketo acids by transaminase reactions. These are
eliminated in the urine. Note: Although the transaminase
reactions are also B_6-PO_4 dependent, it appears that
kynureninase is more sensitive to B_6 deprivation.

iv) In the absence of B_6 and tryptophan in the diet, normal
synthesis of NAD^+ and $NADP^+$ can be achieved if there is
an adequate dietary supplement of nicotinic acid.
However, a diet which is deficient in protein is also
likely to be deficient in vitamins (including niacin and
nicotinamide). In the nutritional deficiency disease
pellagra, there is a lack of both protein (and hence
tryptophan) and the vitamins niacin and nicotinamide.

b. Serotonin

i) Serotonin is a biogenic amine and is formed by the
decarboxylation of 5-hydroxytryptophan.   It is a powerful
vasoconstrictor and stimulator of smooth muscle
contraction. Serotonin has also been shown to have effects
on cerebral activity and has been implicated as a
neurohormonal substance.

ii) Serotonin undergoes oxidative deamination as follows:

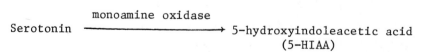

5-HIAA, the detoxified product of serotonin, is excreted in the urine. Other derivatives of 5-HIAA, such as glycine, acetyl and glucuronide conjugates, have been identified in the urine, particularly in individuals with carcinoid disease.

iii) Many inhibitors of monoamine oxidase have been reported, an example of which is the isopropyl derivative of isonicotinic acid hydrazide (iproniazid). These inhibitors lengthen the action of serotonin by stopping or delaying its inactivation, thus presumably giving rise to a stimulation of cerebral activity (see also comments on catecholamine biosynthesis, page 491).

c. Melatonin is a hormone of the pineal gland whose synthesis is shown in Figure 61. Its presumed function has been mentioned under albinism (phenylalanine-tyrosine metabolism).

3. The conversion of tryptophan to skatole and indole takes place predominantly in the large intestines due to the action of bacterial enzymes. The odor of the feces is partly attributed to these molecules.

4. Hartnup Disease

a. This condition is characterized by a defect in the intestinal and renal transport of tryptophan and other amino acids which are excreted in large quantities in urine and feces.

b. In this hereditary disorder, the nature of the biochemical lesions are not known, but a deficiency of tryptophan pyrrolase may be involved.

c. Individuals with this deficiency also excrete indolylacetic acid and indican (3-(β-glucosides)indole). These individuals also have decreased ability to produce kynurenine and nicotinic acid. The major clinical manifestations of this disease include a pellagra-like rash, mental retardation, and other aberrations.

5. Pathways for the synthesis of melatonin, serotonin, and nicotinamide, as well as other reactions of tryptophan metabolism, are summarized in Figure 61. Note that tryptophan is both ketogenic and glucogenic since products of its catabolism include acetyl-CoA (which is the precursor of the ketone bodies) and alanine (which transaminates to pyruvate). Pyruvate can either contribute to gluconeogenesis or be used to form acetyl-CoA.

Figure 61.   Summary of Some Pathways in Tryptophan Metabolism

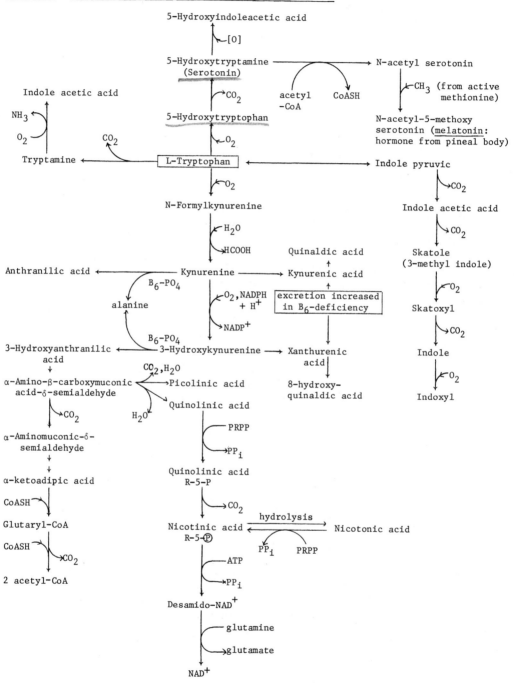

7

Hemoglobin and Porphyrin Metabolism

with John H. Bloor, M.S.

Hemoglobin (Hb) is the protein which, in vertebrates, functions as the principal carrier of oxygen from the lungs to the tissues. It is normally found within the erythrocytes, annucleate cells which are present in the plasma of the circulatory system. The presence of Hb increases the oxygen-carrying capacity of the blood 70-fold.

Some aspects of erythrocytes have already been discussed (see the chapter on carbohydrate metabolism). Hemoglobin will be considered here in detail. To start with, hemoglobin contains iron and in order to understand its structure and function, a knowledge of the chemistry and metabolism of iron is necessary.

A Brief Review of the Coordination Chemistry of Iron Complexes

1. In the formation of a coordinate covalent bond, both the electrons of the bonding pair are supplied by one of the atoms. Transition metal ions (for example, iron ions) accept electrons and form bonds to give rise to complex ions. In a general sense, the transition metal ions may be considered as Lewis acids. In the Lewis sense, acids are electron pair acceptors and bases are electron pair donors. The atom, ion, or molecule that donates the electron pair to form a coordinate covalent bond with a transition metal ion is known as a <u>ligand</u>. Thus, the ligand is a Lewis base and the metal is a Lewis acid.

2. Metal atoms (or ions), in forming complex ions, try to reach a noble-gas electron configuration by accepting electron pairs from ligands. The iron atom, which has 26 electrons, can attain the noble-gas configuration of krypton by accepting 10 electrons.

Examples:

Form of Metal	Number of e^-	Complex molecule or ion	Number of e^-
Fe atom	26	$Fe(CO)_5$	$26 + 10 = 36$
Fe^{+2}	24	$Fe(CN)_6^{-4}$	$24 + 12 = 36$
Fe^{+3}	23	$Fe(CN)_6^{-3}$	$23 + 12 = 35$

Note that when Fe^{+3} forms $Fe(CN)_6^{-3}$, the complex ion does not have the noble-gas configuration of krypton.

3. It is assumed that the electron pairs donated by the ligands enter hybrid orbitals associated with the iron atom. Iron has a coordination number of 6 (can form a complex with six ligands). The orbital hybridization on the iron is d^2sp^3 which gives rise to octahedral complex ions having six bonds. The bonding orbitals are formed from two 3d, one 4s, and three 4p orbitals. These orbitals all have about the same energy, this being necessary for the hybridization to take place. Distribution and arrangement of electrons in the d^2sp^3 orbitals of iron is shown in Table 18. Note that in each atom of iron in the table, there is an inner core of 18 electrons having the configuration $1s^2$, $2s^2$, $2p^6$, $3s^2$, $3p^6$. Thus, the total number of electrons in each iron is 18 plus the number indicated by the small, vertical arrows in the table.

Note that in Table 18, the ferrocyanide ion ($Fe(CN)_6^{-4}$, containing Fe^{+2}) is diamagnetic (all electrons are paired) and that ferricyanide ($Fe(CN)_6^{-3}$, containing Fe^{+3}) is paramagnetic (having an unpaired electron).

The octahedral complex $Fe(H_2O)_6^{+2}$ contains four unpaired electrons and presumably forms sp^3d^2 hybrid orbitals (known as an outer complex; involves 4d orbitals) instead of d^2sp^3 hybrid orbitals (known as an inner complex; uses 3d orbitals), as in the case of $Fe(CN)_6^{-4}$. Apparently the type of hybrid orbitals formed depends upon the strength of the Lewis base (the ligand). For example, CN^- and O_2 are stronger Lewis bases than H_2O. Strong Lewis bases tend to repel electrons more strongly. This causes the electrons to pair up (when possible) and results in inner complex formation. Because of this, outer complexes are also known as high-spin complexes and inner complexes are called low-spin complexes. These terms both refer to the number of unpaired electron spins.

Metabolism of Iron

1. An adult human (70 kg) contains about 4.0-4.5 g of iron which is distributed approximately as follows:

    in hemoglobin--about 2.6-3.0 g

    in ferritin and hemosiderin------------------------1.0-1.5 g
    (storage forms of iron)

    in myoglobin, transferrin (specific ------------0.1-0.2 g
    iron-binding plasma β_1-globulin),
    cytochromes, peroxidase, catalase,
    tryptophan pyrrolase, etc.

Table 18. Electronic Configurations of Some Oxidation States and Complexes of Iron

	3d	4s	4p	4d
Fe atom	(↑↓)(↑)(↑)(↑)(↑)	(↑↓)	( )( )( )	( )( )( )( )
Fe⁺² (gas; high-spin complexes)	(↑↓)(↑)(↑)(↑)(↑)	( )	( )( )( )	( )( )( )( )
Fe⁺² (low-spin complexes)	(↑↓)(↑↓)(↑↓)( )( )	( )	( )( )( )	( )( )( )( )
Fe(CN)₆⁻⁴	(↑↓)(↑↓)(↑↓)[(↑↓)(↑↓)	(↑↓)	(↑↓)(↑↓)(↑↓)] d²sp³ hybrid orbitals	( )( )( )( )
Fe⁺³ (gas; high-spin complexes)	(↑)(↑)(↑)(↑)(↑)	( )	( )( )( )	( )( )( )( )
Fe⁺³ (low-spin complexes)	(↑↓)(↑↓)(↑)( )( )	( )	( )( )( )	( )( )( )( )
Fe(CN)₆⁻³	(↑↓)(↑↓)(↑)[(↑↓)(↑↓) ←unpaired electron	(↑↓)	(↑↓)(↑↓)(↑↓)] d²sp³ hybrid orbitals	( )( )( )( )
Fe(H₂O)₆⁺²	(↑↓)(↑)(↑)(↑)( )	[(↑↓)	(↑↓)(↑↓)(↑↓) sp³d² hybrid orbitals	(↑↓)(↑↓)]( )( )
*Fe(His)(Py)₄(O₂)⁺²	(↑↓)(↑↓)(↑↓)[(↑↓)(↑↓)	(↑↓)	(↑↓)(↑↓)(↑↓)] d²sp³ hybrid orbitals	( )( )( )( )

*This is the configuration of iron in oxyhemoglobin. His = histidyl nitrogen; Py = pyrrole ring nitrogen (part of porphyrin).

2. The amount of iron ordinarily lost from the body is about 1 mg/day, with a variation between 0.7 and 2.4 mg/day. This occurs due to the normal sloughing of the mucosal cells of the duodenum and upper jejunum, exfoliation of the skin, and loss of blood due to gastrointestinal bleeding. Only insignificant amounts of iron are lost in the bile and urine. Iron in the body is combined with proteins and the iron-protein complexes are too large to pass through the kidney glomerular membrane, hence they do not appear in the urine. Iron in the feces is primarily unabsorbed dietary iron.

3. The daily iron loss is made up by ingesting about 12-15 mg of iron in the diet each day. This corresponds to roughly 8% absorption of the metal, a value which is increased in conditions in which there is a low level of plasma or body iron. Such states of enhanced absorption are associated with the growing years, the latter half of pregnancy, female menstruation, extensive loss of blood, iron-deficiency anemias, and hypoxia. In the last three conditions, the low tissue oxygen tension may facilitate the reduction of ferric iron to the ferrous state and hence its transport into the blood. Some typical daily iron requirements are: children (1 mg); adult males (1 mg); menstruating women (2 mg); and pregnant women (2.5 mg).

4. Since there is more iron present in the diet than is required by the body to maintain normal levels, intestinal absorption of iron must be regulated, a process which is not well understood. One reason why the fraction absorbed is so small may be the formation of iron-protein complexes in the diet and in the gastrointestinal tract. Such complexes may not be readily broken down and are too large to be absorbed intact. Iron is recycled continuously in the body. During the breakdown of hemoglobin and other iron-containing proteins, iron is salvaged and reutilized. The body can accumulate abnormal amounts of the metal if there are large amounts of iron in the diet (Bantu siderosis), through repeated blood transfusions, and by excessive parenteral administration of iron compounds.

5. The presence of phosphates and phytates (salts of inositol hexaphosphoric acid) interfere with iron absorption because they frequently form complexes with iron which are insoluble and undissociable.

Figure 62. A Schematic Diagram of Iron Absorption and Distribution in the Body

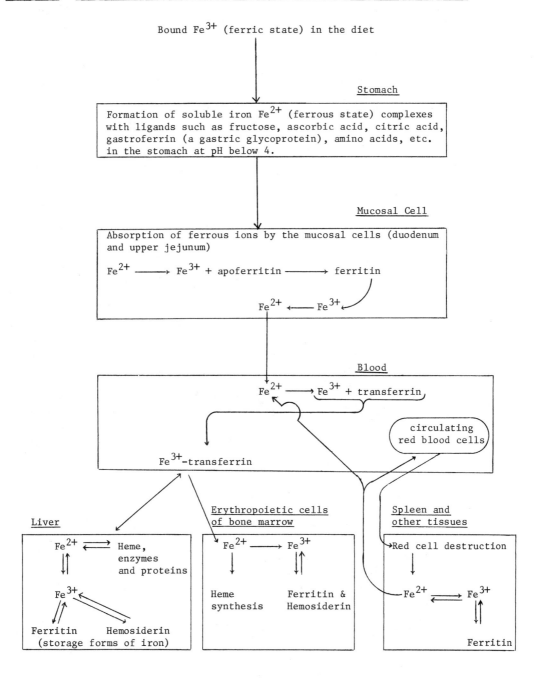

6. Iron storage disorders. Iron can be accumulated in the body due to
    excessive amounts of iron available either in the diet or through
    systemic administration (multiple transfusions). A general increase
    in iron levels in the tissues without any parenchymal cell
    (functional cells of a tissue) damage is known as <u>hemosiderosis</u>.
    (Note: Hemosiderin is a storage form of iron in which ferric
    hydroxide is present as micelles.) Dietary hemosiderosis is seen
    in the African Bantu population, who ingest a great deal of high
    iron (due to cooking in iron utensils) and little phosphate, a diet
    well suited for maximum iron absorption. Hemosiderosis can lead
    to <u>hemochromatosis</u> which is associated with cirrhosis, diabetes
    mellitus, and brown pigmentation. There is also a genetically
    determined disorder known as <u>primary hemochromatosis</u>, where iron
    accumulates in the body far in excess of need. The biochemical
    lesion of this disease is not understood.

7. Laboratory evaluation of serum iron and iron-binding capacity.

    a. Since iron is transported in the blood complexed with transferrin
        (iron-binding protein), it is essential to determine the total
        iron as well as the total iron binding capacity (TIBC) of the
        serum.

    b. The principle used in one method of serum iron determination
        consists of: precipitation of the iron-protein complex and release
        of iron from the complex (accomplished by an acid such as hot
        trichloroacetic acid or acetate buffer pH 5.0), reduction of
        iron from the ferric to the ferrous state (hydroxylamine or
        ascorbic acid), reaction of the iron with a chromogen
        (α,α'-dipyridyl, $2,2',2''$-tripyridine, etc.), and measurement
        of absorbance of the iron-chromogen complex (at 552 nm when
        the chromogen used is tripyridine). Normal range of serum iron

        adult males: 60-150 µg/100 ml

        adult females: 50-130 µg/100 ml

    c. Transferrin in the serum is only partly saturated with iron.
        When it is completely bound with iron and iron is then
        determined, the value obtained will give the total iron binding
        capacity (TIBC). The normal values for TIBC in adults vary
        between 270 and 380 µg/100 ml. These aspects can be illustrated
        in the following diagram.

Total iron-binding capacity (TIBC)

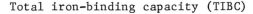

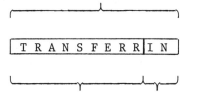

amount of iron bound   unsaturated (or latent)
in a normal serum   iron-binding capacity
                           (UIBC)

TIBC is determined by adding a known excess of iron to a serum
sample, <u>separating</u> the unbound iron and measuring either
unbound iron or iron in the serum by the procedure described
before.

8. Treatment of iron storage diseases

    a. Removal of iron from the body in various types of iron storage
        disorders (and acute iron poisoning) is possible by
        administration of a chelating agent known as <u>deferoxamine</u>,
        isolated from <u>Streptomyces pilosus</u>.

    b. Deferoxamine has the structure shown below.

$$H_2N-(CH_2)_5-\underset{HO}{\overset{}{N}}-\underset{O}{\overset{}{C}}-(CH_2)_5-\underset{O}{\overset{}{C}}-\underset{H}{\overset{}{N}}-(CH_2)_5-\underset{HO}{\overset{}{N}}-\underset{O}{\overset{}{C}}-(CH_2)_2-\underset{O}{\overset{}{C}}-\underset{H}{\overset{}{N}}-(CH_2)_5-\underset{HO}{\overset{}{N}}-\underset{O}{\overset{}{C}}-CH_3$$

        Notice that this molecule has six nitrogens separated by
        fairly long, flexible stretches of methylene groups. Since
        iron forms octahedral complexes (i.e., each iron atom has six
        ligands), one molecule of deferoxamine is probably capable of
        occupying all six coordination sites. This would produce an
        iron-deferoxamine complex having a stoichiometry of 1:1. On
        this basis, the chelater might be compared to a non-cyclic
        stretched-out porphyrin with two extra nitrogens.

    c. Deferoxamine is a highly selective complexing agent. For
        ferric iron, K_{assoc} is about 10^{30}, while K_{assoc} for Ca^{+2} is
        only 10^2. Iron present in hemoproteins is not affected by
        this agent. Such selectivity makes the compound very useful in
        the treatment of iron problems, as its administration does not
        affect the vital functioning of hemoglobin, the cytochromes,
        and calcium metabolism. In addition, the ferric iron of
        ferritin and hemosiderin is chelated in preference to that
        found in transferrin.

d. When deferoxamine is given orally, it complexes with dietary iron, making both the iron and the deferoxamine unavailable for absorption. As a therapeutic agent, the preferred route of administration is parenteral. The compound is excreted in the urine, although its detailed metabolism is unclear at this time. Side-effects of deferoxamine treatment include hypotension (in I.V. administration), pain at the site of administration (intramuscular), and cataract formation during chronic usage.

Structural Aspects of Hemoglobin

1. Mammalian hemoglobins are tetramers, made up of four polypeptide subunits: two α-subunits and two other subunits (usually β, γ or δ). These polypeptide chains differ in the type of amino acids present at specific positions (i.e., in their sequence or primary structure). The normal hemoglobin types are: HbA_1: $\alpha_2\beta_2$ (adult human hemoglobin; most common); HbF: $\alpha_2\gamma_2$ (human fetal hemoglobin); HbA_2: $\alpha_2\delta_2$ (a minor component of normal adult hemoglobin). They are discussed later in somewhat greater detail. The structural and functional aspects of hemoglobin have been worked out almost entirely through studies of HbA_1 and its mutants.

2. Hemoglobin has a molecular weight of about 64,500 daltons and is spherical, due to the remarkable fit of the four subunits. The α-subunits have a molecular weight of about 15,750 daltons and that of the β-subunits is about 16,500 daltons. This arrangement of the subunits is known as the quaternary structure. A useful way of visualizing their orientation is to consider the four polypeptides as being at the four corners of a regular tetrahedron. One should recognize that this is somewhat idealized, since the subunits are not truly spherical and the tetrahedron which their centers describe is not exactly regular. In addition, during binding and release of O_2, the subunits move about, as is discussed later.

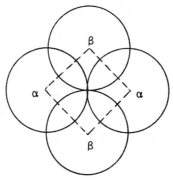

3. In the assembled hemoglobin, contacts between like subunits ($\alpha\alpha$ and $\beta\beta$) are few and limited primarily to six polar interactions between the C-terminal residues and other amino acids in the peptides. These interactions are mediated by salt bridges which can be pictured as (short) chains of salt (e.g., Na^+Cl^-, $(NH_4^+)_2SO_4^=$) molecules (really, ion pairs) between two groups of opposite charge. For example:

$$-NH_3^+ \overbrace{(Cl^-Na^+)}\overline{O}OC-$$

(This is an oversimplification, since there will be many other ions present in the solvent which will also interact with the charged groups. A more accurate view might be that of a <u>charged</u> <u>"atmosphere"</u> of ions surrounding both groups with the charge cloud polarized by the charges on the polypeptide.) Despite this paucity of interactions, the importance of those which do exist will be seen later, in the discussion of the mechanism by which hemoglobin functions.

In deoxyhemoglobin, there is no β-β contact (this is significant in explaining the role of 2,3-DPG). Unlike chains are more strongly joined, mostly via van der Waal's (hydrophobic) interactions. There are probably a few hydrogen bonds which also contribute to these attachments.

These facts are derived principally from the x-ray crystallographic results of Perutz, Kendrew and their coworkers at the Medical Research Council, Laboratory of Molecular Biology in Cambridge, England. They are also, however, supported by other data. Under certain conditions (for example, dilute solutions), normal hemoglobin dissociates in the following way:

There is no evidence for either $\alpha\alpha$ or $\beta\beta$ forming although, in the complete absence of α-chains, β_4 is known. Hemoglobin does not dissociate in red blood cells due to its high concentration.

(Note: M.F. Perutz and J.C. Kendrew received the 1962 Nobel Prize in Chemistry for this work.)

4. In order to understand the mechanism of oxygenation, it is important
    to realize that hemoglobin is not as symmetrical as it appears to
    be. If the two α-and two β-subunits are distinguished from each
    other:

The asymmetry can be described by saying that the $\alpha_1\beta_1$ contact
differs from the $\alpha_1\beta_2$ contact. The contacts $\alpha_1\beta_1$ and $\alpha_2\beta_2$ are the
same, however, as are the contacts $\alpha_1\beta_2$ and $\alpha_2\beta_1$. It has been
shown that when hemoglobin dissociates into dimers, the contact
broken is of the type $\alpha_1\beta_2$ and the one remaining in the dimer is
of the $\alpha_1\beta_1$ sort. This is a mildly confusing notation since $\alpha_2\beta_2$
in one context means HbA_1 (consisting of two α-and two β-subunits),
while in this situation it could mean the second α subunit plus
the second β subunit (although it is not usually used in this way).
The best way to keep this straight is by carefully watching the
context in which the notation is used.

5. a. Figures 63A and B are schematic diagrams of the α-and β-chains,
        respectively, of hemoglobin. Each peptide chain (subunit) is
        made up of helical and non-helical segments, and each surrounds
        a heme group. Hemoglobin is unusual among globular, soluble
        proteins in that almost 80% of its amino acids exist in an
        α-helical conformation. There are two notations used to
        specify the locations of amino acids in the hemoglobin peptides.
        One numbers the residues from the N-terminus of each chain.
        For example, residue α42 of normal adult human hemoglobin is
        tyrosine, while β42 is phenylalanine. The other, devised by
        Kendrew, numbers residues within α-helices from the N-terminal
        amino acid of each of the helices. This is also known as the
        "helical notation". Residues in non-helical regions are
        designated by the letters of the helices at either end of the
        region. Thus, residue α42 would be C7 and β42 is CD1. These
        notations are often combined, yielding α42(C7) or Tyrα42(C7) and
        β42(CD1) or Pheβ42(CD1).

    b. Heme consists of a porphyrin ring system with an Fe^{+2} fixed in
        the center through complexation to the nitrogens of the four
        pyrrole rings. A porphyrin is a planar, resonating (aromatic)
        ring formed from four pyrrole rings linked by =CH-(methene)
        groups. The pyrrole rings may have sidechains, the different
        porphyrins being distinguished on the basis of variations in
        these sidechains. The structure and properties of porphyrins
        are discussed more fully later in this chapter.

Figure 63A. <u>Secondary Structure of the α-Chain of Human Hemoglobin</u>

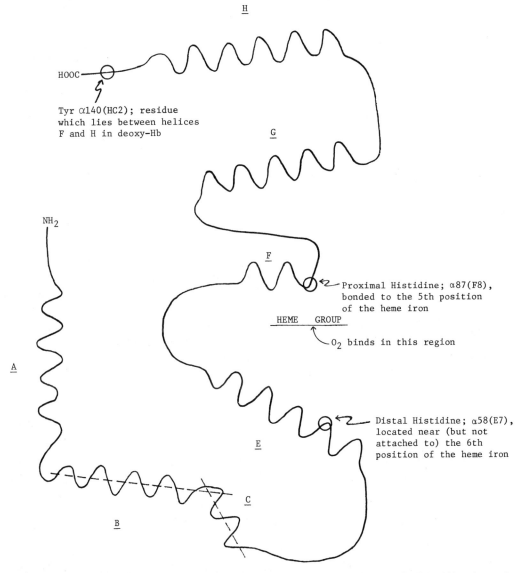

<u>H</u>

HOOC

Tyr α140(HC2); residue
which lies between helices
F and H in deoxy-Hb

<u>G</u>

NH$_2$

<u>F</u>

Proximal Histidine; α87(F8),
bonded to the 5th position
of the heme iron

HEME   GROUP

O$_2$ binds in this region

<u>A</u>

Distal Histidine; α58(E7),
located near (but not
attached to) the 6th
position of the heme iron

<u>E</u>

<u>C</u>

<u>B</u>

The helical regions (labeled A–H, after Kendrew), N– and C-termini, and the histidine located
near the heme group are indicated. The axes of the B and C helices are indicated by
dashed lines.

Figure 63b. Secondary Structure of the β-Chain of Human Hemoglobin

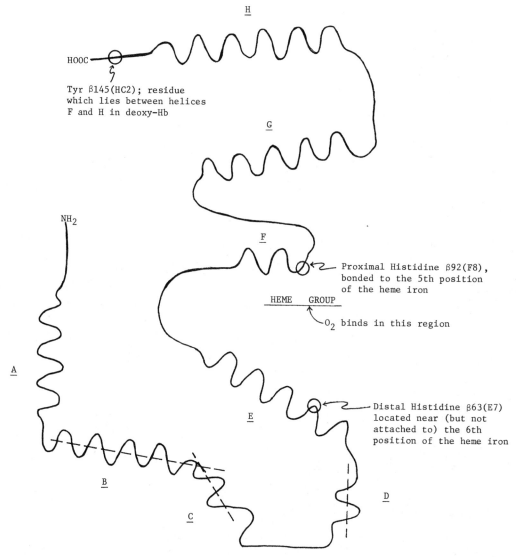

H

HOOC

Tyr β145(HC2); residue
which lies between helices
F and H in deoxy-Hb

G

NH₂

F

Proximal Histidine β92(F8),
bonded to the 5th position
of the heme iron

HEME   GROUP

O₂ binds in this region

A

E

Distal Histidine β63(E7)
located near (but not
attached to) the 6th
position of the heme iron

B

D

C

The helical regions (labeled A-H, after Kendrew), N- and C-termini, and the histidines located
near the heme group are indicated. The axes of the B, C, and D helices are indicated by
dashed lines.

c. Iron in the ferrous state has a coordination number of six (each atom of iron can bind with six electron pairs). Therefore, in the heme molecule, two of iron's coordination positions are still unoccupied. When heme is associated with the peptide, a histidyl nitrogen (from the so-called proximal histidine) of the peptide bonds with the fifth coordination position of iron, leaving the sixth position open for combination with oxygen, water, carbon monoxide, or other ligands.

6. a. The peptide chain folds around and protects the heme groups, which are located in a crevice near the surface of the subunit structure. As indicated in Figures 63A and B, the heme lies between helices E and F. One side of the heme group (the distal side) is open for reversible combination with oxygen. This open part of the molecule is surrounded by hydrophobic groups, resulting in a microchemical environment of low dielectric constant. This prevents O_2 from permanently changing Fe^{+2} to Fe^{+3} and is responsible for the ability of hemoglobin to bind oxygen reversibly. Normally, when Fe^{+2} combines with O_2, the Fe^{+2} is oxidized to Fe^{+3} and Fe^{+3} is not capable of binding reversibly with O_2. There is some evidence (from electron paramagnetic resonance) that, in oxygemoglobin, the oxygen actually does oxidize the Fe^{+2} to Fe^{+3} <u>as long as the O_2 is bound</u>. When the O_2 dissociates, it takes all of its electrons with it, causing the iron to return at once to Fe^{+2}. This transient oxidation does not seriously affect Perutz's model (discussed later) for Hb function, since the dimensions for the Fe^{+2} radius changes were experimental values. What is unique about hemoglobin is not the ability to bind oxygen; it is the ability to bind oxygen <u>reversibly</u>. If the Fe^{+2} in hemoglobin is permanently oxidized to the ferric state (Fe^{+3}), the Fe^{+3} becomes bound tightly to an OH^- group and O_2 will not bind.

b. The heme groups can be removed from hemoglobin by dialysis. Hence, they are not covalently attached. Not counting the proximal histidine (Hisβ92(F8) or α87(F8)), about 60 amino acids contact (come within 4 Å of) one or more of the atoms of the heme group. This is roughly the maximum length of an effective hydrogen bond or hydrophobic interaction. Of these sixty contacts, all but one in the α-subunit and two in the β-subunit are non-polar. This emphasizes the highly hydrophobic environment of the hemes. These polar interactions all involve the carboxyl groups of the propionic acid sidechains on the hemes.

Functional Aspects of Hemoglobin

1. The primary function of Hb is to transport oxygen from the lungs
    to the tissues where it is utilized. Hemoglobin forms a
    dissociable hemoglobin-oxygen complex which can be written as
    follows:

    $$\text{deoxyhemoglobin} + O_2 \rightleftharpoons \text{oxyhemoglobin}$$

    This relationship follows the law of mass action in that the
    reaction is shifted to the right when there is an increase in
    oxygen pressure (as in the lungs) and to the left when there is
    a decrease in oxygen pressure (as in the tissues). The sketch
    below shows the flow of oxygen and some approximate partial
    oxygen pressures. Recall that a torr (named for Torricelli, the
    inventor of the barometer) is a unit of pressure equal to 1 mm
    of Hg.

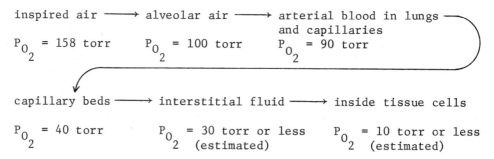

    inspired air $\longrightarrow$ alveolar air $\longrightarrow$ arterial blood in lungs
    and capillaries

    P_{O_2} = 158 torr P_{O_2} = 100 torr P_{O_2} = 90 torr

    capillary beds $\longrightarrow$ interstitial fluid $\longrightarrow$ inside tissue cells

    P_{O_2} = 40 torr P_{O_2} = 30 torr or less P_{O_2} = 10 torr or less
    (estimated)   (estimated)

    It is useful to consider this as the flow of oxygen down a
    continuous pressure (concentration) gradient. One function of
    hemoglobin, then, is to increase the solubility of oxygen in the
    blood, which it does 70-fold. Table 19 summarizes some facts
    concerning erythrocytes and hemoglobin.

2. a. In Figure 64, the sequential change from curve C to B to A
        (termed a "rightward shift" of the oxygen dissociation curve)
        could have been caused by an increase in temperature (as in
        strenuous exercise), a decrease in pH (as in acidosis or
        exercise), or an increase in the 2,3-DPG concentration
        (discussed later), as well as by an increase in P_{CO_2}. As a
        result of such a change, at a fixed P_{O_2} the per cent saturation
        of the hemoglobin in the blood stream decreases. If the
        blood had initially been saturated to the same extent in all
        three cases (for example, by oxygen at a pressure of 95 torr,
        indicated in the figure as P_{O_2} (arterial)), such a shift would
        mean, physiologically, that the tissues would receive an
        increasing amount of oxygen in terms of number of moles

Table 19.   Some Statistics Concerning Human Erythrocytes and Hemoglobin
            in Normal Adults

Property	Male	Female
Hematocrit (% of whole blood volume occupied by erythrocytes)	40–54	37–47
Red cell concentration (millions of cells/microliter of whole blood)	4.5–6.0	4.0–5.5
Mean corpuscular hemoglobin concentration (g Hb/100 ml cells)	32–36	32–36
Mean corpuscular volume (microns^3)	80–94	80–94
	Arterial Blood	Venous Blood
Dissolved oxygen (ml* O_2/100 ml of blood containing 15 g of Hb)	0.29	0.12
Combined oxygen (HbO$_2$) (ml* O_2/100 ml of blood containing 15 g of Hb)	19.5	15.1

* Temperature, pressure, and partial pressure of H_2O under which these
  gas volumes were measured was not given.

Figure 64.   Dissociation Curves of Hemoglobin at Several CO_2 Pressures and of Myoglobin

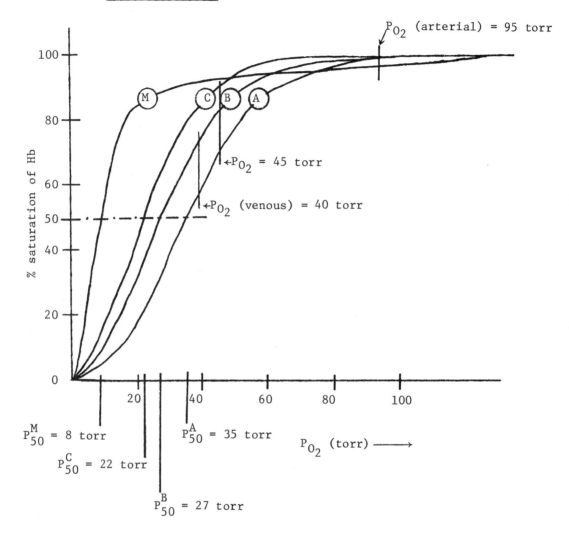

Curve A: P_{CO_2} = 80 torr; Curve B: P_{CO_2} = 40 torr; Curve C: P_{CO_2} = 20 torr; Curve M is for Myoglobin. P_{O_2} (arterial) and P_{O_2} (venous) are normal physiological values for, respectively, arterial and venous oxygen tensions. During exercise, P_{O_2} (venous) will normally be lower, around 20–25 torr.

delivered in a given time. On the other hand, if the arterial pressure was only about 45 torr (indicated in the figure), a rightward shift would decrease both the initial and final per cent saturations to roughly the same extent and, on a molar basis, the tissues would receive almost the same amount of oxygen regardless of the location of the curve. It is important to consider these factors when evaluating the significance with respect to altered oxygen transport (i.e., oxygen delivery to the tissues) of any shifts of the oxygen dissociation curve.

b. For example, it was observed that in going from a low altitude (10 meters above sea level) to a higher altitude (4509 meters above sea level), the 2,3-DPG concentration within the erythrocytes increased by 25-30% within 24 hours after the altitude change and the oxygen dissociation curve shifted to the right (P_{50} increased). About 50% of the change in 2,3-DPG concentration and P_{50} occurred within 15 hours of the change in altitude. This was interpreted as evidence that 2,3-DPG plays a central role in the body's adaption to the hypoxic hypoxia of high altitude. Further evaluation of the experiment revealed, however, that the arterial P_{O_2} also changed, from about 94 torr at 10 meters to about 45 torr at 4509 meters, due to a decrease in atmospheric oxygen pressure from 149 torr to 83 torr. Thus, due to the altitude increase, the arterial pressure decreased to a point where the supply of oxygen to the tissues was largely independent of the position of the curve. Once this was recognized, it was shown that the principal adaptive factor was an increased respiratory rate. Increases in hematocrit and mean corpuscular hemoglobin concentration were also observed.

c. There are, however, situations where a shift in the curve may prove useful. For example, in anemia or cardiac insufficiency, where the supply of oxygen to the lungs is normal but the ability of the blood to deliver it to the tissues is impaired, the intraerythrocytic 2,3-DPG concentration has been observed to increase. A rightward shift of the dissociation curve will not affect the per cent saturation during loading in the lungs (since the upper part of the curve is quite flat), but it will decrease the per cent saturation in the tissues, thereby providing an increase in the number of moles of oxygen available per unit time for metabolism. Respiration in a normal individual provides another example. Since P_{CO_2} is higher in the tissues than it is in the lungs (and, consequently, pH is lower in the tissues than in the lungs), oxygen unloading is facilitated at the tissue level and loading occurs more readily in the lungs. The decrease in hemoglobin saturation with increasing P_{CO_2} (and hence,

decreasing pH) is known as the alkaline Bohr effect (since it
occurs maximally at pH 7.4) and the molecular basis of it
will be discussed later. The acid Bohr effect, occurring
below pH 6.0, involves the increase in Hb saturation with
decreasing pH. It perhaps involves a major conformational
change of the Hb subunits which occurs about pH 5.9. It is
not important in vivo.

3. There are several other important features of Figure 64.

a. The pressure at which the blood is 50% saturated with oxygen,
designated P_{50}, depends on the position of the oxygen
dissociation curve of the hemoglobin. Since the shape of these
curves depends only on the molecular nature of the hemoglobin,
all curves for a given hemoglobin belong to the same family
of curves and they cannot cross each other. Consequently,
P_{50} values are a useful measure of the position of the
dissociation curve for a particular hemoglobin: the larger the
P_{50} value, the further "right" the curve is.

b. Notice that the hemoglobin curves are all sigmoid (S-shaped)
while that for myoglobin is a rectangular hyperbola. The
relationship of hemoglobin to myoglobin is the same as that
of an allosteric enzyme to a "normal" (non-allosteric) enzyme
(see the section on enzyme kinetics). In fact, although
hemoglobin is not truly an enzyme, it is the classic and
best understood example of allosteric control.

i) At P_{O_2} values found in tissues, the myoglobin curve lies
to the left of the hemoglobin curves. Because of this,
hemoglobin can oxygenate myoglobin quite readily. This
can also be expressed by saying that in the equation

$$Hb(O_2)_4 + 4Mb \xleftarrow{\hspace{2cm}} Hb + 4MbO_2$$

the equilibrium lies far to the right. Myoglobin is a
heme protein which is quite similar to a single subunit
of hemoglobin. It occurs at high concentrations in
muscle, where it functions as a storage site for oxygen.
As will be seen later in this section, the difference in
its oxygen dissociation curve is due primarily to the
fact that it has but a single subunit, compared to the
four found in hemoglobin.

ii) The sigmoidicity of the Hb curve can be interpreted in
exactly the same way as were the curves for other
allosteric enzymes. The binding of one molecule of oxygen
to a hemoglobin tetramer makes the binding of successive

molecules to the same tetramer easier, up to the limit of four molecules of oxygen per molecule of hemoglobin. This "cooperative binding" of oxygen by hemoglobin is the basis, as indicated above, for the regulation of oxygen (and indirectly carbon dioxide) levels in the body. The description of a molecular basis for this phenomenon is one of the major triumphs of x-ray crystallographic studies of protein structure.

4. The mechanism of oxygenation of hemoglobin (as proposed by Perutz) may be explained as follows.

   a. Hemoglobin appears to have two quaternary structures corresponding to the deoxygenated (deoxy; five-coordinate iron) and the oxygenated (oxy; six-coordinate iron) forms. X-ray crystallographic results have demonstrated that the number of ligands attached to (coordination number of) the iron is the significant difference between these forms. Thus, all six-coordinate hemoglobins (including oxy- and methemoglobin) have one form and deoxyhemoglobin has the other.

   b. Binding of one or more oxygens causes a change in the tertiary structure of the binding subunits, resulting in a quaternary structure alteration which enhances the ability of oxygen to cause tertiary changes in the remaining, unoxygenated (unliganded) subunits. This makes it much easier to bind more oxygen (up to 4 molecules O_2/molecule Hb) after the first one is attached. The best available data indicate that there are two quaternary structures (deoxy and oxy; T and R in the notation of Monod, Wyman, and Changeux) for the tetramer and that, in addition, each subunit can exist in one of two tertiary structures (again deoxy or oxy; sometimes designated t and r). The transition ($t \rightarrow r$) is caused by O_2 binding to a subunit and the transition ($T \rightarrow R$) occurs when one or more (the exact number is not clear) $t \rightarrow r$ transitions have occurred. Thus, there may be structures in which a subunit in the t-state exists as part of a tetramer which is in the R-state, and vice-versa. These may be the partially oxygenated "intermediates" detected by certain studies. It appears clear, however, that the T $\rightleftarrows$ R equilibrium controls the position of the individual t $\rightleftarrows$ r equilibria. That is, the quaternary (tetrameric) structure determines which tertiary (monomeric) structure is preferred.

   c. This mechanism has some features of both the Monod, Wyman, and Changeux (MWC) and the Koshland, Nemethy, and Filmer (KNF) models (see the section on allosterism).

i) The tertiary structure of the subunits is altered only upon binding oxygen, as in the KNF model. Unlike this model, however, the liganding of one subunit does not appear to <u>directly</u> affect the ability of other subunits to bind oxygen (i.e., no "induced fit").

ii) The change in subunit affinity occurs instead by the initial ligands causing a <u>quaternary</u> structure change of the sort described by the <u>MWC</u> model. Unlike the MWC model though, oxygen apparently binds to Hb molecules in the T-state as well as to ones in the R-state. In their original paper, Monod, Wyman, and Changeux state that their assumption of total non-reactivity of the T-state may have been an oversimplification.

5. Although some aspects are still not rigorously proven, a number of molecular details are becoming clear. Perutz has assembled the pertinent information into several articles, reference to which is given at the end of this book. They are very interesting and provide an excellent view into the way in which molecular data can be used to deduce the mechanism by which an enzyme (or protein) functions.

a. The trigger for the tertiary structure change (and ultimately the quaternary alteration) appears to be the iron atoms in the heme groups. Upon binding oxygen, the iron changes from a high-spin, 5-coordinate ferrous ion having a radius of about 2.24 Å to a low-spin, 6-coordinate ferrous ion with a radius of about 1.99 Å. The radius change occurs because the electrons are nearer the nucleus in the low-spin state. As is indicated in Figure 65, this permits the Fe^{+2} to fit more closely into the hole in the porphyrin ring, resulting in a movement which brings the nitrogen of the proximal histidine (the one bound to the fifth position of the iron) closer to the plane of the porphyrin ring by about 0.85 Å. As was indicated before, the porphyrin is in contact with about sixty residues in the globin chain. Such a change in the position of either the histidine, bound rigidly in the peptide backbone, the heme group itself, or both, causes drastic modifications in the positions of the residues to which they are attached. In particular, certain salt bridges which are instrumental in maintaining the deoxy (T) conformation are broken.

Figure 65. Comparison of the Fe^{+2}-Porphyrin Relationship in Oxy- and
Deoxy-Hemoglobin. The large circle represents the iron
atom in the heme. The drawings below are roughly to scale,
although the iron atom is drawn to the same diameter in
both cases.

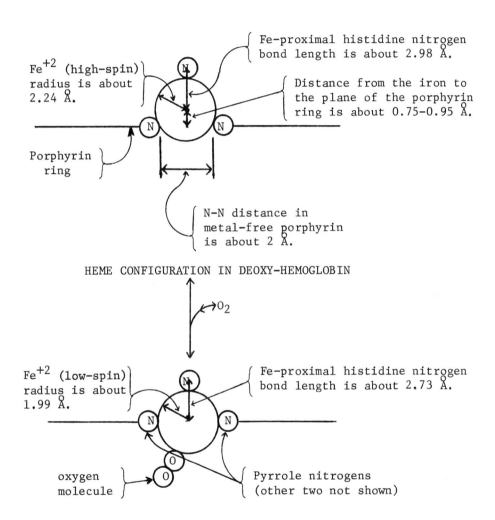

Fe^{+2} (high-spin) radius is about 2.24 Å.

Fe-proximal histidine nitrogen bond length is about 2.98 Å.

Distance from the iron to the plane of the porphyrin ring is about 0.75-0.95 Å.

Porphyrin ring

N-N distance in metal-free porphyrin is about 2 Å.

HEME CONFIGURATION IN DEOXY-HEMOGLOBIN

O_2

Fe^{+2} (low-spin) radius is about 1.99 Å.

Fe-proximal histidine nitrogen bond length is about 2.73 Å.

oxygen molecule

Pyrrole nitrogens (other two not shown)

HEME CONFIGURATION IN OXY-HEMOGLOBIN

b. The sequence of events triggered by the high-spin → low-spin
    change is as follows:

i)   The quaternary deoxy structure is maintained by six salt
     bridges, all involving the C-terminal residues of each of
     the four subunits. ArgHC3(141)α_1 forms two salt bridges
     with residues in subunit α_2; ArgHC3(141)α_2 forms
     corresponding links to residues in subunit α_1; and the
     two HisHC3(146)β residues interact with the opposite
     α-chains (β_1 to α_2; β_2 to α_1). In the absence of these
     salt bridges, the deoxy conformation is unstable and
     spontaneously reverts to the oxy structure (e.g., in
     hemoglobins which lack the C-terminal residues, either
     through mutation or chemical modification). In each
     chain, the penultimate (next to last) residue at the
     C-terminus is tyrosine which in the deoxy form lies in a
     hydrophobic pocket between helices F and H.

ii) When a subunit binds oxygen, the consequent heme movement
     causes helices F and H to approach each other, narrow the
     pocket between them, and expel the penultimate tyrosine.
     The tyrosine drags the C-terminal residue along with it,
     breaking the salt bridges (two if it is an α-subunit, one
     if a β-subunit). In addition, the strain increases in
     the $\alpha_1\beta_2$-type contacts along which the ultimate movement
     will occur in the quaternary change. The word "strain"
     refers to "long" hydrogen bonds, salt bridges, and
     van der Waal's contacts. These long interactions are of
     higher potential energy than the normal length, minimum
     energy ones. Thus, when possible, they shorten to lower
     their energy and thereby bring about movement within the
     molecule. Another tertiary change is apparent in the
     β-subunits. In deoxy-hemoglobin, helix E in these subunits
     is too close to the sixth position of the iron to permit
     the entry of oxygen. Consequently, at some point during
     oxygenation, this pocket opens or is opened by the
     movement of helix E away from the heme and helix F.
     This does not occur in the α-subunits since the distance
     therein between helices E and F is greater.

iii) When a sufficient number of salt bridges have broken and
     the $\alpha_1\beta_2$ strain has built up, the remaining salt bridges
     are inadequate to maintain the deoxy conformation. All
     of the restraining salt bridges now break and the tetramer
     "clicks over" into the oxy form. The exact number of salt
     bridges which must be initially broken by oxygen binding
     is not known and probably depends on factors such as pH,
     temperature, CO_2 concentration, and concentration of

organic phosphates in the solution. This change is
illustrated in Figure 66. The motion can be considered
more or less to be a rotation of $\alpha\beta$ pairs, joined by
$\alpha_1\beta_1$-type interactions, around the α-α axis. The effect
is to move the β-subunits closer to each other in a
pincer-like motion, narrowing the gap between them. It
is clear then why tetramers are <u>required</u> for this
cooperative mechanism to work. Dimers do not possess
the $\alpha_1\beta_2$-type contacts which undergo the change during
quaternary structure alteration. Notice that a change
in <u>tertiary</u> structure brings about a <u>quaternary</u> structure
change.

<u>Figure 66.</u> <u>Diagram of the Gross Quaternary Differences Between Oxy- and</u>
<u>Deoxy-Hemoglobin</u>

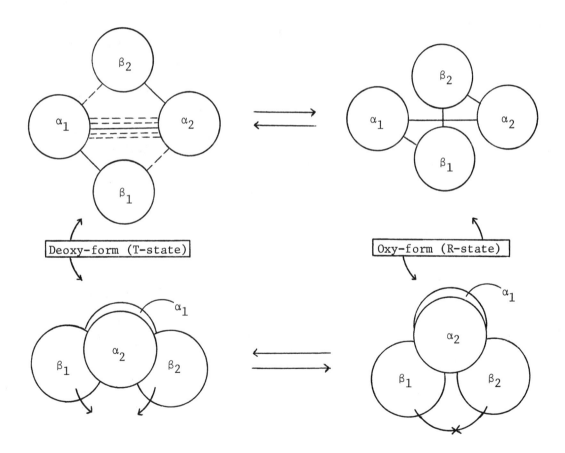

In the upper drawing of Figure 66, the dashed lines indicate the six salt bridges which break during quaternary change, while the solid lines indicate "permanent" (non-covalent) interactions between subunits. Note that in the oxy-form, there is probably a β-β link. The upper pair of drawings is more or less a top view of the tetramer, while the lower pair is a perspective, looking roughly along the α-α axis. It should be recognized that these drawings are highly schematized and that the changes are considerably more detailed and complex than indicated.

iv) The tyrosines in unoxygenated (t) subunits are still held in position in the R-state, but only about half as strongly as they were in the T-state. Consequently, the binding of oxygen to these subunits requires about half the original (deoxy) activation energy and occurs with greatly increased ease. This is the source of the "cooperativity" exhibited by hemoglobin.

v) For some time it has been known that at pH 7.4, hemoglobin releases 0.7 moles of H^+ for each mole of O_2 it binds and that the process is reversible. Since this is considered to be the mechanism by which CO_2 decreases the O_2-affinity of Hb (alkaline Bohr effect), these are referred to as the Bohr protons. This is equivalent to saying that oxy-hemoglobin is more acid than deoxy-hemoglobin. The pK_a for oxy-Hb is 6.62 and the pK_a for deoxy-Hb is 8.18. It appears likely that this change in pK results from the breaking of the six salt bridges and that the groups which release the Bohr protons are those which were involved in the bridges in oxy-hemoglobin. In this way, an increase in P_{CO_2} which causes a decrease in pH, can favor the deoxy structure by stabilizing the salt bridges which maintain that conformation.

vi) Deoxy-hemoglobin is able to bind one mole of organic phosphate (in man, notably 2,3-diphosphoglyceric acid (2,3-DPG) and ATP) per mole of Hb, while oxy-hemoglobin cannot. As was indicated earlier, this is significant in controlling the degree of oxygenation of Hb. From a number of studies, evidence is quite good that at least 2,3-DPG can bind to a group of six positively charged residues on the β-chains located in the central cavity between the β-subunits. Upon oxygenation, some of the positive groups move away and the cavity closes, so that the 2,3-DPG is expelled. (The role of 2,3-DPG is discussed in more detail later in this chapter.) Because the

binding of 2,3-DPG is abolished by the high salt
concentrations generally needed to form single crystals
of Hb for x-ray studies, the actual location of its
interaction has not yet been determined. There may,
in fact, be more than one site. It is significant that
despite its apparent interaction with protonated
nitrogens (Arg, His, Lys, and α-amino groups), 2,3-DPG
does not have any influence over the number of Bohr
protons released during oxygenation.

c. Perutz himself succinctly summarizes these steps (Nature, <u>228</u>,
726 (1970)):

"... The haem group is so constructed that it amplifies
a small change in atomic radius, undergone by the iron atom
in the transition from the high spin to the low spin state,
into a large movement of the haem-linked histidine
relative to the porphyrin ring. This movement, and the
widening of the haem pocket required in the β-chain,
triggers small changes in the tertiary structure of the
reacting subunits, which include a movement of helix F
towards helix H and the centre of the molecule. ... this
movement can be observed to narrow the pocket between
helices F and H so that the penultimate tyrosine is
expelled. The expelled tyrosine pulls the C-terminal
residue with it, rupturing the salt-bridges that had held
the reacting subunit to its neighbours in the deoxy
tetramer. The rupture of each salt-bridge removes one of
the constraints holding the molecule in the deoxy
conformation and tips the equilibrium between the two
alternative quaternary structures some way in favour of
the oxy conformation. In this conformation the oxygen
affinity is raised because the subunits are no longer
constrained by the salt-bridges to maintain the tertiary
deoxy structure ... [this] also leads to the liberation
of the Bohr protons, in agreement with the finding that
their release is exactly proportional to the amount of
oxygen taken up, and that the two reactions are
synchronous. I have suggested that the subunits react in
the order α_1, α_2, β_1, β_2, but this is not certain and my
cooperative mechanism would work equally well in whichever
sequence the individual subunits react."

6. Since the discovery in the late 1960's that 2,3-DPG, ATP, inositol
hexaphosphate (IHP; the principal intraerythrocytic organic
phosphate in birds), and, to a much lesser extent other organic
phosphates, bind to Hb and decrease its affinity for oxygen, a
burgeoning literature on this topic has developed.

a.   The ATP level in erythrocytes is unusually high for a cell
     which does not carry on oxidative phosphorylation. 2,3-
     diphosphoglycerate (2,3-DPG) is present at even higher
     concentrations (about 1:1 mole ratio with Hb) than ATP. The
     2,3-DPG is formed by the rearrangement of 1,3-DPG catalyzed by
     the enzyme diphosphoglycerate mutase. 3-phosphoglyceric acid
     is required as a cofactor and the reaction is allosterically
     stimulated by 2-phosphoglyceric acid. Inorganic phosphate
     appears to be a negative allosteric modifier. 2,3-DPG also
     functions as a cofactor for monophosphoglycerate mutase, an
     enzyme in the glycolytic pathway. In this reaction, it is
     needed only in catalytic amounts. The enzyme 2,3-DPG
     phosphatase catalyzes the reaction

$$\text{2,3-DPG} + \text{H}_2\text{O} \longrightarrow \text{3-PG} + \text{P}_i$$

     which is the degradative route for 2,3-DPG.

b.   The effect of 2,3-DPG on the oxygen dissociation curve of Hb
     and a possible binding site for 2,3-DPG to Hb were discussed
     earlier. It is appropriate to reemphasize here the many
     factors (pH, hematocrit, amount of Hb per red cell,
     respiratory rate, P_{O_2} in the inspired air, etc.), in addition
     to the concentration of 2,3-DPG, work together to control
     oxygen delivery to the tissues.

     Regarding the proposed binding site on Hb for the organic
     phosphates, it appears that 2,3-DPG can be readily accommodated
     in the cleft between the β-chains. There is some doubt,
     however, as to whether inositol hexaphosphate (IHP; much more
     effective than 2,3-DPG) and ATP (whose binding constant is
     an order of magnitude less than that for 2,3-DPG) will fit
     into the same area. Recent studies have indicated, moreover,
     that two binding sites on hemoglobin may be involved. At one
     site, on deoxy-hemoglobin, 2,3-DPG and ATP compete for binding
     while at the other site, on oxy-hemoglobin, ATP enhances the
     binding of 2,3-DPG. This also emphasizes the finding that
     oxy-hemoglobin does apparently bind 2,3-DPG, contrary to earlier
     reports. Clearly there is more to be learned about the
     interactions of organic phosphates with hemoglobin molecules.

c.   Another area of investigation has been the mechanism by which
     2,3-DPG concentration can be altered inside the erythrocyte.
     One or more of these controls must be able to act within a
     period of less than twelve hours (perhaps much shorter; see
     later discussion), since significant (15-25%) changes in
     2,3-DPG concentration have been noted in such a timespan.
     The most probable effects involved are summarized below.

i) As was mentioned in the section on thermodynamics, the binding of 2,3-DPG to Hb decreases the amount of 2,3-DPG which can participate in other reactions, notably the 1,3-DPG $\rightleftharpoons$ 2,3-DPG equilibrium. This causes increased 2,3-DPG synthesis at the expense of 1,3-DPG.

ii) The intraerythrocytic pH appears to have several effects on the 2,3-DPG concentration. First, a decrease in pH increases the binding of 2,3-DPG to Hb, bringing the mechanism above into operation. Second, an increase in pH is known to stimulate glycolysis, which tends to increase the levels of all glycolytic intermediates such as 2,3-DPG. Finally, pH changes affect the activities of diphosphoglyceromutase and phosphatase, the enzymes responsible for 2,3-DPG synthesis and degradation, respectively.

iii) The age of the erythrocytes may or may not have some influence on intracellular organic phosphate levels. Although reports differ somewhat, it appears that either the age has no effect or increasing age corresponds to decreasing concentration.

iv) There may also be genetic control over the 2,3-DPG levels. There is evidence that ATP concentrations within the erythrocyte are under hereditary influence and in the hooded strain of rats, levels of both of these organic phosphates appear to be genetically influenced. Obviously this mechanism is quite different from the previous three in being of no importance in rapid, short-term adaptions.

Whatever the actual regulator(s) is, it is apparent that red cell metabolism is intimately involved in the transport of oxygen by hemoglobin.

d. The processes described in the preceding section might be termed final-step controls. The primary stimuli which trigger these terminal steps are many. A few of them are

i) Decreased O_2-delivery to the tissues due to anemia, altitude, cardiac insufficiency, etc. (mentioned before).

ii) The presence of certain hormones, such as thyroxine (which may directly stimulate diphosphoglycerate mutase) and androgens (which also increase erythropoiesis).

iii) Polycythemia (which decreases the intraerythrocytic concentration of 2,3-DPG).

iv) Some cases of pulmonary disease.

Whether the shift in the O_2-dissociation curve which accompanies 2,3-DPG changes is beneficial or not depends largely on the partial pressure of oxygen in the lungs (see previously). There is wide variation in the degree of change in the 2,3-DPG concentration exhibited by patients having the same disease. For example, in severe pulmonary disease the increase may be from 0-100%; in leukemia (a non-hemolytic anemia), rises from 20-150% have been observed; and patients with iron deficiency display increases ranging from 40-75%.

e. Two other aspects of 2,3-DPG are important.

   i) It has been reported that the 2,3-DPG level is about 45% higher in venous blood than in arterial blood, corresponding to the decreased oxygen saturation in venous blood. This is important because it demonstrates the extreme rapidity with which the 2,3-DPG regulatory mechanism can act. It also indicates a possible mechanism by which elevation of the 2,3-DPG (useful to unload oxygen) might be prevented from impairing oxygen uptake by the pulmonary blood.

   ii) Ion transport across red cell membranes is stimulated by deoxygenation of the Hb in the cell. This may be due to the binding of 2,3-DPG to the deoxy-Hb, causing enhanced glycolysis, glucose uptake, and hence ion transport.

f. An important clinical problem involving 2,3-DPG is the observation that 2,3-DPG and, to a lesser extent, ATP concentrations decrease rapidly in blood which is stored for even a few days in the acid-citrate-dextrose (ACD) medium used for some time by many blood banks. The result of this is that oxygen affinity is increased and the ability to supply oxygen to the tissues is thereby decreased when such blood is used for transfusions. In patients who receive this blood, there is an initial increase in oxygen affinity which does not decrease to normal for six hours or more. Traditionally, red cell survival has been used as the main criterion of the quality of stored blood. Survival does not necessarily correspond to maintenance of adequate organic phosphate (OP) levels, however, so this must be considered as a supplementary criterion.

A number of studies have investigated how best to vary the composition of the storage medium to prevent this loss of organic phosphates. (Actually, these compounds do not really leave the cell since red cell membranes are generally quite impermeable to organic phosphates. Instead, they are metabolized to other forms.) Some of these findings are summarized below.

i) Citrate-phosphate-dextrose (CPD) medium is somewhat better than ACD for maintaining OP levels.

ii) Storage in the deoxy-form at a pH of 7.2 is more satisfactory than oxygenation and lower pH. However, at an alkaline pH, ATP is lost, which decreases cell survival time.

iii) ACD supplemented with the nucleoside adenosine maintains ATP but not 2,3-DPG. In fact, the rate of oxygen affinity elevation is mildly increased in this medium.

iv) ACD supplemented with the nucleoside inosine seems to be the most satisfactory medium at present. Inosine maintains both ATP and 2,3-DPG <u>in vitro</u>, as well as retarding significantly the increase in oxygen affinity. It is not known exactly how inosine brings about this change, but it may partly function by supplying ribose for 2,3-DPG formation (via the HMP shunt).

So far, the actual therapeutic significance of these changes is not clear. The greatest effects should probably be looked for in patients who have received repeated, massive transfusions over a period of, say, six hours, so that a significant fraction of their circulating erythrocytes have increased O_2 affinity.

g. As more is learned about the functioning and control of organic phosphates within the red cell, ways of altering this level artificially in particular situations can be developed. The role of hormones (e.g., androgens and thyroxine) in controlling tissue oxygen supply may be better understood. The physician should also be aware that some drugs may, as a side-effect, change the metabolism of the red cell in a manner which will affect the oxygen-carrying ability of the blood.

7. Although hemoglobin is worth understanding in its own right, one must remember that most regulatory enzymes are also allosteric, multisubunit proteins. That they are allosteric means that the rate at which they catalyze a reaction or bind substrate to one subunit is controlled by the binding of another molecule (a positive or negative allosteric effector) either on the same or another subunit. By understanding how hemoglobin changes its oxygen affinity through quaternary structure alterations, other (regulatory) enzymes can be better understood. Four especially important concepts to arise from the hemoglobin work are

a. The change in Fe^{+2} from high spin to low spin acting as a trigger;

b. The use of salt bridges in maintaining one structure or another;

c. The indication that hemoglobin's allostery can best be described by the Eigen model, i.e., there are features intermediate between the MWC and KNF models). This opportunity to directly test the allosteric models is very important;

d. The description of how an allosteric effector (O_2 and organic phosphates) may function in terms of specific molecular changes.

The approach to the elucidation of hemoglobin's structure and function has been interdisciplinary, involving physical chemistry (x-ray crystallography, magnetic resonance), physiology (respiratory studies, etc.), biochemistry (Bohr effect, Hb derivatives, amino acid sequence), and genetics and clinical medicine (hemoglobinopathies, discussed later). Without all of these contributions, the problem would have been solved much more slowly or not at all.

8. Another function of hemoglobin is to participate in the transport of CO_2 from the tissues to the lungs with very little change in pH. This buffering action of hemoglobin was already discussed in Chapter 1 and is illustrated in Figure 67. Note that it is primarily the Bohr protons which participate in this.

In Figure 67, note:

a. the chloride shift which accompanies the movement of HCO_3^-;

b. the role of H^+ and hence CO_2 and carbonic anhydrase, in unloading oxygen from oxy-hemoglobin;

c. the intermediary role of the plasma as a carrier of CO_2 (as HCO_3^-);

d. that, in addition to the O_2 bound to Hb, there is always some oxygen physically dissolved in the plasma; ε

e. that some CO_2 is chemically combined with Hb, as carbamino compounds.

9. Hemoglobin concentrations are controlled by levels of erythropoietin (page 247). Conditions which give rise to a situation (genetic or otherwise) with a hemoglobin having an increased oxygen affinity (and hence decreased oxygen availability to the tissues), lead to elevated levels of hemoglobin and vice versa.

Figure 67. The Interchange of Carbon Dioxide and Oxygen Between the Tissues and the Lungs

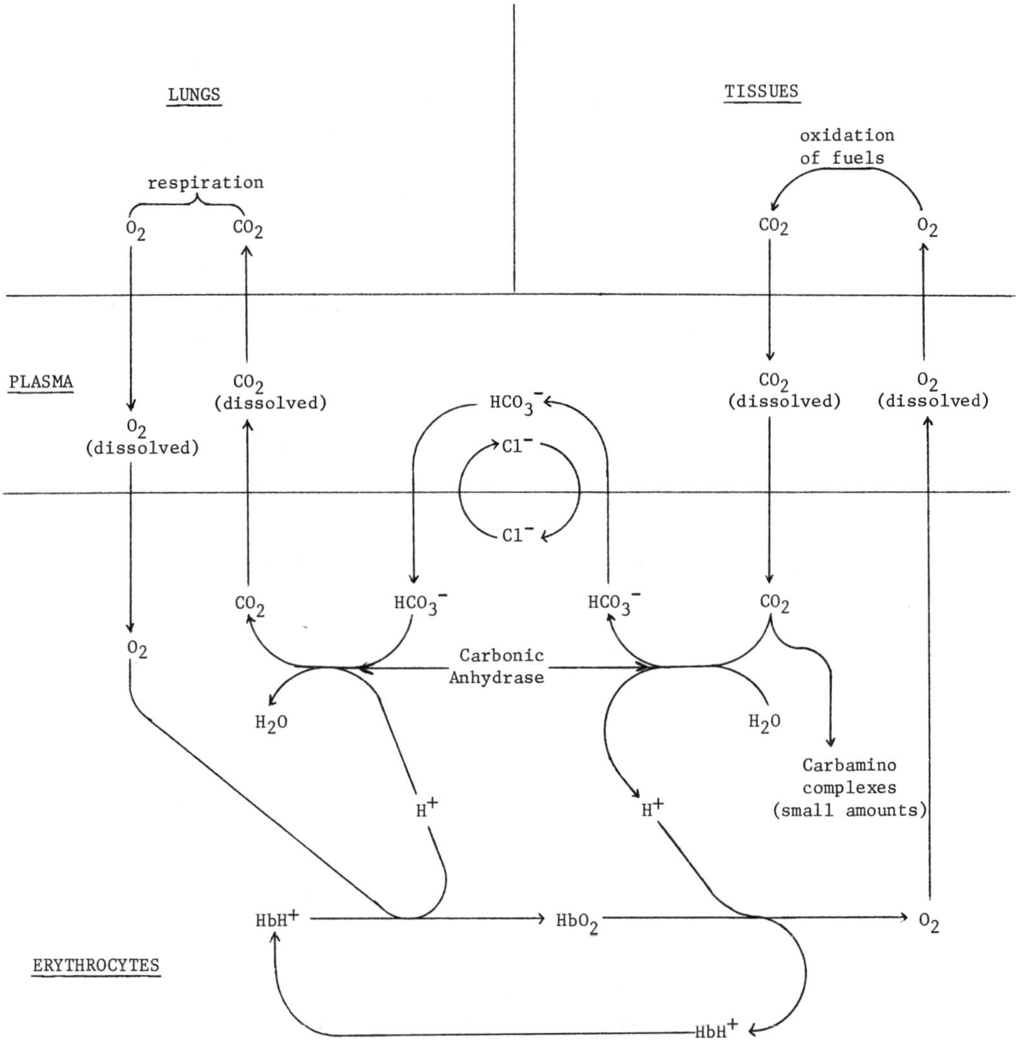

A similar figure is presented in Chapter 1 of this book. Note that the reactions on the <u>right</u> side of this diagram occur in the plasma and erythrocytes in the capillaries, during oxygen delivery to the tissues; while the reactions on the <u>left</u> side occur in the lungs, during CO_2 release and O_2 uptake. The cyclic nature of these changes, in going from lungs to capillaries back to the lungs should be recognized.

10. If the hemoglobin mutant is such that it has an increased or decreased affinity for oxygen, the expected decrease or increase (respectively) in tissue level oxygen tensions does not always occur. This may be partly due to a change in erythrocyte DPG levels (increase or decrease respectively). The hematocrit can also change, however, being higher or lower (respectively) than normal. Studies on such mutants have shown that these abnormal hematocrits occur despite normal erythropoietin levels. This indicates that the body is primarily sensitive to tissue oxygen level and not to red cell mass. Such abnormal hemoglobins frequently can only be detected in the laboratory. Hemoglobin mutants are discussed in the next section.

Diseases of Hemoglobin (Hemoglobinopathies)

More than 150 abnormal hemoglobins have been described (as of about August 1971), a number of which cause no physiological abnormalities. These hemoglobinopathies, of which there are two general types, appear to be the result of various genetic alterations.

1. The thalassemias result from abnormalities in the quantities of the different chains synthesized. All the chains which occur are normal in structure.

2. Abnormal polypeptide chains are synthesized in the second group. The errors in sequence can be a change in the residue at one specific site in the chains, the insertion or deletion of one or more residues, or the combination of different pieces of two normal chains resulting in "hybrid" polypeptides. The hybrid peptide variants (known as the hemoglobins Lepore) are discussed with the thalassemias because of their clinical pattern.

The number of known hemoglobin mutants is still increasing and they continue to provide an ideal system to investigate genetic mechanisms in eucaryotes and to explore the effect on the functioning of Hb of various alterations in its amino acid sequence.

Normal Hemoglobins

As a prelude to discussing the abnormal hemoglobins, a brief survey of the composition and evolution of the normal hemoglobins is useful. The term "normal" should be used with caution since the presence of certain of these Hb's in large quantities in adults can be indicative of quite abnormal conditions. Moreover, there are hemoglobins whose amino acid sequence differs from the normal one but which display no symptoms at all.

1. Each molecule of human hemoglobin is made up of four polypeptide chains. In Hb A_1, A_2 and F, two of the chains are α-chains and the second pair is β for A_1, δ for A_2, and γ for F. These are summarized below.

    a. Hemoglobin A_1: $HbA_1 = \alpha_2\beta_2$ (normal adult hemoglobin:
        (More frequently designated   constitutes over 85% of
              as Hb A)   the total hemoglobin)

    b. Hemoglobin A_2: $HbA_2 = \alpha_2\delta_2$ (about 3% of the total
                                                  hemoglobin in a normal adult)

    c. Hemoglobin F : $HbF = \alpha_2\gamma_2$ (predominant form present at
                                              birth but almost totally
                                              replaced by HbA_1 within the
                                              first few months after birth;
                                              present in adults in amounts
                                              up to 2% of the total Hb;
                                              elevated in some thalassemias)

    In apparently all populations there are two types of γ-chains, differing in only one amino acid residue. Residue 136 is glycine in G_γ-chains and alanine in A_γ-chains. Consequently, these two polypeptides cannot be separated by electrophoresis or chromatography and must be distinguished on the basis of amino acid composition and sequence. HbF has a higher affinity for O_2 than does HbA_1, facilitating O_2 transfer from the mother to the fetus. This is due to a weaker binding of 2,3-DPG to β-chains than to γ-chains. Another minor HbF fraction has been shown to be an N-acetylated derivative of HbF.

2. Hemoglobin A_3 (also called $HbA_1\alpha$) is a complex between HbA_1 and glutathione which occurs in older erythrocytes. It can be separated from other hemoglobins by electrophoresis. HbA_{1c} is a Schiff base between the N-terminal residues of both β-chains and a mole of hexose (making this Hb a glycoprotein). In normal adults, it is present as about 4-6% of the Hb. It is decreased in iron deficiency anemia and in diabetes, either HbA_{1c} or a similar component is elevated twofold.

3. Two types of embryonic hemoglobin have also been observed. These are designated Gower I and Gower II and have the subunit compositions ε_4 and $\alpha_2\varepsilon_2$, respectively. Synthesis of the ε-chains ceases by the end of the third month of embryonic development, being replaced by HbF (γ-chains).

4. Another type of hemoglobin chain, designated ζ (zeta), has been
    observed in a number of cases of severe α-thalassemia, primarily
    among Chinese neonates. It occurs as part of a tetrameric Hb of
    formula $\gamma_2\zeta_2$, which is somewhat unusual in containing two different
    types of subunits, neither of which is α. Slight precedence for
    this occurs in the detection in vitro (and perhaps in vivo) of
    what appears to be $\beta_2\delta_2$ and perhaps $\beta_2\gamma_2$. This could occur (as in
    the case of Hb Portland) in situations where α-chains are either
    totally absent or in extremely short supply. It is postulated
    that the zeta-chains are another example (in addition to ε-chains)
    of an embryonic hemoglobin.

5. The α-chain contains 141 amino acid residues and the rest of the
    chains (β, δ, and γ) contain 146 residues each. It is essential
    to point out each chain has a heme group and in the α-chain,
    the heme group is inserted between the histidine residues at
    positions 58 and 87. On the β-chain, the heme is present between
    histidine residues 63 and 97. Any changes (due to mutation) in
    these histidine residues have a particularly deleterious effect
    on the functioning of the hemoglobin due to problems in maintaining
    iron in the ferrous state (discussed later; see methemoglobin).

6. Production of the different hemoglobin chains is controlled by
    at least four different genes, one coding for each of the four
    possible (α, β, γ, and δ) polypeptides. (Presumably the ε- and
    ζ-chains are also coded for separately.) It is speculated that
    these evolved from one another by gene duplication followed by
    mutation. The α-chain gene is considered to be the oldest. At
    some time long ago, duplication of the α-gene occurred. One copy
    continued to code for α-chains and the other mutated gradually
    into what is today the coding for the γ-chain. Some time after
    this initial duplication event, but still quite a while ago, the
    γ-chain gene was duplicated. The extra gene began diverging
    from the γ-chain until it now codes for the β-chain. The δ-chain
    presumably arose in a similar fashion from the β-gene.

The principal evidence for this developmental path comes from a
comparison of the amino acid sequences of these chains. The
further apart two chains are in evolution, the fewer identical
residues (homologies) will presumably occur between them. This
sort of study shows that there is 39% homology between α-and γ-
chains, 42% homology between α-and β-chains, 71% homology between
β-and γ-chains, and 96% homology between β-and δ-chains. These
evolutionary relationships would be very difficult to actually
prove and should be considered quite speculative.

The Thalassemias

1. This is a group of genetic disorders resulting from unbalanced globin chain synthesis. The unequal synthesis leads to decreased amounts of functional hemoglobin in the maturing erythrocytes (hypochromic erythrocytes). Note that in a normal person, the ratio of production of α-to β-chains is 1:1. These diseases are characterized by the presence of an excess of <u>unused</u> chains which are precipitated intracellularly, appearing as inclusions called <u>Heinz bodies</u>. Such precipitation also occurs in a number of abnormal hemoglobins, especially the unstable ones. This gives rise to premature erythrocyte destruction by the spleen, leading to moderate to severe forms of anemia. Treatment of this anemia with multiple transfusions can cause hemosiderosis and eventually hemochromatosis (discussed earlier in this chapter).

2. Classification of thalassemias

    a. A deficiency of α-chains is known as $\underline{\alpha\text{-thalassemia}}$. Due to the lack of α-chains, the formation of Hb A_1, A_2, and F is severely limited. Death can occur <u>in utero</u> with homozygotes for α-thalassemia. In heterozygous individuals, moderate to severe anemia persists throughout their lifetime. Some of these patients, due to the synthesis of surplus β-, γ-, and δ-chains, show the presence of abnormal hemoglobins. These include: β_4 (four β-chains, known as Hb-H); γ_4 (four γ-chains, known as Hb-Bart's); and δ_4 (four δ-chains). Hb-H has also been observed to occur with some forms of leukemia, usually erythroleukemia or Di Guglielmo's syndrome.

    b. A deficiency of β-chains is known as $\underline{\beta\text{-thalassemia}}$ and the occurrence of abnormal hemoglobins in these disorders has not been observed. In this condition, Hb A_1 is decreased or absent and increased levels of Hb F and Hb A_2 are found. Homozygotes for β-thalassemia (thalassemia major; Cooley's anemia) suffer from a severe form of microcytic anemia. The heterozygotes have thalassemia "minor" or trait, where Hb A_1 levels are reduced and Hb A_2 and Hb F levels are increased. The presence of microcytosis (less than 79 μ^3) and elevated Hb A_2 (greater than 3.5% of the total Hb) are considered by some to be diagnostic for β-thalassemia trait.

    c. Deficiency of δ-and β-chains is known as $\underline{\beta\delta\text{-thalassemia}}$ (Type II β-thalassemia; F-thalassemia). Homozygous individuals for this condition have <u>no</u> Hb A_1 or A_2 and their total hemoglobin complement is made up entirely of hemoglobin F. They show clinical features similar to those of β-thalassemia. Heterozygotes contain lesser amounts of Hb A_1 and A_2 and elevated amounts of Hb F. No abnormal hemoglobin is present.

d. A group of Hb variants known as the <u>hemoglobins Lepore</u> may
    be considered as a variation of $\beta\delta$-thalassemia in which there
    is an abnormal hemoglobin. The abnormal hemoglobin is a
    tetramer containing two normal α-chains and two abnormal,
    hybrid chains. These hybrid chains each contain 146 residues
    but their sequence is partly that of a normal β-chain and
    partly that of a normal δ-chain. These apparently result
    from non-homologous cross-over events in the chromosomes
    which are known to occur in bacteria. The term non-homologous
    indicates that after the crossing over, one hybrid chromosome
    (the duplication chromosome) is longer than the other (the
    deletion chromosome). An example of such an event is shown
    in Figure 68.

    Mitosis or meiosis following a crossing over of this sort will
    give one daughter chromosome with loci for the normal δ- and
    β-chains as well as the abnormal $\beta\delta$-gene, and one daughter
    chromosome with only the abnormal $\delta\beta$-gene. Zygotes receiving
    the former chromosome should develop normally, since they are
    perfectly capable of making normal δ- and β-peptides (left
    side of Figure 68). Hb-Miyada is just such a case, having
    normal δ- and β-synthesis in addition to the $\beta\delta$-subunits.

    The other three Hb_5-Lepore (Washington, Hollandia, and Baltimore)
    all show symptoms similar to those of β-thalassemia. All
    exhibit a complete absence of δ- and β-chains (hence their
    classification as a $\beta\delta$-thalassemia) and the presence of
    $\delta\beta$-hybrids (right side of Figure 68). They differ in the
    specific crossing-over point. Note that, due to the apparent
    order of the δ- and β-genes in the chromosome ($\delta \rightarrow \beta$ is the
    same direction as $5' \rightarrow 3'$), the duplication chromosome
    receives the N-terminal sequence of the β-chain and the C-terminal
    sequence of the δ-chain while the deletion chromosome has the
    N-terminus (5'-end) of the δ-gene and C-terminus (3'-end) of
    the β-peptide. Because of the high degree of homology (96%)
    between the δ- and β-chains, it is difficult to pinpoint the
    exact cross-over point.

e. Cases of β-thalassemia have been reported which show complete
    absence of Hb A_2. They have no clinical symptoms and appear
    to be free of any other red cell abnormalities. A preliminary
    report of γ-thalassemia has also appeared but it has yet to
    be confirmed. Such a disorder would probably result in a
    mild neonatal jaundice which disappears when γ-chain production
    ceases and the γ-chains are replaced by δ- and β-chains.

Figure 68.   Schematic Diagram of a Non-Homologous Crossing-Over Event Involving the
              Hb β-Chain and δ-Chain Loci

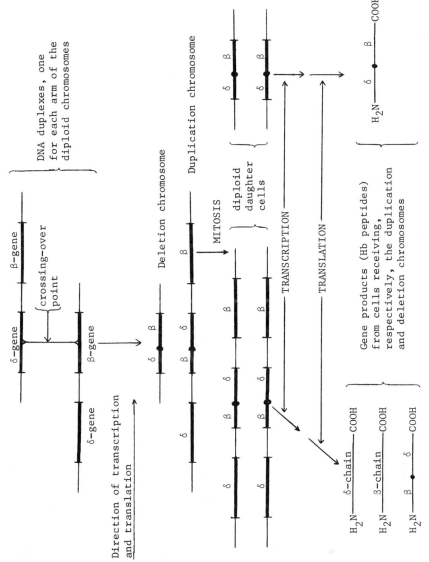

The gene products formed by the cells which receive the two chromosomes from such an event
are also indicated. Note that, in reality, the gene product (peptide) will be shorter than
the gene which codes for it. The dots in recombination genes and their peptide products
represent the point at which the sequence for one chain meets the sequence for the other.

3. The actual biochemical defect(s) which causes thalassemia is not
    known. Most studies in this area have been performed on cells
    from β-thalassemic patients and it has been shown that in these
    cells, decreased β-chain synthesis is <u>not</u> due to defects in
    ribosome function, in chain initiation or elongation factors, or
    in the "supernatant" factors such as tRNA and necessary enzymes.
    Recent work has shown that the <u>quantity</u> of β-chain mRNA synthesized
    by β-thalassemic cells is less than in normal cells. This <u>may</u>
    account for the inadequate β-chain production. There may also be
    a defect in the mRNA which slows down or inhibits translation.

Hemoglobinopathies Caused by Abnormal Subunits

As was indicated previously, these can be divided into two groups

   a. <u>point mutations</u>, involving substitution of one amino acid
       residue for another;

   b. <u>addition and deletion mutants</u>, which have, respectively, more
       or fewer residues than the normal chain from which they are
       derived.

1. Table 20 summarizes the clinical and molecular characteristics
    of a number of point-mutated hemoglobins. The mechanisms by which
    the codon change could occur were discussed in the chapter on
    nucleic acids. Note that the mRNA codon is listed in the table
    rather than the DNA sequence. Mutatory influences function at
    the DNA level.

2. Mutations of this sort (single residue mutations) are the most
    common and important ones. The occurrence or severity of a
    clinical disorder associated with an amino acid substitution
    generally can be said to depend on the location of the residue in
    the chain, the most critical regions seeming to be those near the
    heme group or ones involved in $\alpha_1\beta_1$ and $\alpha_1\beta_2$ contacts.

   a. Mutants which involve the heme contacts or any other aspect of
       the heme pocket can cause weak or absent heme binding, resulting
       in loss of the heme group from the Hb (Hb Sydney, Hb Hammersmith);
       or stabilization of the heme iron in the ferric (Fe^{+3}) state, so
       that O_2 cannot be reversibly bound (the hemoglobins M).
       Methemoglobin (Hb M) also loses its heme group more readily
       than normal Hb.

   b. Mutants in the $\alpha_1\beta_1$ contacts are known which weaken the contacts
       and thereby increase the subunit dissociation (Hb E), and which
       alter the oxygen affinity, either increasing it (Hb Tacoma) or
       decreasing it (Hb Yoshizuka; Hb E).

Table 20.   Molecular and Clinical Aspects of Some Abnormal Hemoglobins

Entries in this table are all mentioned in the text.   This is not intended to be anywhere near a complete listing of all hemoglobin variants which are known.

Hemoglobin	Chain and Position*	Amino Acid Change	mRNA Codon Alteration**	Remarks
I-Philadelphia	α16(A14)	Lys → Glu	AA(A,G) → GA(A,G)	Increased alkali resistance
G-Honolulu	α30(B11)	Glu → Gln	GA(A,G) → CA(A,G)	Also called G-Chinese; benign
Chesapeake	α92(FG4)	Arg → Leu	CGZ → CUZ	Increased O_2 affinity; $\alpha_1\beta_2$-contact mutant
G-Georgia	α95(G2)	Pro → Leu	CCZ → CUZ	Oxy-form dissoc. to dimers; incr. O_2 aff.; $\alpha_1\beta_2$-contact mutant
S	β6(A3)	Glu → Val	GA(A,G) → GU(A,G)	See sickle-cell anemia
G-Makassar	β6(A3)	Glu → Ala	GA(A,G) → GC(A,G)	Compare to HbS; benign
C	β6(A3)	Glu → Lys	GA(A,G) → AA(A,G)	Xtalliz. in RBC; mild hemolytic anemia; compare to HbS
C-Harlem	β6(A3), β73(E17)	Glu → Val, Asp → Asn	GA(A,G) → GU(A,G), GA(U,C) → AA(U,C)	Double mutant; see HbS and Hb Korle-Bu
G-San Jose	β7(A4)	Glu → Gly	GA(A,G) → GG(A,G)	Compare to HbS; benign

Table 20 (continued)

Hemoglobin	Chain and Position*	Amino Acid Change	mRNA Codon Alteration**	Remarks
E	β26(B8)	Glu → Lys	GA(A,G) → AA(A,G)	Mild hemolytic anemia; incr. subunit dissoc.; decr. O_2 affinity; $\alpha_1\beta_1$-contact mutant
Genova	β28(B10)	Leu → Pro	CUZ → CCZ	B helix disrupted; unstable
Tacoma	β30(B12)	Arg → Ser	AG(A,G) → AG(U,C)	Increased O_2 aff.; unstable; $\alpha_1\beta_1$-contact mutant
Hammersmith	β42(CD1)	Phe → Ser	UU(U,C) → UC(U,C)	Decr. O_2 aff.; cyanosis; poor heme binding; inclusion bodies; unstable
Zürich	β63(E7)	His → Arg	CA(U,C) → CG(U,C)	Compare to Hb–M–Saskatoon; incr. O_2 aff.; sulfa drugs cause hemolysis
Sydney	β67(E11)	Val → Ala	GUZ → GCZ	Compare to Hb Hammersmith; inclusion bodies, unstable; cyanosis; poor heme binding
Korle Bu	β73(E17)	Asp → Asn	GA(U,C) → AA(U,C)	Compare to Hb–C–Harlem; benign
Köln	β98(FG5)	Val → Met	GUG → AUG	Unstable; incr. O_2 aff.; $\alpha_1\beta_2$-contact mutant

Table 20 (continued)

Hemoglobin	Chain and Position*	Amino Acid Change	mRNA Codon Alteration**	Remarks
Kansas	β102(G2)	Asn → Thr	AA(U,C) → AC(U,C)	Unstable; cyanosis; incr. O_2 aff.; $\alpha_1\beta_2$-contact mutant
Yoshizuka	β108(G9)	Asn → Asp	AA(U,C) → GA(U,C)	Unstable; incr. O_2 aff.; $\alpha_1\beta_1$-contact mutant
D-Los Angeles	β121(GH4)	Glu → Gln	GA(A,G) → CA(A,G)	Benign; can occur with HbS; also called D-Punjab
M-Boston	α58(E7)	His → Tyr	CA(U,C) → UA(U,C)	Methemoglobins; cause cyanosis
M-Iwate	α87(F8)	His → Tyr	CA(U,C) → UA(U,C)	
M-Saskatoon	β63(E7)	His → Tyr	CA(U,C) → UA(U,C)	
M-Hyde Park	β92(F8)	His → Tyr	CA(U,C) → UA(U,C)	
M-Freiburg	β23(B5)	Val → (deleted)	GUZ GUZ → ---	
M-Milwaukee	β67(E11)	Val → Glu	GU(Z,G) → GA(A,G)	

* Kendrew's helical notation is given in parentheses.
** Because of redundancy in the genetic code, the observed amino acid in the normal and mutant peptides can be coded for by more than one codon (the only exception is the change in Hb Köln, since Met has only one codon). For example, Gln can be coded for by GAA or GAG. When it is replaced by Val (as in HbS), the valine can be coded for by GUA or GUG where the mutation was A→U in the second base of the codon. A Z occurring in a codon (such as CGZ for Arg in Hb Chesapeake) means that the third letter of the codon can be any one of the four bases A, U, C, or G.

Glossary: benign means no observable symptoms.
dissoc. means dissociates.
aff. means affinity.
xtalliz. means crystallization.
incr., decr. mean increased or decreased.

c. The $\alpha_1\beta_2$ contacts are involved in combining the $\alpha\beta$ dimers to form tetramers and these are the contacts which undergo the greatest change in the quaternary structure (T $\overset{\rightarrow}{\leftarrow}$ R) transformation. Some of these show increased (Hb Chesapeake) or decreased (Hb Kansas) oxygen affinity. Others, such as Hb-G-Georgia, completely dissociate into dimers upon oxygenation, then reassociate to tetramers when oxygen is removed. The degree of cooperativity (heme-heme interaction) is decreased in most of the $\alpha_1\beta_2$-contact mutants, presumably due to the tendency to form dimers upon oxygenation.

d. A great number of mutations lead to hemolytic anemia (increased rate of lysis (shortening of the lifespan of erythrocytes)). These are called <u>unstable hemoglobin variants</u> and are usually associated with some change which decreases the solubility of the hemoglobin in the red cell. The lowered solubility results in intraerythrocytic precipitation of hemoglobin (Heinz body formation; Hb appears as intracellular inclusion bodies) which apparently triggers any of several lytic processes. Other factors, such as increased interaction between the sulfhydryl groups of Hb and membrane proteins resulting in weakening of the cell membrane, may also cause increased hemolysis.

Sickle-cell hemoglobin (Hb S) is the best-known example of this and was the first mutant hemoglobin to be described in molecular detail (by L. Pauling, H. Itano, S. Singer, and I. Wells, in 1949). Hb S is discussed more fully later on. Since the presence of heme greatly stabilizes the hemoglobin, mutations which result in impaired heme binding lead to hemolysis due to precipitation of the heme-free globin chains. Similarly, α-subunits by themselves tend to precipitate, so that mutations which decrease intersubunit binding can also result in lytic anemias. This probably also occurs in β-thalassemia. Deletion mutants (Hb Gun Hill, and others; discussed later), which distort the chain conformation have a similar effect. Hemoglobin Köln is an $\alpha_1\beta_2$-contact mutant with an increased positive charge relative to normal Hb. It is unstable and shows increased O_2 affinity. Disruption of a helical region not in contact with the heme group can also apparently distort the molecule sufficiently to cause major changes in its properties. For example, in Hb Genova, a proline is inserted into the B helix in place of a leucine, causing hemolytic anemia even in heterozygotes.

e. Most hemoglobin mutants are very rare, since they tend to confer negative adaptive advantage to the bearer and his progeny. Exceptions to this are Hb S, Hb C, Hb-D-Los Angeles,

Hb E, and some of the thalassemias, which occur in certain populations with a relatively high frequency. Since the appearance of even one mutation is quite uncommon, the occurrence of two abnormalities in one individual is exceedingly rare in most cases. The only known instance of two amino acid substitutions in one chain is Hb-C-Harlem. Since one change is the same as that in Hb S and the other change is the one occurring in Hb Korle-Bu, the double mutant may have resulted from a second point mutation in a carrier of either allele, but most likely in a Hb S heterozygote. It could also have occurred by homologous crossing over in a person heterozygous for both Hb Korle-Bu and Hb S. It is significant that one of the mutations (Hb S) is known to have a high frequency of occurrence. Another double mutant (Hb-C-Georgetown) has been reported but it may be identical to Hb-C-Harlem.

f. Much more common than the doubly mutant hemoglobins is the occurrence of heterozygotes for two abnormal alleles. Examples of this include Hb S-Hb D disease (similar to sickle-cell trait; mild hemolytic anemia) and Hb S-Hb C disease. It is also not uncommon for an abnormal globin chain to be inherited along with the gene for a β-thalassemia. Reports have appeared of Hb S, Hb C, Hb D, Hb E, Hb J, and Hb G occurring in such conjunction. It is important to note that, in general, these "double inborn errors" only happen when one or both of the separate defects occurs with high frequency. In Hb S-Hb C disease, it is significant that not only are both alleles fairly common, but they both appear primarily in the same population.

g. A number of mutants are known which are either completely benign, even in the homozygote, or which result in very mild symptoms of little or no clinical significance. Examples of this include Hb San Jose, Hb-G-Honolulu (also called Hb-G-Chinese; an $\alpha_1\beta_1$-contact mutant), Hb Korle-Bu, and Hb-D-Los Angeles (also called Hb-D-Punjab).

h. Seemingly minor variations between mutations can result in very different clinical and molecular pictures. For example, sickle-cell anemia is a serious disease, brought about by the mutation $\beta6(\text{Glu} \rightarrow \text{Val})$, while Hb San Jose, which is $\beta7(\text{Glu} \rightarrow \text{Gly})$, is a harmless variant. Also compare Hb Zürich ($\alpha63$ His $\rightarrow$ Arg) and Hb-M-Saskatoon ($\alpha63$ His $\rightarrow$ Tyr); and Hb S, Hb C ($\beta6$ Glu $\rightarrow$ Lys), and Hb-G-Makassar ($\beta6$ Glu $\rightarrow$ Ala).

i. Throughout the preceding discussion, reference has been occasionally made to homozygotes for a mutation versus heterozygotes. Persons heterozygous for a trait are usually termed <u>carriers</u> since they frequently display few if any of the symptoms of the homozygous disease. This is probably due to the presence of the normal gene on the other chromosome which can supply normal gene products for use of the cell. (This is similar to complementation in bacterial genetics.) If one carrier has offspring by another carrier however, the children are likely to be homozygous for the trait and have the characteristic symptoms. In some cases, the homozygous state is lethal neonatally and only the heterozygote is even seen.

3. One of the most common, most severe, and best studied hemoglobin mutants is Hb S, which causes sickle-cell anemia (or disease) in homozygotes and sickle-cell trait in heterozygotes. The name arises from the characteristic change of shape of the patient's erythrocytes (from the normal biconcave disc to a curved, sickle-like morphology) under anaerobic conditions. Possessors of the trait appear to be largely asymptomatic, although there are some reports of death occurring among carriers when they are subjected to severe tissue hypoxia (high altitude, strenuous exercise). The oxygen pressure probably must be below about 10 torr before sickling occurs in heterozygotes. This could occur during the administration of an anesthetic or during recovery from anesthesia.

a. As indicated in Table 20, the mutation is a replacement of Glu (a polar amino acid) by Val (a non-polar residue) at position six of the beta chains. The substitution decreases the net negative charge (at pH 8.6) of the tetramer by two. Exactly how this causes the aggregation is not clear despite active speculation. One theory is that the substitution of a neutral residue for a charged residue permits a hydrophobic interaction between tetramers which results in a non-covalent polymerization. Oxy-Hb S is not affected but the solubility of deoxy-Hb S is greatly decreased as indicated below.

$$\frac{\text{Solubility of oxy-Hb A}}{\text{Solubility of deoxy-Hb A}} = 2; \qquad \frac{\text{Solubility of oxy-Hb S}}{\text{Solubility of deoxy-Hb S}} = 50$$

This is explained by the need for a binding site which is present in deoxy-Hb S but absent in oxy-Hb S. The correctness of this hypothesis will not be known until the 3-dimensional structure of deoxy-Hb S is elucidated.

As a consequence of this change, upon deoxygenation Hb S
precipitates within the cell forming a semisolid gel. Under
the microscope, tactoids (a type of liquid crystal) are
visible, approximately 1-15 microns long. This precipitation
causes the characteristic shape-change of the erythrocytes.
The mildness of the heterozygous condition can be explained by
the presence in the erythrocytes of roughly a 1:1 mixture of
Hb S and normal Hb A. Thus, it requires much more complete
deoxygenation to form enough solid to cause sickling. The
homozygote has only Hb S within his erythrocytes.

b. As implied above, almost all of the clinical findings in sickle-
cell anemia can be explained by the precipitation and cell
shape-change. The sickled cells are recognized as abnormal by
the reticuloendothelial system and phagocytized by the RE
cells, resulting in a hemolytic anemia. The "crises" which
occur in homozygotes (and perhaps occasionally in heterozygotes)
are caused by the inability of sickled cells to pass through
the finest capillaries. Once a capillary is blocked and the
blood flow through it ceases, deoxygenation of the static
erythrocytes increases, more cells sickle, and the blockage
becomes larger. An infarct of this sort can occur anywhere
within the body. The normal circulatory rate does not keep
cells deoxygenated for the 15 seconds required for sickling
to begin however, and crises occur only when circulation
slows or hypoxia is present.

c. The sickle-cell allele is transmitted as an autosomal recessive
characteristic. The trait is present in about 8% of the
Afro-American population and to a much greater extent in some
Black African populations. The homozygous condition causes
about 60,000-80,000 deaths per year among African children and
incapacitates many more. Hb S also occurs in India (primarily
among primitive tribes) and in the Mediterranean area.
Although there is some disagreement, it seems probable that
the gene has remained in these populations because of
resistance to the type of malaria caused by the parasite
Plasmodium falciparum which heterozygosity confers. This
"immunity" is present even in young children (less than two
years) who have not developed more conventional forms of
resistance and to whom the disease is quite lethal. A possible
mechanism for this resistance involves infestation by
P. falciparum and growth of the parasite within the red cells.
Erythrocytes so infected tend to adhere to the vessel walls
where they become deoxyganated, sickled, and phagocytized by
the RE cells. Another explanation is that for some reason,
P. falciparum does not thrive on Hb S. This appears unlikely
since sickle cells are as capable as normal cells of supporting
P. falciparum in culture.

d. Recent attempts at treating sickle-cell disease with urea and
    cyanate (CNO^-) have been highly controversial. Particularly
    regarding the use of urea, published results have been
    contradictory. Urea, being a well-known protein denaturant,
    might be able to decrease the presumed hydrophobic interactions
    which maintain deoxy Hb S in the semisolid state. However,
    the in vivo levels which appear to be necessary to produce such
    an effect are difficult to attain clinically and produce
    severe diuresis. It has also been reported that certain
    African tribes imbibe cow's urine (presumably high in urea) to
    ameliorate sickle cell disease. Cyanate, through its ability
    to carbamylate free amino groups, could also decrease this
    sort of interaction. Urea in solution is known to be in
    equilibrium with small amounts of cyanate

$$H_2N-\overset{\overset{\textstyle O}{\|}}{C}-NH_2 \quad \rightleftharpoons \quad NH_4^{\oplus} + N\equiv C-O^{(-)},$$

    raising the possibility that the two substances act by the
    same mechanism.

    Alternatively, Perutz's x-ray work has demonstrated that two
    of the six salt bridges needed to maintain the deoxy-conformation
    are between the α-carboxyl of Arg α141(HC3) and the α-amino
    group of Val α1(NA1) on the other α-subunit. Any α-amino group
    carbamylation should decrease the stability of the deoxy- form,
    thereby increasing O_2-affinity and making it more difficult to
    deoxygenate the cells. Carbamylation of these N-terminal
    valines has been shown to occur with oral and intravenous
    cyanate administration and the oxygenation curve is shifted to
    the right. Studies are currently being conducted to determine
    the usefulness of cyanate therapy in the chronic treatment of
    sickle-cell anemia. The N-terminal carbamylation also prevents
    the formation of carbamino compounds (discussed later) but
    does not cause any significant drop in pH which might be expected
    to result from an increase in free CO_2. This appears to be
    caused by the buffering provided by a slight increase in plasma
    bicarbonate which also occurs. There are still many problems
    to be solved and much yet to be understood about this disease.

e. Hemoglobin C is present primarily in American Blacks and in
    inhabitants of Accra in Ghana, West Africa. The frequency of
    Hb C and Hb S and their occurrence in the same population
    probably contributes to the frequency of Hb S-Hb C heterozygotes
    (mentioned previously). Hb C-Hb A heterozygotes are
    asymptomatic, but Hb C homozygotes show a decreased erythrocyte

lifespan and splenomegaly. The Hb C mutation involves β6(A3) Glu → Lys and shows a decreased Hb solubility (not nearly as severe as in Hb S disease) which results in the appearance of Hb crystals in blood smears from these patients.

4. Although α- and β-mutants have been discussed so far and are much more widely known, γ- and δ-chain mutants have also been found. These are much harder to study because of poorer availability of Hb F and Hb A_2. Another reason that such abnormalities are rarely seen is because these are minor hemoglobins postnatally. Consequently, the occurrence of overt clinical symptoms associated with γ- and δ-chain variants is rare or non-existent. At least six δ-chain mutants are known, all of which are stable. None involve residues which contact the heme or are involved in interchain contacts. In addition to the known heterogeneity of fetal Hb (G_γ-and A_γ-chains, mentioned earlier), six mutant γ-sequences have been determined (involving both A-and G-chains) and several others have been reported but not sequenced.

5. Deletion and addition mutants of hemoglobin are known. They are listed in Table 21. Note that since Hb-M-Freiburg is one of the methemoglobins, it is also listed in Table 21. Mutants of these types can result from unequal crossing over within the same gene producing a deletion or between different genes producing either a deletion or addition; mutation of a termination codon to one which codes for an amino acid (causing an increase in chain length); and perhaps in other ways. Hb Constant Spring (an α-chain extension mutant) probably resulted from the mutation

$$\text{UA(A,G)} \longrightarrow \text{CA(A,G)}$$
$$\text{Term} \qquad\qquad \text{Gln}$$

since residue α142 (the first one after the normal C-terminus) is Gln. Recall that "Term" designates a termination codon (see Table 11, page 330). It could also be the product of a cross-over. It is difficult to explain Hb Tak in this way, however, because residue β147 (the first new one added to the normal C-terminus) is Thr. None of the codons for Thr can be obtained from the three usual Term codons by a single base change.

$$\text{Term codons} \begin{cases} \text{UAA} \\ \text{UAG} \\ \text{UGA} \end{cases} \quad \begin{matrix} \text{ACU} \\ \text{ACC} \\ \text{ACA} \\ \text{ACG} \end{matrix} \Big\} \ \text{Thr codons}$$

Table 21. Deletion and Addition Mutants of Hemoglobin

Hemoglobin	Mutation	Remarks
Hb–Gun Hill	Deletion of residues 93–97 in the β-chain	Deletion near the heme position; no heme binding to β-chains
Hb–Leiden	Deletion of residue 6 or 7 (both Glu) in the β-chain	No clear clinical symptoms; mildly abnormal red cell morphology
Hb–Tochigi	Deletion of residues 56–59 in the β-chain	Unstable hemoglobin
Hb–M–Freiburg	Deletion of residue 23 in the β-chain	One of the methemo-globins, discussed later
Hb–Constant Spring	Addition of 31 residues to the C-terminus of the α-chain	Frequently inherited with α-thalassemia; occurs mostly in persons of Cantonese extraction; symptomless by itself
Hb–Tak	Addition of 10 residues to the C-terminus of the β-chain	Asymptomatic

6. In the chapter on carbohydrate metabolism, the sensitivity to oxidizing substances, especially primaquine, of persons lacking red cell glucose-6-phosphate dehydrogenase (G6PD) was discussed. In those patients, the drugs can trigger hemolytic episodes because of a lack of reducing power (i.e., glutathione) in the erythrocytes. Many, perhaps all, of the unstable Hb variants are similarly unstable to oxidants. In particular, sulfonamides precipitate hemolytic episodes in persons with Hb Zürich. The α-thalassemia known as hemoglobin-H disease confers sensitivity to sulfisoxazole, another drug which affects persons deficient in G6PD. These and other situations in which an inherited characteristic causes persons having the characteristic to respond adversely to a medication are part of the field of pharmacogenetics (see also the chapter on drug metabolism).

Laboratory Evaluation of Hemoglobinopathies

The differential diagnosis of hemoglobinopathies is frequently made on the basis of laboratory findings. Some of the methods used are described below.

1. a. Historically, the most important method of identifying different hemoglobins has been electrophoresis at a pH of 8.6, using paper, cellulose acetate membranes, and starch and polyacrylamide gels as supports. The separation of different hemoglobins occurs because of differences in pI (isoelectric point; see pages 15-18) between proteins having non-identical amino acid sequences. At a particular pH, the differing isoelectric points cause the molecules to have unequal net charges and, hence, to migrate at different rates in an electric field. This is exactly the same as the electrophoretic separation of isoenzymes (see page 164). At pH = 8.6, the great majority of hemoglobin derivatives have a net negative charge and migrate toward the positive (anodic) pole. Most of the differences in charge at this pH between the several normal hemoglobins and the mutant molecules are due to changes in the number of Glu and Asp residues (negatively charged) and Lys and Arg residues (positively charged; protonated) in the peptide chains. Additionally, the same **substitution** in different parts of the hemoglobin can result in different effects on the pI value, since the ionization constant (pK) of a group is affected by its local environment. Examples of these are shown in Figure 69 and Table 22.

Figure 69. Relative Electrophoretic Mobilities of Some Normal and Abnormal Hemoglobins (not drawn to scale)

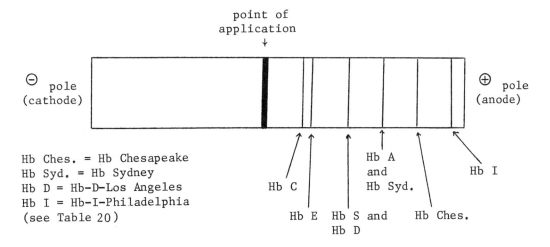

Hb Ches. = Hb Chesapeake
Hb Syd. = Hb Sydney
Hb D = Hb-D-Los Angeles
Hb I = Hb-I-Philadelphia
(see Table 20)

Table 22. Amino Acid Substitutions and Net Charge Alterations in
          Seven Hemoglobins Whose Relative Mobilities are Shown in
          Figure 69.

Hemoglobin	Amino Acid Change*	Charge Alteration in Tetramer*
C	Glu → Lys	+4
E	Glu → Lys	+4
S	Glu → Val	+2
D-Los Angeles	Glu → Gln	+2
Sydney	Val → Ala	0
Chesapeake	Arg → Leu	-2
I-Philadelphia	Lys → Glu	-4

* Normal for amino acid sequence and net charge is Hb A.

---

b. As was implied above, the pH has a great deal to do with the
    mobility. A higher pH will make the molecules more negative
    while a lower pH will make them more positive. At pH's
    around six, histidyl residues begin to contribute to the net
    protein charge. This may be a contributing factor to the
    separation of Hb A and Hb F at pH 6.5. These two hemoglobins
    migrate together at pH 8.6 on cellulose acetate. The Hb A-
    Hb F separation provides an example of another phenomenon
    which affects electrophoretic mobility: interaction of the
    molecules with the support material. On starch gels, at pH 8.6,
    Hb A and Hb F move as clearly separate bands, while they are
    indistinguishable on cellulose acetate at this pH. Hb F is
    discussed again later in this section. Note also that Hb C
    and Hb E do not migrate quite together even though they have
    the same net charge at pH 8.6. This could be due to one of
    the effects discussed above. Care must be taken in varying
    the pH since it is possible to denature hemoglobin and
    precipitate it. Abnormal hemoglobins are especially unstable
    in this regard and, in fact, this provides one way of

qualitatively testing for their presence (discussed below). Mutants also dissociate into dimers and monomers much more readily than does Hb A. Since the α- and β-subunits show up as separate bands during electrophoresis, their appearance is suggestive of an unstable hemoglobin. Abnormal combinations of normal chains also can be detected by electrophoresis. For example, Hb H (β_4, found in one of the alpha-thalassemias) migrates faster than Hb A at pH 8.6 on a cellulose acetate support. Hb H can also be detected by staining (see below).

2. Two modifications of classical electrophoresis have also proved useful in studying hemoglobin.

    a. Isoelectric focusing is based on the observation that in electrophoresis along a pH gradient, a molecule will migrate to the pH which corresponds to its isoelectric point (pI = the pH at which a zwitterion has zero mobility in an electric field; see page 16). This method requires a pI difference of 0.02 pH units between two molecules in order to separate them and is reported to separate minor hemoglobins which are too similar to be resolved by ordinary electrophoresis.

    b. Microelectrophoresis is capable of utilizing as a sample a quantity of hemoglobin as small as that contained within a single erythrocyte. Successful separations have been made of Hb A from Hb F, Hb A_2, Hb S, Hb C, and Hb J.

3. As is apparent from the above discussion, electrophoresis is not adequate to detect nearly all of the abnormal hemoglobins. Other methods have also been very useful in this regard.

    a. It is interesting that Hb A and Hb F migrate very close together on cellulose acetate at pH 8.6, although the β- and γ-subunits have only 39% homology. The most useful method of routinely distinguishing between these two Hb's is on the basis of their relative stability toward alkali. At pH 12.8, Hb A denatures 100 times as rapidly as Hb F. Consequently, any material which migrates in the position of Hb F and Hb A after such treatment must be Hb F.

    b. As indicated in Table 22, Hb S and Hb D have the same net charge and are not separable by electrophoresis at pH 8.6. Upon deoxygenation, Hb S forms a gel while Hb D does not, thus providing a basis for differentiation. Also, these two mutants do not, in general, occur in the same population.

    c. Cytological techniques have likewise proven useful, one example being the observation of cell morphology alteration in sickle-cell disease. Heinz (Heinz-Ehrlich) bodies and other "Heinz-like" inclusion bodies are frequently observed in erythrocytes containing abnormal hemoglobin. They consist of denatured, precipitated hemoglobin resulting from the instability of the mutant. The Heinz-like bodies found in Hb Zürich and Hb H (four beta-subunits (β_4), found primarily in one of the alpha-thalassemias) are stained by brilliant cresyl blue. The Hb H inclusion bodies appear as small, irregularly shaped blue dots within the cell whereas those from patients with Hb Zürich are seen as single, unusually large, blue bodies eccentrically placed and lying close to the cell membrane. Hb H bodies are not visible in Wright's stain preparations but can be stained with crystal violet following the addition of sodium metabisulfite or other agents capable of oxidizing hemoglobin. The Hb Zürich inclusion bodies are visible in Wright's stain. Hb H is also readily distinguished from Hb A and Hb Zürich by its electrophoretic behavior (see above).

d. In keeping with the general instability of the abnormal hemoglobins, they are much more readily denatured by heat than is Hb A. Heating a mutant for 30 minutes at 60°C usually causes almost complete denaturation, while Hb A is hardly precipitated under the same conditions. The sulfhydryl groups of the mutant chains are frequently more exposed than those in normal Hb and they are therefore more reactive towards parachlormercuribenzoate (PCMB; see the section on enzyme inhibitors). Treatment with PCMB for several hours precipitates many mutant peptides, whereas it only causes dissociation of Hb A.

4. One of the most useful methods of studying mutant hemoglobins in which the change does not show up in electrophoresis is peptide mapping or fingerprinting. In this technique, the chains are separated and partially digested by one of several proteolytic processes (see the section on protein sequence determination), yielding smaller oligopeptides. These are then chromatographed by standard methods on paper or silica gel. After the migration (development) is finished, the support is dried and the peptides are located by spraying the chromatogram with ninhydrin or some other reagent which reacts with the peptides to give a color. A given amino acid sequence, when digested and chromatographed under specified conditions, will always give the same peptide map. By comparison of a known map (for, say, the β-subunits present in Hb A) with the map of an unknown variant, the difference in spot pattern (usually amounting to the absence of one of the standard spots in the mutant chromatogram and the appearance of a new spot) can be used to locate the altered residue. In addition, since the composition of the peptides which cause the spots in the standard are known, the specific residue changed can be pinpointed. Even when such results seem to clearly indicate the mutation however, a new hemoglobin variant cannot be described as unequivocally characterized until its entire amino acid sequence is known.

5. The growing knowledge of hemoglobin mutations and their importance
    in the pathology of certain diseases has led to the development of
    screening programs to detect some of the most common and most
    harmful alleles. The primary aim of such screening programs is to
    locate carriers (heterozygotes). There is increasing evidence,
    for example, that heterozygotes for Hb S can, under severely
    hypoxic circumstances, suffer acute sickling crises. Such screening
    not only will aid these persons in avoiding this danger, it should
    provide a broader base for genetic counseling and thereby
    decrease the number of heterozygotes conceived. One such program
    which has been established makes use of electrophoresis and
    differentiates between **sickle-cell** trait, disease, and sickle cell-
    Hb C disease, as well as detecting patients with thalassemia minor,
    hereditary persistence of Hb F, and hemoglobin types AC, CC, and B_2
    (a δ-chain variant of no apparent clinical significance). A method
    has also been reported for detecting Hb S homozygosity in fetuses
    as early as the first trimester of gestation. Another screening
    program for the detection of thalassemia has been proposed. After
    sickle-cell disease, β-thalassemia is probably the most serious
    hereditary hemoglobinopathy in the United States.

Derivatives of Hemoglobin

1. Carboxyhemoglobin (carbon monoxide hemoglobin)

    a. Carbon monoxide (CO) is a relatively inert (chemically
        non-reactive), odorless gas. The affinity of Hb for CO is
        210 times as great as the affinity of Hb for O_2. This means
        that   in the equilibrium reaction

$$HbO_2 + CO \;\; \underset{\longleftarrow}{\overrightarrow{\hspace{2cm}}} \;\; HbCO + O_2 \qquad\qquad K_{eq} = 2.1 \times 10^2$$

    the equilibrium lies far to the right. The type of bond between
    Hb and CO and the binding site, are the same as those for the
    Hb-oxygen complex. That is, CO binds to the sixth position of
    the heme iron with a coordinate-covalent bond.

    b. Because of the identical binding sites and the greater binding
        strength of CO, when both CO and O_2 are present in appreciable
        quantities, the CO is bound preferentially and O_2 is excluded.

    c. CO poisoning may occur in the presence of automobile exhaust,
        poorly oxygenated coal fires in stoves and furnaces, or any
        other situation where incomplete combustion of a carbon-
        containing compound occurs.

    d. For clinical diagnosis, history, unconsciousness, cherry-red
        discoloration of the nailbeds and mucous membranes, and
        spectrophotometric analysis of the blood are used.

e. Treatment for carbon monoxide poisoning consists of purging
    the body of CO. Although the equilibrium

$$CO + HbO_2 \; \underset{\longleftarrow}{\overset{\longrightarrow}{}} \; HbCO + O_2$$

has a large equilibrium constant (is predominantly to the right
when CO and O_2 are present in similar amounts), a large excess
of O_2 will favor the formation of HbO_2 and the release
of CO by the law of mass action. Breathing an atmosphere of
95% O_2 and 5% CO_2 will usually eliminate CO from the body in
30-90 minutes.

f. It is interesting that the oxidation of heme to biliverdin
    produces CO in amounts equimolar to the amount of biliverdin
    (and, hence, bilirubin) formed (see later, under Hb catabolism).
    The CO is transported via the blood to the lungs, where it is
    released. Although no cases are known wherein endogenous
    CO proved toxic, it has been suggested that this carbon
    monoxide may contribute significantly to air pollution.

2. Carbaminohemoglobin

a. As was mentioned previously (see Chapter 1 and earlier in this
    chapter), some of the CO_2 in the blood stream is carried as
    carbamino compounds. These form spontaneously, in a readily
    reversible reaction, with the free α-amino groups of the
    N-terminal amino acids of the Hb chains

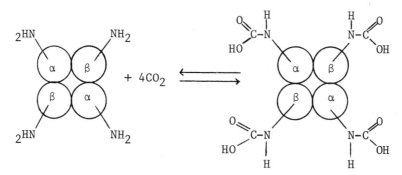

Hemoglobin   Carbaminohemoglobin

b. Although this means that Hb can directly carry as many moles of
    CO_2 as it can O_2, the HCO_3^- system seems to be a much more
    important way of transporting CO_2 in the blood. Evidence for
    this comes from the absence of any marked degree of acidosis
    when the N-terminal amino groups are blocked by carbamylation
    with CNO^- (see sickle-cell anemia), forming carbamylhemoglobin.

c. The presence of carbamino groups decreases the affinity of Hb for O_2. This is a mechanism whereby the P_{CO_2} can affect oxygen affinity independent of any effect which the carbon dioxide may have on the pH and it provides an additional process which may influence the position of the oxygen dissociation curve under various conditions.

3. Methemoglobin (see also the section on red cell metabolism in Chapter 4)

a. Hemoglobin in solution, on standing in the presence of oxygen, is slowly oxidized to methemoglobin, a derivative of hemoglobin in which the iron is present in the ferric (Fe^{+3}) state. In metHb, the ferric ion cannot accept oxygen because the iron is bound tightly to a hydroxyl group or to some other anion. The heme porphyrin (protoporphyrin III) containing an Fe^{+3} ion (instead of Fe^{+2}) is known as a hemin.

b. In the body, there is a small but constant amount of methemoglobin produced. It is re-reduced by specific enzymes (methemoglobin reductases). The electrons are furnished primarily by NADH which is generated in glycolysis (refer to carbohydrate metabolism). The reductases isolated from human red cells are also capable of using NADPH, but to a lesser extent. Because of this continuous formation of metHb, an inability to re-reduce it is pathological. The resultant condition is called <u>methemoglobinemia</u> and causes tissue anoxia.

c. There are two basic types of hereditary methemoglobinemia.

i) Congenital methemoglobinemia may be due to a deficiency in the erythrocytes of one or more of the reducing enzymes which convert metHb back to normal Hb. This is usually a recessive trait. MetHb values may range from 10 to 40% of the total Hb (normal = 0.5%). Treatment involves administering an agent that will reduce the metHb. This is usually ascorbic acid or in severe cases, methylene blue. The reactions involved are shown later.

ii) There may also be a defect in the hemoglobin molecule, making it resistant to re-reduction by both the metHb reductases and exogenous agents such as ascorbate and methylene blue. These abnormal hemoglobins are collectively called the hemoglobins M (HbM). The defects of the six known types are summarized in Table 20. The disease is inherited as a dominant trait. In the four types involving a His → Tyr mutation, the problem seems to be that the phenolic hydroxyl group of Tyr forms a stable complex with Fe^{+3}, making it resistant to conversion back to Fe^{+2}. Hb-M-Milwaukee has a similar problem, having a glutamic acid residue substituted for a valine at position β67(E11) near the distal histidine. The γ-carboxyl group of the Glu is able to form a stable complex with Fe^{+3}. Although a brownish cyanosis is characteristic of blood containing Hb M, diagnosis should be confirmed by examination of the absorption spectrum, either of the Hb or of its CN^- derivative. The cyanide derivative spectrum is preferred, since, of the two, it differs from the Hb A spectrum to a greater extent. Some of the M hemoglobins are unable to form CN^- derivatives, presumably due to a Tyr or Glu blocking the sixth iron positions where CN^- must bind. In these cases, the absorption spectrum of the Hb M itself must be examined.

d. A third type of hereditary methemoglobinemia could occur, in which the hemoglobin is mutated in the region to which the methemoglobin reductase binds to perform its function. In the mutant, the reductase could not attach to the Hb, thereby leaving the iron in a ferric state. This type has not been reported. When more is known about the mechanism of interaction between metHb and metHb reductase, perhaps reasons will become apparent why this is so.

e. Acquired Acute Methemoglobinemia

    i)   This is a relatively common condition caused by introduction into the body of a variety of drugs (all oxidizing agents or capable of being converted to oxidizing agents in the blood) such as phenacetin, aniline, nitrophenol, aminophenol, sulfanilamide, and inorganic as well as organic nitrites and nitrates. The condition is less commonly produced by chlorates, ferricyanide, pyrogallol, sulfonal, and hydrogen peroxide. These compounds appear to act as catalysts for the oxidation of Hb by oxygen, producing metHb.

    ii) The symptoms of this condition include brownish cyanosis, headache, vertigo, and somnolence (sleepiness; unnatural drowsiness).

    iii) Diagnosis is based on history, occurrence of the brownish cyanosis, and the presence of excessive amounts of metHb (measured spectrophotometrically).

iv) Treatment is usually not difficult, consisting of removal of the offending substance and the administration of ascorbic acid (in mild cases) or methylene blue (in severe cases). These reducing agents function according to the reactions below (MB = methylene blue, oxidized; MBH_2 = methylene blue, reduced).

$$MBH_2 \rightleftharpoons MB + H_2 \text{ (gas)}$$

$$2 H_{2 \text{(gas)}} + Hb(Fe^{+3})_4 \longrightarrow Hb(Fe^{+2})_4 + 4H^+$$

(The release of H_2 by MBH_2 is spontaneous since methylene blue is autoxidizable.)

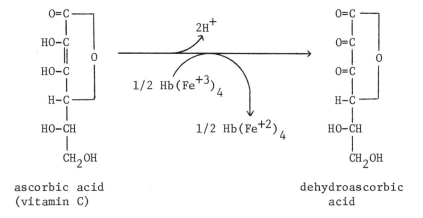

ascorbic acid   dehydroascorbic
(vitamin C)   acid

4. Sulfhemoglobin

   a. Many of the drugs that are responsible for formation of methemoglobin also produce sulf-Hb, a greenish Hb derivative apparently formed only by oxy-Hb. The two appear together in poisoning by sulfanilamide, acetanilid, and sulfanol.

   b. The structure of sulf-Hb is unknown, but the sulfur appears associated with the pyrrole nucleus rather than with the iron. Symptoms are the same as those for metHb.

   c. There is no specific treatment. The sulfhemoglobin will eventually disappear (in mild cases) as the cells containing it are destroyed.

5.   Cyanmethemoglobin

   a.   Cyanide poisoning <u>does not</u> produce cyanohemoglobinemia, nor
        does it produce cyanosis. Cyanosis (bluish coloration,
        especially of the skin and mucous membranes) is caused by an
        excessive concentration (over 5 gm/100 ml) of deoxygenated
        (sometimes called reduced) hemoglobin in the capillary blood.
        Cyanosis may occur:

      i)   when there is a diminished uptake of oxygen in the lungs
           as in extensive pulmonary disease, strangulation,
           pulmonary arterial constriction and certain types of
           congenital heart disease;

      ii) when there is a circulatory inefficiency as in heart
           failure, shock, and localized impairment of venous return;

      iii) when the blood is abnormal (e.g., Hb M) and cannot carry
           enough oxygen.

   b.   Cyanide poisoning produces histotoxic anoxia by poisoning
        cytochrome oxidase and other respiratory enzymes. This
        prevents the utilization of O_2 by tissues. It is detected by
        the characteristic odor of HCN gas (bitter almonds) and by
        laboratory tests (absorption spectra, tests for CN^-).

   c.   Treatment of cyanide poisoning (discussed with enzyme inhibition
        in Chapter 3) provides an example of selective toxicity. It
        consists of diverting the cyanide into the production of
        cyanmethemoglobin. First, some of the normal hemoglobin is
        converted to methemoglobin by intravenous infusion of a solution
        of $NaNO_2$ or inhalation of amyl nitrite. Once the methemoglobin
        is formed, CN^- can replace the OH^- at position 6 of the iron.

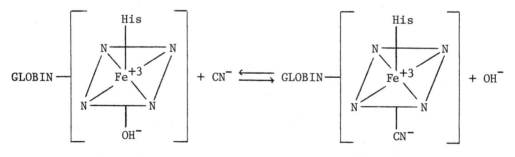

           Methemoglobin   Cyanmethemoglobin

The four nitrogens complexed to the iron in the drawing above are from the pyrrole rings of the porphyrin. The histidine in position 5 of the iron complex is the proximal His, $\alpha 87$(F8) or $\beta 92$(F8). Cyanmethemoglobin is no more toxic than methemoglobin and cells containing it can be eliminated by normal body processes. The cyanide bound to the metHb is always in equilibrium with free CN^- and this uncomplexed cyanide is converted to thiocyanate (SCN^-; non-toxic) by administration of thiosulfate (see also pages 105-106).

Synthesis of Porphyrin and Heme

1. Porphyrin consists of four pyrrole molecules joined by methene
    (-CH=) groups. If methylene (-CH$_2$-) bridges are used instead, the
    ring system is called a porphyrinogen. Specific porphyrins and
    porphyrinogens differ from each other in the groups attached to
    the numbered positions (1-8) below. The bridges are designated
    α-δ and the pyrrole rings are A-D. Porphin, the example given,
    has hydrogens at all 8 substituent positions.

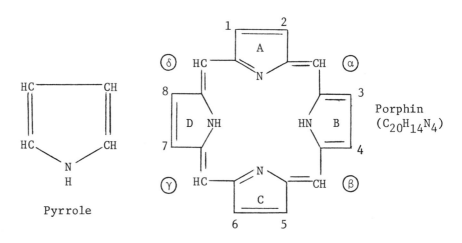

Pyrrole

Porphin
(C$_{20}$H$_{14}$N$_4$)

2. Porphyrins can form complexes with metal ions. This property is
    very important in their functioning in biological systems.
    Examples: Heme is an iron porphyrin; chlorophyll is a magnesium
    porphyrin; vitamin B$_{12}$ is a cobalt porphyrin.

3. These <u>metalloporphyrins</u> are conjugated with proteins to form a
    number of biologically important macromolecules. Some of these are

    a. Hemoglobin: iron porphyrin + globin; functions as a reversible
        oxygen carrier; molecular weight 64,500; contains four Fe^{+2}
        atoms.

    b. Erythrocruorins: iron porphyrin + protein; their function in
        some invertebrates is similar to that of hemoglobin in higher
        animals.

    c. Myoglobins: iron porphyrin + protein; present in muscle cells
        of vertebrates and invertebrates; participate in respiration.

    d. Cytochromes: iron porphyrin + protein; participate in electron
        transfer reactions (oxidation and reduction reactions).

e. Catalases and Peroxidases: iron porphyrin enzymes; catalyze
    the reactions

$$2\ H_2O_2 \xrightarrow{\text{Catalase}} 2\ H_2O + O_2$$

$$H_2O_2 + XH \xrightarrow{\text{Peroxidase}} 2\ H_2O + X$$

f. Tryptophan pyrrolase: iron porphyrin enzyme; catalyzes the
    reaction

$$\text{Tryptophan} \xrightarrow{[O]} \text{Formyl Kynurenine}$$

Biosynthesis of Porphyrins

1. Formation of δ-aminolevulinic acid (ALA)

Note: B_6-PO_4 = pyridoxal phosphate; both of the above steps are
catalyzed by the same enzyme.

2.   Formation of Porphobilinogen (PBG)

2 molecules of ALA   Porphobilinogen (PBG)

Note: A = acetate; P = propionate.   PBG is more conveniently
written as:

PBG

3.   Synthesis of Uroporphyrinogens Types I and III

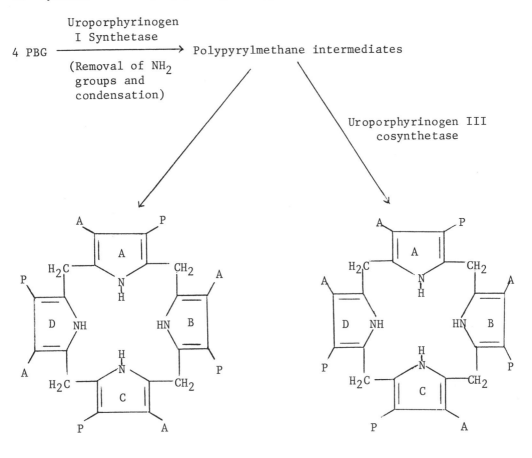

                                  Uroporphyrinogen
                                  I Synthetase
4 PBG ───→ Polypyrylmethane intermediates
                                  (Removal of NH$_2$
                                  groups and
                                  condensation)

                                              Uroporphyrinogen III
                                              cosynthetase

Uroporphyrinogen I   Uroporphyrinogen III

Note: The only difference between types I and III is the
orientation of the substituents on ring D.

4. Synthesis of Protoporphyrin III and Heme

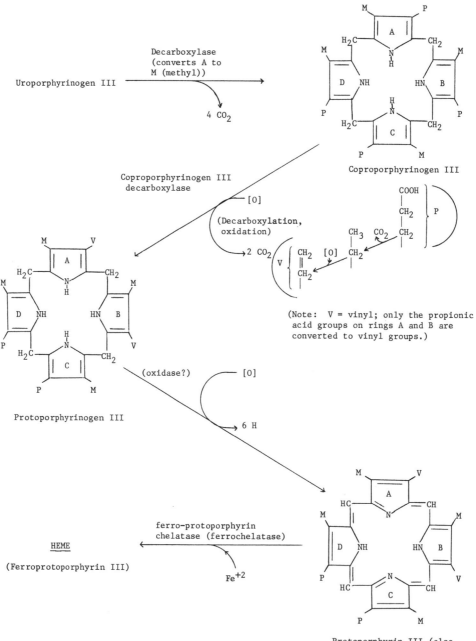

Uroporphyrinogen III → Decarboxylase (converts A to M (methyl)) → 4 CO₂ → Coproporphyrinogen III

Coproporphyrinogen III decarboxylase [O] (Decarboxylation, oxidation) → 2 CO₂

(Note: V = vinyl; only the propionic acid groups on rings A and B are converted to vinyl groups.)

Protoporphyrinogen III → (oxidase?) [O] → 6 H → Protoporphyrin III (also called protoporphyrin IXα)

ferro-protoporphyrin chelatase (ferrochelatase) + Fe⁺² → HEME (Ferroprotoporphyrin III)

Figure 70. Summary of the Biosynthesis of Porphyrins and Heme

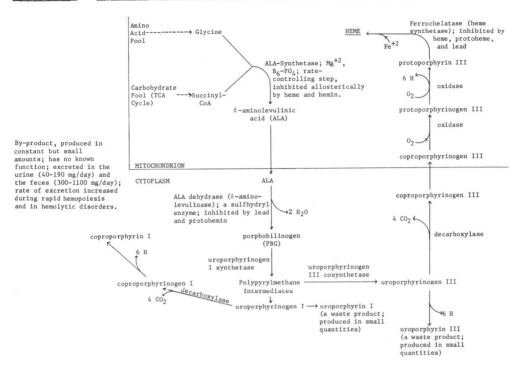

Comments on Porphyrin and Heme Synthesis

1. The ALA-Synthetase catalyzed reaction

    a. The first reaction unique to porphyrin synthesis is the
        formation of δ-aminolevulinic acid (ALA) from one mole each of
        succinyl-CoA (a TCA cycle intermediate) and glycine. It
        requires B_6-PO_4 (pyridoxal phosphate) and Mg^{+2} as cofactors.
        The pyridoxal phosphate is apparently tightly bound to the
        enzyme. The indicated intermediate, α-amino-β-keto-adipic
        acid, has never been isolated. It is perhaps not a true
        intermediate as the condensation and decarboxylation steps may
        occur simultaneously. Since ALA synthetase is found in the
        mitochondrial matrix, it does not occur in mature erythrocytes
        which lack functional mitochondria. In rat hepatocytes,
        although the enzyme is found in the cytosol, it appears to
        be functional only when present within the mitochondria.

    b. In experimental animals, a deficiency of pantothenic acid (needed
        for CoASH and, hence, succinyl-CoA synthesis) can prevent
        heme synthesis and cause anemia. Similarly, a lack of
        vitamin B_6 or the presence of compounds which block the
        functioning of this cofactor (isonicotinic hydrazide, used in
        the treatment of tuberculosis; L-penicillamine, a chelating
        agent used in the treatment of Wilson's disease, discussed
        under copper metabolism) causes decreased production of ALA
        and heme, leading eventually to anemia. Heme synthesis also
        requires that the TCA cycle be functional and that an oxygen
        supply be available (active respiration is needed).

    c. The primary regulatory step (in the liver) is apparently the
        synthesis of ALA from succinyl-CoA and glycine, catalyzed by
        the enzyme ALA synthetase. The normal end product, heme,
        when accumulated in excess and not utilized in the synthesis
        of conjugated proteins (the principal one being hemoglobin),
        is oxidized to hematin. Hematin contains a hydroxyl group
        attached to the trivalent iron. When the hydroxyl group is
        replaced by a chloride ion, the product is known as hemin
        (or hemin chloride). Hemin and heme allosterically inhibit
        the ALA synthetase reaction (an example of hepatic feedback
        or end-product control of an enzyme). It has also been reported
        that heme, functioning as a corepressor, blocks the formation
        of ALA-synthetase protein at the DNA level. The repressor,
        which binds to the DNA, is made of a protein part (aporepressor)
        and heme. This mechanism controls the rate of structural gene
        transcription and hence the rate of synthesis of the enzyme
        protein. It and other aspects of the control of heme synthesis
        are discussed in greater detail in the chapter on nucleic acids.

Presumably, individuals with a defect (loss of control) in this system can show the condition of porphyria (discussed later in this section) or can develop hepatic porphyria when certain drugs (such as barbiturates) are administered. The postulated mechanism is that these persons produce defective repressor molecules which combine with barbiturates (or sex steroids) at the corepressor site, giving rise to an inactive complex. The result is the repressor does not function and, consequently, overproduction of ALA synthetase occurs, resulting in porphyria.

2. The ALA-Dehydrase catalyzed reaction

    a. In this step, two molecules of ALA are condensed to form one molecule of porphobilinogen with the release of two molecules of water. Since ALA dehydrase is present in the cytosol, the ALA must leave the mitochondria prior to this reaction.

    b. This enzyme may play a role in the regulation of heme synthesis. ALA dehydrase isolated from <u>Rhodopseudomonas spheroides</u> appears to be an allosteric enzyme, activated by K^+ and inhibited by heme.

    c. ALA dehydrase is a sulfhydryl enzyme and reduced glutathione is essential for its activity. Agents which bind to -SH groups, such as iodoacetamide, p-chloromercuribenzoate, heavy metal ions, etc. inhibit the reaction (see also page 109). ALA dehydrase is apparently very sensitive to lead and in lead poisoning, there is a significant increase in the urinary excretion of ALA (normal is 2 mg/24 hr). The decrease in ALA-dehydrase activity in blood has been used as a measure of the extent of lead poisoning in clinical situations.

3. Synthesis of Uroporphyrinogens I and III

    a. The nature of these reactions is still not completely clear but certain things are becoming known. There appear to be two enzymes involved; uroporphyrinogen I synthetase and uroporphyrinogen III cosynthetase. Good evidence for this is when an enzyme preparation originally capable of making both types I and III is heated to 60°C, it ceases to make type III but continues synthesis of type I. This indicates that the cosynthetase is heat labile.

b. A likely sequence of events for this synthesis is indicated
below.

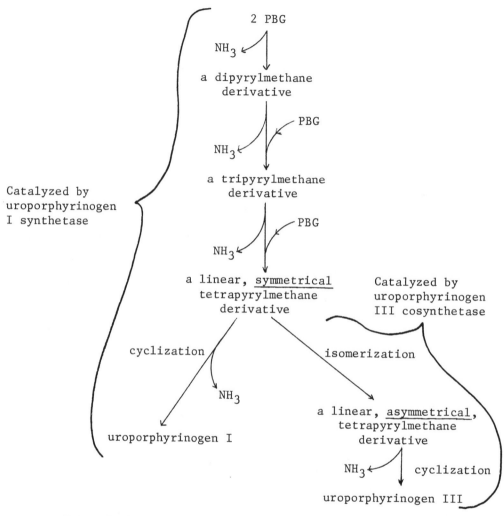

Although these steps remain to be proven, a dipyrrylmethane
derivative has been tentatively identified. It has also been
demonstrated that only four molecules of PBG are needed for
the synthesis of type III isomers, that formaldehyde is neither
consumed nor produced in the reaction, and that dipyrylmethenes
do not participate as reactants.

4. Formation of Heme

    a. The **end-product** of uroporphyrinogen I is coproporphyrin I which
        has no known function and is eliminated (see Figure 70). The
        type III uroporphyrinogen is ultimately converted to heme.

    b. Uroporphyrinogen III is converted in the cytosol to
        coproporphyrinogen III in a reaction catalyzed by
        uroporphyrinogen decarboxylase. The enzyme is highly
        specific for uroporphyrinogen and is inhibited strongly by
        mercury, copper, manganese, and oxygen. The oxygen probably
        functions by promoting the oxidation of uroporphyrinogen III
        to uroporphyrin III, a waste product.

    c. The coproporphyrinogen then moves into the mitochondria for
        further conversions. It is first decarboxylated to
        protoporphyrinogen III by coproporphyrinogen III decarboxylase,
        then oxidized to protoporphyrin III (also called IXα). It is
        not clear whether both steps are catalyzed by the decarboxylase
        or if another enzyme is required for the oxidation. The
        decarboxylase **is** highly specific for type III uroporphyrinogen
        and the process requires oxygen.

    d. The terminal step in heme synthesis is catalyzed by
        ferrochelatase (also known as ferro-protoporphyrin chelatase
        or heme synthetase) which incorporates ferrous iron into
        protoporphyrin III. Reducing substances such as ascorbic acid,
        cysteine, or glutathione are required by the enzyme. The
        reaction is inhibited by heme as well as by lead.

5. Of the many possible porphyrin isomers, only types I and III are
    found in biological systems. Type I isomers have no known useful
    function and are eliminated in the urine and feces. The type III
    isomer, when chelated with iron to form heme or hemin, serves as
    the prosthetic group in hemoglobin, myoglobin, the cytochromes c,
    and other enzymes. The magnesium-porphyrin known as chlorophyll a,
    found in many photosynthesizing plants, is formed by a pathway
    very similar to that of heme up to the synthesis of magnesium-
    protoporphyrin III. It is then further modified for its role in
    photosynthesis. The synthesis of vitamin B_{12}, a chelate of cobalt
    with a porphyrin-derived nucleus, is also similar to heme synthesis
    up to a certain point. It too is finally modified by pathways
    unique to the bacteria in which it is made.

6. Biosynthesis of Hemoglobin

    The protein portion of hemoglobin is called <u>globin</u>. It is
    synthesized primarily in the nucleated erythrocytes of the bone
    marrow and in the reticulocytes, using the labile amino acid pool
    of the body. The synthesis occurs via the regular protein
    biosynthetic reactions involving DNA, mRNA, ribosomes, tRNA, amino
    acid activating enzymes, and so on (see the section on protein
    synthesis in Chapter 5). The α-and β-<u>chains</u> of hemoglobin are
    apparently under the control of different genes and yet are
    synthesized at the same rate. It is also of interest that hemin
    and protoporphyrin III, while inhibiting or decreasing the rate
    of synthesis of heme (discussed above), promote the synthesis of
    globin in an apparant attempt to maintain the proper ratio (1:1)
    of heme to globin. There are genetic defects in the globin
    production which give rise to abnormal hemoglobins. These were
    discussed earlier in this chapter.

7. Some Chemical Properties of Porphyrins

    a. The tertiary nitrogens of the pyrrole rings function as weak
        bases. The carboxyl groups function as acids. The
        iso-electric pH range is 3.0-4.5.

    b. Porphyri<u>nogens</u> are colorless while porphy<u>rins</u> are colored
        compounds, due to the conjugation introduced when the methylene
        bridges are converted to methenes. Porphyrins have
        characteristic absorption spectra in both UV and visible regions.
        All porphyrins show an absorption band at about 400 nm known
        as the Soret band.

    c. When porphyrin complexes with iron or other metals, its
        spectrum in the visible region changes. For example, in
        alkaline solutions, protoporphyrin gives rise to several
        sharp absorption bands at 540, 591, and 645 nm, while heme
        shows only a broad band with a plateau extending from 540 to
        580 nm.

    d. Heme can complex with Lewis bases. A hemochrome (hemochromogen)
        is a compound of heme with any two other ligands to give
        hexacoordinate Fe^{+2}.

8. As we shall see later in this section, breakdown (catabolism) of
    heme does not yield porphyrins but gives rise to products (bile
    pigments) which are toxic to the body and need to be eliminated.
    In general, ALA, PBG, and uroporphyrin are excreted in the urine
    while coproporphyrin (to a large extent) and protoporphyrin (solely)
    are excreted in the bile after appropriate transformations (discussed
    later).

The Porphyrias

1. Accumulation of coproporphyrins and uroporphyrins in the blood
with their excessive elimination in urine and feces is called
porphyria. Porphyrinuria refers to an excess of coproporphyrin
in the urine. Uroporphyrin is not present in the urine in
porphyrinuria.

2. Since the genetic control of porphyrin synthesis differs in the
erythroid tissue and the liver, a defect in one system may occur
independently of the other, giving rise to either erythropoietic
porphyrias or hepatic porphyrias.

3. There are two known types of hereditary hepatic porphyria.

   a. Acute intermittent porphyria (AIP; also called Swedish type
   porphyria) is associated with excessive urinary excretion of
   δ-aminolevulinic acid (ALA) and porphobilinogen (PBG) with
   little or no corresponding rise in either type I or type III
   porphyrinogens. Although ALA-synthetase activity is reportedly
   elevated in these patients, the lack of PBG polymerization
   suggests that the primary enzymic defect may lie in a lack of
   uroporphyrinogen I synthetase. It has now been observed that,
   in fact, the uroporphyrinogen I synthetase activity is
   depressed in the liver and erythrocytes from AIP patients.
   A family study has indicated this is probably the primary
   defect and that the ALA-synthetase activity elevation either
   reflects the lack of feedback control by heme and hemin or is
   an associated enzymic error. Clinical symptoms of this
   disorder include central and peripheral nervous disorder,
   psychosis, and hypertension, although these can vary a great
   deal. Death can occur due to respiratory paralysis. Since
   these individuals are unable to make porphyrins to any great
   extent, they are not photosensitive. This disorder is
   inherited as an autosomal dominant trait.

   b. A second type of hepatic porphyria is called South African
   porphyria or variegate porphyria. Patients with this disease
   excrete excessive amounts of both uroporphyrin and
   coproporphyrin (but not porphobilinogen). These individuals
   are photosensitive (the presence of porphyrins near the
   surface of the body results in light sensitization because
   porphyrins have the ability to concentrate radiant energy) but
   other symptoms vary widely, hence the cognomen variegate.
   This disease appears to be inherited as a dominant trait. The
   primary lesion is possibly overactivity of ALA synthetase.
   A block in heme synthesis distal to this enzyme seems unlikely.

4. A defect in the synthesis of type III isomers from polypyrrylmethane intermediates, possibly due to a deficiency of uroporphyrinogen III cosynthetase, produces <u>congenital erythropoietic porphyria</u> (Günther's disease). In this case, the type I porphyrin (principally uroporphyrin I) is formed, accumulates in the tissues, and is excreted in the urine. This deficiency in the production of the normal (type III) isomer gives rise to further increase in the levels of type I isomers by reducing the regulatory action on ALA synthetase. Red cells may also respond to excessive amounts of porphyrins by episodes of hemolysis. The compensatory increase in hemoglobin formation can then exaggerate the already increased production of type I porphyrins. The accumulation of type I porphyrins produces a pink to dark red color in the teeth, bones, and urine. These individuals are also sensitive to exposure to long wave ultraviolet light and sunlight. This disease appears to be inherited as an autosomal recessive trait.

5. The induced (non-hereditary) porphyrias and porphyrinurias can be caused by the following conditions. The chief urinary porphyrin appearing in each type is mentioned in the parenthesis. Toxic agents: (coproporphyrin III in some cases of lead poisoning, 8000 mg/24 hrs) chemicals, heavy metals, acute alcoholism, cirrhosis of the liver in alcoholics. Liver diseases: (coproporphyrin I, 100-600 mg/24 hrs) infectious hepatitis, mononucleosis, cirrhosis of the liver. Certain blood diseases: (coproporphyrin I) hemolytic anemias, pernicious anemias, leukemias. Miscellaneous: (coproporphyrin III) poliomyelitis, aplastic anemias, Hodgkin's disease.

Bilirubin Metabolism

Since iron and protein metabolism have already been covered, bilirubin is the only hemoglobin-derived material still to be discussed.

1. Sources and Formation of Bilirubin

    a. The heme moiety of hemoglobin is the predominant source of bilirubin, supplying about 80% of the total pigment produced (normally 250-300 mg of bilirubin is formed/24 hrs). The remaining 20% comes from the catabolism of other heme proteins listed previously.

    b. Bilirubin is produced in the reticuloendothelial cells of the liver, spleen, and bone marrow. These cells engulf older red blood cells, cause them to lyse and release hemoglobin, then catabolize the hemoglobin and form the pigment.

c. Although red blood cells contribute 80% of the bilirubin formed, the metabolic turnover time of hemoglobin (about 120 days) is slow compared to that of some of the hepatic hemoproteins. For example, the microsomal cytochromes, P-450 and b_5, that participate in the detoxication reactions of hormones, toxins, drugs, and of bilirubin itself, have a turnover time of between one and two days.

d. Epithelial cells of the renal tubules and the liver parenchymal cells play a role in the conversion of hemoglobin heme to bilirubin when the hemoglobin is produced due to intravascular hemolysis. Bile pigment can also be formed in the macrophages when they catabolize hemoglobin from the phagocytized red blood cells appearing due to blood extravasations.

e. The conversion of heme to bilirubin consists of two steps: the microsomal heme oxygenase system and the biliverdin reductase reaction (see Figure 71). The substrate, heme, effects its own catabolism by activating the heme oxygenase system, presumably by enzyme induction. In hemoglobinuria, the hemoglobin that is present in the glomerular filtrate is taken up by the kidney epithelial cells and stimulates the heme oxygenase system.

2. Some Properties of Bilirubin

a. Bilirubin is a yellow pigment with an absorption maximum at 450 nm. This property can be utilized in the direct measurement of bilirubin in the serum. However, carotene also absorbs near the same wavelength (440 nm), so that the ingestion of carotene (which is present in carrots, tomatoes, and mangoes) seriously limits the direct spectrophotometric measurement of bilirubin in adult sera. In neonatal sera, the direct measurement is of value in emergency situations. The more reliable methods involve coupling of the pigment with diazotized sulfanilic acid (discussed later).

b. Bilirubin is soluble in lipids (and in organic solvents) and consequently can diffuse freely across cell membranes. Inside the cell, bilirubin interferes with many crucial metabolic functions and most of the specific interactions are not understood. The brain damage which it causes has been ascribed to the ability of the pigment to uncouple oxidative phosphorylation on the basis of in vitro experiments. In vivo results, with levels of bilirubin which actually occur in certain diseases, do not support this mechanism.

Figure 71.   Microsomal Oxidation of Heme to Biliverdin and the Reduction of Biliverdin to Bilirubin

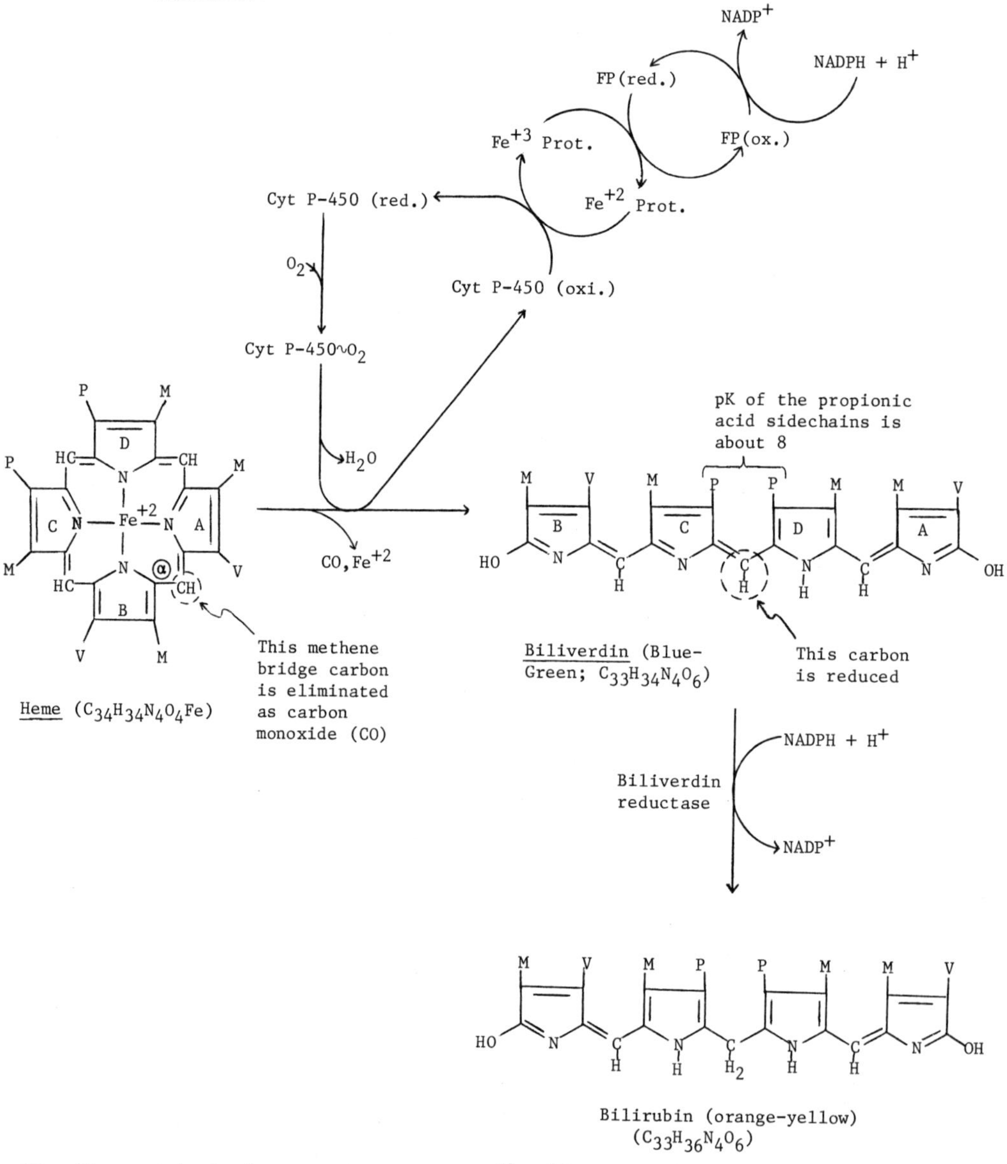

FP = flavoprotein (perhaps containing FAD); Fe^{+2}/Fe^{+3} Prot. is a non-heme iron protein; M = methyl; V = vinyl; P = propionic acid. A similar (probably identical) microsomal oxidase system is described later under the metabolism of drugs.

c. Bilirubin is <u>insoluble</u> in aqueous solutions, particularly at
    physiological pH. It is soluble in a basic solution because
    of the weak acidity conferred by the presence of two propionic
    groups whose pK is around 8.

3. Transport and Uptake of Bilirubin by the Hepatocytes

a. Since bilirubin is insoluble in aqueous systems, it is
    carried in the blood complexed to albumin. This complex
    formation prevents the indiscriminate passage of bilirubin
    into tissue cells other than hepatocytes. The unique ability
    of hepatocytes to take up, concentrate, and eliminate bilirubin
    is presumably due to (i) the presence of a carrier mechanism
    available for the transport of bilirubin; (ii) the presence
    of bilirubin-binding proteins in the cytoplasm of hepatocytes;
    and (iii) conversion of bilirubin within the cells to water-
    soluble conjugates which normally cannot reenter the blood but
    are eliminated through the bile. Two bilirubin-binding
    proteins, named Y and Z, have been isolated from the
    hepatocytes and characterized. Bilirubin is not eliminated
    by the kidney because it will not pass through the glomerular
    capillary walls, in either its free form or bound to protein.

b. Since bilirubin is an anion (due to the presence of the two
    propionic acid residues) and is transported bound to albumin
    (1 molecule of albumin binds more than 2 molecules of
    bilirubin), other anions such as sulfonamides, thyroxine and
    triiodothyronine, fatty acids, and acetyl-salicylate can
    compete for the bilirubin-binding sites on albumin and
    displace bilirubin. If the plasma concentration of displaced
    (free, unbound) bilirubin increases due to a reduction in
    albumin-binding capacity (normal binding capacity: 20-25 mg/
    100 ml), it can enter the brain (giving rise to brain
    encephalopathy) and other tissues, producing cytotoxic effects.
    This aspect (cytotoxicity) assumes a particular significance
    in the newborn because the liver is not yet mature at birth
    (apparently about 10-15 days are required for the liver to
    attain its full biochemical potential) and is unable to
    metabolize bilirubin. As a result, neonatal bilirubin levels
    can surpass the binding capacity of the plasma proteins and
    produce cytotoxic effects if the situation is not corrected.
    This problem is accentuated in erythroblastosis.

c. As mentioned above, hepatocytes take up unbound bilirubin
    from the sinusoidal plasma, presumably in a carrier-mediated
    step. Once the bilirubin enters the cell, it is immediately
    bound by the acceptor proteins so that it cannot escape back
    across the membrane. The uptake of bilirubin by the
    hepatocytes is shown next.

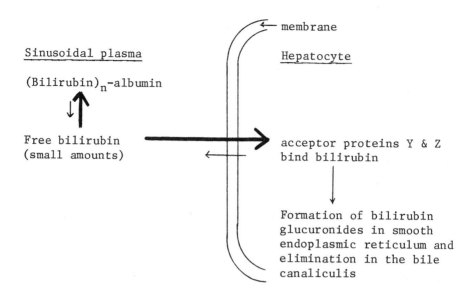

d. Compounds such as phenobarbital and other anionic drugs
    undergo metabolic transformations similar to that of bilirubin
    before they are eliminated. Studies with **phenobarbital** have
    shown that the drug enhances the hepatic uptake, conjugation,
    and excretion of bilirubin, presumably by stimulating the
    overall process (see page 630).

4. Formation and Excretion of Bilirubin Glucuronides

a. The general reaction for glucuronide formation is shown below.

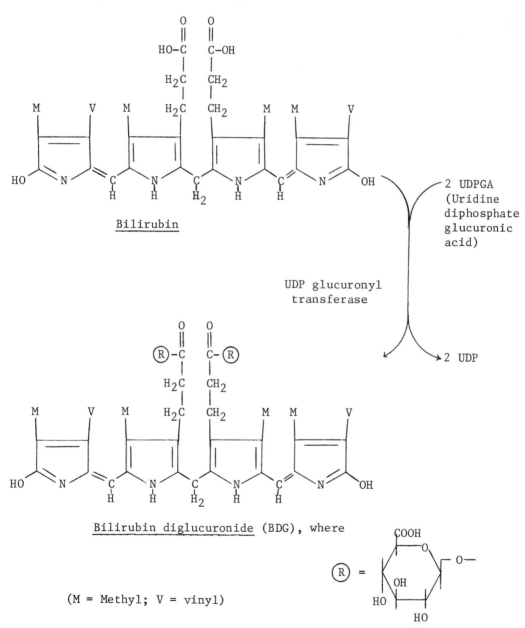

Bilirubin

2 UDPGA
(Uridine
diphosphate
glucuronic
acid)

UDP glucuronyl
transferase

2 UDP

Bilirubin diglucuronide (BDG), where

(M = Methyl; V = vinyl)

R =

glucuronic acid

b. Bilirubin diglucuronide (BDG) is a water-soluble conjugate
    which has a molecular weight of 937 daltons while bilirubin
    has a molecular weight of only 585. These properties make
    BDG unable to diffuse back into the blood once it is formed
    in the hepatocyte and it is eliminated via the bile excretory
    system. This is the reason for the absence of bilirubin
    glucuronides from the blood under <u>normal</u> circumstances. Both
    conjugated and unconjugated bilirubin (bile pigments) are
    present as salts in the bile. In this form, they reach the
    intestine by way of the common duct.

c. In the clinical laboratory, bilirubin and its glucuronides
    are distinguished on the basis of the conditions under which
    they react with diazotized sulfanilic acid (van den Bergh
    reaction; discussed later). The **water-soluble** bilirubin
    glucuronides react <u>directly</u>, without any need for an
    "accelerator" and are known as <u>direct-reacting bilirubin.</u>
    Free bilirubin, on the other hand, requires an accelerator
    such as alcohol or a caffeine-sodium benzoate reagent to make
    it react under ordinary conditions. It is therefore called
    <u>indirect-reacting bilirubin</u>. These properties will be
    discussed more, both in terms of diseases and chemical
    analysis, later in this section.

d. The conjugation of bilirubin with UDPGA is catalyzed by the
    enzyme UDP-glucuronyl transferase, present on the smooth
    endoplasmic reticulum of the hepatocytes. The UDPGA (also
    called UDPG) is synthesized in the uronic acid pathway.
    Although the predominant conjugate synthesized is the
    diglucuronide, monoglucuronide formation has been reported.
    It is possible that the "monoglucuronide" is actually a 1:1
    complex between BDG and free bilirubin. Glucuronide conjugation
    is <u>essential</u> for bilirubin elimination through the biliary
    system. It appears that the excretion involves the Golgi
    apparatus and an energy-requiring, carrier-mediated step.

e. The rate of bilirubin removal from the blood is dependent
    upon its concentration therein and upon the rate of conjugation
    in the hepatocytes, provided everything else is functioning
    normally.

5. Metabolism of Bilirubin in the Gastrointestinal Tract

a. Bilirubin glucuronides, because of their water solubility and
    molecular size, are not absorbed to any significant extent
    by the intestinal cells. Instead, in the intestine the
    glucuronides undergo deconjugation (catalyzed by β-glucuronidase)
    and reduction to a group of colorless compounds known as

urobilinogens. These reactions are catalyzed by enzymes provided by the intestinal bacteria. Trace reabsorption of urobilinogen into the portal blood occurs and the normally functioning hepatocyte reabsorbs most of the urobilinogen and excretes it into the bile. Part of the urobilinogen gets into the systemic blood circulation, however, and is filtered in the kidneys, eventually appearing in the urine (up to about 4 mg/24 hrs). In abnormal conditions, the urobilinogen levels are altered. For example, in obstructive jaundice where the normal entry of bile into the intestines is interrupted and urobilinogen production is thereby diminished, there is an almost total absence of urobilinogen in the urine as well as in the stool. In this condition, the stool does not have the color normally associated with it. In the hemolytic states, urobilinogen levels (fecal and urinary) are increased due to enhanced overall production of pigments. In hepatic disorders, urinary urobilinogen values may be increased due to inability of the hepatocytes to reabsorb urobilinogens from the portal blood and return them to the bile. This gives rise to elevated urobilinogen levels in the systemic circulation and increases urinary excretion of these compounds.

b. Laboratory evaluation of urinary urobilinogen is frequently made on a **two-hour** urine specimen (collected particularly in the early afternoon because, during this period, the urobilinogen clearance has been observed to be maximal and relatively constant). Urobilinogen is quantitated by treating the sample with p-dimethylaminobenzaldehyde in HCl (one of two "Ehrlich's reagents"; the other is diazosulfanilic acid - Ehrlich's diazo reagent - used in bilirubin determinations) and measuring the pink color produced. The color is stabilized and intensified by adding sodium acetate. Sodium acetate appears to provide the specificity for the reaction too because it inhibits the color formed by indole and skatole which also react with Ehrlich's reagent. A reducing agent such as ascorbate is needed in this reaction in order to maintain urobilinogens in the reduced state. The results are expressed in Ehrlich units, where each unit represents the amount of color produced by 1 mg of urobilinogen.

c. Urobilinogens in the intestine are oxidized to urobilins in reactions catalyzed by bacterial enzymes. Other decomposition products include dipyrroles (e.g., **bilifuscins**). These are the compounds which give the characteristic color to the stools.

6. Disposition of Bilirubin by Other Than Hepatic Routes

    a. Under normal circumstances, the liver is capable of handling
        the excretion of bilirubin. Under abnormal conditions, when
        bilirubin glucuronide accumulates, it can be excreted through
        the urine.

    b. Bilirubin is photolabile, that is, it can be decomposed upon
        exposure to light. The reaction requires oxygen and gives
        water-soluble products of unknown structure which can be
        eliminated in the urine. In neonatal jaundice and unconjugated
        hyperbilirubinemia in adults, serum levels of bilirubin have
        been reduced by exposure of the patient to intense light.
        Presumably, bilirubin in the skin and in the cutaneous
        capillary blood is destroyed.

7. Several methods are available for detecting and quantitating
    bilirubin. These are discussed below.

    a. The direct spectrophotometric measurement of bilirubin at
        450 nm has already been described. As was pointed out, this
        is strongly interfered with by carotenes, present in many fruits
        and vegetables; and by hemolysis and turbidity in the sample.
        The <u>icterus index</u> (from icteric = jaundiced) is a modification
        of the direct reading method and, as such, is subject to
        the same interferences. The procedure involves addition of
        10 ml of 5% (W/V) sodium citrate to 1 ml of unhemolysed,
        non-turbid serum. The color intensity is read at 460 nm and
        expressed in icterus units, defined below.

$$\text{icterus units} = \frac{\text{absorbance of sample x 10}}{\text{absorbance of standard } 0.0157\% \text{ } K_2Cr_2O_7 \text{ solution}}$$

        The normal range is 3-8 icterus units. The method is not very
        useful other than as a qualitative measurement or, as indicated,
        in neonatal jaundice where pigments other than bilirubin are
        absent.

    b. By far the most important methods for quantitating bilirubin
        in the serum are based on the van den Bergh reaction. This
        reaction involves the coupling of bilirubin with diazotized
        sulfanilic acid (one of two "Ehrlich's reagents"; the other is
        p-dimethylaminobenzaldehyde, used in urobilinogen determinations)
        to give an "azobilirubin" which is red or blue, depending on
        pH, and which can be measured spectrophotometrically. The
        conversions involved are shown below. They proceed as indicated
        with either free (indirect reacting) or conjugated (direct
        reacting) bilirubin. Thus, in this diagram, R- can be either
        a hydrogen or a glucuronic acid residue.

Diazotization of Sulfanilic Acid

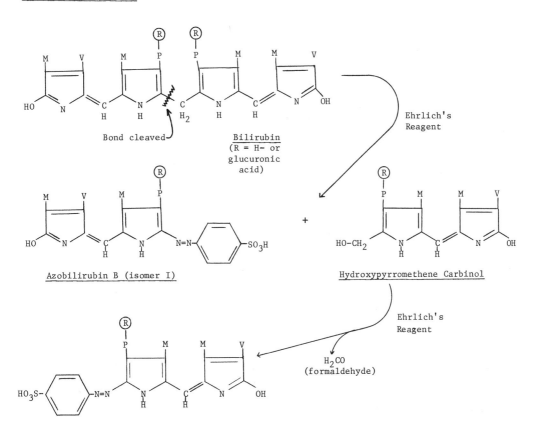

sulfanilic acid

diazosulfanilic acid
(Ehrlich's reagent or
Ehrlich's diazo reagent)

Van den Bergh Reaction

Bond cleaved

Bilirubin
(R = H- or
glucuronic
acid)

Ehrlich's
Reagent

Azobilirubin B (isomer I)

+

Hydroxypyrromethene Carbinol

Ehrlich's
Reagent

H_2CO
(formaldehyde)

Azobilirubin B (isomer II)

i)   In the method of Malloy and Evelyn, a pH of 2-2.5
     (weakly acid) is used for color development. At this pH,
     azobilirubin has a reddish hue which has an absorption
     maximum near 540 nm. The principal drawback to this
     method is the pH sensitivity of the color in this pH
     range. A number of modifications to this method have
     still not been able to avoid this basic trouble. Despite
     this, it is still widely used in clinical laboratories.

ii) At alkaline and strongly acid pH's, the azobilirubin is
     blue, with an absorption maximum near 600 nm. This color
     shift - from blue to red to blue with increasing pH - makes
     the pigment a **double-range indicator**. Jendrasic and
     Grof, and Nosslin, developed a method using an alkaline
     pH. It has the advantages of greater color intensity and
     greatly decreased pH sensitivity, compared to the Malloy-
     Evelyn procedure.

iii) In his original study of these reactions, van den Bergh
     described two types of bilirubin: one which reacted
     rapidly with Ehrlich's reagent, the other which reacted
     quite slowly unless an "accelerator" (methanol, ethanol,
     caffeine-sodium benzoate, and other materials) was added.
     As was described above, the rapidly (direct) reacting
     material is the **water-soluble** glucuronide while the
     compound which required an accelerator (indirect reacting)
     is the lipid and organic solvent-soluble free (unconjugated)
     bilirubin. It is important to distinguish between these
     compounds in an analysis, since their differential
     elevation varies with the disorder. A number of methods
     determine direct-reacting bilirubin (without an accelerator)
     and total bilirubin (with an accelerator), calling the
     difference between the values indirect bilirubin. This
     may underestimate the indirect and overestimate the
     direct, due to the partial reaction of the former even
     without the accelerator. If accurate values are needed,
     it is best to perform a preliminary separation of the
     two compounds, probably by one of several extractive
     procedures.

iv) There are two other important considerations in the use
     of the van den Bergh reaction for bilirubin. First, none
     of the usual methods are capable of distinguishing between
     direct and indirect bilirubin at low concentrations.
     A careful separation of the two must be performed prior
     to analysis if low level results are to have any meaning.
     Second, bilirubin standards are notoriously unstable

toward light, heat, and solvent. Great care must be taken in standard preparation and storage. The College of American Pathologists, the American Association of Clinical Chemists, and the American Academy of Pediatrics have jointly recommended that bilirubin standards be prepared by the addition of acceptable bilirubin preparations to pooled, normal human serum, followed by lyophilization.

c. Bilirubin is not normally present in the urine and, when it is, only conjugated bilirubin is found. Qualitative methods are generally used due to the instability of the pigment toward O_2 and because any bilirubinuria is considered abnormal. These tests are based on oxidation of yellow bilirubin to green biliverdin or blue bilicyanin. The oldest one is the Gmelin or iodine-ring test where I_2 is the oxidizing agent. It is now considered relatively insensitive and can fail to reveal bilirubinuria. The best tests are modifications of the Harrison test, which uses Fouchet's reagent (1% $FeCl_3$ in 25% trichloroacetic acid solution). Addition of this reagent to urine containing barium chloride will produce a noticeable green color if as little as 0.05 mg of bilirubin/100 ml is present. The Huppert-Cole method is also used.

Hyperbilirubinemias

1. As the name implies, these disorders are characterized by elevated serum levels of bilirubin, bilirubin conjugates, and other bile pigment metabolites. When the level of the bile pigments exceeds the capacity of albumin to bind them, the pigments escape into the tissues where they accumulate. The presence of quantities of these pigments, particularly in the skin and mucous membranes, gives rise to a yellow (if bilirubin is deposited) or green (if biliverdin predominates) coloration. Such a syndrome is known as jaundice. It should be noted that jaundice is really a set of symptoms (a syndrome) with multiple possible etiologies. A similar situation is encountered in gout, and perhaps in diabetes and certain other clinical disorders. Laboratory tests provide a powerful tool for distinguishing between the possible causes in such syndromes.

2. Unconjugated hyperbilirubinemia (indirect-reacting bilirubin) can result from three principal causes outlined below. It is important to note that in these disorders, bilirubinuria does not occur.

a. Excessive hemolysis from a variety of disorders. In these situations, the liver is presented with a much greater than normal load of bilirubin. If it is functioning normally, the liver is quite capable of handling the increase. A new equilibrium is apparently established across the hepatocyte membrane, however, with a consequent elevation in circulating bilirubin. Regardless of the reason for red cell destruction, this is known as hemolytic jaundice.

b. Defects in the uptake of bilirubin by the hepatocytes. In Gilbert's syndrome, the hyperbilirubinemia is presumably due to a deficiency or defect(s) in the intrahepatic bilirubin binding proteins (Y and Z). The compound flavaspidic acid, an active constituent of male-fern extract which is used in the treatment of tapeworm infestation, competes with bilirubin- for binding with intrahepatic acceptor proteins and thus causes increased levels of unconjugated bilirubin in the blood.

c. Defects in the conjugation reaction. There are two types. The Crigler-Najjar syndrome (constitutional non-hemolytic hyperbilirubinemia) is a genetic disorder characterized by a complete lack of the conjugating enzyme glucuronyl transferase. Severe jaundice and brain encephalopathy are present and death often occurs at a young age. The second type is a milder form where the conjugation takes place to a lesser degree giving rise to moderate elevation of unconjugated bilirubin levels. Administration of phenobarbital (also used in the treatment of cholestasis) to these patients produces beneficial results by decreasing blood bilirubin levels. As discussed earlier, phenobarbital stimulates hepatic uptake, conjugation, and secretion of bilirubin. It is important to point out that phenobarbital serves no useful purpose when administered to individuals completely lacking the conjugating enzyme (see page 630). Conjugation is an obligatory prerequisite for bilirubin excretion via the biliary system.

3. Types of conjugated hyperbilirubinemia (direct-reacting bilirubin)

a. In these disorders, the defects lie in the hepatic biliary secretory mechanisms. The defects in the normal flow of bile may be due to mechanical obstruction occurring either intra- or extrahepatically (obstructive jaundice) or it can be caused by non-mechanical factors. This syndrome is known as cholestasis. Extrahepatic biliary tract obstruction can be caused by a variety of conditions such as tumors, stones, stricture, pancreatitis, or cholecystitis. Non-mechanical cholestasis is observed in bacterial infections; after

administration of steroids, oral contraceptives or other drugs;
in pregnancy (rarely); and in familial chronic idiopathic
jaundice (Dubin-Johnson syndrome).

b. The Dubin-Johnson syndrome is characterized by deposition of
    a brown pigment (possibly related to melanin) in the liver
    and by a fairly mild, chronic, non-hemolytic hyperbilirubinemia.
    Roughly equal amounts of conjugated and unconjugated pigment
    appear in the plasma with the direct usually predominating
    slightly. The bilirubin levels and a number of other features
    of Rotor's syndrome are quite similar to those of the Dubin-
    Johnson syndrome. Rotor, however, did not observe any liver
    pigmentation in his patients and, unlike Dubin and Johnson,
    was able to visualize the gallbladder by cholecystography.
    Because Rotor described only three patients, it is difficult
    to draw any real conclusions. There is a great deal of
    variation in the severity of symptoms in Dubin-Johnson patients,
    however, and it seems possible that the two syndromes are
    actually different manifestations of one biochemical lesion.
    It is interesting that in the Dubin-Johnson disorder, the ratio
    of coproporphyrinogen I to coproporphyrinogen III excreted in
    the urine is elevated over that of normal individuals and
    bromosulphalein retention is 10-20% (variable) over normal
    (see liver function tests). The disease appears to be
    inherited as an autosomal recessive trait.

c. In the above disorders, the mechanism by which conjugated
    bilirubin appears in the plasma is not clear. It has been
    suggested that there may be injury to the endothelial lining
    of the bile ductules (cholangioles) leading to the leakage
    of the conjugated bilirubin into the plasma. In addition,
    reverse pinocytosis has been implicated as a mechanism of
    transfer of bilirubin glucuronides to sinusoidal plasma.
    Electron microscopic studies have shown a multiplicity of
    structural derangements suggesting that intrahepatic
    cholestasis may be due to several independent factors.

d. It is important to point out-when conjugated glucuronide
    begins to have access to the blood, its appearance in the urine
    also becomes apparent. In liver disorders (viral, hepatitis,
    cirrhosis, etc.), plasma levels of both conjugated and
    unconjugated bilirubin levels are often increased. This is
    indicative of general damage to the hepatocytes.

4. Occurrence of some degree of jaundice in infants (neonatal jaundice)
    is not uncommon. It is usually due to the immaturity of some
    aspect of bilirubin conjugation in the liver. Apparently the
    liver is not "mature" at birth and requires one to two weeks
    postnatally to complete development of the conjugating system.
    Ordinarily, there is a rise in bilirubin levels on the first day
    of life, reaching a maximum on the third or fourth day (at which
    time it rarely exceeds 10 mg/100 ml; see Table 23). In some cases
    (e.g., those complicated by erythroblastosis), however, prolonged
    high levels of bilirubin do occur, giving rise to generalized
    jaundice, brain jaundice, disturbances of the nervous system, and
    death. One cause of erythroblastosis is Rh incompatibility.

    Prolonged elevation of bilirubin has also been reported in
    association with the breast feeding of infants. It is thought that
    pregnanediol, present in human milk but absent in cow's milk,
    may inhibit the glucuronyl transferase enzyme and thereby prevent
    bilirubin conjugation.

Table 23. Serum Bilirubin Levels in Some Normal and Abnormal Conditions

Condition	Bilirubin Values in mg/100 ml of Plasma or Serum		
	Total	Unconjugated (indirect reacting)	Conjugated (direct reacting)
Newborn, cord blood	0.2 - 2.9		
1 day	0 - 6.0		
3 days	0.25-11		
7 days	0.14-9.9		
Normal (adult)	0.2 - 1.67	0.2 - 1.4	0 - 0.27
Hemolytic Disorders (in adults)	2.2 - 3.4	2.0 - 3.0	0.2 - 0.4
Crigler-Najjar Syndrome (glucuronyl transferase deficiency; plasma deeply icteric)	15 -48	15 -48	trace
Dubin-Johnson Syndrome (perhaps identical to Rotor's syndrome; defect in the bile secretory mechanism)	2.4 -19 (ave. of 46 cases = 5.6)	2.2 (ave. of 34 cases; ranged from 74%-14% of total)	3.4 (ave. of 34 cases; ranged from 26%-86% of total)
Gilbert's Syndrome (defects in the hepatic bilirubin uptake)	2.2	2.0	0.2
Cholestasis (severe form)	10.0	1.0	9.0
Cirrhosis (severe)	11.0	5.0	6.0
Hepatitis (acute-severe)	10.0	1.5	8.5

8
Lipids

1. Lipids are insoluble in water but are soluble in fat solvents such as chloroform, ether, and benzene. The list of these compounds includes fats, oils (which are liquid at room temperature), waxes, and related compounds.

2. Fats serve as

    a. Insulating material (in subcutaneous tissue and as "padding" around certain organs).

    b. A major source of caloric energy, used directly or stored. Fat is an efficient and concentrated form of stored energy. Oxidation of 1 gram of fat yields 9.3 kcal, while 1 gram of carbohydrate releases only 4.2 kcal and 1 gram of protein gives 5.6 kcal. Fat is associated with very little water.

    c. Combinations of fats with proteins (lipoproteins) are important cellular constituents. They occur in cell membranes and in membranes of intracellular organelles such as nuclei, microsomes, mitochondria, etc. Lipid derivatives are part of the components of the electron-transport system in mitochondria.

    d. Unsaturated fatty acids containing two or more double bonds, (linoleic, linolenic, and arachidonic) are not synthesized in the body and, hence, must be furnished in the diet. The functions of these essential fatty acids are not well understood, but they are required for normal growth and function. One recently described role may be in the synthesis of prostaglandins.

3. Metabolism of Fats   (a few general statements)

    a. The body fat is in a dynamic state.

    b. Much of the carbohydrate of the diet is converted to fat before it is utilized to provide energy.

    c. Fat may serve as a major source of energy in many tissues. Except for the brain, tissues may use free fatty acids (FFA) as a fuel in preference to carbohydrates.

d. Fat's caloric value is 9 kcal/gm as mentioned above and it is associated with less water in storage than either protein or carbohydrate.

e. Fats are the major source of calories in the postabsorptive state.

Types of Lipids

1. **Simple lipids**: esters of fatty acids with various alcohols.

   a. Fats: esters of fatty acids with glycerol (glycerides).

   b. Waxes: esters of fatty acids with monohydroxy aliphatic alcohols.

      i) True waxes: esters of cetyl alcohol ($CH_3(CH_2)_{14}CH_2OH$) or other higher straight-chain alcohols with palmitic, stearic, oleic or other higher fatty acids.

      ii) Cholesterol esters: esters of cholesterol with fatty acids.

      iii) Vitamin A esters: esters of vitamin A with palmitic or stearic acid.

      iv) Vitamin D esters.

2. **Compound lipids**: esters of alcohols with fatty acids which also contain other groups.

   a. Phospholipids: contain fatty acids, glycerol, phosphoric acid, and in most cases, a nitrogenous base.

   b. Glycolipids or cerebrosides: contain both carbohydrate and nitrogen but not phosphate or glycerol.

   c. Sulfolipids: contain sulfate groups.

   d. Lipoproteins: lipid and protein complexes.

   e. Lipopolysaccharides: lipid and polysaccharide complexes.

3. Derived lipids: substances derived from the above group by hydrolysis; they include

    a. Fatty acids (saturated & unsaturated). Any aliphatic carboxylic acid of about six carbons or more is generally considered a fatty acid. This definition is flexible.

    b. Mono-and diglycerides (derived from triglycerides by saponification of, respectively, two or one of the ester linkages).

    c. Glycerol, sterols and other steroids (vitamin D), and alcohols containing β-ionone rings (vitamin A).

    d. Fatty aldehydes.

    e. The protein portion of lipoproteins.

4. Miscellaneous lipids: aliphatic hydrocarbons (iso-octadecane), carotenoids, squalene (an intermediate in the biosynthesis of cholesterol) and other terpenes (named for turpentine), vitamins E and K, glycerol ethers (diesters of these compounds occur in tissues from various sources including human neoplasms), glycosylglycerols (in plants), etc.

Fatty Acids

1. Saturated fatty acids   General Formula: $C_nH_{2n+1}COOH$
    (No double bonds)

    acetic acid CH_3COOH

    butyric acid C_3H_7COOH – found in fats from butter

    palmitic acid $C_{15}H_{31}COOH$ ⎫ present in all
                                                            animal and plant
    stearic acid $C_{17}H_{35}COOH$ ⎭ fats

    lignoceric acid $C_{23}H_{47}COOH$ – present in cerebrosides

2. Unsaturated fatty acids (general formula: $C_nH_{2n+1-m}COOH$ where m = number of double bonds + number of rings in molecules).

a. One double bond: $C_nH_{2n-1}COOH$

Examples: Palmitoleic - $C_{15}H_{29}COOH$ (C_{16}; $\Delta 9$)

Oleic - $C_{17}H_{33}COOH$ (C_{18}; $\Delta 9$).

Note: The Δ indicates the location of the double bond. For example, $\Delta 9$ means that the double bond is between carbons 9 and 10. The carboxyl carbon is carbon   1.

b. Two double bonds: $(C)_n(H)_{2n-3}COOH$ - linoleic (C_{18}; 9,12).

c. Three double bonds: $(C)_n(H)_{2n-5}COOH$ - linolenic (C_{18}; 9,12,15).

d. Four double bonds: $(C)_n(H)_{2n-7}COOH$ - arachidonic (C_{20}; 5,8,11,14).

3. a. Quantitatively, the largest group of fatty acids in biological systems is composed of acyclic, unbranched, non-hydroxylated molecules containing an even number of carbon atoms. Appreciable amounts of other types of fatty acids do occur in certain tissues, however, and may perform important roles in the life of various organisms including man. The prostaglandins, discussed later, provide one example. Another is the membrane phospholipids, many of which have an unsaturated fatty acyl group as one of their sidechains. The metabolism of some of these compounds is discussed later, along with the more common types of fatty acid metabolism.

b. Some of these "unusual" fatty acids are produced only by plants or microorganisms. Others are synthesized by mammals including man. By means of gas chromatography and mass spectroscopy, human earwax (cerumen) has been shown to contain varying amounts of over fifty different types of fatty acids of all of the kinds mentioned above. Sebum (the secretion of the sebaceous glands in the skin) also contains a wide variety of fatty acids in both free and esterified forms. Skin microorganisms and spontaneous oxidative and hydrolytic reactions may be partly responsible for the diversity which is found.

c. Some specific examples of these compounds include

i) Tuberculostearic acid, part of the lipids of the tubercle bacillus, is 10-methylstearic acid. Thus it is both an odd-chain-length and a branched acid.

ii) Cerebronic acid, mentioned later under cerebrosides, is 2-hydroxylignoceric acid, an α-hydroxy fatty acid.

iii) Ricinoleic acid (12-hydroxyoleic acid) is a hydroxy-unsaturated compound characteristic of castor oil.

iv) Chaulmoogric acid, isolated from chaulmoogra oil and used in the treatment of Hansen's disease (leprosy), has the structure

$$-(CH_2)_{12}-COOH$$

Chaulmoogric Acid

v) Phytanic acid (3,7,11,15-tetramethylhexadecanoic acid; note that hexadecanoic acid is palmitic acid) is the oxidation product of phytol, an open-chain terpene derived from chlorophyll. It is present in foods derived from herbivorous animals, notably cow's milk and animal fat. Because of the 3-methyl group, it is resistant to β-oxidation and is normally catabolized by the phytanic acid α-oxidase system to pristanic acid which can be β-oxidized. In persons with Refsum's syndrome, a nervous disorder, up to about 20% of the serum fatty acids and 50% of the hepatic fatty acids are composed of phytanic acid, whereas the normal amount in serum is less than 1 μg/ml. This is due to the (heritable) absence of phytanic acid α-hydroxylase in affected individuals. This enzyme catalyzes one of the initial reactions in the α-oxidation (see also the formation of β-hydroxy-acyl-CoA during β-oxidation, page 637).

vi) Bacteria are able to synthesize cyclopropyl fatty acids of general formula

$$CH_3-(CH_2)_y-\overset{\displaystyle CH_2}{\underset{}{CH-CH}}-(CH_2)_x-COOH$$

where x and y are integers. The substrate is a cis-unsaturated fatty acid, usually in the β-position of phosphatidylethanolamine. The extra methylene group is supplied by S-adenosylmethionine.

4. There is a group of fatty acids that mammals need but are unable to synthesize. These essential fatty acids must be supplied in the diet. Linoleic, linolenic, and arachidonic acids are generally considered to comprise this group although man can convert linoleic acid to arachidonic acid. These are essential because, in each of

them, <u>one or more of the double bonds lies somewhere between the</u> <u>terminal methyl group and the seventh carbon from that group.</u> In this region, indicated in the drawing below, mammals are unable to introduce a double bond.

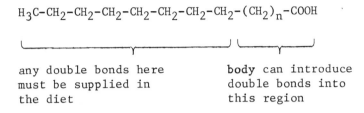

$$H_3C-CH_2-CH_2-CH_2-CH_2-CH_2-CH_2-CH_2-(CH_2)_n-COOH$$

any double bonds here   **body** can introduce
must be supplied in   double bonds into
the diet   this region

a.   They are found in the structural lipids (present in phospholipids) which are essential in maintaining the structural integrity of membranes. They also occur in high concentrations in the reproductive organs.

b.   These compounds are necessary for fat mobilization. Deficiencies of essential fatty acids give rise to a fatty liver in experimental animals. They are also involved in cholesterol metabolism and serve as precursors of prostaglandins. This last may be one of the most important roles.

c.   Palmitoleic acid and oleic acid are not essential fatty acids because the double bond in each of these compounds can be introduced by enzymes which occur in the mammalian liver. The substrates for these enzymes are the appropriate saturated fatty acids.

5.   The roles of free fatty acids in the plasma, as well as methods for their analysis, are discussed in the section on plasma lipids. They are also important constituents of the acylglycerides, phospholipids, and cholesterol esters. The synthesis and catabolism of the fatty acids are also covered later in this chapter.

Prostaglandins (PG)

1.   Prostaglandins, discovered initially in the seminal plasma, are synthesized in every tissue of the body. The seminal plasma PG's are formed in the prostate gland from which the name prostaglandins is derived. PG's are rapidly synthesized as well as inactivated and the catabolizing enzymes are distributed throughout the body.

2.   a.   The parent compound of PG's is prostanoic acid. It is a C20 fatty acid containing a five-membered (cyclopentane) ring. The dotted bond between carbons 7 and 8 is aimed back into the page from carbon 8.

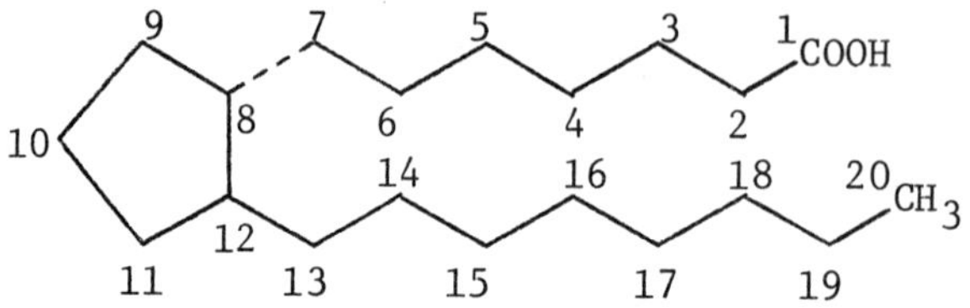

Prostanoic
Acid

b. Differences between various PG's are due to number and position of the double bonds and the presence of hydroxyl and oxy side groups. The names given to the various PG's in order to differentiate them from one another consists of PG followed by a letter (for example, PGE, PGF, etc.) and a numerical subscript (for example, PGE_1, PGF_2) which indicates the number of double bonds. In addition, the subscript α or β is sometimes appended (for example $PGE_{2\alpha}$). The letter E in PGE represents ether solubility and the F in PGF indicates the solubility of this group of compounds in phosphate buffer solution. (F is used to designate P because in Sweden where these compounds were initially isolated, the word for phosphate is spelled with an initial F.) The primary naturally occurring prostaglandins are considered to be PGE_1, PGE_2, PGE_3, $PGF_{1\alpha}$, $PGF_{2\alpha}$, and $PGF_{3\alpha}$, although E_3 and $E_{3\alpha}$ are only rarely encountered <u>in vivo</u>. The A and B families are secondary types.

3. Biosynthesis

a. PG's are synthesized from unsaturated fatty acids such as linoleic, linolenic, and arachidonic acids. The structural relationships between the precursors and the PG's synthesized are shown in the following examples:

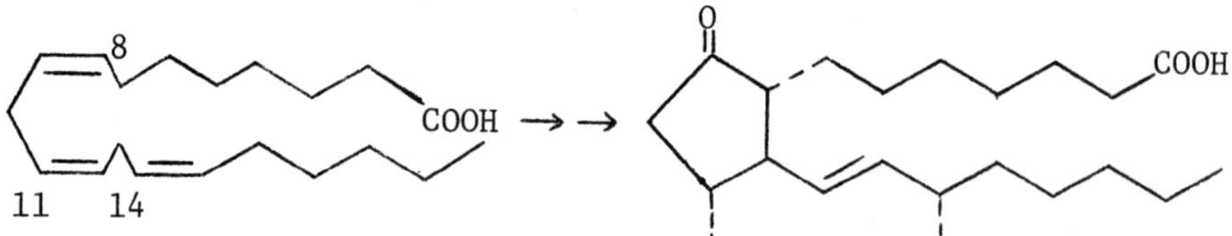

Bishomo-γ-Linolenic Acid
(8,11,14-Eicosatrienoic Acid)

PGE_1

Arachidonic Acid
(5,8,11,14-Eicosatetraenoic Acid)

$PGF_{2\alpha}$

b. An enzyme system obtained from sheep seminal vesicles as an insoluble complex associated with microsomes catalyzes the synthesis of the six primary PG's, starting from essential fatty acid precursors. A given essential fatty acid precursor can give rise to either E or F compounds, but E-type and F-type PG's are <u>not</u> found to be interconvertible. This is significant because it appears that the <u>ratio</u> between E and F compounds may be critical in initiating and/or regulating a physiological process (e.g., fertilization). In the sheep seminal vesicle PG-synthetase system, it has been observed that addition of GSH (reduced glutathione) stimulates the synthesis of PGE at the expense of PGF synthesis and, similarly, Cu^{2+} in the presence of dihydrolipoamide gives rise to increased formation of F compounds at the cost of E synthesis.

c. E and F compounds appear to be the principal PG's. It is not clear whether the A and B compounds have functional roles in physiological systems or whether they are merely artifacts of isolation procedures. Removal of an element of water from the five-membered ring (dehydration) of the PGE's yields the PGA's, and the PGB's are produced by isomerization of the ring double bond of the corresponding PGA's.

4. Catabolism. The catabolism of PG's occurs throughout the body but one of the most efficient processes is the uptake and catabolism of the circulating E and F compounds by the lungs. In fact, the lungs are capable of removing PG's during a single circulatory cycle. Despite this rapid removal, the PG's have adequate access to the target organs to bring about their physiologic effects. The catabolizing processes start with reactions catalyzed by the enzymes PG 15-hydroxy dehydrogenase (oxidation of allylic -OH group at C_{15}) and PG reductase (reduction of the Δ^{13} double bond). These transformations are followed by β-oxidation, ω-oxidation of the alkyl sidechain, ω-hydroxylations, and the eventual elimination of the products of these reactions. (The ω-carbon is the one farthest from the number 1 (carboxyl) carbon in a fatty acid.)

5. Biological Actions of Prostaglandins

    a. This is a rapidly growing area of research and results continue
        to pour in. A complete listing of all the actions of PG's in
        various animal systems is beyond the scope of this work. Some
        of the problems in delineating the primary actions of PG's are
        that they give rise to opposing effects and that it is
        difficult to distinguish between normal physiologic functions
        and pharmacological actions. As mentioned earlier, the ratio
        between E and F compounds may be the crucial factor involved
        in controlling the dynamic level of a given physiological
        response in the body.

    b. Cardiovascular Effects

        i)   PG's cause increases in the cardiac output and enhance
             myocardial contractility. PGE causes reduction in
             arterial blood pressure accompanied by dilation of blood
             vessels and increased vascular permeability. This is
             part of the inflammatory response (wheal and flare
             response) brought about by PG's which can be blocked by
             anti-inflammatory agents such as aspirin (acetyl-salicylic
             acid). This anti-inflammatory action may be due to
             aspirin's inhibition of the enzyme dioxygenase, which
             catalyzes one step in prostaglandin synthesis in the
             sheep vesicular system.

        ii) The effects of PGF on the cardiovascular system are
             opposite to those described above for PGE. Thus PGF
             causes hypertension, constricts blood vessels, and
             decreases the capillary permeability.

        iii) PGE_1 inhibits platelet clumping by stimulation of cAMP
             synthesis.

    c. Renal Effects. PGE_1 (and PGA) increases plasma flow, urinary
        volume, and the excretion of Na^+, K^+, and Cl^-.

    d. Nervous System Effects. PGE in the cerebral cortex produces
        sedation and tranquilizing effects. In the medulla, PGE and
        PGF can produce either excitation or inhibition. PGE and PGF
        can inhibit spinal reflexes in the spinal cord. A wide variety
        of more specific actions in the nervous system have also been
        attributed to prostaglandins.

    e. Effects on Reproductive Glands. PGE and PGF cause increased
        synthesis of steroids. PGE in low concentrations causes
        lowering of uterine peristaltic action but, in higher

concentrations, PGE and PGF can stimulate uterine activity
leading to the expulsion of the fetus and placenta. This
property makes the prostaglandins potentially useful as
abortifacients and labor inducers. PG's also may play a role
in fertility by their action on the Fallopian tubes. PG's
that are present in the human semen can contract the proximal
end of the Fallopian tube while dilating the distal end. This
leads to a trapping of the ovum, making it more accessible
for fertilization by spermatozoa. Since PG's have been shown
to produce opposing effects in the fertilization process, it
may be the concentration ratio of E over F which decides
whether or not fertilization will occur.

f. In the lungs, PGE produces relaxation of the bronchial muscle
while PGF has the opposite effect.

g. PGE inhibits gastric secretions and both forms (E and F) cause
increased peristaltic action in the intestines, with a
corresponding decrease in transit time. They also cause
inhibition of Na^+ uptake (by stimulation of adenyl cyclase)
by mucosal cells accompanied by increased water loss, thereby
producing diarrhea.

h. Metabolic Effects. Many of PG's actions are linked with the
production of cAMP by adenyl cyclase or its degradation by
phosphodiesterase. In many tissues, PG's stimulate the
formation of cAMP. In the thyroid gland, PGE has a similar
action to that produced by TSH. In the adipose tissue cells,
however, PGE_1 prevents lipolysis (discussed later in this
chapter) by decreasing cAMP levels. Insulin also has an
antilipolytic effect in the adipose tissue cells. It has been
observed that PGE_1 produces insulin-like responses in the
peripheral tissues.

The potential possibilities of clinical use of PG's may lie
in the area of treatment of peptic ulcer, hypertension, and
in obstetrics and gynecology (fertility control, abortion,
and induction of labor). At the present time, there is a
great need for a drawing together of the proliferating
observations concerning the multitudinous actions of the
prostaglandins. Until PG research enters fully into the analytic
stage, it is difficult to understand the true role of these
ubiquitous compounds.

Steroids

This group includes cholesterol, ergosterol, the bile acids, the sex
hormones, the D-vitamins, and the cardiac glycosides. The biological
properties of these respective groups are discussed in appropriate
sections of this text.

1. Steroids are often found in association with fat.

2. They occur in the "unsaponifiable residue" following saponification.

3. The basic steroid nucleus, cyclopentanoperhydrophenanthrene,
    consists of four fused rings.

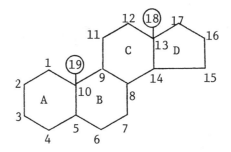

a. Carbons 18 and 19 are angular methyl groups attached at positions
    10 and 13, respectively. They are not part of the steroid
    nucleus but occur in a number of derivatives.

b. A sidechain at position 17 is common.

c. If the compound has one or more OH groups and no $\diagup\!\!\!\overset{}{C}{=}O$ or −COOH
    group, it is ster<u>ol</u>. If it has one or more $\diagup\!\!\!\overset{}{C}{=}O$ or −COOH
    groups, it is ster<u>oid</u>.

Plasma Lipids

1. This is a heterogeneous group of compounds which is seldom
    strongly polar, despite the frequent occurrence of phosphorous and
    to a lesser extent nitrogen. When they are extracted with suitable
    lipid solvents and subsequently separated into the various lipid
    classes, the presence of the following compounds can be demonstrated.

    a. <u>Phospholipids</u> are the largest lipid component of plasma but
        their clinical significance is not understood. One role is as
        a detergent, to increase the solubility of the other lipids.
        This is possible because many of the phospholipids have both

polar and non-polar regions. The importance of phospholipids in amniotic fluid is discussed later.

   b. <u>Cholesterol</u> is the second largest fraction of the plasma lipids.

   c. The <u>triglycerides</u> are third quantitatively among lipids in the plasma.

   d. <u>Unesterified (free) long-chain fatty acids</u> make up most or all of the remaining material.

2. <u>Free fatty acids</u> (FFA) are also called NEFA or non-esterified fatty acids. Quantitatively, the straight-chain, saturated fatty acids of formula $CH_3(CH_2)_nCOOH$ are the most important, with stearic (C_{18}) and palmitic (C_{16}) comprising at least 85% of the total.

   a. Although the FFA account for less than 5% of the total fatty acid present in the plasma, they are the most metabolically active of the plasma lipids.

   b. FFA is released from the adipose tissue and is transported in the plasma bound to albumin (see also bilirubin).

   c. Plasma FFA levels are increased in diabetes mellitus and are often elevated in obese individuals.

   d. Measurement of free fatty acids in serum

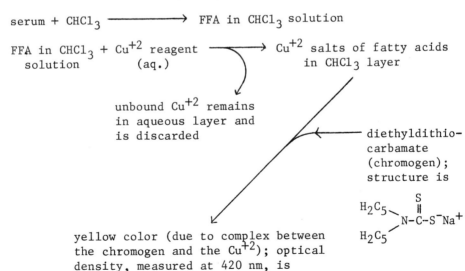

serum + $CHCl_3$ $\longrightarrow$ FFA in $CHCl_3$ solution

FFA in $CHCl_3$ + Cu^{+2} reagent $\longrightarrow$ Cu^{+2} salts of fatty acids
  solution   (aq.)   in $CHCl_3$ layer

unbound Cu^{+2} remains in aqueous layer and is discarded

diethyldithiocarbamate (chromogen); structure is

yellow color (due to complex between the chromogen and the Cu^{+2}); optical density, measured at 420 nm, is proportional to free fatty acid concentration

Diethyldithiocarbamate is also used in quantitation of copper in body fluids (see copper metabolism).

Another procedure for free fatty acids in serum and plasma involves an initial extraction of these molecules from the sample with a 40:10:1 (V/V/V) mixture of isopropanol, heptane, and 1 N H_2SO_4. Water and more heptane are then added to separate the unionized acids (non-aqueous layer) from the other serum components (aqueous layer). The extracted NEFA are titrated with dilute alkali to a thymolpthalein end-point.

3. Triglycerides are triple esters of glycerol with three molecules of fatty acid. The fatty acids may or may not be identical to each other. The general formula is indicated below.

$$
\begin{array}{c}
\qquad\qquad\qquad \overset{\displaystyle O}{\overset{\displaystyle \|}{}} \\
\overset{\displaystyle O}{\overset{\displaystyle \|}{}} \quad CH_2\text{-}O\text{-}C\text{-}R_1 \\
R_2\text{-}C\text{-}O\text{-}CH \qquad \overset{\displaystyle O}{} \\
\qquad\qquad\quad | \qquad \overset{\displaystyle \|}{} \\
\qquad CH_2\text{-}O\text{-}C\text{-}R_3
\end{array}
$$

a. Triglycerides derived from the intestinal absorption (details discussed later) of fat are transported in the blood as chylomicra. Those triglycerides synthesized in the liver and intestines from carbohydrates and FFA are carried as pre-betalipoproteins.

b. Triglycerides are stored in adipose tissue. They function as an energy reservoir and participate in the transport of FFA.

c. Triglycerides (TG's) are usually quantitated by saponifying them to release glycerol and free fatty acids, then measuring the amount of glycerol formed. The triglycerides are isolated from the serum by an extractive and/or chromatographic procedure. One way of obtaining the glycerol is indicated next.

i)

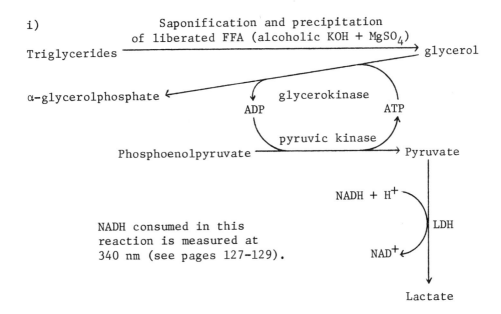

Saponification and precipitation
of liberated FFA (alcoholic KOH + MgSO$_4$)

Triglycerides $\longrightarrow$ glycerol

α-glycerolphosphate

ADP   glycerokinase   ATP

pyruvic kinase

Phosphoenolpyruvate $\longrightarrow$ Pyruvate

NADH + H$^+$

LDH

NADH consumed in this
reaction is measured at
340 nm (see pages 127-129).

NAD$^+$

Lactate

ii) Alternatively, the α-glycerolphosphate can generate NADH
    in the reaction

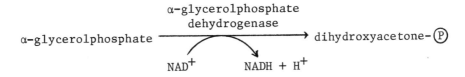

α-glycerolphosphate
dehydrogenase

α-glycerolphosphate $\longrightarrow$ dihydroxyacetone-Ⓟ

NAD$^+$ NADH + H$^+$

The increase in optical density at 340 nm is proportional
to the α-glycerolphosphate concentration and, hence,
to the triglyceride concentration (see pages 127-129).

iii) A chemical method for glycerol involves its oxidation to
     formaldehyde with sodium periodate followed by
     colorimetric determination of the formaldehyde at
     570 nm with chromotropic acid (4,5-dihydroxy-2,7-
     naphthalene sulfonic acid).

4. Cholesterol

This compound is distributed in all cells of the body, especially in the nervous tissue. It occurs in animal fats but not in plant fats. Structure of cholesterol

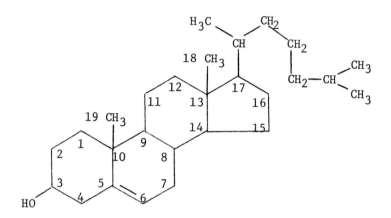

In this molecule, note in particular the hydroxyl in the 3 position, the double bond between carbons 5 and 6, and the alkyl sidechain at position 17.

a. Color Reactions to Detect Cholesterol

   i) The Liebermann–Burchard reaction and the Salkowski test are based on the colors produced (green and red, respectively) when cholesterol is reacted with H_2SO_4 under specified conditions. A proposed mechanism is outlined below.

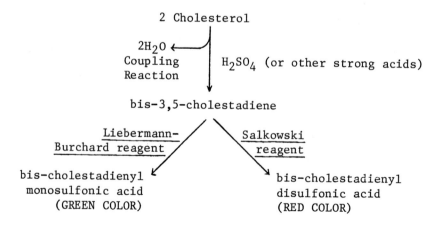

ii) The Liebermann-Burchard procedure consists of adding
acetic anhydride and H_2SO_4 to a chloroform solution of
the sterol. This test has been used to measure blood
cholesterol. A major disadvantage is that various other
sterols also react to produce a green color. The
Salkowski reagent contains a relatively higher
concentration of H_2SO_4, thereby favoring disulfonate
formation. It is apparent that the purity of the green
or red color is difficult to control and, consequently,
the use of these reactions is not highly reproducible.
Despite this, well over 150 different procedures for
cholesterol determination have been published based
on these reactions. Since a number of them differ with
respect to the extent to which the cholesterol is
purified prior to measurement, they are sometimes
classified as one-, two-, three-, or four-step procedures.

iii) The Zak reaction of cholesterol with $FeCl_3$, glacial
acetic acid, and concentrated H_2SO_4, produces a pink to
purple color which is much more stable than the colors
formed in the preceding methods. This procedure is also
more sensitive and reproducible and has been adapted for
and widely used in auto-analyzers in the clinical
laboratory.

b. Free cholesterol has an -OH group in position 3 and is
precipitable by digitonin (a glycoside) to produce digitonides.
This is the form in which cholesterol is biosynthesized and
is the form present as a structural unit of all membranes
(discussed later).

c. Esterified cholesterol has a fatty acid attached to the hydroxyl
group in position 3 through an ester linkage. This type of
cholesterol is not digitonin precipitable. The role of the
cholesterol esters is not known but they are an important
catabolic product of free cholesterol. Cholesterol esters
are formed by at least three processes.

i) In the liver, a fatty acid can be transferred from fatty
acyl CoA to free cholesterol.

ii) Free fatty acids can be esterified with cholesterol in
the intestinal mucosa in the presence of bile salts.

iii) The enzyme lecithin-cholesterol acyl transferase (LCAT),
a plasma-specific enzyme (see uses of enzymes in the
clinical laboratory), catalyzes the reaction

lecithin + free cholesterol ⟶

lysolecithin + cholesterol ester

This is a transesterification process involving transfer of an unsaturated fatty acyl group from the β-carbon of lecithin to the 3-position of cholesterol. A deficiency of LCAT (sometimes called Norum's disease), has been described in seven members of three Scandinavian families. These individuals also show high plasma levels of lecithin and unesterified cholesterol, a complete absence of α_1- and pre-β-lipoproteins, proteinuria, anemia, corneal infiltration, and the presence of foam cells in the glomerular tufts. Blood smears of the peripheral blood show the presence of target cells and the erythrocytes contain increased amounts of cholesterol and lecithin. The disease appears to be transmitted as an autosomal recessive trait.

d. In <u>Wolman's disease</u>, cholesteryl esters (discussed below) and triglycerides (all neutral lipids) accumulate in a number of organs and tissue levels of di- and monoglycerides and free fatty acids are usually elevated, accompanied by steatorrhea. Hepatosplenomegaly and calcification of the adrenals are always present, due to the excessive lipid storage in these organs. The disease usually manifests itself within the first week of life and is invariably fatal, normally by the age of six months. Although the disorder is clearly inherited, probably as an autosomal recessive trait, it remains to be clearly shown whether the biochemical lesion in Wolman's disease is one of synthesis, transport, or degradation. The best evidence, obtained from three patients, is the total deficiency of acid lipase activity in the liver and spleen. This enzyme hydrolyzes triglyceride and cholesteryl esters.

Phospholipids

Phospholipids, the fourth group of plasma lipids, are present in all cells as well as in the plasma. They are also known as phosphorized fats, phospholipins, and phosphatides.

1. All but the sphingomyelins contain a molecule of glycerol as a backbone. The middle (β) carbon of the glycerol, as part of the phospholipid molecule, is asymmetric. It usually has an L-configuration.

α'   CH$_2$-O-C-R$_1$
            ‖
            O

β   HC-O-C-R$_2$
          ‖
          O

α   CH$_2$-O-P-OH
            ‖
            O
            |
            OH

D-α-phosphatidic acid

CH$_2$-O-C-R$_1$
          ‖
          O

R$_2$-C-O-CH
     ‖
     O

CH$_2$-O-P-OH
          ‖
          O
          |
          OH

L-α-phosphatidic acid

2. There are five types of phospholipids.

   a. Phosphatidic acid is an important intermediate. A mole of it
       contains one mole of phosphate, one mole of glycerol, and two
       moles of one type of fatty acid or one mole each of two
       different types. The general structure of these compounds is
       given above.

   b. Lecithins are phosphatidyl cholines (esters of choline with
       phosphatidic acid). They are zwitterions over a wide pH range
       (including physiological pH) due to the presence of the choline
       (+ charge) and the phosphoric acid (- charge; strong acid).
       The fatty acids present usually are palmitic, stearic, oleic,
       linoleic, and arachidonic acids. Lecithins are the most
       abundant phosphoglycerides in animal tissues and are soluble
       in all of the usual fat solvents except for acetone.

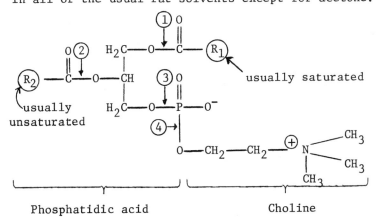

Enzymes that degrade lecithins cleave the molecule at the numbered (1-4) bonds. They are

i)   Phosphalipase A (lecithinase A): hydrolyzes bonds of type 1.

ii) Phospholipase A_2: hydrolyzes bonds of type 2; releases a fatty acid and lysolecithin or lysophosphatidylcholine; secreted by the pancreas as a proenzyme; some snake venoms contain a Ca^{+2}-requiring phospholipase of this type; lysolecithins are powerful detergents and hemolytic agents (due to their ability to solubilize membrane components).

iii) Lysophospholipase: hydrolyzes type 1 bonds in lysolecithins; releases a free fatty acid and glycerylphosphorylcholine.

iv) Phospholipase B (lecithinase B): hydrolyzes bonds of types 1 and 2; secreted by the pancreas.

v)   Phospholipase C: hydrolyzes bonds of type 3, releasing a diglyceride and phosphoryl choline; also acts on phosphatidyl serine and ethanolamine (cephalins); found in the α-toxin of Clostridium welchii and certain other bacteria and in some plant and animal tissues.

vi) Phospholipase D: hydrolyzes bonds of type 4, releasing phosphatidic acid and free choline; can simultaneously remove the choline and add an alcohol in its place (transesterification); found only in plants.

c. Cephalins are mixtures of phosphatides containing ethanolamine, serine, or inositol in place of choline. They are found in all tissues but are particularly abundant in brain and other nerve tissues. They are probably involved in blood clotting. Cephalins are separated on the basis of their differential solubilities in mixtures of chloroform and alcohol. The structures of these compounds are indicated next.

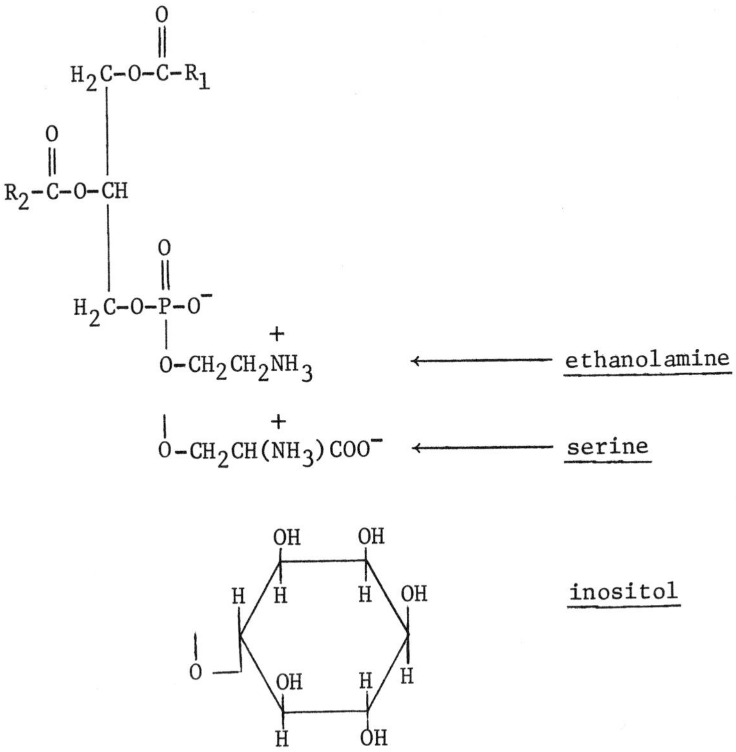

d. Plasmalogens (phosphatidyl ethanolamines, serines, and cholines) are found in cardiac and skeletal muscle, brain, and liver. Their function is not known. The general structure is shown below.

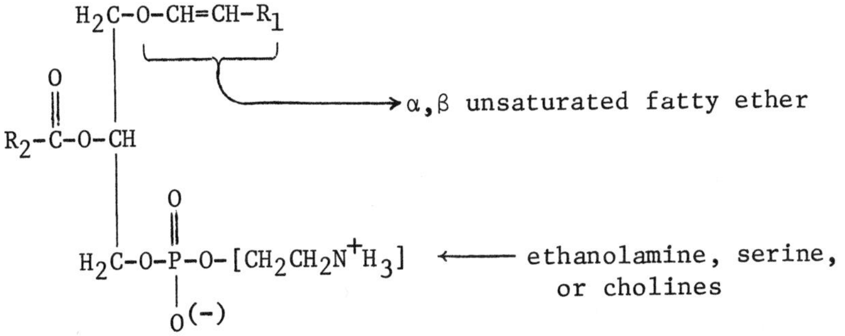

The α,β-unsaturated fatty ether is sometimes called an aldehydogenic group since hydrolysis of the ester releases an α,β-unsaturated primary alcohol which readily tautomerizes to an aldehyde.

e.   <u>Sphingomyelins</u> are found in all tissues, especially in brain
     and other nervous tissue. On complete hydrolysis they yield

1 mole of fatty acid

1 mole of H_3PO_4

1 mole of choline

1 mole of sphingosine

The structure of a **sphingomyelin is shown below**

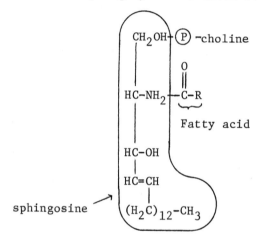

OR

known as a <u>ceramide</u>

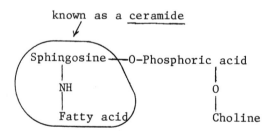

In <u>Niemann-Pick</u> disease, large amounts of sphingomyelins
accumulate in brain, liver, and spleen, accompanied by mental
retardation. The probable biochemical lesion is a deficiency
of sphingomyelinase, the enzyme which catalyzes the cleavage
of the molecule to phosphorylcholine and ceramide. There is
no known treatment.

Measurement of Lecithin and Sphingomyelin in Amniotic Fluid

1. One of the principal causes of death in premature infants is
    respiratory distress syndrome (RDS) with hyaline membrane disease.
    This occasionally occurs even in full-term babies. The syndrome
    is apparently due to immaturity of the lung, specifically with
    respect to its ability to synthesize lecithin and excrete it into
    the alveolar air spaces. Consequently, ways of measuring fetal
    lung maturity are helpful in judging whether a fetus is capable
    of surviving postnatally.

2. In a normal pregnancy, the lung is adequately developed by about
    the 36th or 37th week. There are several changes occurring at
    about this age which can be used to indirectly evaluate fetal lung
    maturity. A recently developed, more direct method which has
    proved useful in this regard is the measurement of lecithin
    (or more commonly, the lecithin to sphingomyelin (L/S) ratio,
    to correct for changes in fluid volume) in the amniotic fluid.
    In normal pregnancies, the L/S ratio is less than 1 prior to about
    the 31st week, rises to about 2 by the 34th week, 4 at the 36th
    week, and 8 at term (39 weeks). This results from an increase in
    lecithin synthesis rather than a decrease in sphingomyelin
    formation. There is some variation in these values for normal
    gestation and, in abnormal pregnancies (due to many causes
    including maternal diseases and fetal and placental conditions),
    the L/S ratio may be elevated or reduced without regard to
    gestational age. It has also been reported that a low L/S ratio
    is not <u>inevitably</u> associated with RDS and hyaline membrane disease.

3. Lecithin is a powerful detergent (surface active agent, capable
    of reducing the surface tension of water) which presumably
    functions by stabilizing the alveolar air spaces, preventing
    them from closing due to surface tension upon exhalation.
    Although the formation of a lecithin by trimethylation of
    α-palmityl, β-myristyl phosphatidyl ethanolamine is detectable
    from week 22-24 onward, this particular lecithin has only marginal
    ability to stabilize the alveoli. In the mature mammalian lung,
    the principal pathway for lecithin synthesis is a salvage reaction.

    CDP-choline + D-1,2-diacylglycerol ⟶ Phosphatidyl choline
                                                   (a lecithin)
                              ↓
                             CMP

    Apparently the lecithins formed by this route are much more
    effective at stabilizing the alveolar air spaces.

4. In one method of L/S ratio determination, chloroform is used to extract the phospholipids from the amniotic fluid. The phospholipids are then separated by thin-layer chromatography of a concentrated sample of the chloroform extract. The phospholipids are located by spraying with sulfuric acid and are identified by comparison with known standards. The spots are quantitated by scanning the plates with a suitable densitometer.

Other Complex Lipids

1. Polyglycerol phospholipids are present in large amounts in heart muscle, liver, and brain (in decreasing order). They also occur in plants.

    a. An important example of this group of compounds in animal tissue is cardiolipin (diphosphatidylglycerol). Its structure is shown below.

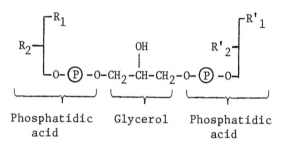

    b. In the mitochondrial membrane, 10% or more of the lipid is cardiolipin. It is apparently linked in the manner shown below.

        cytochrome —cardiolipin— structural protein

    c. Cardiolipin is the only phosphatidate which is known to be antigenic. This property is used in the serologic test for syphillis.

2. Cerebrosides (glycolipids) are <u>not</u> phospholipids. On hydrolysis they yield

    1 mole of sphingosine

    1 mole of fatty acid

    1 mole of a hexose (usually D-galactose)

The general structure is

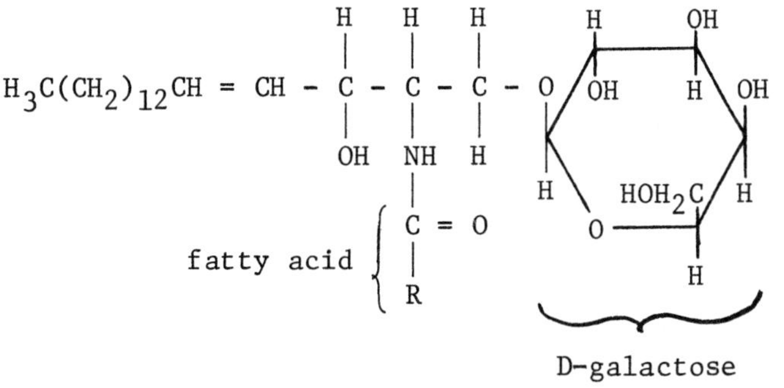

$$H_3C(CH_2)_{12}CH = CH - \overset{\overset{\displaystyle H}{|}}{\underset{\underset{\displaystyle OH}{|}}{C}} - \overset{\overset{\displaystyle H}{|}}{\underset{\underset{\displaystyle NH}{|}}{C}} - \overset{\overset{\displaystyle H}{|}}{\underset{\underset{\displaystyle H}{|}}{C}} - O$$

fatty acid $\left\{ \begin{array}{l} \\ C = O \\ | \\ R \end{array} \right.$

D-galactose

They are found in brain and other tissues and can be considered as monosaccharide derivatives of ceramides.

In addition to gluco- and galactocerebrosides as general names, different fatty acyl groups confer specific names to the cerebrosides.

Cerebroside	Fatty Acid	Formula of Fatty Acid
kerasin	lignoceric acid	$CH_3(CH_2)_{22}COOH$
phrenosin (cerebron)	cerebronic acid	$CH_3(CH_2)_{21}CHOHCOOH$
nervon	nervonic acid	$CH_3(CH_2)_7CH=CH(CH_2)_{13}COOH$

3. Sulfatides (or sulfatidates) are cerebrosides having a sulfate group attached in an ester linkage to the galactosyl residue. In the white matter of the brain, an analogue of phrenosin, having a sulfate ester at C-3 of the galactose, is present in large quantities.

4. Gangliosides (also called ceramide oligosaccharides), the most complex glycosphingolipids, are found in ganglion cells, spleen, and erythrocytes. They generally consist of a ceramide (N-acyl sphingosine in an ester linkage to a hexose (usually galactose or glucose). To this cerebroside nucleus are commonly attached at least one mole of N-acetylglucosamine or N-acetylgalactosamine and at least one mole of N-acetylneuraminic acid (see page 149).

a. One notation for gangliosides, which is used later in the
    discussion of sphingolipidoses, is composed of an initial
    capital G (for ganglioside); a subscripted M, D, or T
    (indicating one, two, or three residues, respectively, of
    N-acetyl neuraminic acid in the molecule); and a further
    subscript, either a number or a number and a lower case letter.
    Thus G_{M1}, G_{D1a}, and G_{T1} designate, respectively, members of
    the class of gangliosides containing one, two, or three
    residues of N-acetyl neuraminic acid per molecule of ganglioside.

b. Globoside, the most abundant glycosphingolipid in erythrocyte
    stroma, is given below as an example.

    ceramide-($\leftarrow$1β)-glucose-(4$\leftarrow$1β)-galactose-(4$\leftarrow$1α)

                          galactose-(3$\leftarrow$1β)-N-acetyl-galactosamine

    Another important ganglioside is ganglioside G_{M1}, a degradation
    product of which (G_{M2}) accumulates in Tay-Sachs disease.
    The structure of G_{M1} is shown below.

    ceramide-($\leftarrow$1β)-glucose-(4$\leftarrow$1β)-galactose-(4$\leftarrow$1β)

                                        N-acetylgalactosamine

                                        (3$\leftarrow$1β)-galactose

            (3$\leftarrow$2)-N-acetyl neuraminic acid

Comments on the Biosynthesis of Phospholipids and Triglycerides

1. The importance of cytidine diphosphate (CDP) and coenzyme A (CoASH)
    as carriers of intermediates in lipid synthesis is apparent from
    Figure 72. Further examples will be given throughout this chapter.
    The structure of CDP is given in the chapter on nucleic acids,
    while that of CoASH is with glycolysis. The structure of CDP-
    diacylglyceride is shown below. These biosynthetic reactions take
    place largely on the endoplasmic reticulum.

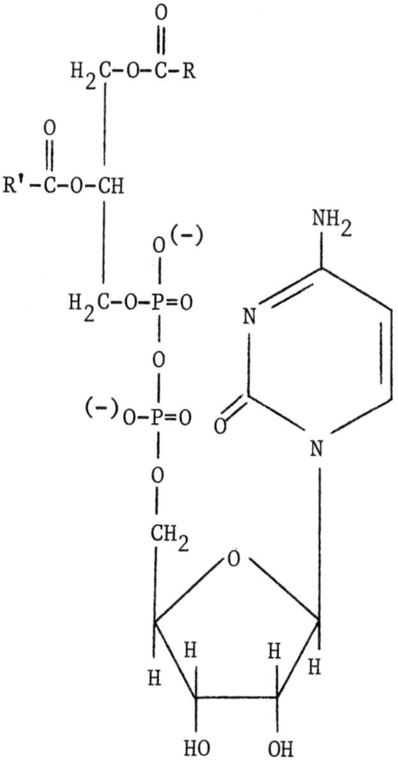

$$H_2C-O-C(=O)-R$$

$$R'-C(=O)-O-CH$$

$$H_2C-O-P(=O)-O^{(-)}$$

Cytidine
diphosphate
diacylglyceride

2. In Figure 72, due to the specificities of the enzyme involved,
    the pathway from dihydroxyacetone phosphate through lysophosphatidyl-
    choline produces phosphatidic acids in which the terminal acyl group
    is saturated and the middle (β) acyl group is usually unsaturated.
    Triglycerides and phosphatides from many sources show this pattern.
    The direct acylation of α-glycerolphosphate, on the other hand,
    produces a random mixture of saturated and unsaturated acyl groups
    in phosphatidic acid. This could produce the lecithins of
    lung tissue which contain predominantly saturated acyl groups in
    both positions.

Figure 72.   Biosynthesis of Di- and Triglycerides, Cephalins, Cardiolipin (A Polyglycerol Phospholipid), and Lecithin

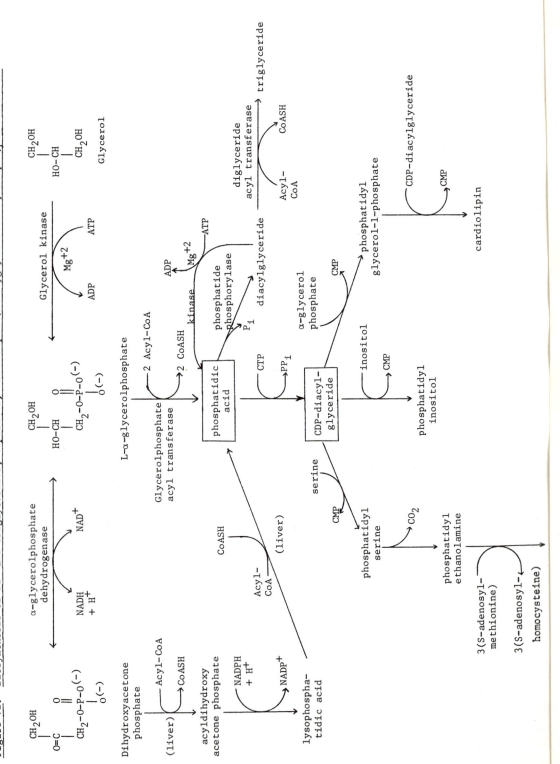

3. In addition to the <u>de novo</u> pathways outlined in Figure 72, there
    exists a salvage pathway which uses choline and ethanolamine
    obtained from the diet and from the catabolism of other
    phospholipids.

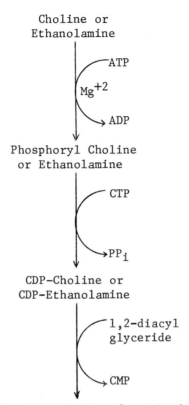

Choline or
Ethanolamine

ATP

Mg^{+2}

ADP

Phosphoryl Choline
or Ethanolamine

CTP

PP_i

CDP-Choline or
CDP-Ethanolamine

1,2-diacyl
glyceride

CMP

Phosphatidyl Choline (Lecithin)
or Phosphatidyl Ethanolamine

4. Plasmalogens are synthesized by a pathway similar to the
    lysophosphatidic acid route shown in Figure 72. The α,β-unsaturated
    ether is derived from the long-chain alcohol initially coupled
    to dihydroxyacetone phosphate. The reactions are indicated below.

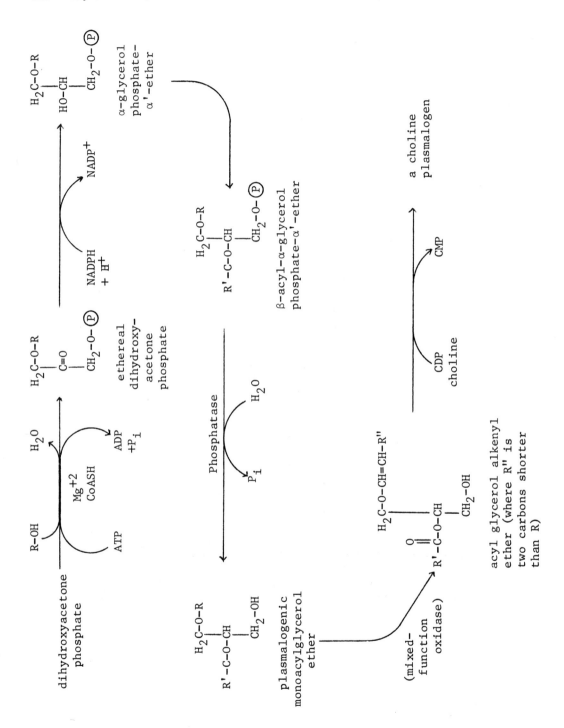

5. Biosynthesis of the sphingolipids (ceramides, sphingomyelins, cerebrosides, sulfatides, and gangliosides) begins with the formation of sphingosine (also called sphingenine) from palmitoyl-CoA (in mouse brain) or some other 16-carbon precursor, and L-serine. These reactions require Mn^{+2}, pyridoxal phosphate, NADPH, and probably a flavin coenzyme. Successive steps are shown below. Note in particular the central role of UDP as a carrier of carbohydrate residues. This was pointed out earlier in Chapter 4.

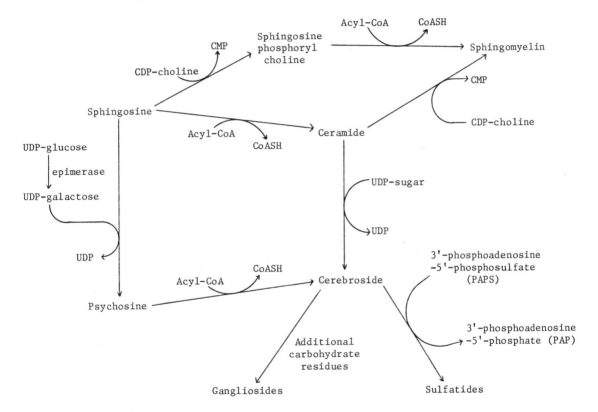

Catabolism and Storage Disorders of Glycosphingolipids

1. An important group of inherited diseases exists in which large
    quantities of different glycosphingolipids accumulate within the
    lysosomes of various tissues. These belong to the group of inborn
    metabolic errors known as lysosomal disorders and all involve
    decreased or absent activity of catabolic enzymes rather than
    overactivity of synthetic pathways. Lysosomes contain over forty
    different hydrolases with specificities for different types of
    molecules. Although there are several criteria for classifying
    a disorder as lysosomal, it is significant that in the majority
    of such diseases, the absence of a specific hydrolase can be
    demonstrated. Other lysosomal disorders include the
    mucopolysaccharidoses (lysosomal enzyme defects not yet shown);
    mucolipidoses (deposition of both mucopolysaccharide and lipid;
    most genetic defects not known); Niemann-Pick disease; I-cell
    disease (may be due to a "leaky" lysosomal membrane); cystinosis;
    acid phosphatase deficiency; Wolman's disease; and others.

2. The particular group of glycosphingolipids to be discussed here
    all contain sphingosine in the form of a ceramide. The catabolism
    of these compounds involves removal of successive glycosyl residues
    until finally the ceramide is released. The errors usually
    involve the enzymes which catalyze these glycosidations. The one
    exception to this is metachromatic leukodystrophy, due to a
    deficiency in a sulfatidase enzyme.

3. The catabolic pathways for the glycosphingolipids are given in
    Figure 73 and their disorders are summarized in Table 24. Certain
    aspects of these diseases warrant further comment and this is given
    in succeeding paragraphs. Accumulation of a specific lipid in
    these disorders is frequently accompanied by deposition of one or
    more polysaccharides which are structurally related to the lipid.
    Their clinical treatment is generally palliative or non-existent.
    Recently, however, enzyme replacement therapy (administration of
    an exogenous preparation of the missing enzyme) has proven useful
    in the treatment of Fabry's disease. Because the exogenous enzymes
    are apparently unable to cross the blood-brain barrier, the
    efficacy of this type of treatment in the majority of the
    glycosphingolipidoses (which have neurological involvement) is
    doubtful. Attempts are being made to modify the enzymes to
    overcome this difficulty. Considerable progress has also been
    made in identifying carriers of these diseases and in prenatal
    diagnosis of homozygotes. This will, presumably, improve the
    prospects for genetic counseling.

Figure 73.   Degradative Pathways for Glycosphingolipids

Note: G_{M1} = Gal-NAc Gal-Gal-Glc-Cer
$$\underset{NANA}{\text{|}}$$

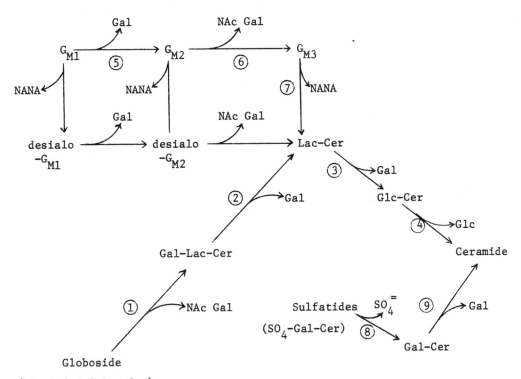

$G_{M1,2,3}$ are gangliosides. The structures of ganglioside G_{M1} and globoside were given previously in the section on gangliosides. The other compounds in this diagram are derived from them by successive removal of the indicated carbohydrate residues. Abbreviations:

    Gal = galactose
    Glc = glucose
    NAc Gal = N-Acetyl-Galactose-2-Amine
    NANA = N-Acetyl Neuraminic Acid (a sialic acid)
    Lac = lactose (galactosyl(β-1$\rightarrow$4) glucose)
    Cer = ceramide
    desialo = without a sialic acid (NANA) residue

The circled numbers correspond to the metabolic lesions listed in Table 24.

Table 24.   Characteristics of Glycosphingolipid Storage Disorders

Disorders	Major Lipids Accumulated	Other Compounds Affected	Enzyme Lacking*	REMARKS
G_{M2} gangliosidosis, Type II (Tay-Sachs Variant; Sandhoff's disease)	Globoside and G_{M2} ganglioside both accumulate		Hexosaminidases A and B (①) and ⑥)	Same clinical picture as Tay-Sachs disease but progresses more rapidly
Fabry's disease (Glycosphingolipid Lipidosis)	Gal-(4←1α)-Lac-Cer	Gal-(4←1α)-Gal-Cer accumulates	α-galactosidase (②)	X-linked recessive; hemizygous males have a characteristic skin lesion usually lacking in heterozygous females; pain in the extremities; death usually results from renal failure or cerebral or cardiovascular disease
Ceramide Lactoside Lipidosis	Gal-(4←1β)-Glc-Cer		Ceramide Lactoside β-galactosidase (③)	Liver and spleen enlargement; slowly progressive brain damage; neurological impairment
Gaucher's disease (glucosyl ceramide lipidoses; three types; see text)	Glc-Cer	Accumulation of G_{M3} ganglioside most frequently; other compounds occasionally	β-glucosidase (glucocerebrosidase; ④)	Hepatosplenomegaly; frequently fatal; no known treatment; occurrence of "Gaucher cells" (RE cells which contain accumulations of erythrocyte-derived glucocerebroside)
G_{M1} gangliosidosis (two types; see text)	G_{M1}- and desialo-G_{M1}-gangliosides	Keratin-sulfate-related polysaccharide accumulates	G_{M1}-β-galactosidase (⑤)	Mental and motor deterioration; accumulation of mucopolysaccharides is as significant as accumulation of gangliosides; invariably fatal; autosomal recessive

Table 24. (continued)

Disorders	Major Lipids Accumulated	Other Compounds Affected	Enzyme Lacking*	REMARKS
G_{M2} gangliosidosis, Type I (Tay-Sachs disease; see text)	G_{M2}- and desialo-G_{M2}-gangliosides	Other desialo hexosyl ceramides; occasionally other compounds accumulate	Hexosaminidase A (6)	Red spot in retina; mental retardation; invariably fatal; autosomal recessive; panracial but especially prevalent among Northern European Jews
G_{M3} gangliosidosis	G_{M3}- and desialo-G_{M3}-ganglioside		Not known (7?)	Sample long stored in formalin prior to analysis; lipid changes could be artifactual; included here primarily for completeness
Metachromatic leukodystrophy (MLD; sulfatide lipidoses; at least three types; see text)	3-sulfato-galacto-sylcerebroside	Cerebrosides other than sulfatides are decreased; ceramide dihexoside sulfate accumulates	Sulfatidases (8; arylsulfatases)	Demyelination; progressive paralysis and dementia; death usually occurs within the first decade; autosomal recessive inheritance
Krabbe's disease (Globoid cell leukodystrophy; galactosyl ceramide lipidosis)	Galactocerebroside	Sulfatides are also greatly decreased, probably as a secondary feature	Galactocerebroside-β-galactosidase (9)	Mental retardation; demyelination; globoid cells in brain white matter; invariably fatal; autosomal recessive inheritance

* The circled numbers following the enzyme names refer to reactions indicated in Figure 73. The abbreviations used here are also the same as those in that figure.

4. Patients exhibiting glucosyl ceramide lipidosis (Gaucher's disease) fall into one of three groups. The genetic defects in these three types appear to be similar errors in the same or related genetic loci.

   a. Chronic non-neuronopathic (adult) A somewhat heterogeneous group of patients characterized by lack of cerebral involvement and presence of hematological abnormalities. Despite the name, it often develops in childhood and may produce death at an early age.

   b. Acute neuronopathic. Usually appears prior to six months of age and is fatal by two years. Mental damage is a primary characteristic and the disease progresses rapidly from its onset.

   c. Subacute neuronopathic (juvenile). Another heterogeneous group in which death occurs anywhere from infancy to about 30 years of age. The cerebral abnormalities usually appear at least two years postnatally.

5. Studies of a number of cases of G_{M1} gangliosidosis have revealed two distinct types.

   a. In generalized gangliosidosis, G_{M1} and desialo-G_{M1}-gangliosides accumulate in both the brain and viscera. Three β-galactosidase activities have been isolated from normal human liver and all three are absent in this disorder. The progress of the disease is rapid, beginning at or near birth and terminating fatally, usually by two years of age.

   b. Juvenile G_{M1} gangliosidosis progresses much more slowly. Psychomotor abnormalities usually begin at about one year and death ensues at 3-10 years. Only two of the three liver β-galactosidase activities are absent in this variant, presumably accounting for the lack of lipid accumulation in this organ. This enzymatic finding also supports the genetic separation of the two forms.

6. Although the name amaurotic familial idiocy (AFI) originally meant specifically Tay-Sachs disease, there are now six subdivisions of AFI. Of these six, infantile amaurotic idiocy is synonymous with Tay-Sachs disease. The only other AFI in which gangliosidosis has been demonstrated is systemic late infantile AFI (a G_{M1} gangliosidosis). The terms amaurotic idiocy and gangliosidosis are not interchangeable, despite common usage.

7. The reason for the accumulation of desialo-G_{M1} and desialo-G_{M2} gangliosides in G_{M1} gangliosidosis and Tay-Sachs disease (TSD), respectively, is not clear. It has been demonstrated that, in vitro, tissue from patients with TSD is capable of metabolizing desialo-G_{M2} ganglioside. The reaction is inhibited by G_{M2} ganglioside but it remains to be demonstrated that this mechanism operates in vivo. Although similar studies have apparently not been done with G_{M1} gangliosidosis material, a parallel explanation seems plausible.

8. The name metachromatic leukodystrophy (MLD) refers to a dysfunction of the white matter of the brain which is characterized by metachromasia of the nervous tissue. Metachromasia is the ability of certain compounds (notably organic anions such as sulfatides which either have a high molecular weight or are capable of polymerizing or forming micelles) to shift the absorption wavelength of some cationic dyes to shorter wavelengths. This causes toluidine blue, for example, to appear red or pink. Other dyes affected include cresyl violet (stains MLD tissue brown), methylene blue, and thionine.

    It is important to distinguish between MLD and other disorders of the white matter. One common disease of this sort is multiple sclerosis, which probably has a viral or autoimmune etiology.

9. As in several other of the diseases discussed here, metachromatic leukodystrophy can be subdivided into several disorders which show distinct clinical patterns.

    a. Late infantile form. Most common type; usually appears 1-4 years postnatally; associated with loss of motor and mental functions and ultimately death; due to lack of arylsulfatase A and cerebroside sulfatase deficiency.

    b. Adult. Rare; onset is not until age 21 years or later but difficult to diagnose; dementia and psychosis; later, loss of motor function usually followed by death; deficiency (but not absence) of arylsulfatase A.

    c. MLD variant with multiple sulfatase deficiency. Rare; appears at age 1-3 years; progresses slowly with psychomotor deterioration and other changes; development of these patients is retarded compared to those with late infantile MLD despite similar age of onset; the disorder is fatal and is apparently due to a lack of arylsulfatases A and C and of steroid sulfatase, and decreased activity of arylsulfatase B.

10. One other disorder of lipid catabolism is <u>alpha-fucosidosis</u>.
In this disease, a glycoceramide having the antigenicity of
H-isoantigen (see the section on ABO blood groups in the chapter
on immunochemistry) and mucopolysaccharides having an
α-L-fucose residue at the reducing end accumulate in liver, brain,
and perhaps other tissues.

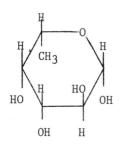

α-L-fucose
(6-deoxy-L-galactose)

Cer-(←1β)-Glc-(4←1β)-Gal-(4←1β)-NAcGlcam
                                          |
                                    (3←1β)-Gal
                                          |
                                    (2←1α)-Fuc

Structure of H-isoantigenic substance which
accumulates in α-fucosidosis. Cer = ceramide;
Glc = glucose; Gal = galactose; NAcGlcam =
N-acetyl-glucosamine; Fuc = α-L-fucose.

Because of the mucopolysaccharide deposition, this is also
classified as a mucolipidosis or even a mucopolysaccharidosis.
Clinical features include cerebral degeneration, thick skin,
electrolyte abnormalities, and muscle spasticity. In the livers
of the patients studied, α-fucosidase was found to be absent.
This enzyme hydrolyzes glycosidic bonds between α-fucose and
other sugars. This enzyme deficiency is apparently the cause
of this disorder.

Biological Membranes

1. As was pointed out earlier (Chapter 3), one of the factors which
distinguishes living matter (cells) from non-living material is
the ability to employ external energy to maintain chemical
reactions in a non-equilibrium steady-state. This would be
practically impossible without the various types of membranes which
permit the cell to establish a "local" (intracellular) environment
which differs from whatever surrounds the cell regarding
concentrations and even the particular molecules present. In
addition to this important physical isolation, membranes provide
a support for many metabolic enzymes and an electrical barrier
(as in nerve conduction). Membranes cannot be totally
impermeable, though, since nutrients must enter the cell and
waste products have to be eliminated. Specific carrier enzymes
(permeases), apparently mounted in the membrane, perform important
roles in the transport of various molecules. Other compounds can
freely diffuse through the membranes. The particular function of

the cell dictates which molecules can and cannot cross the membrane barrier and under what conditions they can do so. Additionally, membranes must be physically flexible (plastic). This is evident from observations of cell packing and motility and from the swelling and shrinking which occurs when cells are placed in a medium having an osmolarity different from that within the cell.

2. The preceding discussion also applies to the use of membranes for intracellular localization seen in eucaryotic cells. Examples of such <u>subcellular organelles</u> include mitochondria, chloroplasts, the nucleus, the endoplasmic reticulum and the Golgi apparatus, lysosomes, peroxisomes, and others. Some of these are discussed elsewhere in this book. Recall the role of conformational change in one theory of the mechanism by which electron flow is coupled to phosphorylation in the mitochondria (page 212).

3. Biological membranes are generally composed of proteins, neutral lipids (especially cholesterol), and phospholipids. The specific ratios of these compounds vary considerably among cell types and between plasma membranes and membranes of subcellular organelles. This is to be expected, since many of the properties which make different cells and organelles unique reside in their membranes. For example, in the myelin nerve sheath membrane, the ratio of lipid (grams) to protein (grams) is 4, the molar ratio of cholesterol to phospholipid is 0.7-1.2, and the principal phospholipids present are phosphatidyl ethanolamine and choline, and cerebrosides. For a liver cell membrane (plasma membrane, plasmalemma), the corresponding values are 0.7-1.0 (lipid/protein), 0.3-0.5 (cholesterol/polar lipids), while phosphatidyl ethanolamine, choline, and serine predominate among the phospholipids. A third example is the ribosome-free endoplasmic reticulum, having values of 0.8-1.4 (lipid/protein), 0.03-0.08 (cholesterol/phospholipid), with phosphatidyl choline and serine, and sphingomyelin as the principal phospholipids. The endoplasmic reticulum is typical of the membranes of other subcellular organelles in containing much smaller amounts of cholesterol than the plasma membranes. Animal cell membranes also contain oligosaccharides, composed principally of fucose, galactose, amino sugars, and (in some cell membranes) sialic acid. Some of these apparently provide the antigenic specificity possessed by most cells.

4. In addition to varying considerably in composition from one cell to another, biological membranes apparently are constantly gaining and losing molecules (plasticity in time). Thus, different membranes are constantly exchanging proteins and lipids and as a cell matures, the protein and (to a lesser extent) lipid content of its membranes alters. This may affect composition and

structural studies of, say, erythrocyte membranes, where any sample
obtained from fresh blood will contain cells of all ages. Exogenous
substances can change membrane composition. In the rat,
phenobarbital causes proliferation of smooth endoplasmic reticulum
(SER) membranes in which the activities of certain enzymes
(including the P-450 cytochromes involves in phenobarbital
detoxication) are ten times higher than the activity in normal SER
membranes. These "abnormal" membranes persist after the
phenobarbital has been metabolized and may explain why this drug
is capable of stimulating heme catabolism.

5. Despite intensive study and much speculation, the structure of
    biological membranes is still not well understood. The "lipid
    bilayer" model was originally proposed by Danielli and Davson in
    1935. It has been modified extensively since then but remains
    an integral part of most theories of membrane structure. In its
    simplest form, the Danielli-Davson model appears as shown in
    Figure 74. Cholesterol, which is an important part of animal
    membranes, probably lies in the hydrophobic region with the
    plane of its ring parallel to the fatty acids of the phospholipids.

<u>Figure 74.</u>   <u>Danielli-Davson Lipid Bilayer Model for Biological Membranes</u>

polar, glycerylphos- 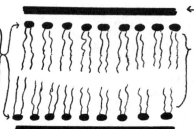 non-polar, hydrophobic
phatidate "heads" of   interior containing
diacylphosphatidate   the fatty acid "tails"
molecules such as   of the phosphatidates
phosphatidyl choline

protein layers coating
the outsides of the
lipid leaflet bilayer

In this figure, note that the phospholipids could be any of the
phosphatidates (e.g., phosphatidyl choline, ethanolamine, or serine)
or even some of the sphingolipids. In modern membrane models, it
is thought that regions of this sort of structure may occur
interspersed with other types of structure (e.g., proteins, see below).

6. Most of the modifications to this basic model deal with the location of the proteins. A number of studies indicate that certain proteins are attached to the inner surfaces of membranes, others are on the outer surfaces, and some penetrate the membrane, so that they are both inside and outside. The forces involved in the protein-lipid interaction are probably non-covalent in most cases. Ionic groups of the proteins may associate with the polar surfaces of the membrane lipids while hydrophobic portions contact the lipid-like interior of the bilayer.

    This leads to the concept of mosaicism or inhomogeneity in biological membranes. The nature and arrangement of the proteins in a particular membrane may well depend on the region of the membrane being studied. There is still much work to be done before these crucial structures can be completely described and understood.

7. The central role of membranes in enzyme function is one of the most important ideas to arise from membrane studies. As was indicated previously, membranes are far more than passive envelopes for the cytoplasm or intraorganellular fluid. A partial list of enzyme systems and the membranes with which they are associated is given below. Some of these systems (or members of them) have a demonstrated functional requirement for lipid.

    a. Electron transport and oxidative phosphorylation enzymes (bound to the inner mitochondrial membrane).

    b. Protein synthesis (polysomes are frequently attached to the rough endoplasmic reticulum and perhaps penetrate the membrane).

    c. Microsomal oxidase system (involved in heme catabolism and detoxication of other compounds; smooth endoplasmic reticulum).

    d. Enzyme system which synthesizes the polysaccharide portion of the mucopolysaccharides (smooth endoplasmic reticulum and Golgi apparatus).

    e. Miscellaneous

        i) Permeases (also called translocases or transporter proteins; used by the cell to transport molecules and ions across membrane permeability barriers; occur in most membranes; when energy for the transport is supplied by ATP hydrolysis, these are sometimes called ATPases; an example is the Na^+-K^+ ATPase or pump in the cell membranes of erythrocytes and other cells).

ii) <u>Adenyl cyclase</u> (located on the inside of a number of cell membranes; it is in communication with a <u>hormone receptor protein</u> attached to the outside of the cell membrane).

Fat Absorption

1. The steps involved in the absorption of dietary fat are outlined in Figure 75. Cholesterol is absorbed directly into mucosal cells where most of it is first esterified with fatty acids then incorporated into chylomicra or (to a small extent) absorbed directly into the lymphatic system. The uptake of cholesterol from the intestine is dependent upon the presence of the bile salts (discussed later) which are presumably needed for emulsification of the sterol. Phospholipids are hydrolyzed by the phospholipases (see page 610) to free fatty acids and glycerylphosphorylcholine. The fatty acids are absorbed and the glycerylphosphorylcholine is either excreted in the feces or further degraded and absorbed.

2. The triglycerides and the lower glycerides (a minor component of dietary fats) are hydrolyzed by pancreatic lipase. As the name implies, it is secreted into the intestine by the pancreas, along with the phospholipases, proteases, amylase, and other compounds. Anything which prevents this material from reaching the intestine can result in fatty stools (steatorrhea; discussed later in this section). The action of pancreatic lipase hydrolyzes about 40% of the dietary triglycerides to glycerol and free fatty acids and another 3-10% is absorbed unchanged as triglycerides. The remainder (50-57%) is degraded to free fatty acids and mono- and diglycerides with β-monoglycerides predominating. This lipase acts at the interface between water and an insoluble substrate (e.g., triglyceride). Gastrointestinal motility helps to increase the available surface by mixing and breaking up the chyme. The fatty acids released are mostly in their ionized form, as Ca^{+2} salts. Such compounds (of an ionized fatty acid and a metal ion) are known as soaps.

3. Following hydrolysis and aided by the churning action of intestinal peristalsis, a micellar emulsion of bile salts, free fatty acids, mono-, di-, and triglycerides, and cholesterol is formed in the intestine and absorbed into the mucosal cells of the intestinal villae. The particles of this emulsion are less than 1 μ in diameter. Oils with a greater percentage of unsaturated triglycerides (e.g., olive oil) are absorbed bettwe than saturated long-chain TG's, as are fatty acids containing less than about 16-18 carbons. Related to this is the observation that lipids and fats which are solids at temperatures much above that of the body are absorbed poorly by the intestine.

Figure 75. Intestinal Absorption of Dietary Fats

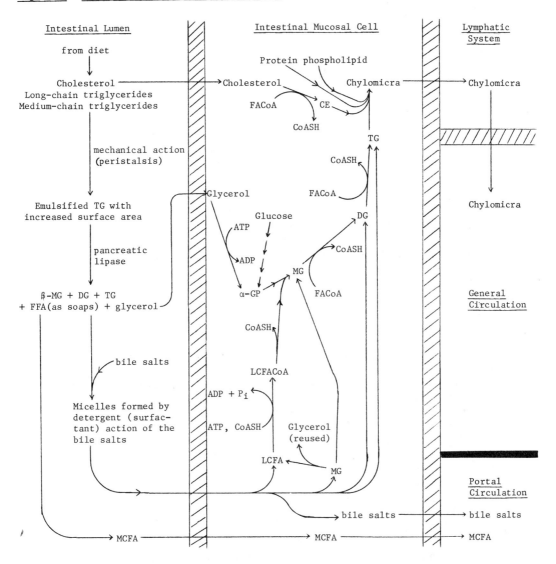

Abbreviations: MG = monoglyceride
                DG = diglyceride
                TG = triglyceride
                FFA = free fatty acid
                LCFA = long-chain fatty acid (14 carbons or more)
                MCFA = medium-chain fatty acid (10 carbons or less)
                α-GP = α-glycerolphosphate
                CE = cholesterol ester
                FACoA = fatty acyl-CoA

4. Inside the mucosal cells, some of the absorbed mono-, di-, and triglycerides are further hydrolyzed by intestinal lipase. More important, though, is the net resynthesis of triglyceride in these cells and the formation of chylomicra (plural of chylomicron; see page 663ff). These are complex lipoproteins, about 1 μ in diameter, which serve as the transport form of exogenous triglyceride. Their composition is approximately

    87% Triglyceride

    4% Free and esterified cholesterol

    8% Phospholipid

    1% Protein

Chylomicron synthesis occurs in the endoplasmic reticulum. A rate-limiting step in the transport of fat from the mucosal cells into the lymphatic system may be the synthesis of β-lipoprotein, the protein used in chylomicron synthesis. In fact, in congenital β-lipoproteinemia, triglycerides accumulate in the mucosal cells.

5. Material which has entered the mucosal cells can leave them by two routes.

    a. Water-soluble substances such as medium- and short-chain fatty acids (MCFA and SCFA; less than 10-12 carbon atoms) and bile salts are absorbed directly into the enterohepatic portal circulation. The MCFA are transported as free (unesterified) fatty acids. The bile salts are removed from this blood by the liver and reinjected, along with the bile, into the duodenum. Because of this recycling, very little bile acid appears in the peripheral circulation and only about 200 mg/day are lost in the feces. The direct absorption of the smaller fatty acids is useful in the maintenance of patients who exhibit fatty acid malabsorption and steatorrhea due to bile salt deficiency. Medium-chain-length triglycerides (MCT) are provided in the diet in place of a large fraction of the more usual long-chain triglycerides. Since neither bile nor micellar formation is directly necessary for lipase activity, the MCT can be hydrolyzed and the MCFA absorbed despite this lack. Special MCT diets are commercially available.

b. The chylomicra and any remaining cholesterol are absorbed into the intestinal lacteals (lymphatics originating in the villi of the small intestine). Following a meal, the presence of large amounts of chylomicra in the lymph give it a milky appearance and endow it with the name <u>chyle</u>. From the lymph, these materials enter the blood by way of the thoracic duct and, to a lesser extent, accessory channels.

6. Lipid malabsorption, leading to the excessive loss of fat and other lipids in the stool (i.e., steatorrhea), can be caused by several things. It is important to note that while the bile salts do not activate the lipase, they do prevent its inactivation. Because albumin can partially replace the bile salts in this function, it is presumed that the solubilization of the fatty acids produced by lipase action stops fatty acid inhibition of the enzyme. In the laboratory assay of serum lipase (see Chapter 3), a much higher value for lipase activity is obtained if sodium deoxycholate (a bile salt) is added to the reaction mixture to emulsify the fatty acids produced.

a. Defective lipolysis in the lumen of the small intestine

i) Bile salt deficiency: due to impaired liver function, obstruction of bile duct or excessive loss of bile salt because of poor reabsorption.

ii) Pancreatic lipase deficiency: due to malfunction or damage of pancreatic tissue, obstruction of the pancreatic duct (as in Cystic Fibrosis of the pancreas; see under protein digestion), or a genetic defect in enzyme synthesis.

b. Defective mucosal cell function

i) Impaired resynthesis of triglycerides: in non-tropical sprue and adrenocortical hormone deficiency.

ii) Impaired protein synthesis of α-and β-lipoproteins (genetic defects).

iii) Blockage of the synthesis of the above proteins by certain drugs that inhibit protein synthesis (e.g., puromycin).

c. Abnormal connections between organs

i) Chyluria: abnormal connection between the urinary tract and the lymphatic drainage system of the small intestine causing the excretion of milky urine.

ii) Chylothorax: abnormal connection between the pleural space and lymphatic drainage system of the small intestine giving rise to the accumulation of milky pleural fluid.

d. Defective fat absorption also can impair the uptake of other nutrients, notably the fat-soluble vitamins (A, D, E, and K). Since vitamin K is essential for blood clotting, care should be taken to avoid its deficiency in patients prior to gall-bladder surgery.

Fatty Acid Oxidation

1. The principal route for fatty acid catabolism, outlined in Figure 76, is β-oxidation. The name is derived from the oxidation of the activated fatty acid (fatty acyl-CoA) at the β-carbon, followed by removal of two carbon fragments as acetyl-CoA. These reactions are catalyzed by a group of mitochondrial enzymes, collectively called fatty acid oxidase, associated with the mitochondrial cytochrome system. Entry of fatty acids into the mitochondria and release of acetyl-CoA to the cytoplasm is discussed later under fatty acid synthesis.

2. From Figure 76, it can be seen that one molecule of ATP is hydrolyzed to AMP and pyrophosphate for each fatty acid molecule oxidized (Reaction 1). This is the only ATP consumed in these reactions. The energy from its hydrolysis is used to form the high-energy thioester bond in acyl-CoA.

   a. Each time a mole of fatty acyl-CoA cycles through reactions 2-5, one mole of reduced flavoprotein (FPH_2), one mole of NADH, and at least one mole of acetyl-CoA are produced. On the last pass of an even-chain-length fatty acid, two moles of acetyl CoA are formed; and the final pass of an odd-chain-length molecule releases one mole of propionyl-CoA.

   b. Based on the preceding paragraph, the amount of ATP formed in the β-oxidation of hexanoic acid can be calculated. In doing so, the hydrolysis of ATP to AMP + PP_i is considered as equivalent to two moles of ATP, since two high-energy phosphate bonds are lost. This calculation is shown on page 638.

Figure 76. Fatty Acid β-Oxidation

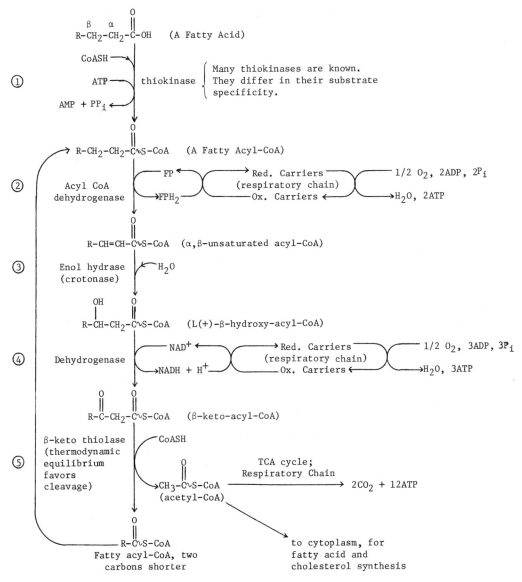

Note that the shortened fatty acyl-CoA from one cycle is further oxidized in successive passes until it is entirely converted to acetyl-CoA. Odd-chain fatty acids, discussed later, produce one molecule of propionyl-CoA. Ox. = oxidized, Red. = reduced; respiratory chain = oxidative phosphorylation and electron transport; "$\sim$" = a high-energy bond (see Chapter 3); FP = flavoprotein.

Reaction	Direct Consequences of the Reaction	Moles of ATP Gained or Lost Per Mole of Hexanoic Acid
1	hexanoic acid → hexanoyl-CoA	-2
2	dehydrogenation of acyl-CoA; $2(FP \rightarrow FPH_2)$	+4
3	hydration of α,β-unsaturated fatty acyl-CoA	0
4	dehydrogenation of β-hydroxy-acyl-CoA; $2(NAD^+ \rightarrow NADH + H^+)$	+6
5	formation of 3 moles of acetyl-CoA, followed by their oxidation to CO_2 and H_2O in the TCA cycle and electron-transport system	+(3x12) = +36

Total ATP = +44

c. As was pointed out earlier (Chapter 4), fatty acid oxidation produces more moles of ATP per mole of CO_2 formed than does carbohydrate oxidation. In this case, oxidation of one mole of glucose (a six-carbon moiety as is hexanoic acid) produces, at most (assuming malate shuttle operation exclusively), 38 moles of ATP. This is six less than obtained from hexanoic acid.

3. The propionyl-CoA produced by β-oxidation of **odd-chain-length** fatty acids is metabolized as shown next.

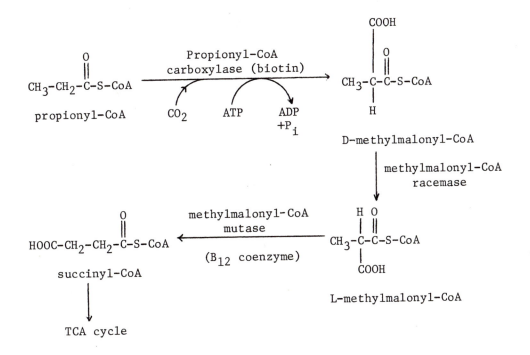

a. Propionyl-CoA is also produced in the oxidation of isoleucine, one of the branched-chain amino acids (see page 467). The fermentation products of certain bacteria (especially strains of the genera Propionibacterium and Veillonella) contain propionic acid. This can be converted to propionyl-CoA by acetyl-CoA synthetase. Methylmalonyl-CoA is the chief product of valine oxidation (see page 467).

b. There are two inheritable types of methylmalonic acidemia and aciduria. In young children, they are associated with mental retardation and failure to thrive. In one type (unresponsive to vitamin B_{12}), there is a lack of methylmalonyl-CoA mutase. The other type probably involves an inability to readily convert vitamin B_{12} to B_{12} coenzyme. It responds to large doses of vitamin B_{12}. Propionic and methylmalonic aciduria results from vitamin B_{12} deficiency in otherwise normal individuals.

4. At least two other oxidative pathways are significant in fatty acid metabolism.

a. Alpha-oxidation (already mentioned in connection with Refsum's syndrome, page 596) is important in the catabolism of branched-chain and odd-chain-length fatty acids. The general reaction, shown below, is catalyzed by a monooxygenase and requires O_2, Fe^{+2}, and either ascorbate or tetrahydropteridine (see page 479). It has been demonstrated in plants and in microsomes from brain and other tissues.

$$R-CH_2-CH_2-COOH \xrightarrow{\text{α-ox.}} R-CH_2-\overset{\displaystyle OH}{\underset{\displaystyle |}{C}}H-COOH$$

This is also one route to hydroxy fatty acids. The α-hydroxy fatty acids can be further converted, in succeeding reactions, to a fatty acid one carbon shorter than the original one. Thus, if an odd-chain-length compound is used initially, an even-chain-length acid is produced which can be further oxidized by β-oxidation.

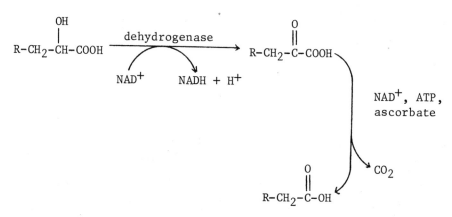

b. Omega-oxidation refers to oxidation of the carbon most remote from the carboxyl group in any fatty acid. The basic reaction, catalyzed by a monooxygenase which requires NADPH, O_2, and cytochrome P-450, is shown below. It has been demonstrated in liver microsomes and in some bacteria.

$$H_3C-(CH_2)_n-COOH \xrightarrow{\text{ω-ox.}} HO-CH_2-(CH_2)_n-COOH$$

Further oxidation of the ω-hydroxy acids can produce the corresponding dicarboxylic acids which can be β-oxidized from either end.

Fatty Acid Synthesis

1. The reactions of fatty acid synthesis are outlined in Figure 77 as they have been elucidated in E. coli. Although equivalent reactions probably occur in this pathway in all organisms, details of carriers, enzyme organization, etc. are known to differ. Several important features of these reactions are

    a. Although the initial and rate-controlling step requires CO_2 fixation, CO_2 is not incorporated into fatty acids. That is, no net CO_2 fixation occurs during fatty acid synthesis.

    b. In the presence of excess malonyl-CoA, acetyl-Coa is incorporated only into carbons-15 and -16 of palmitate. Thus, it is successive residues of malonate which are used in the reactions. In the absence of malonyl-Coa, acetyl-CoA is incorporated, but each residue must be carboxylated first. The Initial acetyl-CoA can be viewed as a "primer" acyl group for the polymerization.

    c. The final release of the finished fatty acid depends on chain length. That is, when the acyl group has grown to the proper size, it is transferred from the acyl carrier protein to either water or a CoASH. It is not known which recipient is of more importance and the exact mechanism of chain length specificity is not clear.

    d. In some organisms, propionyl-CoA can replace acetyl-CoA to some extent in the initial condensation. This provides a route for the synthesis of odd-chain-length fatty acids.

    e. The overall reaction for palmitate synthesis can be written

$$8 \text{ acetyl-CoA} + 14 \text{ NADPH} + 14 \text{ H}^+ + 7 \text{ ATP} \longrightarrow$$

$$\text{Palmitic acid} + 8 \text{ CoASH} + 14 \text{ NADP}^+ + 7 \text{ ADP} + 7 \text{ P}_i$$

2. Fatty acid synthetase can be isolated from yeast as a single particulate complex of molecular weight about 2,300,000 daltons containing at least seven proteins. The system in birds and mammals appears similar. In bacteria and plants, the synthetase enzymes are all soluble and can be fractionated by appropriate techniques. The acyl group remains attached to an acyl carrier protein (ACP) as it grows. One of the seven components of the

Figure 77.   Fatty Acid (Specifically, Palmitic Acid) Synthesis in E. Coli.

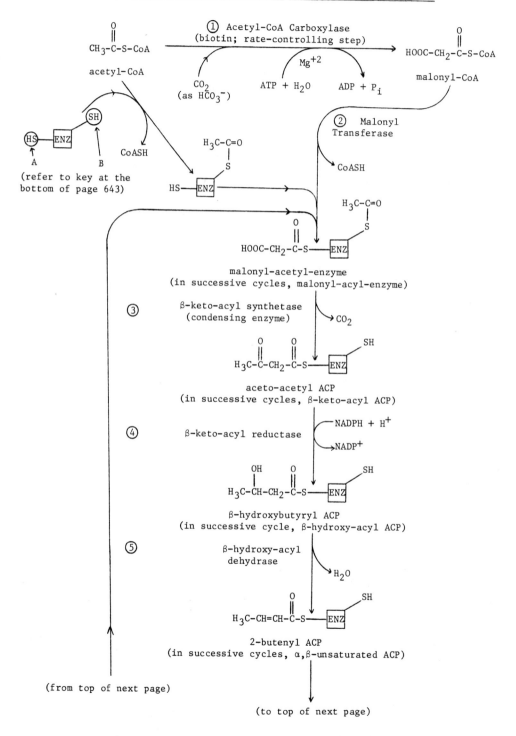

β-keto-acyl synthetase
(condensing enzyme)

aceto-acetyl ACP
(in successive cycles, β-keto-acyl ACP)

β-keto-acyl reductase

β-hydroxybutyryl ACP
(in successive cycle, β-hydroxy-acyl ACP)

β-hydroxy-acyl
dehydrase

2-butenyl ACP
(in successive cycles, α,β-unsaturated ACP)

(from top of next page)

(to top of next page)

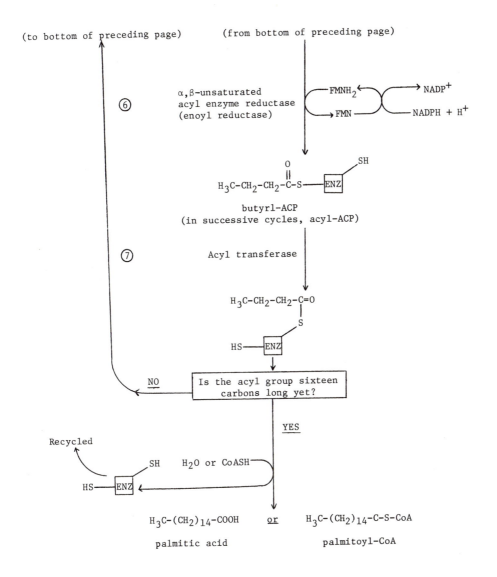

(to bottom of preceding page)   (from bottom of preceding page)

⑥ α,β-unsaturated acyl enzyme reductase (enoyl reductase)

FMNH$_2$ → NADP$^+$

FMN → NADPH + H$^+$

H$_3$C-CH$_2$-CH$_2$-C-S—[ENZ]—SH
(with C=O above the C)

butyrl-ACP
(in successive cycles, acyl-ACP)

⑦ Acyl transferase

H$_3$C-CH$_2$-CH$_2$-C=O
S
HS—[ENZ]

NO ← Is the acyl group sixteen carbons long yet?

YES

Recycled

SH   H$_2$O or CoASH
HS—[ENZ]

H$_3$C-(CH$_2$)$_{14}$-COOH   or   H$_3$C-(CH$_2$)$_{14}$-C-S-CoA

palmitic acid   palmitoyl-CoA

ACP = acyl carrier protein

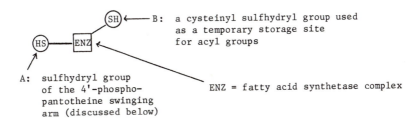

SH ← B: a cysteinyl sulfhydryl group used as a temporary storage site for acyl groups

HS [ENZ]

A: sulfhydryl group of the 4'-phosphopantotheine swinging arm (discussed below)

ENZ = fatty acid synthetase complex

synthetase complexes in higher organisms appears to be similar to ACP. A schematic diagram of fatty acid synthetase is shown below. Each of the outer circles represents an enzyme which catalyzes the indicated reaction. The center circle is ACP or its equivalent. The clashed line with an -SH group is the swinging arm shown in each of its successive positions; and the other -SH group, on the acyl transferase, is the cysteinyl "storage" site on acyl transferase.

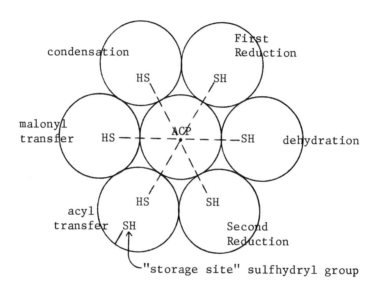

3. The sulfhydryl swinging arm of the enzyme is 20.2 Å long. It is a 4'-phosphopantotheine group, similar in structure to coenzyme-A (see page 174). Linkage to ACP is through a phosphate ester to serine.

4. The reducing agent for fatty acid synthesis is NADPH. It is
    formed primarily by four reactions of carbohydrate metabolism
    which have already been covered.

    a. HMP Shunt (see pages 241ff)

        i)

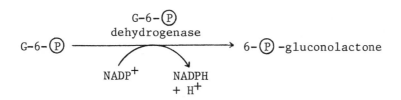

        ii)

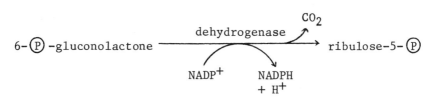

    b. TCA Cycle (see page 170)

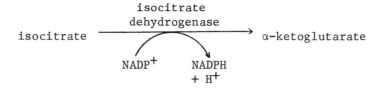

    c. Malic Enzyme

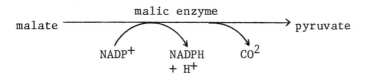

d. The HMP-shunt enzymes are present in the cytoplasm, the TCA
    cycle is strictly intramitochondrial, and malic enzyme is
    found in both places in different tissues. Since the enzymes
    of fatty acid synthesis are cytoplasmic, the most important
    sources of NADPH are probably the HMP shunt and malic enzyme.
    Active lipogenesis takes place in the liver, adipose tissue,
    and lactating mammary gland. All of these tissues show a
    correspondingly high rate of operation of the HMP shunt.

e. As is suggested by the reactions which supply NADPH,
    lipogenesis is closely linked to carbohydrate oxidation. The
    rate of lipogenesis is high in well-fed humans whose diet
    contains a great deal of carbohydrate. Restricted caloric
    intake, a high fat diet, or an insulin deficiency (diabetes)
    decreases fatty acid synthesis.

5. Source and transport of acetyl-CoA. Acetyl-CoA is produced in the
    mitochondria by the breakdown of carbohydrates, fatty acids, and
    amino acids. It cannot pass through the mitochondrial membrane,
    and hence has to be transported out of the mitochondria by means
    of a transport mechanism.

a. Carnitine (β-hydroxy-γ-(trimethyl ammonium) butyrate)

$$(CH_3)_3 \overset{\oplus}{N} - CH_2 - \underset{\underset{OH}{|}}{CH} - CH_2 - COO^{\ominus}$$

is known to stimulate fatty acid oxidation by mitochondria and
fatty acid synthesis from pyruvate. Acetyl and fatty acyl
carnitines move readily across mitochondrial membranes. The
mechanism is illustrated in Figure 78.

The ester bond is a high-energy bond equivalent to a thioester
linkage. Consequently, little or no energy is needed for the
transacylations between carnitine and CoASH. Note the probable
interaction between the positive and negative ends of the
molecule. In rat liver and nemospora, carnitine has been
shown to arise from lysine by trimethylation of the epsilon
nitrogen, deamination, and hydroxylation in that order. It is
interesting that the same carrier is used to bring fatty acids
into the mitochondria for oxidation and to carry acetyl groups
out into the cytoplasm for fatty acid synthesis.

Figure 78. The Role of Carnitine in Mediating Acetyl and Acyl Transfers Across Mitochondrial Membranes

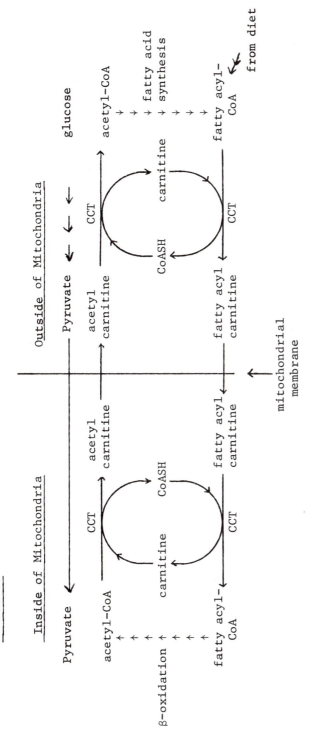

CCT represents carnitine-CoASH transferase. A number of these have been identified in both intramitochondrial and extramitochondrial spaces. The acetyl and fatty acyl groups are attached to the carnitine through an ester linkage to the β-hydroxyl group.

b. Cytoplasmic generation of acetyl-CoA ("citrate shuttle")

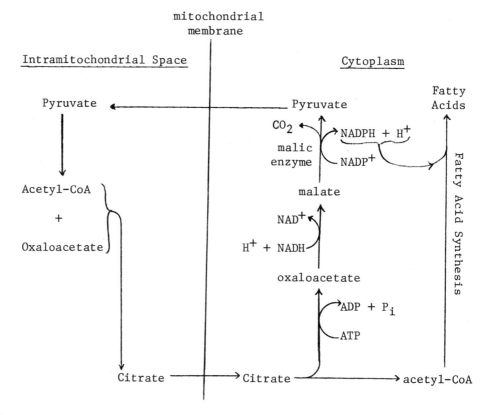

6. The rate-controlling step in fatty acid synthesis is catalyzed by acetyl-CoA carboxylase (reaction 1 in Figure 77)

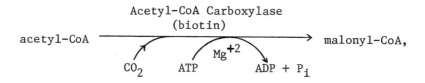

This reaction is allosterically inhibited by long-chain acyl-CoA and allosterically stimulated by citrate, isocitrate, and α-ketoglutarate (see pages 190–192).

7. Although the pathway described above is the only route for <u>de novo</u> fatty acid synthesis, other routes exist for fatty acid <u>elongation</u>.

   a. In the mitochondria, a slightly modified reversal of β-oxidation adds acetyl groups to saturated and unsaturated C_{12}, C_{14}, and C_{16} fatty acids to form primarily C_{18}, C_{20}, C_{22}, and C_{24} fatty acids. This system does not require CO_2, malonyl-CoA, or biotin.

   b. Microsomes can elongate C_{10}-C_{16} saturated and C_{18} unsaturated fatty acids by successive addition of two carbon groups derived from malonyl-CoA. This route appears similar to <u>de novo</u> synthesis except that the intermediates are not bound to an acyl carrier protein. Recall that mammals are <u>unable</u> to synthesize fatty acids with multiple double bonds so that any multiple unsaturated fatty acids used by this pathway must be preformed.

8. Synthesis of branched, odd-chain-length, unsaturated, hydroxy, and dicarboxylic fatty acids has been mentioned a number of times. There are several other reactions which can be used to form such compounds that have not yet been described. They will be briefly summarized here.

   a. Short, branched-chain acyl-CoA's derived from catabolism of valine, leucine, and isoleucine can be elongated, producing branched-chain fatty acyl-CoA's.

   b. In addition to alpha- and omega-oxidation, hydroxy fatty acids can be formed by hydration of singly unsaturated fatty acids. These can then be further elongated by addition of acetyl residues.

   c. As was pointed out under essential fatty acids, there are certain unsaturated fatty acids which the body cannot synthesize. The two ways in which mammals can satisfy their need for such compounds are shown below. These reactions occur primarily in the liver.

      i) The essential fatty acids (linoleic, linolenic, and arachidonic) can be lengthened by adding two-carbon units to them. The elongations may be performed by the microsomal-malonyl-CoA system described earlier. Additional double bonds can be introduced into this added material in a manner similar, but not identical, to the one described below for forming singly unsaturated acids.

ii) Palmitic acid can be oxidized to palmitoleic acid
containing one double bond and stearic acid can be
oxidized to the singly unsaturated oleic acid. Both
reactions simply remove one hydrogen each from carbons
9 and 10. Palmitoleic and oleic acids can be further
elongated and oleic acid can have additional unsaturation
added to the carbons used for this elongation.

The Biosynthesis of Cholesterol consists of the following five stages

1. Synthesis of mevalonate (containing 6 carbons) from acetyl-CoA
    (2 carbons).

2. Loss of CO_2 from the mevalonate to form isoprenoid (5 carbon) units.

3. Condensation of six isoprenoid units to form squalene (30 carbons).

4. Conversion of squalene to lanosterol (30 carbons).

5. Transformation of lanosterol to cholesterol (27 carbons).

2 $CH_3-\overset{O}{\overset{\|}{C}}-S-CoA$ $\xrightarrow{\text{thiolase}}$ $CH_3-\overset{O}{\overset{\|}{C}}-CH_2-\overset{O}{\overset{\|}{C}}-S-CoA$ $\longrightarrow$ $CH_3-\overset{O}{\overset{\|}{C}}-S-CoA + H_2O$

acetyl-CoA   CoASH   aceto-acetyl-CoA

$\rightarrow$ CoASH

2 NADPH + 2 H^+

2 $NADP^+$

$HOOC-CH_2-\overset{\overset{CH_3}{|}}{\underset{\underset{OH}{|}}{C}}-CH_2-\overset{O}{\overset{\|}{C}}-S-CoA$ $\longleftarrow$

HMG-CoA Reductase; Rate-limiting step; dietary cholesterol inhibits this reaction

β-hydroxy-β-methylglutaryl-CoA (HMG-CoA)

$\rightarrow$ CoASH

$HOOC-CH_2-\overset{\overset{CH_3}{|}}{\underset{\underset{OH}{|}}{C}}-CH_2-CH_2-OH$ $\xrightarrow[\text{ATP}\quad\text{ADP}]{\text{mevalonate kinase}}$ $HOOC-CH_2-\overset{\overset{CH_3}{|}}{\underset{\underset{HO}{|}}{C}}-CH_2-CH_2-O-\textcircled{P}$

mevalonic acid

mevalonic acid-5-$\textcircled{P}$

ATP

ADP

phosphomevalonic kinase

$HOOC-CH_2-\overset{\overset{CH_3}{|}}{\underset{\underset{OH}{|}}{C}}-CH_2-CH_2-O-\textcircled{P}-O-\textcircled{P}$

mevalonic acid-5-pyrophosphate

pyrophosphoryl mevalonic kinase

ATP

ADP

$HOOC-CH_2-\overset{\overset{CH_3}{|}}{\underset{\underset{O-\textcircled{P}}{|}}{C}}-CH_2-CH_2-O-\textcircled{P}-O-\textcircled{P}$ $\longrightarrow$ CO_2, P_i

mevalonic acid-3-phosphate-5-pyrophosphate

$\overset{H_3C}{\underset{H_2C}{\diagdown}}C-CH_2-CH_2-O-\textcircled{P}-O-\textcircled{P}$ $\underset{\xrightarrow{\hspace{2cm}}}{\overset{\text{isomerase}}{\xleftarrow{\hspace{2cm}}}}$ $\overset{H_3C}{\underset{H_3C}{\diagdown}}C=CH-CH_2-O-\textcircled{P}-O-\textcircled{P}$

isopentenyl**pyr**ophosphate
(IPPP)

dimethylallylpyrophosphate (DMAPP)

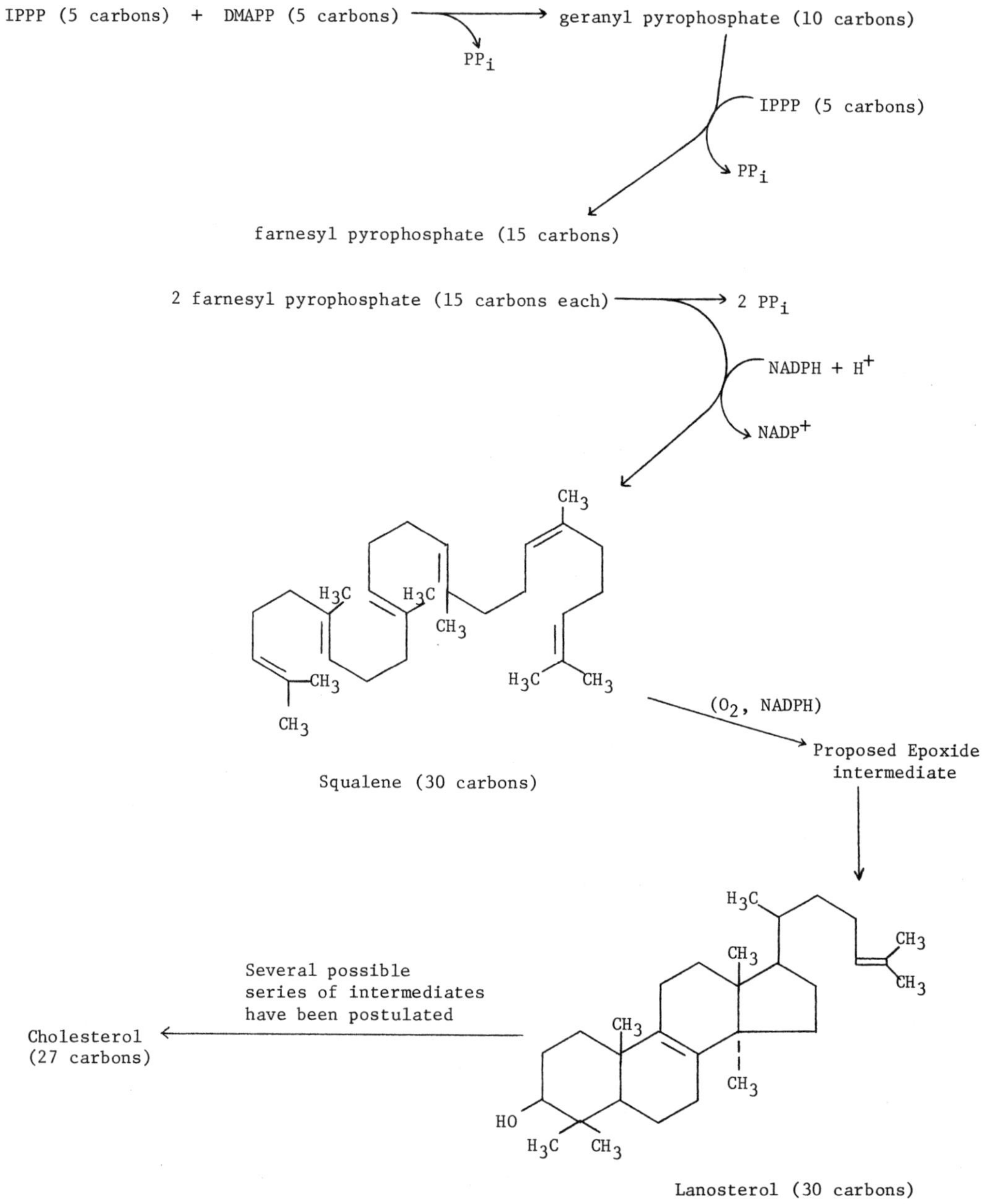

IPPP (5 carbons)   +   DMAPP (5 carbons) ⟶ geranyl pyrophosphate (10 carbons)

PP$_i$

IPPP (5 carbons)

PP$_i$

farnesyl pyrophosphate (15 carbons)

2 farnesyl pyrophosphate (15 carbons each) ⟶ 2 PP$_i$

NADPH + H$^+$

NADP$^+$

Squalene (30 carbons)

(O$_2$, NADPH)

Proposed Epoxide
intermediate

Several possible
series of intermediates
have been postulated

Cholesterol
(27 carbons)

Lanosterol (30 carbons)

Comments on Cholesterol Metabolism

1. A normal adult synthesizes 1.5-2.0 gm of cholesterol each day and
    consumes another 0.3 gm/day in the diet. Cholesterol is a product
    of animal metabolism and occurs primarily in food of animal origin.
    Egg yolk is rich in cholesterol and meat, liver, brain, and
    shellfish contain appreciable amounts. As was mentioned earlier,
    cholesterol is absorbed from the intestines along with other
    lipids. In the mucosal cells, most of it is esterified with fatty
    acids and incorporated into chylomicra which are transported to
    the blood by the lymphatics. Bile salts and pancreatic juice
    (containing cholesterol esterase) are needed for cholesterol
    absorption.

2. The liver is the main site of cholesterol synthesis (1-1.5 gm/day)
    with the adrenal cortex, skin, intestine, testes, aorta, and
    perhaps other tissues forming the remainder of it. All of the
    carbon atoms of cholesterol are derived from acetyl-CoA, as can
    be seen in Figure 79.

Figure 79. Origins of the Carbon Atoms in Cholesterol

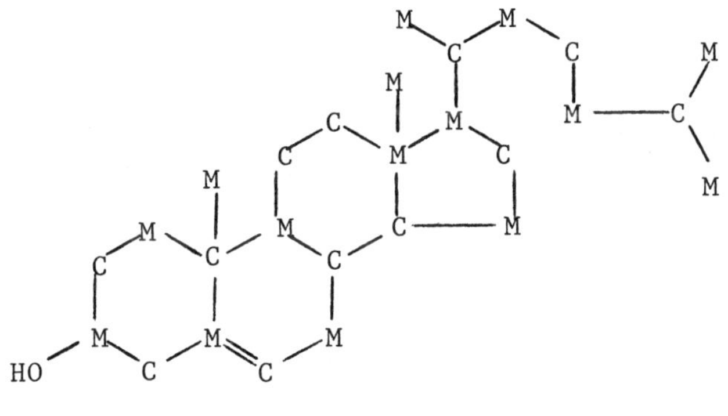

M = derived from the methyl carbon of the acetyl group; C = derived
from the carbonyl carbon of acetyl. The oxygen comes from molecular
oxygen during the epoxidation of squalene.

a.   The cyclization of squalene to cholesterol is complex and not well understood. It probably involves squalene 2,3-epoxide and perhaps other compounds as intermediates. An epoxide has the general formula

$$R_1 \diagdown \underset{\diagup}{\overset{\diagup}{C}} \overset{O}{\diagdown\diagup} \underset{\diagdown}{\overset{}{C}} \diagup R_3$$

Enzymes probably involved in this reaction are squalene epoxidase and squalene oxide cyclase.

b.   Lanosterol is converted to cholesterol by the removal of three methyl groups (two at position 4 and one at position 14); saturation of the sidechain and of the 8,9-double bond; and creation of a new double bond between atoms 5 and 6. As might be expected, these alterations may proceed in more than one order. There are probably at least two, and perhaps more, series of intermediates between lanosterol and cholesterol. It is not yet clear which route is most important if, in fact, any one of them is.

3.   Control of biosynthesis of cholesterol. The primary rate-limiting step is the conversion of HMG-CoA to mevalonic acid, catalyzed by the enzyme HMG-CoA reductase. High cholesterol levels inhibit this reaction by a negative feedback mechanism. It appears that cholesterol itself is not the inhibitor. Instead, a cholesterol-containing lipoprotein formed during the absorption and transport of cholesterol may actually perform this function. In this regard, it should be pointed out that mevalonic acid is also a precursor of coenzyme Q and its formation is not affected by cholesterol feeding. A second rate-limiting reaction may be the cyclization of squalene into lanosterol. Fasting inhibits cholesterol biosynthesis by diverting HMG-CoA to ketone body production, while high-fat diets seem to accelerate the production of cholesterol. The thyroid gland also plays a role in the regulation of cholesterol metabolism. Thyroid hormone possibly increases the catabolism of cholesterol and hypothyroidism is associated with hypercholesterolemia. Administration of thyroid hormone to both normal and hypothyroid individuals reduces the serum cholesterol level.

4. The routes of elimination of cholesterol are indicated below. The
    role of the bile salts in lipid absorption from the intestine has
    already been mentioned and this and other aspects of bile salt
    metabolism will be discussed later in more detail.

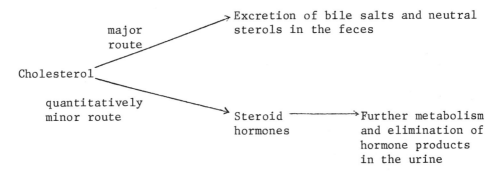

5. Other than as a precursor for bile salts and steroid hormones,
    the exact biochemical function of cholesterol is conjectural.
    Cholesterol esters may aid in the transport of fatty acid,
    particularly unsaturated ones. Brain tissue is rich in cholesterol.
    Whether it is involved in the production and propagation of impulses
    or serves merely as an insulating material is not known.
    Introduction of another double bond to cholesterol forms vitamin D_3
    (7,8-dehydrocholesterol) which is converted to its active form by
    ultraviolet light. The role of cholesterol in steroid hormone
    synthesis is covered in the chapter on hormones.

Role of Cholesterol in Arteriosclerosis and Atherosclerosis

1. <u>Arteriosclerosis</u> is a degeneration and hardening of the walls of
    the arteries, capillaries, and veins due to chronic inflammation
    which results in fibrous tissue formation. It can have several
    etiologies, including defects in the basement membrane (see
    pages 279-280). <u>Atherosclerosis</u> is one type of arteriosclerosis,
    specifically involving an accumulation in the arterial intima of
    plaque-like lipids which undergo calcification. In atherosclerosis,
    cholesterol esters and other lipids are deposited in the connective
    tissue of the arterial walls. It is important to notice that
    atherosclerosis is a subtype of arteriosclerosis and that the two
    words are <u>not</u> synonymous.

2. Diabetes mellitus, lipid nephrosis, hypothyroidism, and other
    hyperlipemic conditions, in which prolonged high levels of low
    density lipoproteins (LDL) and very low density lipoproteins (VLDL)
    occur in the blood, lead to atherosclerosis. High fluctuating

levels of free fatty acids in the blood can be due to emotional
stress, nicotin (from smoking), coffee drinking, and big meals
instead of frequent, smaller ones. High blood pressure, obesity,
lack of exercise, and family history (genetics) also play a role
in the development of atherosclerosis. Normal dietary fluctuation
of cholesterol appears not to cause this disease, although feeding
of greater than usual amounts of cholesterol can induce it.

3. There is a good correlation between high serum lipid levels and
    both atherosclerosis and arteriosclerosis. Serum cholesterol levels
    have been used most extensively in such correlations but serum
    triglycerides, the cholesterol/phospholipid ratio, and the
    concentration of serum lipoproteins with S_f values of 12-400 (the
    cholesterol-rich fraction) have also been employed. S_f is defined
    later in Table 25 in the section on lipoproteins, page 663.

4. Heparin, a polysaccharide normally found in serum (see Chapter 4),
    has a clearing effect <u>in vivo</u> on hyperlipemic blood. This may be
    due to the stimulation of lipolytic activity by heparin and the
    rapid removal of the solubilized lipid from the blood-stream. There
    may be a deficiency of heparin in atherosclerosis. The ability
    of heparin, dextran sulfate, and other polyfunctional ions to
    complex and, in the presence of Mg^{+2} or Mn^{+2}, precipitate all of
    the lipoproteins except the high density ones provides the basis
    of one technique for the separation of this class of lipoproteins.

5. The correlation of cholesterol levels with atherosclerosis suggests
    that the control of these levels may be useful in management or
    prevention of the disease. Blood cholesterol is susceptible to
    dietary influences and to the hypocholesterolemic drugs.

    a. The blood cholesterol can be lowered by replacing some of the
        saturated fatty acids in the diet by polyunsaturated fatty acids.
        There is some evidence that atherosclerosis is due to a
        <u>relative</u> deficiency of unsaturated fatty acids. On this basis,
        safflower, corn, peanut, and cottonseed oils (in order of
        decreasing efficacy) lower blood cholesterol while butter fat
        and coconut oil raise it. The polyunsaturates probably exert
        their effect by

        i)   stimulating cholesterol excretion into the intestine;

        ii) stimulating the oxidation of cholesterol to bile acids;

        iii) increasing the rate of metabolism of cholesterol esters,
             since esters of cholesterol with unsaturated fatty acids
             are more rapidly metabolized in the liver and other
             tissues than are the esters with saturated fatty acids;

iv) shifting the distribution of cholesterol (decreasing it in the plasma and increasing it in the tissues).

b. The hypocholesterolemic drugs function in several ways to lower serum cholesterol levels. Some examples of such drugs and their actions are given below.

  i) One group of drugs acts on cholesterol levels by blocking the various stages of cholesterol synthesis. Clofibrate, shown below, is such a compound.

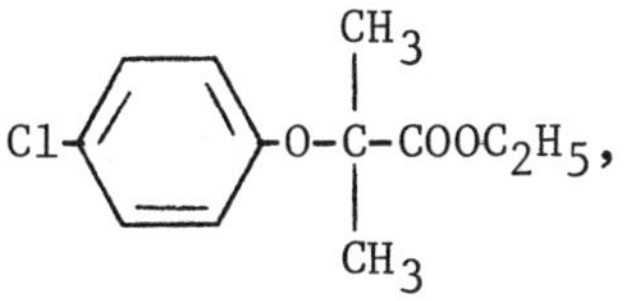

$$Cl-\underset{\underset{CH_3}{|}}{\overset{\overset{CH_3}{|}}{\bigcirc}}-O-C-COOC_2H_5, \text{ Clofibrate}$$

  Although the exact inhibitory mechanism is not understood, clofibrate is known to inhibit hepatic cholesterol synthesis. Clofibrate also lowers the VLDL, presumably by inhibiting triglyceride and lipoprotein synthesis. This compound is used in treating types III, IV, and V hyperlipoproteinemias, but it is not of value in the treatment of type I and II hyperlipoproteinemias (see later). Side-effects of clofibrate include enhanced sensitivity to coumarin anticoagulants and, in some cases, myositis with weakness, tenderness of muscle (with an accompanying increase in CPK levels), transient elevation of SGOT and SGPT, leukopenia, etc.

 ii) D-thyroxine lowers the cholesterol concentration by accelerating the catabolism of cholesterol and LDL. Note that the normal thyroid hormone is L-thyroxine. The D-isomer is used to reduce cholesterol level because it has the highest ratio of hypocholesterolemic to calorigenic activity of any of the thyroid-like compounds tested. One side-effect of D-thyroxine is enhanced sensitivity to coumarin anticoagulants. This drug should not be used in individuals with organic heart disease or abnormal glucose tolerance, and possibly those with an elevated metabolic rate.

iii) Cholestyramine, a basic anionic exchange resin, prevents reabsorption of bile acids and increases their fecal loss. The $Resin^+Cl^-$ exchanges its Cl^- ion for (bile acid)$^-$ ion, giving rise to a Resin-bile acid complex which is eliminated in the feces. This interrupts the enterohepatic recycling and leads to increased catabolism of cholesterol to replace the lost bile acids. This compound has been found to be quite useful in type II hyperlipoproteinemia. Its side-effects include nausea, constipation, possibly hyperchloremic acidosis in children, interference with absorption of fat soluble vitamins (A, D, E, and K), etc.

iv) <u>Neomycin</u> blocks the intestinal absorption of cholesterol.

v) Other drugs

<u>Nicotinic acid</u> decreases the plasma lipids (both cholesterol and triglycerides) by interfering with the mobilization of FFA. It is useful in all of the hyperlipoproteinemias except type I. Side-effects include flushing of the skin, pruritus, G.I. disturbances, hyperglycemia, hyperuricemia, and abnormal liver function.

<u>Estrogen</u> decreases the LDL (the major cholesterol carrier) while increasing the VLDL, thus bringing about the hypocholesterolemic effect. <u>Androgens</u> show an opposite effect to that of estrogens.

Structures of Bile Acids

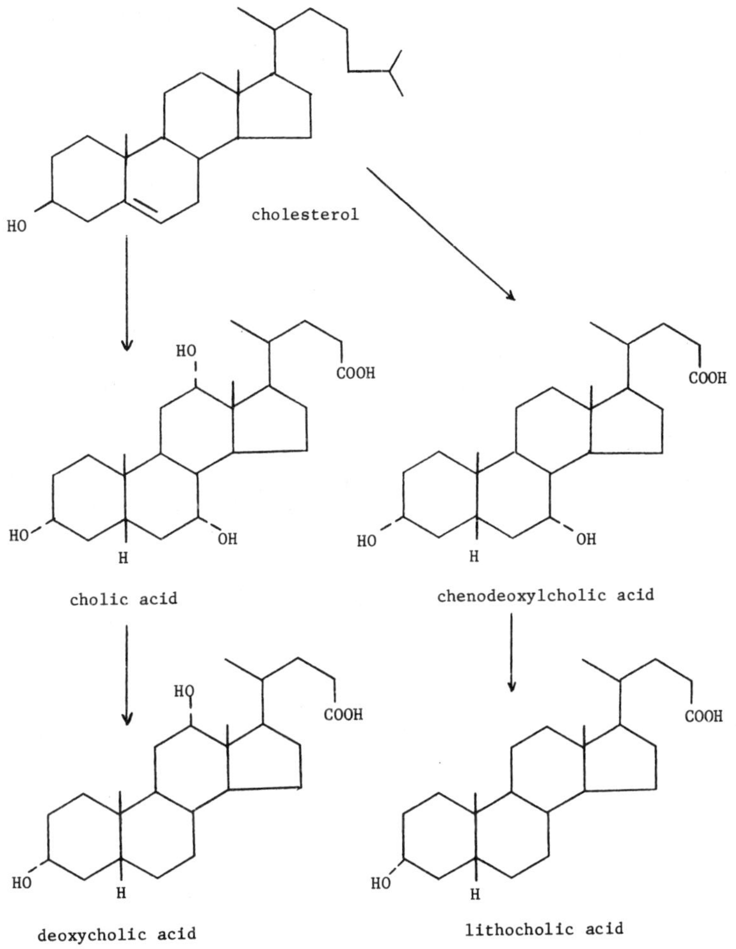

cholesterol

cholic acid

chenodeoxylcholic acid

deoxycholic acid

lithocholic acid

Comments on the Metabolism of Bile Acids

1. The primary bile acids (cholic and chenodeoxycholic) are produced
    in the liver from cholesterol by a mechanism similar to the
    β-oxidation of fatty acids. Cholic acid is the most abundant of
    these compounds in humans. The conjugation of bile acids with
    glycine and taurine makes them soluble at the pH of the intestinal
    contents and thus able to perform their functions. Therefore,
    the conjugation process in the liver is obligatory for optimal
    lipid absorption. At the pH of the bile, these acids are
    actually present as salts, although the names bile acids and bile
    salts are frequently used interchangeably.

2. Bile is formed continuously in the liver. Its chief constituents
   are the bile pigments (bilirubin, biliverdin), bile salts, and
   cholesterol secreted into the gallbladder. When stimulated by
   the hormone cholecystokinin-pancreozymin (see page 433), the
   gallbladder empties its contents into the small intestine.
   Interruption of this process due to malfunction of liver or
   gallbladder or obstruction of any of the ducts leads to poor
   absorption of intestinal lipids and lipid soluble vitamins, such
   as A and K. Lithocholic acid is relatively toxic to the liver and
   its formation following cholestasis may contribute to the liver
   damage in this condition.

3. As pointed out earlier, the bile salts are powerful emulsifying
   agents which solubilize lipid substances. This helps in
   intestinal lipid absorption and aids the action of enzymes, as
   well as stimulating intestinal motility. (Also see lipid
   absorption.)

4. Enzymes of the intestinal bacteria convert the primary bile acids
   to secondary bile acids (deoxycholic and lithocholic acids). The
   bile acids (salts) are largely reabsorbed into the enterohepatic
   circulation. The reabsorbed salts stimulate the liver to secrete
   bile and partially control bile acid synthesis in the liver.
   Apparently bile acids regulate cholesterol synthesis in the liver
   by curtailing the production of mevalonic acid.

5. About 95% of the bile salts secreted into the intestine are
   reabsorbed into the enterohepatic portal blood in the various
   segment of intestine: jejunum (passive absorption; fat absorption
   also takes place here); ileum (active absorption); and colon
   (passive absorption).

6. It has been reproted that the intestinal microorganism may convert
   the bile acids to certain carcinogenic compounds (such as methyl
   cholanthrene). This aspect is receiving considerable attention at
   the present time in the studies on the etiology of cancer of the
   colon.

7. The most common form of gallbladder stones (calculi) contains about
   80% cholesterol. They apparently form when cholesterol precipitates
   due to the inability of the bile salts to adequeately solubilize the
   cholesterol excreted in the bile. Generally, if the mole ratio of
   bile salts + phospholipids) to cholesterol is less than 10:1, the
   bile is considered lithogenic (stone-forming), but this figure is not
   an absolute one in all individuals. Causes of lithogenic bile includes elevated cholesterol excretion, (due to obesity, administration of hypocholesterolemic drugs (see pages 657-658), etc),
   excessive loss of bile salts in the feces, and decreased synthesis
   of bile salts in the liver.

Figure 80. Formation and Disposition of the Bile Salts (Bile Acids)

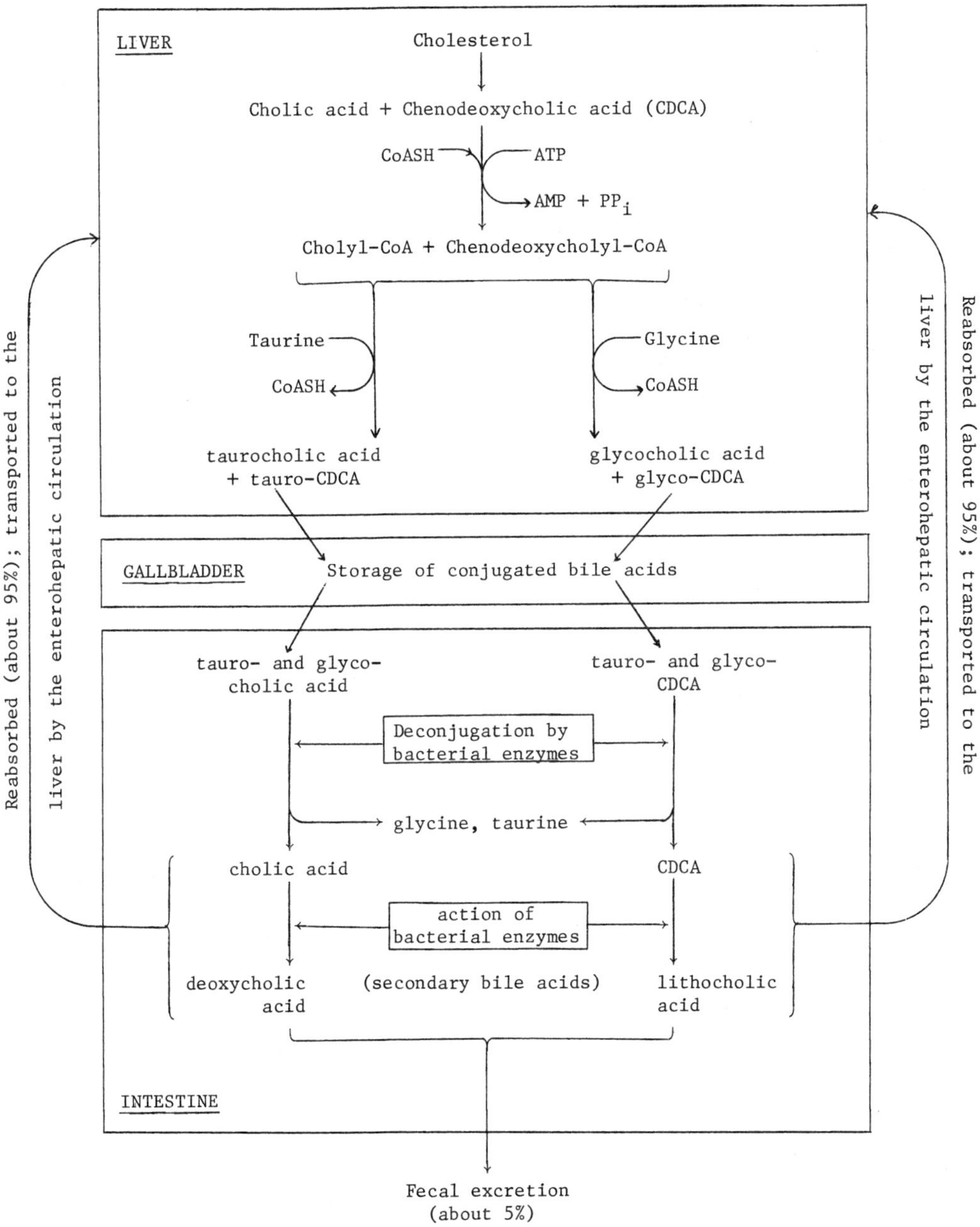

Lipoproteins

1. The lipids, which are generally insoluble in water, are largely
    transported within the body by the blood, an aqueous medium. To
    accomplish this and thereby make the lipids available for metabolism,
    the lipids are first complexed with various specific proteins. The
    resultant lipoproteins (LP's), containing varying amounts of
    triglyceride, phospholipid, cholesterol, and protein, are soluble
    in the blood because of the hydrophilic nature of the protein
    moiety. One example already mentioned is the transport of free
    fatty acids by albumin. There are other, more specific proteins
    involved in carrying other lipids. These are discussed in this
    section. Practically all of the plasma lipids exist as protein-
    lipid complexes.

2. Four major groups of lipoproteins have been separated and identified
    by electrophoretic and ultracentrifugal techniques. These are
    sometimes further subdivided but this is beyond the scope of this
    section. Two systems of nomenclature have been created based on
    the results from the two techniques. In the ultracentrifuge, the
    densities of the lipoproteins are measured and the four major
    classes are designated chylomicra (lowest density), very low, low,
    and high density. As is seen in Table 25, an increase in density
    corresponds to an increase in the protein to lipid ratio in the
    lipoprotein. The serum globulins are used as reference points in
    paper or agarose gel electrophoresis resulting in the names
    pre-β-, β-, and α-lipoprotein (corresponding to very low, low,
    and high density fractions). The chylomicra remain at the origin.
    This nomenclature is summarized in Table 25. The composition of
    each of these four classes is given in Table 26. Because
    electrophoresis is more readily and frequently used in the
    clinical laboratory than ultracentrifugation, the electrophoretic
    nomenclature is the one usually seen in clinical situations.

    Other abnormal lipoproteins also occur, usually in association
    with some disease. Several of these will be discussed later in
    this section. It should be pointed out that although
    polyacrylamide and starch gels can also be used as supports for
    the electrophoresis, on these materials the pre-β-lipoproteins
    migrate behind the β-lipoproteins rather than ahead of them
    (i.e., the order of the pre-β- and β-bands is switched).

Table 25. Nomenclature and Some Physical Data on the Four Major Lipoprotein Types

Lipoprotein	Density (gm/ml)	S_f*	Position in paper or agarose electrophoresis
(1) Chylomicra	d < 0.94	> 400	origin
(2) Pre-β (very low density lipoprotein; VLDL)	0.94 < d < 1.006	20-400	pre-β (i.e., ahead of β-globulin)
(3) β-lipoprotein (low density lipoprotein; LDL)	1.006 < d < 1.063	0-20	β (with β-globulin)
(4) α-lipoprotein (high density lipoprotein, HDL)	1.063 < d < 1.21	---	with α-globulin

* (The rate at which lipoprotein floats up through a solution of NaCl of specific gravity 1.063 is expressed in Svedberg, S_f units: one S_f unit = 10^{-13} cm/second/dyne/gm at 26°C. S_f can be thought of as a negative sedimentation constant.)

Table 26. Average Composition of the Four Major Groups of Serum Lipoproteins

Lipoprotein	Protein	Tri-glyceride	Phos-pholipid	Cholesterol*	Lipoprotein functions principally as a carrier of
Chylomicra	1	87	8	4	Exogenous (dietary) triglyceride
VLDL (pre-β)	7	52	19	22	Endogenous triglyceride
LDL (β)	16	18	23	43	Cholesterol
HDL (α)	45	8	25	22	Phospholipid

Although the amounts of each class present in the blood are subject to considerable variation in different populations, the compositions, as given above, are relatively constant.

* Cholesterol = free cholesterol + cholesterol esters.

3. Although the sites of synthesis of the various lipoproteins have not been fully elucidated, some general observations are possible. The VLDL are formed primarily in the liver while the chylomicra and small amounts of VLDL are made in the mucosal cells of the intestine. The location of LDL and HDL synthesis is not known, but they may be metabolic products of chylomicra and VLDL.

4. The proteins which are used for lipoprotein synthesis appear to be relatively homogeneous, consisting of only a few different peptides. On this basis, and for other reasons, the lipid-protein interaction is probably a fairly specific one. These proteins are called apolipoproteins or simply apoproteins. No uniform nomenclature for these apoproteins has yet been accepted. In one terminology (to be used here), they are identified by their carboxy-terminal amino acid. In addition to the usual chemical methods of protein identification (amino acid analysis, peptide fingerprinting, electrophoresis, etc.), the apoproteins show some degree of immunological specificity. This has been used in some studies both for detection and quantitation of these molecules. This area is still rather new and subject to change. Some important findings are summarized in Table 27. In this table, note that there is some overlap, with a given apoprotein being present in more than one lipoprotein.

Table 27. Apoproteins Present in the Various Lipoprotein Classes

Lipoprotein	Apolipoprotein
VLDL (very low density lipoproteins) or pre-β-lipoproteins	apoLP-Val apoLP-Glu apoLP-Ala (about 50% of the total apoprotein) apoLP-Ser
HDL (high density lipoproteins) or α-lipoprotein	apoLP-Thr ⎫ apoLP-Gln ⎬ major constituents apoLP-Val ⎧ apoLP-Glu ⎨ minor constituents apoLP-Ala ⎩
LDL (low density lipoproteins) or β-lipoproteins	apoLP-Ser ... major component
Chylomicra	Although no specific proteins have been isolated, there is evidence that HDL and LDL apoproteins are present.

5. There are three inherited types of hypolipoproteinemia, all of
    which are usually detected initially by the presence of
    hypocholesterolemia. Their differentiation is based on a more
    detailed analysis of the serum lipoprotein pattern.

    a. Tangier disease (familial HDL deficiency; α-lipoprotein
        deficiency) is characterized by hypocholesterolemia and high
        or (occasionally) normal serum triglyceride. Normal
        α-lipoproteins (HDL) are absent from the serum and an abnormal
        HDL (designated HDL$_T$), containing decreased quantities of
        apoLP-Thr and an excessive amount of apoLP-Gln, is present
        instead. The ratio of apoLP-Thr to apoLP-Gln is 1:11 in
        these complexes, whereas the value in normal HDL is 3:1. The
        HDL$_T$ also differs immunologically from normal HDL and is
        present to the extent of only about 1-2% of normal. Thus,
        this disease apparently involves a true mutant LP in addition
        to a disturbance in plasma LP levels. Clinical manifestations
        of this disorder include cholesteryl ester deposition in
        reticuloendothelial tissues (notably resulting in splenomegaly
        and yellow-orange, enlarged tonsils) and, in a number of
        patients, neurological abnormalities. The mode of inheritance
        is autosomal recessive. The disease does not appear to be
        generally fatal.

    b. In abetalipoproteinemia (acanthocytosis; Bassen-Kornzweig
        syndrome), as the name implies, LDL, VLDL, and chylomicra are
        absent. The important clinical findings are retinitis
        pigmentosa, malabsorption of fat, and atoxic neuropathic
        disease. The erythrocytes are distorted with a "thorny"
        appearance due to protoplasmic projections of varying sizes
        and shapes (hence the term, acanthocytosis, from the Greek
        akantha, meaning thorn). Serum triglyceride levels are low.
        The genetic defect is probably one which completely shuts off
        apoLDL synthesis, although other possibilities do exist. It
        is probably transmitted as an autosomal recessive characteristic.

    c. In hypobetalipoproteinemia, unlike abetalipoproteinemia, LDL
        is low (10-20% of normal) but not absent, HDL is normal, and
        VLDL is mildly lowered. In twenty-three affected individuals
        from the four known affected families, only one subject had
        central nervous system dysfunction and fat malabsorption. The
        other twenty-two known cases had only mild pathologic changes,
        if any were present at all. The symptoms and signs of
        abetalipoproteinemia are generally not shown in this disorder.
        The disease is inherited as an autosomal dominant trait. It
        is interesting to note the benignity of this reduction in
        β-lipoprotein compared to the relative seriousness of
        hyperbetalipoproteinemia (see page 671). In the latter, LDL
        concentrations are 2-6 times normal and such persons are
        predisposed to premature atherosclerosis.

Hyperlipidemias (Hyper L's) and Hyperlipoproteinemias (Hyper LP's)

1. The hyper L's and hyper LP's are, quantitatively, a much more significant group of lipoprotein disorders than the hypolipoproteinemias. The two are considered together because, as was pointed out earlier, all serum lipids are present as lipoproteins. Consequently, an increase in serum lipids (hyperlipidemia) must be accompanied by an elevation of the proteins needed to carry the lipid (hyperlipoproteinemia). There may be exceptions to this but certainly, at present, the most rational approach to these disorders is to treat them as being synonymous. Values of serum lipid levels are useful, however, for aiding in evaluation of the lipoproteins which are abnormal and in the diagnosis of these diseases. This is especially true with respect to determinations of serum triglyceride and cholesterol levels as will be seen later. Most hyper L's are initially detected by observation of elevated triglyceride, cholesterol, or both.

2. The detection of hyperlipidemias is important because numerous studies have shown a correlation between elevated levels of lipids (in particular, cholesterol and triglycerides) and the development of coronary heart disease. Persons with hyperlipidemias also run an increased risk of contracting atherosclerosis, regardless of the cause of the elevated lipid levels. Detection of the hyper L is only the beginning, however. In order to diagnose the specific disorder and to prescribe the most efficacious treatment to correct it, the hyperlipidemias have been divided into five (six, counting two subclasses of one type) classes or phenotypes and a systematic procedure has been developed to aid in the proper classification of hyperlipemic patients.

3. Although the term "phenotyping" implies a genetic origin of these diseases, hyperlipidemia can also be caused by other things. For example, hyper L's are secondary to certain other diseases, including hypothyroidism, uncontrolled insulin-dependent diabetes mellitus, pancreatitis, nephrotic syndrome, dysglobulinemia, and biliary obstruction. Some of the secondary causes for each hyper LP are indicated later in the descriptions of the phenotypes. Environmental factors, such as diet, alcohol consumption, and the use of certain drugs (e.g., estrogens, notably in birth control pills) can also result in elevated serum lipids. Thus, in addition to the phenotype, it is extremely important to consider the etiology of each hyperlipidemia. One factor which is useful in establishing the presence of a primary (genetic) hyperlipidemia is the identification of the same disorder in other members of the family. It is interesting that, unlike the hypolipoproteinemias, no abnormal (mutated) apolipoproteins have been implicated as causative agents (or even detected) in the hyperlipoproteinemias, although the possibility of such an occurrence does exist.

4. Although the values of serum cholesterol and triglyceride which occur in a patient under his normal living conditions are useful in diagnosis, accurate phenotyping requires a standard "preparation" prior to the taking of the blood sample to be used.

    a. The individual should consume a normal diet, with no gain or loss of weight for the two weeks prior to sample collection. No medications which are known to affect plasma lipid or lipoprotein concentrations should be administered to the patient during this period.

    b. Blood should be taken 12-14 hours postprandially (after a meal) and the serum should be separated promptly. Plasma can be used but a correction must be applied to account for the dilution caused by the addition of anticoagulants. The terms plasma and serum will be used interchangeably in this section. The fasting sample precludes the occurrence of chylomicra unless they are present due to an abnormality.

    c. The plasma or serum sample should not be frozen.

5. Some of the techniques which are used in lipoprotein phenotyping are discussed below.

    a. If normal plasma, collected as described above, is stored at 4°C for 18 hours, it should show no chylomicra. The occurrence of a creamy, upper layer (supernatant) and/or uniform turbidity throughout the plasma indicates the presence of chylomicra.

    b. The accurate measurement of serum cholesterol (pages 606-607) and serum triglycerides (pages 604-605) has already been discussed. Normal values for these quantities vary with age, sex, and diet. While the "normal" for one population may be associated with, say, a high risk of coronary disease, the "normal" for another population may be quite benign. This difference between the statistical normal for a population and the biological normal (i.e., that associated with good health) is important to recognize. The values below should be viewed as broad guidelines.

	Cholesterol (mg/100 ml)	Triglyceride (mg/100 ml)
Adults	130-250	10-150
Infants	45-170 (< 10 at birth)	10-140
Children	120-230	10-140

c. Post-heparin lipolyptic activity (PHLA) is a measure of the
    ability of the body to release lipoprotein lipase
    in response to intravenous heparin
    administration. This enzyme is apparently required for the
    hydrolysis of lipids in chylomicra, a necessary process for
    fatty acid utilization. The test is performed simply by doing
    two serum lipase assays (Chapter 3), one before and the other
    following heparin administration.

d. The actual evaluation of the lipoprotein pattern is usually
    done by electrophoresis although phenotyping by this technique
    alone is not feasible. The analytic ultracentrifuge can also
    be used for this purpose and results from it are generally
    more accurate than those obtained by electrophoresis. The
    ultracentrifuge is an expensive instrument, however, and is
    therefore not economically feasible for most laboratories to
    use.

    Agarose electrophoresis (particularly the use of thin agarose
    films) appears to be the method of choice because this
    procedure gives clean separation of lipoprotein bands in a
    short period of time. Because other supports will sometimes
    improve visualization of certain LP's, it is frequently a
    good idea to run LP patterns on several different supports
    (see especially hyper LP Type III, below). The bands are
    stained with a lipid stain (oil red O, fat red 7B (specific for
    unsaturated fatty acids and acyl groups), Sudan black B, and
    others). Following is a schematic diagram indicating the
    relative positions of various lipoprotein bands.

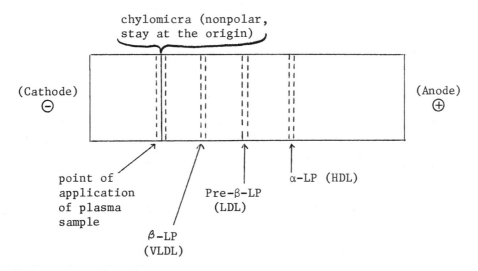

The designations indicated in parentheses are based upon ultracentrifugal studies. The electrophoretic pattern of fasted normal plasma shows essentially only two bands, those corresponding to α- and β-lipoproteins.

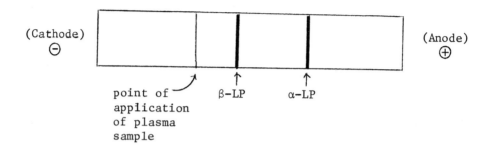

e. It is important to obtain some estimate of VLDL and LDL independent of electrophoresis. For practical purposes, serum triglyceride (TG) concentration directly reflects VLDL levels when chylomicra are absent and abnormal TG values can be interpreted as abnormal VLDL values. When VLDL and HDL are normal, serum cholesterol is approximately equal to LDL concentration.

f. Other techniques which may be useful in lipoprotein analysis include selective precipitation (heparin, polyvinylpyrrolidine, dextran sulfate, etc.), nephelometry (large molecules and particles, such as chylomicra and VLDL scatter light), and immunochemistry. It is frequently helpful to use more than one method in evaluating abnormal conditions.

Classification (Phenotyping) of Hyperlipoproteinemias

1. The classification system for hyper LP's was developed by Fredrickson and his coworkers at the National Heart and Lung Institute, National Institutes of Health, Bethesda, Maryland. The material presented here is condensed from the Bulletin of the World Health Organization, 43, 891-915 (1970), and from CRC Critical Reviews in Clinical Laboratory Science, 2(3), 461-472 (1971). The stepwise procedure for phenotyping using methods described in the preceding section is summarized in Figure 81.

2. Type I (familial fat-induced hyperlipemia; hyperchylomicronemia). Characteristics are

   a. Plasma, collected 14 hours postprandially from a patient maintained on a normal diet, shows a creamy supernatant phase (due to excessive chylomicra) following storage at 4°C for 18 hours.

b. Triglyceride levels are increased with a modest rise in cholesterol, resulting in a ratio of cholesterol/triglyceride of less than 0.2. When the ratio declines to less than 0.1, it is characteristic of <u>only</u> Type I.

c. The electrophoretic pattern shows the presence of a thick chylomicron band at the origin, reduced beta and alpha bands, and the absence of a pre-beta band.

d. Clinical manifestations of this type include eruptive xanthomas, colic (abdominal pain), hepatosplenomegaly and lipemia retinalis, all of which appear at an early age.

e. Type I is rare and is expressed as an autosomal recessive trait. Secondary (non-genetic) causes include uncontrolled diabetes mellitus, acute alcoholism, dysglobulinemia, pancreatitis, hypothyroidism, and the use of oral contraceptives.

f. The biochemical lesion appears to be an absence of heparin-activated lipoprotein lipase. To confirm this type, it is essential to measure PHLA (post-heparin lipolytic activity) which is reduced in these individuals. This lipase aids in clearing the chylomicra from the plasma. PHLA is normal in all of the rest of the types.

g. One study reported an acquired form of the disease in which the affected individual showed decreased PHLA. Apparently an abnormal protein was present which bound to and inactivated the heparin, preventing it from activating the lipoprotein lipase and thereby preventing the clearing of chylomicra from the blood.

h. This type responds to low-fat diets supplemented with medium-chain triglycerides. No particular drug therapy has been found to be of any value.

3. <u>Type II</u> (familial hyperbetalipoproteinemia; familial hypercholesterolemia)

a. This is the most common of the familial LP's and it is subdivided into types IIa and IIb. Both show increased β-LP (LDL) while, in addition, IIb exhibits an elevated pre-β (VLDL) concentration which does not occur in IIa. Both types II's are apparently inherited as autosomal dominant traits and both can occur in different members of the same family, suggesting that they may be genetically related to each other. Hypothyroidism, porphyria, dysglobulinemia, biliary obstruction, and dietary imprudence are non-genetic causes of this disorder which must be ruled out.

b. The clinical manifestations include xanthelasma, tendon and tuberous xanthomas, corneal arcus (juvenile), and premature and accelerated atherosclerosis. The treatment consists of restricted intake of cholesterol (less than 300 mg/day) and saturated fats, enhanced intake of polyunsaturated fats, and regulation of body weight by controlling caloric intake. Drugs such as cholestyramine, dextrothyroxine (D-thyroxine), and nicotinic acid are also used when needed.

c. <u>Characteristics of Type IIa</u>

   i)   Plasma maintained at $4^\circ C$ for 18 hours is clear (no visible chylomicra).

   ii) Total plasma cholesterol values are elevated in association with a normal triglyceride level and a ratio of cholesterol/triglyceride greater than 1.5.

   iii) In this phenotype, the excess cholesterol is <u>associated with the LDL fraction.</u> In other situations, it is possible to have elevated total plasma cholesterol without an increase in the LDL cholesterol. For example, the total plasma cholesterol is elevated in women taking oral contraceptives but the cholesterol is associated with the HDL rather than the LDL fraction. An estimate of LDL cholesterol can be made using the relationship

   $$LDL\ Cholesterol = PC - (PTG/5 + AC)$$

   where: PC = total plasma cholesterol (mg/100 ml)
   PTG = total plasma triglycerides (mg/100 ml)
   AC = α-lipoprotein (HDL) cholesterol, estimated by analysis of the supernatant obtained from the precipitation of the non-HDL lipoproteins with heparin-Mg^{+2}. A value of 45 is commonly used.

   This formula should <u>only</u> be used when PTG is less than 400 mg/100 ml and following exclusion of type III hyper LP as a possible cause. The upper limits for normal LDL cholesterol values (in mg/100 ml) are: 150 (0-29 years of age); 170 (30-49 years of age); and 190 (50-69 years of age).

   iv) The electrophoretic pattern of type IIa plasma shows an absent or normal (not elevated) pre-β-band. The β-band is heavily stained, chylomicra are not visible, and the α-LP is normal.

d. Characteristics of Type IIb

    i) On maintaining plasma from a type IIb patient at $4^{\circ}C$ for 18 hours, it is either clear or uniformly turbid (lipemic) in contrast to the clarity always seen with type IIa plasma treated similarly. No creamy supernatant of chylomicra is present.

    ii) Plasma triglyceride and LDL cholesterol are always increased and total plasma cholesterol is usually elevated.

    iii) Both the beta- and pre-beta-lipoprotein electrophoretic bands are increased in intensity, while the α-LP band is usually normal and chylomicra are not visible.

4. Type III (floating-β-LP; broad-β-LP; familial hypercholesterolemia with lipemia)

a. Plasma maintained at $4^{\circ}C$ for 18 hours is usually turbid with a creamy layer above the plasma.

b. Both plasma triglyceride and cholesterol values are nearly always elevated. The cholesterol/triglyceride ratio is most commonly around 1 but it can vary from 0.3 to greater than 2.

c. In paper, agarose, or cellulose acetate electrophoresis, the plasma reveals a broad beta band encompassing both the beta and pre-beta regions. Sometimes a distinct pre-β-band of abnormal intensity is seen. A faint chylomicron band is also frequently observed, even in a fasting sample. The α-lipoproteins are generally normal. Certain other aspects of the LP pattern are visible when polyacrylamide and starch are used as the electrophoretic support. This is helpful in the diagnosis of this disease.

d. The definitive diagnosis of type III is made by demonstrating the presence of an abnormal lipoprotein with a density of less than 1.006 g/ml ("floating-β" LP). A decrease in one LDL subclass (density 0.010–0.163, S_f = 0–12) is seen with an increase in VLDL (100–400) and of the LDL having S_f = 12–20. The low density fraction also contains an unusually large amount of cholesterol.

e. Type III is relatively rare and is presumably inherited as an autosomal recessive trait. Secondary causes of type III are rare but myxedema and dysglobulinemia must be excluded. The biochemical lesion may be the presence of an abnormal protein.

The clinical manifestations of this type are tendon and palmar xanthomas, accelerated atherosclerosis of the coronary and peripheral vessels, and abnormal glucose tolerance curves.

f. The treatment includes weight control by means of a restricted diet which is low in carbohydrate and cholesterol and which contains increased amounts of polyunsaturated fat. The drugs used in the treatment include clofibrate, D-thyroxine, and nicotinic acid.

5. Type IV (hyper-pre-beta-lipoproteinemia)

a. Plasma maintained at 4°C for 18 hours is either clear or turbid throughout, but no creamy, chylomicral layer is present.

b. Plasma cholesterol is normal or slightly elevated. Triglycerides are elevated.

c. On electrophoresis, the plasma reveals a normal beta band and a pre-beta band with an increased intensity. The α-band appears normal and there is an absence of a chylomicral band.

d. Type IV is a common disorder and appears to be inherited as an autosomal dominant trait. The triglyceridemia is frequently associated with an abnormal glucose tolerance. Estrogens, either alone or in oral contraceptives, seem to often cause triglyceridemia.

The LP abnormalities seen in this disorder can also be due to a number of secondary causes, including pancreatitis, hypothyroidism, the nephrotic syndrome, alcoholism, diabetes mellitus, glycogen storage disease, dysglobulinemia, progestin administration, Werner's Syndrome, Gaucher's disease, Niemann-Pick disease, and others.

Xanthomatosis and hepatosplenomegaly are characteristic clinical signs of this disorder. Type IV hyper LP also carries with it a high risk of developing coronary disease.

e. The biochemical lesions leading to the hypertriglyceridemia are not clear although defects in the endogenous synthesis and clearance of triglycerides are possible causes.

f. The management of this type involves attaining an ideal body weight, the restriction of carbohydrates, and enhanced intake of polyunsaturated fats. Clofibrate and nicotinic acid are drugs which are sometimes useful in treatment.

6. <u>Type V</u> (familial hyperprebetalipoproteinemia with hyperchylomicronemia)

   a. When a plasma sample is maintained at 4°C for 18 hours, it shows chylomicra as a creamy supernatant layer. The underlying plasma is turbid. This plasma turbidity distinguishes type V from type I.

   b. Elevated levels of both cholesterol and triglycerides occur. The plasma cholesterol/triglyceride ratio is usually greater than 0.15 and less than 0.6.

   c. On electrophoresis, the pattern reveals a chylomicron band (at the point of application) and an intense pre-beta band while the α-and β-bands are generally decreased. This type does <u>not</u> show the presence of a "broad-beta" band encompassing both the beta and pre-beta regions as is observed in type III. This can be confirmed by the absence of any "floating-β" LP upon ultracentrifugation. Type V and type I are also distinguished from each other by the normal PHLA observed in the former type compared to the absence of PHLA in type I.

   d. This is an uncommon disorder and the inheritance pattern is not known.

   e. The clinical manifestations of this type are abnormal pain, eruptive xanthomas, hepatosplenomegaly, hyperglycemia, hyperuricemia, and an abnormal glucose tolerance profile.

   f. Other disorders which can show the characteristics of type V include chronic pancreatitis, severe diabetes mellitus (insulin-dependent), alcoholism, nephrosis, myeloma, and dysglobulinemia.

   g. The management of type V involves attaining an optimal body weight, enhanced intake of protein, and diminished intake of fat (less than 70 grams per day) and carbohydrate. Drugs used are nicotinic acid and clofibrate.

Figure 81.   <u>Algorithm for the Differentiation of Lipoprotein Phenotypes</u>*

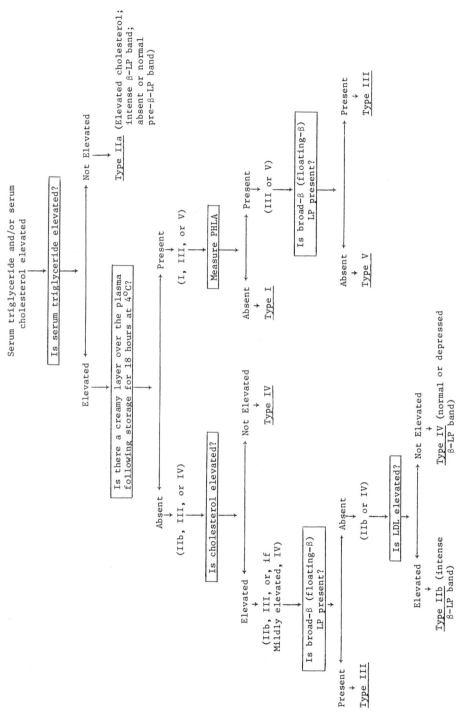

PHLA = postheparin lipolytic activity

*Modified from A.M. Gotto and L. Scott, J. Am. Dietetic Association, <u>62</u>, 617, (1973), with permission.

Some of the Other Disturbances of Fat Metabolism

1. In obesity (abnormal increase in body weight due to fat deposition),
    the rate of lipid deposition is greater than the rate of
    mobilization. All other metabolic factors being equal about one
    gram of fat is deposited for 9 calories ingested in excess of that
    used by the body.

    a. Obesity most commonly results from simply eating more calories
        than are needed for the energy requirements of the body. This
        can occur because of failure to adjust caloric intake to
        decreased metabolic rate or a change from an active to a more
        sedentary life-style.

    b. Endocrine disturbances can also cause obesity. These include
        hypothyroidism (rate of food oxidation decreases) and
        castration. Certain pituitary deficiencies may lead to
        abnormal fat distributions without changing the total quantity
        of stored lipid. Hyperinsulinism sometimes causes obesity
        by stimulation of appetite. The hypothalamus is concerned
        with the regulation of appetite. Injury to certain regions of
        this gland in animals can cause obesity (but see anorexia,
        below).

    c. In general, the reactions involved in fat metabolism appear to
        be normal in obese individuals.

2. Xanthomatosus is the accumulation of excess lipids (often rich in
    cholesterol) in multiple, fatty tumors. The name is derived from
    the yellow-orange color of the deposits (xanthos = yellow). This
    is frequently accompanied by lipemia and hypercholesterolemia.
    Hand-Schüller-Christian disease (chronic idiopathic xanthomatosis)
    is a cholesterol lipidosis associated with diabetes insipidus and
    lipid deposits in the liver, spleen, and the flat bones of the skull.

3. Cachexia is the failure to ingest enough calories to maintain normal
    lipid storage. A high rate of lipid mobilization from fat depots
    occurs relative to the rate of deposition. In severe cases,
    adipose tissue may completely disappear. This may be caused by
    anorexia (loss of appetite) due to damage to certain portions of
    the hypothalamus (see above, under obesity) or to a nervous
    disorder (anorexia nervosa). Cachexia can also be secondary to
    carcinoma, malnutrition, certain chronic infectious diseases,
    hyperthyroidism, and uncontrolled diabetes. In the last disorder,
    fat is oxidized to provide energy and the capacity to synthesize
    fat for storage is drastically reduced. This gives rise to the
    depletion of fat depots.

Functions of Some Tissues in Lipid Metabolism

Liver

1. The liver is actively concerned with lipid metabolism. Liver takes up plasma-free fatty acids very actively and utilizes them for the synthesis of triglycerides and phospholipids, which are either stored or released to the circulation. Note that the liver does not utilize FFA for energy purposes as much as muscle tissue does. About 30-60% of the circulating dietary triglycerides are also taken up by the liver.

2. The liver plays a major role in cholesterol metabolism by regulating the plasma cholesterol levels (maintaining cholesterol homeostasis), carrying on the biosynthesis of cholesterol and its conversion to the bile acids, etc.

3. The normal liver is 5% lipid, this figure being dependent on a number of factors. In fatty livers, due to certain pathological and physiological disturbances, the lipid content increases to 25-30%. In chronic lipid accumulation, the liver cells become fibrotic, leading to cirrhosis and impaired liver function. Lipids are normally found in the Kupffer cells in the form of droplets. In fatty livers, fatty acid droplets may replace the entire cytoplasm of the hepatic cell. In samples of fatty livers, chemical and histological findings correlate well.

4. Increased fat in the liver may result from decreased oxidation of fat, increased formation of fat, increased movement of fat from the depots to the liver, impaired removal of fat from the liver, or a combination of these factors. There are several etiologies for these four abnormal processes.

    a. Nutritional factors. Diets which are high in fat and low in carbohydrate; diets deficient in methionine and choline (lipotropic factors, needed for phospholipid synthesis); excessive amounts of thiamine and biotin; excessive alcohol ingestion (oxidation of the alcohol produces large amounts of NADH and acetyl-CoA. The NADH stimulates α-glycerol-phosphate synthesis from dihydroxyacetone phosphate and the acetyl-CoA contributes to fatty acid and cholesterol biosynthesis).

    b. Endocrine factors. Hormones which promote hepatic lipid storage are ACTH, adipokinin, adrenal cortical hormones, sex steroids, insulin, and the thyroid hormones.

c. <u>Miscellaneous factors</u> include chemicals which inhibit protein
    synthesis (puromycin, ethionine) and other, less specific ones,
    such as CCl_4, $CHCl_3$, phosphorous, and bacterial toxins. Anoxia,
    due to anemia and respiratory or vascular congestion, has a
    similar effect.

5. Steps which are necessary for the transport of fat from the liver
    are summarized in Figure 82. The steps which are inhibited by
    some of the above agents are indicated.

Adipose Tissue

1. Adipose tissue functions as an area of caloric storage. When
    there is excess caloric intake, calories are stored there as
    triglycerides, and when there is stress or other energy deficit,
    free fatty acids are released. Glucose by itself can provide all
    the atoms needed for a triglyceride (TG) molecule. The reactions
    related to lipid metabolism which occur in the adipose cells are
    indicated in Figure 83.

2. The capillary lipoprotein lipase breaks down the TG in chylomicra
    and VLDL to FFA and glycerol. The FFA then enters the adipose
    cells to become adipose FFA and glycerol returns to the blood.
    Adipose tissue cannot utilize glycerol due to the absence of the
    enzyme glycerol kinase. In the formation of triglycerides in the
    adipose tissue, α-glycerolphosphate is obtained from G-6-P.
    Thus a supply of glucose must be available for TG synthesis.

3. Insulin plays several important roles in fatty acid metabolism
    in the adipose tissue: (a) It stimulates the uptake of blood
    glucose, and (b) It increases both fatty acid and triglyceride
    synthesis.

4. A number of other hormones and compounds also affect the adipose
    tissue. These are illustrated in Figure 84. This is another
    important example of hormonal effects mediated by cyclic AMP.

Ketosis

1. Under certain metabolic conditions (starvation, very high fat diets,
    severe diabetes and other states in which carbohydrate reserves
    are depleted), there is a high rate of fatty acid oxidation in the
    liver. This produces considerable quantities of the <u>ketone bodies</u>:
    acetoacetate, D(-)β-hydroxy butyrate, and acetone which are released
    into the circulation, causing <u>ketosis</u>. The name "ketone bodies"
    for these compounds is traditional but somewhat inaccurate.

Figure 82.   Synthesis of VLDL Lipoprotein in the Liver

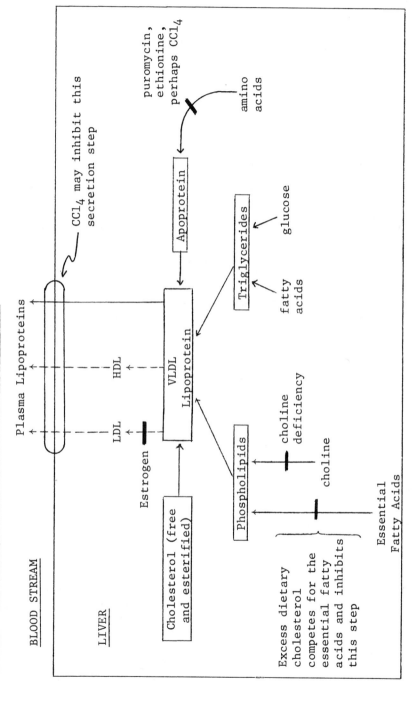

Solid lines indicate known processes while dashed lines are steps which are not known for certain. Heavy bars across some lines indicate inhibition of that step by the specified materials.

Figure 83. Lipid Metabolism in the Adipose Tissue Cells

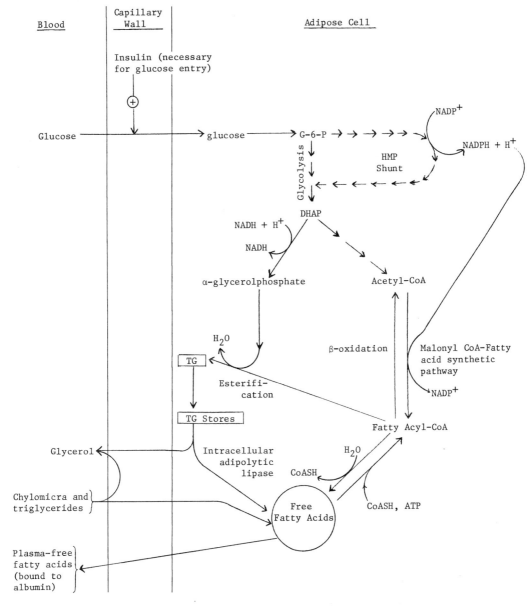

DHAP = dihydroxyacetone phosphate
TG = triglyceride
G-6-P = glucose-6-phosphate
HMP Shunt = hexose monophosphate shunt

Figure 84.   Hormonal Regulation of Lipolysis in Adipose Cells

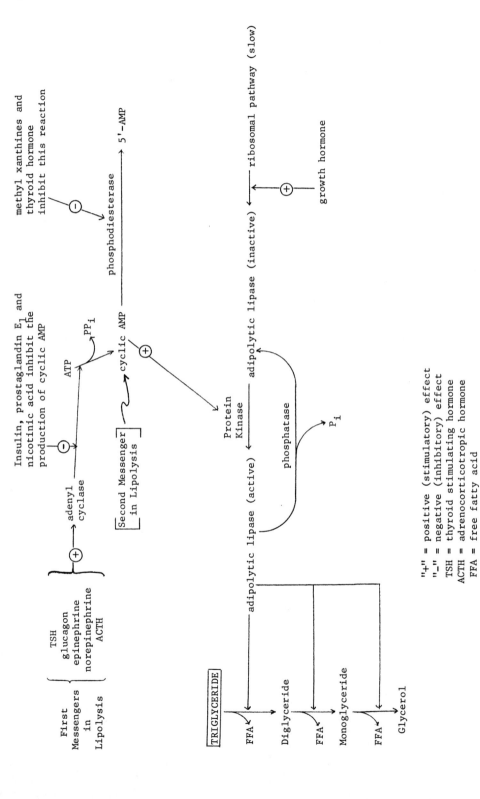

2. Under normal conditions, only small amounts of the ketone bodies
    are formed. They pass from the liver to the blood by diffusion
    and are oxidized in the peripheral tissues (kidney, muscle, heart,
    brain and testes) by way of acetyl-CoA and the citric acid cycle.
    Only when the rate of production of ketone bodies increases much
    more than their utilization do the levels build up.

3. Acetoacetate and β-OH butyrate are moderately strong acids. They
    have to be buffered and, hence, deplete the alkali reserves,
    giving rise to ketoacidosis. This can be fatal, for example, in
    uncontrolled diabetes.

4. The synthesis of ketone bodies in the liver proceeds by the
    pathways outlined in Figure 85. The major pathway for acetoacetic
    acid synthesis seems to be the one from HMG-CoA, catalyzed by
    HMG-CoA lyase. In addition to the fatty acids, three amino acids
    (tyrosine, phenylalanine, and leucine; see Chapter 6) are important
    contributors. The decarboxylation of acetoacetate to form acetone
    appears to be spontaneous in mammals. In bacteria, however, a
    specific enzyme, acetoacetate decarboxylase, has been shown to
    catalyze the reaction. The contribution of acetone to the total
    ketone bodies is minor.

5. The catabolism of acetoacetate and β-hydroxybutyrate, the two major
    ketone bodies, is essentially a reversal of the reactions in
    Figure 85, giving acetyl-CoA as an end-product. The key enzyme
    in the process is β-ketoacid CoA transferase, needed to "activate"
    the acetoacetate by forming aceto-acetyl-CoA. The reactions
    involved are shown below.

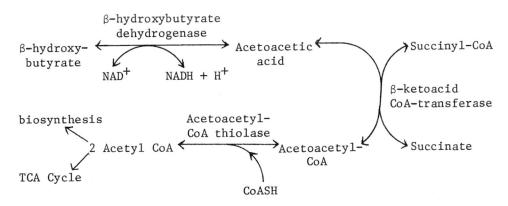

β-hydroxybutyrate dehydrogenase is an enzyme of the inner
mitochondrial membrane.

Figure 85. Hepatic Synthesis of the Ketone Bodies

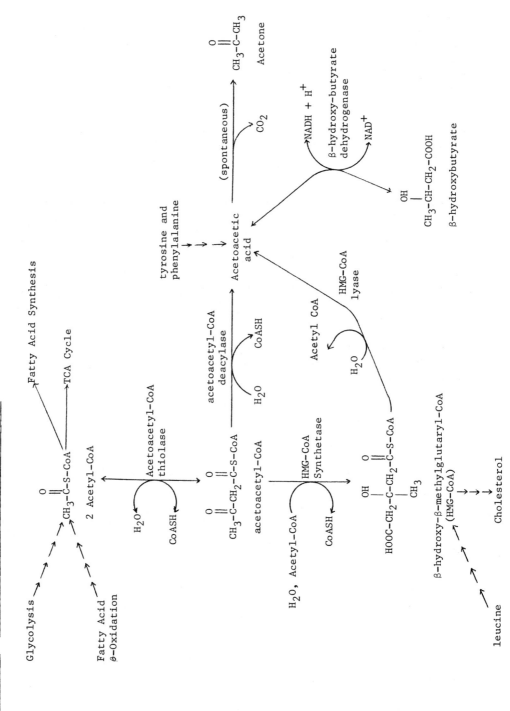

6. β-ketoacid CoA-transferase is absent in the liver which is one
    reason why the ketone bodies must be released into the blood.
    Tissues which can metabolize the ketone bodies (muscle, brain,
    testes, heart, kidney) have this enzyme. As was indicated earlier
    (pages 268-269), the ability of the brain to metabolize the
    ketone bodies as an energy source is especially important in
    starvation. For some time it was thought that the brain had to
    have <u>glucose</u> and it is only fairly recently that the enzymes
    necessary for ketone body catabolism have been demonstrated to
    be present in this tissue. This capacity for ketone body
    utilization in the brain is also of significance in rat, canine,
    and human neonates. In these (and other) species, mother's milk
    is very rich in fats, producing a transient ketosis in suckling
    infants which lasts until weaning. In the rat, levels of most of
    the requisite enzymes increase following birth then decline again
    after weaning. The acetoacetyl-CoA thiolase follows a somewhat
    different pattern, with the final adult activity remaining
    fairly high, probably due to its role in other pathways. The
    brain cannot survive <u>entirely</u> without glucose, however, perhaps
    due to the need for succinyl-CoA in the β-ketoacid CoA-transferase
    reaction. Normally, succinyl-CoA hydrolysis (in the TCA cycle)
    is coupled to the formation of GTP, a compound needed for several
    tissue functions including protein synthesis and gluconeogenesis.
    The glucose may be necessary to supplement the succinyl-CoA
    supply to permit GTP synthesis.

Figure 86. Some Interactions of the Liver, Adipose Tissue, and Other Extrahepatic Tissues in Lipid Metabolism

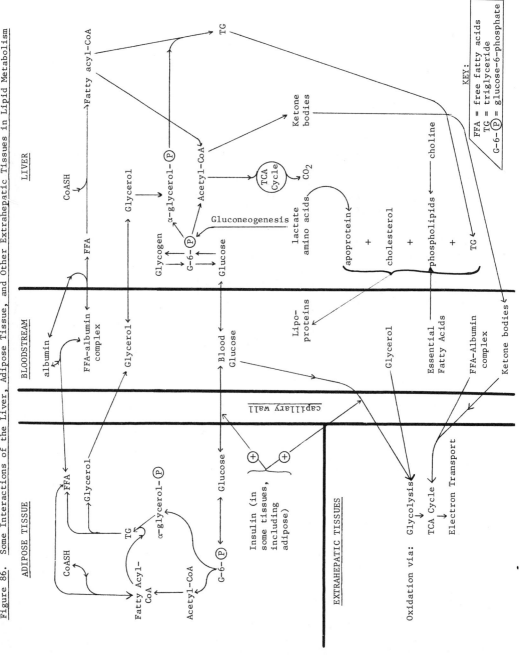

9

Mineral Metabolism

*with Howard F. Mower, Ph.D.**

Calcium Metabolism

1. Functions of Calcium in the Human Body

   a. Membrane Phenomena

   Ca^{+2} is a component of many membranes, presumably being bound
   to the membrane proteins. Hormone binding to the plasma
   membrane receptor sites or depolarization of the membrane by
   electrical stimulation causes release of the bound Ca^{+2} into
   the cytoplasm by an unknown mechanism. This is accompanied by
   a local increase in membrane permeability. Passage of Ca^{+2}
   into the cell creates an unstable situation since, under
   resting conditions, the intracellular Ca^{+2} concentration is
   1/1000 of the extracellular value. This gradient is normally
   maintained by a Ca^{+2} pump located in the plasma membrane.
   Many of the enzymes activated by hormone interaction at the
   receptor site require Ca^{+2} for activity and frequently involve
   the formation of cAMP by the membrane-bound adenyl cyclase
   (see page 226). Examples of processes which require Ca^{+2} and
   which produce a cAMP include the release of neurotransmitter
   substances by the synapse in response to electrical stimuli,
   the release of growth hormone by the anterior pituitary in
   response to growth hormone releasing factor, the synthesis
   and release of glucose by the liver in response to glucagon,
   HCl secretion by the stomach in response to histamine, and
   a number of others.

   A low plasma Ca^{+2} concentration due to dietary deficiency,
   inadequate adsorption, vitamin D deficiency, hypoparathyroidism,
   or a thyroid tumor (of the parafollicular cells) has the
   immediate effect of perturbing several of these processes.
   Nerve fibers fire spontaneously, muscle cells show lowered
   threshold stimuli, and cardiac contractility is increased while
   relaxation is retarded, resulting in muscle spasms. This

---

* Professor of Biochemistry, University of Hawaii School of Medicine.

syndrome is known as tetany and it may be either latent
(triggered by stress) or overt. If uncorrected, it can lead
to death.

b. Blood Clotting

Ca^{+2} is required in the blood clotting process. Collected
blood will not clot if EDTA, citrate, or oxalate (chelating
agents) is added to bind Ca^{+2}. In vivo, however, tetany occurs
before Ca^{+2} levels become low enough to affect clotting.

c. Muscle Contraction

Ca^{+2} plays an important role in muscle contraction related to
a, above. The muscle cell contains an internal store of
Ca^{+2} in the sarcoplasmic reticulum. This Ca^{+2} is released as
a depolarization phenomena initiated by neural stimulation
of the muscle. It is resequestered into the sarcoplasmic
reticulum during the relaxation phase of the muscle
contraction cycle (see also pages 287-288).

d. Serum Ca^{+2}

Total serum calcium is about 8.5-10.5 mg/100 ml (5 mEq/liter
or 2.5 mM/liter), distributed about equally between
non-diffusible (protein-bound) and diffusible forms. The
diffusible fraction is composed of complexed calcium and
ionized calcium. The ionized calcium is less than half of
the total calcium and is the only form which is
physiologically active. A decrease in this fraction is
responsible for the production of tetany, regardless of the
changes in total calcium values. The non-diffusible fraction
is only slowly released to replenish the free metal,should the
levels of the free-form fall. Because most analytical methods
measure total Ca^{+2}, it is important to also measure the
protein concentration of the serum when evaluating serum calcium.
A knowledge of serum protein concentration at pH 7.35 and
25°C permits estimation of the ionized calcium present.
Ionized calcium can also be measured by first removing the
protein-bound material with sephadex gel filtration or
ultrafiltration. It can also be estimated by measuring
calcium in the cerebrospinal fluid (CSF), because CSF may be
considered as an ultrafiltrate of plasma. Normal CSF calcium
values are 4.2-5.8 mg/100 ml.

Tetany can occur at higher levels of total Ca^{+2} if the serum
protein concentration is high, or in alkalosis, when the
ability of serum proteins to bind Ca^{+2} is increased. Symptoms
of tetany also occur when the total serum calcium falls below 15%.

e. Relationship Between Ca^{+2} and Phosphate

Calcium phosphate is a rather insoluble substance and if the value of "total calcium" times serum phosphate exceeds about 35, precipitation of calcium phosphate will occur. As a consequence, high levels of Ca^{+2} result in low levels of PO_4^{-3} and can cause deposition of $Ca_3(PO_4)_2$ within the kidney, liver, etc. The immediate effects of hypercalcemia are muscular weakness, gastrointestinal distress, giddiness, lassitude, and thirst.

f. Bone Structure

Bone may be thought of as having three elements:

i) A fibrous, proteinaceous material with three components:
(a) collagen (quantitatively the largest fraction);
(b) a mucoprotein which has chondroitin sulfate groups attached to a protein backbone; and (c) elastin. In this regard, bone is very much like connective tissue, skin, tendon, etc. These substances contain the same three proteins.

ii) The ground substance of bone is the mineral hydroxyapatite which has the formula $Ca_{10}(PO_4)_6(OH)_2$. It is a solid, embedded in the network of protein fibers. In the more pliable connective tissues, the ground substance is a gel of varying consistency containing different amounts of water, salts, protein, and large polysaccharides (hyaluronic acid, chondroitin-4- and chondroitin-6-sulfates, dermatin, heparin, and keratin sulfates; see pages 147-150).

iii) Bone cells of two types: osteoclasts and osteoblasts. The osteoblasts resemble the chondroblasts of cartilage and the fibroblasts of skin. In each case, they are responsible for the fibrous protein formation and ground substance deposition. The osteoclasts are responsible for bone breakdown by a process of "resorption". The bone is a dynamic, well-vascularized living tissue. The hydroxyapatite is actively turning over and the bone is constantly being resorbed and formed again.

Hydroxyapatite formation during bone growth is thought to be accomplished by the osteoblasts. These cells achieve a high local concentration of PO_4^{-3} by alkaline phosphatase action on the sugar phosphates of glycolysis (see later under vitamin D deficiency and under

osteoporosis). A high local Ca^{+2} concentration is
brought about by the binding of Ca^{+2} to the sulfate groups
of chondroitin sulfate.

2. Control of Ca^{+2} Levels in the Serum

A 15% change of the Ca^{+2} from the normal range of 8.5-10.5 mg/100 ml
can bring about severe clinical symptoms which can lead quickly
to death. It is therefore not surprising to find Ca^{+2}
concentrations within the body under the control of a multicomponent
system. The agents functioning in this system are summarized in
Table 28. Some of them are discussed below while others
(e.g., androgens and the thyroid hormones) are covered elsewhere
in this text.

Vitamin D and Its Effects on Calcium Metabolism

1. Vitamin D is an unusual vitamin in that it can be produced entirely
within the body, provided sufficient light is present. The
pathway to cholesterol synthesis (see pages 650-652) is used to produce
7-dehydrocholesterol, the immediate precursor to cholesterol.

Radiant energy striking the skin catalyzes the opening of the "B"-
ring of 7-dehydrocholesterol, producing cholecalciferol (vitamin D_3).
Ergosterol (24-methyl-21-dehydro-7-dehydrocholesterol), a common
plant steroid which is present in the diet, may also serve as a
substrate in the light reaction, producing ergocalciferol
(vitamin D_2; also called calciferol). This is an important
commercial method for making vitamin D_2. Both D_2 and D_3 evoke
the same physiological response.

Table 28.   Substances Which Participate in the Homeostatic Control of Ca^{+2}
Some of these are discussed elsewhere in this chapter and in other parts of the book.

Substance	Source	Chemical Nature	Site of Action	Effect
Vitamin D	a) Diet b) UV irradiation of cholesta-dienol, a precursor of cholesterol; or of ergosterol, a plant sterol	Steroid derivative	Intestinal mucosa; bone	Increases Ca^{+2} adsorption from intestine by increasing synthesis of Ca^{+2}-binding protein and Ca^{+2}-dependent ATPase in intestinal mucosa. Also acts on bone to cause reabsorption
Parathyroid Hormone (hypercalcemic factor)	Parathyroid	Small protein; molecular weight is 8,447	1) Bone: activates osteoclasts, causing removal of Ca^{+2} from bone into plasma (bone resorption) 2) Kidney; increases reabsorption of Ca^{+2} (decreases urinary excretion of Ca^{+2}); decreases PO_4^{-3} readsorption, thereby increasing urinary PO_4^{-3} excretion 3) Intestinal mucosa: aids vitamin D action	All three effects increase serum Ca^{+2} and lower Mg^{+2} and PO_4^{-3} increase urinary PO_4^{-3}; hydroxyproline in the urine increases, due to increased collagenase action.
Calcitonin (hypocalcemic factor)	Thyroid parafollicular cells	Small proteins; molecular weight is 3,500–4,000.	Bone: increase deposition of Ca^{+2}	Decreases serum Ca^{+2}, and urinary PO_4^{-3} and hydroxyproline (decreased collagenase action)
Growth Hormone	Anterior Pituitary	Protein; molecular weight is 21,500	Bone: activates osteoblasts to increase synthesis of collagen and chondroitin sulfate	Increased utilization of Ca^{+2} during growth
Estradiol	Ovary	Steroid	Bone: affects osteoblasts	Specific action on epiphysial cartilage of bone during puberty; causes ends of epiphyses to close and thereby stops bone elongation
Thyroid Hormone	Thyroid	Iododerivatives of tyrosine	Bone: affects osteoblasts	In growing animal it is synergic to growth hormone in cartilage growth In adults, prevents bone resorption
Cortisone	Adrenal Glands	Steroid	Bone	Aids in Ca^{+2} and PO_4^{-3} deposition in bone; depresses osteolytic action of osteoclasts
Androgens	Testes	Steroid	Bone	Aids in Ca^{+2} and PO_4^{-3} deposition in bone

Primary Ca^{+2} Homeostatic Agents

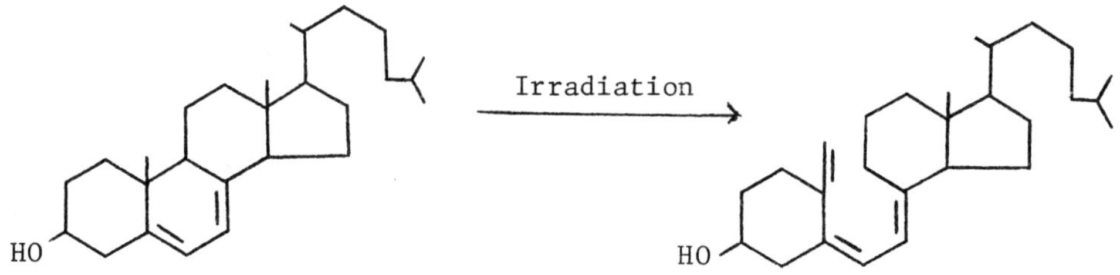

7-dehydrocholesterol

cholecalciferol
(vitamin D$_3$)

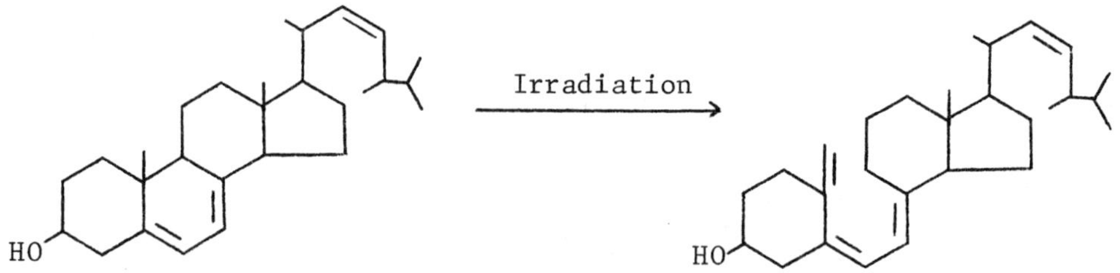

ergosterol

ergocalciferol
(vitamin D$_2$)

2. Neither vitamin D$_2$ nor D$_3$ is the active form of the vitamin. Both
    substances must undergo hydroxylation at carbon 25, becoming,
    respectively, 25-hydroxyergocalciferol (25OHD$_2$) and
    25-hydroxycholecalciferol (25OHD$_3$) before they are biologically
    active. This reaction takes place in the liver and is feedback
    inhibited by 25OHD$_2$ and 25OHD$_3$. A second hydroxylation, at
    carbon one, converts 25OHD$_3$ to 1,25-dihydroxycholecalciferol
    (1,25(OH)$_2$D$_3$), currently thought to be the final active metabolite
    of vitamin D$_3$. This compound has ten times the activity of 25OHD$_3$.
    The conversion is catalyzed by a renal enzyme which requires
    oxygen and reduced pyridine nucleotide. Calcium inhibits the
    reaction. It is thought, but not proven, that 25OHD$_2$ undergoes
    a similar reaction. This sequence is shown below for vitamin D$_3$.

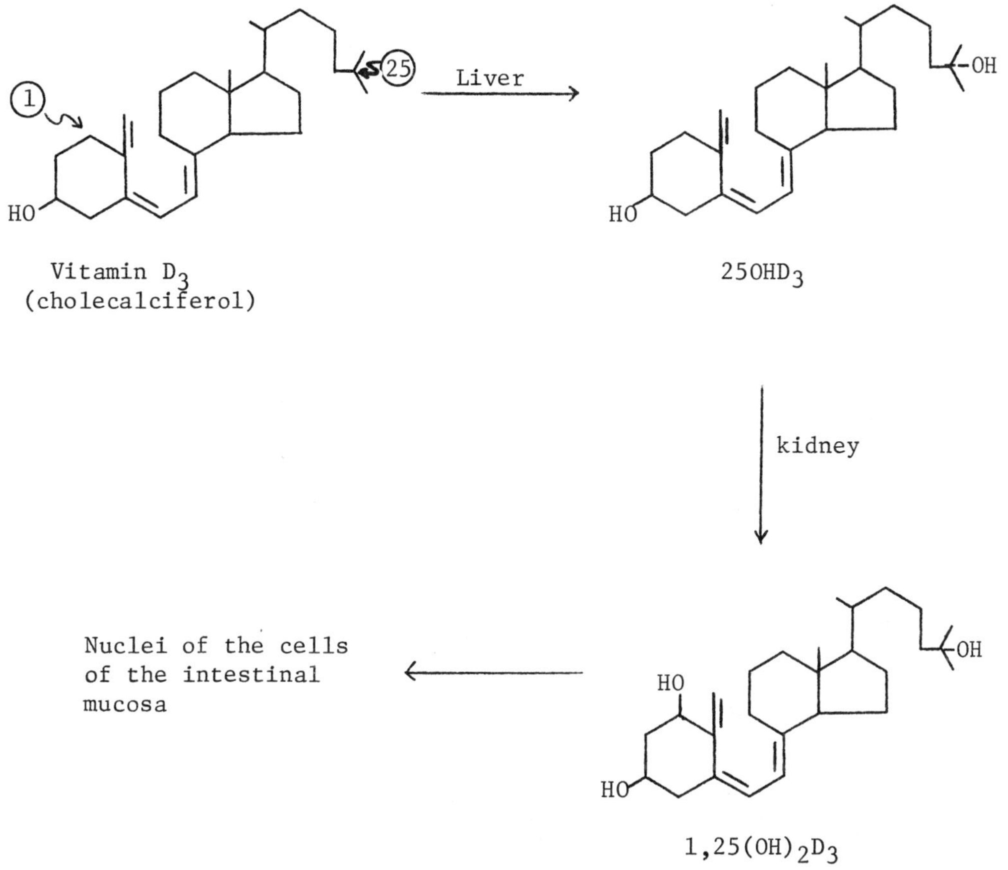
1
25
HO
Liver
OH
HO
Vitamin D3
(cholecalciferol)
25OHD3
kidney
Nuclei of the cells
of the intestinal
mucosa
OH
HO
HO
1,25(OH)2D3

Because of the mode of action of $1,25(OH)_2D_3$ and the interrelationship between serum Ca^{+2} levels and $1,25(OH)_2D_3$ synthesis, this form of the vitamin is sometimes classified as a hormone and the kidney as an endocrine gland.

Dihydrotachysterol, shown at the top of the next page, is a synthetic derivative of ergosterol which possesses antirachitic activity. Although it is less effective than D_3 when low dosages of both compounds are compared, it is clinically valuable because it is <u>more</u> effective than D_3 when <u>high</u> dosages of both are compared.

dihydrotachysterol (DHT)

This makes dihydrotachysterol very useful for treating diseases which require large amounts of vitamin D, such as hypoparathyroidism. Apparently DHT is 25-hydroxylated by the same system as D_2 and D_3 but, unlike $250HD_2$ and $250HD_3$, 25-hydroxyDHT exerts no feedback inhibition on this process.

3. The active forms of vitamin D are transported to the cytoplasm of the cells of the intestinal mucosa. Once there, as in the case of the steroids, they enter the cell nucleus where they stimulate mRNA formation and thereby cause an increase in the synthesis of a variety of proteins. One of the proteins which is made in this process has been identified as a calcium binding protein, while another is a Ca^{+2}-dependent ATPase. These proteins presumably function in the energy-dependent process of Ca^{+2} absorption across the intestinal epithelium. The mechanism by which vitamin D acts on the bone, causing Ca^{+2} readsorption, is not known.

Vitamin D also influences phosphate absorption, in two ways. First, since phosphate is usually the ion accompanying Ca^{+2} during its absorption in the intestine (its counterion), any increase in Ca^{+2} uptake usually results in a simultaneous rise in the rate of phosphate absorption. In a second, more direct fashion, current studies suggest that vitamin D may influence the transport of phosphate.

4. A vitamin D deficiency results in hypocalcemia due to poor Ca^{+2} absorption. In children, this results in rickets, characterized by demineralization (softening) of the skeleton with deformation of the weight bearing bones. In adults, the corresponding disease is osteomalacia. This is in contrast to osteoporosis, where the bones become brittle. In both rickets and osteomalacia, serum alkaline phosphatase is elevated while it is normal in osteoporosis. An increase in this enzyme is also seen in long-standing cases of hyperparathyroidism and may occur in other instances of marked osteoblastic activity. Diseases of the liver or kidney can also result in hypocalcemia.

It was because of disturbances in calcium metabolism during hemodialysis and advanced kidney failure that the role of the kidney in vitamin D chemistry was detected. Such patients show hypocalcemia, impaired intestinal absorption of calcium, and osteomalacia. They fail to respond to the administration of parathyroid hormone. A recent study indicates that it may be possible to treat these symptoms in patients with advanced renal disease by administration of 1,25-dihydroxycholecalciferol.

5. Administration of phenobarbital and diphenylhydantoin, both anticonvulsants used in the treatment of epilepsy, can cause hypocalcemia (termed drug induced hypocalcemia). As pointed out elsewhere (see pages 578-580), phenobarbital (and diphenylhydantoin) stimulate the P-450 microsomal oxidase system in the liver. Once activated, these enzymes apparently increase the rate of oxidation of vitamin D to inactive, polar metabolites.

Parathyroid Hormone (Parathormone)

1. Parathyroid hormone (PTH) is the single most important regulator of Ca^{+2} in humans. Bovine PTH has been completely characterized and shown to be a peptide containing 84 amino acids. The human hormone is a peptide of 9,000 molecular weight secreted by the chief cells of the parathyroid glands. It is initially synthesized as a prohormone which becomes PTH when activated, a process apparently similar to the activation of proinsulin. There are usually four parathyroid glands in humans, two embedded in the superior poles of the thyroid and two in its inferior poles. PTH is secreted in response to a decrease in ionized serum Ca^{+2} concentration. The half-life of PTH in the circulation is about 20-30 minutes, typical of other peptide hormones of similar size (e.g., insulin).

2. All actions of PTH promote and increase the movement of Ca^{+2} into the extracellular fluid from the bone, by increased resorption in the kidney, and by absorption from the intestine. The hormone

secondarily causes plasma Ca^{+2} to rise by enhancing the urinary secretion of phosphate. Recall that, as the phosphate level decreases, the concentration of Ca^{+2} is increased to satisfy the solubility product relationship:

$$[Ca^{+2}]^3 [PO_4^{-3}]^2 = 35.$$

The ability of PTH to increase serum Ca^{+2} by stimulating bone resorption is mediated by its effects on the osteoclasts. It increases collagenase activity in these cells, as well as lysosomal activity. It also appears to inhibit the activity of the osteoblasts and to decrease the rate of collagen or matrix synthesis by these cells.

3. In the absence of PTH, rapid loss of Ca^{+2} in the urine occurs, a situation which can be reversed by administration of the hormone. Under normal conditions, 97% to 99% of the Ca^{+2} filtered by the glomerulus is actively reabsorbed along the entire length of the nephron, with 65% to 80% of the reabsorption occurring in the proximal tubule. Under the same conditions, 80% to 90% of the PO_4^{-3} is also actively reabsorbed. Thus, there is normally a slight continual PO_4^{-3} loss but almost complete Ca^{+2} retention. Any increase in the PTH level beyond the normal range results in progressively smaller amounts of PO_4^{-3} reabsorption. The action of PTH on the intestine is not detected until several hours or days following administration. The vitamins D appear to be the most important control agents of Ca^{+2} absorption from the gut.

4. Recent work suggests that there exists a complex relationship among serum Ca^{+2} and phosphate, PTH, and $1,25(OH)_2D_3$, which functions to maintain calcium homeostasis. In a typical cycle, PTH is released in response to low serum Ca^{+2} (less than 10 mg/100 ml). This hormone mobilizes calcium by direct action on the bone and by increasing the synthesis of the "renal hormone", $1,25(OH)_2D_3$. The vitamin D derivative proceeds to the bone (further mobilizing skeletal Ca^{+2}) and to the intestine (causing increased Ca^{+2} absorption). The feedback loop is completed when the serum Ca^{+2} rises sufficiently to shut off PTH secretion. Phosphate levels are also related to this process. Hypophosphatemia, by itself with no increase in PTH, can stimulate $1,25(OH)_2D_3$ synthesis. This also suggests a mechanism whereby PTH may exert its influences on the endocrine function of the kidney.

5. The mechanism of PTH action in the bone, kidney, and gut appears to involve cAMP elevation within the cells of these tissues. Agents which lower intracellular cAMP concentrations, either by activation of the phosphodiesterase (as does imidazole) or by inhibition of the adenyl cyclase (the mechanism of action of 2-thiophene carboxylic acid), cause a fall in plasma Ca^{+2} even in the presence of PTH. Agents which raise cAMP levels (such as dibutyryl cAMP, or caffeine and theophylline, both phosphodiesterase inhibitors) will raise plasma Ca^{+2}.

6. Under normal circumstances, the urine contains micromolar quantities of cAMP, one hundred times the levels found within cells or in plasma. This urinary cAMP is formed by the action of PTH on the Ca^{+2} pumping mechanism of the cells of the collecting tubules. It is not known how the cAMP is secreted into the urine in this unusual manner. In some cases of hypocalcemia, calcium resorption from the collecting tubules is impaired because the tubule cells do not respond to PTH. In this situation, called pseudohypoparathyroidism, serum levels of PTH are high but the urine contains very little cAMP. Presumably, PTH receptor sites are lacking in the cells of such patients or some other defect in the cAMP mechanism is present.

7. Two common causes of hypercalcemia are parathyroid tumors and so-called "ectopic tumors", particularly of the lung. The latter are spontaneous tumors of some tissue other than the parathyroid which, for some unknown reason, secrete PTH. These cells arise as a result of a neoplastic process in which DNA is transcribed in an uncontrolled and sometimes random way. Bone tumors and prolonged periods of immobilization or exposure to zero gravity also cause hypercalcemia.

Calcitonin

1. Calcitonin is a peptide hormone of molecular weight 3,590 which is produced by the parafollicular cells scattered throughout the thyroid gland. Its presence has also been demonstrated in the parathyroid and thymus glands of humans. Total thyroidectomy may, therefore, not remove all cells capable of producing this hormone. Normal serum Ca^{+2} levels appear to cause continuous calcitonin secretion and the level of this hormone is directly responsive to the concentration of Ca^{+2} in the blood. Glucagon also stimulates calcitonin release and gastrin is reported to act similarly. The half-life of calcitonin in the blood is about 2-15 minutes, somewhat shorter than that of PTH.

2. The only site of action of calcitonin appears to be in the bone, where it can cause deposition of Ca^{+2} and PO_4^{-3} (as hydroxyapatite) and thereby rapidly reduce plasma concentrations of <u>both</u> ions. The effect is most profound in young or growing animals and in other conditions where bone remodelling is active. The effect decreases in the postadolescent human and its homeostatic role in the adult man remains unclear. It is conceivable that the main function of calcitonin may be to prevent the occurrence of hypercalcemia during childhood.

3. Calcitonin appears to inhibit bone resorption by regulating both the number and activity of osteoclasts, making this hormone directly antagonistic to the action of PTH. The mechanism of action of calcitonin is unclear, but it does involve either protein synthesis or adenyl cyclase. The parafollicular cells are capable of storing large amounts of calcitonin, in marked contrast to the parathyroid cells which store very little PTH. Both hormones are present in the plasma of adults (despite a marked decrease in the effect of calcitonin in the adult) and both have a linear response to changes in Ca^{+2} concentration. Calcitonin varies <u>directly</u> with Ca^{+2}, while PTH varies <u>inversely</u> with Ca^{+2}. An <u>increase</u> in the concentration of this ion <u>elevates</u> the calcitonin level while <u>decreasing</u> the PTH concentration.

Adrenocorticosteroids and Thyroxine

The effect of cortisone and thyroxine on the bone is rather perplexing. Both are required in low concentrations for normal bone growth. Additionally, low levels of cortisone are clinically useful in correcting hypercalcemia and the hyperosteolytic activity of malignant bone disease and thyrotoxicosis. Yet both agents in large amounts cause bone deterioration, an effect seen in Cushing's syndrome and thyrotoxicosis.

Osteoporosis

In this condition, normal bone material (both matrix elements and hydroxyapatite) are lost. This can proceed to a state where frequent, multiple fractures occur. There are multiple causes for this disorder, giving rise to several different patterns of bone loss. The most common cause is the loss of anabolic hormones with advancing age. While osteoporosis does occur in old men, it is more common in postmemopausal women where decreased levels of estrogens have a bone-wasting effect. The normalcy of the serum alkaline phosphate levels is useful in distinguishing this disease from rickets and osteomalacia.

Magnesium Metabolism

1. Magnesium is the second most abundant cation (after K^+) within cellular fluids. Although it is indispensable to normal health, its exact daily requirement is not known. The recommended adult dietary allowance is about 350 mg/day and an average 70 kg adult contains about 2000 mEq (20 gm) of magnesium. More than 50% of this magnesium is in the bone, complexed with calcium and phosphorous, the rest being present in the soft tissues (skeletal and cardiac muscle, liver, kidney, brain, and interstitial fluid). The plasma levels of magnesium vary between 1.5 and 2 mEq per liter of which 30% is protein bound. The mechanism for the control of magnesium levels are not understood, but PTH and aldosterone may play a role in this process. Once absorbed, magnesium is excreted in the urine. Unabsorbed fecal magnesium, however, represents the major portion of the metal excreted daily.

2. Magnesium is required in many enzymatic reactions, particularly those in which ATP participates. As was indicated on page 157, the complex $ATP^{-4}-Mg^{+2}$ serves as the substrate in these reactions. In some instances, it can be replaced by Mn^{+2} (see pages 156-157). That magnesium is an intracellular cation is shown by the observation that the magnesium concentration within a muscle cell is about ten times greater than the value for plasma magnesium. This unequal distribution between extracellular and intracellular compartments is comparable to that seen for K^+. The relationship between Mg^{+2} and Ca^{+2} is somewhat analogous to that between K^+ and Na^+. Aldosterone regulates the excretion of both K^+ and Mg^{+2} in the kidney.

3. Antagonism between magnesium and calcium can be observed when they are administered in higher concentrations. The occurrence of a rare, genetic, selective non-absorption of magnesium suggests that they are taken into the body by different mechanisms. When the plasma magnesium level is increased to about 20 mg/100 ml, animals become anesthetized with a paralysis of peripheral neuromuscular activity. This action can be reversed by the intravenous administration of a suitable amount of calcium. There is no biochemical explanation available for this reversal of magnesium narcosis by calcium. When the magnesium levels fall below their normal values (hypomagnesemia), tetany, a functional disturbance also seen in hypocalcemia, can result. In humans, magnesium deficiency can result from renal failure or severe diarrhea. It becomes evident if the fluids administered to such patients for rehydration do not contain magnesium. Hypokalemia is associated with hypomagnesemia.

Copper Metabolism

1. Copper is an element which occurs widely. The suggested daily
    requirement for copper is 2.5 mg/day and the average daily diet
    contains anywhere from 2.5 to 5.0 mg of it. An adult individual
    contains a total of 100-150 mg of copper. This is distributed
    so that about 50% of the copper is in muscles, about 20% is in
    the bones, about 18% is in the liver, and the remainder is
    distributed among the red blood cells, the plasma, and other
    tissues. As in the case of iron, most of the copper in the diet
    is lost in the feces and only a small part of it reaches the blood.
    Copper metabolism is also similar to that of iron in that once
    copper is absorbed, it is almost completely retained by the body.

2. The exact role of copper in human metabolism is not clearly
    understood.

    a. It is a constituent of a number of proteins, some of which
        are important enzymes. These include cytochrome oxidase,
        catalase, tyrosinase, monoamine oxidase, ascorbic acid oxidase,
        uricase, and ceruloplasmin. Ceruloplasmin, the serum copper
        binding protein, exhibits oxidase activity toward some
        polyphenols and toward amines, such as epinephrine and
        serotonin.

    b. Animals maintained on a copper deficient diet develop severe
        anemia which resembles the iron deficiency anemias. Although
        copper is normally present in red blood cells, its exact role
        is not yet delineated. About 80% of it is bound to a protein
        known as erythrocuprein. This protein is colorless, contains
        2 atoms of copper per molecule, and has a molecular weight
        of 33,000 daltons. One hundred milliliters of packed
        erythrocytes from an adult human contain about 30-36 mg of
        erythrocuprein to which is bound most of the red blood cell
        copper (93-114 µg of copper/100 ml of packed red blood cells).

    c. It has been postulated that copper may play a role in the
        formation of bone and in maintaining the normal function of
        myelin. Copper is distributed widely in the brain which
        contains copper-binding proteins known as cerebrocupreins.

    The liver, which contains a major fraction of the total body
    copper, possesses a copper-binding protein known as hepatocuprein.

3. In the <u>serum</u>, 98% of the copper is bound to ceruloplasmin, an
    α_2-globulin having a molecular weight of 160,000 daltons. It
    contains 8 atoms of copper per molecule (about half in the form
    of Cu^{+2}) bound probably to histidyl and either lysyl or tyrosyl
    residues. The normal plasma level of ceruloplasmin is about 30 mg
    per 100 ml. The remainder of the copper present in the serum
    is loosely bound to the histidyl and perhaps the carboxyl groups
    of albumin. This may be a metabolically available source of
    copper for tissue use. The albumin-bound copper is known as
    direct-reacting copper because it is readily available for
    reaction with chromogens. This is in contrast to the
    ceruloplasmin-bound metal which requires prior treatment with
    HCl for its release.

    The colorimetric methods for serum copper determination are based
    on treatment of the sample to release protein-bound copper,
    precipitation of the proteins, and reaction of the copper in the
    filtrate with a chromogen. In order of increasing sensitivity,
    diethyldithiocarbamate (yellow complex with Cu^{+2}), cuprizone
    (blue complex with Cu^{+2}), and oxalyldihydrazide (lavender color
    with Cu^{+2} in the presence of NH_4OH and acetyladehyde) are used as
    chromogens. Copper can also be determined by atomic absorption
    spectrophotometry, following its separation from the serum
    proteins. The principle of this method is the selective absorption
    of light by a flame containing ions of a particular metal. A
    solution of the metal being measured is drawn into a flame, where
    it is ionized. Light passing through this flame will have one
    or several wavelengths (energies) absorbed, corresponding to the
    electronic transitions which can occur in the metal ions. This
    method is both sensitive and selective (though subject to some
    interferences), permitting accurate measurement of a number of
    metals including most or all of the alkali metals (Na^+, K^+, Li^+,
    etc.), alkaline earths (Ca^{+2}, Mg^{+2}, etc.), and transition metals
    (copper, zinc, iron, cobalt, etc.).

4. a. <u>Wilson's disease</u> (also known as hepatolenticular degeneration)
        is a rare, autosomal recessive disturbance of copper metabolism.
        In this disease, there are increased amounts of copper in
        various tissues, particularly the liver, brain, kidneys, and
        the cornea of the eye.

    b. The disease is characterized by abnormalities in the brain,
        cirrhosis, excessive excretion of copper in the urine, and
        aminoaciduria. These patients also show greenish-brown
        Kayser-Fleischer rings at the limbus of the cornea.

c. The nature of the biochemical lesion is not completely clear. Decreased synthesis of functionally normal ceruloplasmin appears to be the major defect, however, and the majority of the patients with Wilson's disease have lowered serum ceruloplasmin levels (less than 20 mg/100 ml). The lack of functional ceruloplasmin leads to larger amounts of unbound copper in the plasma which are picked up by the tissues in excessive amounts and accumulated there. There is also excessive urinary elimination of copper.

d. Penicillamine (β,β-dimethylcysteine) is the drug of choice for treatment of Wilson's disease. The structure of penicillamine is

$$\begin{array}{c} CH_3 \\ | \\ H_3C-C-CH-COOH \\ |\quad| \\ HS\ \ NH_2 \end{array}$$

It is a chelating agent for copper, (as well as for mercury, zinc, and lead) which solubilizes these metals and promotes their elimination from the body in the urine.

Other Inorganic Ions of Biological Importance

1. The importance of a number of inorganic substances (minerals) in normal metabolism has already been discussed, both in this chapter (Ca^{+2}, PO_4^{-3}, Mg^{+2}, and Cu^{+2}) and elsewhere throughout the book. These include iron (Chapter 7); iodine and ammonia (Chapter 6); sodium, potassium, and chloride (Chapter 12); bicarbonate (Chapter 12; pages 6-9); and the role of transition metal ions in enzyme function (pages 123-124). Deficiency diseases of metals (other than perhaps iron, copper, the alkali metals, and the alkaline earths) are uncommon, however, because the small amounts required are present in the great majority of diets.

The toxicity of lead, mercury, and other heavy metals was previously indicated (pages 109; 571) as was that of cyanide, azide, and sulfide (pages 105-106; 205; Chapter 7). Thiocyanate ion (SCN^-) was mentioned as a detoxication product of cyanide (pages 105-106; 563) while cyanate (CNO^-) is a drug used in the treatment of sickle-cell disease (page 550). Many other examples of the biological and clinical importance of inorganic compounds and ions are known.

2. There are several other elements which are required for normal body functions, are toxic in some way, or are used as drugs. Several examples are cited below.

   a. Lithium, in the form of its salts, has found particular use in the control of severe mania. This alkali metal has many similarities to Na^+ and K^+, being almost completely absorbed from the G.I. tract and eliminated by renal excretion. Lithium is not treated entirely like sodium and potassium, however, being much more evenly distributed between the intracellular and extracellular fluids. Some of its diverse effects on different organ systems are probably related to this partial replacement of Na^+ and K^+ in bodily processes. Because Li^+ apparently can move across cell membranes more freely than can other electrolytes, it may also act by alteration of the chemical microenvironment of the cell. Its effects probably include increasing the osmolarity and thereby altering the water distribution, and modifying the pH and the ionic atmosphere of proteins and other biomolecules. Cyclic AMP-mediated processes which are regulated by peptide hormones also seem to be particularly disturbed by lithium, perhaps by one of the mechanisms described above. This suggests the usefulness of lithium salts in the management of thyrotoxicosis and other disorders in which there is excessive activation of adenyl cyclase by peptide hormones. Cases of lithium-induced diabetes insipidus have been reported where the metal apparently inhibits the ability of antidiuretic hormone (vasopressin) to synthesize cAMP, needed as a mediator in the control of water metabolism. There have been reports of lithium intoxication, some resulting in death.

   b. Fluorine (as fluoride) is necessary for the prevention of dental caries (decalcification of the tooth enamel resulting from the action of microorganisms on carbohydrates during formation of the teeth). A fluoride concentration in the drinking water of about 0.9-1.0 mg/liter appears to be effective as a prophylactic. On the other hand, mottled enamel (dull, chalky patches on the teeth with pitting, corrosion, and deficient calcification) results from the consumption of water having fluoride in excess of 1.5 mg/liter during the years of tooth development. Fluoride has also been shown to stimulate adenyl cyclase by an unknown mechanism.

c. Selenium is an analogue of sulfur and can replace it in
    cysteine, methionine, etc. Plants grown in soil which has a
    high selenium content concentrate this element and can cause
    selenium toxicity when used as food. Other actions attributed
    to selenium include carcinogenesis in laboratory animals,
    enhancement of tooth decay (if ingested during tooth
    development), and antagonism of magnesium, copper, and
    manganese levels in rats. This last effect was magnified
    when cobalt was also present in the diet. On the other hand,
    smaller amounts of selenium (as selenate, SeO_4^{-2}) appear to
    be necessary for prevention of hepatic necrosis in laboratory
    mice. This duality (requirement of low levels for normal
    growth but toxicity with slightly greater amounts) seems
    relatively common among a number of "trace nutrients".

d. Depending on the local environment, at least 43 different
    elements are normally incorporated into developing teeth and
    another 25 are seen in some instances. The remaining elements
    (notably the heavy metals) have never been detected. At least
    14 different trace elements are demonstrably necessary for
    normal human health. The list of elements which are toxic
    in some circumstances is also quite long. Cadmium, a waste
    product from certain industrial processes, was implicated as
    the cause of itai-itai ("ouch-ouch") disease, a syndrome
    characterized by severe, painful decalcification of bone and
    other tissues. In the 1950's, it afflicted many residents
    of Japan's Jinzu River basin. Lead poisoning has resulted
    from the ingestion of alcohol prepared illegally in vessels
    partly fabricated from lead. Air pollution and the consumption
    of flakes of lead-based paint by children have been implicated
    as causes of lead poisoning in the poorer sections of several
    cities.

    A number of plants concentrate trace elements from the soil
    and water. These include selenium (see above), strontium
    (mesquite beans), sulfate (cabbage), and lithium (wolfberries,
    used by Indians in the Southwest United States for making jam).
    Patients with certain types of heart disease are reputed to
    benefit from treatment with medicinal herbs containing high
    concentrations of trace elements.

    The application of trace metal analysis to clinical problems
    has generally been very limited, but several observations
    have been made which may prove useful in this regard. The
    serum zinc concentration is elevated by physical trauma such
    as operations and burns. The degree of elevation correlates

well with the severity of the injury and the extent of recovery, returning to normal when the stress abates. Serum manganese and nickel reportedly rise just prior to the onset of a myocardial infarct, suggesting a potential diagnostic tool for early stages of this disease.

e.   As levels of metals in the environment continue to rise and as more examples of metal toxicity present themselves to physicians, a greater knowledge of the metabolism of the trace elements will develop. Additionally, there is the problem of radioactive isotopes which are also increasing in the environment.

10

Vitamins

with John H. Bloor, M.S.

General Considerations

1. The name <u>vitamin</u> arose from vita (= life, in Greek) + amine because
    thiamine (antiberiberi factor; vitamin B_1), the first essential food
    factor to be prepared in relatively pure form, was an amine. The
    word is now used to describe any of a rather loosely-defined
    collection of molecules which are needed by the body but which it
    are unable to synthesize in adequate amounts (see also page 379).
    These include the <u>fat-soluble</u> vitamins (A, D, E, and K), the <u>water-</u>
    <u>soluble</u> vitamins (the B-complex and vitamin C), and sometimes the
    essential amino acids (page 428), the essential fatty acids
    ("vitamin F"; see pages 596-597), and the trace elements (see
    Chapter 9). More commonly "vitamins", in human biochemistry (and in
    this book), refer only to the fat-and water-soluble groups as defined
    above. A number of other substances have been found to be essential
    food factors in various non-human species and a group of compounds
    exists which are classified as "vitamin-like". The latter
    substances will be discussed again later in this chapter.

    a. The B-complex vitamins are a group of about twelve chemically
        unrelated compounds which commonly occur together in dietary
        sources. Many are also synthesized by the intestinal flora.
        Of these, eight are distinct compounds (or groups of closely
        related compounds) known to be of dietary significance in man.
        They are thiamine, riboflavin, niacin, pyridoxine, biotin,
        pantothenic acid, folic acid, and cyanocobalomin. Deficiencies
        of the first six of these generally cause disorders of the
        nervous system, mucous membranes and skin, while folate and B_{12}
        deficiencies result in megaloblastic anemia. Synonyms and
        properties of these compounds have been indicated throughout
        the book and will be summarized later in this chapter.

    b. Despite the fact that individual members of the fat- and water-
        soluble vitamins are chemically dissimilar, these two general
        classifications have been retained because the compounds in each
        group have several other properties in common. The fat-soluble
        vitamins are usually absorbed in the intestine, along with
        other lipids, while the water-soluble ones are taken up poorly
        in that region. Significant amounts of the fat-soluble compounds
        are stored in the liver (perhaps as a consequence of their lipid
        solubility), while water-soluble vitamins are not generally
        stored in the body to any great extent.

c. It is sometimes convenient to use one name to designate a group of closely related substances. "Vitamin D" refers to ergocalciferol (D_2), cholecalciferol (D_3), their 25-hydroxy derivatives (25OHD$_2$ and 25OHD$_3$), and 1,25-dihydroxycholecalciferol (see pages 690-695). Similarly, pyridoxine (vitamin B_6) is actually a mixture of pyridoxol (for which pyridoxine is an accepted synonym), pyridoxal, and pyridoxamine while niacin (nicotinic acid) is really a combination of nicotinic acid and nicotinamide. Frequently, there is only one (or at most two or three) <u>active</u> form, however, to which the several members of the group are converted by the body. Thus, 1,25-dihydroxycholecalciferol is considered to be one of the final active metabolites of vitamin D, while the coenzyme forms of vitamin B_6 and niacin are, respectively, pyridoxal phosphate and the two related dinucleotides NADH and NADPH.

2. Many vitamins were discovered through the observation of patients who were suffering from vitamin deficiency diseases. A substance was generally considered to be a vitamin if it was shown to be a normal dietary constituent absent in the diet of the afflicted individual which, when administered in pure form, would relieve the symptoms seen in the patient.

a. For example, <u>scurvy</u> (scorbutus) used to be observed frequently among sailors who ate no fresh fruit or vegetables during long sea voyages. Citrus fruits (for example, limes, hence the name "limey" or "lime-juicer" for British sailors) were known to relieve this disorder. It was later shown that large amounts of ascorbic acid (vitamin C) were present in these foods and that pure ascorbate was effective as an antiscorbutic (hence the name ascorbic acid). This compound will be discussed again later (see also pages 256 and 257). It is rare to encounter the deficiency of a single vitamin, except in instances such as those discussed below, where the dietary supply is adequate but utilization of the vitamin is impaired. Frequently, if one vitamin is lacking, others are also lacking and the intake of protein, trace elements, and other nutrients is probably insufficient. This is particularly true of the members of the B-complex.

b. Vitamin deficiencies are not always caused by dietary insufficiency. Malabsorption syndromes are known where the vitamins are ingested but excreted in the feces. In disorders which cause steatorrhea (e.g.,biliary obstruction), the fat-soluble vitamins are generally poorly absorbed. Cases of congenital vitamin B_{12} deficiency are known which are due to the absence of a protein factor needed for B_{12} uptake in the intestine (discussed later). The intestinal bacteria are capable

of synthesizing some vitamins (notably K and those of the
B-complex). Although the B-complex vitamins are only poorly
absorbed in the intestine, in the absence of a dietary supply
of vitamin K the intestinal flora seem to be capable of providing
a sufficient amount of vitamin K for normal health. Consequently,
loss of these organisms (as in sulfonamide treatment) may cause
the symptoms of vitamin K deficiency disease. Certain drugs
may antagonize vitamin metabolism, resulting in an inadequate
supply for the body's needs. For example, isoniazid (used in
the treatment of tuberculosis) is reported to produce the
symptoms of pyridoxine deficiency, presumably due to
interference with the metabolism of this vitamin (e.g., its
conversion to pyridoxal phosphate, the active coenzyme form).
Another example, already discussed, is the use of folate
antagonists (methotrexate, etc.) in the treatment of neoplasia
(see Chapter 5).

c. To plan diets which contain adequate supplies of the vitamins
and to aid in diagnosing vitamin deficiency diseases, it is
necessary to know how much of these materials are needed by
members of the population. For this purpose, two sets of
standards have been established.

  i) The Minimum Daily Requirement (MDR) is defined as the
      smallest amount of a substance needed by a person to prevent
      a deficiency syndrome. This is also considered to represent
      the basic physiological requirement of the material by the
      body. These values are established by the United States
      Food and Drug Administration. The MDR is not known for
      all vitamins.

  ii) The Recommended Daily Allowance (RDA) is the daily amount
      of a compound needed to maintain good nutrition in most
      healthy people. These values are intended to serve as
      nutritional goals, not as dietary requirements. They are
      defined by the Food and Nutrition Board of the National
      Academy of Science of the United States.

RDA values have also been established for mineral elements,
energy (calories, carbohydrates, and fats), protein, and
electrolytes and water. The MDR for some of these substances
has also been defined. Of these two, the RDA is used more
widely today than is the MDR. The RDA's for the vitamins are
given later, with the individual compounds. Many of these
values are given in International Units (I.U.), which are
empirically defined for each substance. It is appropriate to
note here that, in general, vitamins are needed only in small

quantities. As was pointed out earlier (page 120), many vitamins serve as coenzymes or coenzyme-precursors and, as such, play a catalytic role in body function. Many examples of this have already been noted (pages 121 and 122) among members of the B-complex.

d. In the case of vitamins A and D, overdosage produces hypervitaminosis. In recognition of this, the United States Food and Drug Administration recently ruled that vitamin A in doses greater than 10,000 I.U. and vitamin D in doses greater than 400 I.U. will be available on a prescription-only basis. Additionally, all products which contain quantities of a vitamin in excess of 150 percent of its RDA will be classified as drugs. This allows closer regulation of the vitamins and will permit a more rapid response to any future cases of toxicity which may result from chronic or acute overdosage. A and D hypervitaminosis will be discussed in the sections on these vitamins.

3. As was implied above, certain vitamins can be synthesized by man in limited quantities.

a. Niacin (nicotinamide), one of the B-complex vitamins, can be formed within the body from tryptophan (see pages 500-503). This pathway is not active enough to satisfy all of the body's needs but, in calculating the recommended allowance for niacin, 60 mg of dietary tryptophan can be considered equivalent to one mg of dietary niacin. In Hartnup's disease (see page 503), a rare hereditary disorder in the transport of monoamino monocarboxylic acids (e.g., tryptophan), a pellagra-like rash may appear. This suggests that, over a long period of time, dietary intake of niacin is insufficient for metabolic needs. This pattern also occurs in carcinoid syndrome in which much tryptophan is shunted into the synthesis of 5-hydroxytryptamine (serotonin; see pages 502-503).

b. As described in Chapter 9, the active form of vitamin D (1,25-dihydroxycholecalciferol) can be synthesized by the body provided radiant energy is available for the conversion.

$$\text{7-dehydrocholesterol} \xrightarrow{\text{photons}} \text{cholecalciferol (vitamin } D_3).$$

Since dietary vitamin D is necessary, the amount supplied by this pathway is apparently not adequate for the body's needs.

Fat-Soluble Vitamins

1. Vitamin A

    a. This vitamin is needed for vision, normal growth, reproduction
        (in both males and females), and normal embryonic development.

        i) The earliest sign of vitamin A deficiency is night blindness
            (discussed later). Keratinization and cornification of the
            epithelial cells of all tissues occur, the best-known changes
            in humans being the development of xerophthalmia and "toad
            skin". The latter lesion may also involve B-complex
            deficiencies. The two visual disorders occur independently
            of each other. Males become sterile and, although deficient
            females are able to conceive, the offspring seldom survive
            to term due to placental defects. Failure of skeletal
            growth is also seen early in vitamin A deficiency. Because
            the most characteristic of these symptoms manifest
            themselves only during growth and because of significant
            liver storage of this vitamin, it is less common to see
            the deficiency disease in adults than in children.
            Additionally, several years' supply of vitamin A are stored
            in the liver of a normal adult. This supply is absent in
            neonates and very young children.

        ii) Although its role in vision is fairly well understood, the
            mechanisms by which vitamin A performs its other functions
            remain obscure. The RDA for vitamin A is 5,000 I.U.,
            one I.U. of vitamin A being equivalent to 0.3 µg of
            all-<u>trans</u> β-carotene.

    b. Eggs, butter, cod liver oil, and the livers of other fish are
        the principal dietary sources of retinol (vitamin A_1). In these
        materials, it is present as the free alcohol (shown below) and
        as retinyl esters of fatty acids (primarily palmitic acid).

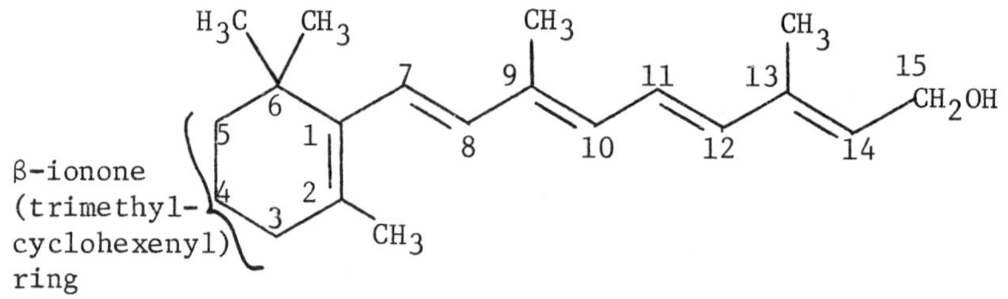

All-<u>trans</u> retinol (Vitamin A_1)

If the alcohol at carbon 15 is oxidized to an aldehyde or an acid, the resulting compounds are, respectively, retinal and retinoic acid. Retinal is the metabolically active form of vitamin A in some processes and can be converted to forms which entirely fulfill the need for the vitamin. The acid, while perhaps being the form needed for growth, is probably a catabolic product, unable to entirely replace the vitamin.

i) In most of the dietary retinol, the four sidechain double bonds are in the <u>trans</u> conformation as shown above. <u>Trans</u> refers to the fact that bonds to adjacent atoms of the chain are on opposite sides of the planar double bond. This type of isomerism is important in vitamin A chemistry. It is further illustrated below, with 2-butene as an example.

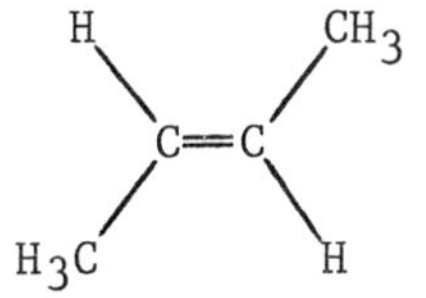

trans-2-butene 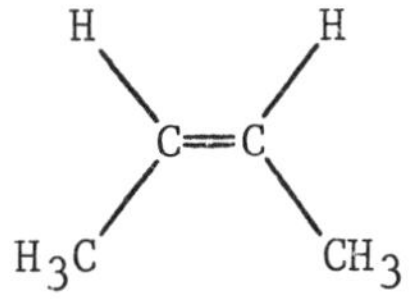   cis-2-butene

In this example, the methyl groups can be on <u>opposite</u> sides of the planar double bond (<u>trans</u> to each other) or on the <u>same</u> side (<u>cis</u> to each other). The double bonds of retinol are readily oxidized by atmospheric oxygen, inactivating the vitamin, unless protected by the presence of antioxidants such as vitamin E.

ii) The retinyl esters are hydrolyzed by one or more specific hydrolases in the lumen and brush border of the intestine prior to their absorption. In the mucosal cells, they are re-esterified and transferred to the lymph. From the lymph, they enter the plasma (probably by the same route as the chylomicra) where they are transported to the liver by a specific retinol-binding protein, one of the α_1-globulins. Retinyl esters are the hepatic storage form of vitamin A and, in an adult human, a year's supply or more may be retained in the liver.

c. Vitamin A can also be derived from a group of plant pigments known as <u>carotenes</u>. These are present in carrots, mangoes, cantaloupe, and other fruits and vegetables (see page 577). One-half cup of canned carrot juice, for example, contains about 20,000 I.U. of vitamin A (four times the RDA). The structure of all-<u>trans</u> β-carotene, the most important of these is shown on the next page.

β–carotene is cleaved
here to produce retinol

All-<u>trans</u> β-carotene

This molecule is cleaved at the indicated bond by β-carotene-15,15'-dioxygenase, a soluble enzyme present in the intestinal mucosa, producing two molecules of retinal. This retinal is then reduced to retinol by NADH or NADPH and absorbed by the route described previously.

d. As was indicated above, retinoic acid is at least partly a catabolic product of retinol and retinal. It can be converted to its β-glucuronide and excreted in the urine or it can be decarboxylated. It has also been shown that retinol is excreted in the urine as its β-glucuronide. These and other metabolic routes of vitamin A are shown in Figure 87. Although it is known that the Δ^{11}-<u>cis</u> isomers of both retinol and retinal are converted to the all-<u>trans</u> forms in the liver, the isomerase enzymes have not actually been studied. Retinal reductase is also found in the retina. The conversion of all-<u>trans</u> retinal to the Δ^{11}-<u>cis</u> isomer can also be accomplished by 380 nm radiation, but the process is slow and inefficient, requiring high levels of illumination.

e. Retinal is the form of vitamin A involved in the visual process. <u>Rhodopsin</u> (visual purple) is a complex of <u>opsin</u> (a protein) with Δ^{11}-<u>cis</u>-retinal. It has an absorption maximum at 498 nm. When rhodopsin absorbs light, the Δ^{11}-<u>cis</u>-retinal is transformed to the all-<u>trans</u> isomer which dissociates from the opsin. In the dark, presumably by a route similar to that shown in Figure 87, the Δ^{11}-<u>trans</u> isomer is regenerated and rhodopsin forms again. These reactions are summarized in Figure 88. Rhodopsin is localized in the <u>retinal rods</u> of the eye which are responsible for the basic detection of light (as opposed to color vision). Night blindness (nyctalopia), due to a vitamin A deficiency, is thus detected by noting an increase in the minimum amount of illumination which can be sensed by the eye (the visual threshold).

Figure 87.   Sources and Metabolism of Vitamin A in the Human Body

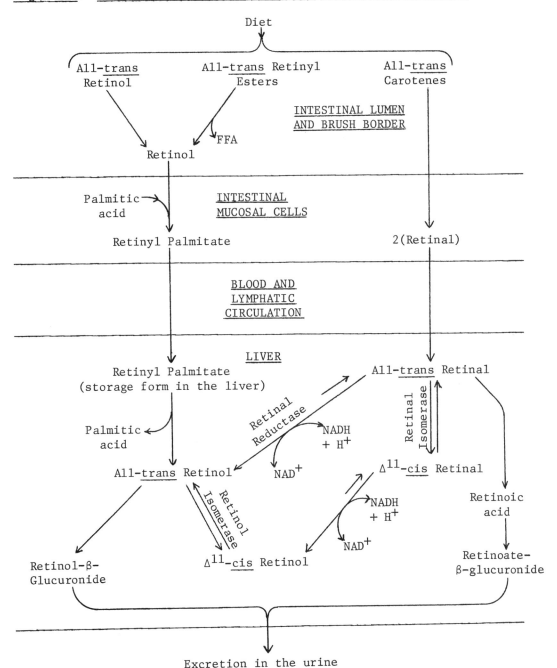

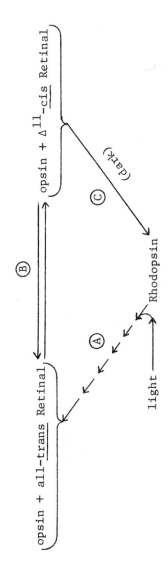

Figure 88.   Chemical Conversions Which Occur During Rod Vision

The equilibrium ((B)) between the two retinal isomers may have retinol as an intermediate (see Figure 87). In vivo, under constant illumination, a steady state is established so that the rate of (A) is equal to the rate of regeneration (reactions (B) and (C)). As indicated by the multiple arrows, process (A) involves several intermediates.

There are three different pigments present in the retinal cones
which are responsible for color vision ("cone vision"). Each
has a different absorption maximum: 430 nm (blue), 540 nm
(green), and 575 nm (red), in accord with the trichromatic
theory of color vision. The pigments all contain Δ^{11}-cis
retinal but appear to differ with respect to the opsin (protein
portion) which they possess. Each of the three types of
hereditary color blindness is due to the absence of a specific
one of these opsins, presumably due to mutation of one of the
three genes coding for these proteins.

f. Hypervitaminosis A, mentioned earlier, is characterized by
    skeletal decalcification, tenderness over the long bones, and
    membrane disruption. In more severe overdosage, the cerebro-
    spinal fluid pressure increases, with severe headache, anorexia,
    nosebleeds, and dermatitis. Since vitamin A esters accumulate
    in the liver, a chronic dosage of 100,000 I.U. per day is
    reportedly able to produce this disorder.

2. Vitamin D

    a. The structure, function, and metabolism of this vitamin are
        discussed in Chapter 9 (Mineral Metabolism). It is needed for
        the proper intestinal absorption of calcium and is thereby
        intimately involved in the actions of parathyroid hormone (PTH)
        and calcitonin (see Chapter 9). It also aids in the regulation
        of phosphate levels by its phosphaturic activity. In the
        Fanconi Syndrome (see page 476), a renal tubular reabsorption
        disorder, the calcium and phosphate losses are treated with
        large doses of Ca^{+2} and vitamin D.

        Several antirachitic agents (compounds which have vitamin D
        activity and are thereby able to prevent rickets) are known,
        especially ergocalciferol, cholecalciferol, and dihydrotachysterol
        which are described in Chapter 9. Fish liver oils (e.g., cod liver
        oil) contain a ketonic sterol, derived from an unknown source,
        which has about twice the potency of the calciferols. The RDA
        for vitamin D is 400 I.U. per day. One international unit is
        the activity present in 0.01 ml of average medicinal cod liver
        oil and is equivalent to approximately 0.05 µg of ergocalciferol.
        The principal dietary sources in the United States are milk
        (which is usually supplemented with vitamin D) and fish liver.
        An unknown but significant amount of cholecalciferol is also
        synthesized within the body from cholesterol. There is
        appreciable storage of vitamin D in the liver.

b. <u>Hypervitaminosis D</u>, mentioned earlier, presents clinically with demineralization and increased fragility of the bones. Serum Ca^{+2} is elevated, resulting in soft-tissue calcification and formation of renal calculi. These symptoms are seen in patients who have received, on a prolonged basis, ten times the amount of vitamin normally required. All forms of vitamin D are toxic in this regard. The adult MDR is not actually known for this vitamin, however, partly due to the difficulty in measuring the amount of active vitamin formed within the body by the action of light striking the skin.

3. <u>Vitamin E</u> (tocopherol)

a. The name tocopherol is derived from the Greek <u>tokos</u> (childbirth) + <u>pherein</u> (to bear) + <u>ol</u> (alcohol) and the classic manifestation of vitamin E deficiency is infertility of both male and female animals.

  i) Seven tocopherols, all derivatives of <u>tocol</u>, have been isolated from natural sources. The most common and most active of these is <u>α-tocopherol</u> (5,7,8-trimethyltocol), shown below.

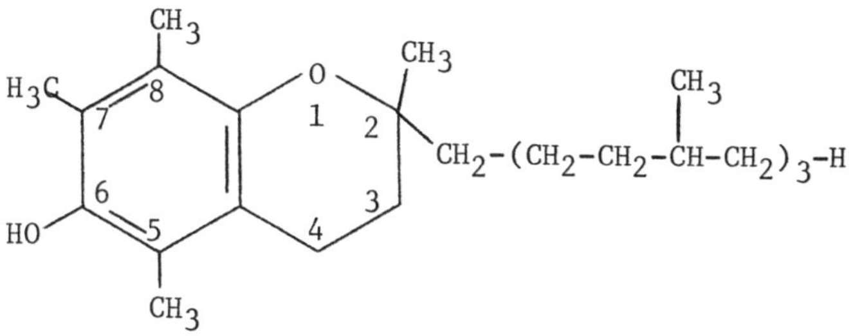

α-tocopherol

  Other tocopherols (β, γ, δ) differ in the number of methyl groups on the aromatic ring.

  ii) The principal sources of the tocopherols are vegetable oils, especially wheat germ oil, legumes, nuts, cereals, and green vegetables are also important sources. Although rich in vitamins A and D, the fish liver oils are devoid of vitamin E. The RDA for vitamin E is 30 I.U.. One I.U. is equivalent to 1 mg of synthetic d,l-α-tocopherol acetate. The potency of d-α-tocopherol acetate, the natural form, is 1.36 I.U./mg while d-α-tocopherol, the free alcohol, has 1.49 I.U. per mg.

iii) Although the metabolism of vitamin E is not well understood, it can be oxidized in man to the product shown below.

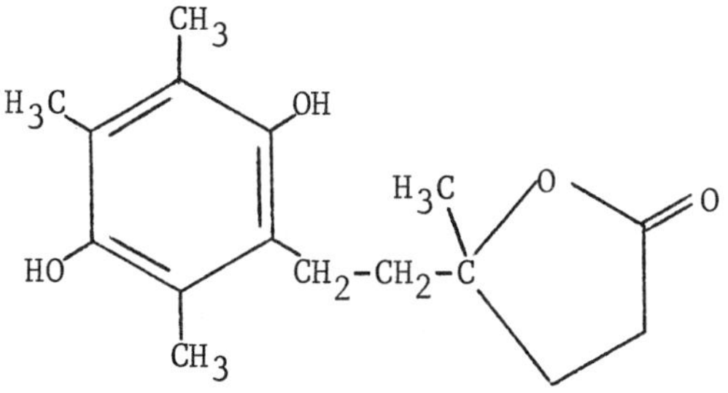

This material has been found in human bile secretions conjugated with two moles of glucuronic acid via the two hydroxyl groups.

b. Despite intensive investigation, the biological role(s) of vitamin E remains unknown. This is at least partly due to the multiplicity of disorders observed in vitamin E deficiency (in animals) and to the way in which various non-vitamin agents can apparently correct some of them.

   i) The role of vitamin E as an antioxidant was referred to in the discussion of vitamin A. The latter material is readily oxidized to an inactive form, a process which can be prevented by the presence of vitamin E. Similarly, E protects dietary essential fatty acids (see page 597) from oxidative degradation, the amount of the vitamin needed being proportional to the quantity of unsaturated fatty acid in the diet. Synthetic antioxidants will replace vitamin E in these processes. It is also interesting that the tissues of vitamin E-deficient animals consume oxygen at a greater rate than do those of normal animals.

   ii) Certain disorders which are prevented only by the tocopherols can be produced in experimental animals by vitamin E deficiency. These include nutritional muscular dystrophy, irreversible degeneration of rat testicular tissue, and anemia in monkeys.

   iii) Selenium is apparently involved in some aspects of vitamin E metabolism, perhaps in its absorption from the intestine. This has been studied in several species of animals (including the rat, the lamb, and the calf) but not in humans.

iv) Membrane changes which are reversed by tocopherol administration have been noted in cases of vitamin E deficiency. These include increased fragility of the erythrocytes in the presence of peroxides and dialuric acid (5-hydroxy barbituric acid) and changes in membranes of the intestinal mucosal cells. In the latter situation, several types of cellular and subcellular membranes lost their ability to be stained by osmium tetroxide. This compound presumably binds to sites of unsaturation (e.g., unsaturated fatty acids) in organic molecules.

v) Several cases of vitamin E deficiency in humans have been reported. Some of these were associated with familial abeta-lipoproteinemia (acanthocytosis; see page 666). A deficiency syndrome has recently been described in low-birth-weight infants who are fed an artificial formula in which the ratio of vitamin E to unsaturated fatty acids is low (milligrams of α-tocopherol to grams of polyunsaturated fatty acid is less than 0.6).

c. There seem to be two major hypotheses regarding the part which vitamin E plays in nutrition. One maintains that it is largely a non-specific antioxidant. The other concedes the vitamin's role as an antioxidant but believes that another, more specific function (perhaps as a coenzyme) exists.

4. <u>Vitamin K</u>

a. This group of compounds (whose name derives from the initial letter of the German word Koagulation) is needed for normal blood clotting and no characteristics other than impaired blood clotting have been noted in vitamin K deficiency. Although they are related to coenzyme Q and have been shown in the mycobacteria to play a role similar to that of CoQ in man (see page 200), the K vitamins do not appear to be involved in oxidative phosphorylation in higher organisms.

b. There are three series of K vitamins; the phyloquinones (exemplified by K_1); the menaquinones (for example K_2); and the menadiones (e.g., K_3). All are derivatives of 1, 4-naphthoquinone, shown below, which itself has some vitamin K activity.

1,4-naphthoquinone

i) The <u>phyloquinones</u> all have the basic structure

Vitamin K_1, found in alfalfa, spinach, cabbage, and other
green vegetables, has n=3. In this compound, the
unsaturated sidechain is known as a phytyl group (see page 596).

ii) The <u>menaquinones</u> are variations on the structure

Vitamin K_2 (n= 4) is synthesized by intestinal bacteria and
other microorganisms and is found in animal tissue. The
intestinal flora are an important source of this vitamin
(see earlier in this chapter) and may be instrumental in
preventing deficiency diseases caused by low dietary
intake. Human milk contains 15 µg/liter while cow's milk
has 60 µg/liter, making milk an important source of this
vitamin, especially in infants.

iii) The menadiones are synthetic compounds. Of these,
vitamin K_3 (menadione), shown below, is the most important.
It is three to four times as potent as vitamin K_1.

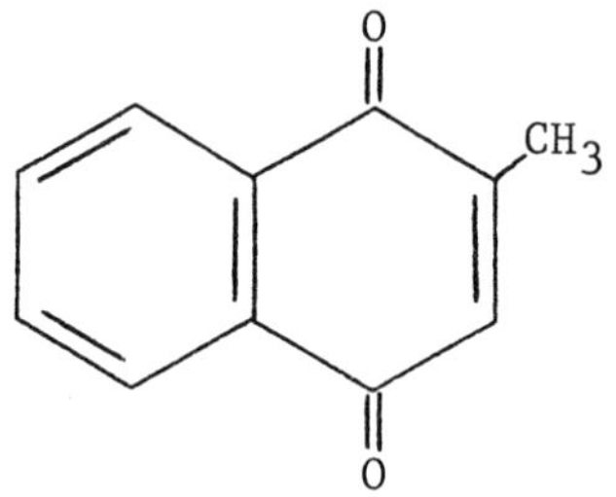

menadione

Note that all of the vitamins K actually contain
menadione as a nucleus.

c. An RDA for vitamin K has not been established because of
inadequate information regarding daily human intake and the
contribution of the intestinal bacteria to this supply. As
with the other fat-soluble vitamins, anything which causes
steatorrhea or poor lipid absorption in the intestine may also
cause some degree of vitamin K deficiency, but few cases of this
have been reported in humans. The symptoms center around
clotting abnormalities, notably hypoprothrombinemia and
hemorrhagic disorders. Menadione and its derivatives, given
in excess, reportedly produced kernicterus in premature infants
and hemolytic anemia in rats. The mechanism for this is not
understood. Consequently, vitamin K_1 is the preferred form for
neonates and pregnant women.

d. Vitamin K is required for the normal formation of clotting
factors II (prothrombin), VII (proconvertin), IX (Christmas
factor), and X (Stuart factor). It is apparently involved in
synthesis of the protein portions of these factors at the
ribosomes (either in translation or chain completion) or in
the post-ribosomal attachment of the carbohydrate groups (see
Chapter 5). Vitamin K is not involved directly in the
clotting process. Two human kindreds have been reported which
have an increased affinity for vitamin K at the liver site for
promoting clotting factor synthesis. This results in an overproduction of the substances in these individuals.

e. A number of compounds antagonize the action of vitamin K,
    thereby acting as anti-coagulants. The most important of these
    are the coumarin-related materials, including warfarin (an
    important rodenticide) and dicumarol. The structures of these
    compounds are shown below.

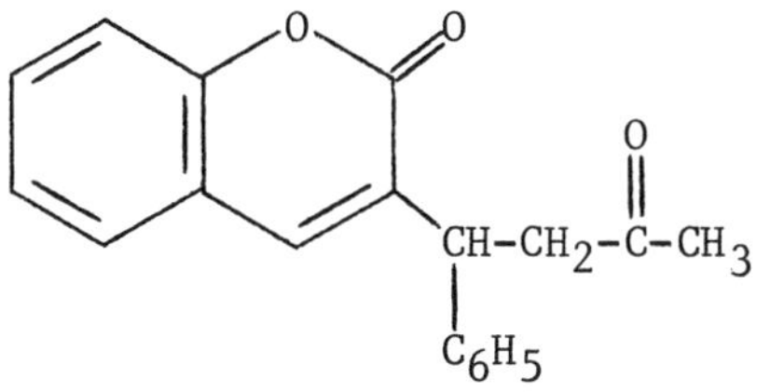

Warfarin

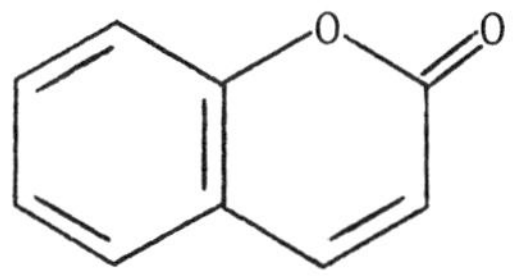

Coumarin

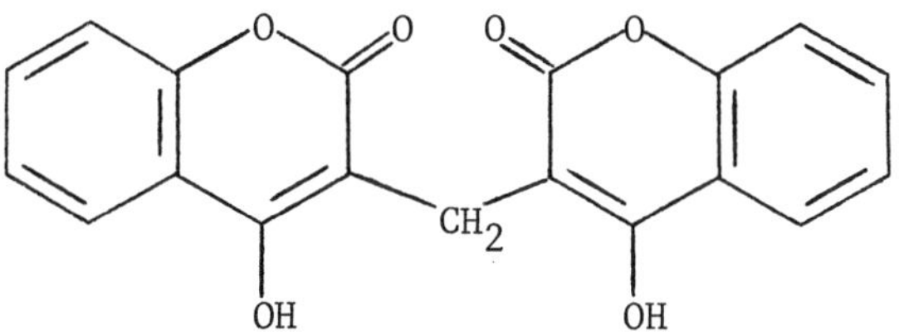

Dicumarol (Bishydroxycoumarin)

Dicumarol is the toxic substance in "spoiled sweet clover"
disease, a hemorrhagic condition which occurs in cows.

i) The mechanism of action of the coumarins is not known.
    Some evidence suggests that they act by decreasing the
    binding of vitamin K to albumin, thereby preventing its
    transport in the blood. Other workers have detected
    abnormal forms of prothrombin and factor IX (two of the
    clotting factors; see Chapter 11) in the blood of
    vitamin K-deficient individuals and in persons treated
    with anticoagulants. It was concluded that these
    compounds function by disturbing the vitamin-mediated
    synthesis of the factors. Whatever the mechanism,
    factor X is depressed most severely while II, VII, and IX
    are decreased to a lesser extent. The effect of the
    coumarins can be totally reversed by sufficiently high
    doses of vitamin K, suggesting a direct competition
    for some active site or other binding site.

ii) Many drugs apparently modify the activity of the coumarins
and, in turn, the coumarins can alter the metabolism of
other compounds. Substances which <u>potentiate</u> the
anticoagulant effect may decrease the albumin binding of
the coumarins, thereby increasing the concentration of
the free molecules in the plasma (e.g., phenylbutazone);
decrease the rate of coumarin metabolism by inhibiting
the liver microsomal oxidase system (e.g., chloramphenicol
and phenyramidol) and perhaps, in some cases, decrease
the synthesis of normal clotting factors. The last
mechanism has yet to be clearly demonstrated. Compounds
which <u>antagonize</u> the coumarins, thereby enhancing
coagulation, generally seem to function by accelerating
coumarin metabolism. Most notable and potent among this
group are the barbiturates, which apparently stimulate
the hepatic microsomal oxidase system, thereby increasing
the rate of coumarin hydroxylation. Other coumarin
antagonists are glutethimide (a hypnotic) and griseofulvin
(an antifungal agent). Coumarins potentiate the activity
of diabinese (chlorpropamide, a hypoglycemic agent),
dilantin (diphenylhydantoin, an anticonvulsant), and
orinase (tolbutamide, a hypoglycemic). They are also
reported to potentiate the activity of phenobarbital,
although both compounds are catabolized by the microsomal
oxidase system in the liver. An unknown mechanism of
interaction is seemingly involved in this instance.

iii) The area of drug-drug interactions is of great practical
importance in medicine. It is discussed generally in
Chapter 14 and other examples are cited in appropriate
places in this book.

Water-Soluble Vitamins

The majority of these compounds have already been covered in appropriate
places throughout the book. Reference to these discussions will be
given below with each vitamin. All apparently serve as coenzymes or
coenzyme-precursors in one or more biological reactions. It is
important to notice that the B-complex vitamins are generally obtained
from the same sources (whole grain cereals, meats, yeast, wheat germ)
and that deficiency of any one of these is usually accompanied by
deficiency of the entire group as well as of protein.

1. Vitamin B_1 (<u>thiamine</u>; aneurine; cocarboxylase) is the antiberiberi
(antineuritic) factor referred to earlier in this chapter. The
active coenzyme form is thiamine pyrophosphate (TPP; see pages
171-173).

a. It is required for a variety of reactions such as

   i)   oxidative decarboxylations of α-keto acids:
        pyruvic acid $\longrightarrow$ acetyl-CoA (pages 171-172)
        α-ketoglutarate $\longrightarrow$ succinyl-CoA (page 177)
        branched-chain amino acid catabolism (pages 465 and 468).

   ii) transketolase reaction (page 244) and related processes.

   iii) non-oxidative decarboxylations of α-keto acids:
        pyruvate $\longrightarrow$ acetaldehyde + CO_2 (in yeast).

b. Principal dietary sources of this vitamin are fish, lean meat
    (especially pork), milk, poultry, dried yeast, and whole grain
    cereals. Bread, cereals, and flour-based products are frequently
    enriched with this material. The RDA is 1.5 milligrams and
    much of the American population is reported to consume less than
    this amount.

c. Beriberi, the deficiency disease in man, is characterized by
    weakness and muscle wasting. "Dry" and "wet" beriberi are
    distinguished by the occurrence of generalized edema and prompt
    response to thiamine administration in the "wet" form.
    Additionally, the dry type is accompanied by neurological
    disorders and the wet form exhibits acute cardiac symptoms.
    Clear-cut thiamine deficiency is seldom seen in Western man
    except in chronic alcoholics (Wernicke's Syndrome). Certain
    thiamine analogues such as pyrithiamine and oxythiamine can
    antagonize the functioning of thiamine, thereby causing
    symptoms of thiamine deficiency.

2. Vitamin B_2 (<u>riboflavin</u>) is incorporated into flavin mononucleotide
    (FMN) and flavin adenine dinucleotide (FAD). Both of these
    function, as part of flavoproteins, in electron-transfer (redox)
    reactions. They are discussed on pages 175 and 178. As was
    pointed out there, these are not truly flavin nucleotides, since the
    bond between the sugar and the flavin ring is not a glycosidic
    one. Enzymes which contain FAD as a prosthetic group include
    D-amino acid oxidase, glucose oxidase, xanthine oxidase, NADH and
    NADPH dehydrogenases (such as NADPH-glutathione reductase),
    succinate dehydrogenase, and acyl-CoA dehydrogenase. FMN is part
    of the L-amino acid oxidase found in the kidney, "old yellow enzyme"
    (involved in glucose-6-phosphate oxidation; first flavoprotein
    discovered), and other dehydrogenases. FAD appears to be more
    commonly used than FMN as a cofactor.

The riboflavins are ubiquitous in plant and animal tissues. Particularly good dietary sources are liver, yeast, and wheat germ; while milk and eggs are important when consumed in sufficient quantities. Leafy green vegetables also provide this vitamin, and cereals and bread are frequently enriched with B_2. The RDA is 1.7 milligrams.

No specific symptoms are associated with riboflavin deficiency (ariboflavinosis), partly because it is difficult to produce uncomplicated B_2 deficiency. Although cheilosis (red, swollen, cracked lips), a magenta-colored tongue, seborrheic dermatitis, and congestion of the conjunctional blood vessels have been observed apparently as a result of riboflavin deficiency, some or all of these are also caused by other dietary inadequacies, notably of niacin and iron. The most sensitive bodily indicators for B_2 deficiency are the erythrocytes; and a normal person contains 20 µg/100 ml of the vitamin in the blood.

3. Vitamin B_6 (pyridoxine) is actually three closely related compounds.

Pyridoxol	Pyridoxal	Pyridoxamine
(for the alcohol group in position four)	(for the aldehyde group in position four)	(for the amine group in position four)

Pyridoxine designates the entire group as well as being synonymous with pyridoxol. The active form in the body is pyridoxal-5-phosphate, which seems to be transiently converted to pyridoxamine-5-phosphate in functioning as a coenzyme (see pages 444-445). This coenzyme is generally important in reactions involving α-amino acids. Examples of this include the formation of N^5,N^{10}-methylene FH_4 from FH_4 and serine (pages 383-384), nicotinic acid synthesis (tryptophan metabolism, pages 501-502), cysteine metabolism (page 472), and heme synthesis (formation of δ-aminolevulinic acid from glycine and succinyl-CoA, page 565, 570). Its role in transaminase reactions is discussed on pages 444-445. Pyridoxol is the form usually present in plants while the aldehyde and amine are animal products. The three compounds are readily interconverted by the body.

a. The RDA for vitamin B_6 is 2.0 milligrams, although the actual requirement increases with an increase in dietary protein and perhaps with age. The principal dietary sources are whole grain cereals, wheat germ, yeast, meat, and egg yolk. Although no deficiency syndrome specifically due to a lack of pyridoxine is known in man, rats and monkeys maintained on a B_6-free diet develop dermatitis and exhibit neuropathological changes. Infants having an inherited requirement for increased amounts of the vitamin have convulsions if the supply is inadequate. This symptom can be relieved by either pyridoxine or γ-aminobutyric acid (GABA) suggesting that such persons have an abnormal glutamate decarboxylase which requires high B_6 concentrations for normal activity (see page 462). The reaction involved is

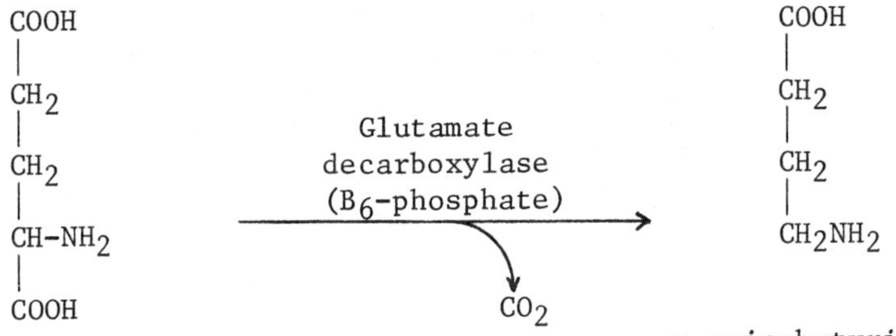

glutamic acid

Glutamate
decarboxylase
(B_6-phosphate)

CO_2

γ-aminobutyric acid
(a neurotransmitter)

b. Pyridoxine antagonists such as deoxypyridoxine and isonicotinoyl hydrazide (isoniazid, a tuberculostatic drug) are capable of producing B_6-deficiency symptoms when administered to humans.

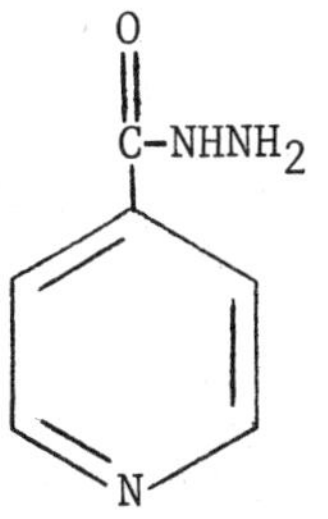

isoniazid

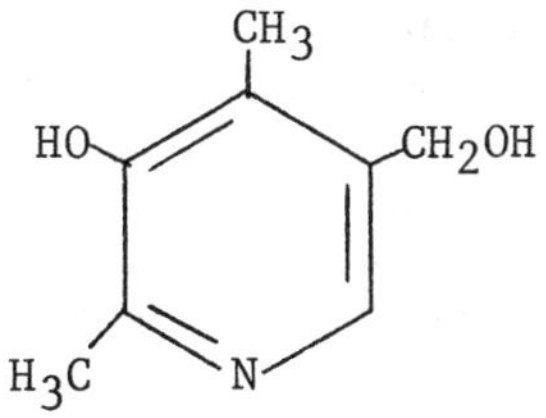

deoxypyridoxine

Although most of these symptoms (e.g., nausea, seborrheic dermatitis, glossitis, and polyneuritis) are associated with other disorders, deoxypyridoxine also causes oxaluria (due to impairment of glycine metabolism, page 456) and excretion of xanthurenic acid (caused by a lack of kynureninase activity, page 501). Roughening of the skin and a pyridoxine-responsive anemia have also been reported in some cases of pyridoxine deficiency.

4. Niacin (the generic name for nicotinic acid and nicotinamide as well as a synonym for nicotinic acid) is a B-vitamin by virtue of its occurrence in meat, eggs, yeast, and whole grain cereals in conjunction with other members of the B-complex. The RDA is 20 milligrams but, since a limited amount can be synthesized from tryptophan by the human body (see pages 500-503), the actual requirements depend on the amount and type of dietary protein. Its principal known role in the body is as part of NADH and NADPH, both dinucleotide coenzymes involved in glycolysis, fatty acid metabolism, and a number of other oxidation-reduction reactions. The structure and functions of these compounds are given on pages 158-160 and their synthesis from nicotinic acid is outlined on page 501. The structures of nicotinic acid and nicotinamide are shown below.

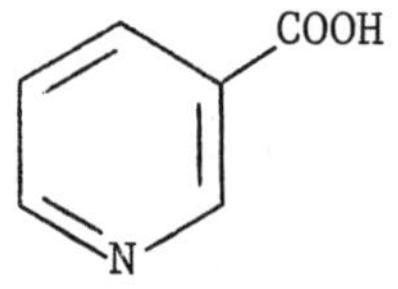

niacin 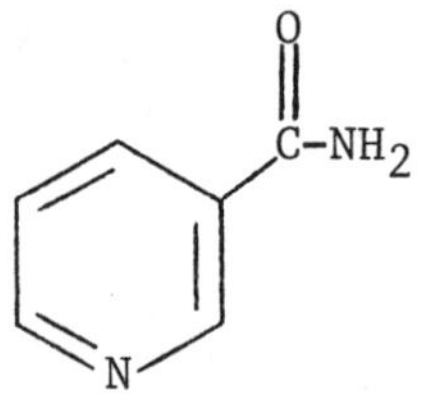   nicotinamide

As with most of the other B-vitamins, it is difficult to pinpoint symptoms due specifically to a deficiency of niacin. Pellagra, generally attributed to a lack of this vitamin, probably reflects an insufficiency of several members of the B-complex. Symptoms include dermatitis (pellagra rash), stomatitis, abnormalities of the tongue and of digestion, and diarrhea. These are difficult to correlate with the known function of the vitamin as a precursor to redox coenzymes, suggesting some other roles for niacin. In Hartnup disease, an inherited disorder of amino acid transport (see page 503), a pellagra-like rash may be seen. Presumably this occurs because the dietary supply of niacin is inadequate for the body's needs in the absence of tryptophan. Carcinoid syndrome, in which tryptophan is diverted to the formation of large quantities of 5-hydroxytryptamine (see pages 502-503), presents a similar clinical picture.

5. <u>Pantothenic acid</u> (Pantoyl-β-alanine) is ubiquitous in plant and animal tissues, being especially abundant in those materials which are rich in other B-vitamins. Its RDA in man is 10 milligrams. In the body, it is an essential precursor to the synthesis of coenzyme A (CoA, CoASH; see page 174), used to carry acetyl groups (acetyl-CoA) in a wide variety of two-carbon-requiring reactions. CoASH is also the carrier of fatty acyl groups (Chapter 8), succinyl groups (pages 177 and 570), and other carboxyl-containing materials (as in amino acid metabolism, pages 465-468). Pantothenic acid is also part of the "swinging sulfhydryl arm" of the fatty acid synthetase complex (page 644). The structure of pantothenic acid is shown below.

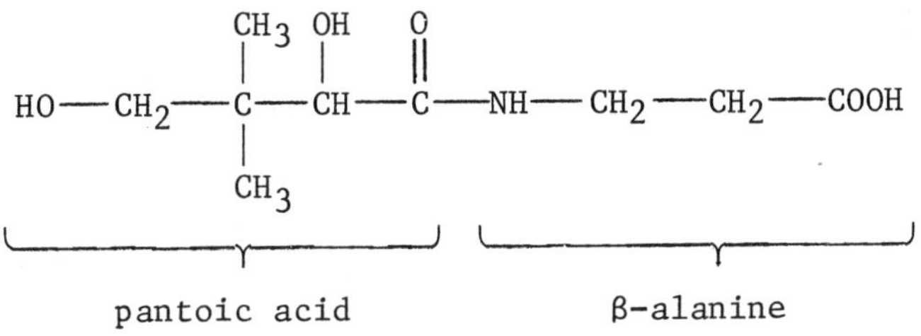

$$HO-CH_2-\underset{\underset{CH_3}{|}}{\overset{\overset{CH_3}{|}}{C}}-\underset{\overset{|}{OH}}{CH}-\overset{\overset{O}{\|}}{C}-NH-CH_2-CH_2-COOH$$

pantoic acid   β-alanine

Pantothenic Acid (Pantoyl-β-Alanine)

Uncomplicated pantothenic acid deficiency due to dietary restriction alone is probably not known in man. When the antivitamin ω-methylpantothenic acid is administered, symptoms similar to those seen in animals occur. The significant biochemical changes are insensitivity of the adrenal cortex to ACTH (resulting in adrenal insufficiency), increased insulin sensitivity, and the acceleration of erythrocyte sedimentation. Neuromuscular degeneration is also observed and antibody synthesis in response to an antigen to which the individual is known to be immune is sluggish. If the person is deficient in both pyridoxine and pantothenic acid, antibody formation is completely abolished.

6. <u>Biotin</u> is another member of the B-complex. It is widely distributed in natural products with beef liver, yeast, peanuts, kidney, chocolate, and egg yolk being especially rich in it. The RDA is 0.3 milligrams although no MDR has been established. As with the other B-vitamins, the intestinal flora synthesize biotin and, in fact, the total daily urinary and fecal excretion of the vitamin usually exceeds the daily dietary intake.

a. Biotin is the coenzyme in a number of carbon dioxide fixation
    reactions. Examples of this are given below.

   i) $CO_2 + NH_3 + 2$ ATP $\longrightarrow$ carbamyl phosphate $+ 2$ ADP $+ P_i$
       (pages 412, 416-417, and 453).

  ii) Acetyl-CoA $+ CO_2 \longrightarrow$ malonyl-CoA   (pages 641-642).

 iii) Propionyl-CoA $+ CO_2 \longrightarrow$ methylmalonyl-CoA   (pages 639
                                                              and 467).

  iv) Pyruvate $+ CO_2 \longrightarrow$ oxaloacetate   (page 79).

The mechanism by which biotin is believed to mediate the CO_2
attachment is shown on page 185. An unusual carboxylation,
for which biotin is apparently <u>not</u> required, is the addition
of carbon 6 to the purine ring (see page 389).

b. The structure of biotin is shown below. In <u>oxybiotin</u>, which is
    capable of substituting for biotin in most species; the sulfur
    is replaced by oxygen.

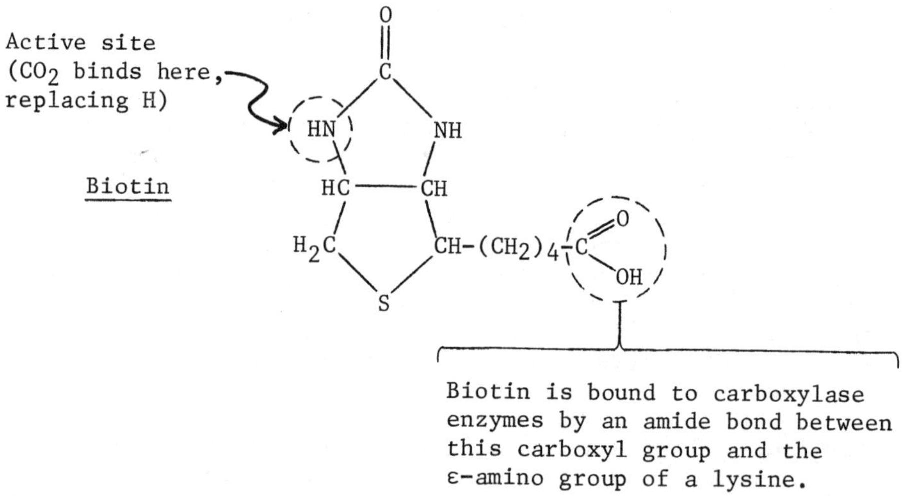

Active site
(CO_2 binds here,
replacing H)

Biotin

Biotin is bound to carboxylase
enzymes by an amide bond between
this carboxyl group and the
ε-amino group of a lysine.

The biotin-carboxylase (biotin-enzyme) is formed in two steps:

   biotin $+$ ATP $\longrightarrow$ biotinyl-5'-adenylate $+ PP_i$

   biotinyl-5'-adenylate $+$ apocarboxylase $\longrightarrow$ holocarboxylase
                                                             $+$ AMP

Proteolysis (in vivo and in vitro) releases biocytin, also known as ε-biotinyllysine. Biocytinase, in the blood and liver, cleaves this molecule to biotin and lysine.

c. Biotin deficiency occurs in man when large amounts of raw egg white are consumed. Avidin, a protein of about 70,000 molecular weight is present in egg white and binds strongly ($K_{assoc.} = 10^{15}$) to biotin, preventing its absorption by the body. Avidin is readily inactivated by heating. It contains four identical subunits, each having 128 amino acid residues and each capable of binding one molecule of biotin. This binding is abolished by heat and other denaturing influences. Sterilization of the intestine (as with sulfonamide treatment) has also been known to cause the biotin-deficiency syndrome and biotin antimetabolites have a similar effect. The symptoms of biotin deficiency initially include a scaling dermatitis, lassitude, anorexia, muscle pains, and depression. As the condition progresses, nausea, anemia, hypercholesterolemia, and changes in the electrocardiogram are observed.

7. Folic acid was discussed extensively in the section on one-carbon metabolism (pages 371-386). Its active coenzymic form is folacin (FH_4, tetrahydrofolic acid). It is found in many foods, especially the glandular meats, leafy green vegetables, and yeast. Although folic acid is stored in the liver (unlike most of the water-soluble vitamins), the RDA for folate is 400 milligrams. This reflects the fact that, while the MDR is probably about 50 milligrams, the amount of utilizable vitamin in food varies, absorption from the G.I. tract may be incomplete, and some of the folic acid is destroyed by cooking.

a. Many dietary sources contain pteroylpolyglutamic acid which is hydrolyzed to pteroylglutamic acid (folic acid) and free glutamic acid by the body. Failure of the body to perform this hydrolysis may produce folate deficiency. Other causes of the deficiency state may be increased demands by the body (as during pregnancy, hemolytic anemia, leukemia, and Hodgkin's disease) or excessive excretion of the vitamin. Folacin antimetabolites (e.g., methotrexate, pages 380 and 395) are also well known. Sulfonamides (pages 380-381 and 397) function as antibacterials by interfering with folate synthesis in susceptible organisms. Sprue, a syndrome characterized by a sore mouth and gastrointestinal disturbances including periodic diarrhea and steatorrhea, is associated with folate deficiency. Megaloblastic anemia (appearance of the red cell precursor in the bone marrow changes) frequently results from either folate or vitamin B_{12} deficiency, but the relationship between these two vitamins is not clear. This point is further discussed in the next section, on vitamin B_{12}.

b. As was suggested previously, folacin functions as a carrier of one-carbon fragments in biosynthetic reactions. Some important examples are summarized below.

i) Glycinamide ribosyl-5-$\circled{P}$ + N^5,N^{10}-methenyl FH_4

FH_4

N-formyl-glycinamide ribonucleotide ⟵

(see page 389)

ii) 5-Aminoimidazole-4-carboxamide ribonucleotide + N^{10}-formyl FH_4

FH_4

5-formamidimidazole-4-carboxamide ribonucleotide

(see page 389)

iii) dUDP + N^5,N^{10}-methylene FH_4 ⟶ Thymidylic Acid

FH_4

(see page 414)

iv) Homocysteine + N^5-methyl FH_4 ⟶ Methionine

FH_4

(see page 471)

8. Vitamin B_{12} (cyanocobalamin) is the last member of the B-vitamin
    complex. Animals and higher plants cannot synthesize B_{12}, although
    a number of animal tissues are able to concentrate it, making lean
    meat, liver, sea food, and milk important dietary sources of this
    vitamin. Microorganisms are responsible for the de novo synthesis
    of the B_{12}. The RDA for cyanocobalamin is 6.0 micrograms.

    a. The structure of cobalamin is shown on page 732. The corrin ring
        system, composed of four pyrrole rings linked by three methene
        groups and one direct carbon-carbon bond, is similar to the
        porphyrin system discussed in Chapters 4 and 7. In the
        commercially available form of the vitamin, the R-group
        (indicated in the structure by (R)) is usually CN^-, hence the
        name cyanocobalamin. This is probably an artifact of the
        isolation procedure, however, and cyanocobalamin does not occur
        naturally in microorganisms. In B_{12}-coenzyme (cobamide
        coenzymes), (R) is 5'-deoxyadenosine (adenosyl B_{12}) or a methyl
        group (methyl B_{12}), depending on the reaction in which the
        coenzyme participates. Both of these compounds are unusual among
        biomolecules because they contain carbon bonded to a metal (in
        this case cobalt). This is termed an organo-metallic bond
        (see page 122). In the coenzymes, the adenosyl group is
        donated by ATP while the methyl group probably comes from
        N^5-methyl FH_4. The 5,6-dimethylbenzimidazole ring is sometimes
        replaced by 5-hydroxybenzimidazole, adenine, or other similar
        groups. Although the cobalt is shown in a +1 oxidation state,
        it is readily oxidized to +2 and +3. Hydroxocobalamine
        ((R) = OH^-; has been isolated from mammalian tissue) contains
        cobalt +3, while methyl, adenosyl, and, cyanocobalamin all have
        cobalt +1. B_{12s}, the precursor to the coenzymic forms, also
        contains coablt +1. In Clostridium tetanomorphum, each of the
        two reduction steps ($Co^{+3} \rightarrow Co^{+2}$ and $Co^{+2} \rightarrow Co^{+1}$) is catalyzed
        by a separate, NADH-dependent reductase system.

    b. The B_{12} coenzymes are involved in at least two and probably
        three or more reactions in mammalian systems and at least two
        reactions in bacterial systems.

        i) Methylmalonyl CoA-mutase requires adenosyl B_{12} as a
            coenzyme in catalyzing the reaction.

$$COOH \qquad\qquad\qquad\qquad COOH$$
$$| \qquad\qquad\qquad\qquad\qquad |$$
$$H_3C-CH \qquad \longrightarrow \qquad H_2C-CH_2$$
$$| \qquad\qquad\qquad\qquad\qquad |$$
$$O=C-S-CoA \qquad\qquad\qquad O=C-S-CoA$$

L-methylmalonyl-CoA   Succinyl-CoA

5,6-dimethyl-benzimidazole

Cobalamin Structure.   If R= -CN, then the compound is cyanocobalamin (Vitamin B$_{12}$).

The L-methylmalonyl-CoA is derived from propionyl CoA (see pages 638-639) in the reactions

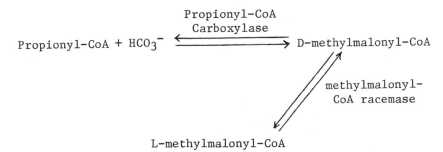

It also arises from the catabolism of valine (see page 467).

ii)   Although it has not been conclusively demonstrated, it is probable that methyl B_{12} is required as a coenzyme by thymidylate synthetase. This enzyme catalyzes the reaction (see also pages 414 and 416-417).

$$dUMP + N^5,N^{10}\text{-methylene-FH}_4 \longrightarrow TMP + FH_2$$

iii) The remethylation of homocysteine to methionine (page 471) requires methyl-B_{12} as a coenzyme. The reaction uses N^5-methyl FH_4 as the methyl donor.

iv)   Adenosyl B_{12} was originally identified as a cofactor for the enzyme catalyzing the conversion of glutamic acid to B-methylaspartic acid in C. tetanomorphum. This reaction is similar to the methylmalonyl-CoA mutase reaction but it has not been shown to occur in mammals.

v)   In various species of Lactobacillus, the deoxynucleoside triphosphates are formed by reduction of the nucleoside triphosphates in a reaction catalyzed by an adenosyl-B_{12} coenzyme-requiring enzyme. This is different from the nucleoside diphosphate reductase system (see page 422) which does not require a B_{12} coenzyme. The triphosphate reduction system has been demonstrated only in the lactobacilli.

c. Although B_{12} deficiency in man is not known to be caused by inadequate dietary intake of the vitamin, several disorders have been described which seem to result from its malabsorption in the alimentary tract or from an inability of the body to utilize the vitamin once it has been absorbed.

The most characteristic result of B_{12} (and folate) deficiency is <u>megaloblastic anemia</u>. In this disorder, the red cell precursors in the bone marrow are mildly enlarged and the nuclear chromatin is altered in appearance (megaloblastic change). Erythroid hyperplasia is also apparent in the bone marrow. Advanced cases also show leukopenia and thrombocytopenia of the peripheral blood. Vitamin B_{12} deficiency also results in neurological damage which is not seen in persons lacking only folate.

Megaloblastic anemia is diagnosed from a bone marrow aspirate. When this disorder is caused specifically by a deficiency of gastric intrinsic factor (IF, needed for B_{12} absorption; discussed below), it is termed <u>pernicious anemia</u>. In classic pernicious anemia, the megaloblastic changes and IF absence are accompanied by a complete lack of gastric HCl (achlorhydria) and the occurrence of atrophic gastritis. This name is somewhat misleading since "pernicious" is a general term meaning "tending to a fatal conclusion". As more is understood about the mechanism of B_{12} absorption, it now appears that some cases of pernicious anemia may also result from a defective intrinsic factor or other heritable errors in the absorption process (see below).

Pernicious anemia is diagnosed by the <u>Schilling test</u> and by analysis of gastric aspirates. In the Schilling test, vitamin B_{12} containing radioactive cobalt is administered orally and the amount of radioactivity appearing in a 24-hour urine specimen is determined. A normal person will excrete about 1/3 of the label in this period while less than 8% is found in the urine of a patient with classic pernicious anemia.(a positive Schilling test). The test is repeated several days later, this time including intrinsic factor with the B_{12}. If the defect is one of factor synthesis or an abnormal IF, the B_{12} uptake should be normal in the second test. The megaloblastic anemia of folate deficiency is usually indistinguishable from that of B_{12} deficiency except with respect to the Schilling test.

i) Two types of <u>ethylmalonic aciduria</u> (see page 468) have been described. One type, responsive to the administration of adenosyl-B_{12} or large doses of vitamin B_{12}, is believed to be caused by a defect in deoxyadenosyl transferase, the enzyme which transfers a deoxyadenosyl group from ATP to cobalamin during adenosyl-B_{12} synthesis. The other type, which is <u>not</u> relieved by vitamin B_{12}, is probably due to an abnormal methylmalonyl-CoA mutase enzyme. Both defects are apparently inherited as autosomal recessive

traits but there are insufficient data to state this
definitely. Propionicacidemia (misleadingly called
ketonic hyperglycinemia because of its clinical picture)
is a defect of propionate metabolism distinct from those
producing methylmalonic aciduria. It is caused by a
defect in propionyl-CoA carboxylase.

ii)   Another case of methylmalonic aciduria, accompanied by
      defects in sulfur-containing amino acid metabolism, has
      been described. Although definitive data are lacking,
      this could be due to an abnormality in an enzyme catalyzing
      the formation of a precursor common to both methyl-B_{12}
      and adenosyl-B_{12}. The reductase enzymes have been
      suggested as possible sites for the mutation. An
      alternative hypothesis which has not been ruled out is
      that the error is one in folate metabolism, since these
      two coenzymes are closely related (see below).

iii) Several hereditary B_{12} malabsorption disorders are also
      known. These are characterized by megaloblastic anemia,
      an inability to take up orally-administered vitamin B_{12},
      and responsiveness to parenteral B_{12}. The normal route
      of B_{12} absorption is shown in Figure 89.

      Gastric juice intrinsic factor (IF), isolated from humans,
      is a mucoprotein of molecular weight about 50,000. It
      usually contains about 13% carbohydrate and is secreted by
      the gastric parietal cells along with HCl. Transcobalamin
      II is a β-globulin of about 35,000 molecular weight. Its
      primary function appears to be the transport of newly
      absorbed B_{12} into the circulation. Transcobalamin I has
      a molecular weight of roughly 121,000 and migrates with
      the α-globulins during serum electrophoresis. Most of
      the circulating B_{12} is bound to this protein, suggesting
      that transcobalamin I is principally a carrier molecule
      for B_{12} within the body. The binding constant of
      transcobalamin I for B_{12} is also greater than that of
      transcobalamin II for the vitamin. The transport of
      B_{12} across the ileal mucosal membranes is an active,
      energy-requiring process.

      Defects have been reported in which no IF is secreted into
      the gastric juice (pernicious anemia; see above).
      Recently, another type of pernicious anemia, caused by
      secretion of a defective IF, has been described. The IF
      in this patient will bind B_{12} normally but it is unable,
      even when administered to normal individuals, to attach

Figure 89.   Normal Route for the Absorption of Vitamin B_{12} from the Diet

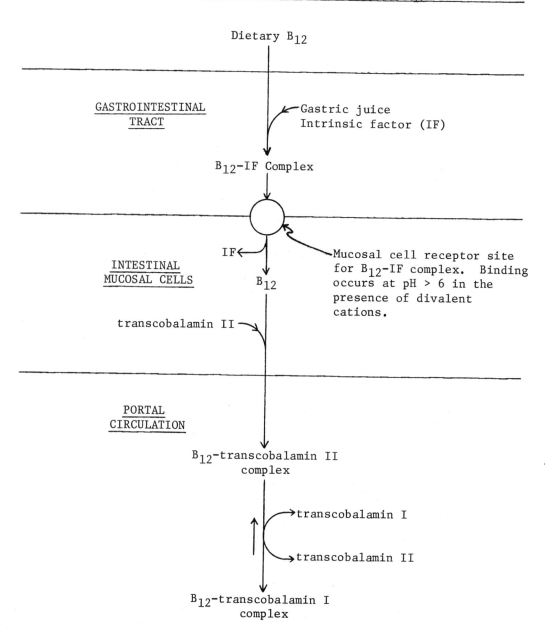

to the ileal mucosal receptor site and transfer the B_{12}
to the circulation. Still a third etiology for B_{12}-
malabsorption pernicious anemia appears to involve
defective mucosal cells. In these cases, the B_{12}-IF
complex is formed and binds to the mucosal cell receptor
sites but the B_{12} cannot pass through the cells and into
the circulation. Certain disorders and blockages of the
terminal ileum can also cause B_{12}-malabsorption and lead
to the symptoms of pernicious anemia. These can usually
be distinguished from a true lack of IF occurrence of a
positive Schilling test even with added factor. This
is not absolute, however, and a Schilling test should
not be used alone for diagnosis.

d. Folate and B_{12} metabolism appear closely related, although just
    how this is so is not clear. A deficiency of either vitamin
    produces megaloblastic anemia and the symptoms of B_{12}-deficiency
    anemia are reversed by large doses of folate while B_{12} ameliorates
    folate-deficiency anemia. (Folate does not reverse the
    degeneration of the long tracts of the spinal cord which occurs
    in B_{12} deficiency, however, and may even worsen it. Consequently,
    folate should not be used clinically to treat B_{12} deficiency and
    care should be taken to accurately determine the cause of
    megaloblastic anemia prior to the initiation of treatment.)

    Additionally

    i)   A lack of either folate or B_{12} causes increased urinary
         excretion of formiminoglutamic acid (FIGLU; see pages 384
         and 464) in the histidine-loading test.

    ii) Patients with a B_{12} deficiency have elevated plasma
         levels of N^5-methyl FH_4.

    iii) Liver stores of B_{12} are decreased in individuals
         who are lacking in folic acid.

    Although the increase in plasma N^5-methyl FH_4 could be explained
    by blockage of the thymidylate synthetase pathway, the other
    effects are not so readily understood. Much remains to be
    learned about the functioning of these coenzymes in the body.

9. Vitamin C (ascorbic acid) was discussed earlier in this chapter.
    Its structure is shown on page 561. It is a vitamin in man, other
    primates, and guinea pigs because these species lack the enzyme which
    converts L-gulonolactone to 2-ketogulonolactone (see pages 256-257).
    All other species investigated are able to synthesize ascorbate.

The principal dietary sources of this vitamin are fresh fruits
(notably citrus fruits), tomatoes, leafy green vegetables, and
new potatoes. Cooking (but not freezing) destroys (by oxidation)
a portion of the ascorbate in these foods. The RDA for vitamin C
is 60 milligrams, an amount which is probably well in excess of
the MDR.

The vitamin C-deficiency disease, scurvy, presents a variety of
symptoms, almost all of which can be related to the formation of
defective collagen. The anemia of scurvy may result from a
defect in iron or folate metabolism as a result of the vitamin C
deficiency. The role of ascorbate as a cofactor for protocollagen
hydroxylase is discussed on pages 360-361 and the collagen formed
in scorbutic patients is reportedly low in hydroxyproline. Despite
this, it has yet to be definitively shown that ascorbate is
absolutely required as a cofactor in this or any other reactions
in the human body. Vitamin C is also involved in the conversion
of p-hydroxyphenylpyruvic acid to homogentisic acid (phenylalanine
metabolism; see page 478) and probably in other hydroxylations and
in the microsomal electron-transport system.

Vitamin-Like Substances

As was mentioned early in this chapter, several compounds are apparently
required in the diet despite the fact that pathways for their synthesis
are known in the body. Such a situation could conceivably arise if
the pathways do not provide an adequate supply for the needs of the
cell (due to low substrate concentrations, rapid degradative processes,
or an inefficient enzyme system) or if the material cannot be readily
transported from the site of synthesis to the place of action.

   a.   The structure of choline (N,N,N-trimethyl-β-hydroxyethylamine)
        is given on page 609. It is an important constituent of
        phospholipids (lecithin is phosphatidyl choline) and of acetyl-
        choline, a neurotransmitter substance. It can be completely
        synthesized from serine in a pathway shown on page 617
        (Figure 72). Note, however, that this synthesis only occurs
        when the serine is present as phosphatidyl serine and is
        dependent on an adequate supply of amino acids (and, hence,
        of protein). Betaine (N,N,N-trimethylglycine) readily replaces
        choline in the diet for all species. Choline is utilized by a
        salvage pathway (page 619). In the lung, this salvage route
        is the principal one for the synthesis of the lecithin needed
        as a surfactant to prevent alveolar collapse (see page 613).
        It has not yet been demonstrated that choline is an essential
        dietary nutrient in man although it is necessary in some other
        species.

b. <u>Inositol</u> (1,2,3,4,5,6-hexahydroxycyclohexane) can occur in a
    number of different isomeric forms. <u>Myo</u>-inositol (<u>meso</u>-inositol),
    depicted on page 611 (see also page 132) is an important
    constituent of phospholipids. It is the only one of the
    inositols with biological activity. Inositol hexaphosphate
    (phytic acid) is found in avian erythrocytes where it binds to
    hemoglobin, thereby regulating the oxygen capacity of the blood
    (see pages 529-533). Although inositol deficiency has been
    shown in laboratory animals, the dietary importance of this
    compound in humans has not been demonstrated.

c. <u>Lipoic acid</u> (pages 171, 174-175; other reactions involving
    thiamine pyrophosphate) and <u>ubiquinone</u> (coenzyme Q; page 198)
    are sometimes classified as <u>vitamin-like</u> materials. In man,
    however, it appears that synthesis of these compounds by the
    body is adequate to meet all normal needs of the cells.

11

Immunochemistry

*with Yoshitsugi Hokama, Ph.D.**

This section on the immunological system of host defense is an integral
part of human biochemistry and physiology. Its consideration in this
text will be limited, and more extensive discussions are available in
any textbook of immunology. This chapter will encompass the
immunoglobulins of serum proteins and the cells and tissue systems
associated with host defense or surveillance against deleterious agents,
particularly those of an infectious nature. It includes

    a.   The immune system: the central and peripheral lymphoid systems,
including the thymus-dependent (T) and thymus-independent (B)
lymphocytes, spleen, lymph nodes and their relationship to
T- and B-lymphocytes (the cell-mediated and humoral systems,
respectively);

    b.   Immunoglobulins (IgA, IgE, IgM, IgG, IgD): isolation and
properties, structure, nomenclature, and functions in disease;

    c.   Phagocytes of the peripheral blood and the reticulo-endothelial
system;

    d.   The complement and kinin systems of serum and their functions
in host reponses.

The Immune System

1.   The central lymphoid system

Currently, the immune system is thought to consist of two major
central lymphoid systems. In mammals, these are the thymus-dependent
(cell-mediated) system and the thymus-independent or bursa-equivalent
(humoral) system. Unlike birds (Aves), which have the bursa of
Fabricius (located near the rectal region of the gastrointestinal
tract and composed of B-cell lymphocytes associated with
immunoglobulin synthesis), man has no such specific organ. Instead,
in mammalian systems, one speaks of thymus-independent or bursa-
equivalent lymphocytes. Since these cells perform the same functions
in mammals as do the bursal lymphocytes in birds, the term
B-lymphocyte is retained. The bone marrow, appendix, tonsils, and

---

* Professor of Pathology, University of Hawaii School of Medicine.

lymphoid tissues of the intestines (GALT: gut associated lymphoid tissues) have been suggested as possible bursa-equivalent tissues (B-lymphocyte regulators) in mammals.

The origin and development of immunocompetent T- and B-lymphocytes, and their relationship to each other and to various facts of the body are outlined in Figure 90. As indicated, both T- and B-cells originate from the same sources (yolk sac, fetal liver, and bone marrow) during embryogenesis. The functions of these two types of cells are summarized in Table 29.

Table 29. Functions of the Plasma Cells (Derived from the B-Cells) and T-Cells, The Members of the Two-Component Lymphoid System

Plasma Cells	T-Cells
Humoral immunity	Cell-mediated immunity
Immunoglobulin synthesis	Homograft rejection
Production of the wide diversity of antibodies	Delayed hypersensitivity reactions
Immediate hypersensitivity reactions	Graft versus host (GVH) reactions

a. T-Lymphocytes

T-lymphocyte precursors from bone marrow, known as pre-thymic cells, seed the thymus. There they become thymocytes and trigger the development of the thymus. The thymocytes proliferate from the cortex of the thymus at a rapid rate and move inward to the medullary region where they undergo maturation. These mature thymocytes enter the circulating pool of lymphocytes and are then called T-lymphocytes. They have a relatively long half-life. In addition to comprising 70-80% of the circulating lymphocytes of the blood, T-cells are found in the perivascular region of the white matter of the spleen (Malpighian corpuscles), the parafollicular and deep cortical regions of the lymph node, the thoracic duct, and the bone marrow. The T-lymphocytes are a heterogeneous group of cells with diverse functions. They differ in turnover rates, circulating routes, and surface antigenic markers. In the blood and the thoracic duct lymph, there are two populations of small lymphocytes. One population is formed at a slower

Figure 90.   Genesis of the Lymphoid Immune System

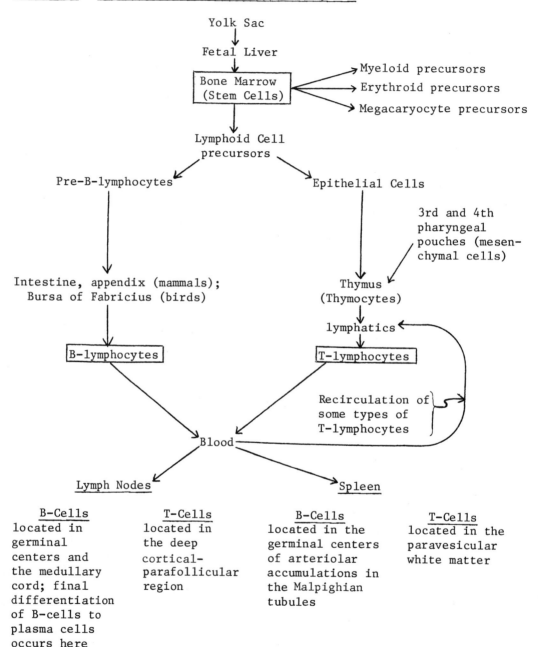

rate, is long-lived (having a half-life of 100-200 days), and moves back and forth between blood and lymph. The other is produced rapidly (2 to 3 mitotic divisions per day) and has a circulating life span of less than two weeks ($t_{1/2}$ of 3 to 4 days). Although the thymus and bone marrow are the major sites for these short-lived lymphocytes, other lymphoid centers can also produce them.

Mesenchymal cells derived from the 3rd and 4th pharyngeal pouches are also present in the thymus. These cells may contribute to the maturation of the bone marrow pre-thymic precursors of T-lymphocytes into mature T-lymphocytes. The products of this maturation are ultimately distributed to the spleen, lymph nodes and other lymphoid tissues which serve as T-cell sources in mammals. This action of the mesenchymal cells may occur by hormonal mediation.

b. B-Lymphocytes

B-lymphocyte precursors are also of bone marrow origin. They are either distributed directly to spleen, lymph nodes, and other lymphoid tissues or may first traverse or be regulated by the appendix, intestines (GALT), or tonsils in order to be functional B-lymphocytes. Evidence for B-cell control similar to that shown for birds is not as convincing in the mammalian system as yet, although the intestine, appendix, tonsils, and bone marrow have been suggested as the possible bursa-equivalent tissues. In addition to the B-lymphocytes in the peripheral blood (20 to 30% of circulating lymphocytes of varying sizes), these cells are also found in the germinal centers of lymph nodes, the lamina propria of the gastro-intestinal tract, secretory glands, bone marrow, appendix, and other lymphoid tissues associated with the reticulo-endothelial system.

2. The peripheral lymphoid system

a. B-Lymphocytes

   i) The peripheral lymphoid system which is associated with the immune system in the mature individual (capable of immune responses) resides primarily in the peripheral lymphoid tissues. These consist of the lymphatics, the spleen, lymph nodes, lymphoid tissues of the gastrointestinal tract, the respiratory tract, secretory glands, and the lymphocytes of the blood.

ii)   As indicated earlier, the B-lymphocytes are situated in
      the germinal centers of lymphoid follicles within the
      lymph nodes and spleen and the plasma
      cells in the medullary regions of the lymph nodes and red
      pulp of the spleen. The distribution of B- and T-cells
      within a lymph node is illustrated in Figure 91. Plasma
      cells, the final cellular product of B-lymphocyte
      development, are involved in antibody (Ab) or specific
      immunoglobulin synthesis. These cells may be found
      scattered throughout the lymphoid systems, including the
      bone marrow. The sequence of B-cell transformation to
      plasma cell via lymphoblast, following antigenic
      stimulation, is shown later, in section 3. Plasma cells
      found in endothelial areas of the respiratory tract,
      nasal cavities, secretory glands, and gastro-intestinal
      tract are associated generally with the secretion of
      IgA and IgE. These immunoglobulins are consequently
      referred to as secretory immunoglobulins.

iii) B-lymphocytes and the process of their conversion to plasma
      cells play a significant role in humoral immunity.
      B-lymphocytes contain the light chains of the
      immunoglobulins bound to their membrane surfaces, while
      the plasma cells, in turn, ultimately secrete the
      immunoglobulin. This system provides defenses against
      viral and bacterial infections by synthesis of specific
      antibodies (immunoglobulins) against the antigens (Ag) of
      the infectious agents. The antibodies interact with
      their specific antigens (a homologous reaction) to give
      antigen-antibody complexes. The nature of these complexes
      varies according to the size and shape of the antigen
      molecules. Visibility (in vitro) of the complexes formed
      is dependent also on the concentrations of the reacting
      antigen and antibody. Extreme concentrations of antigen
      or antibody may obscure visible aggregation, even though
      complexes (soluble) have formed or antigen-antibody
      interactions have occurred. Specific antibodies reacting
      with soluble antigens such as proteins, carbohydrates, and
      viruses, form visible complexes. Such reactions are
      referred to as precipitation, the antibody is referred
      to as a precipitin, and the antigen as a precipitinogen.
      On the other hand, specific antibodies reacting with
      antigens in suspensions (for example, bacteria or
      red blood cells), results in an agglutination reaction.
      The antibody is termed an agglutinin and the antigen
      is an agglutinogen.

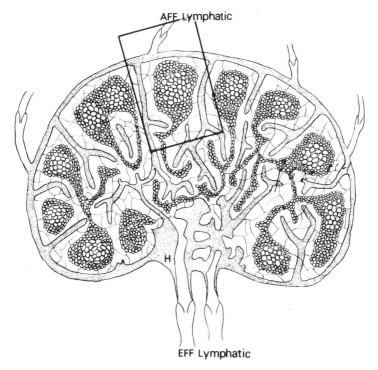

AFF Lymphatic

EFF Lymphatic

Figure 91. <u>DETAILED DIAGRAM OF LYMPH NODE</u>
F(G)C-Follicular (Germinal Center) Cells, H-Hilus;
LS-Lymphatic Sinus; MC-Medullary Cord; P-Para-
follicular Cells; RC-Reticular Cells; T-Trabeculae; AFF = afferent;
EFF = efferent

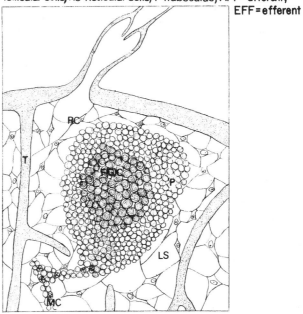

iv)   The humoral immune system has been associated too with
immediate hypersensitivity in man. In many instances,
this type of response to an exogenous or endogenous
antigen is deleterious to the host. One type of immediate
hypersensitivity, referred to below as an <u>atopic</u>
allergy, has been attributed to the immunoglobulin named
IgE. This immunoglobulin has a high affinity for mast
cells and basophils of blood. When an antigen interacts
with an IgE molecule which is bound to a mast cell, lysis
of the cell occurs with the release of granules which
contain histamine and slow-reacting substance of
anaphylaxis (SRS-A). SRS-A initiates a classical
systemic anaphylaxis or an immediate skin reaction. The
immediate type of hypersensitivity associated with IgG
and exogenous antigens (resulting in antigen-antibody
complexes) can also trigger the release of histamine,
acetylcholine, serotonin, anaphylatoxin, and bradykinin.
These pharmacologically active agents affect the smooth
muscles of blood vessels, collagen-containing tissues,
leukocytes, and various other tissues, producing
histological patterns characteristic of early acute
inflammation.

v)   A variety of factors, including both immunological (atopic)
and non-immunological (non-atopic) stimuli, appear to
activate attacks of <u>bronchial asthma.</u> To explain this
clinical entity, Szentivanyi proposed the beta-adrenergic
theory, depicted in Figure 92.

The bronchial system is affected by the catecholamines.
Those affecting the α-receptor, such as norepinephrine,
induce bronchoconstriction and those affecting the
β-receptor (e.g., epinephrine) induce bronchial relaxation.
Under normal circumstances, homeostasis exists between
these two influences. The β-receptor is linked to adenyl
cyclase, an enzyme which catalyzes synthesis of cyclic-
3',5'-AMP (cAMP) from ATP in the presence of Mg^{+2} ions
(see pages 223-228). It has been postulated that patients
with bronchial asthma, eczema, rhinitis, etc., have
deficiencies in adenyl cyclase in the cells of these
various tissues, since the key to suppression or release
of histamine and SRS-A is cAMP. Diminished levels of
cAMP due to insufficient synthesis via adenyl cyclase or
hyperactive degradation (to AMP) via the enzyme
phosphodiesterase, would increase the amount of histamine
(H) and SRS-A released in response to the presence of
IgE-Ag complexes, by infections, or by other agents.

Figure 92. Postulated Mechanism for the Allergic Release of Histamine
and Slow-Reacting Substance of Anaphylaxis (SRS-A)
and the Influence of the Catecholamines Epinephrine and
Norepinephrine

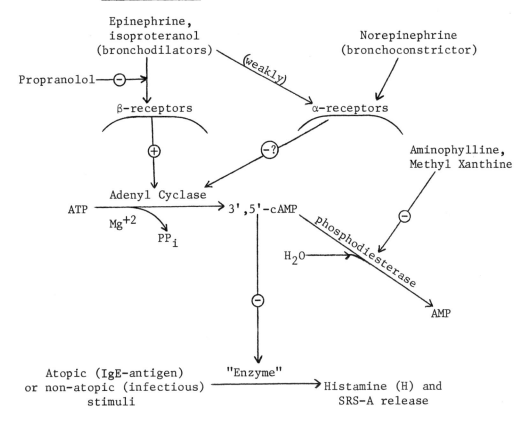

Note: ⊖ = inhibitory (suppressive) effect on the indicated process.

⊕ = stimulating effect on the indicated process.

Norepinephrine is known to have an effect opposite to that of
epinephrine but it remains to be shown (indicated by ⊖?) that
it acts directly on adenyl cyclase. Cyclic AMP inhibits a
postulated enzyme in this process (see also cAMP in Chapter 4).

Experimentally, suppression of adenyl cyclase by
propranolol (a β-adrenergic blocking agent; antagonizes
the action of epinephrine) along with an increase in
α-receptor stimulation will suppress cAMP production,
leading to increased histamine and SRS-A release following
either atopic or non-atopic stimuli. Isoproterenol, which
stimulates β-receptors, together with aminophylline or
methylxanthine (which inhibit phosphodiesterase) would
increase the levels of cAMP and thus cause suppression
of histamine and SRS-A release. Cyclic AMP appears to
suppress the activation of an "enzyme" which initiates
the release of H and SRS-A. Evidently, IgE-Ag complexes
and non-atopic stimuli cause H and SRS-A release by
activating this "enzyme".

vi)   The role of cAMP in immunobiology is presently being
investigated. It appears that cAMP suppresses antibody
synthesis when given in large doses prior to antigen
administration, whereas antibody stimulation is seen when
cAMP is administered after antigen injection.

vii) The B-lymphocytes of the lymph nodes of mammals and the
peripheral blood of man can be functionally differentiated
from T-lymphocytes by their response to non-specific
mitogens (substances which cause mitosis). This is
summarized below. A positive response (indicated by a +)
consists of an increased rate of mitosis and, consequently,
of DNA synthesis.

Mitogens		B-Lymphocytes	T-Lymphocytes
Pokeweed mitogen	(PWM)	+	±
Phytohemagglutinin	(PHA)	−	+
Concanavalin A	(ConA)	−	+
Purified Lipo- polysaccharide	(LPS)	+	−

T-lymphocytes respond to PHA (enhanced by adherent cells --
accessory cells -- or macrophages) and ConA, with no
response to LPS and questionable response to PWM.
B-lymphocytes respond only to PWM and LPS stimulation.
These responses are generally assayed by determining the
amount of ^{3}H-thymidine incorporated into the new DNA
synthesized in the lymphocyte culture following stimulation
by the mitogen in question. This is done using cells in
short-term cultures (generally 2 to 7 days). The ^{3}H-
thymidine uptake is determined in the scintillation

spectrometer following precipitation of the cellular material with trichloroacetic acid, appropriate washing with alcohol and ethyl ether, and solubilization of the precipitate in the scintillation solution.

b.   T-Lymphocytes

   i)   T-lymphocyte function is associated with <u>cell-mediated-immunity</u> (CMI). The distribution of these cells has been discussed in earlier paragraphs. They have been implicated in primary transplantation rejection, delayed hypersensitivity, surveillance against neoplastic transformation, and protection against viruses and bacterial infections. T-lymphocytes do not transform to plasma cells nor do they elaborate immunoglobulins, but they do have specific antigens on their membrane surfaces (the θ antigens of mouse T-lymphocytes, for example) and HLA antigens (histocompatibility lymphocyte antigens).

  ii)   Sensitized T-lymphocytes must be in direct contact with target antigens for their activation. This interaction initiates the release of a variety of factors including

    (1)   <u>mitogenic factor</u>: a non-specific factor which presumably recruits previously quiescent lymphocytes in the circulation;

    (2)   <u>migration-inhibition-factor (MIF)</u>: stabilizes or retains macrophages in the inflamed area (this observation is currently used widely for the <u>in vitro</u> determination of cell-mediated-immunity);

    (3)   <u>lymphotoxin</u>: a factor released by sensitized lymphocytes for the destruction of target cells such as those of a tumor or bacteria;

    (4)   <u>transfer factor (TF;</u> molecular weight 20,000 daltons): presumably recruits lymphocytes and appears to be specific; recently, TF has been utilized for immunotherapy in patients with lepromatous leprosy, mucocandidiasis, and in specific immune deficiency diseases such as Wiskott-Aldrich Syndrome;

    (5)   <u>chemotactic factor</u>: attracts macrophages and polymorphonuclear cells to the area of inflammation;

    (6)   <u>interferon</u>: an agent reacting against viruses;

(7) a factor associated with DNA synthesis.

These lymphocytic factors have collectively been called
lymphokines. Examples of the classical type of delayed
hypersensitivity are the tuberculin reaction and the
contact dermititis caused by poison ivy. With the possible
exception of the transfer factor and interferon, most of
the factors released in the cell-mediated immune responses
have not been chemically characterized. In part, this is
due to the small amounts of these factors released at any
one time. Evidence for their occurrence is derived
primarily from biological assays and observations. Thus,
whether a factor contributes to one or more of the
observed functions of the T-lymphocytes remains to be
determined.

3. A-cell-T-cell-B-cell Interactions

a. Recent evidence suggests that, for some antigens, antibody
synthesis occurs only after the occurrence of T-B lymphocyte
interactions. Aggregated human IgG (AHGG) is an example of a
natural antigen which requires T-cell intervention. This can
be depicted as follows:

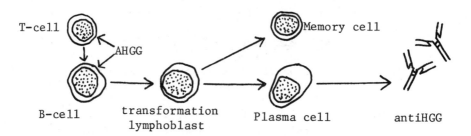

The memory cells stores for future reference the knowledge of
how to make antiHGG. On the other hand, antigens such as
lipopolysaccharide (LPS) of gram-negative organisms like
Escherichia coli require no T-cell intervention. They can act
directly on B-cells resulting in plasma cell formation with
subsequent antiLPS synthesis.

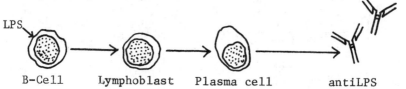

Pneumococcal type-specific capsular polysaccharides behave in a manner similar to LPS. Whether any interaction of T-lymphocytes with other cells are required for cell-mediated responses has not been thoroughly examined.

b. Particulate antigens, such as sheep erythrocytes (SRBC) and bacteria, must initially be digested to appropriate antigenic subunits ("activated" antigens) before T-B-lymphocyte interaction can occur for antibody synthesis. This requires the intervention of a third cell type called the accessory-cell (A-cell). These cells are also referred to as adherent cells because of their ability to attach themselves to surfaces such as glass. They are macrophages and have been shown to enhance antibody synthesis by providing "activated" antigens. This can be depicted as follows:

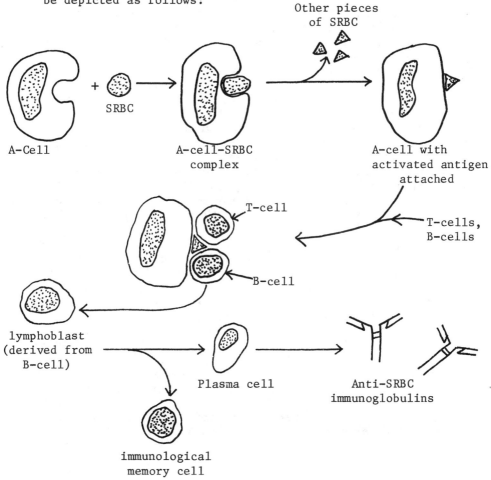

Chemistry of the Immunoglobulins

Immunoglobulins are elaborated by plasma cells following the
transformation of antigen-stimulated B-lymphocytes. As indicated
earlier, this constitutes the humoral immune system. The genesis
of the humoral immune system and the lymphoid tissues involved in this
development have been presented earlier in the chapter. Discussions
here will be confined to (1) the isolation and properties of
immunoglobulins, (2) their biological functions, (3) nomenclature and
classification based on the recommendations of the WHO (World Health
Organization), and (4) the structures of these molecules and the
theories which account for antibody diversity.

1. The isolation and physicochemical characterization of immunoglobulins

  a. The solubility properties of the immunoglobulins in salt
      solutions (especially the salting-out effect; see Chapter 2)
      were utilized in early procedures for their isolation from serum
      proteins. The solubility of any non-electrolyte solute in an
      aqueous solvent is decreased by the addition of a neutral salt.
      This effect is observed only when the dielectric constant of
      the solute is less than that of water. Ammonium sulfate has
      been most frequently used, though other salts such as sodium
      sulfate, sodium chloride, potassium chloride, potassium
      phosphate, and calcium chloride have also been employed. Two
      factors account for the popularity of ammonium sulfate: (1) its
      salting-in action at low ionic strength is less than that of
      calcium, potassium, and sodium chloride, and (2) the salting-out
      effect of the sulfate ions is far greater than that of chloride
      ions. In addition, the greater solubility of ammonium sulfate
      over broad ionic concentrations affords a wider range of
      salting-out conditions than is attainable with the other salts.
      The disadvantages of ammonium sulfate are: (1) ammonium ions
      interfere with direct nitrogen analysis and hence must be
      removed by exhaustive dialysis; (2) it is not a good buffer and
      hence the control of pH below 8.0 is difficult; and (3) its
      concentration cannot be determined accurately, since a saturating
      concentration of ammonium sulfate is strongly dependent on
      temperature. For temperatures of 20 to 25°C, a half-saturated
      ammonium sulfate solution may be taken as equivalent to a
      concentration of 2.05 molal. Sodium sulfate gives a better
      salting-out effect than ammonium sulfate at comparable ionic
      concentrations, but is less popular than the latter because it
      is less soluble in water. This limits the range of conditions
      available for serum protein fractionations.

Immunoglobulins of all classes have been precipitated by one-third saturated ammonium sulfate at room temperature. At best, these are mixtures of IgG, IgA, IgM, and at times other globulins. Nonetheless, with careful reprecipitation and washing with one-third ammonium sulfate solution at pH 7.4, immunoglobulins of reasonable purity (85 to 90%) which show a major peak in the γ-region in cellulose acetate electrophoresis (γ-globulins) may be obtained.

b. During World War II when serum protein fractions were in great demand, Cohn and his coworkers devised a procedure for the low temperature, low ionic strength fractionation of serum in the presence of a water-miscible organic solvent such as ethyl alcohol. This involved a complex series of steps regulating temperature from 0 to -5°C with the pH decreasing from 7.4 to 4.8, and the ethyl alcohol concentrations increasing from 0.027 to 0.163 molar. The principal fractions obtained with their corresponding major serum components are shown below.

| Cohn Fraction | pH | Conditions | | | | Major serum components |
		ionic strength	$^{\circ}$C	ETOH M-F*		
I	7.2	0.14	-3	0.027		Fibrinogen
II & III	6.8	0.09	-5	0.091		gamma- & beta-globulins
IV-I	5.2	0.09	-5	0.062		alpha-globulins
IV-4	5.8	0.09	-5	0.163		alpha, beta-globulins; albumin
V	4.8	0.11	-5	0.163		albumin
VI	4.8	0.11	-5	0.163		alpha-glycoprotein; albumin

*M-F = Mole fraction; fraction VI is obtained by further evaporative concentration at low temperature of the supernatant remaining following precipitation of fraction V.

Fractions II and III were further separated in a series of five steps using 20 to 25% ethyl alcohol with temperatures of -5° to -6°C and ionic strength varying from 0.005 to 0.05. A highly purified gamma-globulin fraction was obtained and designated fraction II (Cohn FII). Fraction III was separated into three subfractions; III-1, containing the isohemagglutinins; III-2, containing prothrombin; and III-3, containing plasminogen.

c. With the advent of gel filtration and chromatography using anionic and cationic adsorbents conjugated to cellulose or other inert polymers and gels (agarose, polyacrylamide), immunoglobulins of high purity have been obtained. For example, chromatography of serum proteins equilibrated with 0.0175 M phosphate buffer, pH 6.3, on diethylaminoethanol-cellulose (DEAE-cellulose), an anionic exchanger, will give relatively homogeneous immuno-globulin-G in the first major eluate fraction. IgA and IgM appear in later fractions with increasing ionic concentration, but these fractions also contain other serum proteins. A typical fractionation pattern on DEAE-cellulose, with the distribution of the immunoglobulins and other serum proteins, is shown in Figure 93.

Isolation of immunoglobulins has also been achieved by gel filtration chromatography. In this technique, the separation is based on the molecular size and shape of the proteins. Gel filtration chromatography has been most useful in the separation of IgM from IgG and of heavy chains (H-chain) from light chains (L-chain), discussed later. These chromatographic methods have been widely used for immunoglobulin isolation and characterization because of their simplicity and less severe denaturation or alteration of the immunoglobulins.

d. Another method utilized in the characterization and, in many instances, isolation of purified immunoglobulin products is zonal electrophoresis. This utilizes a semi-solid medium that serves as a bridge between the electrodes and as a support on which the sample is separated. The migration of the proteins depends on their net electrical charge at the pH being used, and, hence, on their isoelectric point (pI).

Zonal electrophoresis is not the recommended procedure for isoelectric point determination with most support materials. At the alkaline pH used (8.6), the usual supports possess a negative charge. Since the support is a solid and therefore unable to migrate when the electric field is applied, the solvent (water) will move instead, in a direction opposite to that which the support would take if it were able to. Thus, for a negatively charged support (attracted to the anode), water will migrate toward the cathode, creating a decreasing concentration gradient from the anode to the cathode. This partially counters the electrical force and thereby alters the mobility of the serum proteins. This phenomenon of water

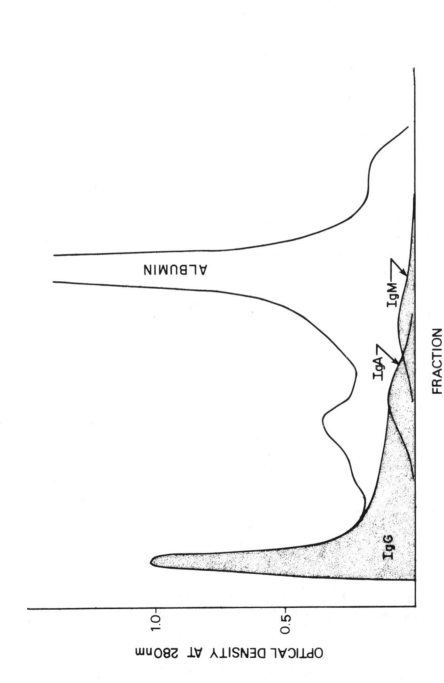

Figure 93. DISTRIBUTION OF IgG, IgA, AND IgM ON DEAE-CELLULOSE COLUMN CHROMATOGRAPHY. THE UPPER LINE OUTLINES THE FRACTIONATION PROFILE OF WHOLE SERUM WHILE THE STIPPLED REGIONS INDICATE WHERE THE THREE MAJOR IMMUNOGLOBULIN FRACTIONS OCCUR WITHIN THIS PROFILE. IgE AND IgD ARE SO SMALL QUANTITATIVELY THAT THEY CANNOT READILY BE SHOWN IN THIS DIAGRAM.

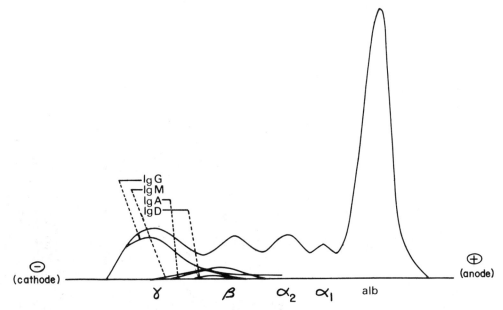

Figure 94. CELLULOSE ACETATE ELECTROPHORESIS
OF SERUM PROTEINS AND OF THE IMMUNOGLOBULINS

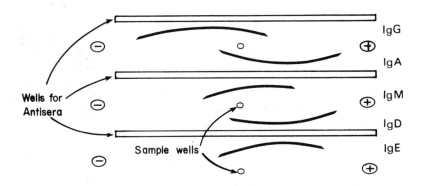

Figure 95. IMMUNOELECTROPHORESIS OF THE IMMUNOGLOBULINS

migration is termed the <u>electro-osmotic effect.</u> It is discussed
in more detail in most books on physical chemistry. The
influence of electro-osmosis is especially significant with the
serum proteins which have isoelectric points ranging from
4.8 (albumin) to 7.4 (γ-globulin). <u>All</u> of these proteins bear
a negative charge at pH 8.6 and will therefore migrate <u>toward</u>
the anode, <u>against</u> the water migration. The isoelectric points
of serum proteins have generally been determined by the free
boundary electrophoresis method of Tiselius, utilizing the
mobilities of the proteins in buffer systems at different pH's.
Here, the problem is one of convection and, hence, temperature
control is a critical factor. To minimize this effect, the
electrophoresis is carried out at 0°C.

Some of the more common media used in zonal electrophoresis are
cellulose acetate, agar, agarose, gelatin, starch, and
polyacrylamide. Zonal electrophoresis is performed at low ionic
concentrations (0.05 to 0.075) and the buffer systems used
(at pH 8.6) include sodium barbital, borate, and tris-glycine.
Starch and polyacrylamide media add a second dimension to
electrophoresis in that proteins are separated on the basis of
molecular shape and size in addition to their net electrical
charges. This is called the sieving effect and is controlled
by altering the concentration of the gels and thus the sizes
of the pores in them. Typical zonal electrophoretic patterns
of normal human serum carried out on cellulose acetate are shown
in Figure 95. Five major components can be seen: near the
anode (+) are the albumin peak, followed by α_1-, α_2-, and β-
globulin peaks and finally the γ-globulin or immunoglobulin peak.
The contributions of the individual immunoglobulin are also
indicated.

e. Immunoelectrophoresis is a useful procedure for the
    characterization of immunoglobulins with specific antibodies
    which are prepared against the Igs (see later). This procedure
    combines the principles of zonal electrophoresis and
    immunodiffusion for the analysis of antigens. The method
    consists of first separating the immunoglobulins in serum by
    zonal electrophoresis in agar, agarose or gelatin, and
    subsequently adding specific antiserum to the electrophoresed
    antigens in wells cut parallel to the electrophoretic migration.
    Figure 95 illustrates how this procedure is used in recognizing
    the five major classes of immunoglobulin.

f. Immunoglobulins are found mostly in the γ-globulin fraction and
    to a small extent in the β-globulin fraction of serum.
    Quantitative determination of human immunoglobulins are made by
    radial immunodiffusion procedures. This method consists of

isolating a given human immunoglobulin and making antibodies
against it by injecting the immunoglobulin into either a
rabbit or a goat. The antibodies produced are purified, then
incorporated into a semisolid buffered agar medium. The
fluid containing the immunoglobulin is placed in a well centered
in the immunodiffusion plate. The antigen diffuses through
the plate forming precipitin discs. By running the appropriate
standards simultaneously with the unknown sample and measuring
the diameter of the precipitin discs after a given time, one
can estimate the immunoglobulin concentration of the unknown
specimen.

2. Nomenclature, Classification and Structure

a. Five major classes of immunoglobulins have been recognized.
Prior to 1964, immunoglobulin designations were a conglomerate
of names based primarily on physicochemical properties (see
Table 30 for the common synonyms). IgD and IgE lack synonyms,
since they were discovered after 1964 and their original naming
followed the standardized procedure for classification
established by the WHO. Each of the major immunoglobulin classes
is designated by one of the capital letters G, A, M, D or E.
For example, Immunoglobulin-E is referred to as IgE or γ-E
(γ-E-globulin). Furthermore, each class can be referred to as a
member of the kappa (κ) or lambda (λ) type. Thus, the
designation of the immunoglobulins depends upon the antigenicity
of the polypeptide chains which constitute the complete molecule
(see below).

b. Immunoglobulins are composed of four peptide chains: two light
chains (L-chains) and two heavy chains (H-chains) per molecule.
These chains are linked by disulfide bonds. There are two
classes of light chains, κ and λ, thus creating two series of
immunoglobulin molecules. Each class of immunoglobulin contains
a unique type of heavy chain. These are designated γ, α, μ, δ,
or ϵ chains. These immunoglobulins and their chemical formulae
are represented as follows:

	H-chain	κ-Type	λ-Type
IgG	γ	$\kappa_2\gamma_2$	$\lambda_2\gamma_2$
IgA	α	$\kappa_2\alpha_2$	$\lambda_2\alpha_2$
IgM	μ	$\kappa_2\mu_2$	$\lambda_2\mu_2$
IgD	δ	$\kappa_2\delta_2$	$\lambda_2\delta_2$
IgE	ϵ	$\kappa_2\epsilon_2$	$\lambda_2\epsilon_2$

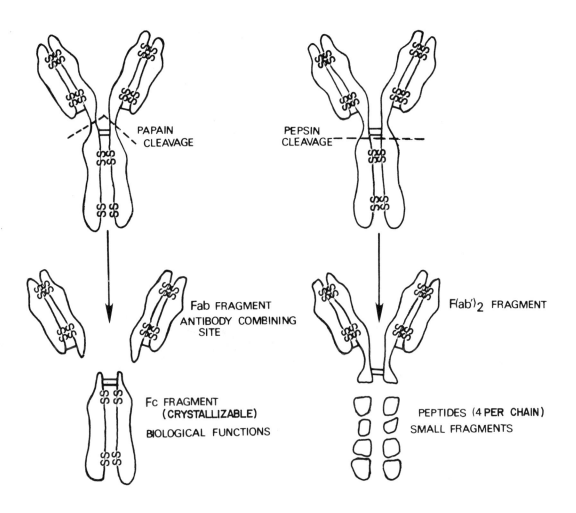

PAPAIN
CLEAVAGE

PEPSIN
CLEAVAGE

Fab FRAGMENT
ANTIBODY COMBINING
SITE

F(ab')₂ FRAGMENT

Fc FRAGMENT
(CRYSTALLIZABLE)

BIOLOGICAL FUNCTIONS

PEPTIDES (4 PER CHAIN)
SMALL FRAGMENTS

Figure 96. Papain and Pepsin Cleavage of IgG   A <u>solid line</u> between chains indicates an <u>interchain</u> disulfide bond while "SS" represents an <u>intrachain</u> cystinyl group.

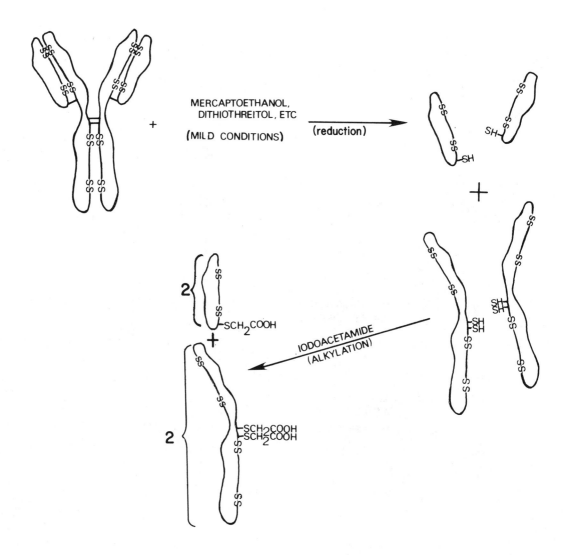

Figure 97. Reduction and Alkylation of IgG

c. The structural analysis of immunoglobulins really began with Porter's early studies using the proteolytic enzyme papain on rabbit immunoglobulins. Since then, in addition to papain, pepsin (another proteolytic enzyme; see page 437) and the reducing agents, mercaptoethanol and dithiothreitol, have been widely used for this work. The reducing agents are necessary for the breakage of the disulfide linkages. Immunoglobulin fragments obtained by proteolysis or reduction have subsequently been analyzed and separated by adsorption and gel filtration chromatography and by zonal electrophoresis on starch or polyacrylamide gel supports. Diagrammatic representations of the fragments obtained following cleavage with papain and pepsin are shown in Figure 96. Papain hydrolysis, under proper pH and time conditions, cleaves IgG into three fragments. The three fragments consist of two similar fragments. Two of the fragments, designated Fab, are alike and each contain the antibody (ab) combining sites. The third fragment is called Fc because it is crystallizable, being obtained as flat, thromboid crystals from aqueous solution. Pepsin, on the other hand, produces one large fragment (designated F(ab')₂), which has both ab sites, and eight smaller peptides (four per chain). The small peptides are derived from the region roughly corresponding to the Fc fragment of papain digestion. The principal site of both papain and pepsin action is in the "hinge" region of the molecule (see Figures 96 and 98). This is a segment of the heavy peptides which is linear (or randomly coiled) due to the presence of three proline residues which prevent helical folding (see Chapter 2). This "openness" makes the hinge region susceptible to enzymic attack.

d. Reduction of IgG with mercaptoethanol or dithiothreitol, followed by alkylation of the exposed sulfhydryls, results in the cleavage of the interchain disulfide linkages. Two L-chains and two H-chains are obtained by gel filtration chromatography on Sephadex G-100 or 150 gels following this treatment. The cleavage and alkylation of the interchain -S-S- bonds is illustrated in Figure 97. Under mild conditions, the intrachain -S-S- bonds remain intact and retain the configuration and antigenic and biological properties of the chains.

These enzymatic cleavages and chemical reductions have led to the understanding of the spatial orientation of the subunits and the structure-function relationships of immunoglobulins, which will be discussed in later paragraphs. In addition, amino acid sequence studies and immunological analysis have contributed much to the elucidation of the heterogeneity, genetics, antibody diversity, and configurational arrangement of the five major classes of immunoglobulins.

e. To reiterate, monomeric immunoglobulins are composed of four
    polypeptide chains (two H-chains and two L-chains), which loop
    and twist to form the three-dimensional configuration. The
    configuration of each Ig is of extreme importance for its
    ultimate biological functions. The three-dimensional structure
    is maintained by two major physical bonding forces:
    (1) noncovalent bonds (non-electron sharing), and (2) disulfide
    bonds formed by two neighboring cysteine (half-cystine)
    residues. These were discussed in Chapter 2. The interchain
    disulfide bridge connecting the light and heavy chains in
    molecules of subclass IgG$_1$ is closer to the bridges between the
    two heavy chains than it is in the other Igs. This is reflected in
    schematic diagrams of the Igs by drawing the L-H bond as a line
    <u>perpendicular</u> to both chains in IgG$_1$ but <u>angled</u> to both chains
    in the other Igs, as shown below.

IgG$_1$ Other Igs

    IgA$_2$ has no disulfide linkages between the L- and H-chains and
    the orientation of the L-chains relative to the H-chains is
    opposite to that normally seen. In this instance, the H-L
    bridges are attributable to hydrogen bonding, salt-
    linkages, and van der Waal's forces. Individually, these are
    weak bonds, but collectively they constitute a strong bonding
    force (see Chapter 2).

f. The light (L)-chain consists of about 214 amino acids. The
    NH$_2$-terminal region, about 107 to 115 amino acids long, is
    called the L-chain <u>variable (VL) region</u>. The remaining
    (carboxyl) portion of the L-chain has about 107 to 110 amino
    acids and is termed the L-chain <u>constant (CL) region</u>. The heavy
    (H)-chain has about 450 amino acid residues. As in the L-chain,
    the 107 to 115 NH$_2$-terminal residues compose the variable
    H-chain (VH) region while the remaining (310 to 330)
    COOH-terminal residues are the H-chain constant (CH) region.
    The CH-region can be further subdivided, on the basis of
    sequence homologies, into the CH$_1$, CH$_2$, and CH$_3$ sequences as
    shown in Figure 98. The "constant" regions (CH$_1$, CH$_2$, and CH$_3$)
    are those portions whose amino acid sequence is relatively

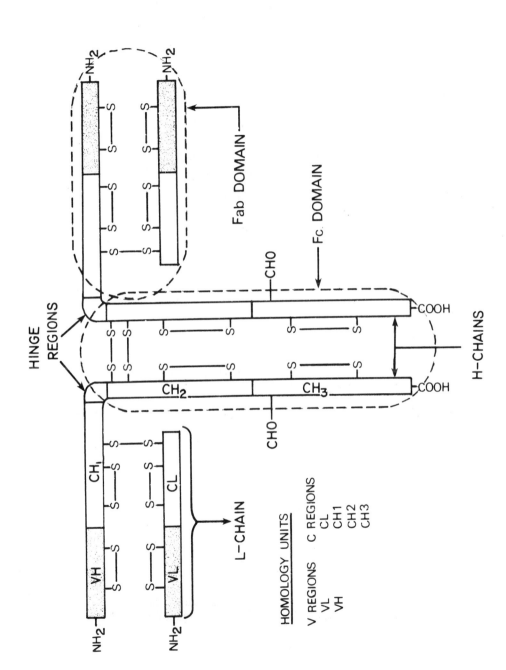

Figure 98 · DIAGRAMMATIC STRUCTURE OF IgG. CHO REPRESENTS CARBOHYDRATE ATTACHED TO THE HEAVY CHAIN CONSTANT REGIONS.

unchanging from one member of a particular class (γ, μ, etc.) of heavy chain to another member of the same class. Likewise, the CL regions of all κ-chains are closely alike as are those of the λ-chains. The "variable" regions (VL and VH) have sequences which depend on the immunological specificity of the immunoglobulin. As might be expected, the antibody binding site is located in the variable sequences.

g. In part, the amino acid sequence regulates the three dimensional configuration of the immunoglobulins. In some cases, the substitution of a single amino acid at a key position may have a significant effect on the immunogenicity or the biological function of the molecule. A classical example is the observation that the immunogenicity of pig and human insulin is not the same. This difference is accounted for by residue 30 of the B-chain which is alanine in the human hormone and threonine in the porcine material. Part of this antigenic variance may be attributable to the three-dimensional configurational differences between pig and human insulin, which would be difficult to assess. Another familiar case in which the changing of a single amino acid profoundly affects the biological function of a molecule is the substitution of valine for glutamate in the 6 position of the β-chain of hemoglobin (see Chapter 7). This represents the only difference in amino acid sequence between normal hemoglobin A and hemoglobin S (sickle cell trait). In this respect, immunoglobulins are similar in behavior to insulin, since antigenic differences result from amino acid substitution without impairment of biological function.

3. Physico-chemical and Biological Properties of the Immunoglobulins

Some of the major physico-chemical and biological properties of the five major classes and several subclasses of immunoglobulins are compiled in Tables 30 and 31.

a. Immunoglobulin-G (IgG; γG)

i) IgG comprises 80 to 85% of the circulating immunoglobulins in the blood. Prior to the standardization of nomenclature of immunoglobulins in 1964 by the WHO, these globulins were designated by a variety of names such as γ, 7sγ, γγ, 6.6γ, γ1, and γss. These designations were derived from electrophoretic mobilities and ultracentrifugal studies. The approximate molecular weights, sedimentation, extinction and diffusion coefficients, isoelectric points, and partial specific

Table 30. Physicochemical Properties and Other Biochemical Characteristics of the Immunoglobulins of Man

Immunoglobulin Class	IgG				IgA		IgM	IgD	IgE
Immunoglobulin Subclasses	IgG$_1$	IgG$_2$	IgG$_3$	IgG$_4$	IgA$_1$	IgA$_2$			
1. Synonyms	γ, 7Sγ, 6.6γ, γ2, γss (all synonyms for IgG)				β$_2$A, γ$_1$A (synonyms for IgA)		γ$_1$M, β$_2$M, 19Sγ γ-macroglobulin	none	none
2. Approximate M.W.*	149,000-153,000				150,000-600,000		900,000	160,000	188,000-209,000
3. Approximate S$_{20,w}$*	6.6 - 7.2				7,9,11,13		18 - 32	7	7 - 8.2
4. Diffusion Coefficient (1 x 10^{-7} cm^2/sec)	4.0				3.0 - 3.6		1.71 - 1.75	---	3.71
5. Extinction Coefficient (E$_{280\ nm}^{1\%}$)	13.8				13.4 - 13.9		13.3	---	15.3
6. Partial Specific Volume ($\overline{V}_{20}$, cm^3/gm)	0.739				0.729 - 0.723		0.723	---	0.713
7. Isoelectric Point (in pH units)	5.8 - 7.3				---		5.1 - 7.8	---	
8. Valence	2				2		5	---	2
9. Total Carbohydrate Content (wt. %)	3.0				12.0		12.0	---	12.0
10. Mean Survival in Serum (T/2 in days)	11-12	11-12	6-7	20-21	5 - 6		5	2.8	1.5
11. Mean Concentration in Serum: (mg/100 ml) (range in parentheses)	Total for IgG: 1240 (900-2000) For Subclasses: 820 / 286 / 90 / 50				280 (200-350)		120 (75-150)	3.0 (<0.3-40)	0.03 (0.007-0.18)
12. Heavy Chain (M.W.)	γ (50,000)				α (65,000)		μ (65,000-70,000)	δ	ε (72,500)
13. Light Chain	κ or λ				κ or λ		κ or λ	κ or λ	κ or λ

Table 30. Physicochemical Properties and Other Biochemical Characteristics of the Immunoglobulins of Man (continued)

Immunoglobulin Class	IgG				IgA		IgM	IgD	IgE
Immunoglobulin Subclasses	IgG$_1$	IgG$_2$	IgG$_3$	IgG$_4$	IgA$_1$	IgA$_2$	IgM	IgD	IgE
14. κ to λ ratio (Range)	1.42 to 2.41	0.96 to 1.10	1.25 to 1.12	7 to 5	1 - 2		1, 1-4	1/20	---
15. Number of Inter-chain Disulfide-Bonds	4	6	6	4	5?, 5?		15	3	3?
16. Allotypes: H-Chain Gm	+	+	+	0	-	-	-	-	-
Am$_2$	-	-	-	-	-	+	-	-	-
L-Chain Inv	+	+	+	+	+	+	+	+	+
Oz	+	+	+	+	+	+	+	+	+
17. Electrophoretic Mobility in Agar	γ_2	γ_2	γ_2	γ_1	$\gamma_1-\beta$		$\gamma_2-\beta_2$	$\gamma-\beta$	γ_1
18. Intrinsic Viscosity	0.06				---		0.162	---	---

* M.W. = molecular weight; S$_{20,w}$ = sedimentation coefficient corrected to water at 20°C; blank spaces = data unknown.

In some cases, data are given for each subclass while in others, average values for the class are indicated.

Table 31. The Biological Properties of Human Immunoglobulins

Immunoglobulin Class	IgG				IgA	IgM	IgD	IgE
Immunoglobulin Subclasses	IgG$_1$	IgG$_2$	IgG$_3$	IgG$_4$	(not distinguished)			
1. Specific antibodies								
a. anti-Rh	+				+	+		+
b. anti-dextran		+	+					
c. anti-levan		+						
d. anti-factor VIII				+				
2. Complement-fixation	+	+$_w$	+	0	0	+		$\pm$
3. Passive cutaneous anaphylaxis (PCA)	+	0	+	+	0	0		0
4. Placental transfer (Pt)	+	+	+	+	0	0		0
5. Reacts with Rheumatoid factor	+	+	0	+	$\pm$			
6. Reacts with Staphylococcus protein A	+	+	0	+				
7. Macrophage receptor (MR)	+	0	+	0	0	0	0	
8. Prausnitz – Küstner P-K reaction	0	0	0	0	0		0	++ (0.1-1.0 ng AbN/ml)
9. Isohemagglutinin	+ (positive, in hyperimmunized individuals)				+	+	0	0
10. Cryoglobulins		+			+ (rare)	+		

+ = reacts (antibody present); 0 = does not react; w = weakly; AbN = antibody nitrogen; $\pm$ = possible reaction; Blanks indicate data not known.

volume are data obtained for IgG without reference to subclass distinction. This may account for the range found, especially in the sedimentation coefficient and isoelectric points. Subclasses IgG_1, IgG_2, and IgG_3 have similar mobilities in agar immunoelectrophoresis (traveling with the γ_2 globulins), while IgG_4 appears to have a slightly faster mobility (moving with the γ_1 fraction). This, in part, may account for the variation in the isoelectric points. The valence or binding capacity (number of binding sites per molecule) for IgG is 2 although, in some instances, a valence of 1 has been reported. This may be attributable to a weaker binding capacity rather than a true valence of 1, since this so-called univalent antibody has been shown to have all four polypeptide chains (two H- and two L-chains). Molecules in all four of the IgG subclasses contain γ-H chains and either κ or λ L-chains. The total carbohydrate content of IgG molecules is approximately one-fourth that of the other Ig classes.

ii)   Kappa/Lambda ratios in the subclasses increase in the order $IgG_2 < IgG_3 < IgG_1 < IgG_4$, while serum concentrations of the subclasses decrease in the order $IgG_1 > IgG_2 > IgG_3 > IgG_4$. The mean total serum concentration of IgG is 1240 mg/100 ml with a normal range of 900–2000 mg/100 ml. The mean half-lives of the IgG subclasses vary from a low of 6–7 days for IgG_3 to a high of 20–21 days for IgG_4. Those for IgG_1 and IgG_2 are intermediate, at 11 to 12 days. A mean of 23 days has been indicated for total IgG. IgG_1 and IgG_4 have four inter-chain disulfide bridges, two between L-H chains and two between H-H chains. IgG_2 and IgG_3 possess six inter-chain disulfide covalent bonds with four bonds between the H-H chains rather than two.

iii) Immunoglobulins are excellent antigens or immunogens and antisera to them can be raised in a variety of animals (monkey, horse, sheep, donkey, goat, and rabbit). Using these antisera, specific antigenic sites on the Ig's can be correlated with structural features and with specific amino acids in the polypeptide chains. On this basis, the polypeptide chain antigens are conveniently categorized into three major groups.

(1)   isotype antigens are present in all normal sera and can be differentiated into the five major classes and their subgroups (types);

(2) <u>allotype</u> antigens are present in some (but not all) normal sera and are governed by allelic genes;

(3) <u>idiotype</u> antigens are unique to one particular antibody molecule, hence by inference, they are related to the V-region or antibody combining site.

iv) The isotype and allotype antigens are related to the C-region of the H-chains. The class-specific antigens of the C-region represent one group of isotypes. Cross-reactions between the major immunoglobulin classes have not been demonstrated, though regions of homology based on sequence studies have been shown. Within subclasses, however, many shared antigens have been found. These analyses have been carried out by precipitation and hemagglutination inhibition tests. Antibodies to some subclass-antigens are difficult to raise in animals.

v) The allotype antigens for the H-chains are referred to as Gm (gamma-markers). The Gm markers are restricted to the CH region of the H-chains of IgG. Gm 1, 22, 4, and 17 are restricted to IgG_1; Gm 24 is specific for IgG_2; and Gm 3, 5, 6, 13, 14, 15, 16, and 21 are specific for IgG_3. No regular Gm for IgG_4 has been demonstrated as yet. IgG_4 has a non-marker, Gm-variant antigen designated 4a and 4b on the CH_3-homology region. Other Gm antigens are shared between the subclasses. There are at present a total of 27 Gm antigens. The allotype marker InV is found on the kappa chain. Valine at residue position 191 on the κ-chain represents a positive test for InV(b+); whereas substitution at this position with leucine results in InV(a+). Amino acid substitution at position 190 in the λ-chain controls the presence or absence of the isotype Oz.

The biological activities attributed to the IgG subclasses are shown in Table 31. The major subclass associated with specific antibodies is IgG_1. Specific antibodies to dextran, levan, Rh, and factor VIII reside in the different subclasses of IgG. IgG_4 does not fix complement, while IgG_2 does so weakly. Passive-cutaneous anaphylaxis (PCA) in guinea pig is negative for IgG_2. IgG_3 does not react with rheumatoid factor and staphylococcal protein A. No reaction with macrophage receptor has been shown for IgG_2 and IgG_4. Isohemagglutinin related to IgG antibody is found in individuals following hyperimmunization with ABO antigens. IgG antibodies are generally related to the later-occurring antibodies following antigenic stimulation. IgG antibodies do not participate in the P-K (Prausnitz-Kützner) reaction. Cryoglobulins and rheumatoid factor in some diseases may be associated with IgG.

b. Immunoglobulin-A (IgA; γA)

    i) IgA has two subclasses designated IgA_1 and IgA_2. There are two major differences between these subclasses: (1) The allotype Am_2 is present in the C region of the H-chain of IgA_2 (Am_2+) and is absent in IgA_1 (Am_2-). (2) The L-chains in IgA_2 are not linked to the H-chains by disulfide bonds. In addition, as was indicated earlier, the orientation of the L-chains in IgA_2 is opposite to that of the L-chains in the other immunoglobulins.

    ii) The significant physicochemical and biological properties of IgA are listed in Tables 30 and 31. Some of these properties are for the secretory IgA which is a dimer of IgA_1 held together by a secretory piece or component (SP,SC) and a polypeptide "J"- (juncture) chain. Properties of both SC and "J"-chain are shown in Table 32.

Table 32. Properties of Secretory Component and J-Chain

Properties	SC	J-Chain
Molecular weight	50,000–90,000	23,000–26,000
Carbohydrate	9.5–15.0%	none
Sedimentation coefficient, $S_{20,w}$	4.2	----
Partial Specific Volume (cm^3/gm)	0.726	----
Electrophoresis: (relative mobility)	β-globulin	fast-moving component in gel

Note that this polymeric IgA contains other subunits in
addition to H- and L-chains. The J-chain has also been
noted in IgM. Skeleton diagrams of each of the major
classes and subclasses of immunoglobulins with inter-chain
disulfide (S-S) linkages are shown below. The locations
of the J-chains and SC in IgM and secretory IgA$_1$ is still
speculative although it is postulated to occur, as shown
below, somewhere near the carboxyl termini of the heavy
chains. The secretory component is antigenic and is
present in its free form in bodily secretions. In IgA$_1$,
IgA$_2$, and IgE the number of disulfide bonds connecting
the two heavy chains is not known for certain.

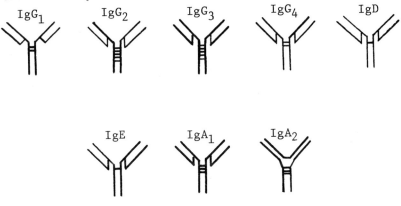

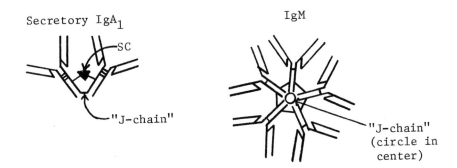

iii)   It appears that there are no chemical differences between circulating forms of IgG, IgD, and IgE and those found in the secretions of mucous membrane cells. As indicated on the previous page, however, a portion of IgM and almost all of IgA have a secretory component (SC) attached to them when they occur in external secretions. An additional connecting glycoprotein (named as J-chain) has also been identified which is associated with dimeric and polymeric forms of IgA and IgM molecules.

In humans, it has been shown that the secretory component is synthesized in the mucosal cells and is joined by disulfide linkages to the H-chains of IgA molecules before they are secreted. IgA molecules themselves are synthesized in response to a proper stimulus in the plasma cells that are present in submucosa. This can be diagrammatically represented as follows:

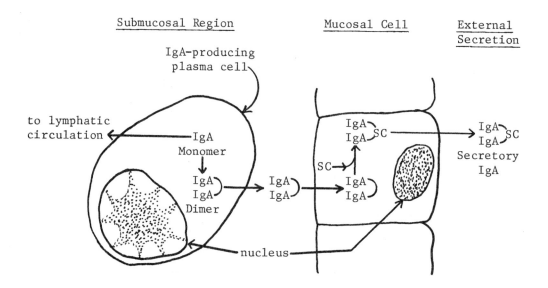

Note: in the above diagram

IgA monomer has a sedimentation coefficient of 7S.
IgA dimer has a sedimentation coefficient of 10S.
IgA dimer plus SC (secretory IgA) has a sedimentation coefficient of 11S.

A question arises as to the need for the SC component that is synthesized and attached to IgA dimers in the mucosal cells. It has been suggested that the SC component may facilitate transport of the dimer through the mucosal cell by protecting the molecule from intracellular degradative processes (e.g., proteolysis). It may also participate in the functioning of IgA as an antibody in an external environment.

iv) The primary function of IgA has not been elucidated. Recent studies in orally vaccinated or spontaneously infected human subjects appear to confirm the protective value of exocrine IgA antibodies. Previous reports on the absence of complement-fixation by IgA antibodies may have to be revised, since it has been observed that colostral-type IgA to Escherichia coli was able to lyse these bacteria in the combined presence of lysozyme and complement. The tendency of IgA to polymerize probably accounts for the various sedimentation coefficients and molecular weights observed. It is likely that this tendency also contributes to the activation of complement via C3-proactivator action in the alternative pathway of complement activation. This will be discussed later, in the section on complement.

It has also been reported that porcine milk IgA enhances the ability of phagocytes to destroy certain types of encapsulated bacteria, such as pneumococcus. This is known as opsonization and substances, such as IgA, which have this ability, are known as opsonins. It has also been suggested that one of the major functions of IgA is to react with various harmful antigens to form unabsorbable antibody-antigen complexes. This would prevent entrance of these deleterious materials via the gut and other endothelial surfaces such as the respiratory tract. This is compatible with the high circulating antibody titers to food antigens found in individuals with IgA deficiency. IgA deficiency also correlates well with a high incidence of autoimmune disease. Nonetheless, complete absence of IgA is not incompatible with good health, since in many instances of secretory IgA deficiency, IgM appears to assume IgA functions. IgA is negative for PCA, placental transfer, and P-K reaction, and several naturally occurring serum isohemagglutinins are IgA molecules. Some IgA cryoglobulins may occur rarely in certain diseases.

Other possible functions of secretory immunoglobulins
(in particular IgA) are as follows:

(1) Prevention of viral and bacterial infections that may
     have access to the external secretions such as in
     the lungs, mouth, intestines, ear canals, etc.;

(2) Participation as opsonins in the phagocytic process
     (involving principally monocytes in the submucosal
     region) in cases where the infective agent has
     penetrated the mucosal barriers.

v) With diseases produced by respiratory viruses such as
    influenza, local immunization appears to augment the
    effectiveness of immunization by parenteral routes. In
    fact, in those viral infections affecting primarily the
    mucosal regions, local immunization may be the preferred
    method. The local immunization obtained, for example, by
    administration of a vaccine via the alimentary route,
    may have an additional advantage in control of viruses
    that produce systemic infections. This advantage is the
    elimination of the carrier state. For example, in
    poliomyelitis immunization by the Salk method (systemic
    introduction of viral antigens), one may develop
    neutralizing antibodies systemically but not in the
    secretory IgA. Therefore, these individuals may be immune
    to the disease, but may become carriers by harboring the
    virus in their intestinal tract. On the other hand, the
    Sabin method of immunization (attenuated virus by oral
    route) not only produces systemic antibodies but also
    secretory antibodies. The latter may prevent local
    infections by neutralizing the virus particles.

c. Immunoglobulin-M (IgM; γM)

i) The IgM's are the largest immunoglobulins with molecular
    weights of 900,000 or greater. Evidence for the occurrence
    of subclasses of different allotypes has not been demonstrated,
    although they have been suggested. The significant
    physicochemical properties of IgM are indicated in Table 30.
    Of particular interest is the valence number. Though 10
    would be anticipated by structural analysis, only 5 binding
    sites have been shown in reactions with antigens. This may,
    in part, be due to steric interferences at binding sites
    and thus may not represent the true potential valence. Like
    IgA and IgE, the total carbohydrate content of the molecule
    is four times that of IgG. Because of its size, the intrinsic
    viscosity and diffusion coefficients are strikingly different
    from the other immunoglobulins.

ii) The significant biological characteristics of IgM are
     shown in Table 31. IgM fixes complement and it is probably
     the major isohemagglutinin (anti-A or anti-B). One of
     the cardinal features of IgM is its early appearance
     following antigenic stimulation. It possesses strong
     agglutination or precipitation properties, perhaps due
     to its pentavalency. It does not react with rheumatoid
     factor but, instead, is considered to be one of the
     factors itself. Rheumatoid factors have also been
     associated with IgG and IgA. It has been suggested that
     IgM is one of the more primitive of the immunoglobulins,
     since it is synthesized early in neonatal life and even,
     in some species, by the fetus in utero. IgM does not
     traverse the placental barrier under normal conditions.
     "J"-chain is also part of the IgM molecule, probably bound
     to some region at or near the carboxyl terminus of the
     H-chains.

d. Immunoglobulin D (IgD; γD)

    Information on IgD is limited and no biological function has
    yet been attributed to these molecules. Of interest is its low
    level in serum (only IgE is lower) and the high lambda/kappa
    ratio. It is not a secretory Ig like IgA or IgE and its P-K
    response is negative. The available physicochemical data for
    IgD is shown in Table 30.

e. Immunoglobulin E (IgE; γE)

    The physicochemical properties of IgE are summarized in
    Table 30. It has the lowest level and shortest survival time
    in serum of any of the immunoglobulins. The affinity of IgE
    for mast cells and its relationship to immediate hypersensitivity
    and to the homocytotropic antibodies of other animals are
    among the most studied aspects of this immunoglobulin. IgE has
    been equated to the reaginic antibody of man associated with
    the release of histamine and SRS-A (slow-reacting-substance of
    anaphylaxis) following interaction with specific antigens.
    The term reagin is used in man to refer to cytotropic antibodies.
    These antibodies are bound to target cells (e.g., mast cells) in
    individuals sensitized to an antigen. When presented again with
    the antigen, the cytotropic antibodies bind it, thereby triggering
    the release of vasoactive amines from the target cells. IgE
    does not bind complement in the classical pathway. IgE can passively

sensitize monkey and human skin (P-K reaction) and leukocytes, but gives no PCA response in the guinea pig. Addition of specific antigen to IgE-sensitized leukocytes induces release of histamine and SRS-A. This has been utilized as a means of detecting allergies in man by examining the morphological changes of the leukocytes or measuring histamine release after addition of the antigen. Similar changes, with release of histamine and SRS-A, occur with monkey lung tissues. The Fc region of the IgE molecule contains the site for binding to basophilic leukocytes and mast cells.

Though much is known of the role of IgE in allergic manifestation of the atopic variety, there is little knowledge of its real function in normal metabolic processes. IgE is considered a secretory immunoglobulin with a distribution similar to that of IgA in areas such as the respiratory and gastro-intestinal tract mucosal tissues.

4. Antibody Diversity

a. Multivalent antigens, such as bacteria, have many antigenic determinants. Since each determinant can, theoretically, cause the synthesis of a different immunoglobulin molecule, the response of the humoral system to even the simplest antigen can be quite heterogeneous. When a B-cell transforms to a plasma cell, each member of the clone which develops from that cell is capable of producing only one molecular species of immunoglobulin. Thus, the number of different Ig molecules reflects the number of B-lymphocyte clones which synthesized them.

b. A typical antibody response of Coturnix coturnix (Japanese quail) to Brucella antigen (containing 1×10^6 bacteria/ml) given intraperitoneally is shown in Figure 99. The initial administration of 0.5 ml of the bacterial suspension initiates a primary response. The following sequence of events accompanies the primary response.

(i) An initial induction period (2 to 4 days for Brucella antigen in Coturnix);

(ii) A rapid synthesis of antibody and its appearance in the serum (note that IgM appears earlier than IgG and has a significantly higher titer);

(iii) Attainment of a maximum on about the seventh day;

(iv) A gradual decline in the level of the antibodies in the serum.

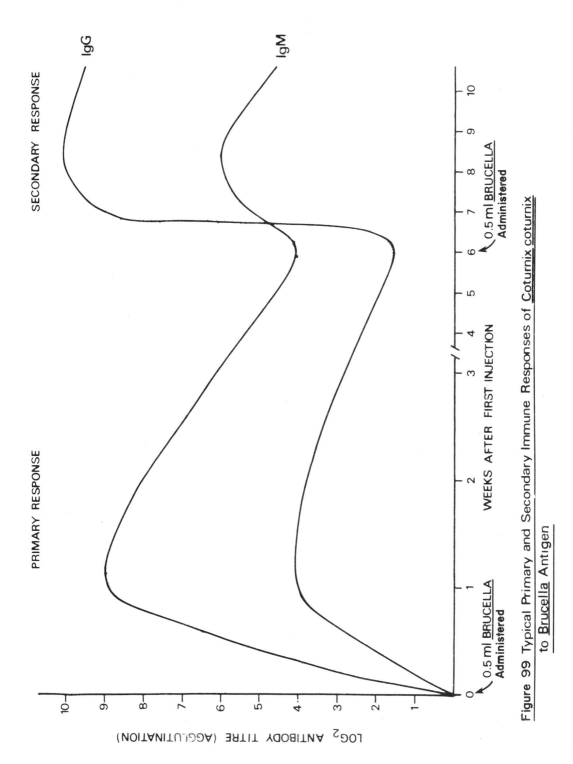

Figure 99 Typical Primary and Secondary Immune Responses of *Coturnix coturnix* to Brucella Antigen

This sequence will vary according to the antigen and animal used. A second administration of the same antigen when antibody levels are decreasing following the primary response results in a secondary response. With acceleration this time, the rate of synthesis is greater, the induction period is shorter, a greater peak titer occurs, and the antibodies disappear less rapidly from the blood. The decreased rate of disappearance could be accounted for by the observations that (1) IgG has a longer serum-half-life than IgM, and (2) in the secondary response, the amount of IgG synthesized is greater than the amount of IgM. The differentiation of IgG and IgM in the agglutination reaction with <u>Brucella</u> antigen <u>in vitro</u> is accomplished by the addition of a dilute solution of mercaptoethanol. This reducing agent inactivates IgM by degrading the large (polymeric) IgM molecules to their monomeric form, which lacks agglutinin activity. IgG is essentially resistant to mild mercaptoethanol treatment. IgM can also be readily separated from IgG by gel filtration chromatography.

c. The biochemistry of protein synthesis, using DNA, mRNA, tRNA, polyribosomes, and the appropriate enzymes was discussed in Chapter 5. Like albumin and hemoglobin, there is a constant rate of immunoglobulin degradation and <u>de novo</u> synthesis. The rates depend on the nature of the immunoglobulins. Specific immunoglobulin (antibody) synthesis occurs when the appropriate clone is stimulated by a specific antigenic determinant as was indicated earlier. Although much is known of the fundamentals of protein synthesis and of the immunoglobulin structures, the manner in which H-chains and L-chains are put together to form the tetrapeptide has not been elucidated. Of interest also is the manner in which a single B-cell switches from IgM to IgG synthesis. Some mechanism of feedback inhibition by IgG has been suggested but information in these areas is very limited.

d. Figure 98 is a schematic diagram of the basic structure of human immunoglobulin. The important features of this molecule are

   (i) Two identical L-chains (about 22,500 daltons each) and two identical H-chains (approximately 53,000 to 70,000 daltons each) are present, linked by disulfide bonds and non-covalent interactions;

  (ii) The molecule is folded into three domains (2 Fab and 1 Fc), which are separated by the H-chain hinge region;

 (iii) The polypeptide chains of both H and L can be divided into an NH_2-terminal (variable) region and a COOH-terminal (constant) portion.

The V-and C-regions are defined by amino acid sequence homology. For example, in the C-region of the predominant L-chain, only one amino acid substitution occurs and it behaves as a single Mendelian allele. In the V-region of the same L-chain, as many as 15 to 40 amino acid substitutions can occur. The V-regions of H- and L-chains from the same Ig show striking sequence homologies with one another. Such homologies would also be anticipated between the V-region of the H-and L-chains of antibodies from different species following stimulation of both species by the same antigenic determinant (i.e., the binding site sequence is relatively independent of species when the Ig's are against the same antigen). Sequence homologies are also found in the C-regions of H-and L-chains.

e. Present evidence strongly supports the concept of a single immunoglobulin polypeptide chain encoded by two germline (heritable) genes, a V-gene and a C-gene, which are joined somatically. Thus, three major families of immunoglobulin polypeptides could be envisioned based on amino acid sequence homology and genetic linkage studies. These are the families of the H-chain and the two L-chains (κ and λ). The constant regions of some of these families can be further divided into classes and subclasses. For example, classes and subclasses of C H-chains are C_{γ_1}, C_{γ_2}, C_{γ_3}, C_{γ_4}, C_{α_1}, C_{α_2}, C_{μ}, C_{δ}, and C_{ε}. Classes and subclasses of C L-chains are C_{arg}, C_{lys} for λ, and C_K for κ. Each of these is encoded by a germline gene. V-regions are also divided into subgroups controlled by separate genes, but they are of a more complex nature.

f. The structural features of the immunoglobulins are intimately associated with their biological features. Thus, antibody diversity is encoded in the V-regions while the general functions of antibody molecules (placental transfer; complement and macrophage binding) are mediated by the various C-regions of the H-chain.

g. Numerous theories have been advanced to explain antibody production in the past. With the tremendous advances over the past decade in the knowledge of antibody molecules, immunologists have now generally accepted Burnet's clonal selection theory. This theory postulates that clones of potential immunologically competent cells, each containing the genetic codes (C-gene and V-gene, one of each for H-and L-chains) of the immunoglobulin complementary to one antigenic determinant, are present from birth. A sufficient number of different clones would provide an immune potential against all existing antigens that an individual might encounter. The immunogen thus selects or affects its specific complementary clone, which is provided with the appropriate, specific receptor site, and stimulates it to proliferate and synthesize the corresponding specific antibody. This concept stresses the role of the immune cells. Since antibody diversity reflects a corresponding immune cell gene diversity, there is the question of how this gene diversity arose in the immune cell population in the first place. Three generally acceptable theories have been formulated for explaining this.

(i) The <u>somatic mutation theory</u> postulates that antibody diversity results from hypermutation during somatic differentiation of the immune cells.

(ii) The <u>germline theory</u> proposes that antibody diversity is encoded by a large number of germline genes (i.e., multiple genes encode the V-regions, while a separate gene encodes the C-region in a given immune cell).

(iii) The <u>somatic recombination theory</u> suggests that somatic recombinations occur in a limited number of antibody genes.

Of these three hypotheses, the germline and somatic mutation theories have greatest support from amino acid sequence analysis of the immunoglobulins from various species.

Immunogen (Antigen)

1. General considerations

   a. It is obvious that any discussion of antibody is intimately associated with reference to antigen. This term, synonymous with immunogen, refers to any substance, exogenous or endogenous, which, when administered to a host, induces or stimulates the production of specific antibodies (any of the five major immunoglobulins discussed earlier). These antibodies can combine specifically with the immunogen as demonstrated by various immunological tests.

   b. There are many varieties of immunogens, including bacteria and their subunits, viruses, proteins, carbohydrates, and in some cases, lipids (see page 614). It is estimated that there are up to 1×10^6 different kinds of antigens. Immunogens are designated complete and incomplete. The former group can induce production of antibody in the host and also react with the specific antibody formed against it. Incomplete immunogens, called <u>haptens</u>, are incapable of antibody stimulation in the host unless coupled with a carrier antigen. Nevertheless, they

can react with antibody as determined by special procedures
such as the blocking test and passive hemagglutination reactions.
Tolerogens are a third class of "antigens" which induce
unresponsiveness in individuals. Persons showing
no response to a tolerogen are said to be tolerant toward it.
The tolerant state can be natural or induced. The latter form
is brought about by introduction of an antigen which is
qualitatively inadequate (as is readily demonstrated in mice,
monomeric γ-globulin is tolerogenic while aggregated γ-globulin
is highly antigenic) or quantitatively insufficient to produce
a true immune response. Tolerogens usually affect the T-cells
so that they lose their ability to interact with the B-cells
to bring about antibody synthesis. However, since tolerance
refers to a general inability of the host to respond to
an antigen via either the cell-mediated or humoral system, an
animal can be rendered tolerant by blocking either B- or
T-cells or both.

c. Proteins of high molecular weights are usually good antigens
capable of stimulating antibody production. Some examples in
order of decreasing molecular weight are viruses, thyroglobulins,
immunoglobulins, and serum albumins. Carbohydrates, which are
basically made up of hexoses, hexosamine, and sialic acid
residues, are generally less satisfactory immunogens than
proteins. Examples of carbohydrate antigens are type-specific
pneumococcal polysaccharides, the lipopolysaccharides of gram-
negative bacteria, and the blood-group mucopolysaccharides of
erythrocytes. Lipids generally act as haptens and are poor
complete antigens. Nucleoproteins have been shown to be
complete immunogens, while nucleic acids and the base residues
of DNA and RNA show good haptenic properties.

2. Isoantigens (homologous antigens)

a. ABO and Rh Blood Groups

All proteins and protein-lipopolysaccharide complexes of human
serous fluids and tissues have immunogenic properties. Of
interest in this section are the isoantigens, which have been
extensively investigated and which play significant roles in
transfusion, transplantation, and certain immunological
disorders (isoimmunization; hemolytic disease of the newborn).
Isoantigens are expressions of different alleles which occur at
one genetic locus within different individuals of one
interbreeding population (genetic polymorphism). They can be
considered as the factors which make a person's tissues and
blood cells immunologically unique (i.e., unlike those found
in another individual).

i) In man, the ABO and Rh red cell isoantigens are medically
     important in transfusions and, to some extent, in tissue
     transplantation. They are found on the outer surface of
     red blood cell membranes and on various cells (especially
     those of epithelial origin) scattered throughout the
     body. Some ABO isoantigens are also related to bacterial
     antigens such as the somatic lipopolysaccharides of gram-
     negative bacteria. Rh isoantigens, on the other hand,
     have been demonstrated only in the erythrocytes of man
     and monkey (Rh is derived from Rhesus monkey). The
     phenotypes and genotypes of the ABO and Rh isoantigens
     are summarized in Tables 33 and 34, respectively. The ABO
     antigens have reciprocal antibodies of isohemagglutinins
     present in the blood. An indicated in the earlier sections,
     these appear soon after birth and are primarily IgM and IgA
     antibodies. Rh antigens have no reciprocal natural
     antibodies, but are so highly immunogenic that improper
     transfusion from an Rh-positive (Rh^+) to an Rh-negative
     (Rh^-) individual results in anti-Rh^+ antibody production in
     the Rh^- recipient. Similarly, transmission of red cells
     during parturition from an Rh^+ infant to an Rh^- mother
     (across the placental barrier, for example) can stimulate
     anti-Rh^+ antibodies in the parent. An Rh^+ fetus conceived
     in subsequent pregnancies can be seriously affected by the
     anti-Rh^+ IgG_1 antibodies carried by the mother since these
     are able to cross the placenta and enter the fetal
     circulation. This disorder is known as erythroblastosis
     fetalis or hemolytic disease of the newborn.

ii) The ABO blood group is genetically controlled by three
     alleles at one genetic locus (three allelic genes). These
     are the A, B, and O alleles, where O is an amorph. The
     O allele is expressed as the absence of the A and B
     alleles and is also involved in the expression of the
     H-isoantigen. Subgroups (notably A_1 and A_2) have been
     demonstrated in the A group. The secretory (Se) genes,
     distinct from the ABO loci (though closely related to them),
     regulate the elaboration of the A, B, and H isoantigens
     in body fluids (especially saliva). Although the ABO
     isoantigens have not been completely characterized
     structurally, the terminal residues are known and are
     indicated in Table 35.

Table 33.   ABO Blood Groups

Genotype	Phenotype	Isoantigen In Red Cells	Isohemagglutinin Present in Plasma	Frequency, United States, (%)		
				Caucasians	Negroes	Chinese
A_1O, A_1A_1, A_1A_2	A_1	A and A_1	Anti-B	41	27	28
A_2A_2, A_2O	A_2	A	Anti-B			
BO, BB	B	B	Anti-A and Anti-A_1	10	21	23
OO	O	H	Anti-A and Anti-B	45	48	36
A_1B	A_1B	A_1 and B	None	4	4	13
A_2B	A_2B	A_2 and B	None			

Table 34. Antigenic Determinants of the ABO and Lewis Blood Groups

Specificity	Allele	Non-Reducing Terminal Reisudes of the Isoantigen
None	"precursor"	Gal$\frac{\beta1,3}{}$-NacGlu$\frac{\beta1,3}{}$-Gal$\frac{\beta1,3}{}$-NacGal·····
Type XIV (pneumococcus)	"precursor"	Gal$\frac{\beta1,4}{}$-NacGlu$\frac{\beta1,3}{}$-Gal$\frac{\beta1,3}{}$-NacGal·····
H	H	$\beta1,3;$ Gal$\frac{\beta1,4}{}$-NacGlu$\frac{\beta1,3}{}$-Gal$\frac{\beta1,3}{}$-NacGal····· |$\alpha1,2$ Fuc
Le^a	Le	Gal$\frac{\beta1,3}{}$-NacGlu$\frac{\beta1,3}{}$-Gal$\frac{\beta1,3}{}$-NacGal····· |$\alpha1,4$ Fuc
Le^b	H & Le	Gal$\frac{\beta1,3}{}$-NacGlu$\frac{\beta1,3}{}$-Gal$\frac{\beta1,3}{}$-NacGal····· |$\alpha1,2$ |$\alpha1,4$ Fuc Fuc
A	A	$\beta1,3;$ NacGal$\frac{\alpha1,3}{}$-Gal$\frac{\beta1,4}{}$-NacGlu$\frac{\beta1,3}{}$-Gal$\frac{\beta1,3}{}$-NacGal····· |$\alpha1,2$ Fuc
B	B	$\beta1,3;$ Gal$\frac{\alpha1,3}{}$-Gal$\frac{\beta1,4}{}$-NacGlu$\frac{\beta1,3}{}$-Gal$\frac{\beta1,3}{}$-NacGal····· |$\alpha1,2$ Fuc

Note: a) The first two isoantigens, labeled as "precursor" are the basic carbohydrate sequences to which are added specific residues to form the ABO antigenic determinants. The first one is non-specific while the second one is, as indicated, related to the antigen of type XIV (pneumococcus).

b) Le^a and Le^b are the Lewis blood group isoantigens. They were discovered after the original ABO grouping was formed but are chemically closely related to the ABO isoantigens.

c) Gal = galactose; NacGlu = N-acetylglucosamine; NacGal = N-acetylgalactosamine; Fuc = fucose (6-deoxy-L-galactose).

d) When more than one type of linkage ($\beta1,3$; $\beta1,4$) is shown, both are known to occur without altering the antigenic specificity.

Table 35. <u>Rh Blood Groups</u>

Genotype*		Phenotype	Blood Factors Present*		Frequency (%) Caucasian
F-R	Wiener		F-R	Wiener	
DCE	R^Z	Rh^+	D,C,E	Rh_o,rh',rh''	85
DCe	R^1		D,C,e	Rh_o,rh',hr''	
DcE	R^2		D,c,E	Rh_o,hr',rh''	
Dce	R^0		D,c,e	Rh_o,hr',hr''	
dCE	r^y	Rh^-	C,E	rh',rh''	15
dCe	r'		C,e	rh',hr''	
dcE	r''		c,E	hr',rh''	
dce	r		c,e	hr',hr''	

*"F-R" refers to the notation (and theory of Rh factor inheritance) of R.A. Fisher and R.R. Race, while "Wiener" indicates the notation (and theory) propounded by A.S. Wiener. Note the correspondence between C and rh', c and hr', E and rh'', e and hr'', and D and Rh_o. The isoantigen corresponding to d (Hr_o) has not yet been recognized. Whether a person is Rh^+ or Rh^- depends on the presence (or absence) of the D (Rh_o) isoantigen in their blood and the corresponding allele in their genome. Further subdivisions of the Rh^+ and Rh^- groups can be made on the basis of the antigenicity of the other four Rh factors (rh', hr', rh'', hr'') in Wiener's notation).

iii)   The isoantigens of the Rh system are genetically complex.
       Two acceptable genetic schema have been suggested for the
       explanation of the expression of the Rh antigens on the
       erythrocyte surfaces. These are the Fisher-Race and the
       Wiener theories of Rh isoantigen inheritance. The
       Fisher-Race theory suggests the presence of five distinct
       determinants designated D, C, E, c, and e, which are
       products of genes situated at three distinct but closely
       linked chromosomal loci. Crossing over is infrequent
       due to the close proximity of the genes to each other.
       It is thought that one set of three alleles is inherited
       from each parent. This is similar to the concept of one
       cistron regulating the product of one antigenic determinant,
       the Rh phenotype, and it can be illustrated as follows.

Chromosomal loci	Antigenic determinant	Rh Phenotype
D	D	
C	C	Rh^+
E	E	
D	D	
C	C	Rh^+
e	e	
d	none	
c	c	Rh^-
e	e	

The gene regulating D is significant since this
expresses the Rh^+ isoantigen important in clinical medicine.
This isoantigen is responsible for Erythroblastosis fetalis
in newborns. It accounts for nearly all (90%) of the
hemolytic diseases of the newborn resulting from maternal
isoimmunization. The alternative allele, small d, is
hypothetical. It represents the absence of D-isoantigen
since the presence of a d-isoantigen has not been shown
to date.

The second theory, that of Wiener, suggests that one chromosomal locus is responsible for the regulation of several antigenic determinants (allele-multiple determinants). This can be shown as follows.

Chromosomal locus	Antigenic determinant	Rh Phenotype
R^1	Rh_o rh' rh''	Rh^+
r'	rh' hr''	Rh^-

In the Wiener scheme R^0, R^1, R^2, and R^z relate to Rh^+ isoantigen and r, r', r'', and r^y represent Rh^-. This is seen in Table 35 in more detail.

   iv) There are approximately twelve other blood group isoantigens of the human erythrocytes. They have been primarily useful in medico-legal and anthropological problems. With the exception of rare transfusion difficulties, the contribution of these groups to diseases has not been ascertained.

b. Histocompatibility Antigens (HLA)

   i) With the advent of tissue transplantation, another major group of isoantigens have become of tremendous importance to clinical medicine. These are the isoantigens found on the surfaces of white blood cells, particularly lymphocytes (HLA stands for histocompatibility lymphocyte antigens). They are also found on platelets and on most of the fixed tissues of the body. These antigens have been utilized for donor selection in tissue transplantation (hence the name histocompatibility).

   ii) The HLA antigenic determinants have not been thoroughly characterized chemically because of the difficulty of separating them from the other components of the cell membrane and because they usually are of a rather heterogeneous molecular composition. Recent work suggests that the HLA alloantigenicity probably resides in a protein (unlike the ABO determinants which are carbohydrates) of

molecular weight about 31,000 daltons. Samples of HLA
antigens have been prepared by methods which remove
non-covalently bound lipids and carbohydrates. These
purified isoantigens are immunogenic proteins containing
less than 1% carbohydrate or lipid indicating that it is
probably not a glyco- or lipoprotein. These findings are
still quite controversial, however, and it may yet be
shown that carbohydrate or lipid is at least partially
necessary for the complete identity of these antigens.

iii)   The genetic control of HLA is accomplished by a complex of
linked genes at two chromosomal loci, termed segregants or
subloci. The HLA system is similar to that of the Rh
system in that the linked genes are transmitted in a group
from one generation to the next. The group of allelic
genes (pair of HLA allelic determinants) contributed by
each parent is referred to as the haplotype. These are
shown schematically as segregated, linked genes according
to Mendelian principles of genetics (see Figure 100).

These genes have many alleles within the population and
are therefore polymorphic. There are approximately 30
known HLA antigens. Phenotypically, no individual can
have more than four major HLA antigens (two from each
segregant series) since there are only four subloci (two
on each member of the chromosomal pair). The segregant
series of HLA antigens and their distribution in the
ethnic groups thus far examined are shown in Table 36.

iv)   The induction of an immune response is noted in vivo as
rejection of the tissue or graft. The primary (or first
set) rejection, which occurs approximately ten days after
transplant, is generally related to the cell-mediated
(T-lymphocyte) system. Secondary (or second set) graft
reaction and hyperacute rejection involve the humoral
system and all the attendant amplification processes
(complement, fibrinolysis, kinins, etc.). In the
secondary response, the graft is generally rejected within
four to five days after transplantation in a sensitized
host. Induction of an immune response can also be
detected in vitro by cytotoxicity and leukoagglutination
tests. Serum from sensitized individuals is incubated
with lymphocytes in the presence of complement and the
viability of the cells is estimated by the dye exclusion
method. Since viable cells do not stain with dyes such
as trypan blue or eosin, the number of surviving cells
may thus be determined. This is the general procedure
utilized in tissue typing. Recipient and donor lymphocytes

Figure 100. Schematic Diagram of the Inheritance of the Histocompatibility Antigens

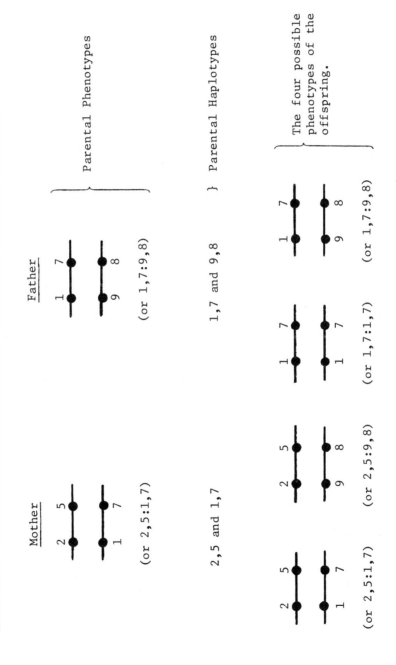

Table 36. Segregant Series and Some Ethnic Distributions of Human Histocompatibility Antigens

Segregant Series		HLA	Ethnic Group	% of the ethnic population having the indicated HLA
First	Second			
HLA-1	HLA-5	HLA-2	Caucasian	46.0
HLA-2	HLA-7	HLA-9	Oriental	56.0
HLA-3	HLA-8	HLA-1,8,3	Oriental	0.0
HLA-9	HLA-12	Te57	Oriental	0.0
HLA-10	HLA-13	Te57	Black	25.0
HLA-11				
Te40	Te50	Te57		
Te59	Te51	Te58		
Te63	Te52	Te60		
Te66	Te54	Te61		
	Te55	Te64		

Note: The Te antigens listed above are tissue antigens found by Terasaki (hence Te) and his coworkers. They have not yet been completely characterized. They cannot be classified as HLA antigens (i.e., members of the HLA series) until it is shown that they map at the same chromosomal loci as do the HLA alleles. Other non-HLA tissue antigens are also known which are not shown here.

are treated, in the presence of complement, with a number
of standard cytotoxic sera capable of reacting with the
major antigens of the two segregant series of HLA (see
earlier) and the percentage of viable cells is determined.
The matching grade of the recipient and donor are classified
as shown in Table 37.

Table 37.   Tissue Typing on the Basis of HLA Antigen

Match Type	Hypothetical Recipient Genotype	Hypothetical Donor Genotype	Remarks
A	2,5:3,8	2,5:3,8	Identical HLA's in both donor and recipient (e.g., identical twins or inbred strains-isologous)
B	2,5:3,8	2,5:3,5	No group mismatches but donor is missing one antigen found in the recipient
C	2,5:3,8	2,5:3,7	One group mismatch; donor has one antigen which recipient does not have (i.e., one unrelated antigen)
D	2,5:3,8	10,12:3,8	Two group mismatches, both on the <u>same</u> chromosome (i.e., one haplotype completely different between donor and recipient)
E	2,5:3,8	2,12:3,7	Two or more group mismatches involving <u>both</u> chromosomes (i.e., both donor haplotypes are at least partially different from recipient haplotypes)
F	---	---	Match is classified as type F if the recipient has an antibody to one of donor's HLA antigens. can occur even in identical twins if one of them has an autoimmunity to one of his own HLA antigens.

v) In any discussion of transplantation immunity, the
terminologies listed in Table 38 are of importance.

Table 38. Transplantation Immunology: Terminology

Nature of Graft	Tissue	Definition
Autograft	autologous	Graft from one part of the body to another in the same individual
Isograft	isologous	Graft from genetically identical individuals (identical twins, inbred strains of animals)
Allograft	homologous	Graft from a genetically dissimilar donor of the same species
Xenograft	heterologous	Graft from a donor of another species

Relative to genetic identity, the terms syngeneic and
isogeneic (isologous), allogeneic (homologous) and
xenogeneic (heterologous) are also used.

vi) In summary, significant properties of the three known
human isoantigen systems are compared in Table 39.
Preliminary evidence suggests that soluble forms of HLA
occur in the sera of normal individuals, in a manner
similar to that of the ABO system. This finding may
have significant implications in the tolerance states
of individuals.

Complement System

This section on complement, designated C, will include discussions
relative to (1) the chemical composition of the 11 components of the
system, (2) the molecular events of complement action leading to immune
cytolysis, (3) a brief mention of complement deficiencies, and (4) the
relationship of this to other inflammatory amplification systems.

Table 39. Comparison of ABO, Rh, and HLA Isoantigens

Characteristics	Blood-Group System		HLA
	ABO	Rh	
Clinical significance	Transfusion reaction Hemolytic disease of the newborn	Erythroblastosis fetalis	Transplantation and Transfusion Reactions
Techniques for the immunosuppression of the humoral and cell-mediated responses	None	Specific anti-Rh antibody given to Rh⁻ mother just prior to Rh exposure	Steroids; antilymphocyte serum (ALS)
Genetic polymorphism	Moderate	Extensive	Extensive
Chromosome loci	Single	Complex	Complex

1. Chemical Characteristics

    a. Complement (C) comprises a group of eleven distinct but
        functionally interrelated plasma proteins. These molecules are
        found in the fresh sera of animals of many species. Complement
        levels tend to fluctuate during certain diseases. In
        immunological diseases, complement consumption occurs so that
        serum levels generally decrease. In some inflammatory states
        where immune systems are not involved, an increase in complement
        activity may occur. Differences in complement activity between
        species, which are attributable to the variation in concentrations
        of one or the other of the eleven components, have been
        demonstrated.

    b. Examination of fresh serum has demonstrated that certain
        components of C are readily destroyed by oxidizing agents and
        by heat. Dialysis of fresh serum against distilled water
        results in the formation of an insoluble protein precipitate
        that can be solubilized in normal physiological saline following
        separation by centrifugation. The aqueous, soluble proteins,
        called the "endpiece", consist of C2-and C4-components, whereas
        the "midpiece" (soluble in saline but insoluble in water)
        contains the C1-and C3-components. Each fraction alone has
        no complement activity but when recombined, C-activity can be
        restored. Heat treatment of fresh serum at 56°C irreversibly
        inactivates C1-and C2-activities, while treatment of fresh
        serum with zymosan (a complex containing mostly protein and
        carbohydrate; prepared from yeast cell walls; also called
        anticomplementary factor) removes C3-component. Dilute ammonium
        hydroxide treatment of fresh serum inactivates C4-component.
        These observations have been known for many years.

    c. With the advent of chromatographic procedures utilizing anionic
        and cationic cellulose derivatives, gel filtration chromatography,
        and zonal electrophoresis, rapid advances in the characterization
        of all of the eleven components have been achieved. The
        significant, known physicochemical characteristics of the eleven
        components of C are shown in Table 40. Molecular weights range
        from a low of 80,000 daltons (C1s and C9) to a high of
        400,000 daltons (C1q). Concentrations of C-components in fresh
        serum range from a low of 2.5 mg/100 ml (C2) to a high of
        125 mg/100 ml (C3). Thus, in immunological assays, C3 and C4
        have been utilized for recognition of complement action in
        immunological diseases. Complement fixed to an antigen-antibody
        complex in tissue has been detected with fluorescence labeled
        anti-C3 in immunofluorescence analysis using fluorescence-
        microscopy. Complement levels have also been estimated by

Table 40. Physicochemical Properties of the Components of Complement

Properties	C1 Clq	C1 Cls	C1 Clr	C4	C2	C3	C5	C6	C7	C8	C9
Molecular Weight (daltons)	40,000	80,000	150,000	250,000	120,000	180,000	160,000	90,000	110,000	150,000	80,000
Sedimentation Coefficient ($S_{20,w}$)	11.1	4.0	7.0	10.0	5.5	9.5	8.7	5-6	5-6	8.0	4.5
Serum Concentration (mg/100 ml)	19.0	22.0	20.0	43.0	2.5	125.0	25.0	10.0	10.0	5-10	5-10
Relative Electrophoretic Mobility	γ	α_1	β_1	β_1	$\beta_1-\alpha_2$	$\beta_2-\beta_1$	β_1	β_2	β_2	γ_1	α_2
Carbohydrate Content (%)	15.0			14.0		2.7	19.0				
Functional Sulfhydryl Content (number of SH groups)					+(2)	+(1-2)	+(?)				

measuring the concentration of C3 in serum by radial immunodiffusion. The majority of the C-components migrate in the β-globulin region in zonal electrophoresis, although Clq, C8, and C9 migrate in the γ, γ_1, and α regions, respectively. Clq is an interesting molecule containing large amounts of carbohydrate, hydroxyproline, hydroxylysine, and glycine. It appears to be chemically related to the collagenic proteins. This is the component which binds to the Fc portion of the antibody in an antibody-antigen complex.

2. Complement Actions, Regulation, and the Alternative Pathway

a. Complement Action

The eleven C-components interact sequentially in a cascade process analogous to the blood-clotting components and their reactions. The C-system is activated by the interaction of an antibody with its specific antigen (present either on a cell surface such as an erythrocyte, parasitic cell, or malignant cell or in an aqueous system as soluble antigens). Classical C-reaction sequence has been thoroughly defined using the lysis of sheep erythrocytes (SRBC-antigen) by their specific antibody (hemolysin) produced in rabbit by administration of SRBC. The general scheme of the complement reactions is shown in Figure 101.

Associated with the sequential activation of the C-system are a series of enzymatic activities and the release of polypeptides with various significant biochemical functions. For example

i) $\overline{C1}$ (where the bar over the 1 indicates <u>activated</u> component 1) possesses an esterase-like activity with C2 and C4 as the primary substrates. Activation of Cl is accomplished by linking Clq (with an affinity for IgG and IgM molecules) to Cls (a protease) and Clr (a protein which converts Cls from a protease to an esterase). The link involved is calcium ion (Ca^{+2}).

ii) Activated C142 ($\overline{C142}$; a complex containing $\overline{C1}$ and active fragments of C2 and C4; also called C3 convertase) cleaves C3, forming large amounts of C3a and C3b. The significant active peptides are C3a and C5a, having anaphylatoxic and chemotactic properties. The physicochemical nature of C3a has been known for some time but C5a has only recently been characterized. C3a (molecular weight = 8700 by gel filtration) is somewhat smaller than C5a (17,500 molecular weight by gel filtration) and is less active on a molal

Figure 101. Diagrammatic Scheme of C-Action

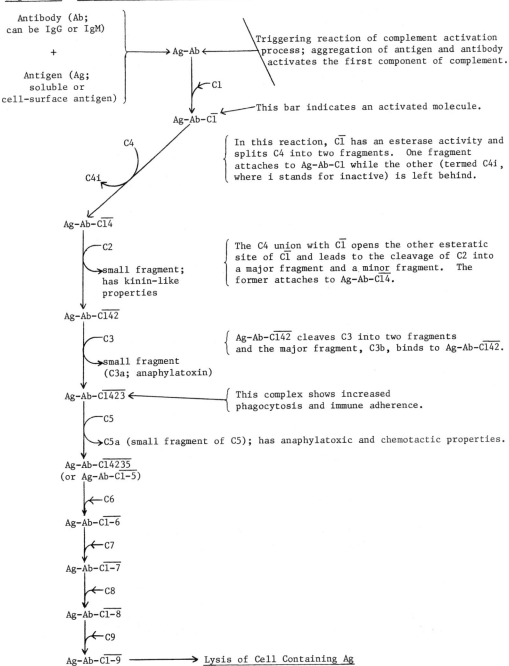

Antibody (Ab; can be IgG or IgM)

+

Antigen (Ag; soluble or cell-surface antigen)

→ Ag-Ab ←

Triggering reaction of complement activation process; aggregation of antigen and antibody activates the first component of complement.

← C1

Ag-Ab-C̄1̄ ← This bar indicates an activated molecule.

C4

C4i

In this reaction, C̄1̄ has an esterase activity and splits C4 into two fragments. One fragment attaches to Ag-Ab-C1 while the other (termed C4i, where i stands for inactive) is left behind.

Ag-Ab-C̄1̄4̄

C2

→ small fragment; has kinin-like properties

The C4 union with C̄1̄ opens the other esteratic site of C̄1̄ and leads to the cleavage of C2 into a major fragment and a minor fragment. The former attaches to Ag-Ab-C̄1̄4̄.

Ag-Ab-C̄1̄4̄2̄

C3

→ small fragment (C3a; anaphylatoxin)

Ag-Ab-C̄1̄4̄2̄ cleaves C3 into two fragments and the major fragment, C3b, binds to Ag-Ab-C̄1̄4̄2̄.

Ag-Ab-C̄1̄4̄2̄3̄ ← This complex shows increased phagocytosis and immune adherence.

C5

→ C5a (small fragment of C5); has anaphylatoxic and chemotactic properties.

Ag-Ab-C̄1̄4̄2̄3̄5̄ (or Ag-Ab-C̄1̄-5̄)

← C6

Ag-Ab-C̄1̄-6̄

← C7

Ag-Ab-C̄1̄-7̄

← C8

Ag-Ab-C̄1̄-8̄

← C9

Ag-Ab-C̄1̄-9̄ ⟶ Lysis of Cell Containing Ag

basis. Using an <u>in vivo</u> assay (wheal and erythema on human skin), C3a was active at a concentration of 2.1×10^{-12} molal while C5a elicited the same response at a concentration of 1×10^{-15} molal. In serum electrophoresis (cellulose acetate, pH 8.5), C5a had a mobility of -1.7 while that of C3a was $+2.1$. The activities of these and other relative peptides is indicated in Figure 101 and Table 41.

b. Regulation of Complement System

This is accomplished by two sets of processes: the presence of <u>specific inhibitors</u> (or inactivators) of the enzymes that participate in the complement system and the <u>instability</u> of some complexes that are produced in the sequential reactions.

   i) C$\overline{1}$-inhibitor (C$\overline{1}$INH) is an α-globulin which inhibits C$\overline{1}$-esterase activity by combining with the enzyme in stoichiometric proportions. A congenital deficiency of C$\overline{1}$INH, known as angioedema, gives rise to recurrent episodes of subepithelial edema of the skin and the upper respiratory and gastro-intestinal tracts. It is inherited as an autosomal dominant trait.

  ii) The inactivation of C3b (the <u>fragment</u> of C3 that attaches to Ag-Ab-C$\overline{142}$ giving Ag-Ab-C$\overline{1423}$) is accomplished by an enzyme that catalyzes the cleavage of C3b to two fragments, C3c and C3d.

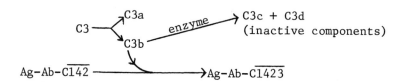

  iii) A substance has been isolated which inactivates C6.

  iv) Inactivation of C3a and C5a (anaphylatoxins) is done by cleavage of the carboxy-terminal arginine from these molecules by carboxypeptidases.

   v) Formation of unstable complexes also provides a control mechanism by preventing the accumulation of intermediates. The <u>unstable</u> complexes are the C2 portion of <u>the</u> Ag-Ab-C$\overline{142}$ complex and the C5 part of Ag-Ab-C$\overline{14235}$.

Table 41. Biologically Active Peptides of Complement and Kinin Systems

Active Peptides	Biological Action
1. Kinin System	
Bradykinin	Increases vascular permeability
"C-kinin"	Increases vascular permeability
Kallikrein	Responsible for chemotaxis of PMN-leukocytes
2. Complement	
C3b	Enhances phagocytosis of PMN leukocytes, with lysosomal release
C3a, C5a, $\overline{C567}$	Increases chemotaxis
C3a, C5a	Anaphylatoxin, (releases histamine from mast cells)
C1-7	Facilitates lymphocytotoxicity
C1-9	Cytotoxic to specific sensitized target cell or adjacent cells; activates tissue lysosomal enzymes

c. Alternative Pathway of Complement Activation

There is an alternative pathway which can activate the later reactions
of the complement system, starting from C3 to C9. An example
of such an activation is the properdin system. This is dependent
on the presence of C3-proactivator (C3PA), a plasma protein with
β-mobility and a molecular weight of 80,000 daltons. C3PA can
be converted to C3-activator, a protein of 60,000 daltons with
γ mobility, by C3PA-convertase which is formed by the action of
various agents (inulin, zymosan, endotoxin, and IgA-Ag
complexes) on a serum precursor. C3PA-convertase transforms
C3PA to C3-activator (C3A) which then activates C3. This
alternative pathway can be shown as follows:

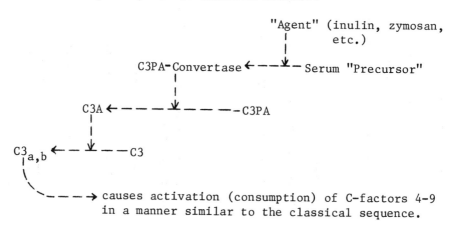

        causes activation (consumption) of C-factors 4-9
        in a manner similar to the classical sequence.

3. Complement Deficiency

Inherited abnormalities of the complement system of man include:
(a) Cl-inhibitor deficiency; (b) deficiency of C2; (c) polymorphism
of C4; (d) deficiency and polymorphism of C3; (e) dysfunction of C5;
(f) abnormalities in the alternative pathway (deficiency of C3PA).
The deficiencies of C2 and C3 have essentially no effect on the
health of the individual. Lack of Cl-inhibitor, on the other hand,
results in hereditary angioedema (HAE), which appears as a
non-pitting, non-pruritic edema of the skin, G.I. tract, and
respiratory mucosa following an allergic reaction. There are two
variants, one due to a complete lack of inhibitor, the other
associated with a defective inhibitor. C3PA deficiency has been
associated with diminished bactericidal activity. C2-and C5-
dysfunctions have also been related to diseases in man. There are
also acquired defects of the human complement system. In lupus
erythematosus and hypocomplementemic nephritis, C3-levels are
depressed. Complement levels may be transiently elevated by
inflammatory diseases and infections.

4. Interrelationships of the Complement, Clotting, Fibrinolysis, and
    Kinin Systems in Inflammation

    a. The inflammatory process is a significant and orderly response
        of living tissue to injury. It comprises a series of biochemical
        and anatomical changes of the terminal vascular bed and of the
        connective tissues. These processes are intended to eliminate
        the injurious agents and to repair the damaged tissues.
        Inflammation involves a great many responses of the host.
        Discussions in this section will be limited to a brief review
        of the biochemical events associated with complement, coagulation,
        fibrinolysis, and kinin generation in their relationships to
        the inflammatory state. Discussions of the inflammatory cells,
        peripheral blood leukocytes, and tissue macrophages will be
        presented later.

    b. The sequence of events resulting from tissue injury which
        make up the inflammatory response are

        i)   Tissue injury results in an increased permeability and
              dilation of blood vessels (erythema) in the injured region.
              Initiation of fibrin deposition via the clotting system
              occurs with the activation of Hageman factor (HF; clotting
              factor XII; see also gout in Chapter 5).

        ii)   The deposited fibrin aids in trapping the deleterious
              agents and platelets within the coagulum. This also
              enhances phagocytosis by neutrophils and macrophages.

        iii) HF subunits (formed by the action of plasmin on activated
              HF) activate prekallikrein by converting it to kallikrein
              (chemotaxis factor). This stimulates the emigration of
              blood leukocytes and plasma cells and these cells
              contribute to phagocytosis and localized antibody synthesis.

        iv)   Increases in blood plasma and tissue fluids at the injured
              site occurs (as shown by swelling-edema) and this elevates
              the local levels of the bactericidal serum factors (C3PA,
              complement, kinin, opsonins, enzymes, etc.). At the
              same time, the increase in fluids contributes to the
              dilution of the toxins formed.

        v)   The appearance of non-specific acute phase proteins such
              as C-reactive protein may occur, which could stimulate
              localized phagocytosis and activation or suppression of
              lymphocytes.

vi) The physiological and biochemical alterations (e.g., an increase in CO_2, decrease in O_2, increase in body temperature, and accumulation of organic acids) which occur at the site of injury may be deleterious to the invading bacteria.

vii) Finally, tissue repair occurs following macrophage action and the appearance of fibroblastic cells and deposition of collagen.

c. Figure 102 is a schematic diagram showing the interrelationship of the coagulation, fibrinolytic, kinin generating, and complement systems and the role of HF. Table 42 summarizes the major factors released in the interactions of inflammation and their biological functions.

d. One of the pivotal plasma proteins in the physiological and biochemical interrelationships of these various systems is HF or clotting factor XII, a γ_1-globulin with a molecular weight approximately of 100,000. Tissue injury activates HF which initiates the intrinsic clotting processes, while tissue thromboplastin (factor III) initiates the extrinsic clotting mechanism. Activated HF together with HF-cofactor converts plasminogen to active plasmin, a proteolytic enzyme. Plasmin acts on the stabilized fibrin clot and fibrinolysis ensues. Plasmin also acts on HF to give HF-subunits. The latter subunits then convert prekallikrein to kallikrein, a chemotactic factor for polymorphonuclear leukocytes. Kallikrein, together with plasmin, converts kininogen, an α_2-globulin, to bradykinin. Kallikrein also acts on C4,2 to give "C-kinin" peptide. Bradykinin and "C-kinin" affect vascular permeability at the injured site. Bradykinin is a non-apeptide (H-Arg-Pro-Pro-Cly-Phe-Ser-Pro-Phe-Arg-OH), various derivatives of which have been synthesized and are available commercially. The action of plasmin on C1, converting it to an active esterase, initiates the C-component reactions in a manner similar to that of the classical pathway, although not as efficient. Table 42 summarizes the blood coagulation proteins and the members of the kinin system, with molecular weights and electrophoretic properties obtained by zonal electrophoresis.

Peripheral Leukocytes and Macrophages

The previous sections dealt with the contribution of the humoral system to the exudation associated with inflammation and of the many biological peptides stimulating emigration of the cells making up the exudates. The present section will briefly describe the pertinent morphological characteristics, functions, and the deficiencies or dysfunctions of these cells in man.

Figure 102.   A Schematic Diagram Showing the Interrelationships of the Coagulative,
Fibrinolytic, Kinin, and Complement Systems

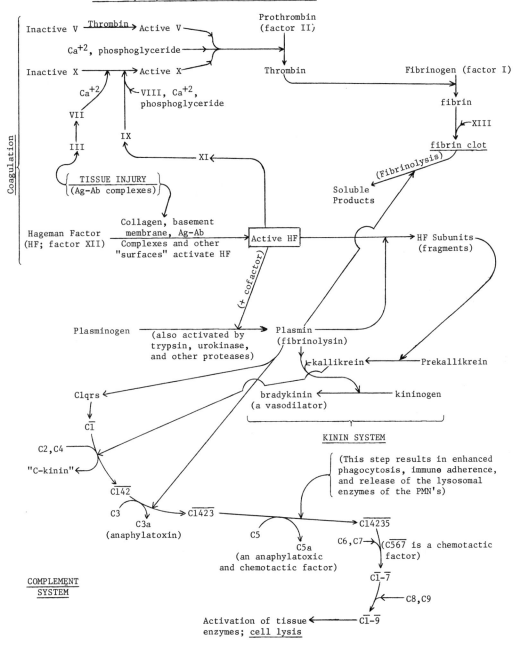

Table 42. Nomenclature and Some Physicochemical Properties of the Blood Coagulation Factors and Members of the Kinin System

International Classification (number)	Synonyms	Molecular Weight		Electrophoretic mobility
		Bovine	Human	
Factor I	Fibrinogen	340,000	341,000	β_1
Factor II	Prothrombin	68,500	69,000	α_2
Factor IIa	Thrombin	33,700	35,000	β
Factor III	Tissue thromboplastin	1.7×10^7	---	---
Factor IV	Calcium	---	---	---
Factor V	Ac-globulin, labile factor, proaccelerin prothrombin accelerator	290,000	70,000-350,000	β_2,β
Factor VII	Stable factor, autoprothrombin I	35,000-63,000	50,000-100,000	β,α
Factor VIII	Antihemophilic globulin	180,000	300,000-400,000	β_2,β
Factor IX	Christmas factor, autoprothrombin II, plasma thromboplastin component	49,900	100,000-200,000	β
Factor X	Stuart-Prower factor	86,000	---	α
Factor XI	Plasma thromboplastin antecedent	---	100,000-200,000	$\beta-\alpha$
Factor XII	Hageman factor	---	60,000-100,000	γ_1
Factor XII (activated)	---	---	100,000	γ_1
Factor XII (fragments; subunits)	---	---	33,000	prealbumin
Factor XIII	Fibrin-stabilizing factor	---	350,000	globulin
Prekallikrein	---	---	127,000	γ_2
Kallikrein	---	---	108,000	γ_2
Plasminogen	pro-fibrinolysin	---	81,000	β
Plasmin	fibrinolysin	---	75,400	β
Kininogen	α_2-globulin	---	70,000	pretransferrin
Bradykinin	Kallidin I	---	1,060	---

1.   Characteristics of Leukocytes

a. Neutrophils

i) Polymorphonuclear neutrophilic leukocytes (PMN-leukocytes;
    see also pages 254-255) are granulocytes which possess a
    number of functional properties important in host
    resistance. These typical inflammatory cells, associated
    with acute inflammation, are essential for the pathogenesis
    of acute, necrotic reactions of immunological diseases,
    such as the vasculitis of the immune-complex disease (serum
    sickness), Arthus' reaction, the severe glomerulonephritis
    of nephritoxic nephritis, and many other immune-complex
    diseases. PMN-leukocytes are, nevertheless, primarily
    involved in the continuous defense against non-immunologically
    related inflammatory infections caused by bacteria and other
    parasites. The major route by which PMN-leukocytes exert
    their defenses is phagocytosis which will be discussed
    in detail later in this section. Neutrophils vary in
    size from 10 to 15 μm diameter.

ii) Neutrophils are the major granulocytes in peripheral blood
    (2 to 3 x 10^{10} cells in the total blood volume and
    1.5 to 3.0 x 10^{12} in bone marrow). A mature neutrophil
    has a half-life of 6.6 hours as measured by use of ^{32}DFP-
    labeled PMN-leukocytes (DFP = diisopropylfluorophosphate;
    see Chapter 3). The nucleus of neutrophils is multilobed
    and in stained smears many pink-staining granules can
    be seen in the cytoplasm. When examined with the electron
    microscope, these granules are of varying density and size
    and have been designated α, β, and γ. They constitute
    the major source of the enzymes used for the killing of
    bacteria following phagocytosis. Some of the enzymes found
    in the cytoplasm of neutrophils are summarized in Table 43.

b. Eosinophils

i) The eosinophilic leukocytes are similar in size and in
    nuclear structure to the neutrophils. The major
    distinguishing characteristics of eosinophils are their
    cytoplasmic granules (0.2 to 1.0 μ), which are much larger
    than neutrophilic granules and which have a strong
    affinity for acidic aniline dyes such as eosin. The
    granules contain large quantities of stable peroxidase,
    lipids, and proteins. Crystalloid structures of various
    patterns can be seen in these granules by electron-
    microscopic examinations. In human eosinophils, the

Table 43.   Some of the Enzymes Associated with the Cytoplasmic Granules of Granulocytes, Macrophages (Monocytic Origin), and Lymphocytes

Cell Type	Organelle	Enzymes
1. Granulocytes		
a. Neutrophil	Lysosome	Acid phosphatase, acid lipase, aryl sulfatases A and B, acid ribonucleases and deoxyribonucleases, cathepsin B,C,D,E, collagenase, phosphoprotein and phosphatidic acid phosphatases, phospholipase, organophosphate-resistant esterase, β-glucuronidase, β-galactosidase, β-N-acetyl-glucosaminidase, α-L-fucosidase, α-1,4-glucosidase, α-mannosidase, α-N-acetylglucosaminidase, α-N-acetyl-galactosaminidase, hyaluronidase, lysozyme
	Peroxisome-related	D-amino acid oxidase, L-α-OH acid oxidase, catalase, myeloperoxidase
b. Eosinophil	Lysosome	Acid protease, β-glucuronidase, aryl sulfatase, nucleases and phosphatase, peroxidase
c. Basophil (mast cell)	Lysosome	Proteases, phospholipase A, glucuronidase, acid and alkaline phosphatases, ATPase
	Granules	Histamine, serotonin, heparin, dopamine, acid mucopoly-saccharides, (dopa, tryptophan and histidine decarboxylases), heparin-synthesizing enzymes of cytoplasm (Note that not all of these substances are enzymes.)
2. Lymphocytes	Peroxisome-related	D-amino acid oxidase, L-α-OH acid oxidase, peroxidase catalase
	Lysosome	Arylsulfatase, β-glucuronidase, β-galactosidase, N-acetyl-β-glucosaminidase, N-acetyl-α-galactosaminidase, α-mannosidase, α-arabinosidase, β-xylosidase, β-cellobiosidase, β-fucosidase, cathepsin D
3. Macrophage	Lysosome and cytoplasmic mitochondria	Acid-phosphatase, β-glucuronidase, cathepsin, esterase, lysozyme, α-glycerolphosphate dehydrogenase, DPN- and TPN-diaphorases (isocitric, lactic, malic, succinic dehydrogenases), uridine diphosphate glucose-glycogen transglycosylase

crystalloid structures have various shapes (oval, squarish,
two or more cores, etc.) and are generally localized in
the central region of the granule. The crystalloids appear
to stain strongly for peroxidase activity when analyzed
by electron-microscope histochemical staining using
3,3'-diaminobenzidine tetrachloride. Eosinophil granules
also contain enzymes similar to these in neutrophils, but
appear to lack lysozyme and phagocytin. Nevertheless,
these granules are considered to be lysosomes. Like the
neutrophils, eosinophils have a short life span in
peripheral circulation. They constitute 2 to 5% of the
total circulating leukocytes of peripheral blood. Bone
marrow contains 200 times more eosinophils than blood,
whereas, the tissues have 500 times more than the blood.
The eosinophils of the tissues are primarily found in the
intestinal walls, skin, external genitalia, and lungs.
Blood eosinophils are regulated by the adrenal
corticosteroids. An increase in these hormones due to
stress or therapeutic administration results in decreased
numbers of eosinophils in the circulating blood.

ii) The functions of eosinophils are unknown, but they do have
limited phagocytic capabilities. When phagocytosis occurs,
the cells can degranulate and in this respect eosinophils
appear to be similar to neutrophils. It has been suggested
that eosinophils contain a factor(s) which neutralizes
histamine, serotonin, and perhaps kinins. This could be
the means by which these mediators of vascular changes
are controlled. This concept helps to explain the
appearance of large numbers of eosinophils during an
allergic manifestation due to an immediate type of
hypersensitivity associated with the release of the
vasoamines. However, evidence for the existence of the
neutralizing factor(s) has not been presented as yet.
The enzymes of eosinophils are summarized in Table 43.

c. Basophils (Mast Cells)

i) Basophils are essentially non-inflammatory cells which are
chemically, structurally, and functionally identical to
the mast cells encountered in connective tissues. These
cells, together with platelets, probably play a significant
role in the release of vasoamines such as histamine and
serotonin. Basophils constitute 0.5% of the circulating
blood leukocytes. They are characterized by the presence
of large electron dense cytoplasmic granules which have
a predilection for basic aniline dyes suggesting that

they contain acid mucopolysaccharides. It has been known for some time that these granules contain heparin which is responsible for the metachromasia. Degranulation of mast cells releases histamine (see Chapter 6) in addition to heparin. In some species, such as the rat, mast cells also contain serotonin (5-hydroxytryptamine) in their cytoplasmic granules. About 30% of the dry weight of mast cell granules is heparin and approximately 35% is basic protein. The histamine content of mast cell granules varies from species to species and from tissue to tissue (7 to 40 µg/cell). Packed peritoneal mast cells of the rat contain 630 to 700 µg serotonin/ml of cells. It has been suggested that mast cell granules are composed of polysaccharide-protein-ion complexes to which vasoamines (histamine, serotonin, dopamine) are bound electrostatically. Other constituents of mast cell granules are shown in Table 43.

ii) Substances which can initiate mast cell or basophil degranulation with the concomitant release of heparin and vasoactive amines, are IgE-Ag complexes, anaphylatoxin (C3A, C5A), basic protein (neutrophil lysosomal protein), and antimast cell antibody plus complement; chemicals such as dextrans, polyvinylpyrrolidine, bee venom, rose thorn; surface active agents such as bile salts, lysolecithin, and Tween 20; and physical agents (heat, ultraviolet radiation, x-rays, and radioisotopes). The mechanisms by which these agents produce degranulation differ, although vasoactive amines are released in all cases. The degranulation process is an active one and requires energy. The primary physiological functions of basophils and mast cells are not clear. Their characteristic perivascular distributions suggest a possible role in the regulation of the permeability of the terminal vascular beds via the release of vasoactive amines. Abnormal release of the vasoactive amines by mast cells contributes to the inflammatory patterns observed in immediate and delayed hypersensitivities and in non-immunological disorders.

d. Platelets

i) The significance of platelets in the intrinsic pathway for coagulation of blood is well known. They are the fundamental formed elements involved in the creation of the hemostatic plug (thrombus) of flowing blood. The brief discussion presented here will pertain to the vasoactive amines of platelets involved in the immunological

reactions. Like mast cells and basophils, platelets
are affected by antigen-antibody complexes in the presence
of complement. In some species, vasoactive amines are
also released in this process. The reaction in rabbit
platelets can be summarized by

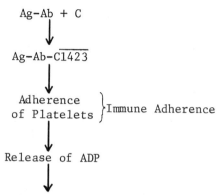

(As mentioned earlier, in immune adherence, neutrophils,
eosinophils, and macrophages also adhere to the Ag-Ab-C1423
complex.) These amines are contained within the platelets
in granules distinct from lysosomes. Energy (presumably
as ATP) is required for the release of the vasoactive
amines and platelet lysis does not usually accompany the
release.

ii) The concentrations of vasoactive amines are lower in
human platelets than in those of the rabbit. Also,
instead of adhering to Ag-Ab-C1423 complexes, they
aggregate directly with the antibody globulin of the
AgAb complexes. Thus, the release of ADP and vasoactive
amines occurs without the intervention of complement.
Platelets of ruminants and pigs are similar to those of
man. However, although the mechanisms are different
in all of the species mentioned here, the end result
(release of vasoactive amines) is the same.

iii) Collagen and several non-immunological materials can also
initiate release of vasoactive amines from platelets.
Other immunological factors which can trigger platelet
release of amines include aggregated IgG, antigen plus
sensitized leukocytes, antiplatelet antibody, and the
enzymes thrombin and trypsin.

e. Lymphocytes in the Plasma

  i) Lymphocytes of the peripheral blood comprise nearly 40 to
      50% of circulating leukocytes and are the major
      immunological cells. They are a heterogeneous population
      in both size and function. Both B-cells and T-cells can
      be found in the circulation. The T-cells constitute the
      major group of lymphocytes in a normal individual (70 to
      80% of the circulating small lymphocytes). B-cell
      lymphocytes constitute the remainder of the lymphocytes
      (20 to 30%) varying in size from small to large
      lymphocytes (8 to 12 μ). The functions of lymphocytes have
      been discussed in earlier sections. These cells are
      the pivotal cells in immunology. The intracellular enzyme
      profile, which is very limited, of these lymphocytes is
      shown in Table 43.

  ii) The major morphologic characteristics of lymphocytes are
      their large homogeneous, non-lobed nuclei, and light-
      staining, non-granular cytoplasm. Plasma cells are rarely
      found in peripheral blood in normal individuals. These
      cells, as indicated earlier, are associated with antibody
      synthesis and are the end cells of B-lymphocyte development.
      Their major morphologic characteristics are their eccentric
      cartwheel-like nuclei. The cytoplasmic area is much larger
      and stains much more intensely with basic dyes (greyish
      blue) than does that of a lymphocyte. Electron micrographs
      show numerous channels of endoplasmic reticulum lined with
      polyribosomes in the cytoplasm. Few or no granules
      characteristic of granulocytes are evident.

f. Monocytes and Macrophages

  i) Monocytes of the peripheral blood measure 12 to 15 μ in
      diameter when examined in stained blood smears. The
      nucleus is generally kidney shaped and the cytoplasm
      stains greyish blue with aniline dyes such as crystal
      violet. The cytoplasm also contains azurophil granules
      which are reddish blue in color. Blood monocytes are
      considered immature macrophages and their primary source
      is the bone marrow. There is general agreement that
      monocytes migrate from the circulation and mature into
      typical tissue and inflammatory macrophages at the site
      of injury. This has been substantiated with radioautographic
      procedures using ^{3}H-thymidine. The source of the
      macrophages in the lung, peritoneal cavity, and inflammatory
      exudate is the bone marrow. Macrophages may also arise

by the mobilization and proliferation of existing tissue
macrophages, migration of Kupffer cells from the liver
into the lungs, the transformation of lymphocytes to
macrophages, and (possibly) differentiation of specialized
endothelial cells to macrophages. Differences and
similarities are evident between macrophage cells from
different tissue sites. This can be shown as follows.

Source of Macrophage	Mitotic Rate	Ability to Adhere and Spread on Glass
1. Sessile or fixed tissue (bone marrow, spleen, and liver)	high	slow
2. Alveolar tissue	high	readily
3. Peritoneal and blood monocytes	low	readily

ii) Tissue macrophages are larger than blood monocytes (15 to
80 µ) and the number of azurophil granules varies. Their
nuclei vary in size and shape and multinucleated forms
may be present. The epithelioid cells of chronic
inflammation (granulomous) are macrophages with ovoid
nuclei resembling epithelial cells. The presence of
diffuse lipids in the cytoplasm gives these cells a pale
appearance. Lipid droplets may also be shown in the
cytoplasm of macrophages. Electron micrographs of
monocytes and unstimulated peritoneal macrophages show
great similarity. Unlike peripheral granulocytes, these
cells have a well-defined Golgi apparatus composed of
flattened sacs, small vesicles, and a few granules;
with a moderate amount of rough endoplasmic reticulum,
varying numbers of electron dense granules, and many
small vesicles in the cytoplasm.

iii) The major function of macrophages, like that of
neutrophils, is the phagocytosis and the degradation of
the ingested material. Significant numbers of neutrophils
are also ingested by macrophages following tissue damage

or an acute infection. The half-life of monocytes has been estimated to be three days. Some of the enzymes found in macrophages are listed in Table 43.

iv) Macrophages, again like <u>neutrophils</u>, are influenced by complement factor ($AgAb\overline{C1423}$ in immune adherence) and by sensitized T-cell lymphokines such as MIF (migration inhibitory factor). The ability of certain antibodies (IgG_1 and IgG_3) to adhere to macrophage membrane (macrophage receptor - MR) permits the IgG to react with its antigen and may play a significant role in macrophage-mediated protection against certain infections. These macrophage-adherent immunoglobulins are known as the cytophilic antibodies. The adherence occurs through the Fc portion of the IgG molecule. Similar binding of IgG to macrophages which react with erythrocytes (sheep RBC) give the typical rosette appearance attributed to macrophages. The role of the macrophage as an accessory or auxiliary cell (A-cell) in the processing of antigens (which is sometimes needed for T-cell--B-cell interactions) has been discussed previously.

2. Phagocytosis

a. Phagocytosis - the process by which a phagocytic cell destroys invading bacteria - can be viewed as a sequence of three events. As discussed earlier, the major phagocytic cells are the neutrophils and the macrophages.

i) <u>Chemotaxis</u>, the attraction of the phagocytes to the site of injury, has been discussed. It is related to the peptides released during complement activation, kallikrein activation, and antigen reaction with specific sensitized T-lymphocytes. In addition, chemotactic factors released by bacteria and from tissue breakdown have been demonstrated. These seem to be small glycoprotein fragments and polysaccharide.

ii) The term <u>phagocytosis</u> is used both for the overall process and for the second step, the engulfment (ingestion) of the bacteria. This occurs when the outer cytoplasmic membrane of the cell surrounds the bacteria, with the formation of a vacuole within the cytoplasm of the phagocyte.

iii) Soon after engulfment, the actual killing of the bacteria begins. Lysosomal (and perhaps peroxisomal) granules fuse with the vacuole and release their enzymatic contents

into it. The resultant packet is now called a
phagolysosome or phagosome, within which the killing
and digestion of the bacteria occurs.

The subunits or fragments formed are egested (eliminated from
the cell) and, in many instances, passed on to the appropriate
lymphocytes for cell-mediated or humoral immune responses.
These subunits or antigens coming from macrophages have been
designated as "super-immunogens" by some investigators.
Figure 103 is a schematic representation of the phagocytic
process.

Phagocytosis is enhanced by naturally occurring opsonins
(immunoglobulins) against bacteria and by specific antibody
immunoglobulins. These effects can be amplified through the
participation of complement and the immune adherence phenomenon
as discussed earlier. Fibrin clots also assist phagocytosis by
creating surfaces and by restricting bacterial migration, both
of which make entrapment easier.

Pinocytosis is used to describe a process similar to
phagocytosis. It involves the intake of soluble material
(proteins) and colloidal suspensions (such as virus particles).

b. Biochemistry of Phagocytosis

  i) The significant biochemical events associated with
      phagocytosis are increases in anaerobic glycolysis
      (Embden-Meyerhof Pathway; pages 155ff) and in the hexose
      monophosphate-shunt pathway (HMP-shunt; pages 241ff).
      These increases are necessary for the ingestion step.
      Increases in glucose utilization and in the formation of
      lactic acid and hydrogen peroxide occur following
      phagocytosis. Energy increases of from 5 to 40% over those
      in a resting leukocyte have been shown to occur in a
      leukocyte which is actively ingesting bacteria. Thus,
      suppression of glycolysis or the HMP will result in
      decreased ingestion of bacteria. The H_2O_2 (which increases
      to four to six times normal levels following active
      phagocytosis) plays a significant role in killing the
      ingested organisms. Hydrogen peroxide plus peroxidase
      and chloride ions or other appropriate hydrogen-donors
      have been shown to play a significant role in killing of
      ingested bacteria. In some instances, halogenation
      (involving Cl^-, hydrogen peroxide, and peroxidase) of the
      bacterial membrane resulted in the death of the organism.
      In other instances, the aldehydes and ketones formed
      through peroxidatic action have been shown to be the
      toxic substances.

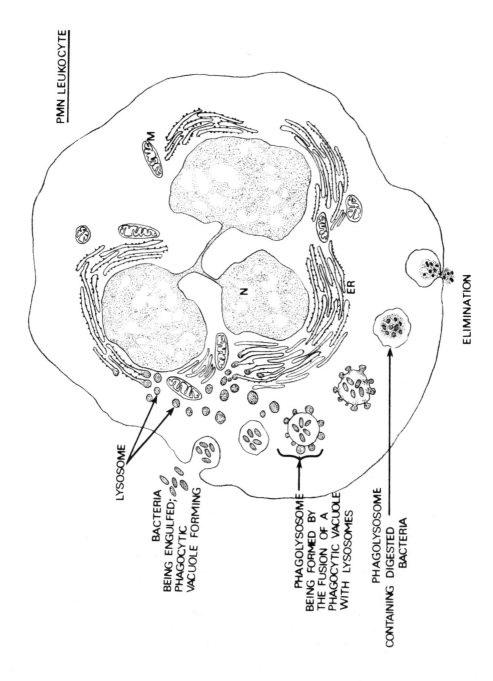

PMN LEUKOCYTE

M

N

ER

LYSOSOME

BACTERIA
BEING ENGULFED;
PHAGOCYTIC
VACUOLE FORMING

PHAGOLYSOSOME
BEING FORMED BY
THE FUSION OF A
PHAGOCYTIC VACUOLE
WITH LYSOSOMES

PHAGOLYSOSOME
CONTAINING DIGESTED
BACTERIA

ELIMINATION

Figure 103 . PHAGOCYTOSIS OF BACTERIA BY PMN LEUKOCYTE
(N-NUCLEUS, M-MITOCHONDRIA, ER-ENDOPLASMIC
RETICULUM)

ii) Hydrogen peroxide, in the presence of ascorbic acid, results in the formation of free radicals which could alter membrane surfaces such as those of the ingested bacteria. This effect would increase the vulnerability of the bacterial surface to the lysosomal enzymes present in the phagolysosome. For example, it has been shown that H_2O_2 plus ascorbic acid mixtures can render <u>Salmonella</u> membrane surfaces susceptible to lysozyme attack by exposing the muramic acid (2-amino-3-0-(1-carboxyethyl)-2-deoxy-D-glucose) residues previously protected by the membrane capsules of the bacteria. Evidently, the free radicals formed depolymerize the protective covering of the plasma membrane, since any reducing chemical, such as thiosulfite, which inhibits free radical formation, reverses this phenomenon. In this context, it is interesting to note that macrophages have a high ascorbic acid content.

iii) Once the ingested bacteria are killed, the complete destruction follows, catalyzed by the numerous hydrolases found in the lysosomes. The enzymes contained in the phagocytes have been summarized in Table 43. A congenital absence of any of these enzymes causes one of the so-called <u>lysosomal disorders</u> discussed on pages 622-628. The synthesis of H_2O_2 as it occurs in PMN leukocytes is presented on pages 254-255. The oxidases (which produce H_2O_2 from O_2) and the peroxidases (which degrade it to H_2O) are probably those enzymes which are found in the peroxisomes. The aldehydes and ketones are perhaps formed by $NADP^+$-dependent dehydrogenases but this process, as it pertains to phagocytosis, is not well understood in man.

3. Leukocyte Deficiencies

Defects in the non-specific immune system associated with neutrophils have been demonstrated. These changes include (1) quantitatively, a decrease in the number of neutrophils (neutropenia); and (2) qualitatively, defects in some metabolic or biochemical function of these cells, but normal morphologic appearance and number.

a. Neutropenias associated with decreased production of myeloid precursors due to bone marrow defects have been observed. There are acquired and inherited forms of neutropenias. The acquired types can be caused by drugs, pollutants, radiation, endotoxin, overwhelming infections, neoplasia, and disorders of the bone marrow and the spleen. The inherited forms are x-linked recessive or autosomal dominant characteristics.

Increased susceptibility to infections is common in both
acquired and inherited neutropenias. <u>Micrococcus pyogenes</u>,
<u>Streptococcus pyogenes</u>, and <u>Diplococcus pneumoniae</u> are the
common causes of infection in the neutropenic individual.

b. Qualitative changes in neutrophils have been associated with
    <u>chronic granulomatous disease</u> (CGD; see page 255). Transmission
    is familial, presumably as an x-linked recessive trait. However,
    both males and females have been affected, although at different
    ages. Defects in the biochemical pathways have been implicated
    in the inability to kill low-grade pathogens such as
    <u>Escherichia coli</u>, <u>Micrococcus pyogenes</u>, <u>Serratia marcescens</u>,
    <u>Paracolon hafnia</u>, and <u>Klebsiella enterobacter</u>. CGD has been
    associated with a deficiency of glucose-6-phosphate dehydrogenase,
    a decrease in NADPH oxidase, a deficient myeloperoxidase, or
    defects in the glutathione peroxidase and reductase systems.
    Since all of these defects have not been found in any one
    patient, the defect in CGD has yet to be completely elucidated.

c. Myeloperoxidase deficiency in leukocytes has been reported which
    is associated with an inability of the leukocytes to kill
    ingested bacteria. The generation of H_2O_2 is severely diminished
    and the ability to break down H_2O_2 is defective in leukocytes
    with this disorder.

Abnormalities of the Immune System

1. Three major categories of immune diseases can be considered.

    a. Diseases characterized by abnormal proliferation of the lymphoid
        cells (B-lymphocytes and plasma cells) which normally synthesize
        immunoglobulins. Abnormal increases in plasma cells are
        associated with excessive immunoglobulin synthesis and the Ig
        produced is characteristic of the disorder;

    b. Diseases characterized by immunodeficiencies, either congenital
        or acquired, with a decrease or lack of Ig synthesis,
        agammaglobulinemia (B-lymphocyte deficiency), or decrease in
        cell-mediated immunity (T-lymphocyte deficiency or defect);

    c. Diseases associated with lymphocytic proliferation with or
        without an increase in immunoglobulin synthesis and with
        general deficiencies in the T-cell system.

2. Diseases in the first and third categories are immunoproliferative
    disorders which include multiple myeloma, macroglobulinemia, heavy
    and light chain diseases. The third also includes lymphomas,

lymphosarcomas, lymphatic leukemias, and Hodgkins disease. These two groups are also synonymous with neoplasia. By immunofluorescence procedures, using fluorescence-labeled antibodies specific for IgM, IgG, IgA, IgD, and IgE and kappa and lambda chains, the specific nature of the lymphoproliferative disorder in lymphatic leukemias has been ascertained in some studies.

3. The multiple myelomas are also called gammopathies and are classified according to the immunoglobulin whose level is increased in the circulating blood. For example, a myeloma with an increase in IgG and decrease in IgM and IgA is called a monoclonal gammopathy-IgG. Increases in two or more immunoglobulins are called polyclonal gammopathies. For example, if IgG and IgA are increased, the disorder is referred to as a polyclonal gammopathy-IgG,A. Figure 104 shows normal immunoglobulin synthesis compared to the situation in a monoclonal gammopathy-IgG.

4. The major criteria for the diagnosis of multiple myelomas are

   a. abnormal quantitative increase in serum immunoglobulin;

   b. changes in the x-ray patterns of the bones, especially the skull, vertebrae, and long bones;

   c. a marked increase in number of pre-plasma and plasma cells in the bone marrow.

   The changes in serum can be detected by qualitative changes of the immunoelectrophoretic pattern and by the presence of an M-protein peak in cellulose acetate electrophoresis. This M-protein is the immunoglobulin (or Ig-chain) which is being synthesized by the abnormally proliferating B-lymphocyte. It can be a myeloma globulin (IgG, A, D, or E), a Bence-Jones protein (free kappa-or free lambda-chains) or a combination of the two. An increase in total serum protein is suggestive of but not diagnostic for a multiple myeloma. Quantitation of the immunoglobulin by radial immunodiffusion with specific anti-Ig is highly important in the diagnosis of multiple myeloma.

5. Because an increase in serum immunoglobulins is normally an expression of stimulation by specific antigens, the immunoproliferative disorders such as multiple myeloma – a classical monoclonal gammopathy – are an enigma, since no specific antigens have been implicated in any of these diseases. Recently, however, the IgM of macroglobulinemia in man and the IgA of multiple myeloma of the experimental mouse have been shown to interact in an antigen-antibody type reaction with, respectively, lecithin and pneumococcal

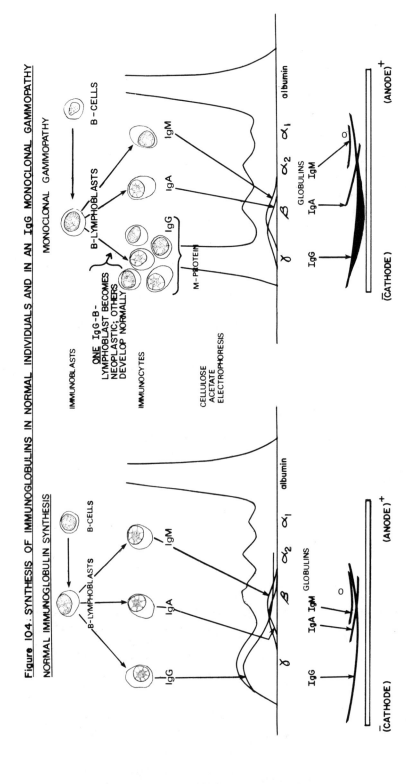

Figure 104 : SYNTHESIS OF IMMUNOGLOBULINS IN NORMAL INDIVIDUALS AND IN AN IgG MONOCLONAL GAMMOPATHY

C-polysaccharide. The antigenic determinant in the C-polysaccharide-IgA reaction was shown to be phosphorylcholine. The immunoglobulins of monoclonal gammopathies have contributed immeasurably to the elucidation of structure and thus to structural-functional relationships of the immunoglobulins.

6. The significant immunoproliferative and immunodeficiency diseases are summarized in Tables 44 and 45. Figure 105 summarizes the areas of the immune system which are affected (indicated by a heavy bar) by the diseases associated with immunodeficiency. Pertinent remarks related to these diseases are also included. More details are available in pathology texts and in specific reviews pertaining to these diseases.

7. A significant increase in the incidence of cancer has been observed in patients with immunodeficiency disorders such as the Wiskott-Aldrich syndrome and ataxia-telangiectasia. A high incidence of cancer is also more common in patients undergoing long periods of immunosuppressive therapy following kidney transplants. These findings strongly support the concept of "immune surveillance" against cancer as expressed by Burnet and others. In addition, patients with lymphoproliferative disorders tend to have depressed T-lymphocyte functions.

Table 44. Immunoproliferative Disorders

Diseases	Immunoglobulin					Characteristics
	IgG	IgA	IgM	IgD	IgE	
1. Multiple Myeloma						Bone marrow changes; osteoporosis; increase in plasma cells; increase in a single immunoglobulin (M-protein in cellulose acetate electrophoresis); general decrease (D) in other Ig; 30 to 40% of the patients have Bence-Jones protein in urine
Monoclonal Gammopathy-G	I	D	D	D	D	
Monoclonal Gammopathy-A	D	I	D	D	D	
Monoclonal Gammopathy-D	D	D	D	I	D	
Monoclonal Gammopathy-E	D	D	D	D	I	
2. Macroglobulinemia						Increase in IgM with general decrease in the other Ig; changes are in lymphoid tissue, hyperplasia, M-protein peak
Monoclonal Gammopathy-M	D	D	I	D	D	
3. Heavy Chain Disease	H-chains in serum and urine; abnormal γ, α, μ H-chains have been reported					Lymphoid cell hyperplasia; other Ig generally decrease; M-protein peak
4. Light Chain Disease	L-chains in serum and in urine					Bone marrow changes; increase in plasma cells; generally no M-protein peak in cellulose acetate electrophoresis
5. Hodgkin's disease, lymphoma, lymphocytic leukemias	Variable Ig levels, dependent on whether T- or B-cell hyperplasia occurs					Changes in lymphoid tissues; hyperplasia; anaplasia; variable sizes of cells; specific cell characteristics (Hodgkins-Reed-Sternberg giant cells)

Table 45. Immune Deficiency Disorders

Nature of Deficiency	Inheritance	Characteristic Defects and Pattern	Susceptibility
1. Cell-Mediated			
a. Congenital thymic aplasia (Di Georges' Syndrome)	---	Defect in embryogenesis; lack thymus and parathyroid; defect of the third and fourth pharyngeal pouches	Viruses, fungi, bacteria
b. Congenital thymic dysplasia (Nezelof Syndrome)	Autosomal recessive	Faulty dysgenesis of thymus; normal immunoglobulin synthesis; defect in cell-mediated immune system	Candida, Herpes, Pneumocystis carinii, Pneumococcus, Streptococcus, Staphylococcus
2. Humoral Immunity			
a. Agammaglobulinemia			
i) Congenital (Bruton Type)	X-linked (males); autosomal recessive (males and females)	Congenital; generally normal until nine months of age or older (some until age 5 to 6 years); immunoglobulins 0 to less than 100 mg/100 ml; plasma cells absent in lymph nodes	Micrococcus pyogenes, Diplococcus pneumoniae, Streptococcus, Pneumocystis carinii
ii) Acquired	---	Acquired; seen at any age, "Primary"--no underlying disease; "Secondary"--associated with disease	---

Table 45.   Immune Deficiency Disorders (continued)

Nature of Deficiency	Inheritance	Characteristic Defects and Pattern*					Susceptibility
			IgG	IgA	IgM	IgE	
iii) Transient hypogamma-globulinemia	---	Gamma-globulin synthesis delayed until 18-30 months of age; normal at three months; cell-mediated response intact; infections of skin, meninges, and respiratory tract					Gram positive bacteria
b. Dysgammaglobulinemia (Selective Ig deficiencies)							
i) Type I	X-linked (males)	(150-1000 mg per 100 ml)	D	D	I	---	Increased pyogenic infections of lung and skin
		Manifest autoimmune disorders					
ii) Type II	---	Nodular lymphoid hyperplasia	N	D	D	---	Increased
iii) Type III	---	Associated with autoimmune disease	N	D	N	I	Recurrent respiratory infections
iv) Type IV	(Familial?)		D	N	N	N	---
v) Type V	(Familial?)	Low levels of isohemagglutinin	N	N	D	---	Gram-negative bacteria

Table 45. Immune Deficiency Disorders (continued)

Nature of Deficiency	Inheritance	Characteristic Defects and Pattern*				Susceptibility
		IgG	IgA	IgM	IgE	
vi) Type VI	---	N	N	N	---	Staphylococcus
		Deficiency in IgG subclass				
vii) Type VII	---	D	I	N	---	Pneumonias
c. Wiskott-Aldrich Syndrome	X-linked recessive (males)	N	N	D	---	Viruses: Vaccinia, and Herpes; Gram- and Gram+ bacteria; Fungi: C. albicans; Protozoa: P. carinii.
		Absence of isohemagglutinin; thrombocytopenia; eczema; inability to process polysaccharide antigens; some reconstitution with transfer factor also occurs in some T-cell deficiencies				
d. Ataxia-telangiectasia	Autosomal recessive	N?	D	N?	D	Increased; respiratory infections.
		T-cell defects also; degeneration of cerebellum with ataxia				
3. Cell-Mediated and Humoral Swiss type, thymic dysplasia, agamma-globulinemia	X-linked recessive, autosomal recessive	Peripheral lymphoid; lack B- and T-cells; defect at lymphoid stem cell; no humoral or cell-mediated responses; lymphopenia and agammaglobulinemia				Increase: viral fungal, bacterial (Gram+ and Gram Protozoan infections.

Table 45.   Immune Deficiency Disorders (continued)

Nature of Deficiency	Inheritance	Characteristic Defects and Patterns*	Susceptibility
4. Reticular dysgenesis	---	Stem cell defect; infants die within the first week of life; thymic dysplasia; lymphopenia; lymphoid depletion; neutropenia; depletion of myeloid cell precursors	Severe sepsis due to Staphylococcus infection

* N = normal; D = decreased; I = increased; all refer to the serum levels of the indicated immunoglobulin.

Figure 105. <u>Summary of the Steps Affected in the Immune Deficiency Diseases</u>

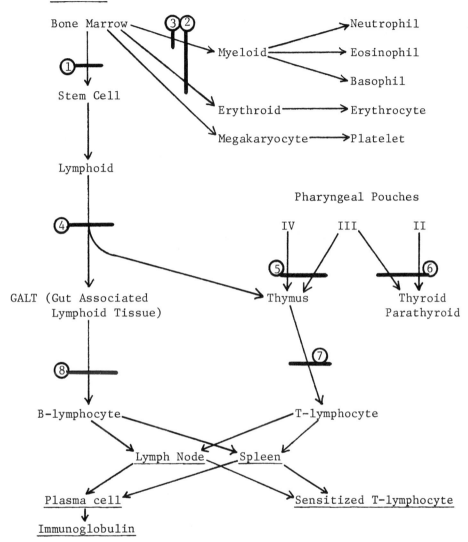

Note: The numbered bars on the diagram indicate the step or steps affected in each of the following disorders:

1. Reticular Dysgenesis
2. Fanconi's Syndrome (page 476)
3. Chronic Granulomatous Disease (page 255) and Myeloperoxidase deficiency
4. Thymic Dysplasia and Agammaglobulinemia (Swiss Type)
5. Thymic Aplasia (III and IV; Di Georges' Syndrome)
6. Thymic Aplasia (II and III)
7. Thymic Dysplasia (Nezelof Syndrome)
8. Agammaglobulinemia (Bruton Type)

12

Water and Electrolyte Balance

with John C. McIntosh, Ph.D.*

Water Metabolism

1. Water is the most abundant body constituent, comprising 45 to 60% of the total body weight (see Figure 106). In a lean person, water accounts for a larger fraction of the body mass than it does in a fat person, accounting for this wide variation. Since biochemical reactions take place in a predominantly aqueous environment, control of water balance is an important requirement for homeostasis.

Figure 106. Proportional Distribution of Solids and Water in a Healthy Adult**

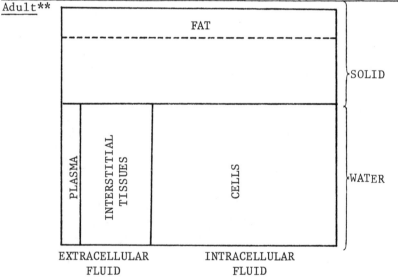

2. Although water permeates freely throughout the body, the various solutes are less mobile due to barriers in the form of the various membrane systems. This gives rise to various fluid pools or compartments of differing composition (see Figure 107). The composition of each compartment is kept constant by a combination of active and passive processes.

---

* Clinical Chemist, Pathology Associates Medical Laboratories, Honolulu, Hawaii.

** From Baron, D.N.: A Short Textbook of Clinical Biochemistry, Philadelphia, 1969, J.B. Lippincott, page 1.

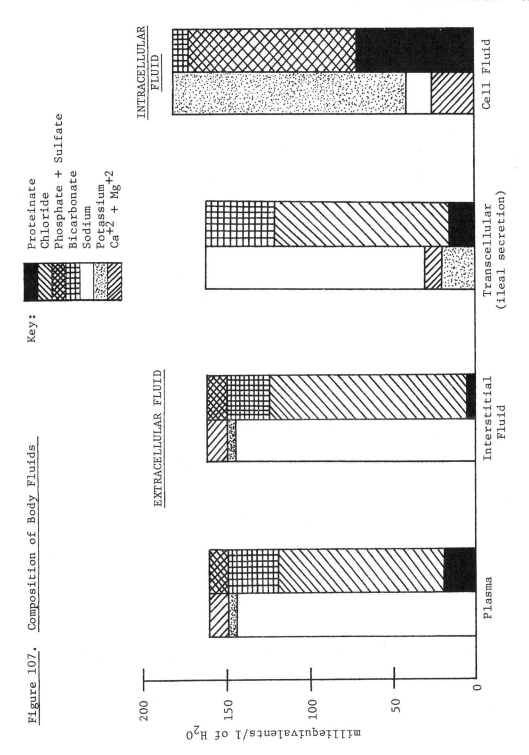

Figure 107. Composition of Body Fluids

Key:

Proteinate
Chloride
Phosphate + Sulfate
Bicarbonate
Sodium
Potassium
$Ca^{+2} + Mg^{+2}$

INTRACELLULAR FLUID

EXTRACELLULAR FLUID

Cell Fluid

Transcellular
(ileal secretion)

Interstitial
Fluid

Plasma

milliequivalents/l of H_2O

200

150

100

50

0

a. Intracellular fluid comprises 30 to 40% of the body weight
    (about two-thirds of the total body water). Potassium is the
    predominant cation with magnesium being the next most common.
    The anions are mainly protein residues and organic phosphates,
    chloride and bicarbonate being low in intracellular fluid.

b. Extracellular fluid has sodium as the predominant cation and
    accounts for 20 to 25% of the body weight (one-third of the
    body water). It can be subdivided into four pools   (1) vascular
    fluid; (2) interstitial fluid; (3) transcellular fluid;
    (4) bone-dense connective tissue fluid.

    i)   The vascular pool is the circulating portion of the
         extracellular fluid. It is characterized by a high
         protein content which does not readily pass the
         endothelial membranes.

    ii) Interstitial fluid is the solution that actually surrounds
         the cells. It accounts for 18 to 20% of the body water.
         It exchanges with the vascular space via the lymph system.

    iii) The transcellular pool includes the fluid present in
         secretions of the body, including the digestive juice,
         intra-ocular fluid, cerebrospinal fluid (CSF) and
         synovial (joint) fluid. These solutions are actively
         secreted by specialized cells and can differ considerably
         from the rest of the extracellular fluid. Under normal
         conditions they rapidly recycle with the main pool.

    iv) Dense connective tissue (bone, cartilage) fluid exchanges
         slowly with the rest of the extracellular fluid so it is
         considered separately. It accounts for 15% of the total
         body water.

3. Water Movements

Unlike ions, water is not actively secreted. Its movements are
due to osmosis and filtration. In osmosis water will move to the
area of highest solute concentration. Thus the active movement
of salts into an area creates a concentration gradient down which
water will flow following the salt movement. In filtration,
hydrostatic pressure in the arterial blood moves water and
non-protein solutes through specialized membranes to give a protein
free filtrate. The formation of the renal glomerular filtrate
is such a process. Filtration also accounts for the transfer of
water from the vascular space into the interstitial compartment.
This fluid movement is opposed by the osmotic pressure (oncotic
pressure) caused by the proteins present in the plasma (Starling's
Law; see page 364).

4. Ionic Pumps

Cells have the capacity to move ions (especially Na^+ and K^+) against a concentration gradient. The postulated "sodium pump" actively transports sodium across cellular membranes, usually from the interior to the exterior of the cell. It is an active process requiring metabolic energy and is stopped by cooling and by metabolic inhibitors. This carrier system may be associated with a magnesium-, sodium-, and potassium-dependent ATPase which is inhibited by ouabain (a digitalis derivative). This enzyme has been isolated from cellular membranes. The sodium pumping action is frequently coupled to the cellular uptake of potassium as shown below. As is indicated, it appears that two potassium ions are pumped into the cell for each three sodium ions secreted.

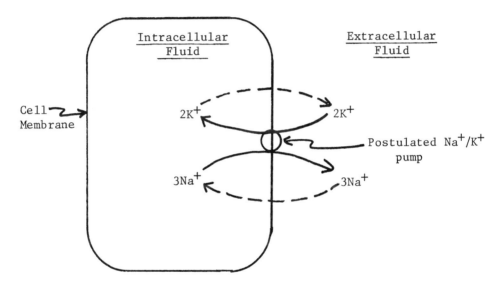

Note: Passive processes (diffusion) are represented by dashed lines while active (energy requiring) processes are shown by solid lines. Diffusion occurs down a concentration gradient (from more concentrated to more dilute); active transport moves ions up a concentration gradient.

This is the postulated mechanism by which the difference in cation concentrations between intra- and extracellular fluid is explained. Other postulated pumps include a K^+-H^+ exchange by cells while the renal tubular cells have an active H^+ pump which results in the formation of an acid urine.

Renal Physiology

As the kidney is the major organ regulating extracellular composition and volume, a brief explanation of renal physiology is needed to understand its function. Further details can be obtained from standard physiology texts.

1. Three main processes occur in the renal nephron. (1) Formation of a virtually protein-free ultrafiltrate at the glomerulus; (2) Active reabsorption (principally in the proximal tubule) of solutes from the glomerular filtrate; (3) Active excretion of substances such as hydrogen ions into the tubular lumen, usually in the distal portion of the tubule (see Figure 108).

2. Under normal circumstances, the amount of sodium reabsorbed from the glomerular filtrate exceeds 99% of that originally filtered by the glomerulus. Most (65-85%) of this reabsorption occurs in the proximal tubule. It is thought that entry of sodium from the filtrate into the tubule cells is a passive process. The cation is then actively transported to intercellular channels which line the tubule. This produces a gradient and results in the passive diffusion of further salt and water from the tubular lumen. It has been calculated that one-third of the sodium is transported actively and the remainder is passively transported. In the loop of Henle and the distal tubule, the remaining sodium is absorbed by active processes, possibly coupled with H^+ and K^+ secretory mechanisms.

3. The amount of sodium delivered to the tubules varies with the glomerular filtration rate (GFR). When the filtered load of sodium is increased, the readsorption of sodium and water increases and when the load falls, the readsorption of sodium falls. This means that, in spite of GFR variation, the excretion of sodium does not change. This phenomenon is called glomerulotubular balance and implies that the nephron responds to a varying sodium load. This is more important in animals like the dog where the GFR is variable, usually rising after a meal due to the increased urea load.

4. K^+ ion is thought to be totally reabsorbed in the proximal tubule. The potassium appearing in the urine is due to additional secretion in the distal tubule.

5. Reabsorption of Water

    a. The normal GFR is 100-120 ml/min which means that about 150 liters of fluid pass through the renal tubules each day. Ninety-nine percent of this is reabsorbed, giving an average

Figure 108.   A Summary of Renal Physiology As It Relates to Electrolytes
$(Na^+$ and $K^+)$ and Water

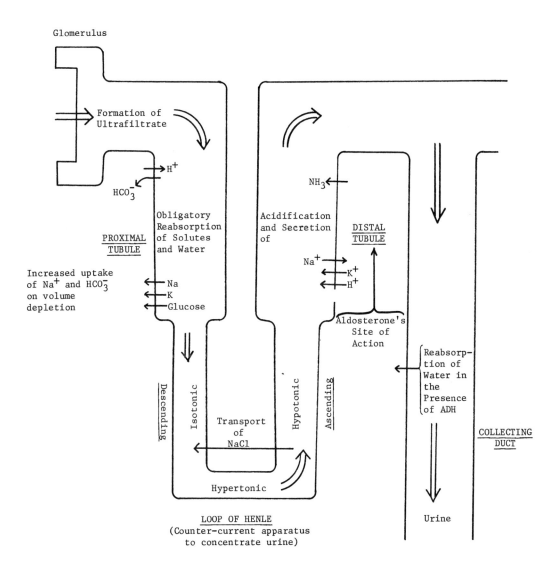

Glomerulus

Formation of
Ultrafiltrate

$\rightarrow H^+$

HCO_3^-

Obligatory
Reabsorption
of Solutes
and Water

PROXIMAL
TUBULE

Increased uptake
of Na^+ and HCO_3^-
on volume
depletion

$\leftarrow$ Na
$\leftarrow$ K
$\leftarrow$ Glucose

Acidification
and Secretion
of

$NH_3 \leftarrow$

DISTAL
TUBULE

Na^+
$\leftarrow K^+$
$\leftarrow H^+$

Aldosterone's
Site of
Action

Reabsorp-
tion of
Water in
the
Presence
of ADH

Descending   Isotonic   Hypotonic   Ascending

Transport
of
NaCl

Hypertonic

COLLECTING
DUCT

Urine

LOOP OF HENLE
(Counter-current apparatus
to concentrate urine)

daily urine volume of 1-1.5 liters. Approximately 80% of the <u>water</u> is reabsorbed in the proximal tubule, a consequence of the active absorption of solutes in this region. The filtrate is still iso-osmotic with the blood, however, and water reabsorption in the rest of the tubule can be varied according to the water balance of the individual as opposed to the <u>obligatory</u> water reabsorption in the proximal tubule.

b. The facultative absorption of water depends on the loop of Henle establishing an osmotic gradient by the secretion of sodium ions from the ascending and the uptake of Na^+ by the descending loop. As a result, the tip of the loop is hyperosmotic (1200 mOsmol) and the distal end is hypo-osmotic with respect to blood. The collecting ducts run through the hyperosmotic region. In the absence of ADH, the membranes of the cells comprising the ducts are relatively impermeable to water. They become permeable in the presence of ADH, water is withdrawn, and the urine becomes hyperosmotic with respect to the blood.

Homeostatic Controls

1. The composition and volume of extracellular fluid is regulated by a number of complex hormonal and nervous mechanisms which are only partially understood. They can be divided into three groups although there is interaction between all three: (1) osmolarity; (2) volume control; (3) pH control.

2. The osmolarity of the body fluid is kept within narrow limits (260-300 mOsmol/kg of body water) through regulation of water intake (via a thirst center?) and water excretion by the kidney under the influence of an antidiuretic hormone (ADH; vasopressin). The osmolarity of extracellular fluid is mainly due to sodium ion and its accompanying anions.

3. The volume of extracellular fluid is kept relatively constant, provided an individual's weight remains constant to within $\pm$ 1 kg. Volume receptors measure the effective circulating blood volume and a decreased blood volume leads to stimulation of the renin-angiotensin-aldosterone system with retention of sodium ions. The increased sodium ion levels lead to rises in osmolarity and stimulation of ADH, resulting in increased water retention. An antagonistic system has been postulated but no definitive evidence has been accumulated. This system would result in increased Na^+ excretion when extracellular fluid volume was increased.

4. The pH of extracellular fluid is kept within very narrow units
    (7.35-7.45) by means of buffering mechanisms and regulation by
    lungs and kidneys. The buffers in the blood were discussed in
    Chapters 1 and 7.

5. The three systems do not act independently. For example, acute
    blood loss results in the release of ADH and aldosterone to restore
    blood volume and renal regulation of the pH leads to ionic shifts
    in potassium and sodium.

Water and Osmolarity Control

1. In spite of considerable variations in fluid intake, an individual
    maintains water balance and a constant composition of body fluids.
    The homeostatic regulation of water is described in this section.
    Figure 109 summarizes the factors which will be covered and their
    interactions.

Figure 109.   Regulation of Osmolarity Within the Body

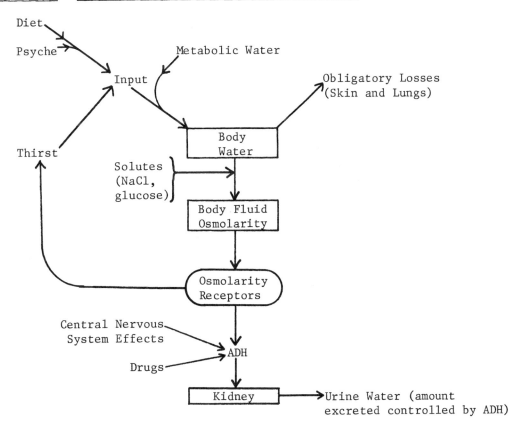

a. Sources of Water

   i) Two to four liters of water a day are consumed in food and
       drink.

   ii) 300 ml of metabolic water are formed each day by
        oxidation of lipids and carbohydrates (see pages 217-218).

b. Loss of Water

   i)   Perspiration⎫ 1 liter/day; increased by fever or
   ii) Lungs   ⎬ hot climatic conditions.

   iii) Gastro-intestinal tract. Although a large volume of fluid
         is secreted during digestion, most of the water is
         reabsorbed and losses in the feces are small. However,
         disturbances in the gastro-intestinal tract can lead to a
         considerable loss of salts and water with profound
         effects on the homeostasis of these materials (see later).

   iv) One to two liters of water are excreted each day in the
        urine. This volume can be regulated by the body to
        maintain homeostasis.

2. The water balance of the body is regulated to maintain constant
    osmolarity of the body fluids.

   a. If two solutions of unequal concentration are separated by a
       semi-permeable membrane, solvent (and any solute particles small
       enough to pass through the membrane) will move through the
       membrane, by osmosis, from the low-concentration side into the
       solution of high concentration. Osmotic pressure is defined as
       the hydrostatic pressure which must be applied (to the solution
       of higher concentration) in order to prevent this solvent flow.
       When one speaks of the osmotic pressure of a solution, it is
       the value relative to pure water (i.e., the pressure differential
       across the membrane when the solution is on one side and water
       is on the other).

   b. Osmotic pressure is directly related to the number of particles
       present per unit weight of solvent. Thus it is a colligative
       property, as are boiling point elevation (and vapor pressure)
       and freezing point depression. A solution containing one mole
       of particles in 22.4 kg of water (22.4 l at 4°C) exerts an
       osmotic pressure of one atmosphere and is said to have an
       osmolality of 0.0446. Conversely, the osmotic pressure of a
       one osmolal solution (one mole of particles in one kilogram

of water) is 22.4 atmospheres. In this sense, "number of
particles" is roughly defined as the quantity of non-interacting
molecular or ionic groups which are present. Since glucose
does not readily dissociate, one mole of this compound
dissolved in one kilogram of water (a one molal solution)
produces one mole of "particles" and the solution has an
osmolality of one. Sodium chloride, on the other hand,
dissociates completely in water, forming two particles from
each molecule of NaCl. Thus a one molal solution of NaCl is a
two osmolal solution. A one molal solution of Na_2SO_4 or $(NH_4)_2SO_4$
is a three osmolal solution, and so on. The number of moles of
a substance which, when added to one kg of water, produces a
one osmolal solution, is referred to as an osmol. One osmol of
glucose is one mole, while one osmol of NaCl is 0.5 moles and
one osmol of Na_2SO_4 is 0.333 moles. In practice, a milliosmol
(mOsmol), equal to one-one thousandth of an osmol, is the unit
usually used.

c. When working with aqueous solutions, osmolarity is sometimes used
    interchangeably with (or instead of) osmolality. Although this
    is not strictly correct (since one is moles of particles per
    liter of solution and the other is moles of particles per
    kilogram of solvent), in water at temperatures of biological
    interest the error is fairly small unless solute concentrations
    are high (i.e., when an appreciable fraction of the solution is
    not water). Thus, with urine the approximation is acceptable
    while with serum it is not, because of the large amount of
    protein present. Although osmolarity is more readily measured,
    it is temperature dependent, unlike osmolality.

d. Osmolality is commonly measured by freezing point or vapor
    pressure depression.

    i) In terms of vapor pressure, (P^V), the osmotic pressure (Π)
        is defined as

$$\Pi = P^V_{pure\ solvent} - P^V_{solution} .$$

As mentioned before,

$$Osmolality = \Pi/22.4$$

where the osmotic pressure is in atmospheres. In one
commercially available instrument, the solution and solvent
vapor pressures are measured by using sensitive thermistors
to detect the difference in temperature decrease caused by

the evaporation of solvent from a drop of pure solvent and a drop of solution. Because the rate of evaporation (vapor pressure) of the solution is lower, the temperature change will be less and the vapor pressure difference can be calculated. This change in temperature due to a difference in vapor pressure is related to the phenomenon of boiling point elevation.

ii)   The freezing point of a solution is always lower than that of the pure solvent. The exact value of the depression depends on the solvent and the osmolality of the solution. For water,

$$\text{Osmolality} = \frac{\Delta T}{1.86}$$

where ΔT is the freezing point depression in degrees centigrade. Instruments which measure the freezing of a sample are used in clinical laboratories to determine serum and urine osmolality.

e.   Since water passes freely through most biological membranes, all fluids in the body are in osmotic equilibrium. The osmolarity of a plasma sample is thus roughly representative of the osmolarity of the other body fluids.

i)   The osmotic pressure of extracellular fluid is due primarily to Na^+ and its accompanying anions, Cl^- and HCO_3^-. Since Na^+ is the principal osmotically-active cation, doubling the sodium concentration provides a good estimation of serum osmolality. Thus, the normal range of plasma Na^+ values is 135–145 mEq/l (about 3.1–3.3 g/l) and the normal plasma osmolality is about 270–290 mOsmol/kg (corresponding to an osmotic pressure of 6.8–7.3 atmospheres and a freezing point depression of 0.50–0.54°C).

ii)   Although glucose is present to the extent of one gram/liter it does not dissociate and so provides only 5–6 mOsmols of particles per kg (contributing about 0.1 atmospheres to the osmotic pressure). For similar reasons, although there are 60–80 grams of protein per liter of plasma, the contributions of these molecules to the osmolarity is small (about 10.8 mOsmols/kg). Because these proteins are large and generally unable to pass through biological membranes, they are important in determining fluid balance between the intravascular and extravascular spaces. Hence, that portion of the osmotic pressure due to proteins is often referred to as the <u>oncotic pressure</u>.

iii) Since many of the molecules in plasma interact with each other, plasma is not an ideal solution. Consequently, the measured osmolarity of a plasma sample is an effective osmolality and is lower than that which would be obtained by adding up the concentrations of all ions and molecules present. A solution which has the same effective osmolality as plasma is said to be isotonic. Examples include 0.9% saline, 5% glucose, and the more complex Ringer's and Locke's solutions. It should be noted that, if a solute is able to permeate a membrane freely, then a solution of that solute will behave as if it were pure water, with respect to the membrane. Thus, a urea solution will cause red cells to swell and burst as does pure water, because urea moves freely across erythrocyte membranes.

iv) The osmolality of urine can differ markedly from that of plasma, as pointed out earlier, because of active (energy-requiring) concentrative processes going on in the renal tubules. Also, the membranes of the collecting ducts show a varying degree of water permeability (see later), permitting the removal of certain solutes from the urine without simultaneous loss of water.

3. Changes in osmolality on an order of 2% or more are detected by osmoreceptors, probably located in the hypothalamic region (see page 863). These receptors have properties similar to a semi-permeable membrane which permits passage of urea but restricts Na^+ and glucose. Thus, changes in the plasma levels of Na^+ and glucose trigger the receptors while urea has no effect.

a. The physiological responses to an increase in osmolality are a sensation of thirst and restriction of urinary water loss (production of hypertonic urine). Under conditions of fluid restriction, urine osmolarity can rise to between 800 and 1200 mOsmol/kg (normal is 390-1090 mOsmol/kg), three to four times the plasma levels of 270-290 mOsmol/kg. A decrease in plasma osmolarity (as in excessive water intake) can cause the excretion of hypotonic urine with an osmolarity as low as 40 to 50 mOsmol/kg. Other water losses, from the skin and lungs, are not subject to any controls of this sort.

b. The biochemical response to hypertonicity of the body fluids is the release of antidiuretic hormone (ADH; also called vasopressin) from the neurohypophysis. As indicated, this is mediated by the osmoreceptors via neuronal impulses. The synthesis and storage of this hormone is discussed in Chapter 13. ADH acts to increase water resorption from the urine.

i)   ADH is a nonapeptide having (in man and most mammals)
the sequence shown below.

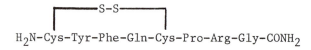

$$H_2N-Cys-Tyr-Phe-Gln-Cys-Pro-Arg-Gly-CONH_2$$

Arginine Vasopressin

The "arginine" in the name refers to the amino acid at
position 7. Lysine vasopressin, with a lysine residue
in place of arginine, is the ADH in pigs. Vasopressin
closely resembles oxytocin, another neurohypophyseal
hormone (see Chapter 13). Both are non-apeptides and have
some overlap in their activities. ADH also has
vasoconstrictive activity (as the synonym vasopressin
implies) but this and its weak oxytocic action are
probably not physiologically important.

ii)   ADH is released into the blood-stream where it is
transported apparently not bound to any proteins. Being
small enough to be filtered by the glomerulus, it is
rapidly cleared by the kidney and excreted in the urine.
It is also destroyed in the liver. One milligram of
vasopressin is defined as having 450 units of biological
activity. Under conditions of normal water intake
(no restrictions) 11 to 30 milliunits (about 70 ng) of
ADH are secreted each day. When water intake is restricted,
this secretion rate can double.

iii)   The site of action of ADH is believed to be the renal
collecting ducts which become permeable to water (due to
increased pore size) in the presence of ADH. A partial
interchange (equilibration) can then occur between the
urine and hypertonic regions of the nephron causing the
urine to become hypertonic relative to the plasma. The
action of ADH is probably mediated by cAMP and the urinary
levels of cAMP are increased in the presence of ADH. The
disulfide ring (cystine group) is necessary for action
and may be involved in the binding of the hormone to the
receptor site.

c.   Although the primary stimulus for ADH secretion is a rise in
the osmolarity of the blood perfusing the receptors, other
factors may also be involved.

i) In acute hemorrhage where there is an abrupt drop in extracellular fluid volume, there is apparent coordination between the two systems controlling blood volume and osmolarity. Under these circumstances, ADH is released to increase the volume even at the expense of a drop in osmolarity. Central nervous stimuli are probably involved in this reaction.

ii) In some situations, ADH is released even in the presence of a water overload and a decline in plasma sodium and osmolarity. This is called <u>inappropriate ADH secretion</u> and results in the formation of a hypertonic urine. It can be caused by fear and pain through cortical influence on the hypothalamus. Various drugs such as morphine, barbiturates, and porphyrins (as in porphyria) cause release of ADH from the pituitary. Ethanol inhibits the release of ADH.

4. Man's intake of water (thirst) is predominantly influenced by psychic and cultural influences. Hyperosmolarity will cause thirst but even under usual conditions man consumes more water than he needs. In certain situations, however, abnormal water loss can occur, resulting in pathological conditions.

a. Excretion of water alone will be accompanied by an increase in osmolarity and a contraction of both the extracellular and intracellular fluid compartments. Cations are usually lost along with water, but this depletion may be more restricted to the extracellular compartment. Unless corrected by water and cation replacement, this loss can lead to circulatory collapse, failure of body heat control, and death.

b. Several factors can increase the amount of urine formed (polyuria), resulting in elevated water excretion by this route.

i) In <u>diabetes insipidus</u> (see page 281) the renal tubules fail to concentrate the urine and recover the water from the glomerular filtrate. This can be due to defective ADH receptors in the nephrons (nephrogenic diabetes insipidus; unresponsiveness of the receptors to ADH); or to diminished or absent ADH secretion.

ii) In <u>osmotic diuresis</u>, as is seen in diabetes mellitus with severe glycosuria (see pages 276-282), a large solute load increases the osmolality of the glomerular filtrate and impairs the ability of the kidney to concentrate the urine. Recall (page 276) that the name diabetes refers to the polyuria which usually accompanies this disorder.

c. Excessive amounts of water may also leave the body during
    sweating, vomiting, and diarrhea. Some of the mechanisms by
    which enteric pathogens cause water and electrolyte losses from
    the intestines have recently been investigated.

    i)   There are two "classes" of pathogens which can cause
         diarrhea. One group, exemplified by Shigella dysenteriae,
         must invade the epithelial cells of the intestinal mucosa
         for its toxins to work. The other type, which includes
         E. coli and Vibrio cholerae, grow in the intestinal lumen
         or on the surface of the epithelial cells. The toxin
         which they secrete is sufficient, by itself, to cause
         the disease.

    ii) Cholera toxin (from Vibrio cholerae) has been studied most
         extensively. It has been obtained in crystalline form and
         shown to be a protein with a molecular weight of 84,000
         daltons. It is heat- and acid-labile and is hydrolyzed
         by pronase but not trypsin.

    iii) The water and electrolyte loss associated with cholera
         seems to result from the ability of the toxin to stimulate
         chloride secretion into and to inhibit sodium absorption
         from the intestinal lumen. This raises the osmolarity
         of the intestinal contents and causes diarrhea. The
         mechanism by which cholera toxin accomplishes this is not
         completely understood but there seems to be an increase
         in the adenyl cyclase activity (and cAMP levels) of the
         epithelial cells of the entire small intestine (duodenum,
         jejunum, and ileum) and a decrease in the activity of
         a Na^+-K^+ ATPase. The elevated cAMP synthesis may be due
         to the formation of additional molecules of adenyl
         cyclase. The rate of lipolysis in isolated fat cells, a
         process mediated by cAMP (see pages 227; 601; 679), is
         also stimulated by the toxin. Cycloheximide, which
         supposedly inhibits protein synthesis (page 351) inhibits
         the fluid loss from the intestine but does not block the
         effect of cholera toxin on fat cells. This somewhat
         obscures just how cholera toxin interacts with the cAMP
         system. Further, the possibility has been raised that
         cholera toxin has a phospholipase activity and that it
         performs its functions by altering the phospholipids in
         the cell membrane where   the cyclase and the ATPase
         are located. Both cyclase and ATPase are known to require
         phospholipids for activity. There is also evidence that
         the membrane binding site for toxin may be a ganglioside
         (see pages 615-616). Much remains to be learned about
         this process.

iv) Currently, the most reliable treatment for cholera is the parenteral and oral administration of solutions containing glucose and electrolytes. The latter fluid replacement therapy is based on the observation that the glucose- and amino acid-stimulated Na^+ and water absorption is not affected by cholera toxin. Antibiotics such as tetracycline and chloramphenicol, used in conjunction with this therapy, reduce the duration of the diarrhea and hence the amount of fluid needed. Adenyl cyclase inhibitors such as ethacrynic acid and cycloheximide have also been tried. Although ethacrynic acid has been successful in relieving diarrhea in dogs with experimentally induced cholera, this approach is not yet ready for use in humans.

v) As indicated above, some strains of E. coli produce an enterotoxin which is immunologically related to cholera toxin and which causes symptoms similar to those of cholera. Enterotoxins of Clostridium perfringens (an organism involved in food infection) has been implicated in the production of a mild diarrhea of short duration in the disease produced by this organism. The toxin has been shown to be heat-labile and stimulates adenyl cyclase activity in the frog erythrocyte membrane.

d. Water excess gives rise to vague cerebral symptoms of headaches, malaise, and, in more severe cases, muscular weakness and cramps. It is associated with low plasma sodium and hypo-osmolarity. Abnormal water intake may be iatrogenic (e.g., administration of excessive I.V. fluids) or may stem from psychological causes. Inappropriate ADH secretion, as is sometimes seen due to postoperative pain, causes increased water retention. Ectopic ADH production, as in bronchiogenic carcinoma, has a similar effect.

Extracellular Volume Control

1. The volume of extracellular fluid (ECF) in a normal adult is kept surprisingly constant. This can be seen from the observation that a person's body weight does not vary by more than a pound per day despite fluctuations in food and fluid intake. Decreases in extracellular fluid result in a lowering of the effective blood volume, and a compromised circulatory system. An increase in extracellular fluid may lead to hypertension, edema, or both. The homeostatic mechanism for volume control centers upon the renal regulation of the sodium balance. When the ECF volume decreases, less sodium is excreted, while increased ECF volume increases sodium

loss. Increased sodium retention leads to an expansion of the extracellular space. This is because sodium is confined to this region and the regulation of extracellular osmolarity (discussed earlier) causes increased water retention in order to keep the sodium concentration constant. Sodium movement in the kidneys is under hormonal influence from the aldosterone-angiotensin-renin system and the postulated natriuretic hormone. These controls are summarized in Figure 110.

2. Volume receptors which perceive changes in extracellular volume and then transduce this information into hormonal changes for ECF regulation must exist, but their nature is controversial. They are probably stretch receptors which monitor the vascular compartment and respond to the degree of filling of some area of the cardiovascular system. Increases in effective blood volume result in increased sodium excretion as is seen in prolonged supinity or exposure to low temperatures. Both conditions cause peripheral vascular constriction which increases the effective blood volume because there is less pooling of blood in the venous part of the circulation. Decreases in effective blood volume lead to sodium retention. This occurs in hemorrhage or prolonged standing with pooling of blood in the limbs.

3. Aldosterone is the principal mineral corticoid in man. It is the hormone which produces increased sodium retention and expansion of the extravascular space. It is formed in the zona glomerulosa of the adrenal cortex. The structure and biosynthesis of aldosterone is given in Chapter 13.

    a. There is no specific binding protein for aldosterone in the plasma and it circulates bound non-specifically to albumin with a certain amount "free" in solution. The plasma levels are 30-130 ng/l with a secretion rate of 75-175 μg/24 hours. The hormone is rapidly inactivated by both liver and kidney, with 80-90% of it being removed in one passage through the liver. The hormone is reduced to the tetrahydro derivative, then conjugated with glucuronic acid and excreted in the urine. 10% of the hormone is removed by the kidney as the 18 oxyglucuronide. A small percentage (less than 1%) of aldosterone is excreted unchanged in the urine, the rest occurring as inactive products.

    b. Aldosterone stimulates the sodium pump causing the retention of sodium ions and an increased excretion of potassium or hydrogen ions. This could be due to a coupled exchange of ions via a common carrier or may be secondary to the increased sodium uptake (i.e., a passive H^+-K^+ loss in response to Na^+ uptake).

Figure 110.   Regulation of Extracellular Fluid (ECF)

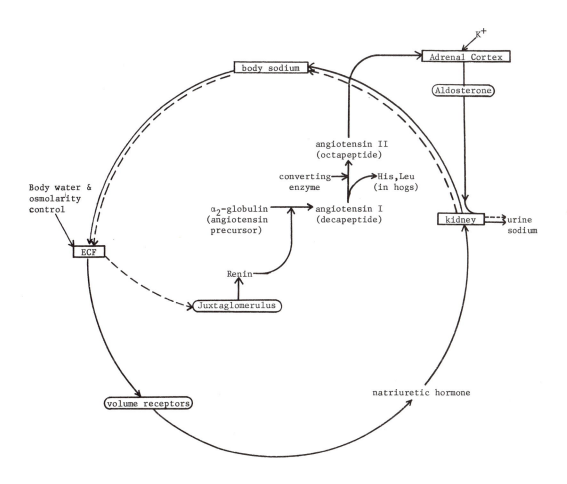

--- decreased or inhibitory

—— increased or stimulatory

The principal site of action is the distal renal tubule, and
aldosterone has no effect on sodium absorption by the proximal
tubules. Other areas which have an active sodium pump that
is sensitive to aldosterone are the gastro-intestinal tract,
sweat glands, and muscle. All show net potassium loss and
net sodium gain under the influence of aldosterone.

c. Aldosterone acts by stimulating the production of messenger RNA,
    thereby increasing the production of specific proteins. The
    exact nature of these proteins is unknown. A sodium permease
    has been proposed as one candidate, but the only definite
    increase noticed has been in citrate synthetase. It is possible
    that the limiting factor in the activity of the sodium pump is
    the availability of energy-providing substrates. Then increased
    levels of enzymes, such as citrate synthetase, that provide
    these substrates will have the effect of stimulating sodium
    transport.

d. Several factors cause the release of aldosterone from the
    adrenal cortex.

    i)   Angiotensin II is a potent stimulator and is the main
         physiological influence.

    ii) Elevated plasma potassium will cause aldosterone levels
         to rise. This is part of the homeostatic regulation of
         potassium levels.

    iii) ACTH plays a "permissive" role. It is necessary for
         the effect of angiotensin or potassium ions.
         Pharmacological doses cause a release of aldosterone but
         its physiological role may be slight.

    iv) The kidney produces several prostaglandins which have a
         hypotensive and natriuretic effect. One PGA causes the
         release of aldosterone independently of angiotensin.

e. Two controlling steps in the biosynthesis of aldosterone exist
    (see Table 50a, page 870 for structures).

    i)   The conversion of cholesterol to pregnenolone is stimulated
         by ACTH, angiotensin, and K^+. This step is common to
         the formation of all steroid hormones.

    ii) The last step in the conversion of corticosterone to
         aldosterone, catalyzed by an 18-ol-dehydrogenase also
         appears to be a control point, although little is known

about it. The enzyme is found only in the zona
glomerulosa. Since this 18-aldehyde formation is unique
to aldosterone synthesis, control at this point is more
specific in regulating this process.

4. <u>Renin</u> is a proteolytic enzyme secreted by the juxtaglomerular
apparatus. The substrate for renin is an alpha-2 globulin, present
in plasma, called angiotensinogen. Renin splits off a decapeptide
(angiotensin I) from angiotensinogen. Angiotensin I is converted
to angiotensin II by loss of a terminal histidylleucine group,
leaving an octapeptide. This reaction is catalyzed by a
"converting enzyme" present in the pulmonary vascular bed.
Angiotensin II is the physiological active form of the hormone,
having a much greater activity than angiotensin I, its precursor.
Renin has also been found in non-renal tissue such as the brain,
submaxillary glands and uterus. Angiotensin has been found in
amniotic fluid, presumably formed by uterine renin. The function
of renin in these tissues is unknown.

a. The release of renin is caused by three stimuli.

    i) A <u>decrease in renal perfusion pressure</u> is a measure of the
effective blood volume. Renin levels <u>increase</u> in
prolonged standing (where pooling of blood in the
extremities occurs) and <u>decrease</u> upon lying down. The
juxtaglomerular apparatus acts as a local volume receptor.

    ii) A <u>decrease in sodium transport across the tubular membrane</u>
(decrease in tubular sodium) also stimulates renin
release, a factor which is of obvious importance in the
regulation of sodium balance. A fall in the filtered
load of sodium reaching the tubule results in an increase
in aldosterone and an increase in sodium retention.

    iii) <u>Increased adrenergic nervous activity</u>, independent of the
other two factors, causes renin release. It is
apparently mediated by cAMP.

    iv) The release of renin is <u>inhibited</u> by angiotensin II, thus
completing the feedback loop. ADH also inhibits renin
release.

b. Angiotensin II is a potent stimulator of the release of aldosterone from the adrenal cortex and this is its main physiological role. It is also a vasoconstrictive agent and may make a significant contribution to blood pressure regulation.

5. Third factor or <u>natriuretic hormone</u> has not been isolated but evidence has accumulated to substantiate its existence. The hormone antagonizes aldosterone and mediates increased excretion of sodium ions. It may act on the proximal tubule portion of the nephron. The stimulus for the hormone's release is an expansion of the extracellular volume. Natriuretic hormone is possibly a polypeptide and it may be identical to melanocyte stimulating hormone from the pituitary.

6. A <u>low extracellular fluid volume</u> (hypovolemia) leads to a decrease in the effective blood volume. This is best detected by measuring <u>postural hypotension</u>. A fall in blood pressure of over 10 torr in moving from a supine to sitting position indicates hypovolemia. The commonest cause of low ECF is the loss of sodium-containing body fluids from the skin by sweating or from the gastro-intestinal tract through vomiting or diarrhea. There are frequently attendant acid-base disorders due to losses of hydrogen or bicarbonate ions at the same time. Deficiency of sodium can also result from increased renal losses as in cases of renal disease or adrenal insufficiency. Treatment consists of the replacement of sodium ion, correction of accompanying potassium deficiency, and correction of the acid-base disorder.

7. An <u>increase in extracellular fluid volume</u> results first in a weight gain followed by edema. There are a number of causes of edema (see pages 363 to 367).

   a. Hypoalbuminemia (nephrosis and cirrhosis); low plasma albumin results in a lower oncotic pressure and an increase in fluid transfer to the interstitial space from the vascular system (see pages 361 to 363).

   b. Interference with lymphatic drainage;

   c. Increased salt retention (as in cardiac failure and hyperactivity of the adrenal gland).

Treatment consists in eliminating (where possible) the cause of the edema. Excess fluid can be removed by diuretics which interfere with the reabsorption of sodium by the kidneys. There is increased loss of potassium in diuretic therapy which can cause a deficiency of this ion.

Some forms of hypertension are associated increased aldosterone levels. It has even been suggested that as many as one quarter of all hypertensive patients suffer from hyperaldosteronism. Primary aldosteronism results in high blood pressure, sodium retention, and low renin levels. This can be ameliorated by spirolactone, an aldosterone antagonist.

Sodium

The average sodium content of the human body in 60 meq/kg of which 50% is in extracellular fluids, 40% is in the bone, and 10% is intracellular. The chief dietary source of sodium is the salt added in cooking. Excess sodium is largely excreted in the urine although some sodium is lost in sweat. Gastro-intestinal losses are small except in diarrhea. Most aspects of sodium metabolism have already been considered in earlier sections of this chapter.

Potassium

1. The average human body contains 40 meq/kg of potassium, distributed mainly in the intracellular space. Potassium is required for carbohydrate metabolism and there is increased cellular uptake of potassium ions during glucose catabolism. Potassium is widely distributed in plant and animals foods. The human requirement is about 4 grams a day. Excess potassium is excreted in the urine, the process being regulated by aldosterone (see previously).

2. Plasma potassium plays a role in the irritability of excitable tissue. A high plasma potassium leads to electrocardiogram (EKG) abnormalities and possibly to cardiac arrhythmia. This is possibly due to a lowering of the membrane potential. Low potassium in the plasma increases the membrane potential, resulting in decreased irritability, EKG abnormalities of another sort, and muscle paralysis.

3. Hyperkalemia (potassium retention with elevated plasma potassium) can occur in renal disease and adrenal insufficiency, where the normal secretory mechanisms are impaired. Metabolic acidosis, particular diabetic acidosis, causes K^+ to be released from the cells as does the catabolism of cellular protein which occurs in starvation (see pages 268 to 269) or fever. Treatment consists of correction of the acidosis and the promotion of cellular uptake of potassium by insulin administration. Insulin enhances glucose uptake and therefore glycolysis, a process which requires K^+. In severe cases, ion exchange resins are given orally to bind potassium in the intestinal secretions.

4. Hypokalemia (potassium deficiency) can be caused by the loss of
   gastro-intestinal secretions which contain significant amounts of
   potassium and by excessive losses of potassium in the urine due to an
   elevated aldosterone level or diuretic therapy. Low potassium
   levels are usually associated with alkalosis because, in potassium
   deficiency, the kidneys tend to secrete a more acid urine with
   consequent increase in the production of bicarbonate.

Acid-Base Balance

1. The blood pH is usually regulated within very narrow limits. The
   normal pH range is 7.35-7.45 (corresponding to 35-45 nmoles of
   H^+/1) and values below 6.80 (160 nmoles of H^+/1) or above 7.70
   (20 nmoles of H^+/1) are seldom compatible with life. At the same
   time, the body generates a large quantity of acid as a by-product
   of metabolism. Fourteen thousand milliequivalents of CO_2 are
   removed by the lungs each day. On a diet of 1-2 grams of protein
   per day, the kidneys remove 40-70 meq of acid each day. This is in
   the form of sulfate (from the oxidation of sulfur-containing amino
   acids), phosphate (from phospholipid and phosphoamino acid
   catabolism), and organic acids (e.g., lactic, β-hydroxybutyric and
   acetoacetic). The organic acids are produced by incomplete
   oxidation of carbohydrate and fats. Under some conditions (e.g.,
   ketosis; see pages 679-685), considerable amounts of these materials
   may form.

2. The major buffer systems of the blood were discussed previously
   (pages 6-9 and 534-535). The most important extracellular buffer is
   the carbonic acids-bicarbonate system, shown below.

$$CO_2 + H_2O \rightleftharpoons H_2CO_3 \rightleftharpoons HCO_3^- + H^+$$

As was shown on pages 6-7, at the normal blood pH of 7.4, the ratio
of HCO_3^- to H_2CO_3 is 20 to 1. Thus the system is not at its maximum
buffering capacity but it is capable of neutralizing a large amount
of acid. The carbonic acid-bicarbonate system is independently
regulated by two organs. The kidneys control the HCO_3^- levels in
the blood while the lungs (via the respiratory rate) regulate the
P_{CO_2} (partial pressure of CO_2). Other buffer systems (proteins,
phosphates) also operate in the plasma and erythrocytes.
Proteins are especially important in the buffering of
the intracellular fluid. Lastly, the hydroxyapatite of the bone also
acts as a buffer.

   a. The respiratory center in the medulla responds to the pH of the
      blood perfusing it and is the source of pulmonary control. P_{CO_2}
      and perhaps P_{O_2} also influence the center along with nervous

impulses from higher centers of the brain. A fall in pH results in an increased rate and deeper breathing with a consequent increase in the respiratory exchange of gases and a lowering of the P_{CO_2}. The lower P_{CO_2} (and hence H_2CO_3) brings about an elevation of the pH. Similarly, a fall in respiratory rate leads to an accumulation of CO_2, a rise in P_{CO_2}, and a fall in pH. The pulmonary responses to pH fluctuations are rapid while the renal compensatory mechanisms (below) are slower to develop.

b. The kidneys actively secrete hydrogen ions derived from the dissociation of H_2CO_3, into the urine. The formation of carbonic acid is a slow reaction whose rate is increased by carbonic anhydrase.

$$H_2O + CO_2 \underset{\text{(slow)}}{\overset{\text{carbonic anhydrase}}{\rightleftharpoons}} H_2CO_3 \underset{\text{(fast)}}{\rightleftharpoons} H^+ + HCO_3^-$$

This is a zinc-containing enzyme found in renal tubular cells and erythrocytes. The dissociation of carbonic acid is rapid with no need for catalysis. The H^+ ion released is available for secretion into the lumen of the nephron while the bicarbonate ion passively diffuses into the blood stream. Thus, every H^+ secreted into the urine is associated with production of a bicarbonate ion in the extracellular fluid.

c. The hydrogen ion secretion is associated with three processes in the nephron: (1) bicarbonate reabsorption, (2) acidification of the urine, and (3) ammonium ion production.

   i) The proximal tubule is responsible for the major reabsorption of the 4500 meq of bicarbonate filtered through the glomerulus each day. Hydrogen ion secreted into the tubule lumen combines with HCO_3^- to form CO_2 and water. The CO_2 diffuses into the tubular cells where it is rehydrated to H_2CO_3 by carbonic anhydrase. It then dissociates to bicarbonate and H^+ again. The HCO_3^- diffuses into the blood stream, resulting in net reabsorption of bicarbonate (see Figure 111). This proximal tubular mechanism becomes saturated at approximately 26 mEq/l. At higher levels HCO_3^- appears in the distal tubule and can be excreted in the urine. Under conditions of elevated P_{CO_2}, H^+ secretion is more active, possibly due to intracellular acidosis.

   In proximal renal tubular acidosis, bicarbonate reabsorption is impaired and saturation occurs at a lower value (about 16-18 meq/l). The plasma bicarbonate is low while the urine pH is frequently high due to the presence of bicarbonate.

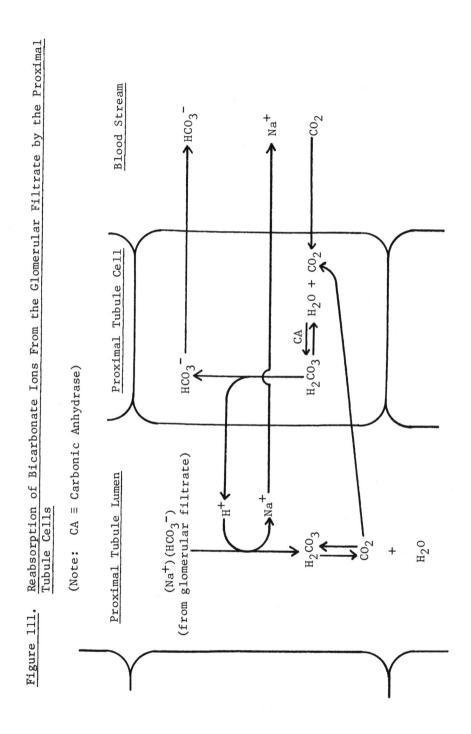

Figure 111. Reabsorption of Bicarbonate Ions From the Glomerular Filtrate by the Proximal Tubule Cells

(Note: CA ≡ Carbonic Anhydrase)

ii) The distal tubule cells can actively secrete hydrogen ions even against a concentration gradient until the urine attains a pH of 4.4. Beyond this point, any additional hydrogen ions that are to be excreted must be buffered as the tubule cells cannot work against a lower pH. This buffering comes from phosphate ions and ammonia. These two processes, described below, are illustrated in Figure 112.

Any phosphate ions in the glomerular filtrate that are not reabsorbed are converted to the dihydrogen form. The amount of $H_2PO_4^-$ present (termed the <u>titratable acidity</u>) can be measured by titrating the urine to pH 7 (pK_2 for H_3PO_4 = 6.15). Normal values for titratable acidity are 16-60 mEq /24 hours, depending on the phosphate load.

Phosphate is a valuable body resource, however, and is mostly reabsorbed, so additional buffering is provided by ammonia. NH_3 readily diffuses into the lumen of the nephron where the addition of a hydrogen ion to it forms an ammonium ion. Because of the positive charge on NH_4^+, it is no longer able to freely pass through the membranes and remains "trapped" in the urine. One hydrogen ion is absorbed for each ammonia molecule produced. The source of the ammonia is the hydrolysis of glutamine catalyzed by <u>glutaminase</u>, a specific renal enzyme. The reaction is shown below.

$$H_2O + glutamine \xrightarrow{\text{glutaminase}} glutamic\ acid + NH_3$$

$$NH_3 + H^+ \longrightarrow NH_4^+$$

This enzyme is induced under acidotic conditions and is responsible for the increased capacity of the kidney to remove hydrogen ions in renal compensation. On a normal protein diet, 0.2 to 1.0 gm of ammonia are excreted each day by this route.

3. In any derangement of acid base balance, compensatory changes occur to restore homeostasis. Acidosis due to respiratory failure will lead to compensatory changes in the kidney which can give rise to a metabolic alkalosis.

Disorders of acid-base balance are classified into four major groups based on their cause and the direction of the pH change. They are (1) respiratory acidosis, (2) metabolic acidosis, (3) respiratory alkalosis, and (4) metabolic alkalosis. Alterations in the plasma electrolytes in each of these conditions is indicated in Table 46.

Figure 112.   Buffering of the Hydrogen Ions Secreted into the Distal Tubule Lumen

(Note: CA ≡ Carbonic Anhydrase)

a.   Excretion of Titratable Acid

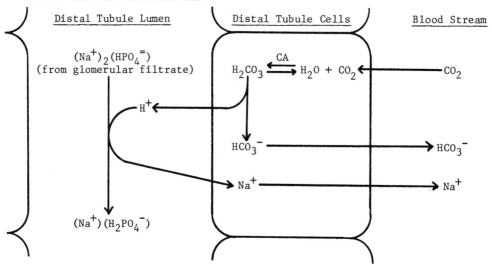

b.   Secretion of Ammonia

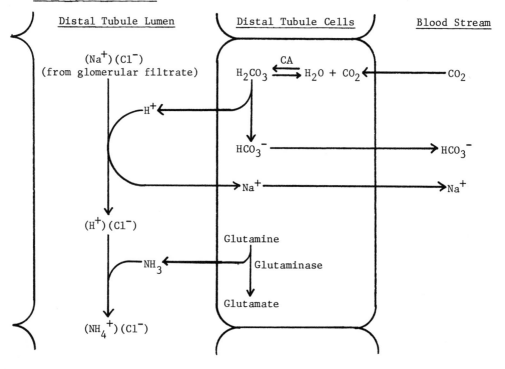

Table 46.   Changes in Plasma Electrolytes During Acid-Base Disorders

Parameter	Normal Values	Acidosis Resp. U.	Acidosis Resp. C.	Acidosis Metabolic U.	Acidosis Metabolic C.	Alkalosis Resp. U.	Alkalosis Resp. C.	Alkalosis Metabolic U.	Alkalosis Metabolic C.
Plasma bicarbonate (mEq/l)	23–33	–	↑	↓	↓	–	↓	↑	↑
P_{CO_2} (torr)	38–42	↑	↑	–	↓	↓	↓	–	↑
pH	7.35–7.45	↓	–	↓	–	↑	–	↑	–
Plasma K^+ (mEq/l)	3.5–5.0	↑	↑	↑	↑	↓	↓	↓	↓

U = uncompensated

C = compensated

↑ = elevated

↓ = lowered

– means no change

a. Respiratory acidosis is defined as a decrease in blood pH caused
    by an accumulation of CO_2 in the lungs with a consequent rise
    in P_{CO_2}. It is associated with emphysema, pneumonia, and other
    disorders which interfere with the pulmonary removal of CO_2.
    The increased carbonic acid in the blood is partly buffered by
    the uptake of H^+ by cells with a corresponding loss of
    intracellular K^+.

    The kidneys respond to the elevated P_{CO_2} by increasing the amount
    of HCO_3^- formed by the carbonic anhydrase system in the renal
    tubules and by excreting more hydrogen ions (producing a more
    acid urine). These renal compensatory mechanisms
    neutralize the respiratory acidosis by producing a metabolic
    alkalosis.

    The usual treatment for this disorder is to improve the
    respiration of the patient.

b. Metabolic acidosis results from the production of abnormal
    amounts of hydrogen ions by metabolic processes (acid-gaining
    acidosis) or by excessive loss of bicarbonate ions (base-losing
    acidosis). In the acid-gaining type, there is also a decrease
    in plasma HCO_3^- caused by the increase in plasma H^+. Cellular
    buffering occurs in both sorts of metabolic acidosis with
    uptake of hydrogen ions by the cells and release of K^+, often
    leading to hyperkalemia. There is also dissolution of bone
    hydroxyapatite (which is basic) and mobilization of calcium.

    i)   Compensation initially consists of the prompt stimulation
         of respiratory rate leading to a decrease in P_{CO_2} and
         respiratory alkalosis. This cannot be sustained, however,
         due to tiring of the respiratory muscles. A slower renal
         compensation also occurs which can be maintained for an
         extended period because of the induction of glutaminase.
         In this process, the excess acid is excreted and the
         production of HCO_3^- is increased, correcting the acidosis.

    ii) Diabetic acidosis, in which ketone bodies accumulate (see
         page 277), provides an example of acid-gaining acidosis.
         Under conditions of poor perfusion, lactic acid may
         accumulate sufficiently to create a similar problem. The
         formic and oxalic acids formed, respectively, from ingested
         methanol and glycols are known to cause acidosis. Ammonium
         salts may lead to acidosis because of the H^+ released when
         NH_4^+ is converted to NH_3, the form in which ammonia is
         metabolized (as, say, in urea synthesis). Failure of the
         renal acidification process also may produce acidosis.
         A base-losing acidosis may occur if excessive quantities
         of HCO_3^- are lost in the digestive secretions.

iii)   Metabolic acidosis is treated by first correcting the cause of the acidosis (as in insulin administration for diabetic ketoacidosis). The acid is then neutralized by treatment with a base such as $NaHCO_3$, sodium lactate, or tris (trishydroxymethylaminomethane) buffer. Potassium replacement therapy is also frequently needed because of the intracellular K^+ loss during cellular buffering.

c.   Respiratory alkalosis occurs when the respiratory rate increases abnormally (hyperventilation) leading to a decrease in P_{CO_2} and a rise in pH.

   i)   Hyperventilation can be caused by hysteria, by pulmonary irritation (as in pulmonary emboli), or by a head injury with damage to the respiratory center. If the alkalosis is the result of hysteria, the usual treatment is simply sedation.

   ii)   The pH increase is buffered by a lowering of plasma HCO_3^- (probably due to the law of mass action) and, to some extent, by the exchange of plasma K^+ for intracellular H^+. Renal compensation seldom occurs because this type of alkalosis is usually transitory.

d.   Metabolic alkalosis is used to describe a decrease in hydrogen ion or an elevated plasma HCO_3^- level. It can be subdivided into three types.

   i)   The administration of excessive amounts of alkali (as during $NaHCO_3$ treatment of a peptic ulcer) increases the $HCO_3^-:H_2CO_3$ ratio with a consequent rise in plasma pH. Acetate, citrate, lactate, and other substrates which can be oxidized to HCO_3^- can also cause alkalosis.

   ii)   Vomiting, with the concomitant loss of H^+ and Cl^-, produces an acid-losing alkalosis.

   iii)   If there is excessive loss of extracellular potassium (and extracellular fluid) in the kidneys, K^+ diffuses out of the cells and is replaced by Na^+ and H^+ from the extracellular fluid. Additionally, K^+ and H^+ are normally secreted by the distal tubule cells to balance Na^+ uptake during Na^+ reabsorption (see Figure 113). If extracellular K^+ is depleted, more H^+ is lost to permit reabsorption of the same amount of Na^+. Loss of H^+ by both routes causes hypokalemic alkalosis. The use of excessive amounts of some diuretics or elevated aldosterone production can cause the hypokalemia which initiates this type of alkalosis.

Figure 113.   Secretion of Potassium and Hydrogen Ions by the Distal Tubule Cells During Sodium Ion Reabsorption

(Note: CA = Carbonic Anhydrase)

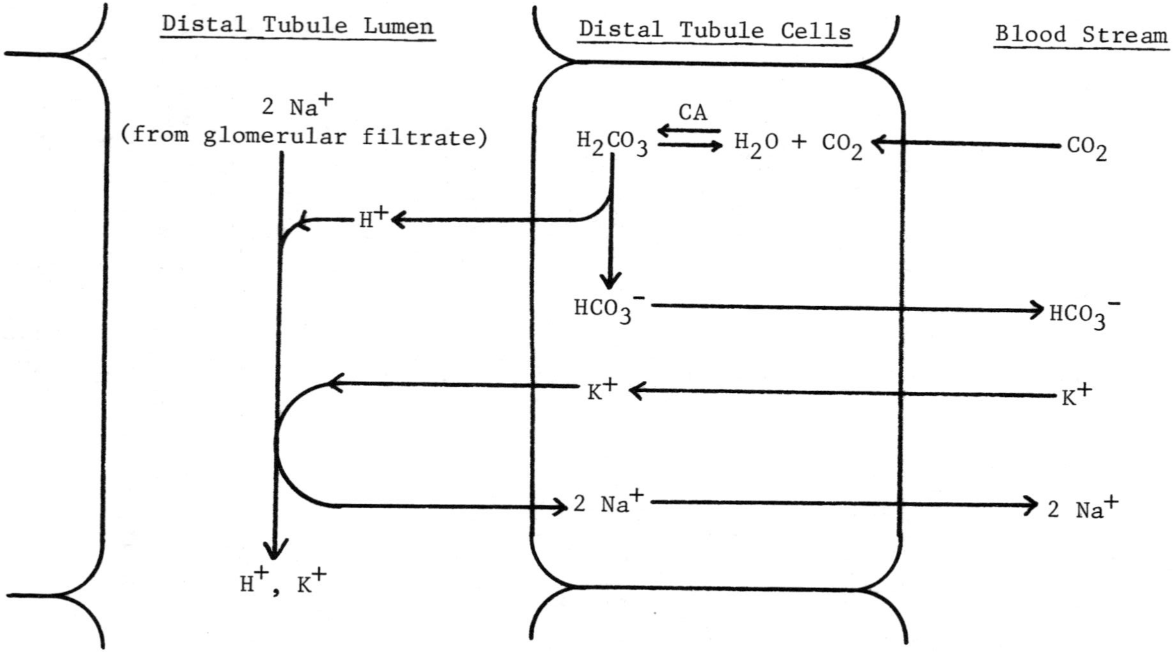

iv) To compensate for a metabolic alkalosis, the respiratory rate decreases, raising P_{CO_2} and lowering the pH. This mechanism is limited, however, because if the respiratory rate falls too low, P_{O_2} decreases to the point where respiration is once again stimulated. Renal compensation involves decreased reabsorption of bicarbonate and the consequent formation of an alkaline urine. Unfortunately, the bicarbonate excreted is accompanied by a loss of cations (Na^+, K^+) and if the alkalosis is accompanied by extracellular fluid depletion, renal compensation by this mechanism may not be possible.

v) The treatment of metabolic alkalosis consists of fluid replacement therapy with NaCl and KCl to correct electrolyte (especially K^+ and Cl^-) losses and NH_4Cl to counteract the alkalosis.

4. The emphasis so far has been in changes in extracellular fluid. Much less is known about the regulation of intracellular pH which occurs independently of extracellular control. The cell interior is probably more acid (about pH 7) than the extracellular environment and it is probably not uniform. Some localized areas, such as the mitochondria, are likely to be more acid (with a pH of about 6.8) than other regions.

13

Hormones

General Aspects of Hormones

1. The roles of hormones in the metabolism of a number of substances
    have been discussed in appropriate places throughout this text.
    A list of the most important instances of this are given below.

This chapter concerns itself with some general aspects of hormones and with those hormones (notably the steroids) which have not already been discussed.

2. Hormones are elaborated and secreted by the <u>endocrine systems</u> of the body under the direction of higher centers of the brain and of chemical stimuli transmitted in the blood stream. The actions of hormones include the regulation of normal levels of various substances within the body (homeostasis), integration of bodily activities with one another, and morphogenesis (both during development and in changes occurring after maturation). Some aspects of the interrelationship of the nervous and endocrine systems (neuroendocrinology) are discussed in the section on the hypothalamus.

3. Hormones produced by specialized cells are released into the circulation in order to reach their target tissues and exert their action. A few hormones (notably those of the hypothalamus) are also transported at least partially by secretory neurons. A given hormone may have a specific effect on a particular target cell (testosterone acts on seminiferous tubule epithelial cells in the formation of sperm) as well as the rest of the body cells (testosterone stimulates biosynthetic processes in non-testicular tissue). It is important to point out that, from an evolutionary point of view, the development of an efficient circulatory system was a prerequisite to the development of an endocrine system. Endocrine (as opposed to exocrine) glands are those tissues which secrete hormones <u>into</u> the body. Note that, although albumin is synthesized in the liver and secreted into the circulation, it is not classified as an endocrine secretion or hormone because it has no specific physiological action or target tissue.

4. Some of the properties of hormones are summarized below.

    a. Hormones are synthesized and released in extremely minute quantities and exert their biological effects at quite low concentrations (10^{-6} g/100 ml for steroids; somewhat less for peptide hormones).

    b. The rate of secretion of a hormone by an endocrine gland is established by the need for the hormone and its rate of inactivation. Several types of self-regulatory feedback mechanisms exist to control the secretion of a given hormone and to thus maintain it at a required physiologic level.

c. Although not all hormones are stored (e.g., some steroid hormones), those which are may be kept in the endocrine cells as inactive precursors in the form of granules (pituitary cells; β-cells of the islets of Langerhans in the pancreas); or as a colloid in the lumen (cavity) of the gland, surrounded by secretory cells (thyroid acini). Examples of inactive precursors and their associated active forms (indicated in parentheses) include: thyroglobulin, (T_3 and T_4); proinsulin, (insulin); testosterone, (dihydrotestosterone); and progesterone, (dehydroprogesterone).

d. The stimuli which cause the release of the stored hormones vary. Insulin, for example, is secreted when blood glucose levels increase above the physiologic "fasting" level. In the case of the adrenal medullary chromaffin cells (cells of neural crest origin which have lost their ability to participate in the conduction of nerve impulses), the hormones epinephrine and norepinephrine (catecholamines) are secreted only in response to neural stimulation. The hypothalamic neurosecretory cells can participate both in nerve impulse conduction and the elaboration and release (in response to nervous and chemical stimuli)of neurohormones.

e. Hormones normally do not exert any significant effects on the cells (tissues) that produce them.

f. Hormones may have very specific effects on a particular target organ (the action of luteinizing hormone on the interstitial cells of the male and female gonads) or may influence several organs (recall the action of insulin on muscle, adipose tissue, etc.). Hormones can also show multiple effects (i.e., different responses to one hormone occur in different tissues).

g. The physiologic response of a tissue to a hormone is dependent upon other variables such as age, genetic makeup, etc.

h. Hormonally initiated responses may persist long after the hormone is no longer measurable by biochemical and biological assays.

Mechanisms of Actions of Hormones

1. The initial encounter of any hormone with a target cell is mediated by a specific receptor molecule. This receptor molecule can be located on the plasma membrane or be present in the cytoplasm of the cell. Direct interaction of steroid hormones with the nucleus has also not been ruled out completely. Some aspects of receptor proteins were discussed in Chapter 4 under cyclic AMP.

2. There are two general types of mechanism which have been described to explain the action of hormones.

    a. A number of (non-steroid) hormones activate the adenyl cyclase system leading to the production of cAMP, which appears to stimulate intracellular metabolism by releasing inhibitory constraints on these pathways. The cAMP produced functions as an allosteric modulator and promotes the phosphorylation of proteins, in addition to altering membrane permeability and calcium flux (see chapters 4 and 9). The locus of specific receptor sites for these hormones is on the cell plasma membrane. The details of cAMP production have already been discussed in Chapter 4.

    b. The action of steroid hormones <u>does not</u> invoke the production of cAMP. They stimulate general and specific protein synthesis in the target cells, acting at the levels of transcription and translation. In the fibroblasts, however, cortisone appears to alter the activity of alkaline phosphatase by changing the kinetic characteristics of this enzyme rather than by increasing the number of phosphatase molecules synthesized.

3. Estrogens, androgens, progesterone, ecdysone (an insect steroid hormone; causes visible puffing of fruit fly salivary cell chromosomes), aldosterone and other steroid hormones have been shown to stimulate messenger RNA synthesis in their respective target tissues. Studies using actinomycin D inhibition of mRNA production and DNA hybridization techniques with newly formed RNA have shown that the steroid hormones are capable of causing transcription of unique portions of the DNA as well as stimulating the generalized production of RNA to enhance overall protein synthesis.

4. Following is a possible sequence of events leading to effects of steroid hormones on transcription.

    a. Entry of steroid hormone into the target cell.

    b. Binding of the hormone or its metabolite with a specific protein receptor molecule in the cytoplasm. These cytoplasmic steroid binding proteins show specificity for particular steroids, as well as serving as carriers to facilitate the entry of the steroids into the nucleus.

    c. The hormone-receptor protein complex enters the nucleus where it becomes bound to the chromosome through a specific acceptor protein associated with chromatin. The chromosomal acceptor protein appears to be an acidic, non-histone protein.

d. The interaction between the three molecules (steroid, receptor protein, and chromosomal acceptor protein) may lead to the derepression of a segment of the chromosome. This results in synthesis of the appropriate mRNA and the subsequent production in the cytoplasm of the protein coded for.

e. In the intestinal cells, 1,25-dihydroxy cholecalciferol, an active metabolite of vitamin D, brings about the absorption of calcium by a similar mechanism (see page 694).

5. Testosterone and progesterone are partly converted to their reduced derivatives, 5α-dihydrotestosterone and 5α-dihydroprogesterone. These mediate some of the actions ascribed to the parent hormones in the respective target tissues. These metabolites bind with specific receptor proteins and then enter the nucleus to initiate the transcriptional changes. The conversion of testosterone to its 5α-dihydro derivative is shown below.

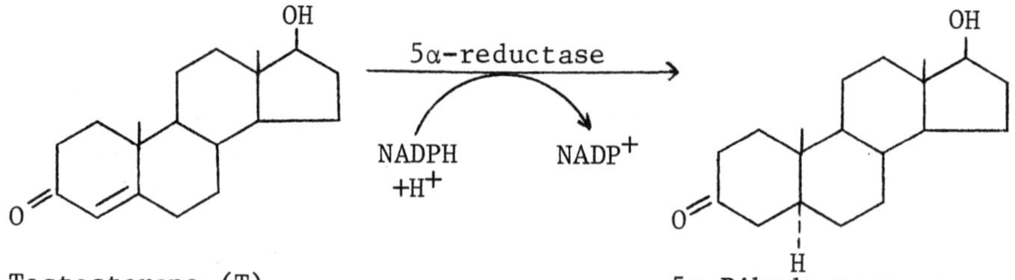

Testosterone (T)   5α-Dihydrotestosterone (DHT)

The 5α-reductase appears to be associated with the nuclear membrane as well as with the endoplasmic reticulum.

6. Pretreatment of a progesterone target tissue (guinea pig uterus, mouse vagina, chick oviduct) with estrogen enhances the uptake of progesterone by cells of these tissues, thereby demonstrating that there can be an interdependence (synergism) between steroid hormones in some tissues. Estrogen has been shown to induce the production of progesterone-receptor protein.

7. Steroid hormones have also been shown to affect protein synthesis at the translational levels but the mechanisms of these processes is not yet clear.

The Hypothalamus

1. This endocrine gland is located below (ventral to) the thalamic region of the forebrain. Under the control of chemical and nervous stimuli from higher centers of the brain, it regulates the secretion of hormones (see Table 48) from the adenohypophysis (and hence many other hormonal secretions in the body) and provides the neurohypophyseal hormones, vasopressin and oxytocin. These processes are summarized in Figure 114. It is interesting to note that, since the nervous impulses are at least partially mediated by neurotransmitter substances, hormones play a role in yet one more step of the control sequence.

2. Vasopressin and oxytocin, the neurohypophyseal hormones (see Table 48), are synthesized by cells in the supraoptic and paraventricular nuclei of the anterior hypothalamus. As granules, in association with the neurophysins, they migrate along secretory neurons and accumulate at the ends of these nerves in the neurohypophysis. The neurophysins are a class of "carrier" proteins (about 10,000 molecular weight) synthesized in the hypothalamus. In the bovine system, neurophysin I is primarily associated with oxytocin in the neurosecretory granules while neurophysin II apparently carries vasopressin, but this is known to differ in other species. Oxytocin is released from the neurohypophysis in response to the suckling of an infant, contraction of the uterus during labor, vaginal stimulation, and cervical dilation. Vasopressin (antidiuretic hormone) secretion is stimulated by small changes (a few percent) in the osmolarity of the blood (an effect mediated by osmoreceptors in the hypothalamus), hemorrhage, adrenalcortical insufficiency, and alterations in the renin-angiotensin system (see Chapter 12). The physiological effects of these hormones are indicated in Figure 114.

3. As is discussed below, the adenohypophysis synthesizes and secretes at least six hormones. The release of these compounds into the blood stream is regulated by at least eight hormones and factors elaborated in the hypothalamus. Two of these substances inhibit adenohypophyseal secretion while the remainder stimulate it. It is not yet clear whether these regulators are all synthesized by different cell-types or if two or more are made by the same sort of cell. Regardless, once formed, these neurosecretions pass down nerve fibers to the primary plexus of the hypothalamic-hypophyseal portal circulation, where they are absorbed and carried by the blood to the adenohypophysis. There they act upon the adenohypophyseal secretory cells, causing them to release the appropriate hormones into the systemic circulation. Some properties of these inhibitory and releasing hormones are summarized in

Figure 114. Hypothalamic and Hypophyseal Hormonal Systems

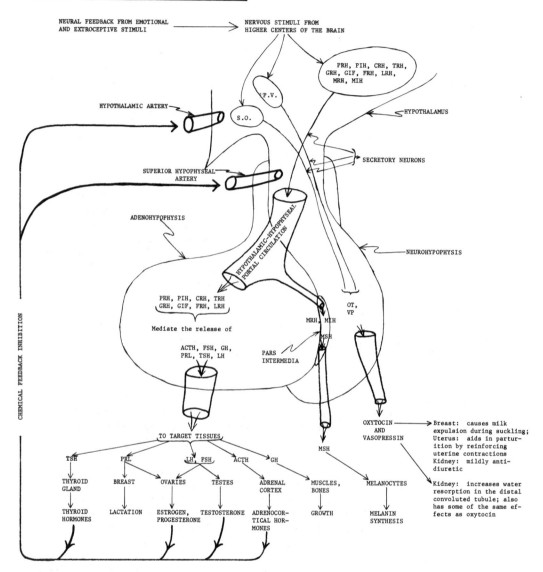

Key: P.V. = paraventricular nucleus; S.O. = supraoptic nucleus; OT = oxytocin; VP = vasopressin; TRH, CRH, PRH, GRH, FRH, LRH,
      MRH, PIH, GIF, MIH: See Table 47, page 865; MSH, ACTH, FSH, GH, TSH, LH: See Table 48, page 867; PRL = prolactin

Table 47. Hypothalamic Releasing and Inhibitory Hormones

	Hormone	Chemical Nature	Target Organ	Effect Produced
A.	Corticotrophin-releasing hormone (CRH)	polypeptide	adenohypophysis	Adrenocorticotrophic hormone (ACTH) release
B.	Follicle-stimulating hormone releasing hormone (FRH) (may be the same as LRH, below)	peptide	adenohypophysis	Follicle-stimulating hormone (FSH) release
C.	Thyrotrophic hormone releasing hormone (TRH) (see page 492)	tripeptide	adenohypophysis	Thyrotrophic hormone (TSH) release
D.	Luteinizing hormone releasing hormone (LRH)	probably a decapeptide	adenohypophysis	Luteinizing hormone (LH) release
E.	Growth hormone releasing hormone (GRH)	polypeptide	adenohypophysis	Growth hormone (GH) release
F.	Growth hormone inhibitory hormone (GIF)	peptide	adenohypophysis	Inhibition of GH release
G.	Prolactin releasing hormone (PRH)	not yet certain	adenohypophysis	Prolactin release
H.	Prolactin release inhibiting hormone (PIH)	not yet certain	adenohypophysis	Inhibition of prolactin release
I.	Melanocyte stimulating hormone releasing hormone (MRH)	not yet certain	intermediate lobe of the pituitary	Melanocyte stimulating hormone (MSH) release
J.	MSH release inhibiting hormone (MIH)	tripeptide	intermediate lobe of the pituitary	Inhibition of MSH release

Table 47. The nervous and chemical stimuli which can initiate this sequence of events are quite diverse. In general, they are associated with the body's need for the physiological response produced by the hormone.

4. The hypothalamus also regulates the release of melanocyte stimulating hormone (MSH) from the pars intermedia of the hypophysis (see below). In this instance also, both releasing and inhibitory factors are secreted by the hypothalamus. This is indicated in Table 47 and Figure 114.

The Hypophysis (Pituitary Gland)

1. The hypophysis (meaning undergrowth) is located below the brain and plays a major role in the endocrine function of the organism. The gland is divided into three parts. They are the anterior lobe (adenohypophysis), the posterior lobe (neurohypophysis), and the intermediate"lobe"(pans intermedia), really just a small collection of cells between the other two lobes. The adenohypophysis and neurohypophysis have different embryological origins and functions. The anterior lobe (of oral epithelial origin) is highly vascularized and its cells are regulated by the hormones elaborated by the hypothalamus and released into the hypothalamic-hypophyseal portal system (see previously). The neurohypophysis (derived from the neural epithelium), on the other hand, is connected with the hypothalamus through many nerve fibers as well as by vascular interaction through the hypophyseal stalk. The neurohypophysis also contains non-neural connective tissue cells (neuroglial cells) known as pituicytes.

2. The adenohypophysis contains two main groups of cells which are histologically distinct. They are designated chromophobes (cells that do not stain very well; small, with granular cytoplasm) and chromophils (cells that stain readily; large with granular epithelium). The chromophils are further subdivided into basophils and acidophils, depending on the chemical nature of the dyes which are capable of staining them. By the use of a variety of dyes, additional divisions of these classes of cells are possible. Functionally, the adenohypophysis contains at least six different cell types, each secreting a specific hormone.

3. Table 48 lists the hormones of the hypophysis along with their target organs and functions. As can be seen, FSH, LH, and TSH are glycoproteins and are secreted by (morphologically distinct) basophilic cells. There is some overlap in the secretions from these cells. Chorionic gonadotrophin (HCG), also a glycoprotein, bears a structural similarity to TSH and LH. All three contain

Table 48.   Actions and Some Characteristics of the Pituitary (Hypophyseal) Hormones

Hormone	Chemical Nature	Target Tissues	Effect Produced
A. Neurohypophysis (Posterior Pituitary)			
i) Oxytocin	cyclic non-apeptide	Mammary glands; uterus	Ejection of milk from the breasts; smooth muscle contraction; pressor efferent
ii) Vasopressin (antidiuretic hormone ADH)	cyclic non-apeptide	Kidneys, arteries	Reabsorption of water; smooth muscle contraction
B. Pars intermedia (intermediary lobe)			
i) Melanocyte-stimulating hormone (MSH)	peptide α-MSH: 13 amino acids β-MSH: 18 amino acids	Skin, melanophores	Dispersal of skin in melanophores, leading to darkening of skin
C. Adenohypophysis (Anterior pituitary)			
i) Prolactin (leuteotropin) secreted by basophil cells	Protein	Mammary gland	Initiation of production of milk in a prepared gland
ii) Adrenocorticotrophin (ACTH; corticotrophin) secreted by chromophobe	44 residue peptide	Adrenal cortex	Synthesis and/or secretion of adrenal cortical steroids
iii) Thyrotrophin(see pages 492-493) (Thyroid-stimulating hormone, TSH) secreted by basophil cells	Glycoprotein; molecular weight about 28,000	Thyroid gland Adipose tissue	Synthesis and secretion of T_3 and T_4 Lipolysis and release of lipid
iv) Growth hormone (somato-trophin; GH); secreted by acidophil cells	188 residue peptide	Adipose tissue; Rest of the tissues of the body	Release of lipids (mobilization of fat); Anabolic hormone; growth of long bones
v) Follicle-stimulating hormone (FSH); secreted by the basophil cells	Glycoprotein; molecular weight about 16,000-17,000	Follicle of ovary in female; seminiferous tubules in males	Maturation of follicles; sperm production
vi) Luteinizing hormone (LH), (interstitial cell-stimulating hormone; ICSH) secreted by basophil cells	Glycoprotein; molecular weight about 16,000-17,000	Interstitial cells of the ovary in females. Interstitial cells of the testis in males	Estrogen and progesterone secretion; maturation of follicle; formation of corpus luteum Androgen secretion; interstitial tissue cell maturation

one each of two types of subunits, designated α and β. The
α-subunits of TSH, LH, and HCG are very similar chemically, are
responsible for the immunological cross-reactivity of these three
hormones, and have no significant hormonal activity. The
β-subunits of these glycoprotein hormones are different and
confer biological specificity and hormonal activity.

4. Several disorders of the pituitary have been described, producing
both under and oversecretion of the hypophyseal hormones.

   a. Overproduction of growth hormone (GH) causes gigantism if it
   occurs in children and acromegaly if it is present in adults.
   Gigantism results from a delay in epiphyseal fusion due to
   hypogonadism resulting from the excessive GH. It may be
   accompanied by mild acromegalic features. In acromegaly, the
   bulk of bone and soft tissue increases causing an increase in
   the size of the hands and other parts of the body. Changes
   in facial appearance and growth of excessive hair (hirsutism)
   also frequently occur. The usual cause is a secreting
   acidophil tumor of the pituitary gland.

   b. Elevated ACTH levels are found in Addison's disease and are
   one of the common causes of Cushing's syndrome. Both of these
   disorders are discussed later.

   c. Panhypopituitarism (see also page 485) is, as the name implies,
   a general undersecretion of the pituitary hormones. It is
   considered to be a chronic condition which poses no threat to
   life except in times of stress. Causes include pituitary
   tumors and infarctions, granulomatous lesions due to
   tuberculosis and other diseases, and pituitary surgery or
   irradiation. The degree to which each hormone is decreased
   depends on the duration of the disorder and on the individual
   patient.

   d. A specific deficiency of growth hormone is the cause of about
   10% of the cases of dwarfism. When GH is demonstrably decreased,
   the disorder is known as pituitary dwarfism. Dwarfism with
   this specific etiology is responsive to treatment with GH.

Adrenals

1. In an adult, the adrenals lie on the superior pole of the kidneys (ad-renal meaning next to the kidney). The glands are triangular and each is enclosed by a capsule of fibrous connective tissue. In a cross section of a fresh gland, the cortex region appears yellow in color and is readily distinguishable from the brownish medullary region.

2. Each adrenal actually consists of two endocrine glands which differ in their embryological origins. The adrenal medulla (the inner region) is of ectodermal (neural crest) origin while the cortex is derived from the cephalic end of the mesodermal region.

3. The cortex and medulla secrete chemically distinct hormones which produce differing effects. The medullary hormones are catecholamines which have already been discussed (see pages 486-491), while the cortical hormones are steroids (adrenocorticosteroids). The medullary hormones indirectly (through the adenohypophysis) affect the secretory activity of the cortex.

4. Over fifty different steroids have been isolated and characterized. They can be grouped into three classes based upon their functional characteristics: mineral corticoids (affect electrolyte (Na^+, K^+, Cl^-) concentrations and water balance), glucocorticoids (affect the metabolism of carbohydrates, proteins, and lipids), and a few sex steroids. The principal mineral corticoid is aldosterone and the principal glucocorticoid is cortisol. There is some overlap in the activities of these hormones, however, as can be seen in Table 49.

Table 49. Activities of Some Natural and Synthetic Corticoids

	Glucocorticoid Activity (units/mg)	Mineralcorticoid Activity (units/mg)
Aldosterone	0.4	3000
Cortisol (hydrocortisone)	1	1
Cortisone	0.8	0.8
Prednisone (Δ^1-cortisone)	3.5	0.8
Dexamethasone (9α-fluoro-16β-methylprednisolone)	25	0
Fluorocortisone (9α-fluorocortisol)	15	125

Cortisol is defined as having unit glucocorticoid and mineralcorticoid activity. The first three compounds occur naturally while the last three are synthetic.

<u>Table 50a.</u> Biosynthesis of Cortisol and Aldosterone
(For explanatory notes, see Table 50b)

Table 50b. **Biosynthesis** of Estrogen, Testosterone, and Other Sex Steroids

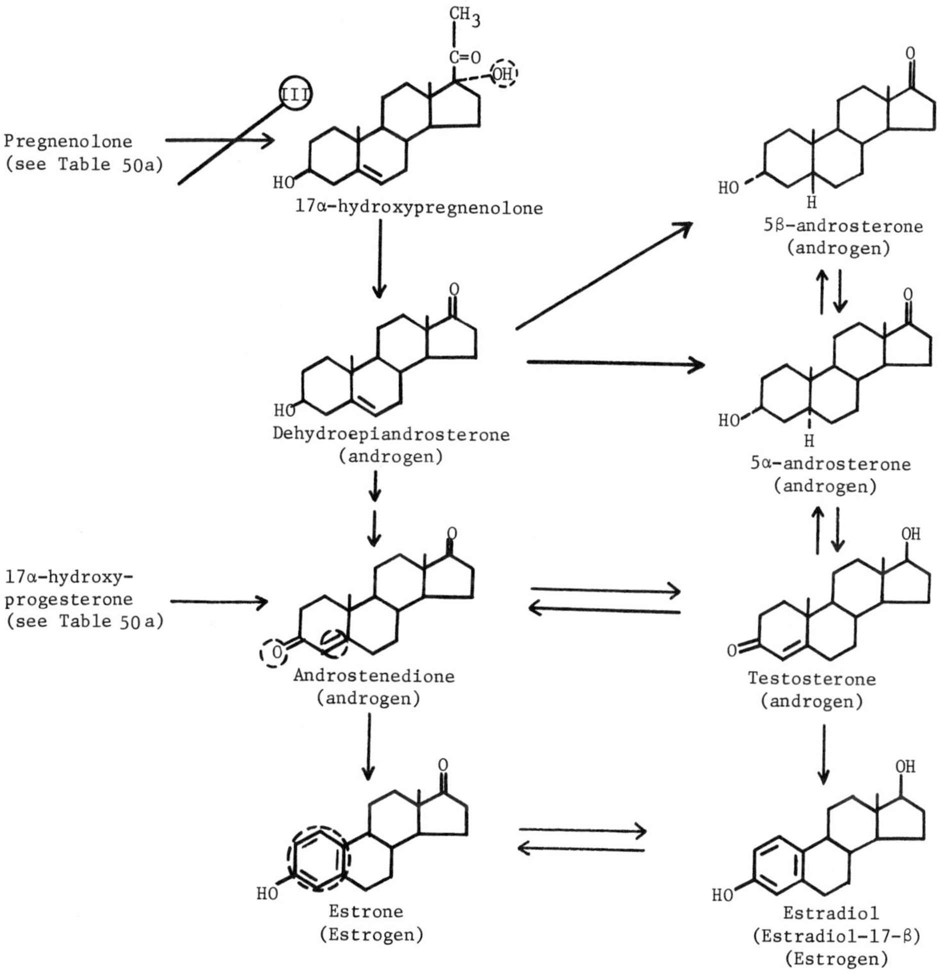

Pregnenolone
(see Table 50a)

17α-hydroxypregnenolone

5β-androsterone
(androgen)

Dehydroepiandrosterone
(androgen)

5α-androsterone
(androgen)

17α-hydroxy-
progesterone
(see Table 50a)

Androstenedione
(androgen)

Testosterone
(androgen)

Estrone
(Estrogen)

Estradiol
(Estradiol-17-β)
(Estrogen)

In both Tables 50a and 50b, the Roman numerals (I-VI) refer to biochemical lesions of steroidogenesis. They are described on page 876. The groups enclosed by dashed lines are the regions of the molecules which are altered in the reaction leading to the structure indicated.

The reason that aldosterone does not exhibit a physiologically noticeable effect as a glucocorticoid, even though it has 40% of the activity of cortisol on a units per milligram basis, is because its normal, physiological blood level (in mg/100 ml of plasma) is low compared to that of the glucocorticoids.

5. The cortex is further differentiated into three concentrically arranged layers or zones of cells. The outermost layer, known as the zona glomerulosa, is thin and is primarily responsible for secretion of aldosterone. The middle layer is the zona fasciculata and is the widest of the three concentric layers. The zona reticularis is the innermost layer and it surrounds the adrenal medulla. The two inner layers are responsible for the synthesis and secretion of glucocorticoids (in particular cortisol) and small amounts of adrenal androgens. Under normal circumstances, in man, the amounts of the principal corticoids secreted during a 24 hour period are as follows: cortisol (10-30 mg); aldosterone (0.3-0.4 mg); and corticosterone (2-4 mg).

6. The zonae fasciculata and reticularis are regulated by ACTH in a negative-feedback type of control system. ACTH maintains the structure and function of these cells and in hypophysectomy the zonae fasciculata and reticularis undergo atrophy. On the other hand, the zona glomerulosa maintains normal structure and function upon hypophysectomy or may even show hypertrophy. Thus the zona glomerulosa is not dependent upon ACTH for its secretion and function, although its activity may be influenced indirectly by ACTH. Recall that ACTH secretion is regulated by ACTH-releasing hormone produced by the hypothalamus. Stress stimuli which bring about hypothalamic secretions originate at higher centers of the central nervous system.

7. Aldosterone secretion is regulated by a variety of factors including the renin-angiotensin system, the plasma electrolyte (Na^+, K^+) levels, growth hormone, etc. The metabolism of aldosterone is discussed in the chapter on water and electrolyte balance (see pages 837-839).

Biosynthesis of Adrenal Cortical Hormones

1. The precursor to the adrenal steroids is cholesterol which in turn is synthesized from acetate (see pages 650-655). More cholesterol is present in the adrenal cortex than in any other structure of the body except for the nervous tissue. The adrenal cortex cholesterol is esterified with long-chain unsaturated fatty acids.

2. In the biosynthetic pathways, a number of specific hydroxylations of the steroid ring system take place. These reactions are catalyzed by specific hydroxylases which are mixed function oxidases requiring O_2, NADPH, cytochrome P-450 and adrenodoxin (see page 205). Except for the 11- and 18-hydroxylating systems (which are mitochondrial), these reactions occur in the microsomes. The cortex also contains large amounts of vitamin C.

3. The synthetic pathways for some of the steroids are shown in Table 50. Also shown in this figure are the biochemical lesions of steroidogenesis which cause various congenital adrenal cortical hyperplasias. In these syndromes, there generally are elevations in levels of testosterone, estrogens (along with 17-ketosteroids), and (except for type II) pregnanetriol.

4. The first step in steroidogenesis is the conversion of cholesterol to pregnenolone, catalyzed by a mixed function oxidase enzyme complex system known as desmolase (cholesterol oxidase). The reaction takes place in the mitochondria and requires NADPH, cytochrome P-450, and either Mg^{+2} or Ca^{+2}. This step is the rate-limiting reaction in steroidogenesis and pregnenolone functions as a feedback inhibitor of this enzyme.

5. The basis for ACTH stimulation of steroidogenesis is its activation of the adrenal cortical adenyl cyclase, causing a rise in cAMP and activating glycogen phosphorylase (see Chapter 4). This enzyme breaks down glycogen to produce glucose-6-phosphate, which then is oxidized via the hexose monophosphate shunt pathway yielding NADPH. The NADPH is a co-substrate in the hydroxylase reactions and an increase in its availability promotes overall steroid synthesis. The action of ACTH also involves stimulation of protein synthesis, presumably by increasing mRNA synthesis and, hence, growth of the cortex.

6. Transport and catabolism of corticosteroids: albumin is a general carrier of steroid hormones, particularly aldosterone. Cortisol, however, is bound to an α_1-globulin known as transcortin or corticosteroid-binding protein. Each molecule of transcortin binds one molecule of cortisol (or corticosterone). Transcortin has a molecular weight of 52,000 daltons and is synthesized in the liver. Estrogen and thyroid hormone increase the transcortin levels while liver diseases and nephrosis decrease them. Catabolism of steroids occurs through the various reduction and conjugation reactions that occur in the liver (see page 205 and Chapter 14).

7. Cortisol has several biological actions

    a. It exerts a catabolic effect on muscle, epidermal, and
        connective tissue leading to negative nitrogen balance, release
        of amino acids, and atrophy of these tissues.

    b. Most of these tissues also exhibit reduced amino acid uptake.
        Hence, any unutilized amino acids released from protein
        breakdown enter the liver where they are converted to glucose
        via gluconeogenesis. The glucose produced is released into
        the blood, causing hyperglycemia. A prolonged hyperglycemia
        due to excess glucocorticoids can lead to diabetes mellitus.

    c. The immediate (acute) effect of cortisol is to transfer amino
        acids to the liver where they are converted to glucose and
        glycogen. Cortisol induces the synthesis of tyrosine
        transaminase, pyruvic-glutamic transaminase, tryptophan
        oxygenase, and ornithine decarboxylase. The long term effect
        also leads to induction of synthesis of enzymes which enhance
        the production of glucose: glucose-6-phosphatase, fructose-
        1,6-diphosphatase, PEP carboxykinase, and pyruvate carboxylase.
        The net effect of all of these is to produce glucose in the
        liver and hyperglycemia in the blood. Recall that insulin
        suppresses the activity of these gluconeogenic enzymes while
        inducing glycolytic and glycogenic ones.

    d. Cortisol causes increased release of glycerol and fatty acids
        from the adipose tissue. The glycerol is converted to glucose
        in the liver while the increased fatty acid levels cause an
        inhibition of key glycolytic enzymes in the liver.
        Administration of cortisol increases peripheral lipogenesis.
        This action is partially explained by the observation that
        cortisol produces hyperglycemia which increases insulin
        secretion and thereby promotes lipid synthesis. The fat
        synthesized has the characteristic distribution seen in
        Cushing's Syndrome (moon face, buffalo hump, pendulous abdomen).

    e. Cortisol has anti-allergic and anti-inflammatory activities.
        These properties are therapeutically useful and are employed
        widely. There is also enhanced lymph node lysis resulting in
        lymphocytopenia and decreased antibody production.
        Erythropoiesis is increased. The antiallergic reactions may
        be due to decreased production of histamine and histamine-like
        substances and protection of the surface of the affected
        cells from the antigen and antibody interaction.

f. Excess cortisol can cause increased gastric acidity due to elevated HCl secretion. Enhanced pepsin secretion is also observed, with production of a watery mucous in the stomach. These can cause ulcers or aggravate an already existent ulcer.

8. The interrelationships between the adrenal cortex, adenohypophysis, the hypothalamus, and the higher centers of the central nervous system are shown in Figure 115.

Figure 115. Interactive Controls Among the Brain, Hypothalamus, Anterior Pituitary, and Adrenal Cortex

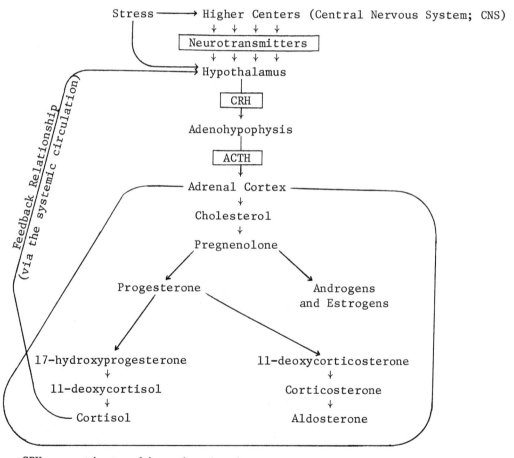

CRH = corticotrophin-releasing hormone.

9. The proper functioning of the feedback mechanisms at various levels (outlined in Figure 115) can be assessed by the following methods.

    a. The <u>administration of ACTH</u> to a normal person should cause the adrenal cortex to respond by releasing cortisol, resulting in an increase in blood cortisol levels. Hypofunctioning of this gland should be suspected when this response does not occur.

    b. <u>Dexamethasone administration</u> (see page 488) will indicate whether or not the hypothalamic feedback centers are capable of turning off CRH release and thus ACTH release. If this does not occur, then cortisol cannot function as a feedback inhibitor of its own synthesis.

    c. The response of the hypothalamic feedback centers to low levels of cortisol is assessed by inhibiting the synthesis of cortisol. This is accomplished by administering metyrapone which inhibits the 11-hydroxylation step (final hydroxylation step) causing a fall in circulating cortisol levels and stimulating the hypothalamic feedback center to release CRH and hence ACTH.

    d. The adrenal response of the patient to stress is assessed by the insulin hypoglycemic test. This consists of reducing the blood glucose level of the patient to 35 mg/100 ml by intravenous administration of insulin. Normal response to this stressful situation is the rapid elevation of plasma cortisol levels.

<u>Disorders of Adrenal Steroidogenesis</u>

Six disorders of steroid synthesis in the adrenal gland are indicated in Table 50. The numbers below (1-6) refer to this diagram.

1. <u>Desmolase deficiency</u> is also known as lipoid adrenal hyperplasia. No steroid hormone synthesis takes place in the adrenals, leading to the excessive accumulation of cholesterol and other lipids. Adrenal insufficiency (salt and water loss) occurs and an early death ensues unless the disease is treated extensively. No urinary steroids are detectable.

2. In <u>3β-ol-dehydrogenase deficiency</u>, progesterone is not formed from pregnenolone resulting in deficient production of mineral and glucocorticoids and elevation of dehydroepiandrosterone and related products. Affected individuals show salt loss, hypospadias (male), and mild virilism (female). High urinary excretion of Δ^5-3β-hydroxysteroids is observed.

3. **17α-hydroxylase deficiency** is a defect in the production of androgens, estrogens, and cortisol with an accompanying elevation of corticosterone and deoxycorticosterone. These individuals show immature sexual development and hypertension. Urinary 17-ketosteroids and estrogens are absent, while elevated levels of pregnanediol and corticosterone are present.

4. **21-hydroxylase deficiency** is the most common of these lesions. It is transmitted as an autosomal recessive trait and is found particularly in Eskimos. It is a deficiency of production of cortisol and aldosterone. Elevated levels of androgens are seen and affected individuals show masculinization with increased urinary excretion of 17-ketosteroids, pregnanediol, and C-21 deoxysteroids.

5. Patients with **11β-hydroxylase deficiency** show depressed production of cortisol and aldosterone, elevated levels of androgens, and to a lesser extent, an increase in 11-deoxycorticosterone and 11-deoxycortisol. These individuals show masculinization and hypertension with elevated urinary 17-ketosteroids, 11-deoxycorticosterone, and 11-deoxycortisol.

6. **18-hydroxylase deficiency** is a defect in the production of aldosterone. There is an increase in corticosterone. Affected individuals show the salt-losing syndrome and urinary levels of corticosterone and related products are elevated.

   As a general consequence of these lesions, the plasma cortisol levels are low and a feedback mechanism is triggered to compensate for this. ACTH production is enhanced until normal levels of cortisol are attained. This situation eventually leads to adrenal hyperplasia.

Structural and Chemical Aspects of Steroid Hormones (see Table 50)

1. Cholesterol (cholestane is the name of the parent ring system) is the precursor of the following five principal classes of steroid hormones. These are listed below with the parent ring system indicated in the parentheses: progestogen (pregnane), estrogen (estrane), androgen (androstane), mineral corticoid (pregnane), and glucocorticoid (pregnane).

2. Glucocorticoids (cortisol), mineral corticoids (aldosterone), and progestogens (progesterone) all contain 21 carbon atoms and are referred to as C-21 steroids. The androgens (testosterone) are C-19 steroids while the estrogens (estradiol) are C-18 steroids. Note that estrogen is unique because it contains an aromatic

(ring A is benzenoid) ring system and does not contain the methyl
group (C-19) attached to C-10. Consequently, in estrogens, the
-OH group attached to the benzene ring is phenolic and weakly
acidic, making these compounds soluble in dilute basic solutions.
This property is employed in the extraction procedures used in
the identification of estrogens.

3. Androgens and estrogens contain a potential keto ($>$C=O) group in
the C-17 position and hence are called 17-ketosteroids. It should
be pointed out that, in man, 17-ketosteroids are not only secreted
by the adrenal cortex, but also by the ovaries and testes. About
one-third of the total 17-ketosteroids are of testicular origin.
Androgens generally appear in the urine as 17-ketosteroids.
Urinary 17-ketosteroids can be separated into three classes:
acidic compounds (derived from the bile acids), phenolic compounds
(which are slightly acidic and are metabolites of estrogens), and
neutral compounds (derived from the androgens of the adrenal
cortex and gonads). These metabolites are present in the urine
as glucuronides. Prior to their separation for purposes of
identification, free steroid derivatives are obtained by hydrolysis
of these glucuronides. This is done by treatment with concentrated
H_2SO_4 or HCl, or by use of the enzyme glucuronidase. The steroids
are extracted into ether, the acidic (bile acid derived) fraction
is separated with $NaHCO_3$, and the phenolic portion with 2N NaOH.

4. The attachment of an -OH group to a saturated ring system can give
rise to two isomers. In the α-configuration, the -OH group extends
below the plane of the ring and the substitution is shown with a
dashed line (---). The β-isomer, in which the hydroxyl bond is
indicated with a solid line, has the -OH group above the plane of
the ring. The 17-OH groups present in testosterone and estradiol
are β-substitutions, while cortisol has an α-OH group (indicated
by ---) attached to C-17.

5. An oxy compound has one additional oxygen atom relative to its
parent compound while a deoxy compound has one less oxygen.
Similarly, a dihydro compound contains two additional hydrogen
atoms while dehydro designates a compound having two less hydrogen
atoms than the molecule from which the name is derived.

Cushing's Syndrome

1. This problem is due to the presence of excess circulating cortisol.
It can result from (a) excessive secretion of ACTH by the
adenohypophysis (the most common form), (b) ectopic ACTH production
by a tumor of non-endocrine origin (as seen in bronchiogenic
carcinoma) with the excess ACTH production leading to adrenal
hyperplasia, (c) carcinoma or adenoma of the adrenal cortex,
(d) administration of cortisone and its synthetic substitutes in
a variety of disorders.

2. The clinical features consist of red cheeks, bruisability, thin skin, obesity with a characteristic distribution of fat (moon face, pendulous abdomen, buffalo hump), hypertension, muscular weakness, osteoporosis, hirsutism, menstrual disturbances, poor wound healing, hyperacidity, peptic ulcers, etc.

3. Laboratory Aspects

   a. The plasma ACTH and, hence, the cortisol level normally shows a diurnal variation with the highest level occurring in the morning and the lowest level around midnight. In Cushing's Syndrome, the cortisol level does <u>not</u> show a diurnal change, remaining constantly elevated.

   b. Urinary 11-hydroxycorticosteroids (cortisol) are elevated. Note any stressful situation can result in both (a) and (b).

   c. Dexamethasone suppression test. Dexamethasone is a synthetic steroid which produces the same effect as cortisol. When this compound is administered to normal individuals, there is a fall in the adenohypophysis ACTH secretion due to an effect on the hypothalamus. This reduction in ACTH output can be monitored by plasma cortisol or urinary 17-hydroxycorticosteroid levels and both should drop in normal individuals. In Cushing's Syndrome, due to adrenal hyperplasia there is an impairment in the suppression of the cortisol level at low dosages of dexamethasone, while with higher levels of the drug, the suppression can be accomplished. In adrenal and ectopic ACTH tumors, however, there is no suppression even after the administration of high doses of dexamethasone.

<u>Addison's Disease</u> (see also page 485)

1. This disease is caused by the <u>hypofunctioning</u> of the adrenal cortex, as a result of tuberculosis (the most common cause), idiopathic atrophy of the cortex (presumably due to an autoimmune mechanism), secondary malignancy, mycotic infections, and amyloidosis.

2. The severity of the disease depends upon the amount of functional tissue which remains available to synthesize and secrete the steroid hormones. The most important consequence of the deficiency is the lack of aldosterone, leading to sodium depletion. The clinical characteristics include dehydration, hemoconcentration, hyponatremia, elevated plasma potassium, metabolic acidosis, low blood pressure, muscular weakness, hypoglycemia, gastro-intestinal disturbances, and increased circulating levels of ACTH due to the deficiency of

cortisol (lack of feedback suppression of ACTH). The excess ACTH leads to brown pigmentation since the ACTH molecule also has some MSH activity. The two are both peptide hormones and they share part of their amino acid sequence in common. An Addisonian crisis, which needs immediate treatment, is a condition in which the patient is in a dehydrational shock.

3. The laboratory investigation shows decreased excretion of 17-hydroxycorticosteroids and aldosterone. The most important diagnostic step is the demonstration of the non-responsiveness of the adrenal cortex to ACTH administration. This can be accomplished by measuring 17-hydroxycorticosteroids before and after the administration of synthetic ACTH peptide (which contains 24 of the 39 amino acid residues yet possesses the same biological activity). A rise in plasma values of these hormones excludes Addison's disease.

Gonadal Hormones

Androgens

1. The name androgen means "producing male characteristics". Any substance which has this effect may be referred to as an androgen. The Leydig cells (interstitial cells) of the testes synthesize two androgens: testosterone and androstenedione. Testosterone is the principal one of these and it is responsible for the observed endocrine effects of testicular activity. The ovaries also secrete small amounts of testosterone and the adrenal cortex in both sexes secretes androgens (mentioned earlier).

2. The gonadotrophic hormones FSH and LH (also called ICSH, see Table 48) stimulate the production of spermatozoa in the seminiferous tubules and of androgens in the Leydig cells, respectively. The circulating level of testosterone regulates the secretion of the gonadotrophic hormones by a feedback mechanism.

3. Biosynthesis of testosterone in the Leydig cells is similar to that described in adrenal cortex (see Table 50b). It has been shown that the active form of testosterone is dihydrotestosterone (see page 862). The major catabolic transformations of the androgens take place in the liver. Testosterone is converted to 5α- and 5β-androsterone (5β-androsterone is also known as etiocholanolone) which is then conjugated with glucuronic acid or sulfate and excreted as a 17-ketosteroid.

4. The basic molecular mechanism by which steroid hormones (including the androgens) function were discussed earlier. The specific effects of androgens include

    a. Development of primary and secondary male sexual characteristics (masculinization). The effects of testosterone are manifested in both sexes. In males, its influence predominates while in females, it has a minimal effect partly because it has to overcome feminization caused by the estrogens.

    b. A major response to androgens is the anabolism of nitrogen and calcium compounds, reflected (especially in males) by an increase in skeletal muscle mass and growth of the long bones prior to closure of the epiphyseal cartilage. Castration results in a delay of epiphyseal closure, while excessive secretion of androgens can cause premature closing of the epiphyses.

    c. Some of the other effects of testosterone are an increase in the basal metabolic rate (BMR), promotion of Na^+, Cl^- and water reabsorption by the kidney tubules, and stimulation of erythrocyte production in the bone marrow.

5. Attempts have been made to synthesize androgens with anabolic activity but without their masculinizing effects. Testosterone is used therapeutically in carcinoma of breast. In this regard, it is highly desirable to have a compound with androgenic activity which does not cause masculinization. Following are some examples of synthetic androgens.

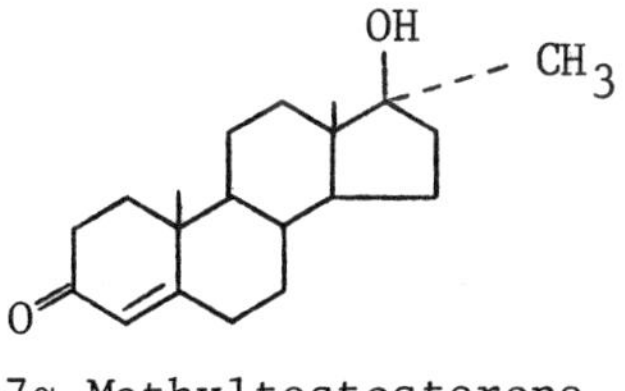

17α-Methyltestosterone
(a potent androgen, orally effective)

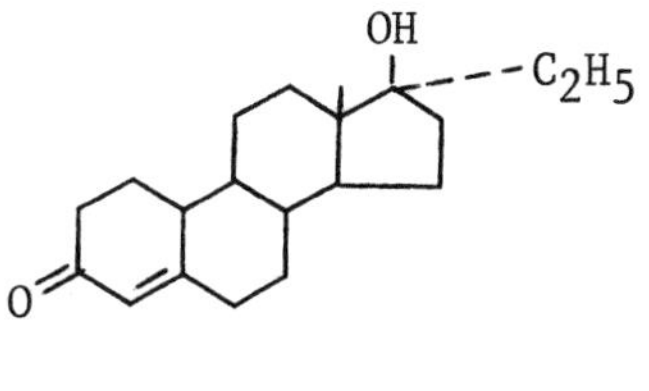

17α-Ethyl-19-nortestosterone

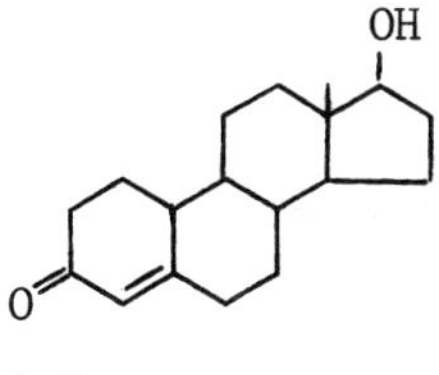

19-Nortestosterone

The above compounds are named as <u>nor</u>steroids because they lack an
angular methyl group (C-19) attached to C-10. Estrogens also lack
this C-19 angular methyl group. These norsteroids have been shown
by bioassay procedures to have an anabolic activity twenty times
greater than their masculinizing activity.

Estrogens

1. The term estrus (or oestrus) from which the word estrogen is derived
    comes from the Greek <u>oistros</u> (gadfly). It refers to the state of
    sexual excitability or receptivity in females of many species which
    occurs with varying periodicity. Apparently, the frenzy caused
    by the imitation of the gadfly is likened to this state. Estrogen
    is now used to refer to any of a group of compounds which cause
    feminization as seen by primary and secondary sexual
    characteristics. The natural estrogens are produced in the
    Graafian follicles of the ovaries under the stimulus of FSH. Three
    types of estrogens have been isolated from ovarian tissue and
    human urine. These are

β-Estradiol   Estrone   Estriol

2. Estradiol is the major hormone of the ovarian secretion and is
    the most potent of the three. The other two are metabolic products
    of estradiol. Estriol is produced mainly in the liver from the
    other two hormones. It undergoes conjugation with glucuronic acid
    and sulfate and is eliminated in the bile and urine. Measurement
    of urinary estriol provides a useful index of estrogen production
    (see below).

3. Estrogens are C-18 steroids, lacking the angular methyl group at
    C-10, and ring A is aromatic in these compounds. Androgens are
    C-19 steroids. The biosynthesis of the estrogens is given in
    Table 50b.

4. The target organs of the estrogens are the uterus, the vagina, the
    adenohypophysis, the hypothalamus, and the mammary gland. At these
    sites, these hormones have the actions summarized next.

a. They promote the development, maturation, and function of the internal and external female reproductive organs. The proliferative phase of the uterine cycle is caused by the cyclic production of estrogen.

b. Their action on the hypothalamus and the adenohypophysis is to suppress FSH secretion and promote LH secretion.

c. Other effects include enhancement of the activity of alkaline phosphatase, an increase in the amounts of glycogen in the endometrium and the vagina, increased RNA and protein synthesis, activation of transhydrogenase, and increased accumulation of lipids in muscle and other tissues.

5. Natural and synthetic estrogens are used clinically in ovarian agenesis, other estrogen deficiencies, and in treating the symptoms associated with menopause. One of the synthetic estrogens is diethylstilbestrol whose structure is shown below.

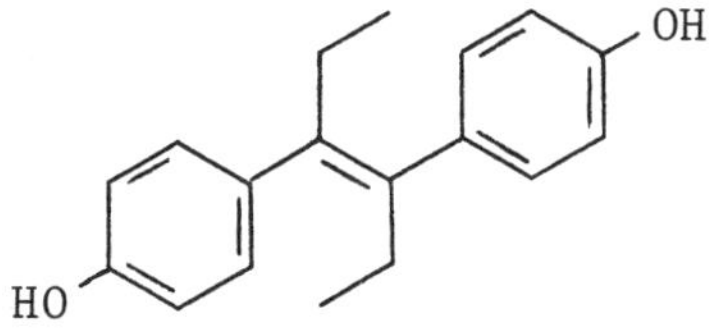

or

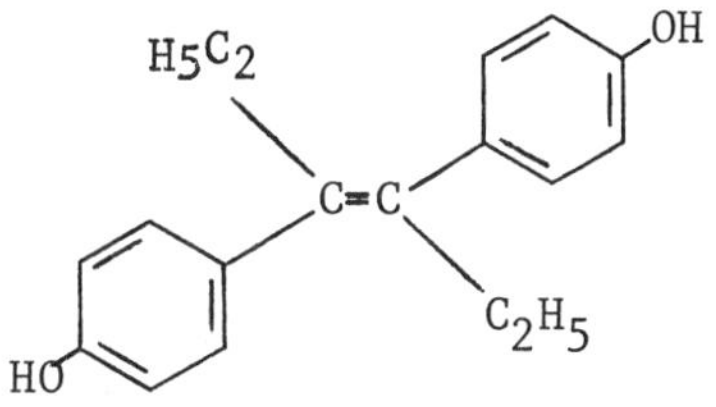

Diethylstilbestrol therapy has been initiated in the first trimester of pregnancy if bleeding occurs or if there is risk of a miscarriage. A tragic side effect of this treatment has recently been reported. The adolescent female offspring of women who had received this non-steroidal estrogen during the indicated period of pregnancy show an increased incidence of adenocarcinoma of the vagina.

6. Measurement of estriol, a metabolite of estrogen, in a 24-hour urine sample has been useful in the assessment of placental and fetal functions in pregnant women. After fertilization, the ovum gets implanted in the uterine wall and the developing placenta secretes chorionic gonadotrophin (CG) which continues the secretion of estrogen and progesterone from the corpus luteum. This maintains the corpus luteum in the luteal phase and prevents the onset of menstruation. The CG secretion reaches a peak at about 13 weeks of gestational life. At this time, the fetus and placenta together take over the production of estrogens and progesterone and chorionic gonadotrophin secretion declines. The estrogens produced by the integrated feto-placental unit appear in the maternal circulation and the estriol formed appears in the urine. Urinary

estriol levels can reach sufficiently high values for accurate laboratory measurements to be made. For a particular individual under normal conditions, urinary estriol increases throughout pregnancy. In monitoring normal feto-placental development, it is essential to perform assays on 24-hour urine specimens and to make periodic determinations. Single determinations are misleading unless they are extremely low (indicating non-secretion), because the normal range varies greatly.

Progesterone

1. Progesterone is an intermediate in the biosynthesis of the adrenal corticosteroids, testesterone, and estradiol (see Table 50). Thus it is present in the adrenal cortex, the placenta, and the ovary. It is also produced in the corpus luteum, and this tissue may therefore be considered as a transitory endocrine gland.

2. In a non-pregnant female, the corpus luteum is the principal source of progresterone, while the placenta provides most of this hormone during the second and third trimesters of pregnancy. During the follicular phase of a normal menstrual cycle, progesterone is present in the blood in very low concentrations. This follicular phase progesterone is of adrenal origin. The progesterone rises in the luteal phase of the cycle if and only if ovulation occurs. Thus, a blood progesterone measurement during the second half of the cycle can be used to determine if ovulation did occur and if, in fact, a corpus luteum capable of secreting progesterone is present in the ovary.

3. The major metabolite of progesterone is pregnanediol which is excreted in the urine as both the glucuronide and the sulfate.

Progesterone   Pregnanediol

4. The major biological effects of progesterone are

    a. Preparation of the uterus for implantation of the ovum to
        occur, a process initiated by estrogens. The actions of
        progesterone occur only <u>after</u> prior estrogen action. The
        preparation consists of developing a secretory type of
        endometrium and of diminishing the muscular contractions of
        the Fallopian tubes and of the uterus.

    b. Stimulation of the glandular development of the breasts by
        inhibition of the effects of estrogens.

    c. Progesterone is reponsible for stimulating the basal metabolic
        rate and thus causing a periodic rise in the body temperature
        (about 0.5 to 1.0°F) during the luteal phase (second half)
        of the normal menstrual cycle.

    d. Suppression of ovulation, hence progesterone and its
        derivatives, are used to prevent conception.

The Menstrual Cycle

The reproductive system of the female, unlike that of the male, shows
regular cyclic changes. During each cycle, the organs in the female
which are needed for reproduction are prepared for fertilization and
pregnancy. Although the duration of a menstrual cycle is extremely
variable, an average figure of 28 days is generally used. The first
day of the cycle corresponds with the first day of vaginal menstrual
bleeding. The period from then until ovulation is the <u>follicular phase</u>
of the cycle, during which time the follicle is developing. At the
end of this phase, the follicle cell bursts open and releases the
ovum. From then until onset of the next menses, the corpus luteum
is developing, hence the name <u>luteal phase</u> for the second half of the
cycle.

The hormonal changes associated with the menstrual cycle are illustrated
in Figure 116. To start the cycle, a hypothalamic-releasing factor (LRH)
stimulates secretion of the pituitary gonadotropins which in turn
stimulate the secretion of estrogen. Rising estrogen levels stimulate
a sharp rise in releasing factor and a rise in LH that triggers
ovulation and formation of the corpus luteum. This structure secretes
high levels of progesterone and estrogen which inhibit further output
of gonadotropins. As the gonadotropin levels decline, the corpus
luteum ceases to function, steroid (estrogen and progesterone) levels
drop, and gonadotropin levels once again begin to rise, starting
the hormonal cycle over again. If pregnancy occurs, chorionic

Figure 116. Composite Diagram of the Variation of Certain Hormones
in the Normal Menstrual Cycle
(note: ng = nanograms; pg = picograms; mIU = milli-interna-
tional units)

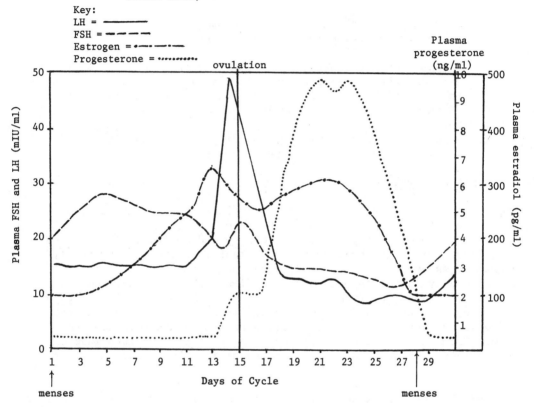

The hypothalamic-releasing factor (LRH) stimulates pituitary
gonadotropins, which in turn stimulate the secretion of estrogen.
Rising estrogen levels stimulate a sharp rise in releasing factor and
a rise in LH that triggers ovulation and the formation of the corpus
luteum. High levels of progesterone and estrogen inhibit further
output of gonadotropins, and the levels decline. The corpus luteum
ceases to function, steroid levels drop.

gonadotropin (CG), secreted by the corion of the placenta, keeps the corpus luteum competent (replacing the pituitary gonadotropins) so that estrogen and progesterone continue to be secreted at high levels. Thus the cycle is broken until the pregnancy terminates. Some birth control pills function in a similar fashion to break the rhythm and prevent ovulation.

For many years, it was thought that the hypothalamic-hypophyseal system contained inherent cyclic capabilities. More recent studies on negative and positive feedback modulations by the ovarian steroids have shown this to be incorrect. Rather, the cyclic gonadotropin released during the reproductive years is a <u>secondary</u> phenomenon dependent on ovarian function. Operationally, the hypothalamic-hypophyseal system can be viewed as a provider and the cyclic ovarian steroid output as a modulator. This output exerts an overlapping, interrelated sequence of negative and positive feedback action and produces a differential effect on the release of FSH and LH.

14

Biochemical Transformation of Drugs

*with Pushkar N. Kaul, Ph.D.**

General Considerations

1. The biochemical transformation of drugs and of other chemicals
    foreign to the body is a function of various enzymatic reaction
    systems analogous to those involved in metabolizing food and
    endogenous substances. It is fortunate that this necessary
    biotransformational capability has developed in terrestrial mammals,
    perhaps as a result of the evolutionary biochemical adaptation
    process. Without it, man would find it difficult to survive the
    toxicity of foreign organic substances which, due to their non-polar
    character and consequent high lipid solubility, would tend to
    accumulate in the lipoidal and lipoprotein compartments of the body.

    In most instances, drugs are metabolized into relatively more
    polar compounds which are subsequently eliminated from the body by
    the usual excretory processes. In a few instances, a drug may
    first be converted into a more non-polar metabolite which is then
    further metabolized by another pathway. In most cases, the length
    of time required for the body to rid itself of a foreign substance
    depends on the nature of the material and the drug-metabolizing
    efficiency of the individual.

2. Following administration, a drug generally attains a steady state
    equilibrium of the type shown in Figure 117. It is the algebraic
    sum of the rates of the various processes involved (absorption,
    distribution, enzymatic biotransformation, and excretion from the
    body) which determines the overall effectiveness and course of
    drug action. Increased consideration of these factors in current
    clinical practice has improved the rationale of therapy and thus
    the quality of patient care.

3. a. Drugs, unlike vitamins and amino acids, are absorbed from the
        gastro-intestinal tract by passive diffusion. Only unionized
        drug molecules are believed to cross the biological membranes,
        except in rare cases when an active or a facilitated transport
        may be involved. Acidic drugs do not ionize in the acidic

---

* Associate Professor of Pharmacology and of Research Medicine and
   Pediatrics, University of Oklahoma, Norman, Oklahoma.

Figure 117.   Processes Which Contribute to the Steady State Equilibrium of a Drug In Vivo

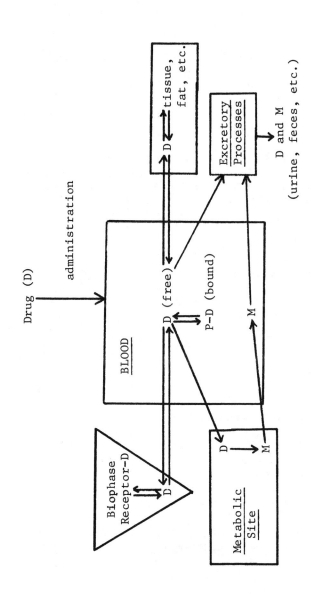

D = drug

P–D = protein-bound drug

M = metabolite of D

medium of the stomach and are therefore absorbed through the
gastric mucosa. Basic drugs ionize in the stomach and are hence
absorbed later, in the intestinal segment.

b.   Not only is the extent of absorption of a drug important but
     also the rate at which its absorption occurs. In general, the
     faster a drug is absorbed, the higher and more rapidly attained
     will be the maximum blood level. It is possible to have two
     dosage forms of the same drug which differ in efficacy simply
     because of unequal rates of absorption. Although both forms
     may be completely absorbed, the one which enters the circulation
     more rapidly will result in a higher peak blood level and, hence,
     a greater clinical effect. This can be critical in the
     administration of antibiotics where it is desirable to attain
     the highest possible peak level.

c.   In the actual process of absorption of a drug from any site of
     administration other than intravascular, the drug molecules
     must cross the capillary membrane to enter the blood. The
     unidirectional flow of drug molecules into the circulation
     continues until no more of the drug remains outside of the
     capillary wall at the site of absorption. Like a dialysis system,
     this entry of molecules can be considerably affected by the rate
     of blood flow inside the capillary lumen. Any changes in the
     blood flow due to normal diurnal variations, spontaneous
     over-exertion, or a pathological state affecting the cardio-
     vascular system can significantly influence the rate of absorp-
     tion of a drug.

4.   Having entered the circulation, a drug molecule has several possible
     fates. It may undergo elimination via the urine or bile, escape
     into the interstitial fluid and then into the cells of the various
     organs, or be metabolically altered while passing through the liver,
     kidney, lungs, or other organs. All these possibilities, however,
     are open only to the free molecules (see Figure 117). Molecules
     which are bound to plasma or cellular proteins are not available for
     these processes unless they first dissociate from their binding site.
     The extent of this dissociation increases (by the law of mass action)
     when the quantity of free drug in the blood decreases as a result
     of the various dispositional processes described above.

5.   The binding of drugs to blood proteins, as well as to the cellular
     or subcellular structures of the biophase (site of action), involves
     an interaction between the drug molecule and a small portion of the
     biological macromolecule known as the receptor. This binding may
     lead to no discernable pharmacological response, as in the case of

plasma protein-binding of imipramine. In other cases, a definite
response may be seen as when acetazolamide binds to and inhibits
carbonic anhydrase in the kidney tubular cells, producing diuresis.
In both situations, however, the binding involves similar types
of forces.

6. There are factors other than protein-binding which influence the
distribution of a drug. Preferential uptake of iodides by the
thyroid, of tetracyclines by bone, of chlorpromazine by hair and
skin, and of thiopental by adipose tissues are only some of the
examples of how specific tissues may affect the distribution and
blood level of a drug following its administration. A drug may be
distributed in all the fluid compartments of the body and consequently
develop only a low circulating concentration. Such a drug is said
to have a large volume of distribution.

7. Those drugs which enter the cerebrospinal fluid (CSF) are believed
to have crossed the so-called blood-brain barrier. This barrier
is not an anatomical structure but rather a concept devised to
explain why some drugs enter the CSF while others do not. From
the evidence available, this property of drugs is probably due to
their physico-chemical characteristics and to the presence of glial
cells in the brain. Generally speaking, highly lipid-soluble drugs
cross the blood-brain barrier whereas highly polar and ionized
drugs cannot enter the CSF.

Protein-Binding of Drugs

1. The bonding forces involved in the drug-protein binding are those
encountered earlier in the discussion of protein structure (see
pages 40-42). They are generally classified into primary and
secondary bonds.   The primary bonds are mostly ionic in nature,
involving coulombic (electrostatic) attractions between opposite
charges on the drug and the receptor molecules. They operate at
relatively long distances (< 4 Å) and aid in the initial
orientation of the drug and the receptor site so that the secondary
bonds can be easily formed. The secondary bonds include hydrogen
bonds and van der Waal's (London) forces.

2. Hydrogen bonds were discussed earlier on page 41. They involve
the sharing of a proton by two electronegative atoms having
unbonded electrons. The two atoms sharing the H^+ are usually
about 3 Å apart. An example of hydrogen bonding is illustrated by
the hypothetical binding of warfarin to albumin shown in
Figure 118.

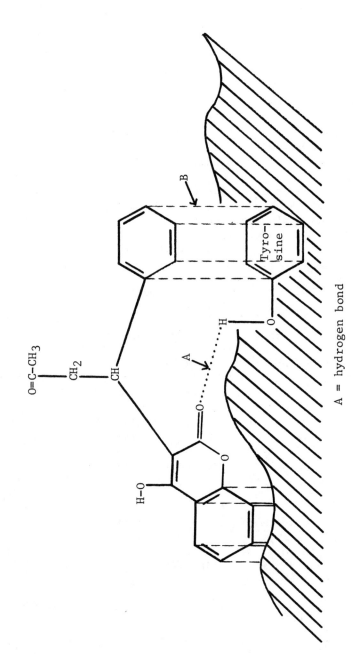

Figure 118.   Postulated Hydrogen and Van der Waal's Bond Formation Between Warfarin (Coumadin) and a Tyrosine Residue of Albumin

A = hydrogen bond

B = van der Waal's bond

3. Another secondary bonding force is the van der Waal's bond or London dispersion force. These are weak (0.1 to 1 kcal/mole), but when several form they can reinforce the binding interaction significantly. These bonds usually occur between carbon atoms of the two interacting molecules (drug and protein) when they approach each other closely enough for the outermost electrons of the carbon of one molecule to be within the range of the electrostatic field of the nucleus belonging to the carbon atom of the other molecule. Such interactions usually occur in multiples and involve either flat aromatic rings or long aliphatic carbon chains of the drug molecule and a comparable structural region of the protein molecule. The sum of the primary and secondary forces determines the strength of the drug-protein interaction.

4. The majority of drugs which undergo plasma-protein binding are attached to albumin. Some drugs may also be bound to erythrocyte membranes or to cytoplasmic proteins of erythrocytes. For example, it is postulated that chlorpromazine is bound to the erythrocyte membrane, whereas its metabolites are most likely bound to sites in the red cell interior (other than carbonic anhydrase). Acetazolamide, on the other hand, is largely bound to carbonic anhydrase.

5. The binding of drugs to blood proteins may have important clinical implications. If the drugs are strongly bound, as are the dicumarol anticoagulants (see pages 721-722) and suramin (a sulfonamide), the circulating level of free drug will be low and the drug will be only slowly removed from the blood stream. As was indicated earlier, the amount of free drug in the blood is crucial because it is only these molecules which are active. The slow elimination of a tightly bound drug is another manifestation of the same effect, since only free drug is actively metabolized and excreted.

    a. Protein-bound drug molecules act as a storage form of the drug. This can be either advantageous or dangerous. Suramin is used to treat African sleeping sickness which is caused by the parasite Trypanosoma gambiense. Because of its binding to plasma proteins, a single injection of the drug provides a parasiticidal effect for up to a week, a distinct advantage.

    b. Warfarin, one of the cumarol anticoagulants, is normally distributed in the blood with 97% of it bound to albumin and only 3% free. Since it is a potent drug, this is clinically adequate for it to function as an anticoagulant. A number of acidic drugs (penicillin, sulfonamides, salicylates, etc.) are able to displace warfarin from albumin, increasing the level of free anticoagulant. A displacement of only 3% would double the free level and the danger of internal hemorrhage is imminent in such a situation. Many drug-drug interactions (discussed later) are based on this type of competitive binding and displacement of bound drugs.

6. Although both acidic and basic drugs are capable of binding to blood proteins, most studies carried out so far deal mainly with the acidic drugs. In general, all drugs possessing one or more homocyclic aromatic rings (benzene, naphthalene, etc.), one or more potential sites for hydrogen bonding, and an atom capable of acquiring a charge at physiological pH, are potential protein binders.

Pathways of Metabolism

1. It is frequently true that the activity of a drug ends when it is metabolized, but it is not correct to believe that drug metabolism is synonymous with detoxication. In some cases, the metabolites have greater activity than the drug itself and may be more toxic or even better therapeutic agents. Phenacetin and acetanilid are both converted in the body to acetaminophen which is also an antipyretic analgesic agent. Recently it was shown that chlordiazepoxide (Librium) yields two metabolites which are at least as active as the parent drug. One of these metabolites is already being used clinically as a minor tranquilizer.

2. Drugs undergo four major types of reactions in the body. These are hydrolysis, reduction, oxidation, and conjugation (synthesis). The major site of these reactions is the liver, but recent evidence suggests that the lungs may also have a rich store of drug-metabolizing enzymes. These enzymes are generally located in the endoplasmic reticulum, a plasma membrane located inside the cell. During disruption of the cell and fractionation of the subcellular components, it disintegrates yielding small particles or fragments known as microsomes. Enzymes which are associated with this cell fraction are termed microsomal enzymes. There are also many drug-metabolizing enzyme systems present in the cytoplasm.

3. Hydrolysis

Drugs which are either esters or amides are usually hydrolyzed. For example, acetylcholine, an endogenous ester, is hydrolyzed by acetylcholinesterase to choline and acetic acid. Several other choline esters are handled similarly by plasma cholinesterases. Local anesthetics such as procaine and analgesics such as aspirin

are also hydrolyzed by the body. These reactions are catalyzed
by hydrolases (esterases and amidases) found in blood plasma and
in the soluble fraction of the cells of tissues such as the liver.
Substrate specificity of the hydrolases may vary from tissue to
tissue and from species to species.

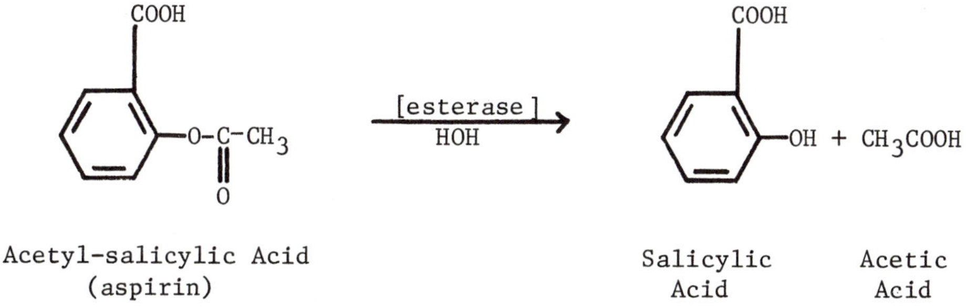

Acetyl-salicylic Acid   Salicylic   Acetic
     (aspirin)   Acid   Acid

Hydrolytic biotransformations are relatively simple in that the
products are usually predictable and can therefore be traced and
studied with ease. Not all ester-containing drugs are hydrolyzed
to a significant extent in man. For example, only about 2% of a
given dose of atropine is degraded in this manner in man. Some
strains of rabbits, however, have atropinesterase, an enzyme not
found in human tissues, which effectively metabolizes this drug by
hydrolysis.

4. Reduction

    a. Only a few drugs have structures capable of being reduced
       in vivo. Azo reduction of prontosil rubrum (sulfamidochrysoidine),
       an azo dye and the precursor of the sulfonamide family
       of drugs, is perhaps the first (1935) and best-known
       example of this type of transformation. Though inactive by
       itself, the dye is hydrolyzed in vivo to an active antibacterial
       metabolite, sulfanilamide, as shown below.

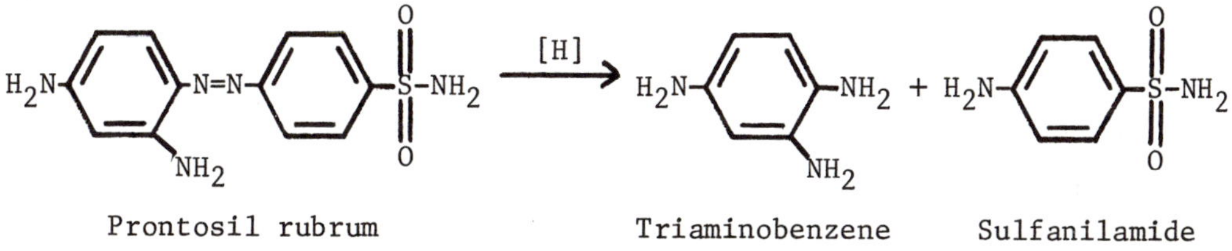

    Prontosil rubrum   Triaminobenzene   Sulfanilamide

Reduction of a nitro group can also occur if one is present in the drug. Chloramphenicol, a broad-spectrum antibiotic, is an example. Both the bacterial and the mammalian nitroreductase systems are capable of catalyzing this metabolic reaction.

b. The azo and nitro reductase activities are present in the light microsomal fraction which sediments more slowly than that containing the drug-oxidizing enzyme systems. Unlike oxidases, the reductases are found not only in the liver but also in other tissues. These enzymes require NADH or NADPH under anaerobic conditions and may be stimulated by FMN and FAD.

c. The reductive dehalogenation of volatile anesthetics such as halothane may also be of importance. In these reactions, again catalyzed by microsomal enzymes, chlorine is removed and replaced by hydrogen. Chlorpromazine, a major tranquilizer, appears likely to undergo a similar transformation, but this has yet to be confirmed experimentally.

d. Non-microsomal reductions of aldehydes are generally catalyzed by alcohol dehydrogenase in the presence of NADH. (Note that this is the reverse of the reaction which occurs when an alcohol is the substrate.) This enzyme, under the name retinal reductase, plays an important role of converting retinal to retinol during the process of vision (see pages 712 to 714). Chloral hydrate, a classical sedative, is another aldehyde which is reduced in the body by alcohol dehydrogenase, in this case to trichloroethanol. It is this metabolite which possesses the sedative property and not the parent drug. Dimethyl sulfoxide, disulfiram (the inhibitor of alcohol dehydrogenase used in alcoholism; see pages 292 and 488), and compounds of pentavalent arsenic and antimony are all reduced by non-microsomal enzymatic systems.

5. Oxidations

a. The greatest proportion of drugs undergo oxidation in the body. The oxidases are present both in the endoplasmic reticulum and in the cytoplasm. These drug-oxidizing systems can be isolated by well-established cellular fractionation techniques and for this reason have been studied fairly extensively.

b. Oxidation usually generates a more polar and reactive group in the molecule which can subsequently undergo glucuronidation, sulfatation, acetylation, methylation, or other synthetic or conjugative reactions.

6. <u>Non-microsomal Oxidations</u> (Cytoplasmic and Mitochondrial)

    a. <u>Alcohols</u> are oxidized by NAD-linked alcohol dehydrogenase (see pages 291-293) found in liver, kidney and lung. Recall that this same enzyme, in the presence of NADH, also catalyzes aldehyde reductions. Certainly the major pathway of metabolism of ethanol is oxidation by this enzyme, though recent evidence suggests that a small proportion may be metabolized by the microsomal ethanol oxidizing system (MEOS; see page 292).

    b. <u>Aldehydes</u> are oxidized by an NAD-dependent aldehyde dehydrogenase and by aldehyde oxidase. Both of these enzymes have been isolated from liver. Purines and xanthines are oxidized by xanthine oxidase (see pages 108-109 and 402) which is very similar to aldehyde oxidase.

    c. <u>Monoamines</u> such as catecholamine, tryptophan, tyramine, and norepinephrine, having alkylamine chains with no α-substitution, are oxidized by MAO present in mitochondria. Alpha-carbon substituted short-chain amines such as amphetamine are metabolized by the microsomal oxidizing system. Non-specific MAO and diamine oxidase, present in plasma, can also oxidize the monoamines mentioned, but the reaction rates may be relatively slow.

    d. <u>Steroidal hormones</u> such as deoxycorticosterone, are oxygenated by a mixed function oxygenase system which, as an exception to the rule, is present in the mitochondria of the adrenal cortex. Unlike the rest of the non-microsomal oxygenase systems, the adreno-cortical mitochondrial oxygenase contains cytochrome P-450, the heme protein otherwise found only in the endoplasmic reticulum of the liver, kidney and lung cells (see page 205).

7. The Microsomal Oxidizing System

    a. The supernatant fraction of a cell homogenate centrifuged at 9000 gravities contains microsomes and several co-factors capable of the oxidative transformation of drugs. The microsomes can be isolated as a pellet by centrifuging at about 110,000 gravities for one hour. When reconstituted, in the presence of oxygen, with the cell supernate, $NADP^+$, Mg^{++}, and nicotinamide, these microsomes regain their ability to perform oxidations. The supernate provides the enzymes required to maintain $NADP^+$ in its reduced (NADPH) form.

b. The microsomal oxygenation system differs from the mitochondrial
    and cytoplasmic systems in that it fixes molecular oxygen
    to the organic substrate (drug) and does not produce any useful
    energy. Briefly, NADPH (or NADH) reduces (activates) molecular
    oxygen. The "active oxygen", [0], is carried by cytochrome P-450
    to the drug molecule where it is attached, usually as a
    hydroxyl group. The microsomal system is referred to as a mixed
    function oxygenase because it reduces O_2 while oxidizing the
    drug or other organic substrate. It is capable of oxidizing
    carbon, sulfur, nitrogen, and other groups.

c. Although the $NADP^+ \leftrightarrow NADPH$ and flavoprotein dependence of the
    oxidation is common to both the mitochondrial and the microsomal
    oxidative systems, there is an established difference between
    them. Unlike mitochondria, microsomes normally contain no
    cytochrome C which is part of the electron-flow pathway in
    the mitochondrial oxidative chain. Instead, this function is
    carried out in a somewhat modified fashion by cytochrome P-450.
    This heme protein, in its reduced form, is capable of combining
    with carbon monoxide to form a complex having a characteristic
    absorption maximum at 450 nm. According to current data,
    cytochrome P-450 appears to be the oxygen-activating enzyme.
    It reduces one atom of oxygen to water and introduces the
    other into the drug molecule. The electron carriers
    involved and their relationship to each other was shown
    earlier in Figure 71, page 578.

d. Because it has been difficult to solubilize and purify
    cytochrome P-450 (the systematic name is B-420; see page 205),
    a vast amount of questionable data has been gathered on this
    hepatic microsomal enzyme. It has been established, however,
    that it is present on the smooth endoplasmic reticulum.

e. Oxidation, frequently followed by conjugation, is the primary
    pathway for the biotransformation of drugs. The oxidation
    usually occurs at a carbon, nitrogen, or sulfur atom and many
    times involves removal of a group, such as the demethylation
    of an alkylated oxygen, ring nitrogen, or amino group. Some
    examples of these reactions are given below. Conjugation
    (synthetic) reactions are discussed in the section following
    this one.

    Oxidized drug may further be transformed into a relatively
    more polar derivative by any one of many synthetic routes
    described in the next section. Since oxidations usually occur

on either a carbon atom of the drug molecule or on nitrogen
or sulfur, the examples will be discussed in that order.   In
addition, many oxidative transformations involve removal of a
group such as a methyl from a nitrogen or an oxygen, or from
an amine group.

i)   Aliphatic (alkyl) sidechain oxidation results in the
attachment of an oxygen atom to the carbon to produce a
hydroxyl group. Pentobarbital in man is transformed
into an alcohol by this sort of reaction.

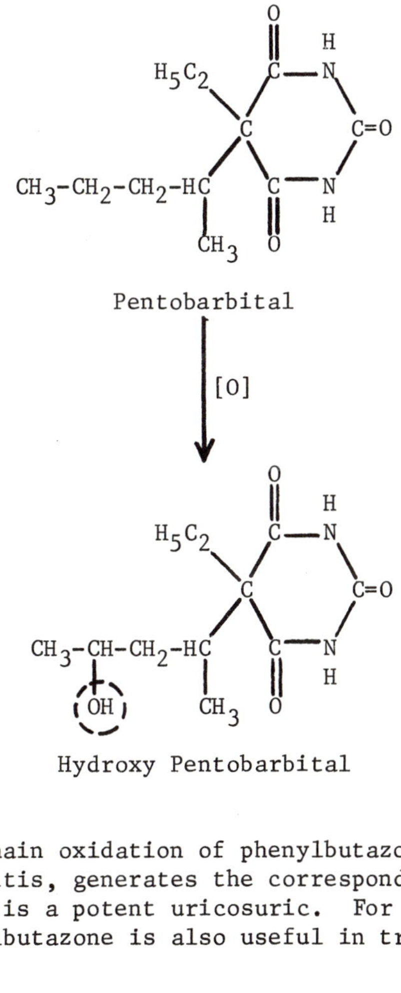

Pentobarbital

[O]

Hydroxy Pentobarbital

Sidechain oxidation of phenylbutazone, a drug used in
arthritis, generates the corresponding hydroxy derivative
which is a potent uricosuric. For this reason,
phenylbutazone is also useful in treating acute gout.

ii) <u>Ring hydroxylation</u> usually occurs with both alicyclic
and aromatic compounds. Phenobarbital, for example, is
hydroxylated on the phenyl ring to <u>p</u>-hydroxyphenobarbital

Phenobarbital   Hydroxy Phenobarbital

iii) <u>Dealkylation</u> by oxidation of a carbon can also be
accomplished by the microsomal oxygenase system. Oxidative
removal of an alkyl group from nitrogen (N-dealkylation),
oxygen (O-dealkylation), or sulfur (S-dealkylation)
generates the corresponding hydrogenated ($>$NH, -OH or -SH)
group in the drug molecule. Each of these groups is then
capable of conjugation or synthesis. During the removal
process, the alkyl functions are oxidized to an aldehyde
and then further to an acid. Methyl groups generally
are finally converted to carbon dioxide. The
N-demethylation of morphine was one of the first reactions
of this type to be studied. N-(^{14}C-methyl)-morphine was
given to subjects and the radioactivity was partially
recovered from their breath as $^{14}CO_2$.

iv) N-dealkylation of primary amines occurs at a faster rate
than that of secondary and tertiary amines. Demethylation
of tertiary amine drugs is of importance, however, because
in several cases the demethylated metabolites have turned
out to be active and have thus led to the discovery of
new drugs. Imipramine, a tertiary amine used as an
antidepressant, is demethylated to desipramine, an active
metabolite now also used as an antidepressant.
Chlorpromazine, a major tranquilizer is also demethylated
to yield monodesmethyl (Nor_1) chlorpromazine which has
many pharmacological activities similar to the parent
drug.

Chlorpromazine   Nor$_1$ Chlorpromazine

v) O-dealkylation, yielding a phenol, usually occurs in drugs
    which have an alkylaryl ether structure. In man, nearly
    10 percent of a dose of codeine is dealkylated to morphine.

Codeine   Morphine

Acetophenetidin (phenacetin) is an analgesic and
antipyretic drug which is O-deethylated to generate an
active antipyretic metabolite, N-acetyl-p-aminophenol.

vi) S-dealkylation of various thioethers has been found to
     occur in vitro with liver, kidney, and spleen microsomal
     preparations in the presence of NADPH and oxygen.
     An in vivo example is the S-demethylation of 6-methyl
     thiopurine to 6-mercaptopurine, an antitumor compound
     (see pages 395-396).

vii) Tertiary amine drugs (e.g., tremarine, imipramine, chlorpromazine) are also oxygenated on the nitrogen to form N-oxides. Primary and secondary amines, on the other hand, undergo N-hydroxylation to form hydroxylamines. In some cases, the N-hydroxylated metabolites are more toxic than the parent drug. Of considerable interest, relative to chlorpromazine metabolism, is the recent identification of large amounts of N-hydroxyl-Nor_1 Chlorpromazine in the red cells of schizophrenic patients who had been receiving daily chlorpromazine doses. Using chlorpromazine as an example, these two types of transformation are shown below.

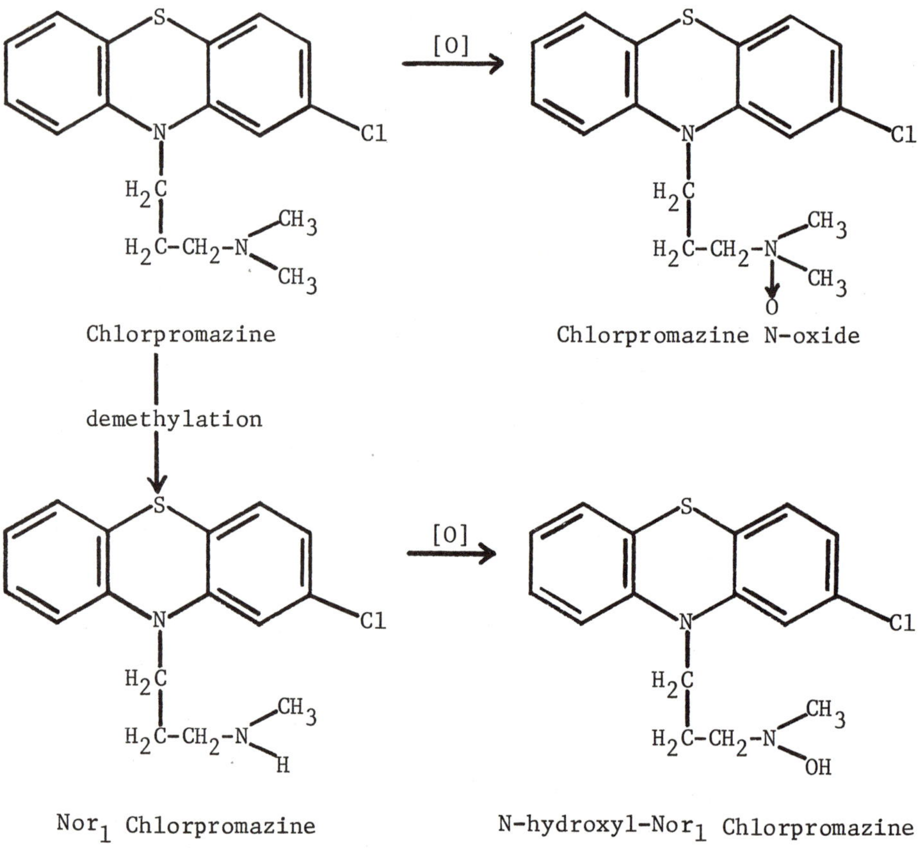

Chlorpromazine

Chlorpromazine N-oxide

demethylation

Nor_1 Chlorpromazine

N-hydroxyl-Nor_1 Chlorpromazine

viii) Oxidative deamination (of a primary amino group) is
analogous to dealkylation. Amphetamine and other alpha
methyl amine drugs are metabolized by oxidative
deamination catalyzed by the mixed function microsomal
oxygenase system of the endoplasmic reticulum. This
reaction is different from the non-microsomal oxidative
deamination of short-chain, non-alpha-substituted amines,
such as norepinephrine. Those reactions are carried out
by mitochondrial monoamine oxidase (MAO). The
deamination of amphetamine to phenylacetone was the
first microsomal oxidation discovered.

Amphetamine   Phenylacetone

ix) The S-oxidations of phenothiazine derivatives such as
chlorpromazine are good examples of this type of reaction.
Sulfoxides of this major tranquilizer as well as of its
various demethylated and ring hydroxylated metabolites
have been recognized in man. The same oxygenase system
may further oxidize the sulfoxides to sulfone.

Chlorpromazine   Chlorpromazine Sulfoxide

x) Oxidative desulfuration of the thiobarbiturates and thioureas can be accomplished by the microsomal oxygenase system, replacing the sulfur with an atom of oxygen. Thiopental is biotransformed into pentobarbital by a reaction wich requires NADH (instead of the usual NADPH) and which is stimulated by Mg^{++}.

Thiopental (enol)   Pentobarbital (enol)

xi) In conclusion, the oxidative pathways of drug metabolism occur frequently and include a wide variety of biochemical transformations. In most of the reactions, both the microsomal and the non-microsomal, the product is usually more polar (water soluble) than the reactant. Often a hydroxyl or a primary or secondary amine is produced, all of which are capable of undergoing further synthetic reactions.

8. Synthesis or Conjugation

These are reactions in which a number of endogenous small molecules react with polar functions of drugs or their metabolites to yield various derivatives. The reactions are catalyzed by specific enzymes and require energy. The conjugations include glucuronidation, O- and N-acetylation, O-, S-, and N-methylation, glycine conjugation, ethereal sulfate formation, and (occasionally) glutathione conjugation. The commonest and probably most-studied conjugations are those producing glucuronides.

a. Glucuronidation

i) In those tissues capable of this type of conjugation, the enzymatic machinery necessary is present in the soluble fraction and the microsomes. The active donor of glucuronic acid, uridine diphosphoglucuronic acid (UDPGA), is synthesized from glucose-1-phosphate and uridine triphosphate (UTP) by the cytoplasmic enzymes of the uronic acid pathway (see pages 256-258). The transfer of the glucuronic acid moiety to the drug molecule containing a hydroxyl, amine, or a carboxyl function is carried out by glucuronyl transferases present in the endoplasmic reticulum (microsomes). Typical reactions of this sort were shown previously on pages 257 and 258.

ii)   Glucuronidation is essentially a nucleophilic substitution in which the drug molecule acts as a nucleophile and launches a back-side attack on the electron-deficient carbon atom 1 of the glucuronic acid part of UDPGA. Inversion of configuration is an inherent characteristic of this type of reaction. Thus, although the configuration of glucuronic acid in UDPGA is alpha, the final reaction product (the glucuronidated drug) has the beta configuration. The glucuronide products of hydroxylated compounds are known as O- or ether glucuronides, those of carboxylic compounds are called ester glucuronides, and the ones formed from amines are termed N-glucuronides.

iii) Although many drugs (and other compounds) are metabolized by glucuronidation, the true significance of these derivatives is still open to question. Previously, it had been believed that it was simply a way of rendering drugs and their metabolites more polar and water soluble, thereby increasing their rate of renal excretion. It may also, however, be an important mechanism for enabling these compounds to cross biological membranes, including the blood-brain barrier. There is one report (yet to be confirmed or refuted) that the glucuronide of a drug is in fact able to enter the brain while the unconjugated drug cannot.

Another consideration is the ubiquitousness in tissues of β-glucuronidase, an enzyme which specifically hydrolyzes the β-glucuronides of drugs and other compounds. A conceivable sequence of events in the action of a drug might be administration, glucuronidation, entry of the glucuronide into the cell (or wherever the site of action for the drug is), and hydrolysis of the glucuronide by β-glucuronidase, releasing the free drug molecule. The biochemical machinery for the active transport of glucuronides is known to be present in the body. Bilirubin glucuronide is actively excreted into the bile by hepatocytes (see page 579) and other glucuronides are generally secreted into the urine from the renal tubules by an active (energy-requiring) process. Further studies may well show that glucuronidation is required for the transport of many drug molecules to their site of action.

iv) Research into the role of glucuronic acid conjugation is hindered by the poor availability of glucuronides. These compounds are difficult to synthesize in vitro and their water solubility makes them hard to isolate from natural sources. Any efforts to generate workable quantities of glucuronides of drugs and their metabolites would certainly be worthwhile.

b. Acylation

i) Various carboxylic acids such as acetic or benzoic acid are activated by coenzyme A (CoA) to form acetyl or benzoyl∿CoA. In this form, they can react with primary amines and amides to form acylated metabolites. Derivatives of aniline (such as the sulfonamide drugs) and hydrazines (e.g., isoniazid, an antitubercular drug) are metabolized in this fashion.

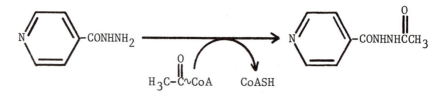

Isoniazid   Acetylated Isoniazid

ii) There are also examples in man where the carboxylic group of the drug is activated by conjugation with CoA. This compound then reacts with an amino acid such as glycine or glutamic acid to form peptide acylates. The ability of the liver to conjugate glycine with benzoic acid reflects the condition of this organ and may be used as a liver function test in humans. Glycine conjugation of the bile acids (see pages 659-661) is an important step in their excretion.

iii) Acetylation can occur in lung, liver, spleen, and many other tissues in the presence of appropriate acyl transferase. All reactions, however, are dependent on CoA. A genetic deficiency of acetyl transferase enzyme has been found to exist in nearly half the Caucasian population of the United States. When administered to this group, the usual therapeutic dose of isoniazid is toxic, producing symptoms such as peripheral neuritis.

c. Methylation

  i) Transmethylation reactions are brought about by a variety of transmethylase enzymes present in microsomal and soluble fractions of the liver cell. Drugs containing aromatic hydroxyl, sulfhydryl, or amine functions undergo methylation. S-adenosylmethionine (SAM), synthesized from methionine (the actual methyl donor) and ATP, serves as the methyl carrier. The synthesis and structure of this compound were shown previously on pages 469-471. The methylation reaction is summarized below.

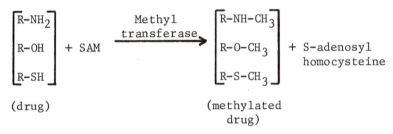

$$\begin{bmatrix} R\text{-}NH_2 \\ R\text{-}OH \\ R\text{-}SH \end{bmatrix} + SAM \xrightarrow[\text{transferase}]{\text{Methyl}} \begin{bmatrix} R\text{-}NH\text{-}CH_3 \\ R\text{-}O\text{-}CH_3 \\ R\text{-}S\text{-}CH_3 \end{bmatrix} + \text{S-adenosyl homocysteine}$$

(drug)   (methylated drug)

  ii) Some of the drugs methylated in the body include norepinephrine which is both N- and O-methylated; norcodeine and normorphine both of which are N-methylated; and dimercaprol (British, anti-lewisite) which is S-methylated. Of these, norepinephrine is one example of a series of catecholamines with a phenylethylamine structure. Such compounds are O-methylated in the presence of catechol O-methyl transferase (COMT; see page 490).

  iii) The adenosylmethionine-methyl transferase system has no relationship to the oxidative demethylation discussed earlier. These two types of biochemical transformations of drugs are not the reverse of each other, but are two distinctly different reactions. Both are, however, catalyzed by microsomal enzymes.

d. Sulfate Ester Conjugation

Various alcoholic and phenolic drugs are sulfated to form ethereal sulfates. Sulfate is first activated by ATP to adenosine-5'-phosphosulfate, then to 3'-phosphoadenosine-5'-

phosphosulfate (PAPS) and is finally transferred to the oxygen function of the drug in the presence of sulfokinases (also known as sulfate transferases) found in the cytoplasm of mammalian livers. The formation of PAPS and the structures of the compounds involved were shown previously on pages 474-475. A typical sulfatation is shown below.

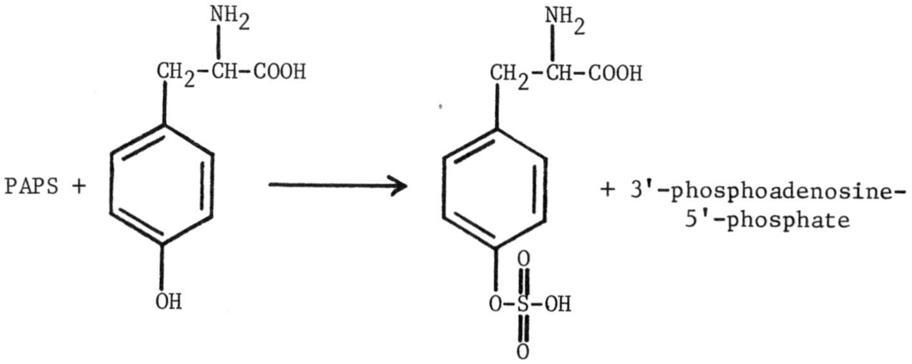

Tyrosine   Tyrosine-O-Sulfate

e. Drug-Induced Alkylations

Several sulfur and nitrogen mustards spontaneously change into an electrophilic intermediate and then attack important endogenous molecules such as RNA and DNA, forming alkylated products within the functioning cells. This interferes with the cellular activities, since the alkylated molecules behave differently than do the normal ones. This is the basis for the antineoplastic and mutagenic activity (see pages 333-334) of the nitrogen mustards. Alkylating agents of this sort are degraded in the body by hydrolysis.

9. Predicting the Metabolic Fate of Drugs

From what has been described about the biotransformations of drugs, it is clear that the route by which a drug is metabolized depends on its chemical structure. Based on this, some predictions concerning the fate of a drug in the body can be made.

a. Aromatic and aliphatic cyclic structures are usually hydroxylated and subsequently conjugated with either glucuronic acid or sulfate. Aliphatic sidechains are also hydroxylated.

b. Aromatic hydroxyl groups are generally glucuronidated or sulfated, but occasionally may be methylated. Aliphatic hydroxyl groups undergo oxidation or glucuronide conjugation.

c. Aliphatic carboxyl groups are oxidized and decarboxylated, or conjugated with glucuronic acid. Aromatic carboxyl groups are conjugated with either glucuronic acid or an amino acid.

d. Primary aliphatic amines are glucuronidated or oxidatively deaminated, whereas aromatic primary amines may be methylated, acetylated, or glucuronidated. Secondary and tertiary amines are dealkylated or sometimes methylated.

Although these are generalities and may not hold in all cases, they may aid in the synthesis of derivatives of drugs which have a desired duration of action on the basis of their ability or lack of it to get metabolized. Highly polar drugs such as strong acids and bases are excreted unchanged while extremely non-polar substances such as liquid paraffin do not get absorbed in the first place.

Clinical Significance of Drug Metabolism

1. One thing common to all biotransformational reactions of drugs is that they involve specific enzymes. In recent years, it has been observed that these drug-metabolizing enzymes can be induced or inhibited by drugs and organic air pollutants which find entry into the body. In the current practice of polypharmacy (simultaneous prescribing of more than one drug), there is always a chance that one of the administered drugs may affect the enzymes involved in metabolizing the other coadministered drugs. Since biotransformation may either terminate or enhance drug action, the implications of the effect of drugs on the drug-metabolizing enzymes are obvious. This is one type of drug-drug interaction (discussed below).

2. A number of compounds can inhibit the oxidation of drugs. An example of a widely studied experimental agent having this effect is beta-diethylaminoethyl diphenylpropylacetate, known commonly as SKF 525A. Monoamine oxidase inhibitors such as iproniazid (see page 503) and pargyline are more familiar examples of agents which have clinical applications. Estrogens are also capable of inhibiting the metabolism of certain drugs.

3. a. Similarly, there are a number of compounds, many used commonly as drugs, which are capable of inducing drug-metabolizing enzymes. Among the most widely investigated of these are phenobarbital, the steroidal hormones, and 3,4-benzpyrene, a

carcinogenic hydrocarbon. These agents induce the microsomal
enzymes which metabolize a large number of drugs and endogenous
waste products (e.g., bilirubin).

b. The induction is believed to involve enhanced enzyme synthesis
via an increased synthesis of mRNA. So far, however, in
experiments designed to prove induction, only the end results
such as the increased rate of metabolism of a drug and increased
synthesis of liver proteins have been measurable. By inference,
it is believed that induction of drug metabolism is a result
of increased protein (enzyme) synthesis.

4. Drug-Drug Interactions

As was pointed out at the start of this chapter, the level of any
compound in the body, including drugs from exogenous sources,
depends on a balance among several factors. Any factor which
alters this balance, including the presence of another drug, is
likely to affect the systemic response to the compound.

a. In the past decade, a large number of adverse drug reactions
or paradoxical drug responses were discovered to have occurred
as a result of inhibition or induction of the metabolic pathway
of one drug by another. One of the simplest clinical examples
of induction is the development of tolerance to a drug. Many
drugs, on repeated administration, induce their own metabolizing
enzymes, with the result that larger doses are required to
produce the desired therapeutic effect as time passes.

i)   Although inhibition of drug metabolism has not been studied
as much as induction, toxicity of a normal therapeutic
dose of one material may be due to the inhibition of its
metabolism by another drug which had been previously
given to the patient for some other purpose. Estrogens are
known to inhibit oxidative and conjugational metabolism
of some drugs. Occasionally, the inability of a female
patient to tolerate a dose of a drug, which has no adverse
effect on a male patient, may be attributed to this.

ii) There are many clinical examples and uses of the phenomenon
of enzyme induction. One well-known clinical application
is the use of phenobarbital in patients suffering from a
deficiency of glucuronyl transferase. In such cases,
the circulating levels of bilirubin, which is normally
glucuronidated and secreted into bile, increase to the
point of producing jaundice. Administration of phenobarbital results in the lowering of bilirubin levels to near
normal.

since this barbiturate induces the enzyme glucuronyl transferase (see also page 580). Another example, mentioned on page 722, is the inhibition of the anticoagulative effect of the coumarins by barbiturates and glutethimide. These compounds both induce the hepatic microsomal oxidase system which catabolizes the coumarins.

b. There are other ways in which one drug may influence the biological activity of another. These include alterations in absorption, transport, and function at the receptor site. Some examples are given below.

   i) Protamines (salmin , clupeine, etc.; see page 305) act as heparin antagonists. These are basic peptides which are thought to neutralize the relatively acidic heparin and prevent it from acting as an anticoagulant.

   ii) Gentamicin is inactivated by carbenicillin or ampicillin when they are combined in the same infusion fluid. Once in the serum, however, this does not occur, so that administration of these antimicrobials by separate routes is desirable.

   iii) Most drugs are carried to their site of action bound to plasma proteins. Low levels of a specific binding protein (e.g., a deficiency of cortisol-binding globulin) or a general decrease in plasma proteins (as in hypoalbuminemia) will thus affect this transport. One compound may compete with another for binding sites on the plasma proteins. In hyperbilirubinemia, the ability of albumin to bind other compounds (drugs) is limited because it is heavily loaded with bilirubin. Aspirin (acetyl-salicylic acid) behaves similarly. The ability of some substances to potentiate the activity of the coumarin anticoagulants by displacing them from protein-binding sites and thereby increasing the level of the free compound was cited previously (page 722). Chloral hydrate, a sedative and hypnotic, has this effect.

   iv) The potentiation of 6-mercaptopurine (6MP) by allopurinol was discussed earlier (page 396). Allopurinol inhibits xanthine oxidase, the enzyme which catabolizes 6MP to 6-thiouric acid, an inactive metabolite. This also increases the toxic side effects (bone marrow degeneration leading to pancytopenia) of 6MP, however.

v) Disulfiram, the inhibitor of aldehyde oxidase (pages 292 and 488) used in many alcoholism treatment programs, greatly increases the toxicity of ethanol due to the accumulation of acetaldehyde. Deaths have been reported due to massive alcohol ingestion among alcoholic patients participating in mandatory disulfiram programs.

c. As was briefly indicated above, drug-drug interactions are potentially useful as well as dangerous. As more is understood about these processes, a more rational approach to the control of the magnitude and duration of drug effects should arise. Also, much of the estimated three billion dollars currently spent each year on the treatment of adverse drug reactions in patients could be put to better use if such reactions were avoided in the first place.

5. Pharmacogenetics is a relatively new term used to refer to the genetically based variation among individuals in their response to drugs. In many instances, these differences are related to a decreased (or sometimes increased) ability of some persons to metabolize a drug. This in turn governs the intensity and duration of action of the compound. Pharmacogenetics is a quite logical extension of the growing interest and awareness of enzyme mutants and inborn metabolic errors. It was mentioned briefly in connection with hemoglobin variants, on page 552.

a. Genetic variants of various esterases including carbonic anhydrase, acetylcholinesterase, and serum cholinesterase have been recognized in man. Not only do these enzymes behave differently in various physico-chemical analytical systems, but they also clearly exhibit varying degrees of specificity and activity toward their substrates.

i) Clinically, this may be witnessed dramatically following the administration of succinylcholine. In normal individuals, this drug is hydrolyzed by the plasma cholinesterase at a rate such that doses of 50 to 80 mg do not produce any toxic symptoms (prolonged paralysis). In individuals with a genetic deficiency of the active enzyme (fluoride resistant or silent variants), these doses produced paralysis lasting from 25 to 300 minutes and requiring extended artificial respiration.

ii) Individuals having the cynthiana serum cholinesterase variant show a resistance to succinylcholine. This seems to be due to an increase in the serum level of cholinesterase, resulting in abnormally rapid hydrolytic inactivation of the molecule.

iii) The significance of esterase variation is not known in terms of the basic physiological function of the body and the genetic variants of esterases have been found in healthy individuals. Whether or not these variants evolved in a certain generation or over several generations as a result of constant exposure to some inhibitor cannot be answered unequivocally, but a possibility for this does exist.

b. Isoniazid, a drug used in the treatment of tuberculosis, is inactivated by acetylation catalyzed by the liver microsomal enzyme N-acetyl transferase. In individuals with a slow acetylation capacity this drug may cause toxic effects such as peripheral neuritis. This can be prevented by administration of excess pyridoxine, which does not interfere with the antitubercular activity.

c. Diphenylhydantoin is inactivated by the liver microsomal hydroxylase system. Three genetic types have been demonstrated with regard to the rate of inactivation: (1) slow metabolizers show toxic effects (nystagmus, ataxia, etc.) with normal doses; (2) normal metabolizers; and (3) rapid metabolizers who require an increased dose of the drug to produce the expected biological response.

d. Individuals who are deficient in liver microsomal dealkylase enzyme develop methemoglobinemia and hemolysis upon the administration of acetophenetidin.

e. G-6-P dehydrogenase deficient individuals show methemoglobinemia and hemolysis upon administration of a variety of drugs (primaquine, etc.; see pages 247-249).

f. Other enzymes involved in drug metabolism which exhibit genetic variation in man include glucuronyl transferase (discussed earlier) and various aromatic hydroxylases and oxidases. A reported lack of sulfite oxidase in children is interesting but may represent an example of developmental enzymology rather than a pharmacogenetic characteristic. This newly evolving concept recognizes that infants and children are deficient in certain enzyme activities which develop as they approach or attain maturity. Another example, discussed earlier (pages 579 and 590) is neonatal jaundice resulting from delayed development of hepatic glucuronyl transferase in neonates.

g. The clinical implications of the genetic differences in man's ability to metabolize drugs is currently evolving with reference to the blood levels of drugs obtained following administration of a fixed dose. A tenfold difference in blood levels of imipramine has been found to occur in different patients receiving the same dose, on a body weight basis, of this drug. Likewise, antipyrine, chlorpromazine, meperidine (demerol) and many other important drugs develop varying blood levels in different individuals. Most of these differences can be attributed to inherent variations in the drug-metabolizing enzymes of the patients. Those who are totally or partially deficient in the enzymes may exhibit very high blood levels of drugs which can at times result in undue toxicity.

h. Practically, it would add to the quality of clinical practice if the physician tested his subjects for pharmacogenetic anomalies before administering certain drugs known to be metabolized by any one of the enzymes which exist as genetic variants. For example, a small test dose of succinylcholine could be injected to see that no paralysis developed, prior to embarking on the usual dosage schedule of this drug.

6. Another topic which is currently being considered is the role of the intestinal microflora in the metabolism of exogenous and endogenous substances.

a. In patients with a decreased NH_3 tolerance due to advanced liver disease, neomycin may be administered to suppress the formation of this compound by the action of bacterial urease.

b. Salicylazosulfopyridine, an anti-inflammatory agent used to treat ulcerative colitis and other bowel inflammations, is cleaved at the diazo bond by intestinal bacteria. The products, 5-aminosalicylate and sulfopyridine, have respectively anti-inflammatory and antibacterial actions. The effectiveness of the parent (uncleaved) compound can be potentiated by the concomitant administration of antibiotics such as neomycin.

c. Caffeic acid, a constituent of most of the normal dietary vegetables, seems to be metabolized exclusively by the intestinal flora. Of the four reactions involved in this process, none of eleven species of human fecal bacteria tested was able to carry out more than one of them. This suggests a way of investigating in situ the functioning of these organisms. Future studies will hopefully reveal other examples of this sort of selective (species-specific) metabolism.

d. L-dihydroxyphenylacetic acid (L-DOPA; see page 488), a drug used to treat Parkinsonism, is apparently converted to m-tyramine and m-hydroxyphenylacetic acid by bacterial enzymes of the intestines. This may contribute to the variability of patients to L-DOPA treatment.

e. Quite little is known about other transformations catalyzed by bacterial enzyme systems in the gut. In vitro studies are difficult because the conditions of pH, osmolarity, and oxygen supply found in this region are hard to reproduce in the laboratory. The use of fecal isolates has only questionable value for understanding how these microorganisms function in their actual environment. Germ-free strains of animals have provided some information but its applicability to human systems is not clear. This is a new and relatively unexplored area of study whose importance may be much greater than was hitherto expected.

15

Liver Function

Anatomical Considerations (with Virgil L. Jacobs, Ph.D.*)

1. The liver represents the largest and heaviest gland in the body
    (3.5 pounds or 1500-1600 grams in the adult). It is situated in
    the upper abdomen directly beneath the diaphragm and can be
    palpated mainly in the right hypochondriac and epigastric regions.

2. The four lobes of the liver are further subdivided into lobules by
    partitions or septa of connective tissue that are continuous with
    the peritoneal covering (Glisson's capsule).

3. The arterial supply is derived from branches of the celiac artery
    called the right and left hepatic arteries. These arteries further
    subdivide and eventually supply oxygenated blood to the liver
    sinusoids.

4. The portal vein is formed from the venous drainage of the intestines,
    spleen, and pancreas. Branches of the portal vein in the liver
    follow the hepatic arterial route and drain into the sinusoids.
    Thus both arterial and venous blood mixes in the liver sinusoids.
    About 75% of the blood supply of the liver is from the portal vein
    and about 25% from the hepatic artery. The function of the arterial
    blood is to raise both the oxygen content and the pressure of the
    sinusoidal blood.

5. Blood in the sinusoids passes slowly through the rows or cords of
    liver cells where the exchange of oxygen, metabolites, and nutrients
    occur (see Figure 119). A group of sinusoids anastomose to
    form the central vein. This vein represents the beginning of the
    venous drainage of the liver. The central veins of many surrounding
    liver lobules join to form larger veins and these further anastomose
    to form the hepatic vein which joins the inferior vena cava.

6. Bile is both a secretory and an excretory product of liver cells.
    The biliary system begins as the bile canaliculi running between
    two apposed rows of liver cells. As more canaliculi anastomose,
    larger bile ducts with a simple cuboidal epithelium are formed.
    Bile ducts from the various liver lobes join to form the common
    bile duct which opens into the duodenum.

---

* Associate Professor of Anatomy, University of Hawaii School of Medicine,
Honolulu, Hawaii.

Figure 119.   Schematic Drawing of the Anatomy and Circulation in the
               Liver

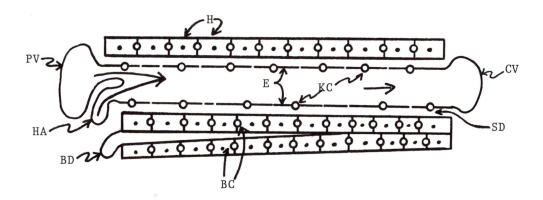

The portal area (triad) is to the left. It consists of a branch of
the portal vein (PV), a branch of the hepatic artery (HA), and a bile
duct (BD). The arterial and venous blood enter the sinusoid which
runs between two rows (cords) of hepatocytes (H). The sinusoid is
lined by endothelial cells (E) which are periodically studded with
Kupffer cells (KC). The endothelium is not continuous but has openings
permitting the fluid to contact the surface of the hepatocytes. The
space of Disse (SD) is between the endothelium and the hepatocyte
surface. After passing through the sinusoids, the blood enters the
central vein (CV) which joins with the general venous drainage
of the liver. Bile is secreted by the hepatocytes into the bile
canaliculi (BC) which collectively drain into the bile ducts (BD).
The bile ducts transport it to the gallbladder where it ultimately
enters the intestinal tract at the duodenum.

7. The parenchyma of the liver consists of one type of cell, the
    hepatic cell. These are arranged in rows or cords, usually as
    doublets, with a great deal of branching. The general histological
    appearance of the cells and sinusoids is that of a sieve.

8. The classical or hepatic liver lobule has the central vein in its
    center from which liver cords and sinusoids radiate peripherally
    like spokes of a wheel. The rim or edge of the lobule contains
    connective tissue and several portal areas, each of which contains
    a portal triad (hepatic artery branch, portal vein branch, and a
    bile duct) and sometimes a lymphatic vessel. The shape of the
    hepatic lobule is usually hexagonal in cross-section (see Figure 120).

9. Alternatively, the liver can be subdivided into portal lobules which
    have the portal canals at their centers. The peripheral boundary
    of this type of lobule is marked by three central veins. This forms
    a triangular lobule with the biliary system draining bile into
    the centrally placed bile duct.

10. A so-called functional unit of the liver can also be defined based
    on the blood supply to the lobules. The unit is diamond shaped
    with two central veins at opposite ends and two portal canals in
    the corners. Pathologically, liver damage is frequently related to
    the blood supply. Normally those parenchymal cells nearest the
    central vein will receive less oxygen and food material than the
    liver cells nearest to the hepatic artery branches.

11. Liver cells are polygonal in shape with one surface related to a
    sinusoid, another applied to other liver cells, and one other
    surface minutely separated from an adjacent liver cell to form a
    bile canaliculus. The nuclei are highly variable in size but
    somewhat spherical in shape. The cytoplasm has a slightly basophilic
    appearance. Glycogen in the liver is first deposited in the more
    peripheral cells of a lobule. As more glucose is brought in
    through the portal system, the liver cells closer to the central
    vein begin to show deposits. The reverse occurs when a glycogen
    demand is placed on the liver (i.e., the cells nearest the central
    vein release their glycogen first).

12. The hepatic cell membrane is thrown into folds or microvilli where
    the cell surface is applied to the sinusoid lining. This cell
    surface is separated from the sinusoid by the space of Disse (a
    potential space between the sinusoidal lining and the cell surface).
    Near the midportion of where two liver cells meet, the two
    membranes of an interface separate slightly to form the bile
    canaliculus. This end of the cell is called the biliary pole.

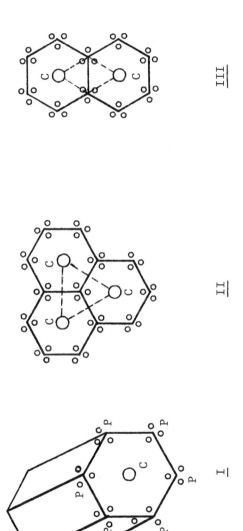

Figure 120. Three Types of Hypothetical Subunits (Lobules) of the Liver

I: Classical hepatic liver lobule (hexagonal)

II: Portal lobule (defined by dashed triangle)

III: Functional unit of the liver (defined by dashed diamond)

Note: C = central vein (large circle)

P = portal triad (portal area containing branches of the portal vein, hepatic artery, and bile ducts, represented by three small circles)

13.   Sinusoids of the liver are larger than capillaries and lined by a reticuloendothelium. The cell types of this lining are an endothelial cell and a phagocytic cell (Cell of Kupffer). These cells do not form a continuous lining of the sinusoid but are interrupted at intervals, so that the blood plasma has direct access to the microvilli on the surface of the liver cell. The phagocytic cells are active in that they show iron-containing granules, degenerating erythrocytes, pigment granules, and droplets of fatty material. The space of Disse is not seen in biopsies of normal liver and is thought to be revealed toward death since autopsied specimens show an edematous expansion of this space.

Functions of the Liver

1.   The liver performs a multitude of diverse functions which are discussed throughout this text at the appropriate places. It is important to stress that this organ is placed in a strategic position where it receives all of the nutrients, with the exception of lipids, from the portal vein which drains the capillary beds of the stomach, intestines, spleen, and pancreas. The portal blood goes through the columns of liver cells and enters the systemic circulation through the hepatic vein. This propitious intervention of the liver between the portal and systemic circulations not only permits the liver to absorb, transform, and store the nutrients resulting for the intestinal digestion and absorption but also protects the rest of the body from toxic metabolites (such as NH_3) produced in the intestines or present in the diet, by converting them into non-toxic substances before they enter the systemic blood.

2.   The numerous functions of liver can be categorized into the following groups. More details on these can be found in other parts of this book.

   a.   Metabolic functions: involving carbohydrates, proteins, lipids, minerals, and vitamins.

   b.   Detoxication and protective functions: conversion of ammonia to urea; formation of bilirubin glucuronides; drug detoxication; steroid inactivation; phagocytic activity of Kupffer cells in removing foreign substances; others.

   c.   Excretory functions: formation of bile and the excretion of bilirubin glucuronides (bile pigments), bile salts (cholesterol metabolites), bromosulfalein (BSP), and other foreign compounds.

   d.   Storage functions: storage of glycogen, iron, several vitamins (A, D, B_{12}), and other materials.

e. Hematological functions: hematopoiesis during embryonic development; formation of clotting factors (VII, IX, X, and prothrombin, all of which are dependent upon vitamin K for their synthesis); synthesis of many plasma proteins except for immunoglobulins (recently, though, it has been shown that hepatocytes do synthesize some immunoglobulins and that this may play an important role in the overall immune mechanism); special carrier proteins for various metabolites; etc.

f. Circulatory functions: participates in blood storage; acts as an intermediary between the portal and systemic blood circulation; reticulo-endothelial activities of Kupffer cells lining the sinusoids.

Tests of Liver Function

1. Since the liver plays a central role in many diverse metabolic processes, no single test or combination of tests will clearly establish the diagnosis of a specific disease. Often the results of a test indicate the nature of the pathological process but not its cause. In order to assess a liver disease, a spectrum of tests, commonly known as a liver profile, is performed.

2. Hepatic disorders can cause parenchymal cell dysfunction or damage and/or cholestasis. Consequently, the most common laboratory tests performed measure the secretory and excretory capabilities of the hepatocytes and liver cell dysfunction and damage.

3. Tests used to evaluate the secretory and excretory functions

   a. Measurements of plasma bilirubin and bilirubin glucuronides

      An increase in unconjugated serum bilirubin is associated with excessive erythrocyte destruction, or defects in the uptake or conjugation of bilirubin by the liver cells. In jaundice, both the conjugated and unconjugated forms are often increased. In cholestasis, conjugated bilirubin is the major species present. In general, the presence of bilirubin glucuronide in the plasma is suggestive of a pathological condition. These diseases are discussed further on pages 576-591.

   b. Measurement of urobilinogen in the urine (see page 583)

      The presence of excess amounts of urinary urobilinogen are associated with excessive erythrocyte destruction and failure of the liver to re-excrete urobilinogen due to liver cell damage, as in hepatitis. Excessive concentration of normal urine can also cause apparently elevated urobilinogen levels. Since urobilinogen gets converted to urobilin on standing, a more accurate way of assessing the amount of urobilinogen involves converting the urobilinogen to urobilin and measuring the level of this compound.

A total absence of urinary urobilinogen is indicative of complete obstruction of bile flow (cholestasis). In cholestasis, the feces have a very pale coloration due to a lack of bile pigments.

c.   The bromosulfalein (BSP) excretion test

When BSP (a dye) is administered intravenously liver cells remove the dye from the circulation, conjugate it with the cysteine portion of a glutathione molecule, and excrete it into the bile. A 45-minute post-BSP injected (in one procedure 5 mg/kg body weight is used) blood sample normally contains about 3% of the total dose administered. Since excretion of the BSP depends upon the integrity of the liver cells, adequacy of the blood circulation, and the patency of the biliary tract, any abnormalities in these cause elevated retention of BSP. Loss of BSP in the urine is negligible and normally is discounted. This is a sensitive test and is only performed when other tests have produced negative results.

d.   Serum alkaline phosphatase measurement

   i)   There are a number of enzymes present in the serum that hydrolyze mono-esters of phosphates. The acid and alkaline phosphatases show a lack of substrate specificity and show optimal activities at alkaline and acid pH's, respectively. 5'-Nucleotidase, which catalyzes the hydrolysis of nucleoside-5'-phosphates (such as AMP), does have a substrate specificity permitting its distinction from the other two phosphatases.

   ii) Serum alkaline phosphatase is derived from several sources including the liver (cells lining the sinusoids and the bile canaliculi), bone (where it is needed for osteoblastic activity; see page 689), kidney, intestine, and placenta (in pregnant women). Therefore, elevated levels of serum alkaline phosphatase are observed in a variety of disorders such as rickets and osteomalacia (see page 695), Paget's disease, hyperparathyroidism, during the healing of fractures, osteoblastic carcinoma, and intra- and extrahepatic cholestasis (hepato-biliary involvement).

The elevation of serum alkaline phosphatase occurs in biliary obstruction because the enzyme is normally eliminated from the plasma by this route. Blockage of bile flow causes accumulation of the enzyme within the liver and impairs its removal from the blood.

iii) The various tissues have different isoenzymes of alkaline phosphatase. These isoenzymes can be distinguished by electrophoretic, chemical, and physical means (the enzyme of each permitting identification of the source of the elevated alkaline phosphatase). Electrophoresis (of lactic acid dehydrogenase) was discussed on pages 163-165. The relative susceptibility of the bone isozyme to heat inactivation permits its distinction from the liver form. The intestinal and placental isozymes are inhibited by L-phenylalanine, unlike the liver and bone forms. A combination of heating and inhibitory methods can be used to quantitate all four isozymes. (Serum <u>acid</u> phosphatase activity, which may also arise from several sources, notably the prostate, can also be fractionated by selective inhibition with L-tartrate which inhibits only prostatic acid phosphatase.)

iv) The level of serum 5'-nucleotidase is not enhanced in individuals with bone diseases but is elevated in hepatic disorders. Serum 5'-nucleotidase measurement is thus useful in confirming hepatic involvement when alkaline phosphatase level are elevated and liver disease is suspected.

4. Tests based on liver cell dysfunctions and damage

a. <u>Measurement of plasma proteins</u>: Since the liver is responsible for the synthesis of albumin (and other plasma proteins), the plasma albumin level is decreased in liver cell dysfunction such as that caused by viral hepatitis. This can be assessed by plasma protein electrophoresis or flocculation tests. Examples of the latter include the thymol turbidity test, cephalin-cholesterol flocculation test, and the zinc sulfate turbidity test.

b. <u>Alterations in the coagulation factors</u>: In the presence of vitamin K, liver parenchymal cells are responsible for the synthesis of clotting factors II (prothrombin), VII, IX, and X. The plasma levels of these molecules can be assessed by measuring prothrombin time. In addition to the above four factors, deficiencies of factors I (fibrinogen) and V occur in severe forms of liver disease. Recall that vitamin K is a fat–soluble vitamin which requires bile salts for its absorption. Consequently, cholestasis results in a lack of absorption of vitamin K giving rise to abnormal clotting time despite the normalcy of the liver parenchymal cells. In order to establish parenchymal damage using tests involving clotting factors, it is essential to administer vitamin K parenterally.

c. As has already been mentioned, liver cells are rich in a wide variety of enzymes, several of which are to some extent tissue-specific (see page 126). An increase in the plasma levels of these enzymes is strongly suggestive of extensive damage of the hepatic cells and the release of their contents into the circulation. Enzymes whose levels are frequently measured for this purpose include the transaminases SGOT and SGPT (see pages 443–448), choline esterase, LDH (isozymes LD-4 and LD-5; see page 163), leucine amino peptidase (see page 437), and 5'-nucleotidase.

d. Recent work indicates that the detection of antimitochondrial antibody in the circulation is indicative of primary biliary cirrhosis (cirrhosis resulting from the chronic retention of bile in the liver). This type of autoimmunity is also found in chronic active hepatitis and cryptogenic cirrhosis, but with a much lower frequency. The assay is performed by the fluorescent antibody technique. This test may be useful in confirming a diagnosis of primary biliary cirrhosis since the other clinical and laboratory findings are not pathognomic.

Suggested Readings

General

   The Metabolic Basis of Inherited Disease, Third Edition; Edited
by John B. Stanbury, James B. Wyngaarden, and Donald S. Fredrickson,
(McGraw-Hill Book Co., New York, 1972).

   Biological Chemistry, Second Edition; by Henry R. Mahler and
Eugene H. Cordes, (Harper and Row, New York, 1971).

   Scientific Tables, Seventh Edition; Edited by K. Diem and
C. Lentner, (Ciba-Geigy Ltd., Basil, 1970).

   Fundamentals of Clinical Chemistry; Edited by Norbert W. Tietz,
(W.B. Saunders Co., Philadelphia, 1970).

Chapter I (Acids, Bases, and Buffers)

   Chemistry: Principles and Properties; by Michell J. Sienko and
Robert A. Plane, (McGraw-Hill Book Co., New York, 1966).

   pH and Dissociation, Second Edition; by Halvor N. Christensen,
(W.B. Saunders, Philadelphia, 1964).

Chapter II (Amino Acids and Proteins)

   Chemistry of the Amino Acids; by Jesse P. Greenstein and
Milton Winitz, (John Wiley and Sons, New York, 1961).

   The Proteins, Second Edition; Edited by Hans Neurath, (Academic
Press, New York).

   A. Marglin and R.B. Merrifield, Annual Review of Biochemistry,
39, 841 (1970); "Chemical Synthesis of Peptides and Proteins".

Chapter III (Enzymes)

   Enzyme-Catalysed Reactions, by C.J. Gray, (Van Nostrand Reinhold
Co., New York, 1971).

   Methods in Enzymology; Sidney P. Colowick and Nathan O. Kaplan,
Editors-in-Chief, (Academic Press, New York).

Structure and Function of Proteins at the Three Dimensional Level, Cold Spring Harbor Symposia on Quantitative Biology, Volume XXXVI, (Cold Spring Harbor Laboratory, Cold Spring Harbor, New York, 1972).

The Enzymes, Third Edition, Edited by Paul D. Boyer, (Academic Press, New York, 1973). Eight Volumes.

Enzymes: Physical Principles; by H. Gutfreund, (Wiley-Interscience, New York, 1972).

Chapter IV (Carbohydrates)

Richard J. Havel, New England Journal of Medicine, 287(23), 1186 (1972). "Caloric Homeostasis and Disorders of Fuel Transport".

Kenneth H. Gabbay, New England Journal of Medicine, 288(16), 831 (1973). "The Sorbitol Pathway and the Complications of Diabetes".

Robert G. Spiro, New England Journal of Medicine, 288(25), 1337 (1973). "Biochemistry of the Renal Glomerular Basement Membrane and Its Alterations in Diabetes Mellitus".

E.W. Taylor, Annual Review of Biochemistry, 41, 577 (1972). "Chemistry of Muscle Contraction".

Chapter V (Nucleic Acids)

W. Eckhart, Annual Review of Biochemistry, 41, 503 (1972). "Oncogenic Viruses".

The Biochemistry of the Nucleic Acids, Seventh Edition; by J.N. Davidson, (Academic Press, New York, 1972).

Asghar Rastegar and Samuel O. Thier, New England Journal of Medicine, 286(9), 470 (1972). "The Physiological Approach to Hyperuricemia".

Lloyd H. Smith, New England Journal of Medicine, 288(15), 764 (1973). "Pyrimidine Metabolism in Man".

Marcus A. Rothschild, Murray Oratz, and Sidney S. Schreiber, New England Journal of Medicine, 286(14), 748; 286(15), 816 (1972). "Albumin Synthesis (in Two Parts)".

Michael E. Grant and Darwin J. Prockop, New England Journal of Medicine, 286(4), 194; 286(5), 242; 286(6), 291 (1972). "The Biosynthesis of Collagen (In Three Parts)".

Gerald Weissmann and Giuseppe A. Rita, Nature (New Biology), 240, 167 (1972). "Molecular Basis of Gouty Inflammation: Interaction of Monosodium Urate Crystals with Lysosomes and Liposomes".

Chapter VI (Amino Acid Metabolism)

P. Truffa-Bachi and G.N. Cohen, Annual Review of Biochemistry, 42, 113 (1973). "Amino Acid Metabolism".

S. Andersson, Annual Review of Physiology, 35, 431 (1973). "Secretion of Gastrointestinal Hormones".

R.J. Baldessarini and M. Karobath, Annual Review of Physiology, 35, 273 (1973). "Biochemical Physiology of Central Synapses".

George W. Frimpter, New England Journal of Medicine, 289(16), 835; 289(17), 895 (1973). "Aminoacidurias Due to Inherited Disorders of Metabolism (In Two Parts)".

Julius Axelrod and Richard Weinshilboum, New England Journal of Medicine, 287(5), 237 (1972). "Catecholamines".

Chapter VII (Hemoglobin and Porphyrin Metabolism)

George B. Segel, Stephen A. Feig, William C. Mentzer, Ronald P. McCaffery, Roe Wells, H. Franklin Bunn, Stephen B. Shohet, and David G. Nathan, New England Journal of Medicine, 287(2), 59 (1972). "Effects of Urea and Cyanate on Sickling In Vitro".

David G. Nathan, New England Journal of Medicine, 286(11), 586 (1972). "Thalassemia".

Rudi Schmid, New England Journal of Medicine, 287(14), 703 (1972). "Bilirubin Metabolism in Man".

Reinhold Benesch and Ruth E. Benesch, Nature, 221, 618 (1969). "Intracellular Organic Phosphates as Regulators of Oxygen Release by Haemoglobin".

George J. Brewer and John W. Eaton, Science, 171(3977), 1205 (1971). "Erythrocyte Metabolism: Interaction with Oxygen Transport".

M.F. Perutz, Nature, <u>228</u>, 726 (1970). "Stereochemistry of Cooperative Effects in Haemoglobin".

G.H. Elder, C.H. Gray and D.C. Nicholson, Journal of Clinical Pathology, <u>25</u>, 1013 (1972). "The Porphyrias: A Review".

<u>New Aspects of the Structure, Function, and Synthesis of Hemoglobins</u>; by T.H.J. Huisman and W.A. Schroeder, (CRC Press, Cleveland, Ohio, 1971).

Chapter VIII (Lipids)

J.W. Hinman, Annual Review of Biochemistry, <u>41</u>, 161 (1972). "Prostaglandins".

W. Stoffel, Annual Review of Biochemistry, <u>40</u>, 57 (1971). "Sphingolipids".

J.J. Volpe and P.R. Vagelos, Annual Review of Biochemistry, <u>42</u>, 21 (1973). "Saturated Fatty Acid Biosynthesis and Its Regulation".

A.D. Bangham, Annual Review of Biochemistry, <u>41</u>, 753 (1972). "Lipid Bilayers and Biomembranes".

P. Siekevitz, Annual Review of Physiology, <u>34</u>, 117 (1972). "Biological Membranes: The Dynamics of Their Organization".

L. Sokoloff, Annual Review of Medicine, <u>24</u>, 271 (1973). "Metabolism of Ketone Bodies by the Brain".

John W.C. Johnson, Jr. and Gerard B. Odell, Annual Review of Medicine, <u>24</u>, 165 (1973). "Estimation of Gestational Age in the Fetal State".

J.L. Beaumont, L.A. Carlson, G.R. Cooper, Z. Fejfar, D.S. Fredrickson, and T. Strasser, Bulletin of the World Health Organization, <u>43</u>, 891 (1970). "Classification of Hyperlipidaemias and Hyperlipoproteinaemias".

Robert I. Levy and Donald S. Fredrickson, Postgraduate Medicine, January 1970, pages 15-21. "The Current Status of Hypolipidemic Drugs".

Chapter IX (Mineral Metabolism)

B.L. Vallee and D.D. Ulmer, Annual Review of Biochemistry, 41, 91 (1972). "Biochemical Effects of Mercury, Cadmium, and Lead".

S. Gershon, Annual Review of Medicine, 23, 439 (1972). "Lithium Salts in the Management of the Manic-Depressive Syndrome".

Jane S. Lin-Fu, New England Journal of Medicine, 286(13), 702 (1972). "Undue Absorption of Lead Among Children--A New Look at an Old Problem".

Hector F. De Lucca, New England Journal of Medicine, 289(7), 359 (1973). "The Kidney as an Endocrine Organ: Production of 1,25-Dihydroxyvitamin D_3".

Jean M. Morgan and Helen B. Burch, Archives of Internal Medicine, 130, 335 (1972). "Comparative Tests for Diagnosis of Lead Poisoning".

Chapter X (Vitamins)

R.H. Wasserman and R.A. Corradino, Annual Review of Biochemistry, 40, 501 (1971). "Metabolic Role of Vitamins A and D".

R.H. Wasserman and A.N. Taylor, Annual Review of Biochemistry, 41, 179 (1972). "Metabolic Roles of Fat-Soluble Vitamins D, E, and K".

Max Katz, Sook K. Lee, and Bernard A. Cooper, New England Journal of Medicine, 287(9), 425 (1972). "Vitamin B_{12} Malabsorption Due to Biologically Inert Intrinsic Factor".

I.L. MacKenzie, R.M. Donaldson, Jr., J.S. Trier, and V.I. Mathan, New England Journal of Medicine, 286(19), 1021 (1972). "Ileal Mucosa in Familial Selective Vitamin B_{12} Malabsorption".

Recommended Dietary Allowances, Seventh Revised Edition, Food and Nutrition Board, National Research Council (Publication 1694, National Academy of Sciences, 1968).

Chapter XI (Immunochemistry)

Immunology;   by J.A. Bellanti, (W.B. Saunders Co.,
Philadelphia, 1971).

Clinical Immunobiology, Volume 1; Edited by F.H. Bach and
R.A. Good, (Academic Press, New York, 1972).

G.P. Smith, L. Hood, and W.M. Fitch, Annual Review of Biochemistry,
40, 1969 (1971). "Antibody Diversity".

T.B. Tomasi, Jr. and J. Bienenstoch, "Secretory Immunoglobulins"
in Advances in Immunology, Volume 9, pp. 1; Edited by F.J. Dixon, Jr.
and H.G. Kunkel, (Academic Press, New York, 1968).

J.B. Natvig and H.G. Kunkel, "Human Immunoglobulin Class,
Subclasses, Genetic Variants and Idiotypes" in Advances in Immunology,
Volume 16, pp. 1; Edited by F.J. Dixon, Jr. and H.G. Kunkel,
(Academic Press, New York, 1973).

H. Metzger, "Structure and Function of γ-M Macroglobulins" in
Advances in Immunology, Volume 57, pp. 12; Edited by F.J. Dixon, Jr.
and H.G. Kunkel, (Academic Press, New York, 1970).

R.P. McCombs, NEJM, 286, 1186 (1972). "Diseases Due to
Immunologic Reactions in the Lungs".

Inflammation, Immunity and Hypersensitivity; Edited by
H.Z. Movat, (Harper and Row Publishers, New York, 1971).

Infectious Agents and Host Reactions; Edited by S. Mudd,
(W.B. Saunders Co., Philadelphia, 1970).

Blood Clotting Enzymology; Edited by W.H. Seegers, (Academic
Press, New York, 1967).

Immunoglobulins, Annals of the New York Academy of Science, Volume 190;
Edited by Kochwa and H.G. Kunkel, (New York Academy of Science, 1971).

Microbiology; Edited by B.D. Davis, R. Dulbecco, H.N. Eisen,
H.S. Ginsberg and W.B. Wood, (Hoober Medical Division, Harper and Row,
Publishers, New York, 1968).

P.I. Terasaki and D.P. Singal, Annual Review of Medicine, 20,
175 (1969). "Human Histocompatability Antigens of Lymphocytes".

Shaun Ruddy, Irma Gigli, and K. Frank Austen, New England
Journal of Medicine, 287(10), 489; 287(11), 545; 287(12), 592;
287(13), 642 (1972). "The Complement System of Man (In Four Parts)".

Chapter XII (Water and Electrolyte Balance)

P.D. Siegel, Medical Clinics of North America, 57, 4 July 1973.
"Physiological Approach to Acid Base Balance".

S. Klahr, S. Wessler and L.V. Avioli, JAHA, 222, 567 (1972).
"Acid-Base Disorders in Health and Disease".

American Journal of Medicine, 53, 529-684 (November 1972).
"Symposium on the Adrenal Cortex".

Archives of Internal Medicine, 131, 779-938 (1973). "Symposium on
Renal Physiology".

Fluid and Electrolyte Balance, Sixth Revision; by B.H. Scribner
and J.H. Burnell, (The University of Washington Press, 1963).

Gerhard Giebisch, New England Journal of Medicine, 287(18), 913
(1972). "Coupled Ion and Fluid Transport in the Kidney".

G.W.G. Sharp, Annual Review of Medicine, 24, 19 (1973).
"Action of Cholera Toxin on Fluid and Electrolyte Movement in the
Small Intestine".

S.B. Formal, H.L. DuPont, and R.B. Hornick, Annual Review of
Medicine, 24, 103 (1973). "Enterotoxic Diarrheal Syndromes".

Chapter XIII (Hormones)

Biochemical Endocrinology of the Vertebrates; by Earl Frieden
and Harry Lipner, (Prentice-Hall, Englewood Cliffs, New Jersey, 1971).

C.C. Gale, Annual Review of Physiology, 35, 391 (1973).
"Neuroendocrine Aspects of Thermoregulation".

Robert J. Lefkowitz, New England Journal of Medicine, 288(20),
1061 (1973). "Isolated Hormone Receptors".

Jean D. Wilson, New England Journal of Medicine, 287(25),
1284 (1972). "Recent Studies on the Mechanism of Action of
Testosterone".

Chapter XIV (Biochemical Transformation of Drugs)

Drill's Pharmacology in Medicine; Edited by J.R. DiPalma, (McGraw-Hill, New York, 1971).

Drug Metabolism and Disposition: The Biological Fate of Chemicals, Proceedings of the Second International Symposium on Microsomes and Drug Oxidations; Edited by K.C. Leibman, (Williams and Wilkins, Baltimore, 1973).

Pharmacogenetics, Annals of the New York Academy of Science, Volume 151; Edited by B.N. LaDu and W. Kalow, (New York Academy of Science, 1968).

Drug Metabolism in Man, Annals of the New York Academy of Science, Volume 179; Edited by E.S. Vessell, (New York Academy of Science, 1971).

Principles of Drug Action; by A. Goldstein, L. Aronow and M. Kalman, (Hoeber, Harper and Row, New York, 1968).

A.H. Conney and J.J. Burns, Advances in Pharmacology, 1, 31 (1962). "Factors Influencing Drug Metabolism".

D.P. Rall and C.G. Zubrod, Annual Review of Pharmacology, 2, 109 (1962). "Mechanisms of Drug Absorption and Excretion".

Metabolic Factors Controlling Duration of Drug Action, Proceedings of the First International Pharmacological Meeting, Volume 6; Edited by B.B. Brodie and E.G. Erdös, (McMillan, New York, 1962).

B.N. LaDu, Annual Review of Medicine, 23, 453 (1972). "Pharmacogenetics: Defective Enzymes in Relation to Reaction to Drugs".

I.H. Raisfeld, Annual Review of Medicine, 24, 385 (1973). "Clinical Pharmacology of Drug Interactions".

Chapter XV (Liver Function)

A.I. Sutnick, I. Millman, W.T. London, and B.S. Blumberg, Annual Review of Medicine, 23, 161 (1972). "The Role of Australia Antigen in Viral Hepatitis and Other Diseases".

Gerald Klatskin and Fred S. Kantor, Annals of Internal Medicine, 77, 533 (1972). "Mitochondrial Antibody in Primary Biliary Cirrhosis and Other Diseases".

Sidney Green, Frederick Cantor, Normal R. Inglis, and William H. Fishman, American Journal of Clinical Pathology, 57, 52 (1972). "Normal Serum Alkaline Phosphatase Isoenzymes Examined by Acrylamide and Starch Gel Electrophoresis and by Isoenzyme Analysis Using Organ-Specific Inhibitors".

INDEX

Numerals in *italics* indicate a figure and a "t" following a page number indicates a table concerning the subject.

O.